AF393804

Classes of
Linear Operators
Vol. II

Israel Gohberg
Seymour Goldberg
Marinus A. Kaashoek

Springer Basel AG

Authors:

I. Gohberg
School of Mathematical Sciences
Raymond and Beverly Sackler
Faculty of Exact Sciences
Tel Aviv University
69978 Tel Aviv
Israel

M.A. Kaashoek
Faculteit Wiskunde en Informatica
Vrije Universiteit
De Boelelaan 1081
1081 HV Amsterdam
The Netherlands

S. Goldberg
Department of Mathematics
University of Maryland
College Park, MD 20742
USA

A CIP catalogue record for this book is available from the Library of Congress, Washington D.C., USA

Deutsche Bibliothek Cataloging-in-Publication Data
Gochberg, Izrail'
Classes of linear operators / Israel Gohberg ; Seymour
Goldberg ; Marinus A. Kaashoek. – Basel ; Boston ; Berlin:
NE: Goldberg, Seymour:; Kaashoek, Marinus A.:
Vol. 2 (1993)
   (Operator theory ; Vol. 63)

NE: GT

© 1993 Springer Basel AG
Originally published by Birkhäuser Verlag, Basel, Switzerland in 1993

Camera-ready copy prepared by the authors
Printed on acid-free paper produced from chlorine-free pulp
Cover design: Heinz Hiltbrunner, Basel

ISBN 978-3-0348-9679-5   ISBN 978-3-0348-8558-4 (eBook)
DOI 10.1007/978-3-0348-8558-4
9 8 7 6 5 4 3 2 1

PREFACE TO VOLUME II

The present book is the second volume of Classes of Linear Operators. The main body of the text was ready by the time the first volume appeared (Summer 1990), and its contents correspond to the plan as it was announced in the first volume. However, to make the exposition more complete an essential number of new sections has been added. Most of the text in this volume can be read independently of that of the first volume.

We are grateful to our colleagues Robert Ellis, Naum Krupnik, Leonid Lerer, and André Ran for valuable remarks about parts of the text. We thank Henno Brandsma, Gilbert Groenewald, Hein van der Holst and Derk Pik for providing us with lists of corrections. The help of Evert Wattel with the pictures is highly appreciated.

As in the preface to Volume I we gratefully acknowledge the support from our home institutions and the financial assistance from the Nathan and Lillian Silver Chair in Mathematical Analysis and Operator Theory.

June 15, 1993The authors

TABLE OF CONTENTS OF VOLUME II

INTRODUCTION

These two volumes constitute texts for graduate courses in linear operator theory. The reader is assumed to have a knowledge of both complex analysis and the first elements of operator theory. The texts are intended to concisely present a variety of classes of linear operators, each with its own character, theory, techniques and tools. For each of the classes, various differential and integral operators motivate or illustrate the main results. Although each class is treated seperately and the first impression may be that of many different theories, interconnections appear frequently and unexpectedly. The result is a beautiful, unified and powerful theory.

The classes we have chosen are representatives of the principal important classes of operators, and we believe that these illustrate the richness of operator theory, both in its theoretical developments and in its applicants. Because we wanted the books to be of reasonable size, we were selective in the classes we chose and restricted our attention to the main features of the corresponding theories. However, these theories have been updated and enhanced by new developments, many of which appear here for the first time in an operator-theory text. In the selection of the material the taste and interest of the authors played an important role.

The books present a wide panorama of modern operator theory. They are not encyclopedic in nature and do not delve too deeply into one particular area. In our opinion it is this combination that will make the books attractive to readers who know basic operator theory.

The exposition is self-contained and has been simplified and polished in an effort to make advanced topics accessible to a wide audience of students and researchers in mathematics, science and engineering.

The classes encompass compact operators, various subclasses of compact operators (such as trace and Hilbert-Schmidt operators), Fredholm operators (bounded and unbounded), Wiener-Hopf and Toeplitz operators, selfadjoint operators (bounded and unbounded), and integral and differential operators on finite and infinite intervals. The two volumes also contain an introduction to the theory of Banach algebras with applications to algebras of Toeplitz operators, the first elements of the theory of operator semigroups with applications to initial value problems, the theory of triangular representation, the method of factorization for general operators and for matrix functions, an introduction to the theory of characteristic operator functions for contractions. Also included are recent developments concerning extension and completion problems for operator matrices and matrix functions.

The present second volume contains Parts V–IX. The material presented in this volume is fundamental in modern operator theory and perhaps less standard than that of the first volume. Triangular representation and factorization are leading themes throughout this volume. Most of the text can be read independently from the first volume.

Part V concerns triangular operators and decomposition of operators into triangular ones. Here the word triangular is understood relative to a chain of projections. The analogue of additive lower-upper decomposition of matrices is discussed. The simplest classes of infinite dimensional triangular operators are analyzed. A major topic is LU-factorization relative to a chain. Formulas for such factorizations are presented together with applications to various classes of integral operators.

Part VI presents an introduction to the general theory of infinite dimensional block Toeplitz and block Laurent operators on Hilbert space. It develops the Fredholm theory and inversion methods based on the theory of Wiener-Hopf factorization. Special attention is paid to block Toeplitz operators defined by rational matrix functions. The latter operators are analyzed by using tools which come from mathematical system theory. Another main topic is the Fredholm theory of block Toeplitz operators defined by piecewise continuous matrix functions.

Part VII concerns the study of unitary invariants of operators, the analysis and construction of invariant subspaces, the theory of operator models, and the theory of characteristic operator functions. These topics are treated for contractive operators and related operators. The theory of unitary systems provides a unifying approach. Triangular representations of operators derived in Part V play an important role, in particular, in the multiplicative decomposition theorems for characteristic operator functions. Block shift operators and their invariant subspaces serve as basic elements. Dilation theory and the commutant lifting theory, together with applications to interpolation problems, are also included.

Part VIII presents an introduction to the theory of Banach algebras and algebras of operators. It contains the basic elements of the general theory. The theory of commutative Banach algebras and that of commutative C^*-algebras are major topics. Special attention is payed to Banach algebras generated by block Toeplitz operators defined by piecewise continuous matrix functions. The notion of a symbol is extended to operators in such algebras. Also included are applications to Wiener-Hopf factorization and to the spectral theory of normal operators.

Part IX presents an operator theoretical method to solve extension and completion problems of various types. It contains an abstract theory, called the band method, which is based on ideas and results of the previous parts. Applications of the general theory to a number of matrix-valued versions of classical interpolation problems are included. One step extension problems for finite matrices serve as an introduction and provide the motivation for the general theory. The analysis of the maximum entropy extension is a major topic.

The first volume consists of Parts I–V. The titles are as follows. Part I: General Spectral Theory; Part II: Classes of Compact Operators; Part III; Fredholm Operators: General Theory and Wiener-Hopf Integral Operators; Part IV: Classes of Unbounded Operators.

In both volumes each part concludes with comments and a series of exercises.

PART V

TRIANGULAR REPRESENTATIONS

In this part the infinite dimensional analogues of upper and lower triangular matrices are studied. The attention is focused on the problem of decomposing an infinite dimensional operator as a sum or a product of triangular operators. The simplest classes of infinite dimensional triangular operators are analysed.

This part begins with the infinite dimensional analogue of lower-upper additive triangular decompositions of operators (Chapter XX). In this case the decomposition is related to a chain of projections (and not necessarily to an orthogonal basis as in the finite dimensional case) and hence in general it has a continuous and not a discrete character. The existence of the decomposition for a given operator relative to a given chain is a difficult problem by itself. A number of illustrative examples for integral operators is presented. Triangular representations of operators are studied in Chapter XXI. A simple unicellular operator (the analogue of a matrix with one Jordan block) is analysed. The last chapter (Chapter XXII) is concerned with multiplicative lower-upper decompositions for operators relative to a chain of projections. It contains applications to Hilbert-Schmidt operators and different classes of integral operators.

CHAPTER XX

ADDITIVE LOWER-UPPER TRIANGULAR DECOMPOSITIONS OF OPERATORS

The decomposition

$$\int_a^b k(t,s)\varphi(s)ds = \int_a^t k(t,s)\varphi(s)ds + \int_t^b k(t,s)\varphi(s)ds$$

may be viewed as the analogue of the additive lower-upper triangular decomposition of a square matrix. In this chapter we study such decompositions for general operators. We begin with the first generalization of the matrix case.

XX.1 ADDITIVE LOWER-UPPER TRIANGULAR DECOMPOSITIONS RELATIVE TO FINITE CHAINS

Throughout this section X is an arbitrary vector space. In the set of all projections on X we introduce a partial ordering by setting $P_1 \leq P_2$ if $\operatorname{Im} P_1 \subset \operatorname{Im} P_2$ and $\operatorname{Ker} P_1 \supset \operatorname{Ker} P_2$. We write $P_1 < P_2$ if, in addition, $P_1 \neq P_2$. Note that

$$(1) \qquad P_1 \leq P_2 \iff P_1 P_2 = P_2 P_1 = P_1.$$

An ordered set of projections $\{P_0, P_1, \ldots, P_n\}$ is called a *finite chain* on X if

$$(2) \qquad 0 = P_0 < P_1 < \cdots < P_n = I.$$

For a finite chain $\{P_j\}_{j=0}^n$ we put $\Delta P_j = P_j - P_{j-1}$. Note that the operators $\Delta P_1, \ldots, \Delta P_n$ are mutually disjoint projections, that is

$$(\Delta P_i)(\Delta P_j) = \begin{cases} 0 & \text{for} \quad i \neq j, \\ \Delta P_j & \text{for} \quad i = j. \end{cases}$$

Further, $\sum_{j=1}^n \Delta P_j = I$. It follows that

$$(3) \qquad X = \operatorname{Im} \Delta P_1 \oplus \operatorname{Im} \Delta P_2 \oplus \cdots \oplus \operatorname{Im} \Delta P_n.$$

Conversely, let $X = X_1 \oplus X_2 \oplus \cdots \oplus X_n$ be a given direct sum decomposition of X with $X_j \neq (0)$ for each j. Define P_j to be the projection of X onto $X_1 \oplus \cdots \oplus X_j$ along the space $X_{j+1} \oplus \cdots \oplus X_n$ $(j = 1, \ldots, n-1)$ and put $P_0 = 0$, $P_n = I$. Then $\{P_j\}_{j=0}^n$ is a finite chain and $\operatorname{Im} \Delta P_j = X_j$ $(j = 1, \ldots, n)$.

THEOREM 1.1. *Let* $\pi = \{P_j\}_{j=0}^n$ *be a finite chain on* X, *and let* A *be a linear operator on* X. *Put*

(i) $A_- = \sum_{j=1}^{n} (\Delta P_j) A P_{j-1}$,

(ii) $A_0 = \sum_{j=1}^{n} (\Delta P_j) A (\Delta P_j)$,

(iii) $A_+ = \sum_{j=1}^{n} P_{j-1} A (\Delta P_j)$.

Then $A = A_- + A_0 + A_+$.

PROOF. We have

$$A = \left(\sum_{j=1}^{n} \Delta P_j \right) A \left(\sum_{k=1}^{n} \Delta P_k \right) = \sum_{j,k=1}^{n} (\Delta P_j) A (\Delta P_k)$$

$$= \sum_{j>k} (\Delta P_j) A (\Delta P_k) + A_0 + \sum_{j<k} (\Delta P_j) A (\Delta P_k)$$

$$= A_- + A_0 + A_+. \quad \square$$

Let us consider A_-, A_0 and A_+ as block operator matrices with respect to the decomposition (3). Then A_0 is a block diagonal matrix and for A_- and A_+ the block diagonal elements are zero. Furthermore, A_- and A_+ are lower and upper block triangular, respectively. For this reason we call the decomposition $A = A_- + A_0 + A_+$ the *additive LU-decomposition* of A with respect to the finite chain π.

The operator A_0 we call the *diagonal* of A with respect to the finite chain π. We shall write

$$A_0 = \operatorname{diag}(A; \pi) = \sum_{j=1}^{n} (\Delta P_j) A (\Delta P_j).$$

XX.2 PRELIMINARIES ABOUT CHAINS

Throughout this section X is a Banach space. By definition a projection of X is assumed to be a continuous projection. A set $\mathbb{P} = \{P\}$ of projections of X is called a *chain* on X if $O, I \in \mathbb{P}$ and $\mathbb{P}$ is linearly ordered with respect to the ordering $\leq$ (which has been introduced in the previous section). Note that the elements in a chain commute with each other (cf. formula (1) in the previous section).

If $\mathbb{P}$ is a chain, the set $\mathbb{P}^c = \{I - P \mid P \in \mathbb{P}\}$ is also a chain, which is called the *complement* of $\mathbb{P}$. A chain $\mathbb{P}$ is said to be *maximal* if $\mathbb{P}$ is not a proper subset of any other chain. First a few examples.

(A) Consider the space ℓ_p $(1 \leq p \leq \infty)$. Define $P_k \colon \ell_p \to \ell_p$ by setting

$$P_k(x_1, x_2, \ldots) = (x_1, \ldots, x_k, 0, 0, \ldots).$$

Put $\mathbb{P} = \{0, P_1, P_2, \ldots, I\}$. Then $\mathbb{P}$ is a maximal chain. We call $\mathbb{P}$ the *standard chain* on ℓ_p.

(B) Consider the space $L_p([0,1])$. For $0 \leq t \leq 1$ let P_t be the projection of $L_p([0,1])$ defined by

$$(P_t f)(s) = \begin{cases} f(s) & \text{for} \quad 0 \leq s \leq t, \\ 0 & \text{for} \quad t < s \leq 1. \end{cases}$$

Then $\mathbb{P} = \{P_t \mid 0 \le t \le 1\}$ is a maximal chain.

There is an important difference between the two examples. The chain in the first example has jumps, while no jumps appear in the chain of the second example. To make this more precise the following notion of discontinuity is useful. Let $\mathbb{P}$ be a chain. A pair (P_1, P_2) with P_1, P_2 in $\mathbb{P}$ is called a *jump* (or *discontinuity*) of $\mathbb{P}$ if

(i) $P_1 < P_2$,

(ii) $P_1 < P < P_2 \implies P \notin \mathbb{P}$.

In that case the rank of the projection $P_2 - P_1$ is called the *dimension* of the jump. A jump in a maximal chain has precisely dimension 1.

A chain $\mathbb{P}$ on X is said to be *invariant* under the operator $A \in \mathcal{L}(X)$ if $AP = PAP$ for each $P \in \mathbb{P}$. Thus $\mathbb{P}$ is invariant under A if and only if $\operatorname{Im} P$ is A-invariant for each $P \in \mathbb{P}$. We say that A is *reduced* by $\mathbb{P}$ (or $\mathbb{P}$ is *reducing* for A) if both $\mathbb{P}$ and its complement $\mathbb{P}^c$ are invariant under A. In other words, A is reduced by $\mathbb{P}$ if and only if $AP = PA$ for all $P \in \mathbb{P}$.

XX.3 DIAGONALS

Roughly speaking the diagonal of an operator A with respect to a chain $\mathbb{P}$ is the limit of $\operatorname{diag}(A; \pi)$, where π runs through all finite subchains of $\mathbb{P}$. To make this more precise we shall use the notion of a partition of a chain. Throughout this section $\mathbb{P}$ is a chain on the Banach space X.

An ordered set $\pi = \{P_0, P_1, \ldots, P_n\}$ is called a *partition* of the chain $\mathbb{P}$ if $P_j \in \mathbb{P}$ $(j = 0, \ldots, n)$ and

$$0 = P_0 < P_1 < \cdots < P_n = I.$$

In other words, a partition π of $\mathbb{P}$ is just a finite subchain of $\mathbb{P}$. A partition π_1 is said to be *finer* than the partition π_2 if $\pi_1 \supset \pi_2$.

An operator A on X is said to have a *diagonal* A_0 with respect to the chain $\mathbb{P}$ if in the operator norm

$$(1) \qquad A_0 = \lim_{\pi \subset \mathbb{P}} \operatorname{diag}(A; \pi),$$

that is, A_0 is a (bounded linear) operator on X and given $\varepsilon > 0$ there exists a partition π_ε of $\mathbb{P}$ such that

$$\left\| A_0 - \sum_{j=1}^{n} (\Delta P_j) A (\Delta P_j) \right\| < \varepsilon$$

for each partition $\pi = \{P_0, P_1, \ldots, P_n\}$ which is finer than π_ε. If the operator A has a diagonal A_0 with respect to $\mathbb{P}$, then A_0 is uniquely determined and we write

$$(2) \qquad A_0 = \int_{\mathbb{P}} (dP) A (dP).$$

Assume the operator A is reduced by the chain $\mathbb{P}$. Then $\operatorname{diag}(A; \pi) = A$ for each partition π, and hence the limit (1) exists and is equal to A. Thus if the chain $\mathbb{P}$ reduces the operator A, then

$$A = \int_{\mathbb{P}} (dP)A(dP).$$

The converse of this statement is also true.

The diagonal of an operator with respect to a chain is a non-trivial notion. For example, consider the backward shift S on ℓ_2, i.e.,

$$(3) \qquad S(x_1, x_2, x_3, \ldots) = (x_2, x_3, x_4, \ldots).$$

Let $\mathbb{P}$ be the standard chain on ℓ_2. Thus P_j is the orthogonal projection onto the first j coordinates. Although the operator S has a simple representation as an infinite matrix, the diagonal of S with respect to $\mathbb{P}$ does not exist. To see this, put $\pi_k = \{0, P_1, \ldots, P_{k-1}, I\}$. Then

$$\| \operatorname{diag}(S; \pi_{k+1}) - \operatorname{diag}(S, \pi_k)\| = 1$$

for each k, and hence for $A = S$ the limit (1) cannot exist.

On the other hand, if (3) is replaced by a weighted shift,

$$(4) \qquad S(x_1, x_2, x_3, \ldots) = (\alpha_1 x_2, \alpha_2 x_3, \alpha_3 x_4, \ldots),$$

with weights $\alpha_n \to 0$ $(n \to \infty)$, then the diagonal of S with respect to $\mathbb{P}$ exists and is equal to the zero operator.

By $A(\mathbb{P})$ we denote the set of all $A \in \mathcal{L}(X)$ such that A has a diagonal with respect to $\mathbb{P}$. Obviously, $A(\mathbb{P})$ is a linear submanifold of $\mathcal{L}(X)$ and the map

$$(5) \qquad A \mapsto \int_{\mathbb{P}} (dP)A(dP)$$

is linear on $A(\mathbb{P})$. To ensure the continuity of the map (5) we introduce the following notion of uniformity.

A chain $\mathbb{P}$ is said to be *uniform* if there exists a constant $\gamma \geq 0$ such that

$$(6) \qquad \| \operatorname{diag}(A; \pi)\| \leq \gamma \|A\|$$

for each operator A on X and any partition π of $\mathbb{P}$.

PROPOSITION 3.1. *If $\mathbb{P}$ is a uniform chain, then $A(\mathbb{P})$ is a closed subspace of $\mathcal{L}(X)$ and for some constant $\gamma \geq 0$*

$$(7) \qquad \left\| \int_{\mathbb{P}} (dP)A(dP)\right\| \leq \gamma \|A\| \qquad (A \in A(\mathbb{P})).$$

PROOF. By taking the limit in (6) we see that (7) holds true. Now, let A be an element in the closure of $A(\mathbb{P})$. So there exists sequence $(A_n)_{n=1}^{\infty}$ in $A(\mathbb{P})$ such that $A_n \to A$ in the operator norm. Put

$$A_{n0} = \int_{\mathbb{P}} (dP) A_n (dP).$$

From (7) it is clear that $\{A_{n0}\}_{n=1}^{\infty}$ is a Cauchy sequence in $\mathcal{L}(X)$, and thus $S = \lim_{n\to\infty} A_{n0}$ exists. We shall prove that S is the diagonal of A. Since $\|A_{n0} - A_{m0}\| \leq \gamma \|A_n - A_m\|$, we have $\|S - A_{m0}\| \leq \gamma \|A - A_m\|$. It follows that

$$\|S - \operatorname{diag}(A; \pi)\| \leq 2\gamma \|A - A_n\| + \|A_{n0} - \operatorname{diag}(A_n; \pi)\|.$$

Let $\varepsilon > 0$ be given. Choose n such that $2\gamma \|A - A_n\| < \frac{1}{2}\varepsilon$. Next, let π_ε be a partition of $\mathbb{P}$ such that

$$\|A_{n0} - \operatorname{diag}(A_n; \pi)\| < \frac{\varepsilon}{2}$$

for $\pi_\varepsilon \subset \pi \subset \mathbb{P}$. But then we have

$$\|S - \operatorname{diag}(A; \pi)\| < \varepsilon, \qquad \pi_\varepsilon \subset \pi \subset \mathbb{P},$$

and the proof is complete. $\square$

Note that $A(\mathbb{P}) = A(\mathbb{P}^c)$ and for $A \in A(\mathbb{P})$ we have

$$\int_{\mathbb{P}} (dP) A (dP) = \int_{\mathbb{P}^c} (dP) A (dP).$$

XX.4 CHAINS ON HILBERT SPACE

In this section H is a separable Hilbert space. On H we shall consider chains with orthogonal projections only. In other words in a chain on a Hilbert space (or in a *Hilbert space chain*) the elements are orthogonal projections by definition. For two orthogonal projections P_1 and P_2 on H the relation $P_1 \leq P_2$ is equivalent to the requirement that $\langle P_1 x, x \rangle \leq \langle P_2 x, x \rangle$ for each $x \in H$.

To study chains on H we shall need the *strong operator topology* on $\mathcal{L}(H)$. This topology is defined as the weakest topology on $\mathcal{L}(H)$ such that for each $x \in H$ the linear map

$$(1) \qquad\qquad S \mapsto Sx, \qquad \mathcal{L}(H) \to H$$

is continuous. Given $T \in \mathcal{L}(H)$, vectors $x_1, \ldots, x_k$ in H and $\varepsilon > 0$, put

$$(2) \qquad \mathcal{N}(T; x_1, \ldots, x_k, \varepsilon) = \bigcap_{j=1}^{k} \{ S \in \mathcal{L}(H) \,|\, \|Sx_j - Tx_j\| < \varepsilon \}.$$

Obviously, $\mathcal{N}(T; x_1, \ldots, x_k, \varepsilon)$ is an open neighborhood of T in the strong operator topology. Conversely, if U is an open neighbourhood of T in the strong operator topology, then there exist $x_1, \ldots, x_k$ in H and $\varepsilon > 0$ such that $\mathcal{N}(T; x_1, \ldots, x_k, \varepsilon) \subset U$. Since H is separable, the strong operator topology is metrizable on bounded subsets of $\mathcal{L}(H)$ (see e.g., Conway [1], page 262).

Let $\mathbb{P}$ be a chain on H. The closure of $\mathbb{P}$ in the strong operator topology is again a chain, which we shall denote by $\overline{\mathbb{P}}$. To see that $\overline{\mathbb{P}}$ is a chain, let us first prove that its elements are orthogonal projections. Take $P \in \overline{\mathbb{P}}$. Fix x and y in H. Then there exists a sequence $Q_1, Q_2, \ldots$ in $\mathbb{P}$ such that

$$(3) \qquad Q_n x \to P x, \quad Q_n y \to P y \qquad (n \to \infty).$$

It follows that

$$\langle P x, y \rangle = \lim_{n \to \infty} \langle Q_n x, y \rangle = \lim_{n \to \infty} \langle x, Q_n y \rangle = \langle x, P y \rangle,$$

and thus $P = P^*$. Next, assume that $y = P x$. Since $\|Q_n\| \leq 1$, the first limit in (3) gives

$$Q_n x - Q_n y = Q_n^2 x - Q_n P x = Q_n (Q_n x - P x) \to 0 \qquad (n \to \infty),$$

and we conclude that $P x = P y = P^2 x$. We have shown now that P is an orthogonal projection. It remains to prove that $\overline{\mathbb{P}}$ is linearly ordered. Take P and Q in $\overline{\mathbb{P}}$, and assume that there exist x and y in H such that

$$(4) \qquad \langle P x, x \rangle < \langle Q x, x \rangle, \qquad \langle P y, y \rangle > \langle Q y, y \rangle.$$

Since P and Q are in strong operator topology closure of $\mathbb{P}$, we can find P_1 and Q_1 in $\mathbb{P}$ such that (4) holds for P replaced by P_1 and Q replaced by Q_1. But this contradicts the linear ordering. So $\overline{\mathbb{P}}$ is a chain.

The chain $\mathbb{P}$ is called *closed* if $\mathbb{P} = \overline{\mathbb{P}}$. If $\mathbb{P}$ is closed and Σ is a non-empty subset of $\mathbb{P}$, then

$$(5) \qquad \inf \Sigma \in \mathbb{P}, \qquad \sup \Sigma \in \mathbb{P},$$

where inf and sup are taken in the set of all orthogonal projections on H. Let us prove the first part of (5). Put

$$M = \bigcap \{ \operatorname{Im} P \mid P \in \Sigma \}, \qquad N = \overline{\bigcup \{ \operatorname{Ker} P \mid P \in \Sigma \}}.$$

Obviously, M and N are closed linear manifolds in H. If $x \perp N$, then $x \perp \operatorname{Ker} P$ for each $P \in \Sigma$, and hence $x \in M$. Thus $N^\perp \subset M$. We shall prove that $N^\perp = M$. To do this it suffices to show that $M \cap N = \{0\}$. Take $x \in M \cap N$. Then there exists sequences $P_1, P_2, \ldots$ in Σ and $x_1, x_2, \ldots$ in N such that $x_n \to x$ if $n \to \infty$ and $x_n \in \operatorname{Ker} P_n$ for each n. The latter property implies that $x \perp x_n$ for each n, and hence also $x \perp x$, which implies that $x = 0$. Let Q be the orthogonal projection onto M. Thus $\operatorname{Ker} Q = N$.

Clearly, $Q = \inf \Sigma$. We want to show that $Q \in \overline{\mathbb{P}}$. Take vectors $x_1, \ldots, x_k$ in H and $\varepsilon > 0$. For $j = 1, \ldots, k$ put $n_j = (I - Q)x_j$ and choose $P_j \in \Sigma$ and $z_j \in \operatorname{Ker} P_j$ such that $\|n_j - z_j\| < \varepsilon$. Since $\mathbb{P}$ is linearly ordered, $P_i \leq P_j$ ($j = 1, \ldots, k$) for some $1 \leq i \leq k$. Put $P = P_i$. Obviously, $Q \leq P$, and thus $PQ = Q$. Furthermore, $z_j \in \operatorname{Ker} P_j \subset \operatorname{Ker} P$ for $j = 1, \ldots, k$. So

$$
\begin{aligned}
\|Px_j - Qx_j\| &= \|PQx_j - Qx_j + P(I - Q)x_j\| \\
&= \|Pn_j\| = \|P(n_j - z_j)\| \\
&\leq \|n_j - z_j\| < \varepsilon \qquad (j = 1, \ldots, k).
\end{aligned}
$$

In other words, $P \in \mathcal{N}(Q; x_1, \ldots, x_k, \varepsilon) \cap \mathbb{P}$, which proves that $Q \in \overline{\mathbb{P}}$.

A chain on the Hilbert space H is called *maximal* if it is not a proper subset of any other chain (with orthogonal projections) on H.

PROPOSITION 4.1. *A Hilbert space chain $\mathbb{P}$ is maximal if and only if $\mathbb{P}$ is closed and every jump of $\mathbb{P}$ has dimension 1.*

PROOF. Let $\mathbb{P}$ be a maximal chain. Since $\mathbb{P} \subset \overline{\mathbb{P}}$ and $\overline{\mathbb{P}}$ is a chain, it follows that $\mathbb{P} = \overline{\mathbb{P}}$, and thus $\mathbb{P}$ is closed. If a jump (P_-, P_+) in $\mathbb{P}$ has dimension at least 2, then there exists an orthogonal projection Q such that $P_- < Q < P_+$ and $Q \notin \mathbb{P}$. In that case $\mathbb{P} \cup \{Q\}$ is a chain strictly larger than $\mathbb{P}$, which contradicts the maximality. So jumps in $\mathbb{P}$ have dimension 1.

To prove the converse statement, assume that $\mathbb{P}$ is a closed chain such that every jump of $\mathbb{P}$ has dimension 1. Suppose $\mathbb{P}$ is not maximal. Then there exists an orthogonal projection Q such that $\mathbb{P} \cup \{Q\}$ is again a chain. Put

$$
P_- = \sup\{P \in \mathbb{P} \mid P < Q\}, \qquad P_+ = \inf\{P \in \mathbb{P} \mid Q < P\}.
$$

Since $\mathbb{P}$ is closed, the orthogonal projections P_- and P_+ belong to $\mathbb{P}$. Obviously, $P_- \leq Q \leq P_+$. Since $Q \notin \mathbb{P}$, it follows that (P_-, P_+) is a jump of dimension at least 2. Contradiction. $\square$

Given a compact operator A on H there exists a maximal chain on H that is invariant under A. To prove this we need the so-called *invariant subspace theorem* for compact operators.

THEOREM 4.2. *A nonzero compact operator on a Banach space of dimension ≥ 2 has a nontrivial invariant subspace.*

PROOF. Let K be a nonzero compact operator acting on the Banach space X, and assume $\dim X \geq 2$. We have to prove that there exists a closed linear manifold M in X with $M \neq \{0\}$ and $M \neq X$ such that $KM \subset M$. If λ is a nonzero point in the spectrum of K, then λ is an eigenvalue of K, and hence there exists $x \neq 0$ in X such that $Kx = \lambda x$. It follows that $M = \operatorname{span}\{x\}$ is a nontrivial subspace of X invariant under K. So it remains to consider the case when the spectral radius (cf., Part I, Exercise 18; also Section XXIX.7)

$$
(6) \qquad\qquad r(K) = \lim_{n \to \infty} \|K^n\|^{1/n} = 0.
$$

Let $\mathcal{A}$ be the set of all $A \in \mathcal{L}(X)$ that commute with K. For each nonzero y in X put

$$M(y) = \overline{\{Ay \mid A \in \mathcal{A}\}}.$$

Since $\mathcal{A}$ is a subalgebra of $\mathcal{L}(X)$ containing the identity on X, the set $M(y)$ is a nonzero subspace of X which is invariant under each operator A in $\mathcal{A}$. In particular, $M(y)$ is invariant under K. It suffices to prove that $M(y) \neq X$ for some $y \neq 0$. Assume not. So, let us suppose that

$$(7) \qquad\qquad M(y) = X, \qquad 0 \neq y \in X.$$

Take $x_0 \in X$ such that $Kx_0 \neq 0$. Obviously, this implies that $x_0 \neq 0$. Since K is continuous, there exists an open ball B centered at x_0 such that $0 \notin \overline{K(B)}$. We may assume that B has been chosen in such a way that $0 \notin \overline{B}$. Take $y \in \overline{K(B)}$. Because B is an open neighbourhood of x_0, the assumption (7) implies that there exists an operator A in $\mathcal{A}$, depending on y, such that $Ay \in B$. Now use that A is continuous and B is open. So there exists an open neighbourhood U_y of y such that $AU_y \subset B$. The open sets U_y, $y \in \overline{K(B)}$, cover the compact set $\overline{K(B)}$, and hence there exists a finite number of U_y covering $\overline{K(B)}$. Let $y(1), \ldots, y(m)$ be the corresponding vectors in $\overline{K(B)}$ and $A_1, \ldots, A_m$ the corresponding operators in $\mathcal{A}$. Since $Kx_0 \in \overline{K(B)}$, there exists $1 \leq i_1 \leq m$ such that $Kx_0 \in U_{y(i_1)}$, and hence $A_{i_1}Kx_0 \in B$. But then $KA_{i_1}Kx_0$ in KB, and so there exists $1 \leq i_2 \leq m$ such that $A_{i_2}KA_{i_1}Kx_0 \in B$. Proceeding in this way, we obtain a sequence $i_1, i_2, \ldots$ in $\{1, \ldots m\}$ such that for each $n \geq 1$

$$x_n := A_{i_n}K \cdots A_{i_1}Kx_0 \in B.$$

Put $\gamma = \max\{\|A_1\|, \ldots, \|A_m\|\}$, and use the fact that the A_i commute with K. We conclude that

$$\|x_n\| \leq \gamma^n \|K^n\| \|x_0\|.$$

But then we can use (6) to show that $\|x_n\|^{1/n} \to 0$ if $n \to \infty$. This implies $x_n \to 0$, and thus $0 \in \overline{B}$, which contradicts the choice of B. So (7) does not hold true, and there exists $y \neq 0$ such that $M(y) \neq X$. $\square$

Note that the proof of Theorem 4.2 also shows that any operator commuting with a nonzero Volterra operator has a nontrivial invariant subspace.

THEOREM 4.3. *A compact operator on H has an invariant maximal chain.*

PROOF. Let A be a compact operator on H, and let $\mathcal{P}$ be the family of all A-invariant chains on H. We order $\mathcal{P}$ by inclusion. Then $\mathcal{P}$ is a nonempty partially ordered set. Let Σ be a linearly ordered subset of $\mathcal{P}$, and let

$$\widetilde{\mathbb{P}} = \bigcup \{\mathbb{P} \mid \mathbb{P} \in \Sigma\}.$$

The set $\widetilde{\mathbb{P}}$ is again an A-invariant chain. So we may apply Zorn's lemma to show that in $\mathcal{P}$ there exists a maximal element $\mathbb{P}$, say.

It is easy to check that the closure $\overline{\mathbb{P}}$ of $\mathbb{P}$ is again an A-invariant chain. So $\overline{\mathbb{P}} \in \mathcal{P}$. Since $\mathbb{P} \subset \overline{\mathbb{P}}$, it follows that $\mathbb{P} = \overline{\mathbb{P}}$. Hence $\mathbb{P}$ is closed.

Next, let (P_-, P_+) be a jump of $\mathbb{P}$, and assume that its dimension is 2 or more. By the invariant subspace theorem for compact operators (Theorem 4.2), applied to $K := A|\operatorname{Im} P_+$, there exists an A-invariant subspace M such that

$$\operatorname{Im} P_- \underset{\neq}{\subseteq} M \underset{\neq}{\subseteq} \operatorname{Im} P_+.$$

Let Q be the orthogonal projection onto M. Then $\widetilde{\mathbb{P}} = \mathbb{P} \cup \{Q\}$ is an A-invariant chain larger than $\mathbb{P}$, which is impossible. So $\mathbb{P}$ is closed and every jump in $\mathbb{P}$ has dimension 1. By Proposition 4.1 this implies that $\mathbb{P}$ is maximal. $\square$

An orthonormal basis $\varphi_1, \varphi_2, \ldots$ of H can be used to parametrize a chain $\mathbb{P}$ on H. Indeed, define

$$(8) \qquad t: \mathbb{P} \to [0,1], \qquad t(P) = \sum_{j=1}^{\infty} 2^{-j} \langle P\varphi_j, \varphi_j \rangle.$$

Since the elements of $\mathbb{P}$ are orthogonal projections, we have $\langle P\varphi_j, \varphi_j \rangle = \|P\varphi_j\|^2 \leq 1$, and thus $0 \leq t(P) \leq 1$. We call $t = t(P)$ the *parameter* of P. Obviously, the map t is order preserving. It is also one-one. To see this, assume $t(P_1) = t(P_2)$ with $P_1 \leq P_2$. Then $\langle (P_2 - P_1)\varphi_j, \varphi_j \rangle \geq 0$ for each j and

$$0 = t(P_2) - t(P_1) = \sum_{j=1}^{\infty} 2^{-j} \langle (P_2 - P_1)\varphi_j, \varphi_j \rangle.$$

Thus $\langle (P_2 - P_1)\varphi_j, \varphi_j \rangle = 0$ for each j, and it follows that $P_2 - P_1$ is zero on the elements of an orthonormal basis, which implies that $P_1 = P_2$. Let Ω be the set of parameters t, and let $\gamma = t^{-1}: \Omega \to \mathbb{P}$ be the inverse map. Then γ is an order preserving parametrization of $\mathbb{P}$.

From the previous remarks it is clear that a chain on H has at most a countable number of jumps. A chain $\mathbb{P}$ on H is called *continuous* if $\mathbb{P}$ is closed and has no jumps. For a continuous chain the set of parameters t (introduced in the previous paragraph) is the full interval $[0,1]$ and the parameterization $\gamma: [0,1] \to \mathbb{P}$ is continuous, assuming $\mathbb{P}$ is endowed with the strong operator topology.

A chain $\mathbb{P}$ on a Hilbert space is automatically a uniform chain. In fact,

$$(9) \qquad \|\operatorname{diag}(A; \pi)\| \leq \|A\|$$

for each partition π of $\mathbb{P}$. Hence, if the diagonal of A with respect to $\mathbb{P}$ exists, then its norm is less than or equal to the norm of A (cf. formula (1) in the previous section).

THEOREM 4.4. *Let $\mathbb{P}$ be a closed chain on H, and let A be a compact operator on H. Then the diagonal of A with respect to $\mathbb{P}$ exists and*

$$(10) \qquad \int_{\mathbb{P}} (dP)A(dP) = \sum_{\nu} (P_\nu^+ - P_\nu^-)A(P_\nu^+ - P_\nu^-),$$

where (P_ν^-, P_ν^+), $\nu = 1, 2, \ldots$, are the jumps of $\mathbb{P}$.

PROOF. For an arbitrary operator A on H the series in the right hand side of (10) converges pointwise and its sum does not depend on the way the jumps are ordered. This follows from the fact that the orthogonal projections $P_\nu^+ - P_\nu^-$, $\nu = 1, 2, \ldots$, are mutually orthogonal. We denote the right hand side of (10) by $\operatorname{diag}(A; \mathbb{P})$. Note that we have to prove

$$(11) \qquad \lim_{\pi \subset \mathbb{P}} \operatorname{diag}(A; \pi) = \operatorname{diag}(A; \mathbb{P}).$$

We begin with two remarks. If $A_1, A_2 \colon H \to H$ are arbitrary operators on H, then

$$(12a) \qquad \| \operatorname{diag}(A_1; \mathbb{P}) - \operatorname{diag}(A_2; \mathbb{P}) \| \le \| A_1 - A_2 \|,$$
$$(12b) \qquad \| \operatorname{diag}(A_1; \pi) - \operatorname{diag}(A_2; \pi) \| \le \| A_1 - A_2 \|.$$

Here π is any partition of $\mathbb{P}$. From (12a) and (12b) it follows that it suffices to prove (11) for operators of finite rank. Next, observe that $\operatorname{diag}(A; \mathbb{P})$ and $\operatorname{diag}(A; \pi)$ are linear in A. Hence to prove (11) we may assume without loss of generality that A is a selfadjoint operator of rank 1.

For $\varphi \in H$ let A_φ be the one-dimensional selfadjoint operator $A_\varphi = \langle \cdot, \varphi \rangle \varphi$. Assume $\| \varphi \| = 1$, and let $\varepsilon > 0$ be given. Choose N such that

$$\left\| \sum_{\nu \ge N+1} (P_\nu^+ - P_\nu^-) \varphi \right\| < \varepsilon/6,$$

and put

$$\psi := \varphi - \sum_{\nu \ge N+1} (P_\nu^+ - P_\nu^-) \varphi.$$

Note that $\| \psi \| \le 1$ and $\| A_\varphi - A_\psi \| \le 2 \| \varphi - \psi \| < \frac{\varepsilon}{3}$.

From the definition of ψ it follows that $P_\nu^- \psi = P_\nu^+ \psi$ for $\nu \ge N + 1$. Since

$$(P_\nu^+ - P_\nu^-) A_\psi (P_\nu^+ - P_\nu^-) = \langle \cdot, (P_\nu^+ - P_\nu^-) \psi \rangle (P_\nu^+ - P_\nu^-) \psi,$$

we may conclude

$$\operatorname{diag}(A_\psi; \mathbb{P}) = \sum_{\nu=1}^{N} (P_\nu^+ - P_\nu^-) A_\psi (P_\nu^+ - P_\nu^-).$$

Next consider the set $\Omega_0 = \{ \| P\psi \|^2 \mid P \in \mathbb{P} \}$. Put $\alpha = \| \psi \|^2$. Obviously, $\Omega_0 \subset [0, \alpha]$. The fact that $\mathbb{P}$ is closed implies that Ω_0 is the complement in $[0, \alpha]$ of the union of all open intervals $(\| P_\nu^- \psi \|^2, \| P_\nu^+ \psi \|^2)$. So, because of the special properties of ψ, we have

$$\Omega_0 = [0, \alpha] \setminus \bigcup_{\nu=1}^{N} (\| P_\nu^- \psi \|^2, \| P_\nu^+ \psi \|^2).$$

Let $\pi_\varepsilon = \{\widetilde{P}_0, \ldots, \widetilde{P}_n\}$ be a partition of $\mathbb{P}$ which contains all projections $P_1^-, P_1^+, \ldots, P_N^-$, P_N^+. Further, we assume that π_ε has been chosen such that

$$\|\widetilde{P}_j \psi\|^2 - \|\widetilde{P}_{j-1} \psi\|^2 < \frac{\varepsilon}{3},$$

whenever $(\widetilde{P}_{j-1}, \widetilde{P}_j)$ is not a jump of $\mathbb{P}$. Let $\pi = \{P_0, \ldots, P_m\}$ be any partition of $\mathbb{P}$ finer than π_ε. Then

$$\| \operatorname{diag}(A_\psi; \mathbb{P}) - \operatorname{diag}(A_\psi; \pi)\| \leq \left\|\sum_{j \in \Lambda}(P_j - P_{j-1})A_\psi(P_j - P_{j-1})\right\|,$$

where Λ is the set of indices j such that (P_{j-1}, P_j) is not a jump of $\mathbb{P}$. It follows that

$$\| \operatorname{diag}(A_\psi; \mathbb{P}) - \operatorname{diag}(A_\psi; \pi)\| \leq \max_{j \in \Lambda}\|(P_j - P_{j-1})A_\psi(P_j - P_{j-1})\|$$

$$= \max_{j \in \Lambda}\|(P_j - P_{j-1})\psi\|^2 < \frac{\varepsilon}{3}.$$

But then, using (12a) and (12b), we may conclude that

$$\| \operatorname{diag}(A_\varphi; \mathbb{P}) - \operatorname{diag}(A_\varphi; \pi)\| \leq 2\|A_\varphi - A_\psi\| + \frac{\varepsilon}{3} < \varepsilon.$$

The theorem is proved. $\square$

COROLLARY 4.5. *The diagonal of a compact operator with respect to a continuous chain on H exists and is equal to the zero operator.*

PROOF. A continuous chain is closed and has no jumps. So the statement is clear from formula (10). $\square$

COROLLARY 4.6. *The diagonal of a compact operator A with respect to a maximal invariant chain $\mathbb{P}$ on H exists and is equal to*

$$(13) \qquad\qquad \sum_j \lambda_j(P_j^+ - P_j^-),$$

where $\lambda_1, \lambda_2, \ldots$ are the nonzero eigenvalues of A repeated according to their algebraic multiplicity, $|\lambda_1| \geq |\lambda_2| \geq \cdots$ and for each j the pair (P_j^-, P_j^+) is a (one dimensional) jump of $\mathbb{P}$.

PROOF. Since $\mathbb{P}$ is maximal, the chain $\mathbb{P}$ is closed, and hence Theorem 4.4 guarantees that A has a diagonal, D say, with respect to $\mathbb{P}$. We know that D is given by the right hand side of (10). By Proposition 4.1 every jump in $\mathbb{P}$ has dimension one. Therefore

$$(P_\nu^+ - P_\nu^-)A(P_\nu^+ - P_\nu^-) = \alpha_\nu(P_\nu^+ - P_\nu^-)$$

for each ν. The sum in the right hand side of (10) does not depend on the way the jumps are ordered. So we may write

$$D = \sum_j \alpha_j(P_j^+ - P_j^-),$$

where $|\alpha_1| \geq |\alpha_2| \geq \cdots > 0$ and (P_j^-, P_j^+) is a one dimensional jump in $\mathbb{P}$ for each j. Relative to the decomposition $\operatorname{Im} P_j^+ = \operatorname{Im} P_j^- \oplus \operatorname{Im}(P_j^+ - P_j^-)$ the operator $A|\operatorname{Im} P_j^+$ admits the following partitioning:

$$A|\operatorname{Im} P_j^+ = \left[\begin{array}{cc} A|\operatorname{Im} P_j^- & 0 \\ * & \alpha \end{array} \right].$$

It follows that α_j is in the spectrum of $A|\operatorname{Im} P_j^+$. The compactness of A implies that α_j is an eigenvalue of A.

Let $\lambda \neq 0$ be a non-zero eigenvalue of A. It remains to show that

$$(14) \qquad m(\lambda; A) = \#\{j \mid \alpha_j = \lambda\}.$$

Here $m(\lambda; T)$ denotes the algebraic multiplicity of λ as an eigenvalue of T. Given a partition π of $\mathbb{P}$, we write D_π for $\operatorname{diag}(A; \pi)$. Since $\mathbb{P}$ is invariant under A, the point λ is also an eigenvalue of D_π and $m(\lambda; A) = m(\lambda; D_\pi)$ for each π. From

$$(15) \qquad D = \lim_{\pi \subset \mathbb{P}} D_\pi$$

and Theorem II.4.1 we conclude that λ belongs to the spectrum of D. Thus $\lambda = \alpha_j$ for some j. According to Theorem II.4.2, formula (15) also implies that $m(\lambda, D) = m(\lambda, D_\pi)$ for some π. Since $m(\lambda; D_\pi) = m(\lambda; A)$ for all π, the corollary is proved. $\square$

XX.5 TRIANGULAR ALGEBRAS

In this section $\mathbb{P}$ is a chain on a Banach space X. Recall that $A(\mathbb{P})$ denotes the set of all operators $A \in \mathcal{L}(X)$ such that A has a diagonal with respect to $\mathbb{P}$. We already know that $A(\mathbb{P})$ is a linear subset of $\mathcal{L}(X)$ which is closed whenever $\mathbb{P}$ is a uniform chain (cf. Proposition 3.1). We consider the following three subsets of $A(\mathbb{P})$:

$$A_+(\mathbb{P}) = \left\{ A \in A(\mathbb{P}) \mid AP = PAP \; (P \in \mathbb{P}), \int_{\mathbb{P}} (dP)A(dP) = 0 \right\};$$

$$A_-(\mathbb{P}) = \left\{ A \in A(\mathbb{P}) \mid PA = PAP \; (P \in \mathbb{P}), \int_{\mathbb{P}} (dP)A(dP) = 0 \right\};$$

$$A_0(\mathbb{P}) = \{ A \in A(\mathbb{P}) \mid AP = PA \; (P \in \mathbb{P}) \}.$$

Note that $A_+(\mathbb{P})$ consists of all operators in $A(\mathbb{P})$ that leave invariant the chain $\mathbb{P}$ and have a zero diagonal. Similarly, $A_-(\mathbb{P})$ consists of all operators A in $A(\mathbb{P})$ that leave invariant the complementary chain $\mathbb{P}^c$ and have a zero diagonal. If $AP = PA$ for all $P \in \mathbb{P}$ or, equivalently, if A is reduced by $\mathbb{P}$, then A has a diagonal with respect to $\mathbb{P}$, and hence $A_0(\mathbb{P})$ consists of all $A \in \mathcal{L}(X)$ that are reduced by $\mathbb{P}$. Furthermore

$$\int_{\mathbb{P}} (dP)A(dP) = A, \qquad A \in A_0(\mathbb{P}).$$

PROPOSITION 5.1. *If $\mathbb{P}$ is a uniform chain, then $A_+(\mathbb{P})$, $A_-(\mathbb{P})$ and $A_0(\mathbb{P})$ are closed subalgebras of $\mathcal{L}(X)$, and $A_+(\mathbb{P})$ and $A_-(\mathbb{P})$ consist of quasi-nilpotent operators, i.e., of operators of which the spectrum consists of the zero element only. Furthermore, if $A_\pm \in A_\pm(\mathbb{P})$, then*

$$(I + A_\pm)^{-1} - I \in A_\pm(\mathbb{P}).$$

PROOF. The fact that $\mathbb{P}$ is uniform implies that $A(\mathbb{P})$ is closed (Proposition 3.1). Furthermore, the map $A \mapsto \int(dP)A(dP)$ is continuous. Since $A \mapsto (AP - PAP)$ is also a continuous map, it is clear that $A_+(\mathbb{P})$ is closed. It is easy to check that $A_+(\mathbb{P})$ is a linear subset of $\mathcal{L}(X)$.

Take $A_1, A_2 \in A_+(\mathbb{P})$. Note that

$$A_1 A_2 P = A_1 P A_2 P = P A_1 P A_2 P = P A_1 A_2 P.$$

Let $\pi = \{P_0, P_1, \ldots, P_n\}$ be a partition of $\mathbb{P}$. With respect to the decomposition

$$(1) \qquad X = \operatorname{Im} \Delta P_1 \oplus \operatorname{Im} \Delta P_2 \oplus \cdots \oplus \operatorname{Im} \Delta P_n$$

the elements of $A_+(\mathbb{P})$ are block upper triangular matrices. It follows that

$$\operatorname{diag}(A_1 A_2; \pi) = \operatorname{diag}(A_1; \pi) \cdot \operatorname{diag}(A_2; \pi).$$

By taking limits over π we see that

$$\int_{\mathbb{P}} (dP)A_1 A_2(dP) = \left(\int_{\mathbb{P}} (dP)A_1(dP)\right)\left(\int_{\mathbb{P}} (dP)A_2(dP)\right) = 0.$$

Thus $A_1 A_2 \in A_+(\mathbb{P})$ and $A_+(\mathbb{P})$ is a subalgebra of $\mathcal{L}(X)$.

Take $A \in A_+(\mathbb{P})$, and let $\pi = \{P_0, P_1, \ldots, P_n\}$ be a partition of $\mathbb{P}$. Write $A = U_\pi + D_\pi$, where $D_\pi = \operatorname{diag}(A, \pi)$. With respect to the decomposition (1) the operator U_π is a block upper triangular operator matrix with zeros on the main diagonal. Further, $D_\pi \to 0$. Fix $\lambda \neq 0$, and take π such that $\|D_\pi\| < |\lambda|$. Then $\lambda I - D_\pi$ is invertible. Note that $(\lambda I - D_\pi)^{-1} U_\pi$ is again a block upper triangular operator with zeros on the main diagonal. In particular, $(\lambda I - D_\pi)^{-1} U_\pi$ is nilpotent, and hence $I - (\lambda I - D_\pi)^{-1} U_\pi$ is invertible. It follows that

$$\lambda I - A = (\lambda I - D_\pi)[I - (\lambda I - D_\pi)^{-1} U_\pi]$$

is invertible. This shows that A is quasi-nilpotent. Furthermore, by the results just proved,

$$(I + A)^{-1} - I = \sum_{n=1}^{\infty} (-1)^n A^n \in A_+(\mathbb{P}).$$

So for $A_+(\mathbb{P})$ the theorem is proved. The statements for $A_-(\mathbb{P})$ and $A_0(\mathbb{P})$ can be proved in a similar way. $\square$

XX.6 RIEMANN-STIELTJES INTEGRATION ALONG CHAINS

Let $\mathbb{P}$ be a chain on a Banach space X, and let $F\colon\mathbb{P} \to \mathcal{L}(X)$ be a given operator function. We shall define the following integrals

$$(1) \qquad \int_{\mathbb{P}} F(P)dP, \qquad \int_{\mathbb{P}} (dP)F(P), \qquad \int_{\mathbb{P}} (dP)F(P)dP.$$

Let $\pi = \{P_0, P_1, \ldots, P_n\}$ be a partition of $\mathbb{P}$, and let $\tau = \{Q_1, Q_2, \ldots, Q_n\}$ be a set of projections in $\mathbb{P}$ such that

$$(2) \qquad P_{j-1} \le Q_j \le P_j, \qquad j = 1, \ldots, n.$$

For π and τ related in this way one considers the Riemann-Stieltjes sum

$$(3) \qquad S_\tau(F; \pi) = \sum_{j=1}^{n} F(Q_j)(P_j - P_{j-1}).$$

We say that the first integral in (1) converges if for some operator $T \in \mathcal{L}(X)$ we have

$$(4) \qquad T = \lim_{\pi \subset \mathbb{P}} S_\tau(F; \pi),$$

that is, given $\varepsilon > 0$ there exists a partition π_ε such that

$$\|T - S_\tau(F, \pi)\| < \varepsilon$$

for any partition π of $\mathbb{P}$ finer than π_ε and every choice of τ (with π and τ related as in (2)). When the limit (4) exists, it is determined uniquely and we write

$$T = \int_{\mathbb{P}} F(P)dP.$$

The integrals $\int_{\mathbb{P}}(dP)F(P)$ and $\int_{\mathbb{P}}(dP)F(P)dP$ are defined in a similar way. They appear, respectively, as the limits of the following Riemann Stieltjes sums:

$$(5a) \qquad S_\tau(\pi; F) = \sum_{j=1}^{n} (P_j - P_{j-1})F(Q_j),$$

$$(5b) \qquad S_{\tau,\pi}(F) = \sum_{j=1}^{n} (P_j - P_{j-1})F(Q_j)(P_j - P_{j-1}).$$

Note that the definition of $\int_{\mathbb{P}}(dP)F(P)dP$ agrees with the way we have introduced the expression $\int_{\mathbb{P}}(dP)AdP$.

In the integrals in (1) we replace $\int_{\mathbb{P}}$ by $\int_{[\mathbb{P}}$ or $\int_{\mathbb{P}]}$ if there is an additional restriction on the choice of the projections Q_j. We shall use the symbol $\int_{[\mathbb{P}}$ if $Q_j = P_{j-1}$ $(j = 1, \ldots, n)$ and $\int_{\mathbb{P}]}$ if $Q_j = P_j$ $(j = 1, \ldots, n)$. So, for example,

$$(6a) \qquad \int_{[\mathbb{P}} F(P)dP = \lim_{\pi \subset \mathbb{P}} \sum_{j=1}^{n} F(P_{j-1})(P_j - P_{j-1}),$$

$$(6b) \qquad \int_{\mathbb{P}]} F(P)dP = \lim_{\pi \subset \mathbb{P}} \sum_{j=1}^{n} F(P_j)(P_j - P_{j-1}).$$

If the integral $\int_{\mathbb{P}} F(P)dP$ converges, then the same is true for the integrals $\int_{[\mathbb{P}} F(P)dP$ and $\int_{\mathbb{P}]} F(P)dP$ and in that case all three integrals are equal.

As usual there are Cauchy conditions which guarantee the convergence of the various integrals. For example, the integral $\int_{\mathbb{P}} F(P)dP$ converges if and only if for every $\delta > 0$ there exists a partition π_δ such that

$$(7) \qquad \|S_{\tau_1}(F, \pi_1) - S_{\tau_2}(F, \pi_2)\| < \delta$$

for all partitions π_1 and π_2 of $\mathbb{P}$ finer than π_δ. Note that in (7) we are free in the choice of the sets τ_1 and τ_2 as long as for $i = 1, 2$ the projections in π_i and τ_i are related as in (2).

PROPOSITION 6.1. *If the integral $\int_{\mathbb{P}} F(P)dP$ exists, then necessarily*

$$(8) \qquad [F(P_+) - F(P_-)](P_+ - P_-) = 0$$

for every jump (P_-, P_+) of $\mathbb{P}$. If the chain $\mathbb{P}$ is finite, then this condition is also sufficient for the existence of the integral.

PROOF. Assume the integral $\int_{\mathbb{P}} F(P)dP$ exists, and let (P_-, P_+) be a jump of $\mathbb{P}$. Take $\delta > 0$ and let $\pi_\delta = \{P_0, \ldots, P_n\}$ be as in the Cauchy condition mentioned above. Without loss of generality we may assume that P_-, P_+ are two consecutive projections in π_δ. Let us say that $P_- = P_{k-1}$ and $P_+ = P_k$. For $\nu = 1, 2$ let $\tau_\nu = \{Q_{\nu 1}, \ldots, Q_{\nu n}\}$ be a subset of $\mathbb{P}$ for which the relation (2) holds. We assume that $Q_{1j} = Q_{2j}$ for $j \neq k$, $Q_{1k} = P_+$ and $Q_{2k} = P_-$. Then

$$S_{\tau_1}(F, \pi_\delta) - S_{\tau_2}(F, \pi_\delta) = [F(P_+) - F(P_-)](P_+ - P_-).$$

So by the Cauchy condition $\|[F(P_+) - F(P_-)](P_+ - P_-)\| < \delta$. Since $\delta > 0$ is arbitrary we have proved (8). The second part of the lemma is trivial. $\square$

With appropriate modifications (interchange in the left hand side of (8) the order of the factors) Proposition 6.1 also holds true for the integral $\int_{\mathbb{P}} (dP)F(P)$.

We conclude this section with two lemmas which we shall use in Section XXI.1. In both lemmas $\mathbb{P}$ is a Hilbert space chain.

LEMMA 6.2. *Let $\mathbb{P}$ be a closed chain on a separable Hilbert space H, and let $A\colon H \to H$ be a compact operator. Then $\int_{\mathbb{P}} PAdP$ converges if and only if $\int_{[\mathbb{P}} PAdP$ converges and $(P_+ - P_-)A(P_+ - P_-) = 0$ for every jump (P_-, P_+) in $\mathbb{P}$. In that case*

$$(9) \qquad \int_{[\mathbb{P}} PAdP = \int_{\mathbb{P}} PAdP.$$

PROOF. Assume $\int_{\mathbb{P}} PAdP$ converges. Then $\int_{[\mathbb{P}} PAdP$ converges trivially and obviously (9) holds true. Further we can apply Proposition 6.1 to show that $(P_+ - P_-)A(P_+ - P_-) = 0$ for every jump (P_-, P_+) of $\mathbb{P}$.

To prove the converse statement, assume that $\int_{[\mathbb{P}} PAdP$ converges and $(P_+ - P_-)A(P_+ - P_-) = 0$ for every jump (P_-, P_+) of $\mathbb{P}$. Let $\pi = \{P_0, \ldots, P_n\}$ be a partition of $\mathbb{P}$. Take $Q_j \in \mathbb{P}$ such that $P_{j-1} \le Q_j \le P_j$, $j = 1, \ldots, n$. We have to prove that

$$(10) \qquad \lim_{\pi \subset \mathbb{P}} \sum_{k=1}^{n} Q_k A(P_k - P_{k-1})$$

exists. Since $\int_{[\mathbb{P}} PAdP$ converges, we know that the limit (10) exists whenever we take $Q_k = P_{k-1}$. Hence, it suffices to show that

$$(11) \qquad \lim_{\pi \subset \mathbb{P}} \sum_{k=1}^{n} (Q_k - P_{k-1})A(P_k - P_{k-1})$$

exists.

To do this take $\varphi \in H$. We have

$$\left\| \sum_{k=1}^{n} (Q_k - P_{k-1})A(P_k - P_{k-1})\varphi \right\|^2 = \sum_{k=1}^{n} \|(Q_k - P_{k-1})A(P_k - P_{k-1})\varphi\|^2$$

$$\le \sum_{k=1}^{n} \|(P_k - P_{k-1})A(P_k - P_{k-1})\varphi\|^2$$

$$= \left\| \sum_{k=1}^{n} (P_k - P_{k-1})A(P_k - P_{k-1})\varphi \right\|^2.$$

Thus

$$(12) \qquad \left\| \sum_{k=1}^{n} (Q_k - P_{k-1})A(P_k - P_{k-1}) \right\| \le \| \operatorname{diag}(A; \pi) \|.$$

Since $(P_+ - P_-)A(P_+ - P_-) = 0$ for every jump (P_-, P_+) of $\mathbb{P}$, we can apply Theorem 4.4 to show that the diagonal of A is the zero operator. Thus $\lim_{\pi \subset \mathbb{P}} \operatorname{diag}(A; \pi) = 0$. But then we see from (12) that the limit (11) exists (and is equal to zero). $\square$

LEMMA 6.3. (*Formula of partial integration*). *Let* $\mathbb{P}$ *be a closed chain on a separable Hilbert space* H, *and let* $A\colon H \to H$ *be a compact operator. Then* $\int_{\mathbb{P}} PA\,dP$ *converges if and only if* $\int_{\mathbb{P}}(dP)AP$ *converges, and in that case*

$$(13) \qquad A = \int_{\mathbb{P}} PA\,dP + \int_{\mathbb{P}}(dP)AP.$$

PROOF. For a partition $\pi = \{P_0, \ldots, P_n\}$ of $\mathbb{P}$ we have

$$\begin{aligned}
A &= \sum_{j,k}(\Delta P_j)A(\Delta P_k) \\
&= \sum_{j<k}(\Delta P_j)A(\Delta P_k) + \sum_{j\geq k}(\Delta P_j)A(\Delta P_k) \\
&= \sum_{k=1}^{n} P_{k-1}A(\Delta P_k) + \sum_{j=1}^{n}(\Delta P_j)AP_j.
\end{aligned}$$

Assume that $\int_{\mathbb{P}}(dP)AP$ converges. From the previous identities it follows that $\int_{[\mathbb{P}} PA(dP)$ converges and

$$(14) \qquad \int_{[\mathbb{P}} PA(dP) = A - \int_{\mathbb{P}}(dP)AP.$$

The fact that $\int_{\mathbb{P}}(dP)AP$ converges also implies (cf. Proposition 6.1) that

$$(P_+ - P_-)A(P_+ - P_-) = 0$$

for each jump (P_-, P_+) in $\mathbb{P}$. But then we can apply the preceding lemma to show that $\int_{\mathbb{P}} PA(dP)$ converges, and hence we see from (9) and (14) that the identity (13) holds true.

In a similar way one proves that the convergence of the integral $\int_{\mathbb{P}} PA(dP)$ implies that $\int_{\mathbb{P}}(dP)AP$ is convergent. $\square$

XX.7 ADDITIVE LOWER-UPPER DECOMPOSITION THEOREM

In this section $\mathbb{P}$ is a uniform chain on a Banach space X. Given an operator $A \in \mathcal{L}(X)$ we call

$$(1) \qquad A = A_- + A_0 + A_+$$

an *additive LU-decomposition* of A with respect to the chain $\mathbb{P}$ if $A_\alpha \in A_\alpha(\mathbb{P})$ for $\alpha = -, 0, +$. Here $A_-(\mathbb{P})$, $A_0(\mathbb{P})$ and $A_+(\mathbb{P})$ are the triangular algebras introduced in Section XX.5.

THEOREM 7.1. *Let $\mathbb{P}$ be a uniform chain. If $A = A_- + A_0 + A_+$ is an additive LU-decomposition of A with respect to the uniform chain $\mathbb{P}$, then*

$$(2) \qquad A_- = \int_{[\mathbb{P}} (dP)AP, \qquad A_0 = \int_{\mathbb{P}} (dP)A(dP), \qquad A_+ = \int_{[\mathbb{P}} PAdP.$$

Conversely, if in (2) any two of the three integrals converge, then the third is convergent and A admits an additive LU-decomposition with respect to $\mathbb{P}$.

PROOF. The following identities hold true:

$$(3a) \qquad \int_{[\mathbb{P}} (dP)TP = \begin{cases} T & \text{for} & T \in A_-(\mathbb{P}), \\ 0 & \text{for} & T \in A_0(\mathbb{P}), \\ 0 & \text{for} & T \in A_+(\mathbb{P}); \end{cases}$$

$$(3b) \qquad \int_{\mathbb{P}} (dP)T(dP) = \begin{cases} 0 & \text{for} & T \in A_-(\mathbb{P}), \\ T & \text{for} & T \in A_0(\mathbb{P}), \\ 0 & \text{for} & T \in A_+(\mathbb{P}); \end{cases}$$

$$(3c) \qquad \int_{[\mathbb{P}} PTdP = \begin{cases} 0 & \text{for} & T \in A_-(\mathbb{P}), \\ 0 & \text{for} & T \in A_0(\mathbb{P}), \\ T & \text{for} & T \in A_+(\mathbb{P}). \end{cases}$$

To see this, let us prove (3a). Let $\pi = \{P_0, \ldots, P_n\}$ be a partition of $\mathbb{P}$. By definition

$$(4) \qquad \int_{[\mathbb{P}} (dP)TP = \lim_{\pi \subset \mathbb{P}} \sum_{j=1}^{n} (\Delta P_j)TP_{j-1}.$$

Take $T \in A_-(\mathbb{P})$. Then $T = \sum_{j=1}^{n} (\Delta P_j)TP_{j-1} + \operatorname{diag}(T;\pi)$ (cf., Theorem 1.1). Further, we know that $\operatorname{diag}(T, \pi) \to 0$. So in this case the integral in (4) converges and is equal to T. If $T \in A_0(\mathbb{P})$ or $A_+(\mathbb{P})$, then

$$\sum_{j=1}^{n} (\Delta P_j)TP_{j-1} = 0,$$

and hence $\int_{[\mathbb{P}} (dP)TP$ is equal to the zero operator too. This proves (3a). The identity (3c) is proved in a similar way and (3b) follows from the definition of the triangular algebras.

Now, assume $A = A_- + A_0 + A_+$ is an additive LU-decomposition. Using (3a) one sees that the integral $\int_{[\mathbb{P}} (dP)AP$ exists and is equal to A_-. In a similar way, using (3b) and (3c), one proves the second and third identity in (2).

To prove the second part of the theorem, let $\pi = \{P_0, P_1, \ldots, P_n\}$ be a partition of $\mathbb{P}$, and let

$$(5) \qquad A = L_\pi + \operatorname{diag}(A; \pi) + U_\pi$$

be the additive LU-decomposition of A with respect to the finite chain π (see Section XX.1). We know (see Theorem 1.1) that

$$L_\pi = \sum_{j=1}^{n} (\Delta P_j) A P_{j-1}, \qquad U_\pi = \sum_{j-1}^{n} P_{j-1} A (\Delta P_j).$$

Consider the limits

$$(6) \qquad \lim_{\pi \subset \mathbb{P}} L_\pi, \qquad \lim_{\pi \subset \mathbb{P}} \operatorname{diag}(A; \pi), \qquad \lim_{\pi \subset \mathbb{P}} U_\pi.$$

From (5) it is clear that the three limits in (6) exist, whenever any two of them exist. Observe that L_π is a Riemann-Stieltjes sum for the first integral in (2), $\operatorname{diag}(A; \pi)$ is an arbitrary Riemann-Stieltjes sum for the second integral in (2) and U_π is an arbitrary Riemann-Stieltjes sum for the third integral in (2). It follows that in (2) all three integrals converge whenever two of them converge.

Assume all three integrals in (2) converge, and let A_-, A_0 and A_+ be as in (2). From (5) and the remarks made in the previous paragraph it is clear that $A = A_- + A_0 + A_+$. It remains to prove that $A_\alpha \in A_\alpha(\mathbb{P})$ for $\alpha = -, 0, +$.

Consider the operator A_-. First we show that $PA_- = PA_-P$ for each $P \in \mathbb{P}$. Fix $P \in \mathbb{P}$, and let π_0 be a partition of $\mathbb{P}$ with $P \in \pi_0$. Then $PL_\pi = PL_\pi P$ for any partition finer than π_0. Recall that $L_\pi \to A_-$. Thus by taking limits we see that $PA_- = PA_-P$.

Next we show that the diagonal of A_- exists and is equal to the zero operator. Take $\varepsilon > 0$. Then there exists a partition π_ε of $\mathbb{P}$ such that $\|A - L_\pi\| < \varepsilon$ for every partition π of $\mathbb{P}$ finer than π_ε. Let $\pi = \{P_0, P_1, \ldots, P_n\}$ be such a partition. Then $\sum_{j=1}^{n} \Delta P_j L_\pi \Delta P_j = 0$. It follows that

$$\left\| \sum_{j=1}^{n} (\Delta P_j) A_- (\Delta P_j) \right\| = \left\| \sum_{j=1}^{n} (\Delta P_j)(A_- - L_\pi) \Delta P_j \right\|$$
$$\leq \gamma \| A - L_\pi \|,$$

where γ is a certain positive constant (cf. formula (6) in Section XX.3). We conclude that $\| \operatorname{diag}(A_-, \pi) \| < \gamma \varepsilon$ for any partition π of $\mathbb{P}$ finer than π_ε. But then we see that the diagonal of A_- exists and is equal to zero. We have proved that $A_- \in A_-(\mathbb{P})$.

Next consider A_0. According to our hypothesis $\operatorname{diag}(A; \pi) \to A_0$. Let P be an element of $\mathbb{P}$. Let π_0 be a partition of $\mathbb{P}$ such that $P \in \pi_0$. For any partition π of $\mathbb{P}$ finer than π_0 we have

$$\operatorname{diag}(A; \pi) P = P \operatorname{diag}(A; \pi).$$

By taking limits we see that $A_0 P = P A_0$. Thus $\mathbb{P}$ is a reducing chain for A_0. It follows that A_0 has a diagonal and $A_0 \in A_0(\mathbb{P})$. To prove that $A_+ \in A_+(\mathbb{P})$ one can use similar arguments as for the corresponding statement about A_-. $\square$

XX.8 ADDITIVE LOWER-UPPER DECOMPOSITION OF A HILBERT-SCHMIDT OPERATOR

In this section $\mathbb{P}$ is a chain of orthogonal projections on a separable Hilbert space H.

THEOREM 8.1. *For a Hilbert-Schmidt operator A the integrals*

$$(1) \qquad \int_{[\mathbb{P}} (dP)AP, \qquad \int_{[\mathbb{P}} PAdP$$

converge in the S_2-norm.

PROOF. To prove the convergence of the integrals in (1) we use the fact that S_2 is a Hilbert space in its own right with $\langle A, B \rangle = \operatorname{tr} AB^*$ as inner product (see Section VIII.2). As usual the norm on S_2 is denoted by $\| \cdot \|_2$.

Let $\pi = \{P_0, P_1, \ldots, P_n\}$ be a partition of $\mathbb{P}$. Put

$$L(\pi) = \sum_{j=1}^{n}(\Delta P_j)AP_{j-1} = \sum_{k<j}(P_j - P_{j-1})A(P_k - P_{k-1}).$$

First note that

$$\begin{aligned}
\langle (P_j - P_{j-1})A(P_k - P_{k-1}), (P_r - P_{r-1})A(P_s - P_{s-1}) \rangle \\
= \operatorname{tr}[(P_j - P_{j-1})A(P_k - P_{k-1})(P_s - P_{s-1})A^*(P_r - P_{r-1})] \\
= \operatorname{tr}[A(P_k - P_{k-1})(P_s - P_{s-1})A^*(P_r - P_{r-1})(P_j - P_{j-1})] \\
= \delta_{jr}\delta_{ks} \operatorname{tr}[A(P_k - P_{k-1})A^*(P_r - P_{r-1})].
\end{aligned}$$

In other words, the operators $(P_j - P_{j-1})A(P_k - P_{k-1})$, $j,k = 1,\ldots,n$, are mutually orthogonal elements in S_2. It follows that

$$\begin{aligned}
\|L(\pi)\|_2^2 &= \sum_{k<j} \|(P_j - P_{j-1})A(P_k - P_{k-1})\|_2^2 \\
&\leq \sum_{j,k=1}^{n} \|(P_j - P_{j-1})A(P_k - P_{k-1})\|_2^2 \\
&= \|A\|_2^2.
\end{aligned}$$

Further, $\|L(\pi_2)\|_2 \geq \|(L(\pi_1)\|_2$ whenever $\pi_1 \subset \pi_2$, because

$$\|L(\pi_2) - L(\pi_1)\|_2^2 = \|L(\pi_2)\|_2^2 - \|L(\pi_1)\|_2^2$$

for π_2 finer than π_1.

Let $m = \sup \|L(\pi)\|_2$, where the supremum is taken over all partitions π of $\mathbb{P}$. Take $\varepsilon > 0$, and choose a partition π_ε of $\mathbb{P}$ such that

$$\|L(\pi_\varepsilon)\|_2^2 > m^2 - \frac{\varepsilon^2}{4}.$$

Let π_1 and π_2 be two partitions of $\mathbb{P}$ finer than π_ε. We have

$$\|L(\pi_1) - L(\pi_2)\|_2^2 \leq 2\big(\|L(\pi_1) - L(\pi_\varepsilon)\|_2^2 + \|L(\pi_2) - L(\pi_\varepsilon)\|_2^2\big)$$
$$= 2\big(\|L(\pi_1)\|_2^2 + \|L(\pi_2)\|_2^2 - 2\|L(\pi_\varepsilon)\|_2^2\big)$$
$$\leq 4(m^2 - \|L(\pi_\varepsilon)\|_2^2) < \varepsilon^2.$$

It follows that $\lim_{\pi \subset \mathbb{P}} L(\pi)$ exists in S_2-norm.

Since $L(\pi)$ is an arbitrary Riemann-Stieltjes sum for the first integral in (1) it follows that this integral converges in the S_2-norm. The convergence in the S_2-norm of the second integral in (1) is proved in a similar way. $\square$

COROLLARY 8.2. *If A is a Hilbert-Schmidt operator, then*

(2)
$$A = \int_{[\mathbb{P}} (dP)AP + \int_{\mathbb{P}} (dP)A(dP) + \int_{[\mathbb{P}} PAdP$$

and this decomposition is the additive LU-decomposition of A with respect to $\mathbb{P}$ with terms that are Hilbert-Schmidt operators.

PROOF. Since the S_2-norm dominates the usual operator norm, it follows from Theorem 8.1 that the integrals in (1) also converge in the usual operator norm. According to Theorem 7.1 this implies that the diagonal $\int_{\mathbb{P}}(dP)A(dP)$ exists. Moreover (2) is the additive LU-decomposition of A with respect to $\mathbb{P}$.

It remains to prove that the factors are in S_2. For $\int_{[\mathbb{P}}(dP)AP$ and $\int_{[\mathbb{P}} PAdP$ this is clear from the previous theorem. Since A is Hilbert-Schmidt, we see from (2) that the diagonal term also must be in S_2. The proof is complete. $\square$

COROLLARY 8.3. *The maps $J_-, J_0, J_+ : S_2 \to S_2$ defined by*

$$J_-(A) = \int_{[\mathbb{P}} (dP)AP, \qquad J_0(A) = \int_{\mathbb{P}} (dP)A(dP), \qquad J_+(A) = \int_{[\mathbb{P}} PAdP$$

are mutually disjoint orthogonal projections.

PROOF. Clearly, J_-, J_0 and J_+ are linear operators. We know that $J_-(A) \in A_-(\mathbb{P})$. Hence (see formulas (3a), (3b) and (3c) in the previous section)

$$J_-\big(J_-(A)\big) = J_-(A), \quad J_0\big(J_-(A)\big) = 0, \quad J_+\big(J_-(A)\big) = 0.$$

Similar identities hold true for $J_0(A)$ and $J_+(A)$. It follows that J_-, J_0 and J_+ are mutually disjoint projections on S_2. To see that J_-, J_0 and J_+ are orthogonal, we go back to the LU-decomposition of A with respect to a finite subchain $\pi = \{P_0, P_1, \ldots, P_n\}$ of $\mathbb{P}$:

$$A = L(\pi) + \mathrm{diag}(A;\pi) + U(\pi).$$

Since the elements $(P_j - P_{j-1})A(P_k - P_{k-1})$, $j,k = 1,\ldots,n$, are mutually orthogonal in S_2, it is clear that $L(\pi)$, $\mathrm{diag}(A;\pi)$ and $U(\pi)$ are mutually orthogonal too. But then this is also true for their limits. Thus $J_-(A)$, $J_0(A)$ and $J_+(A)$ are orthogonal elements in S_2. It follows that J_-, J_0 and J_+ are orthogonal projections. $\square$

XX.9 MULTIPLICATIVE INTEGRATION ALONG CHAINS

Let $\mathbb{P}$ be a chain on a Banach space X, and let $F\colon\mathbb{P} \to \mathcal{L}(X)$ be a given operator function. We define the following *multiplicative integrals*:

$$
(1) \qquad
\overset{\frown}{\int_{\mathbf{P}}} (I + F(P)dP), \qquad
\overset{\frown}{\int_{\mathbf{P}}} ((dP)F(P) + I),
$$

$$
\underset{\mathbf{P}}{\overset{\frown}{\int}} (I + F(P)dP), \qquad
\underset{\mathbf{P}}{\overset{\frown}{\int}} ((dP)F(P) + I).
$$

Let $\pi = \{P_0, P_1, \ldots, P_n\}$ be a partition of $\mathbb{P}$, and $\tau = \{Q_1, Q_2, \ldots, Q_n\}$ be a set of projections in $\mathbb{P}$ such that

$$
(2) \qquad\qquad P_{j-1} \le Q_j \le P_j, \qquad j = 1, \ldots, n.
$$

For π and τ related in this way, one considers the product

$$
(3) \qquad \overset{\frown}{\prod_{j=1}^{n}} (F;\pi,\tau) = \big(I + F(Q_n)\Delta P_n\big)\big(I + F(Q_{n-1})\Delta P_{n-1}\big) \cdots \big(I + F(Q_1)\Delta P_1\big),
$$

where $\Delta P_j = P_j - P_{j-1}$, $j = 1, \ldots, n$. We say that the first integral in (1) *converges* if for some operator $T \in \mathcal{L}(X)$ we have

$$
(4) \qquad T = \lim_{\pi \subset \mathbb{P}} \overset{\frown}{\prod_{j=1}^{n}} (F;\pi,\tau).
$$

The definition of this limit is the same as the one given in Section XX.6 with $S_\tau(F;\pi)$ replaced by the left hand side of (3). When the limit (4) exists, it is determined uniquely, and we write

$$
T = \overset{\frown}{\int_{\mathbf{P}}} (I + F(P)dP).
$$

The remaining integrals in (1) are defined in a similar way. They appear, respectively, as limits of the following products:

$$
(5) \qquad \overset{\frown}{\prod_{j=1}^{n}} (\pi,\tau;F) = \big((\Delta P_n)F(Q_n) + I\big)\big((\Delta P_{n-1})F(Q_{n-1}) + I\big) \cdots \big((\Delta P_1)F(Q_1) + I\big),
$$

$$
(6) \qquad \underset{j=1}{\overset{\frown}{\prod}}{}^{n} (F;\pi,\tau) = \big(I + F(Q_1)\Delta P_1\big)\big(I + F(Q_2)\Delta P_2\big) \cdots \big(I + F(Q_n)\Delta P_n\big),
$$

$$(7) \qquad \overset{\frown}{\underset{j=1}{\prod^{n}}}\, (\pi, \tau; F) = \big(I + (\Delta P_1)F(Q_1)\big)\big(I + (\Delta P_2)F(Q_2)\big) \cdots \big(I + (\Delta P_n)F(Q_n)\big).$$

If in (3), (5)–(7) we take $Q_j = P_{j-1}$, $j = 1, \ldots, n$, and the corresponding products converge, then we replace $\int_{\mathbf{P}}$ by $\int_{[\mathbf{P}}$ in (1). Similarly, if in (3), (5)–(7) we take $Q_j = P_j$, $j = 1, \ldots, n$, and the corresponding products converge, then $\int_{\mathbf{P}}$ is replaced by $\int_{\mathbf{P}]}$.

THEOREM 9.1. *Given a chain* $\mathbb{P}$ *on* X, *let* $G\colon \mathbb{P} \to \mathcal{L}(X)$ *be an operator function with the property that* $PG(P) = G(P)$ *for each* $P \in \mathbb{P}$. *Assume that the integral*

$$C = \int\limits_{\mathbf{P}} G(P)\,dP$$

converges and that the diagonal of C *exists and is equal to zero. Then* C *is in* $A_+(\mathbb{P})$ *and* $(I - C)^{-1}$ *is the multiplicative integral*

$$(I - C)^{-1} = \overset{\frown}{\int\limits_{\mathbf{P}}} \big(I + G(P)\,dP\big).$$

PROOF. Given a partition $\pi = \{P_0, P_1, \ldots, P_n\}$ of $\mathbb{P}$ and $P_{j-1} \le Q_j \le P_j$ with $Q_j \in \mathbb{P}$ for $j = 1, \ldots, n$, put

$$S_\pi = \sum_{j=1}^{n} P_{j-1}G(Q_j)\Delta P_j, \qquad S'_\pi = \sum_{j=1}^{n} G(Q_j)\Delta P_j.$$

To see that C is in $A_+(\mathbb{P})$, we first note that $P_k S'_\pi P_k = S'_\pi P_k$. Indeed,

$$(8) \qquad P_k S'_\pi P_k = \sum_{j=1}^{k} P_j G(Q_j)\Delta P_j = S'_\pi P_k.$$

Here we used that

$$P_j G(Q_j) = P_j Q_j G(Q_j) = Q_j G(Q_j) = G(Q_j),$$

by virtue of our hypotheses on G. Since C is the limit of operators of the form S'_π, we may, without loss of generality, assume that given $P \in \mathbb{P}$, the partition π contains P. Now $\operatorname{Im} P$ is invariant under S'_π if $P \in \pi$, because of (8). It follows that $\operatorname{Im} P$ is invariant under C. Also, the diagonal of C is zero by hypothesis. Hence $C \in A_+(\mathbb{P})$, and therefore $I - C$ is invertible by Proposition 5.1.

Next, note that

$$\|S'_\pi - S_\pi\| = \|\sum_{j=1}^{n} \Delta P_j G(Q_j)\Delta P_j\| = \|\sum_{j=1}^{n} \Delta P_j S'_\pi \Delta P_j\|$$

$$\le \|\sum_{j=1}^{n} \Delta P_j (S'_\pi - C)\Delta P_j\| + \|\sum_{j=1}^{n} \Delta P_j C \Delta P_j\|$$

$$\le \|S'_\pi - C\| + \|\sum_{j=1}^{n} \Delta P_j C \Delta P_j\|.$$

Since S'_π converges in the operator norm to C and the diagonal of C is zero, we have that S_π converges in the operator norm to C. Hence $(I - S_\pi)^{-1}$ converges in the operator norm to $(I - C)^{-1}$. Now $S_\pi^n = 0$. To see this, take $S_j = P_{j-1}G(Q_j)\Delta P_j$, $j = 2, \ldots, n$. Then

$$(9) \qquad S_\pi = \sum_{j=2}^{n} S_j, \quad S_j S_k = 0 \qquad (j \geq k),$$

and an easy induction argument shows that

$$(10) \qquad S_\pi^k = \sum_{2 \leq i_1 < i_2 < \cdots < i_k \leq n} S_{i_1} S_{i_2} \cdots S_{i_k}, \qquad k = 1, \ldots, n-1.$$

Note that for $k = n - 1$, we have $S_{i_k} = S_n$. Thus (9) and (10) imply $S_\pi^n = 0$. Hence

$$(11) \qquad (I - S_\pi)^{-1} = I + S_\pi + S_\pi^2 + \cdots + S_\pi^{n-1}.$$

Since $\lim_\pi \|S'_\pi - S_\pi\| = 0$, it follows that

$$(I - C)^{-1} = \lim_\pi (I - S_\pi)^{-1} = \lim_\pi (I - S_\pi + S'_\pi)(I - S_\pi)^{-1}$$
$$= \lim_\pi \left(I + S'_\pi(I - S_\pi)^{-1}\right).$$

To prove the theorem it suffices to show that

$$(12) \qquad I + S'_\pi(I - S_\pi)^{-1} = \overset{\curvearrowright}{\prod_{j=1}^{n}} \left(I + G(Q_j)\Delta P_j\right).$$

Put $S'_j = G(Q_j)\Delta P_j$, $j = 1, \ldots, n$. Then

$$S'_j S_k = 0 \quad (k \leq j), \qquad S'_j S_k = S'_j S'_k (k > j),$$

and we can apply (10) to show that

$$(13) \qquad \begin{aligned} S'_\pi S_\pi^k &= \left(\sum_{j=1}^{n} S'_j\right) S_\pi^k \\ &= \sum_{1 \leq j_1 < j_2 < \cdots < j_{k+1} \leq n} S'_{j_1} S'_{j_2} \cdots S'_{j_{k+1}}, \qquad k = 1, 2, \ldots, n-1. \end{aligned}$$

Equality (12) now follows from (11) and (13). $\quad\square$

XX.10 BASIC PROPERTIES OF REPRODUCING KERNEL HILBERT SPACES AND CHAINS

Let H be a Hilbert space of complex-valued functions defined on a set Ω. A function $R(\cdot, \cdot)$ defined on $\Omega \times \Omega$ is said to be a *reproducing kernel* for H if for every $t \in \Omega$, the function $R(t, \cdot)$ is in H and

$$(1) \qquad \langle f, R(t, \cdot) \rangle = f(t), \qquad f \in H.$$

The space H is then called a *reproducing kernel Hilbert space* (RKHS).

By considering the vectors in ℓ_2 as functions on the positive integers $\mathbb{Z}_+$, the function $R(m, n) = \delta_{mn}$, defined on $\mathbb{Z}_+ \times \mathbb{Z}_+$, where δ_{mn} is the Kronecker delta, is a reproducing kernel for ℓ_2.

The space $L_2([0, 1])$ is not a RKHS on $[0, 1]$ (because it is not a space of functions), but by applying a unitary operator it may be turned into such a space. Let $\widetilde{H}$ be the space of all complex-valued absolutely continuous functions defined on $[0, 1]$ with $f(0) = 0$ and $f' \in L_2([0, 1])$. Define the inner product on $\widetilde{H}$ by $\langle f, g \rangle = \int_0^1 f'(s)\overline{g'(s)}ds$, and let

$$(2) \qquad R(t, s) = t \wedge s = \min\{t, s\}, \qquad 0 \le t,\ s \le 1.$$

The inner product space $\widetilde{H}$ is complete since the map U defined by

$$(3) \qquad (Uf)(t) = \int\limits_0^t f(s)ds, \qquad 0 \le t \le 1,$$

is a unitary operator mapping $L_2([0, 1])$ onto $\widetilde{H}$. Also

$$\langle f, R(t, \cdot) \rangle = \int\limits_0^t f'(s)ds = f(t), \qquad 0 \le t \le 1,$$

for each $f \in \widetilde{H}$. Thus $\widetilde{H}$ is a RKHS.

A reproducing kernel Hilbert space H has only one reproducing kernel. For if R and R_1 are reproducing kernels for H, then for $t \in \Omega$,

$$\begin{aligned}
\|R(t, \cdot) - R_1(t, \cdot)\|^2 &= \langle R(t, \cdot) - R_1(t, \cdot), R(t, \cdot) - R_1(t, \cdot) \rangle \\
&= \langle R(t, \cdot) - R_1(t, \cdot), R(t, \cdot) \rangle - \langle R(t, \cdot) - R_1(t, \cdot), R_1(t, \cdot) \rangle \\
&= \big[R(t, t) - R_1(t, t)\big] - \big[R(t, t) - R_1(t, t)\big] = 0.
\end{aligned}$$

THEOREM 10.1. *A Hilbert space H of complex-valued functions on a set Ω possesses a reproducing kernel if and only if for each $t \in \Omega$ the functional $F_t(f) = f(t)$ is bounded on H.*

PROOF. Suppose $R(t,s)$ is a reproducing kernel for H. Then

$$(4) \qquad \|R(t,\cdot)\|^2 = \langle R(t,\cdot), R(t,\cdot)\rangle = R(t,t).$$

Hence

$$(5) \qquad |F_t(f)| = |f(t)| = |\langle f, R(t,\cdot)\rangle| \le \|f\|\|R(t,\cdot)\| = \|f\|R(t,t)^{1/2}.$$

Thus F_t is bounded. Now suppose the functional F_t is bounded. Then, by the Riesz representation theorem, there exists a $\varphi_t \in H$ such that $F_t(f) = \langle f, \varphi_t\rangle$, $f \in H$. Define $R(t,s) = \varphi_t(s)$. Then $R(t,\cdot) = \varphi_t$ and

$$\langle f, R(t,\cdot)\rangle = \langle f, \varphi_t\rangle = F_t(f) = f(t), \qquad f \in H,\ t \in \Omega.$$

Thus $R(t,s)$ is a reproducing kernel for H. $\square$

If H is a RKHS and $\|f_n - f\| \to 0$, then $f_n(t) \to f(t)$ for all t. Indeed, by (4),

$$|f_n(t) - f(t)| \le \|f_n - f\|R(t,t)^{1/2} \to 0.$$

The latter inequality also shows that the sequence (f_n) converges uniformly to f on the set of t on which $R(t,t)$ is bounded.

Let H be a RKHS of functions on Ω with reproducing kernel $R(\cdot,\cdot)$. Given $A \in \mathcal{L}(H)$, define $A(t,s) = [A^*R(t,\cdot)](s)$. We call $A(\cdot,\cdot)$ the *defining kernel* for A.

PROPOSITION 10.2. *Let $A(\cdot,\cdot)$ be the defining kernel for $A \in \mathcal{L}(H)$. Then*

$$(6) \qquad A(t,s) = \langle R(t,\cdot), AR(s,\cdot)\rangle, \qquad t,s \in \Omega,$$

$$(7) \qquad \langle f, A(t,\cdot)\rangle = (Af)(t), \qquad t \in \Omega.$$

If $A^*(t,s)$ is the defining kernel for A^*, then

$$(8) \qquad A^*(t,s) = \overline{A(s,t)}, \qquad t,s \in \Omega.$$

For $A,B \in \mathcal{L}(H)$ the defining kernels of the product AB and sum $A+B$ are given by:

$$(9) \qquad (AB)(t,s) = \langle A(t,\cdot), \overline{B(\cdot,s)}\rangle, \qquad t,s \in \Omega,$$

$$(10) \qquad (A+B)(t,s) = A(t,s) + B(t,s), \qquad t,s \in \Omega.$$

If A_n converges in $\mathcal{L}(H)$ to A, then

$$(11) \qquad A_n(t,s) \to A(t,s), \qquad (t,s) \in \Omega \times \Omega.$$

PROOF. Take t,s in Ω. To prove (6), note that

$$A(t,s) = A^*R(t,\cdot)(s) = \langle A^*R(t,\cdot), R(s,\cdot)\rangle = \langle R(t,\cdot), AR(s,\cdot)\rangle.$$

The identity (7) follows from

$$\langle f, A(t, \cdot)\rangle = \langle f, A^* R(t, \cdot)\rangle = \langle Af, R(t, \cdot)\rangle = (Af)(t).$$

Since $A^{**} = A$, we get from (6)

$$A^*(t, s) = \langle R(t, \cdot), A^* R(s, \cdot)\rangle = \overline{\langle A^* R(s, \cdot), R(t, \cdot)\rangle} = \overline{A(s, t)},$$

which proves (8). By (6) and (8)

$$(AB)(t, s) = \langle R(t, \cdot), ABR(s, \cdot)\rangle = \langle A^* R(t, \cdot), BR(s, \cdot)\rangle$$
$$= \langle A(t, \cdot), B^*(s, \cdot)\rangle = \langle A(t, \cdot), \overline{B(\cdot, s)}\rangle.$$

The identity (10) follows directly from (6). To prove (11), note that

$$\|A_n(t, \cdot) - A(t, \cdot)\| = \|(A_n^* - A^*)R(t, \cdot)\| \le \|A_n^* - A^*\|\|R(t, \cdot)\|$$
$$= \|A_n^* - A\|R(t, t)^{1/2} \to 0 \qquad (n \to \infty).$$

The remark in the first paragraph after the proof of Theorem 10.1 now yields (11).

In the remaining part of this section, H is a RKHS of functions defined on an interval $[a, b]$ with reproducing kernel $R(t, s)$. For each $a \le \lambda \le b$ put

$$(12) \qquad R_\lambda = \overline{\operatorname{span}\{R(t, \cdot) \mid a \le t \le \lambda\}}.$$

We shall use the space R_λ to define a chain of orthogonal projections on H.

PROPOSITION 10.3. *Let R_λ be defined by (12), and let P_λ be the orthogonal projection of H onto R_λ. We have:*

(a) *P_b is the identity operator on H;*

(b) *$(P_\lambda f)(t) = f(t)$ on $a \le t \le \lambda$ for all $f \in H$;*

(c) *if $f_1(t) = f_2(t)$ on $a \le t \le \lambda$, where f_1, f_2 are in H, then $P_\lambda f_1 = P_\lambda f_2$;*

(d) *$\|P_\lambda f\| = \min\{\|g\| \mid g(t) = f(t)$ on $a \le t \le \lambda$, $g \in H\}$, where $f \in H$;*

(e) *if $f, g \in H$ and $g(t) = f(t)$ on $a \le t \le \lambda$, then $\|P_\lambda f\| < \|g\|$ whenever $g \ne P_\lambda f$.*

PROOF. If $f \perp R_\lambda$, then

$$0 = \langle f, R(t, \cdot)\rangle = f(t), \qquad a \le t \le \lambda.$$

In particular, $f \perp R_b$ implies that $f = 0$, and thus $P_b = I$, which proves (a). To prove (b), note that

$$(P_\lambda f)(t) = \langle P_\lambda f, R(t, \cdot)\rangle = \langle f, R(t, \cdot)\rangle = f(t), \qquad a \le t \le \lambda.$$

By considering $h = f_1 - f_2$ we see that (c) is an immediate consequence of (b). Next, let $g(t) = f(t)$ on $a \le t \le \lambda$. Then $P_\lambda f = P_\lambda g$ by (c), and so

$$(13) \qquad \|g\|^2 = \|P_\lambda f\|^2 + \|(I - P_\lambda)g\|^2 \ge \|P_\lambda f\|^2,$$

and in the right hand side of (13) we have equality if and only if $g = P_\lambda f$. Hence (d) and (e) hold. $\square$

Let P_λ be the orthogonal projection of H onto the space R_λ. Proposition 10.3 shows that 0 and P_λ, $a \leq \lambda \leq b$, form a chain which we shall call the *chain associated with* H and $R(\cdot,\cdot)$. Let us compute this chain for the RKHS $\widetilde{H}$ introduced in the second paragraph of this section. Note that in this case $[a,b] = [0,1]$. Since $(P_\lambda f)(t) = f(t)$ for $0 \leq t \leq \lambda$, we have

$$\|P_\lambda f\|^2 = \int_0^\lambda |f'(s)|^2 ds + \int_\lambda^1 |(P_\lambda f)'(s)|^2 ds.$$

Now let $\widetilde{f}(t) = f(t)$ on $0 \leq t \leq \lambda$ and $\widetilde{f}(t) = f(\lambda)$ for $\lambda \leq t \leq 1$. Then $\widetilde{f} \in \widetilde{H}$ and

$$(14) \qquad \|\widetilde{f}\|^2 = \int_0^\lambda |f'(t)|^2 dt \leq \|P_\lambda f\|^2.$$

Parts (d) and (e) in Proposition 10.3 and (14) imply that $P_\lambda f = \widetilde{f}$. So we have

$$(15) \qquad (P_\lambda f)(t) = \begin{cases} f(t), & 0 \leq t \leq \lambda, \\ f(\lambda), & \lambda \leq t \leq 1. \end{cases}$$

In this case $P_0 f = 0$, because $f(0) = 0$ for each $f \in \widetilde{H}$. Thus the projections P_λ, $0 \leq \lambda \leq 1$, in (15) define a chain on $\widetilde{H}$. With U as in (3), we have

$$(U^{-1} P_\lambda U f)(t) = \begin{cases} f(t), & 0 \leq t \leq \lambda, \\ 0, & \lambda < t \leq 1, \end{cases}$$

and hence the chain defined by (15) is unitarily equivalent to the usual chain on $L_2([0,1])$.

XX.11 EXAMPLE OF AN ADDITIVE *LU*-DECOMPOSITION IN A RKHS

In this section H denotes the space of absolutely continuous complex-valued functions defined on $[0,1]$ with $f(0) = 0$ and $f' \in L_2([0,1])$. In the previous section we have seen that H is a RKHS with inner product

$$\langle f, g \rangle = \int_0^1 f'(t)\overline{g'(t)}dt,$$

and the reproducing kernel $R(\cdot,\cdot)$ given by

$$R(t,s) = t \wedge s = \min\{t,s\}, \qquad 0 \leq t \leq 1, \ 0 \leq s \leq 1.$$

We shall consider an additive LU-decomposition relative to the chain $\mathbb{P} = \{P_\lambda\}_{0 \le \lambda \le 1}$ associated with the reproducing kernel $R(\cdot, \cdot)$. Thus in the following P_λ is the projection defined by (15) in the previous section.

Let A be a Hilbert-Schmidt operator on H with defining kernel $a(t, s)$. Assume that $a(t, s)$ and the partial derivatives $a_1(t, s) = \frac{\partial}{\partial t} a(t, s)$ and $a_2(t, s) = \frac{\partial}{\partial s} a(t, s)$ are continuous on $[0, 1] \times [0, 1]$. We shall now determine the defining kernels for the upper triangular part A_+ and the lower triangular part A_- of A.

Let $P_\lambda(t, s)$ be the defining kernel for P_λ. Then $P_\lambda(t, s) = [P_\lambda R(t, \cdot)](s) = R(t, s) = t \wedge s$, $s \le \lambda$, $P_\lambda(t, s) = R(t, \lambda) = t \wedge \lambda$, $\lambda \le s$. Hence

$$(1) \qquad P_\lambda(t, s) = \lambda \wedge t \wedge s.$$

Let $(AP_\lambda)(t, s)$ be the defining kernel for AP_λ. By (1) and Proposition (10.2),

$$(2) \qquad \begin{aligned}
(AP_\lambda)(t, s) &= \langle a(t, \cdot), P_\lambda(\cdot, s) \rangle = \int_0^1 \frac{\partial a}{\partial \eta}(t, \eta) \frac{\partial}{\partial \eta}(\lambda \wedge \eta \wedge s) d\eta \\
&= \int_0^{\lambda \wedge s} \frac{\partial a}{\partial \eta}(t, \eta) d\eta = a(t, \lambda \wedge s).
\end{aligned}$$

Hence the defining kernel $(P_\xi A P_\lambda)(t, s)$ for $P_\xi A P_\lambda$ is given by

$$\begin{aligned}
(P_\xi A P_\lambda)(t, s) &= \langle P_\xi(t, \cdot), a(\cdot, \lambda \wedge s) \rangle \\
&= \int_0^1 \frac{\partial}{\partial \eta} P_\xi(t, \eta) \frac{\partial}{\partial \eta} a(\eta, \lambda \wedge s) d\eta \\
&= \int_0^{\xi \wedge t} \frac{\partial}{\partial \eta} a(\eta, \lambda \wedge s) = a(\xi \wedge t, \lambda \wedge s).
\end{aligned}$$

Now Corollary 8.2 implies that $A_- = \int dP_\lambda A P_\lambda$ is a limit in the Hilbert-Schmidt norm of sums of the form

$$\sum_{k=0}^n (P_{\lambda_k} - P_{\lambda_{k-1}}) A P_{\lambda_k}, \qquad 0 = \lambda_0 < \lambda_1 < \cdots < \lambda_n = 1.$$

Hence the defining kernel $a_-(t, s)$ for A_- is the pointwise limit of sums of the form

$$(3) \qquad S = \sum_{k=0}^n a(\lambda_k \wedge t, \lambda_k \wedge s) - a(\lambda_{k-1} \wedge t, \lambda_k \wedge s).$$

This observation is a consequence of the identity (10) in Proposition 10.2. From the assumptions on $a(t, s)$ and the Mean Value Theorem we have that S in (3) is a limit of sums of the form

$$\sum_{k=1}^n a_1(\lambda_k \wedge t, \lambda_k \wedge s)(\lambda_k - \lambda_{k-1}),$$

where $a_1 = \frac{\partial}{\partial t} a$. It now follows that

$$a_-(t,s) = \int_0^t a_1(\lambda, \lambda \wedge s)\,d\lambda. \tag{4}$$

So far we only used that a and $\frac{\partial}{\partial t} a$ are continuous on $[0,1] \times [0,1]$.

Next, we use a duality argument to find the defining kernel $a_+(\cdot, \cdot)$ for A_+. First note that

$$(A_+)^* = \left(\int P_\lambda A \, dP_\lambda\right)^* = \int (dP_\lambda) A^* P_\lambda = (A^*)_-.$$

By (8) in Proposition 10.2 the defining kernel of A^* is $\overline{a(s,t)}$. Thus formula (4) applied to A^* shows that the defining kernel $a_-^*(\cdot, \cdot)$ for $(A^*)_-$ is given by

$$a_-^*(t,s) = \int_0^t \overline{a_2(\lambda \wedge s, \lambda)}\,d\lambda$$

where $a_2 = \frac{\partial}{\partial s} a$. Since $A_+ = \left((A^*)_-\right)^*$, we have

$$a_+(t,s) = \overline{a_-^*(s,t)} = \int_0^s a_2(\lambda \wedge s, \lambda)\,d\lambda. \tag{5}$$

CHAPTER XXI
OPERATORS IN TRIANGULAR FORM

Operators that leave invariant a given maximal chain form the main topic of this chapter. Such operators may be viewed as the infinite dimensional analogues of matrices in upper triangular form. Volterra operators with a one dimensional imaginary part provide the simplest examples and they are identified up to unitary equivalence. In this chapter all operators act on a separable Hilbert space H, and the elements of a chain are assumed to be orthogonal projections.

XXI.1 TRIANGULAR REPRESENTATION

An operator A on $\mathbb{C}^n$ can be recovered from its imaginary part $A_\Im = \frac{1}{2i}(A - A^*)$ if one knows its eigenvalues and one of its invariant maximal chains. Indeed, let $\pi = \{P_0, P_1, \ldots, P_n\}$ be such a chain. Then

$$(1) \qquad A = \sum_{j,k=1}^{n} (\Delta P_j) A (\Delta P_k).$$

The fact that the chain is invariant under A implies that $(\Delta P_j) A (\Delta P_k) = 0$ for $j > k$. By using orthogonality of the projections and by taking adjoints one sees that $(\Delta P_j) A^* (\Delta P_k) = 0$ for $j < k$. Since ΔP_j has rank one, the operator $(\Delta P_j) A (\Delta P_j)$ is a scalar multiple of ΔP_j, i.e., $(\Delta P_j) A (\Delta P_j) = \lambda_j \Delta P_j$. It follows that (1) can be rewritten as

$$(2) \qquad A = 2i \sum_{k=1}^{n} P_{k-1} A_\Im (\Delta P_k) + \sum_{k=1}^{n} \lambda_k \Delta P_k,$$

where $\lambda_1, \ldots, \lambda_n$ are the eigenvalues of A repeated according to their algebraic multiplicity. The infinite dimensional analogue of the representation (2) is the main topic of this section.

THEOREM 1.1. *Let $\mathbb{P}$ be an invariant closed chain for the compact operator A on H. Then the integral $\int_{\mathbb{P}} P A_\Im dP$ converges and*

$$(3) \qquad A = 2i \int_{\mathbb{P}} P A_\Im dP + \sum_{\nu} (P_\nu^+ - P_\nu^-) A (P_\nu^+ - P_\nu^-),$$

where (P_ν^-, P_ν^+), $\nu = 1, 2, \ldots$, are the jumps of $\mathbb{P}$.

PROOF. Since A has a diagonal (see Theorem XX.4.4) and leaves $\mathbb{P}$ invariant, it is clear that A admits an additive LU-decomposition with respect to $\mathbb{P}$ (apply Theorem

XX.7.1 and note that $\int_{[\mathbb{P}}(dP)AP = 0$). In fact, $A = A_+ + A_0$, where A_0 is the diagonal of A and $A_+ \in A_+(\mathbb{P})$ (the A_--term being the zero operator). So we have

$$(4) \qquad A = \int_{[\mathbb{P}} PAdP + \sum_\nu (P_\nu^+ - P_\nu^-)A(P_\nu^+ - P_\nu^-).$$

Next, since $AP = PAP$ for each $P \in \mathbb{P}$, we see that

$$\sum_{j=1}^n P_{j-1}A^*(P_j - P_{j-1}) = 0$$

for each partition $\{P_0, \ldots, P_n\}$ of $\mathbb{P}$. It follows that $\int_{[\mathbb{P}} PA^*dP$ converges and is equal to the zero operator. Since $2iA_\Im = A - A^*$ we obtain from (4) the desired formula (3). $\square$

According to Corollary XX.4.6 the diagonal term in (3) can be specified further for a maximal chain. This yields the following theorem.

THEOREM 1.2. *Let $\mathbb{P}$ be an invariant maximal chain for the compact operator A on H. Then*

$$(5) \qquad A = 2i \int_{[\mathbb{P}} PA_\Im dP + \sum_j \lambda_j(P_j^+ - P_j^-),$$

where $\lambda_1, \lambda_2, \ldots$ are the non-zero eigenvalues of A repeated according to their algebraic multiplicity, $\lambda_n \to 0$ if the sequence of eigenvalues is infinite and for each j the pair (P_j^-, P_j^+) is a jump in $\mathbb{P}$.

The operator $T = 2i \int_{[\mathbb{P}} PA_\Im dP$ appearing in the identities (3) and (5) is compact and belongs to the triangular algebra $A_+(\mathbb{P})$. In particular, the spectrum of T consists of the zero element only, and hence T is a Volterra operator. It follows that the operator A of Theorem 1.1 is Volterra if

$$(P_+ - P_-)A(P_+ - P_-) = 0$$

for each jump (P_-, P_+) in the closed invariant chain $\mathbb{P}$. (In particular, A is Volterra if $\mathbb{P}$ is continuous.) If $\mathbb{P}$ is a maximal chain, then it is clear from the identities (3) and (5) that the converse statement is also true. So we have proved the following corollary

COROLLARY 1.3. *Let $\mathbb{P}$ be an invariant maximal chain for the compact operator A on H. Then A is a Volterra operator if and only if $(P_+ - P_-)A(P_+ - P_-) = 0$ for every jump (P_-, P_+) of $\mathbb{P}$.*

THEOREM 1.4. *Let $\mathbb{P}$ be an invariant closed chain for the compact operator A on H. If $(P_+ - P_-)A(P_+ - P_-) = 0$ for every jump (P_-, P_+) of $\mathbb{P}$, then*

$$(6) \qquad A = 2i \int_{\mathbb{P}} PA_\Im dP.$$

PROOF. In view of Theorem 1.1 it suffices to show that the integral in (6) converges. Let (P_-, P_+) be a jump of $\mathbb{P}$. From our hypotheses and the fact that the projection $P_+ - P_-$ is orthogonal, we see that $(P_+ - P_-)A^*(P_+ - P_-) = 0$. Thus

$$(P_+ - P_-)A_{\mathfrak{J}}(P_+ - P_-) = 0$$

for every jump (P_-, P_+) of $\mathbb{P}$. From Theorem 1.1 we know that $\int_{[\mathbb{P}} PAdP$ converges. So we can apply Lemma XX.6.2 to finish the proof. $\square$

THEOREM 1.5. *Let $\mathbb{P}$ be a maximal invariant chain for the Volterra operator A on H. Then*

$$A = 2i \int_{\mathbb{P}} PA_{\mathfrak{J}}dP.$$

PROOF. Apply Corollary 1.3 and Theorem 1.4. $\square$

We now come to the converse of Theorem 1.4.

THEOREM 1.6. *Let $\mathbb{P}$ be a closed chain on H, and let S be a compact operator on H for which the integral*

$$(7) \qquad\qquad T = 2i \int_{\mathbb{P}} PSdP$$

converges. Then (i) T is a Volterra operator, (ii) $\mathbb{P}$ is an invariant chain for T and (iii) $(P_+ - P_-)T(P_+ - P_-) = 0$ for every jump (P_-, P_+) of $\mathbb{P}$.

If, in addition, S is selfadjoint, then

$$(8) \qquad\qquad T_{\mathfrak{J}} = \frac{1}{2i}(T - T^*) = S,$$

and T is the unique compact operator satisfying (ii) and (iii) such that the imaginary part of T is equal to S.

PROOF. Since T is the limit in the operator norm of a sequence of compact operators, the operator T is compact. From (the one but last two paragraphs of) the proof of Theorem XX.7.1 we know that T belongs to the triangular algebra $A_+(\mathbb{P})$. It follows (cf., Theorem XX.5.1 and Lemma XX.6.2) that T has the properties (i), (ii) and (iii).

Next, assume that S is selfadjoint. Then the formula of partial integration (see Lemma XX.6.3) implies that

$$T_{\mathfrak{J}} = \frac{1}{2i}(T - T^*) = \int_{\mathbb{P}} PSdP + \int_{\mathbb{P}} (dP)SP = S.$$

Finally, let S be selfadjoint, and let A be a compact operator for which (ii) and (iii) are satisfied such that $A_{\mathfrak{J}} = S$. So (6) holds true. But $A_{\mathfrak{J}} = S$. So $A = T$, and the uniqueness is proved. $\square$

Given an arbitrary continuous chain $\mathbb{P}$ on H, any selfadjoint operator on H in the Hilbert-Schmidt class appears as the imaginary part of a Volterra operator which has $\mathbb{P}$ as an invariant chain. To prove this remark, recall (see Theorem XX.8.1) that for a Hilbert-Schmidt operator S the integral $\int_{[\mathbb{P}} PSdP$ converges. Now, if in addition, $\mathbb{P}$ is continuous, then we we know from Lemma XX.6.2 that $\int_{\mathbb{P}} PSdP$ converges, and we can apply Theorem 1.6 to get the desired statement. For an arbitrary compact selfadjoint operator the remark is not true, because for such an operator S the integral in (7) does not have to converge (see Lemma III.4.1 in Gohberg-Kreĭn [4] for examples).

Since a Volterra operator has an invariant maximal chain, the results of this section allow us to study Volterra operators in terms of their imaginary part. In particular, the triangular representation theorems given here can be used to classify Volterra operators via the ranks of their imaginary parts. We shall come back to this topic in the third section where we shall study in detail the simplest class, namely Volterra operators with a one-dimensional imaginary part.

XXI.2 INTERMEZZO ABOUT COMPLETELY NON-SELFADJOINT OPERATORS

Given an operator $A: H \to H$ we denote by $H_T(A)$ (the symbol T stands for trivial) the largest subspace of H invariant under A and A^* on which A and A^* coincide. If $H_T(A) = H$, then A is selfadjoint. The operator A is called *completely non-selfadjoint* (shortly, c.n.s.) if the space $H_T(A) = \{0\}$. In that case the operators A and A^* do not have a common non-zero invariant subspace on which they coincide.

In general the space $H_T(A)$ is equal to the closed linear hull of all subspaces of H invariant under A and A^* on which A and A^* coincide. This implies that the orthogonal complement $H_T(A)^\perp$ of $H_T(A)$ is invariant under A and the restriction of A to $H_T(A)^\perp$ is completely non-selfadjoint. It follows that given an operator $A: H \to H$ there exists an orthogonal decomposition $H = H_0 \oplus H_1$ such that with respect to this decomposition one can write A as a 2×2 operator matrix,

$$(1) \qquad A = \begin{bmatrix} A_0 & 0 \\ 0 & A_1 \end{bmatrix} : H_0 \oplus H_1 \to H_0 \oplus H_1,$$

with A_0 selfadjoint and A_1 completely non-selfadjoint. It is easily seen that there is only one orthogonal decomposition of H with these properties.

Since the restriction of a Volterra operator to an invariant subspace is again a Volterra operator, it follows that for a Volterra operator A the operator A_0 appearing in (1) is again a Volterra operator. But a selfadjoint Volterra operator is equal to the zero operator. So $A_0 = 0$ in this case. We see that a Volterra operator can be written in a unique way as an orthogonal direct sum of a c.n.s. Volterra operator and a zero operator.

Various descriptions of $H_T(A)$ can be given. The following identities are straightforward to check:

$$(2a) \qquad H_T(A) = \{f \in H \mid A^n f = A^{*n} f, n = 1, 2, \ldots\},$$

$$(2b) \qquad H_T(A) = \bigcap_{n=0}^{\infty} \operatorname{Ker} A_{\Im}(A^*)^n.$$

Here $A_{\Im}$ denotes the imaginary part of A. The identity (2a) implies that $H_T(A) = H_T(A^*)$, and thus A is c.n.s. if and only if A^* is c.n.s. From (2b) it is clear that

$$(3) \qquad H_T(A) = \{A^n A_{\Im} f \mid f \in H, n = 0, 1, 2, \ldots\}^{\perp},$$

which yields the following proposition.

PROPOSITION 2.1. *The operator A is completely non-selfadjoint if and only if the linear hull of the set*

$$(4) \qquad \{A^n A_{\Im} f \mid f \in H, n = 0, 1, 2, \ldots\}$$

is dense in H.

We shall say that an operator is *i-dissipative* if $A_{\Im} \geq 0$. In other words, A is *i*-dissipative if and only if iA is dissipative (cf., Section XIX.4). Let A be such an operator, i.e., $A_{\Im} \geq 0$, and let M be an invariant subspace of A on which A acts as a selfadjoint operator. For $f \in M$ we have

$$(5) \qquad \|A_{\Im}^{1/2} f\| = \langle A_{\Im} f, f \rangle = \Im \langle Af, f \rangle = 0.$$

Thus M is invariant under A and A^* and the operators A and A^* coincide on M. Hence $M \subset H_T(A)$, and we see that for an *i*-dissipative operator A the space $H_T(A)$ is equal to the largest A-invariant subspace on which A acts as a selfadjoint operator. It follows that the restriction of a c.n.s. *i*-dissipative operator to an invariant subspace is again c.n.s.

PROPOSITION 2.2. *For an i-dissipative Volterra operator we have $H_T(A) = \operatorname{Ker} A$. In particular, an i-dissipative Volterra operator is completely non-selfadjoint if and only if it has a trivial kernel.*

PROOF. We already know that a Volterra operator A is an orthogonal direct sum of a c.n.s. Volterra operator and a zero operator. Thus $H_T(A) \subset \operatorname{Ker} A$. On the other hand, $\operatorname{Ker} A$ is invariant under A and A acts as a selfadjoint operator on $\operatorname{Ker} A$. So, since A is *i*-dissipative, we also have $\operatorname{Ker} A \subset H_T(A)$. $\square$

PROPOSITION 2.3. *Any invariant maximal chain for an i-dissipative completely non-selfadjoint Volterra operator is continuous.*

PROOF. Let $\mathbb{P}$ be an invariant maximal chain for the c.n.s. *i*-dissipative Volterra operator A, and let (P_-, P_+) be a jump in $\mathbb{P}$. We know (see Corollary 1.3) that

$$(P_+ - P_-)A(P_+ - P_-) = 0,$$

and hence also $(P_+ - P_-)A^*(P_+ - P_-) = 0$. It follows that $(P_+ - P_-)A_{\Im}(P_+ - P_-) = 0$. Since $A_{\Im} \geq 0$, this implies $A_{\Im}(P_+ - P_-) = 0$.

Now write A and A^* as 3×3 operator matrices with respect to the orthogonal decomposition

$$H = \operatorname{Im} P_- \oplus \operatorname{Im}(P_+ - P_-) \oplus (\operatorname{Im} P_+)^{\perp}.$$

We have

$$A = \begin{bmatrix} A_1 & A_2 & * \\ 0 & 0 & * \\ 0 & 0 & * \end{bmatrix}, \qquad A^* = \begin{bmatrix} A_1^* & 0 & 0 \\ A_2^* & 0 & 0 \\ * & * & * \end{bmatrix}.$$

So

$$0 = \frac{1}{2i}(A - A^*)|\operatorname{Im}(P_+ - P_-) = \begin{bmatrix} \frac{1}{2i}A_2 \\ 0 \\ * \end{bmatrix}.$$

It follows that $A_2 = 0$. So $\operatorname{Ker} A \neq (0)$, but this contradicts the fact that A is c.n.s. (see the previous proposition). $\square$

PROPOSITION 2.4. *Let $\mathbb{P}$ be a closed chain on H, and let A be a Volterra operator for which the integral $\int_{\mathbb{P}} PA_\Im dP$ converges. Then*

$$(6) \qquad\qquad H_T(A) \supset \{PA_\Im f \mid f \in H, P \in \mathbb{P}\}^{\perp}.$$

If, in addition, A is i-dissipative, then we have equality in (6).

PROOF. The operator $A_\Im$ is compact and selfadjoint. Thus the fact that the integral $\int_{\mathbb{P}} PA_\Im dP$ converges, implies (see Theorem 1.6) that $A = 2i \int_{\mathbb{P}} PA_\Im dP$. Let $\pi = \{P_0, \ldots, P_n\}$ be a partition of $\mathbb{P}$. We know that

$$(7) \qquad\qquad S(\pi) = 2i \sum_{j=1}^{n} P_{j-1} A_\Im (P_j - P_{j-1}) \to A.$$

Let X be the linear hull of $\{PA_\Im f \mid f \in H, P \in \mathbb{P}\}$. For each $g \in H$ the vector $S(\pi)g$ belongs to X. Thus $Ag \in \overline{X}$, and hence $\operatorname{Im} A \subset \overline{X}$. This implies that $A^* X^{\perp} = \{0\}$. Next, take $h \in X^{\perp}$. Then $A_\Im Ph = 0$ for each $P \in \mathbb{P}$. It follows that $S(\pi)h = 0$, and so $Ah = 0$. We see that $AX^{\perp} = \{0\}$. So $X^{\perp}$ is invariant under A and A^* and the operators A and A^* coincide on $X^{\perp}$, which proves (6).

Next, assume that $A_\Im \geq 0$. Take $g \in H_T(A)$. We have to prove that $A_\Im Pg = 0$ for each $P \in \mathbb{P}$. Since A is i-dissipative, we know that $H_T(A) = \operatorname{Ker} A$ and A and A^* coincide on $H_T(A)$. Thus $Ag = A^*g = 0$. From $AP = PAP$ and $PA^* = PA^*P$ for each $P \in \mathbb{P}$ we see that

$$\langle PA_\Im Pg, g \rangle = \frac{1}{2i}\{\langle PAPg, g \rangle - \langle PA^*Pg, g \rangle\}$$

$$= \frac{1}{2i}\{\langle Pg, A^*g \rangle - \langle A^*g, Pg \rangle\} = 0.$$

So $\langle A_\Im Pg, Pg \rangle = 0$, which implies that $A_\Im Pg = 0$. $\square$

XXI.3 VOLTERRA OPERATORS WITH A ONE-DIMENSIONAL IMAGINARY PART

In this section we study Volterra operators A with a one-dimensional imaginary part. Since a Volterra operator is an orthogonal direct sum of a zero operator and a completely non-selfadjoint operator we restrict ourselves to Volterra operators A that are completely non-selfadjoint. Note that the imaginary part of A is of the form $\pm\langle\cdot,e\rangle e$. So multiplying by -1 if necessary, we may assume that A is of the form

$$(1) \qquad A = A_{\mathfrak{R}} + i\langle\cdot,e\rangle e,$$

where $A_{\mathfrak{R}}$ is the real part of A. From Proposition 2.1 we know that such an operator is c.n.s. if and only if the linear hull of the set $\{A^n e \mid n = 0,1,2,\ldots\}$ is dense in the Hilbert space H. Note that the operator A in (1) is i-dissipative. For the sake of simplicity we shall always assume that the vector e in formula (1) has norm 1. If A is as in (1), then $A_{\mathfrak{I}}x = \langle x,e\rangle e$, and the triangular integrals of Section XXI.1 lead to the following formula:

$$(2) \qquad Ax = 2i\int_{\mathbb{P}}\langle x,(dP)e\rangle Pe, \qquad x \in H,$$

where $\mathbb{P}$ is an invariant maximal chain for A.

The *operator of integration* on $L_2([0,1])$, i.e.,

$$(3) \qquad (Vf)(t) = 2i\int_t^1 f(s)ds,$$

is an example of a c.n.s. Volterra operator with a one-dimensional imaginary part. Indeed, for this operator V we have $(V_{\mathfrak{I}}f)(t) = \int_0^1 f(s)ds$, and thus

$$(4) \qquad V = V_{\mathfrak{R}} + i\langle\cdot,\varphi\rangle\varphi,$$

where $\varphi(t) = 1$ on $0 \le t \le 1$. (Note that also $\|\varphi\| = 1$.) Since V is injective, we know from Proposition 2.2 that V is c.n.s. .

Given Hilbert space operators $A_1 \in \mathcal{L}(H_1)$ and $A_2 \in \mathcal{L}(H_2)$ we say that A_1 is *unitarily equivalent* to A_2 if there exists a unitary operator $U\colon H_1 \to H_2$ such that $A_2 = UA_1U^{-1}$. In this section we shall prove that two c.n.s. Volterra operators of type (1) are unitarily equivalent. In particular, any operator of this kind is unitarily equivalent to the operator of integration. In terms of the latter equivalence, formula (2) is just an abstract version of formula (3). To prove the equivalence our main tools are the existence of invariant maximal chains and the triangular integral representations in terms of imaginary parts. We begin with a lemma.

LEMMA 3.1. *Let $A = A_{\mathfrak{R}} + i\langle\cdot,e\rangle e$ be a completely non-selfadjoint Volterra operator, and let $\mathbb{P}$ be an invariant maximal chain for A. Then the function*

$$(5) \qquad t\colon\mathbb{P} \to [0,1], \qquad t(P):= \langle Pe,e\rangle$$

is bijective and monotonically increasing.

PROOF. From Proposition 2.3 we know that $\mathbb{P}$ is continuous, and hence the map t is surjective. Obviously, t is monotonically increasing. Assume $t(P_\alpha) = t(P_\beta)$ for $P_\alpha < P_\beta$ in $\mathbb{P}$. Then $\langle(P_\beta - P_\alpha)e,e\rangle = 0$ and hence $(P_\beta - P_\alpha)e = 0$. Since

$$A_{\mathfrak{I}}(P_\beta - P_\alpha) = \langle\cdot,(P_\beta - P_\alpha)e\rangle e,$$

it follows that $A_{\Im}(P_\beta - P_\alpha) = 0$. Let $\pi = \{P_0, \ldots, P_n\}$ be a partition of $\mathbb{P}$, and consider the Riemann-Stieltjes sum

$$S(\pi) = 2i \sum_{j=1}^{n} P_{j-1} A_{\Im}(P_j - P_{j-1}).$$

The projections P_α and P_β commute with P_j. So $S(\pi)(P_\beta - P_\alpha) = 0$. Since

$$A = 2i \int_{\mathbb{P}} P A_{\Im} dP = \lim_{\pi \subset \mathbb{P}} S(\pi),$$

it follows that $A(P_\beta - P_\alpha) = 0$. From Proposition 2.2 we know that $\operatorname{Ker} A$ is trivial. So $P_\beta = P_\alpha$. Contradiction. $\square$

THEOREM 3.2. *For $\nu = 1, 2$ let*

$$A_\nu = A_{\nu\Re} + i\langle \cdot, e_\nu \rangle e_\nu, \qquad \|e_\nu\| = 1,$$

be a completely non-selfadjoint Volterra operator on H. Then there exists a unitary operator U on H such that $A_2 = U A_1 U^{-1}$ and $U e_1 = e_2$.

PROOF. For $\nu = 1, 2$ let $\mathbb{P}_\nu$ be an invariant maximal chain for A_ν, which exists according to Theorem XX.4.2. Put

$$t_\nu(P) = \langle P e_\nu, e_\nu \rangle, \quad P \in \mathbb{P}_\nu \qquad (\nu = 1, 2).$$

We know that t_1 and t_2 are both bijective and monotonically increasing. Let

$$t = t_2^{-1} \circ t_1 \colon \mathbb{P}_1 \to \mathbb{P}_2.$$

Then t is again bijective and monotonically increasing. Hence $t(0) = 0$ and $t(I) = I$. From the monotonicity of t and formula (1) in Section XX.1 it is clear that

$$(6) \qquad t(QP) = t(Q)t(P) \qquad (P, Q \in \mathbb{P}_1).$$

It follows that for P and Q in $\mathbb{P}_1$ we have

$$\langle P e_1, Q e_1 \rangle = \langle Q P e_1, e_1 \rangle = \langle t(QP) e_2, e_2 \rangle = \langle t(P) e_2, t(Q) e_2 \rangle.$$

By linearity this extends to

$$(7) \qquad \left\langle \sum_j \alpha_j P_j e_1, \sum_k \beta_k Q_k e_1 \right\rangle = \left\langle \sum_j \alpha_j t(P_j) e_2, \sum_k \beta_k t(Q_k) e_2 \right\rangle$$

for every choice of P_j and Q_k in $\mathbb{P}_1$ and every set of scalars α_j and β_k. In particular,

$$(8) \qquad \left\| \sum_j \alpha_j P_j e_1 \right\| = \left\| \sum_j \alpha_j t(P_j) e_2 \right\|.$$

For $\nu = 1, 2$ let X_ν be the linear hull of $\{Pe_\nu \mid P \in \mathbb{P}_\nu\}$. Since $A_{\nu\Im}Pf = \langle f, Pe_\nu \rangle e_\nu$ for each $P \in \mathbb{P}_\nu$, we can apply Proposition 2.4 to show that X_ν is dense in H ($\nu = 1, 2$). Define $U: X_1 \to X_2$ by setting

$$U\left(\sum_j \alpha_j P_j e_1\right) = \sum_j \alpha_j t(P_j)e_2.$$

According to (8) the map U is well-defined and does not depend on the special representation of the elements of X_1. Since t is bijective, it is clear that U maps X_1 in a one-one manner onto X_2. Also, by (8), the map U is an isometry. So by continuity we may extend U to an isometry, also denoted by U, from H into H. Since $X_2 \subset \mathrm{Im}\, U$ and X_2 is dense in H, it follows that U is unitary.

Clearly, $Ue_1 = e_2$. It remains to prove that $A_2 U = U A_1$. To do this we use the integral representation. Let $\pi = \{P_0, P_1, \ldots, P_n\}$ be a partition of $\mathbb{P}_1$, and put $t(\pi) = \{t(P_0), t(P_1), \ldots, t(P_n)\}$. Then $t(\pi)$ is a partition of $\mathbb{P}_2$. Let

$$S_1(\pi) = \sum_j P_{j-1} A_{1\Im}(P_j - P_{j-1}),$$

$$S_2\big(t(\pi)\big) = \sum_j t(P_{j-1}) A_{2\Im}\big(t(P_j) - t(P_{j-1})\big).$$

Then $S_1(\pi) \to A_1$ and $S_2\big(t(\pi)\big) \to A_2$. Recall that

$$P_{j-1} A_{1\Im}(P_j - P_{j-1})f = \langle f, (P_j - P_{j-1})e_1 \rangle P_{j-1}e_1.$$

So

$$
\begin{aligned}
U S_1(\pi) f &= \sum_j \langle f, (P_j - P_{j-1})e_1 \rangle t(P_{j-1})e_2 \\
&= \sum_j \langle f, U^{-1}\big(t(P_j) - t(P_{j-1})\big)e_2 \rangle t(P_{j-1})e_2 \\
&= \sum_j t(P_{j-1})\big[\langle Uf, \big(t(P_j) - t(P_{j-1})\big)e_2 \rangle e_2\big] \\
&= S_2\big(t(\pi)\big) Uf.
\end{aligned}
$$

By taking limits one obtains the desired result. $\quad\square$

Since the operator of integration (see formula (3)) is an operator of the type considered in the previous theorem, we have the following corollary.

COROLLARY 3.3. *Any completely non-selfadjoint Volterra operator with a one-dimensional imaginary part is (up to a non-zero real factor) unitarily equivalent to the operator of integration.*

For $0 \leq \nu \leq 1$ let P_ν be the orthogonal projection of $L_2([0,1])$ defined by

$$(9) \qquad (P_\nu f)(s) = \begin{cases} f(s) & \text{for} \quad 0 \leq s \leq \nu, \\ 0 & \text{for} \quad t < s \leq 1. \end{cases}$$

The orthogonal projections $\{P_\nu \mid 0 \le \nu \le 1\}$ form a maximal chain which is invariant under the operator of integration V. Let φ be as in (4). Then $\langle P_\nu \varphi, \varphi \rangle = \nu$. It follows that for the operator of integration the function t defined in (5) is just the function $t(P_\nu) = \nu$. Now, let $A = A_\Re + i\langle \cdot, e \rangle e$, $\|e\| = 1$, be a completely non-selfadjoint Volterra operator on H, and let $\mathbb{P}$ be an invariant maximal chain for A. We know that A and V are unitarily equivalent. From the proof of Theorem 3.2 and the remark made above it is clear that the unitary equivalence between A and V is given by the unitary operator $U : H \to L_2([0,1])$ defined by

$$UPe = \chi_{[0,\langle Pe,e \rangle]},$$

where $\chi_{[0,a]}$ is the characteristic function of the interval $[0, a]$, i.e., $\chi_{[0,a]}(t) = 1$ for $0 \le t \le a$ and equal to zero otherwise.

Note that the non-zero real factor referred to in Corollary 3.3 comes from the normalization described in the first paragraph of this section.

XXI.4 UNICELLULAR OPERATORS

Operators on a finite dimensional space with the property that their Jordan normal form consists of one single Jordan block may be viewed as the simplest components from which more complicated operators may build. In fact on a finite dimensional space each operator is a direct sum of these "unicellular" operators. Geometrically unicellular operators are characterized by the fact that given two invariant subspaces, one is always contained in the other. For operators on an infinite dimensional space this last property is taken as a definition.

An operator A on H is called *unicellular* if given two A-invariant subspaces one is always contained in the other. In other words, an operator A is unicellular whenever the A-invariant subspaces are linearly ordered by inclusion or, equivalently, A has only one invariant maximal chain.

THEOREM 4.1.   *A completely non-selfadjoint Volterra operator with a one dimensional imaginary part is unicellular.*

PROOF.   Let A be a c.n.s. Volterra operator with a one dimensional imaginary part. Without loss of generality we may assume that $A = A_\Re + i\langle \cdot, e \rangle e$ with $\|e\| = 1$. Let $\mathbb{P}_1$ and $\mathbb{P}_2$ be A-invariant maximal chains. We have to prove that $\mathbb{P}_1 = \mathbb{P}_2$. From the proof of Theorem 3.2 (with $A_1 = A_2 = A$) we know that there exist a bijective map $t : \mathbb{P}_1 \to \mathbb{P}_2$ and a unitary operator U on H such that $UA = AU$ and

$$(1) \qquad\qquad UPe = t(P)e, \qquad P \in \mathbb{P}_1.$$

Further we know that $t(QP) = t(Q)t(P)$ for all $P, Q \in \mathbb{P}$. Since $Ue = e$ and $UA^n e = A^n Ue = A^n e$, the operator U is the identity on the set $\{A^n e \mid n = 0, 1, 2, \ldots\}$. But the linear hull of the latter set is dense in H (by Proposition 2.1). So U is the identity operator on H. It follows (from (1)) that $Pe = t(P)e$, $P \in \mathbb{P}_1$. But then

$$QPe = t(QP)e = t(Q)t(P)e = t(Q)Pe$$

for every Q and P in $\mathbb{P}_1$. Now we use that the linear hull of $\{Pe \mid P \in \mathbb{P}_1\}$ is dense in H (by Proposition 2.4). It follows that Q and $t(Q)$ coincide on a dense set. Thus $t(Q) = Q$. Hence $\mathbb{P}_1 = \mathbb{P}_2$. $\square$

COROLLARY 4.2. *For the operator of integration V on $L_2([0,1])$, i.e.,*

$$(Vf)(t) = 2i \int_t^1 f(s)ds, \qquad f \in L_2([0,1]),$$

the subspaces

$$M_\nu := \{f \in L_2([0,1]) \mid f(s) = 0 \ (\nu < s \leq 1)\}, \qquad 0 \leq \nu \leq 1$$

are the only invariant subspaces.

PROOF. Let P_ν be the orthogonal projection on M_ν. We know that $\{P_\nu \mid 0 \leq \nu \leq 1\}$ is a maximal chain which is invariant under V. According to Theorem 4.1 the operator V is unicellular. So any invariant subspace of V must be of the form M_ν. $\square$

All unicellular Volterra operators on a fixed finite dimensional space are similar (because they have the same Jordan normal form). For infinite dimensional spaces this statement does not hold true. To see this, let us consider on ℓ_2 the weighted shift $S_\alpha: \ell_2 \to \ell_2$ defined by

$$S_\alpha(x_1, x_2, x_3, \ldots) = (0, \alpha_1 x_1, \alpha_2 x_2, \ldots).$$

We assume that the sequence $\alpha = (\alpha_1, \alpha_2, \ldots)$ consists of positive numbers decreasing monotonically to zero. Obviously, S_α is a Volterra operator. If, in addition, the series $\sum_{j=1}^{\infty} \alpha_j^p$ converges for some p with $0 < p < \infty$, then it can be shown (see Nikol'skiǐ [1]) that S_α is unicellular. Now, take

$$\alpha_\nu = \left(1, \left(\frac{1}{2}\right)^\nu, \left(\frac{1}{3}\right)^\nu, \ldots\right), \qquad \nu > 0.$$

Then the operators S_{α_ν} ($\nu > 0$) are all unicellular, but their similarity classes are mutually disjoint, i.e., if $\nu \neq \mu$, then S_{α_ν} is not similar to S_{α_μ}. So on a separable Hilbert space there is a continuum of non-similar unicellular operators. In order to prove that the operators S_{α_ν} ($\nu > 0$) are mutually non-similar, we first note that the j-th singular value of the operator S_α is equal to α_j. Next, if F is an invertible operator, then, by Proposition VI.1.3, we have the following inequalities:

$$\|F^{-1}\|^{-1}\|F\|^{-1} \leq \frac{s_j(F^{-1}S_\alpha F)}{s_j(S_\alpha)} \leq \|F^{-1}\|\|F\|, \qquad j = 1, 2, \ldots \ .$$

Now

$$s_j(S_{\alpha_\nu})/s_j(S_{\alpha_\mu}) = \left(\frac{1}{j}\right)^{\nu-\mu}, \qquad j = 1, 2, \ldots,$$

and the latter sequence is bounded and bounded away from zero if and only if $\nu = \mu$. It follows that for $\nu \neq \mu$ the operators S_{α_ν} and S_{α_μ} cannot be similar.

CHAPTER XXII
MULTIPLICATIVE LOWER-UPPER TRIANGULAR
DECOMPOSITIONS OF OPERATORS

In Chapter XX we dealt with the problem of *additive LU*-decompositions of operators with respect to a chain. In this chapter we are concerned with the more challenging problem of *multiplicative LU*-decomposition of certain operators. The theorems which are presented encompass classical results from Linear Algebra which show that, under certain conditions, a matrix can be represented as a product of a lower with an upper triangular matrix. Also, the possibility of factoring certain integral operators as the product of integral operators with lower and upper triangular kernel functions is treated as a special case of the general theory which we now develop.

Let $\mathbb{P}$ be a chain in a Banach space X. An operator $T \in \mathcal{L}(X)$ is said to admit a *multiplicative LU-decomposition* (or *LU-factorization* for short) with respect to $\mathbb{P}$ if it can be represented in the form

$$(1) \qquad T = (I + Y_-)D(I + Y_+), \qquad Y_\pm \in A_\pm(\mathbb{P}), \quad D \in A_0(\mathbb{P}).$$

Our aim in this chapter is to determine when such LU-factorizations are possible and to give formulas for the factors.

First we note that *if $\mathbb{P}$ is a uniform chain and T is invertible, then T admits at most one LU-factorization.* To see this, suppose

$$(2) \qquad T = (I + Y_-)D(I + Y_+) = (I + X_-)D_1(I + X_+),$$

where $X_\pm, Y_\pm \in A_\pm(\mathbb{P})$ and $D, D_1 \in A_0(\mathbb{P})$. Now $Y_\pm$, $X_\pm$ are quasinilpotent operators and

$$(I + Y_\pm)^{-1} - I \in A_\pm(\mathbb{P}),$$

by Proposition XX.5.1. Also, $(I + X_\pm)^{-1} - I \in A_\pm(\mathbb{P})$. Since T is invertible, D is invertible and

$$(3) \qquad (I + X_-)^{-1}(I + Y_-) = D_1(I + X_+)^{-1}(I + Y_+)^{-1}D^{-1}.$$

To prove that $X_- = Y_-$ we show that $V_- = (I + X_-)^{-1}(I + Y_-) - I$ is equal to zero. The chains $\mathbb{P}$ and $\mathbb{P}^c$ are invariant under the operators appearing on the right and left, respectively, of (3). Hence $V_- \in A_0(\mathbb{P})$. On the other hand

$$V_- = Z_- + Z_-Y_- + Y_-,$$

where $Z_- = (I + X_-)^{-1} - I \in A_-(\mathbb{P})$. Since $Y_- \in A_-(\mathbb{P})$ and $A_-(\mathbb{P})$ is a subalgebra of $\mathcal{L}(X)$, it follows that V_- is also in $A_-(\mathbb{P})$. Thus $V_- \in A_0(\mathbb{P}) \cap A_-(\mathbb{P})$, and hence $V_- = 0$. The proof that $X_+ = Y_+$ is similar. Therefore, $D = D_1$.

The uniqueness statement mentioned above is of particular interest in the Hilbert space case. Indeed, let $\mathbb{P}$ be a chain of orthogonal projection on the Hilbert space H, and let $T \in \mathcal{L}(H)$ be a selfadjoint invertible operator which admits the *LU*-factorization (1). Taking adjoints yields

$$T = T^* = (I + Y_+^*)D^*(I + Y_-^*).$$

Now $Y_+^* \in A_-(\mathbb{P})$, $Y_-^* \in A_+(\mathbb{P})$ and $D^* \in A_0(\mathbb{P})$. Since a Hilbert chain is always uniform, the *LU*-factorization is unique, and therefore $Y_- = Y_+^*$, $Y_+ = Y_-^*$ and $D^* = D$. Thus the type of factorizations considered here include the so-called Cholesky factorizations of symmetric matrices.

XXII.1 *LU*-FACTORIZATION WITH RESPECT TO A FINITE CHAIN

Before proving our first factorization theorem let us explain one of the main steps in the construction. Suppose $V = V_1 \oplus V_2$, where V_1 and V_2 are subspaces of a vector space V. Let K be an operator on V which, with respect to this decomposition, has the matrix representation

$$K = \begin{bmatrix} A & B \\ C & D \end{bmatrix} : V_1 \oplus V_2 \to V_1 \oplus V_2.$$

If A is invertible on V_1, then

$$\begin{bmatrix} A & B \\ C & D \end{bmatrix} = \begin{bmatrix} I_1 & 0 \\ CA^{-1} & I_2 \end{bmatrix} \begin{bmatrix} A & 0 \\ 0 & D - CA^{-1}B \end{bmatrix} \begin{bmatrix} I_1 & A^{-1}B \\ 0 & I_2 \end{bmatrix},$$

where I_1 and I_2 are the identity operators on V_1 and V_2, respectively. Therefore

$$(1) \qquad K = (I + CA^{-1}P)(AP + (D - CA^{-1}B)Q)(I + A^{-1}BQ),$$

where P is the projection of V onto V_1 along V_2 and $Q = I - P$. By repeatedly applying the above factorization formula more complicated factorizations may be constructed.

THEOREM 1.1. *Let $I - K$ be an invertible operator on a Banach space X. A necessary and sufficient condition in order that $I - K$ admits an LU-factorization with respect to a finite chain $\mathbb{P} = \{P_0, P_1, \ldots, P_n\}$ on X is that $I - P_j K P_j$ be invertible for $j = 1, \ldots, n$. In this case $I - K$ has the LU-factorization*

$$(2) \qquad I - K = (I + Y_-)D(I + Y_+),$$

with

$$I + Y_- = \prod_{j=1}^{\curvearrowleft n} \left[I - \Delta P_j K P_{j-1}(I - P_{j-1}K P_{j-1})^{-1} \right],$$

$$I + Y_+ = \prod_{j=1}^{\curvearrowright n} \left[I - (I - P_{j-1}K P_{j-1})^{-1} P_{j-1}K \Delta P_j \right],$$

$$D = \sum_{j=1}^{n} \Delta P_j \left[I - K - K P_{j-1}(I - P_{j-1}K P_{j-1})^{-1}K \right] \Delta P_j.$$

The inverses of these operators are given by

$$(I + Y_-)^{-1} = I + \sum_{j=1}^{n} (\Delta P_j) K P_{j-1} (I - P_{j-1} K P_{j-1})^{-1},$$

$$(I + Y_+)^{-1} = I + \sum_{j=1}^{n} (I - P_{j-1} K P_{j-1})^{-1} P_{j-1} K (\Delta P_j),$$

$$D^{-1} = \sum_{j=1}^{n} (\Delta P_j)(I - P_j K P_j)^{-1}(\Delta P_j).$$

PROOF. Suppose $I - K$ admits the factorization (2). We have already seen that the factors $I + Y_+$ and $I + Y_-$ are invertible, that the chain $\mathbb{P}$ is invariant under $(I+Y_+)^{\pm 1}$ and that the complementary chain $\mathbb{P}^c$ is invariant under $(I-Y_-)^{\pm 1}$. It follows that for each $P \in \mathbb{P}$

(3) $$P(I - K)P = \{P(I + Y_-)P\} PDP \{P(I + Y_+)P\}$$

and

(4) $$\{P(I + Y_\pm)P\}\{P(I + Y_\pm)^{-1}P\} = P(I + Y_\pm)(I + Y_\pm)^{-1}P = P.$$

The latter equality implies that $P(I + Y_\pm)P$ are invertible on $\operatorname{Im} P$. Note that D is invertible as a consequence of (2). Since D commutes with P, the operator PDP is invertible on $\operatorname{Im} P$. Thus $P(I - K)P$ is invertible on $\operatorname{Im} P$ by (3). But then

$$I - PKP = I - P + P(I - K)P$$

is invertible on X.

Now suppose that $I - P_j K P_j$ is invertible for $j = 1, 2, \ldots, n$. Then $P_j(I - K)P_j$ is invertible on $\operatorname{Im} P_j$. Let $A = I - K$, and define $(P_j A P_j)^{-1}$ to be the inverse of $P_j A P_j$ on $\operatorname{Im} P_j$ and zero on $\operatorname{Ker} P_j$. We now express $(P_j A P_j)^{-1}$ in terms of $(P_{j-1} A P_{j-1})^{-1}$ by using the idea mentioned in the first paragraph of this section.

With respect to the decomposition $\operatorname{Im} P_j = \operatorname{Im} P_{j-1} \oplus \operatorname{Im} \Delta P_j$ we have the matrix representation

$$P_j A P_j = \begin{bmatrix} P_{j-1} A P_{j-1} & P_{j-1} A (\Delta P_j) \\ (\Delta P_j) A P_{j-1} & (\Delta P_j) A (\Delta P_j) \end{bmatrix}.$$

Therefore, by (1), the operator $P_j A P_j$ factorizes as follows:

(5) $$P_j A P_j = (I + G_j)(P_{j-1} A P_{j-1} + D_j)(I + F_j), \qquad j = 1, 2, \ldots, n,$$

where

$$G_j = (\Delta P_j) A P_{j-1} (P_{j-1} A P_{j-1})^{-1} \in A_-(\mathbb{P}),$$
$$D_j = (\Delta P_j) A (\Delta P_j) - (\Delta P_j) A P_{j-1} (P_{j-1} A P_{j-1})^{-1} P_{j-1} A (\Delta P_j) \in A_0(\mathbb{P}),$$
$$F_j = (P_{j-1} A P_{j-1})^{-1} P_{j-1} A (\Delta P_j) \in A_+(\mathbb{P}).$$

Using induction, we shall show that for $j = 1, \ldots, n$

$$(6) \qquad P_j A P_j = \prod_{k=1}^{\curvearrowleft j} (I + G_k) \left(\sum_{k=1}^{j} D_k \right) \prod_{k=1}^{\curvearrowright j} (I + F_k).$$

Suppose (6) holds for $j = m$. By (5),

$$P_{m+1} A P_{m+1} = (I + G_{m+1})[P_m A P_m + D_{m+1}](I + F_{m+1})$$
$$= \prod_{k=1}^{\curvearrowleft m+1} (I + G_k) \left(\sum_{k=1}^{m+1} D_k \right) \prod_{k=1}^{\curvearrowright m+1} (I + F_k).$$

Here we used $G_k D_j = D_j F_k = 0$ for $1 \leq k \leq j \leq n$. Thus (6) holds for $1 \leq k \leq n$. In particular,

$$(7) \qquad A = \prod_{k=1}^{\curvearrowleft n} (I + G_k) \left(\sum_{k=1}^{n} D_k \right) \prod_{k=1}^{\curvearrowright n} (I + F_k).$$

Note that

$$(8) \qquad F_j F_k = 0 \quad (j \geq k), \qquad G_j G_k = 0 \quad (j \leq k).$$

Therefore

$$(9) \qquad (I + G_k)^{-1} = I - G_k, \qquad (I + F_k)^{-1} = I - F_k.$$

From (7), (8) and (9) we get

$$(10) \qquad A^{-1} = \left(I - \sum_{k=1}^{n} F_k \right) D^{-1} \left(I - \sum_{k=1}^{n} G_k \right), \qquad D = \sum_{k=1}^{n} D_k \in A_0(\mathbb{P}).$$

Since $\sum_{k=1}^{n} F_k \in A_+(\mathbb{P})$ and $\sum_{k=1}^{n} G_k \in A_-(\mathbb{P})$, these operators are quasinilpotent. Define

$$Y_+ = \left(I - \sum_{k=1}^{n} F_k \right)^{-1} - I, \qquad Y_- = \left(I - \sum_{k=1}^{n} G_k \right)^{-1} - I.$$

Now $Y_\pm \in A_\pm(\mathbb{P})$, and, by (7), we see that (2) holds for $A = I - K$.

In order to obtain a formula for D^{-1}, we write $A = \widetilde{Y}_- \widetilde{Y}_+$, with $\widetilde{Y}_- = I + Y_-$ and $\widetilde{Y}_+ = D(I + Y_+)$. Then, because of (3),

$$P_j A P_j = (P_j \widetilde{Y}_- P_j)(P_j \widetilde{Y}_+ P_j), \qquad j = 1, \ldots, n,$$

and therefore (see (4))

$$(P_j A P_j)^{-1} = \widetilde{Y}_+^{-1} P_j \widetilde{Y}_-^{-1} P_j, \qquad j = 1, \ldots, n.$$

It follows that

$$(P_j A P_j)^{-1} \Delta P_j = \widetilde{Y}_+^{-1}(\Delta P_j + P_{j-1})\widetilde{Y}_-^{-1}(\Delta P_j)$$
$$= \widetilde{Y}_+^{-1}(\Delta P_j)\widetilde{Y}_-^{-1}(\Delta P_j)$$
$$= \widetilde{Y}_+^{-1}(\Delta P_j) = (I + Y_+)^{-1} D^{-1}(\Delta P_j), \qquad j = 1, \ldots, n,$$

because $\widetilde{Y}_-^{-1} - I \in A_-(\mathbb{P})$ and $\widetilde{Y}_+^{-1} \in A_0(\mathbb{P}) \oplus A_+(\mathbb{P})$. Thus

$$\Delta P_j (P_j A P_j)^{-1} \Delta P_j = \Delta P_j (I + Y_+)^{-1} D^{-1} \Delta P_j = D^{-1} \Delta P_j, \qquad j = 1, \ldots, n,$$

by virtue of the fact that $(I + Y_+)^{-1} - I \in A_+(\mathbb{P})$. Therefore

$$(11) \qquad\qquad D^{-1} = \sum_{j=1}^{n} D^{-1} \Delta P_j = \sum_{j=1}^{n} (\Delta P_j)(P_j A P_j)^{-1} \Delta P_j.$$

The theorem now follows from the identities (7), (10), (11) and the equalities

$$(P_j A P_j)^{-1} = (I - P_j K P_j)^{-1} P_j, \qquad j = 1, \ldots, n. \quad \square$$

If A is a nonsingular $n \times n$ matrix with entries in $\mathbb{C}$, then the well-known result (see, e.g., Strang [1]) concerning the LU-factorization of A is a special case of the theorem above. One need only let K be the operator corresponding to the matrix $I - A$ and the standard basis in $\mathbb{C}^n$. Then define P_j on $\mathbb{C}^n$ by $P_j(\alpha_1, \ldots, \alpha_n) = (\alpha_1, \ldots, \alpha_j, 0, \ldots, 0)$, and take $\mathbb{P} = \{P_0, P_1, \ldots, P_n\}$.

XXII.2 THE *LU*-FACTORIZATION THEOREM

THEOREM 2.1. *Let $I - K$ be an invertible operator on a Banach space X. A necessary and sufficient condition in order that $I - K$ admits a multiplicative LU-factorization with respect to a uniform chain $\mathbb{P}$ on X is that the following two conditions hold:*

(a) *for each $P \in \mathbb{P}$, the operator $I - PKP$ is invertible;*

(b) *the integrals*

$$X_- = \int_{[\mathbb{P}} (dP) K P (I - PKP)^{-1},$$

$$X_+ = \int_{[\mathbb{P}} (I - PKP)^{-1} PK \, dP,$$

$$W = \int_{\mathbb{P}]} (dP)(I - PKP)^{-1} \, dP,$$

converge in $\mathcal{L}(X)$. If (a) *and* (b) *hold, then $I - K$ admits the LU-factorization*

$$(1) \qquad\qquad I - K = (I + Y_-) D (I + Y_+),$$

with

$$I + Y_- = \int_{[\mathbb{P}}^{\frown} I - (dP)KP(I - PKP)^{-1},$$

$$I + Y_+ = \int_{[\mathbb{P}}^{\frown} I - (I - PKP)^{-1}PK\,dP,$$

$$D = \int_{[\mathbb{P}} (dP)\big[I - K - KP(I - PKP)^{-1}K\big]\,dP.$$

The inverses are given by

$$(I + Y_-)^{-1} = I + \int_{[\mathbb{P}} (dP)KP(I - PKP)^{-1},$$

$$(I + Y_+)^{-1} = I + \int_{[\mathbb{P}} (I - PKP)^{-1}PK\,dP,$$

$$D^{-1} = \int_{\mathbb{P}]} (dP)(I - PKP)^{-1}\,dP.$$

PROOF. We split the proof into four parts. In the first three parts we assume that $I - K$ admits the *LU*-factorization (1).

Part (α). We prove that (a) holds. Writing $D = I + M$ and $D(I + Y_+) = I + \widetilde{Y}_+$ gives

$$(2) \qquad\qquad\qquad I - K = (I + Y_-)(I + \widetilde{Y}_+),$$

where $\widetilde{Y}_+ = M + Y_+ + MY_+$ leaves $\mathbb{P}$ invariant. From (2) we get

$$\begin{aligned}
I - PKP &= I + P(Y_- + \widetilde{Y}_+ + Y_-\widetilde{Y}_+)P \\
&= I + PY_-P + P\widetilde{Y}_+P + (PY_-P)(P\widetilde{Y}_+P) \\
&= (I + PY_-P)(I + P\widetilde{Y}_+P).
\end{aligned}$$

Since Y_+ is quasi-nilpotent,

$$(I + \widetilde{Y}_+)^{-1} = (I + Y_+)^{-1}D^{-1} = \sum_{\nu=0}^{\infty} (-1)^\nu Y_+^\nu D^{-1}$$

leaves $\mathbb{P}$ invariant. Here we used Proposition XX.5.1 and the observation that $D \in A_0(\mathbb{P})$ implies $D^{-1} \in A_0(\mathbb{P})$. A direct calculation verifies that $I + P\widetilde{Y}_+P$ and $I + PY_-P$ are invertible with

$$(3) \qquad\qquad (I + P\widetilde{Y}_+P)^{-1} = I - P + (I + \widetilde{Y}_+)^{-1}P,$$

$$(4) \qquad (I + PY_-P)^{-1} = I - P + P(I + Y_-)^{-1}.$$

Hence $I - PKP$ is invertible for every $P \in \mathbb{P}$ and

$$(5) \qquad (I - PKP)^{-1} = (I + P\widetilde{Y}_+P)^{-1}(I + PY_-P)^{-1}.$$

Part (β). We prove the convergence of the second integral in (b) and compute its value. Put $X_- := (I + Y_-)^{-1} - I$. Then (2)–(5) give

$$(6) \qquad \begin{aligned} (I - PKP)^{-1}PK &= \{I - P + (I + \widetilde{Y}_+)^{-1}P\}\{I - P + P(I + Y_-)^{-1}\}PK \\ &= (I + \widetilde{Y}_+)^{-1}P(I + Y_-)^{-1}P\{I - (I + Y_-)(I + \widetilde{Y}_+)\} \\ &= (I + \widetilde{Y}_+)^{-1}P(I + X_-) - (I + \widetilde{Y}_+)^{-1}P(I + \widetilde{Y}_+) \\ &= (I + \widetilde{Y}_+)^{-1}PX_- - (I + \widetilde{Y}_+)^{-1}P\widetilde{Y}_+. \end{aligned}$$

Now let $\pi = \{P_0, P_1, \ldots, P_n\}$ be a partition of $\mathbb{P}$. Then

$$\begin{aligned} (I - P_{j-1}KP_{j-1})^{-1}P_{j-1}K(\Delta P_j) &= (I + \widetilde{Y}_+)^{-1}P_{j-1}X_-(\Delta P_j) - \\ &\quad - (I + \widetilde{Y}_+)^{-1}P_{j-1}\widetilde{Y}_+(\Delta P_j) \\ &= -(I + \widetilde{Y}_+)^{-1}P_{j-1}(-I + D + DY_+)(\Delta P_j) \\ &= -(I + \widetilde{Y}_+)^{-1}DP_{j-1}Y_+(\Delta P_j). \end{aligned}$$

From (3c) in the proof of Theorem XX.7.1 we know that

$$\lim_{\pi \subset \mathbb{P}} -(I + \widetilde{Y}_+)^{-1}DP_{j-1}Y_+(\Delta P_j) = -(I + \widetilde{Y}_+)^{-1}D\int_{\mathbb{P}} PY_+(dP)$$

$$= -(I + Y_+)^{-1}Y_- = -I + (I + Y_+)^{-1},$$

which proves the second integral in (b). In a similar way one shows that

$$\int_{\mathbb{P}} (dP)KP(I - PKP)^{-1} = -I + (I + Y_-)^{-1}.$$

Part (γ). We proceed with the third integral in (b). Let $\pi = \{P_0, P_1, \ldots, P_n\}$ be a partition of $\mathbb{P}$. We know from Theorem 1.1 that

$$(7) \qquad I - K = \big(I + Y_-(\pi)\big)D_\pi\big(I + Y_+(\pi)\big),$$

where

$$\big(I + Y_-(\pi)\big)^{-1} = I + \sum_{j=1}^{n} \Delta P_j K P_{j-1}(I - P_{j-1}KP_{j-1})^{-1},$$

$$\big(I + Y_+(\pi)\big)^{-1} = I + \sum_{j=1}^{n} (I - P_{j-1}KP_{j-1})^{-1}P_{j-1}K\Delta P_j,$$

$$D_\pi^{-1} = \sum_{j=1}^{n} \Delta P_j(I - P_jKP_j)^{-1}\Delta P_j.$$

Since $\left(I + Y_{\pm}(\pi)\right)^{-1}$ converge to $(I + Y_{\pm})^{-1}$, it follows from (7) that

$$\lim_{\pi \subset \mathbb{P}} D_\pi = (I + Y_-)^{-1}(I - K)(I + Y_+)^{-1} = D.$$

Hence D_π^{-1} converges to D^{-1}, and thus the third integral in (b) converges to D^{-1}. We have proved (b).

Part (δ). Now suppose that (a) and (b) hold. Our aim is to establish the factorization (1). By (7)

$$(8) \qquad\qquad D_\pi^{-1}\left(I + Y_-(\pi)\right)^{-1}(I - K) = I + Y_+(\pi).$$

Hence $Y_+(\pi)$ converges in $\mathcal{L}(X)$ to $Y_+ = -I + W(I + X_-)(I - K)$. Since $\mathbb{P}$ is a uniform chain, $A_+(\mathbb{P})$ is closed in $\mathcal{L}(X)$, and hence $Y_+ \in A_+(\mathbb{P})$. A similar argument shows that $Y_-(\pi)$ converges in $\mathcal{L}(X)$ to some $Y_- \in A_-(\mathbb{P})$. Also $D_\pi^{-1} \to W$, and hence $W \in A_0(\mathbb{P})$ (because $A_0(\mathbb{P})$ is closed in $\mathcal{L}(X)$). Now

$$D_\pi^{-1} = \left(I + Y_+(\pi)\right)(I - K)^{-1}\left(I + Y_-(\pi)\right) \to (I + Y_+)(I - K)^{-1}(I + Y_-),$$

and therefore

$$W = (I + Y_+)(I - K)^{-1}(I + Y_-).$$

Let $D = W^{-1}$. Then (1) holds and $D \in A_0(\mathbb{P})$ because $W \in A_0(\mathbb{P})$. Finally, since $I + Y_{\pm}(\pi)$ converges to $I + Y_{\pm}$ and $\left(I + Y_{\pm}(\pi)\right)^{-1}$ converges to $(I + Y_{\pm})^{-1}$, the remaining formulas in the theorem follow from Theorem 1.1. $\square$

For an invertible operator $I - K$ on a Banach space X and a uniform chain $\mathbb{P}$ on X the following holds. *If statement* (a) *in* Theorem 2.1 *is fulfilled and two of the three integrals defined in* (b) *converge in* $\mathcal{L}(X)$, *then the remaining integral also converges in* $\mathcal{L}(X)$. This may be seen by using an argument similar to the one used in part (γ) of the above proof.

For the next theorem we need the following definition. Let $h : [0, 1] \to Z$ be a map from the compact interval $[0, 1]$ to a normed linear space Z. We say that h is of *bounded variation* on $[0, 1]$ if there exists a constant M such that for every partition

$$(9) \qquad\qquad \pi := \{0 = t_0 < t_1 < \cdots < t_n = 1\}$$

the sum

$$\mathrm{Var}(h; \pi) = \sum_{j=1}^{n} \|h(t_j) - h(t_{j-1})\| \le M.$$

In this case we define $\mathrm{Var}(h) = \sup_\pi \mathrm{Var}(h; \pi)$.

THEOREM 2.2. *Let A be a non-negative operator on a Hilbert space with $\|A\| < 1$. Let $\mathbb{P} = \{P(t) \mid 0 \le t \le 1\}$ be a continuous chain of orthogonal projections on H. Assume that the function $t \mapsto A^{1/2}P(t)A^{1/2}$ is continuous and of bounded variation on $[0, 1]$ with respect to the operator norm. Then $I - A$ admits the LU-factorization*

$$(10) \qquad\qquad I - A = (I + Y_+^*)(I + Y_+),$$

where

$$(11) \qquad (I + Y_+)^{-1} = I + X_+, \qquad X_+ = \int_{\mathbb{P}} (I - PAP)^{-1} PA\, dP \in A_+(\mathbb{P}).$$

PROOF. Since $\|A\| < 1$, the operator $I - PAP$ is invertible for each $P \in \mathbb{P}$ (including $P = I$), and hence condition (a) in Theorem 2.1 is fulfilled (for A in place of K). Hence, by the remark made in the paragraph after the proof of Theorem 2.1, in order that $I - A$ admits an LU-factorization it is sufficient to show that the integrals

$$(12) \qquad \int_{\mathbb{P}} (I - PAP)^{-1} PA\, dP, \qquad \int_{\mathbb{P}]} (dP)(I - PAP)^{-1}\, dP$$

converge. Here we use that the convergence of the first integral in (12) implies that this integral also converges when $\mathbb{P}$ is replaced by $[\mathbb{P}$, and in this case the integrals with $\mathbb{P}$ and $[\mathbb{P}$ have the same value. Furthermore, by Theorem 2.1, if both integrals in (12) converge, then

$$(13) \qquad I - A = (I + Y_-)D(I + Y_+),$$

where Y_+ satisfies (11) and $D^{-1} = W$, which is the value of the second integral in (12). Note that Theorem 2.1 is applicable, because a Hilbert space chain is uniform. Since A is also selfadjoint, the uniqueness of the factors in the LU-factorization (13) implies that $Y_- = Y_+^*$. So to prove the theorem we have to show that the two integrals in (12) converge and that the value of the second integral in (12) is equal to I.

To prove the convergence of the first integral in (12), we first note that the functions $t \mapsto A^{1/2}P(t)$ and $t \mapsto P(t)A^{1/2}$ are continuous on $[0,1]$ in the norm of $\mathcal{L}(H)$. This follows from the continuity of the function $t \mapsto A^{1/2}P(t)A^{1/2}$ and the equalities

$$\begin{aligned}
\|A^{1/2}\Delta P\|^2 &= \|(\Delta P)A^{1/2}\|^2 \\
&= \sup_{\|x\|=1} \langle (\Delta P)A^{1/2}x, (\Delta P)A^{1/2}x \rangle \\
&= \sup_{\|x\|=1} \langle A^{1/2}(\Delta P)A^{1/2}x, x \rangle = \|A^{1/2}(\Delta P)A^{1/2}\|,
\end{aligned}$$

where $\Delta P = P(t') - P(t'')$. Consequently, $\Phi(t) = \big(I - P(t)AP(t)\big)^{-1}P(t)A^{1/2}$ is continuous, and therefore uniformly continuous on $[0,1]$. Let τ_1 and π_2 be partitions of the interval $[0,1]$, and let S_{π_1}, S_{π_2} be the corresponding Riemann-Stieltjes sum associated with the first integral in (12). Let π in (9) be the union of the partitions π_1, π_2. Then

$$S_{\pi_1} - S_{\pi_2} = \sum_{i=0}^{n} \big\{ \Phi(t_i'') - \Phi(t_i') \big\} A^{1/2}\Delta P(t_i), \qquad \Delta P(t_i) = P(t_i) - P(t_{i-1}),$$

where t_i' and t_i'' are certain intermediate points related to π_1 and π_2, respectively. It follows that

$$\|S_{\pi_1} - S_{\pi_2}\|^2 = \|(S_{\pi_1} - S_{\pi_2})(S_{\pi_1} - S_{\pi_2})^*\|$$

$$= \left\| \sum_{i=1}^{n} \{\Phi(t_i'') - \Phi(t_i')\} A^{1/2} (\Delta P(t_i)) A^{1/2} \{\Phi(t_i'') - \Phi(t_i')\}^* \right\|$$

$$\leqq \max_i \|\Phi(t_i'') - \Phi(t_i')\|^2 \operatorname{Var}(A^{1/2} P(\cdot) A^{1/2}).$$

The convergence of the first integral in (12) now follows from the above inequality and the uniform continuity of Φ.

It remains to prove that the second integral in (12) converges to I. Let the partition π be as in (9). The corresponding upper Riemann-Stieltjes sum W_π associated with the second integral in (12) is

$$W_\pi = \sum_{j=1}^{n} \Delta P_j \big(I - P(t_j) A P(t_j)\big)^{-1} \Delta P_j, \qquad \Delta P_j = P(t_j) - P(t_{j-1}).$$

Now

$$W_\pi - I = \sum_{j=1}^{n} \Delta P_j \big[\big(I - P(t_j) A P(t_j)\big)^{-1} - I\big] \Delta P_j$$

$$= \sum_{j=1}^{n} \Delta P_j \big(I - P(t_j) A P(t_j)\big)^{-1} P(t_j) A \Delta P_j$$

and

$$\|W_\pi - I\| = \max_{1 \leq j \leq n} \big\|\Delta P_j \big(I - P(t_j) A P(t_j)\big)^{-1} A \Delta P_j\big\|$$

$$\leq \left(\sup_t \big\|\big(I - P(t) A P(t)\big)^{-1} A^{1/2}\big\|\right) \|A^{1/2} \Delta P_j\|.$$

Since $t \to A^{1/2} P(t)$ is uniformly continuous, it follows that $\|W_\pi - I\| \to 0$ and therefore $W = I$. $\square$

Theorem 2.2 remains valid if the condition $\|A\| < 1$ is replaced by the weaker requirement that $I - P(t) A P(t)$ be invertible for each $0 \leq t \leq 1$.

XXII.3 *LU*-FACTORIZATIONS OF COMPACT PERTURBATIONS OF THE IDENTITY

Throughout this section H is a separable Hilbert space and the chains are Hilbert space chains. Our aim is to specify Theorem 2.1 further for the case when K is a compact operator on H. We start with some preliminaries.

Let $\mathbb{P}$ be a closed chain on the Hilbert space H. A function F which maps $\mathbb{P}$ into $\mathcal{L}(H)$ is said to be *continuous* on $\mathbb{P}$ if $\|F(P_n) - F(P)\| \to 0$ $(n \to \infty)$ whenever the sequence $(P_n) \subset \mathbb{P}$ converges strongly to $P \in \mathbb{P}$, i.e., whenever

$$\lim_{n \to \infty} P_n x = P x \qquad (x \in H).$$

A continuous map $F\colon \mathbb{P} \to \mathcal{L}(H)$ is automatically bounded. For if this is not the case, then there exists a sequence $(P_n) \subset \mathbb{P}$ such that $\|F(P_n)\| \geq n$, $n = 1, 2, \ldots$. Now (P_n) has a monotone subsequence $(P_{n'})$ which converges strongly to some $P \in \mathbb{P}$. Therefore $\big(F(P_{n'})\big)$ converges in $\mathcal{L}(H)$, which is impossible.

LEMMA 3.1. *Let K be a compact operator on H, and let $\mathbb{P}$ be a closed chain on H. The following holds:*

(a) *the maps $P \mapsto PK$ and $P \mapsto KP$ are continuous on $\mathbb{P}$;*

(b) *if $I - PKP$ is invertible for every $P \in \mathbb{P}$, then the map $P \mapsto (I - PKP)^{-1}$ is continuous on $\mathbb{P}$.*

PROOF. (a) Suppose $F(P) = PK$ is not continuous on $\mathbb{P}$. Then there exists a sequence $(P_n) \subset \mathbb{P}$ which converges strongly to $P \in \mathbb{P}$, yet $\|P_n K - PK\| \geq \varepsilon$ for some $\varepsilon > 0$. Choose $(x_n) \subset H$ such that $\|x_n\| = 1$ and

$$\|P_n K x_n - PK x_n\| \geq \varepsilon/2, \qquad n = 1, 2, \ldots .$$

Since K is compact, there exists a subsequence $(K x_{n'})$ which converges to some $y \in H$. But then

$$\varepsilon/2 \leq \|P_{n'} K x_{n'} - PK x_{n'}\| \leq \|P_{n'} K x_{n'} - P_{n'} y\| + \|P_{n'} y - Py\| + \|Py - PK x_{n'}\|$$
$$\leq \|K x_{n'} - y\| + \|P_{n'} y - Py\| + \|y - K x_{n'}\| \to 0,$$

which is a contradiction.

Note that $KP = (PK^*)^*$. Thus $P \mapsto KP$ is the composition of two continuous maps, and therefore also continuous.

(b) The map $P \mapsto PKP$ is continuous on H. For suppose $(P_n) \subset P$ converges strongly to $P \in \mathbb{P}$. Then, by the above results,

$$\|P_n K P_n - PKP\| \leq \|P_n K P_n - PK P_n\| + \|PK P_n - PKP\|$$
$$\leq \|P_n K - PK\| + \|K P_n - KP\| \to 0.$$

Hence $\|(I - P_n K P_n)^{-1} - (I - PKP)^{-1}\| \to 0$. $\quad\square$

LEMMA 3.2. *Let $F\colon \mathbb{P} \to \mathcal{L}(H)$ be continuous on the closed chain $\mathbb{P}$. Given $\varepsilon > 0$, there exists a partition π such that for any partition $\{Q_0, Q_1, \ldots, Q_m\} \supset \pi$*

$$\|F(Q_j) - F(Q_{j-1})\| \leq \varepsilon, \qquad 1 \leq j \leq m,$$

provided (Q_{j-1}, Q_j) is not a jump of $\mathbb{P}$ which lies in π.

PROOF. Put $P_0 = 0$. To find $\pi = \{P_0, P_1, \ldots, P_n\}$, we start by defining

$$\Sigma_0 = \big\{P \in \mathbb{P} \mid \|F(P) - F(P_0)\| \geq \varepsilon/2\big\}.$$

If $\Sigma_0 = \emptyset$, take $\pi = \{0, I\}$, and we are done. Suppose $\Sigma_0 \neq \emptyset$. Let $\widetilde{P}_1 = \min \Sigma_0$. If $(P_0, \widetilde{P}_1)$ is a jump of $\mathbb{P}$, take $P_1 = \widetilde{P}_1$; if not, take

$$P_1 = \max\big\{P \in \mathbb{P} \mid P \leq \widetilde{P}_1, \|F(P) - F(P_0)\| \leq \varepsilon/2\big\}.$$

Then $P_0 < P_1 \leq \widetilde{P}_1$ and $\|F(P) - F(Q)\| \leq \varepsilon$ if (P_0, P_1) is not a jump of $\mathbb{P}$ and $P_0 \leq P < Q \leq P_1$. Note that $P_1 = \widetilde{P}_1$ or $(P_1, \widetilde{P}_1)$ is a jump. Continuing, define

$$\Sigma_1 = \{P \in \mathbb{P} \mid P > P_1, \|F(P) - F(P_1)\| \geq \varepsilon/2\}.$$

If $\Sigma_1 = \emptyset$, take $\pi = \{0, P_1, I\}$, and we are done. Now suppose $\Sigma_1 \neq \emptyset$. Let $\widetilde{P}_2 = \min \Sigma_1$. If $(P_1, \widetilde{P}_2)$ is a jump, take $P_2 = \widetilde{P}_2$; if not, take

$$P_2 = \max\{P \in \mathbb{P} \mid P_1 \leq P \leq \widetilde{P}_2, \|F(P) - F(P_1)\| \leq \varepsilon/2\}.$$

Then $P_1 < P_2 \leq \widetilde{P}_2$ and $\|F(P) - F(Q)\| \leq \varepsilon$ if (P_1, P_2) is not a jump and $P_1 \leq P < Q \leq P_2$. Also, $\widetilde{P}_1 \leq P_2$, since $(P_1, \widetilde{P}_1)$ is a jump if $P_1 \neq \widetilde{P}_1$. Furthermore, $P_2 = \widetilde{P}_2$ or $(P_2, \widetilde{P}_2)$ is a jump. Continuing in this manner we obtain $P_0 < P_1 \leq \widetilde{P}_1 \leq P_2 \leq \widetilde{P}_2 \leq \cdots \leq I$, and

$$\Sigma_k = \{P \in \mathbb{P} \mid P > P_k, \|F(P) - F(P_k)\| \geq \varepsilon/2\}.$$

For some k, we must have $\Sigma_k = \emptyset$. Indeed, if $\Sigma_k \neq \emptyset$ for $k = 1, 2, \ldots$, then the sequence $P_1, \widetilde{P}_1, P_2, \widetilde{P}_2, \ldots$ converges strongly. But then the sequence

$$F(P_1), F(\widetilde{P}_1), F(P_2), F(\widetilde{P}_2), \ldots$$

converges in $\mathcal{L}(H)$ by the continuity of F. But this is impossible since $\|F(\widetilde{P}_j) - F(P_{j-1})\| \geq \varepsilon/2$ for all j. So, if $\Sigma_{n-1} = \emptyset$, then the process stops and $\pi = \{P_0, \ldots, P_{n-1}, I\}$ is the desired partition. $\square$

THEOREM 3.3. *Suppose* $T \in \mathcal{L}(H)$ *is compact. Let* $F : \mathbb{P} \to \mathcal{L}(H)$ *be a bounded function on the closed chain* $\mathbb{P}$. *If* (P_ν^-, P_ν^+) *are the jumps in* $\mathbb{P}$, $\nu = 1, 2, \ldots,$ *then*

$$(1) \qquad \int_{\mathbb{P}]} (dP)TF(P)dP = \sum_\nu (P_\nu^+ - P_\nu^-)TF(P_\nu^+)(P_\nu^+ - P_\nu^-).$$

Furthermore, if K *is a Hilbert-Schmidt operator on* H, *then*

$$(2) \qquad \int_{\mathbb{P}]} (dP)TF(P)KdP = \sum_\nu (P_\nu^+ - P_\nu^-)TF(P_\nu^+)K(P_\nu^+ - P_\nu^-),$$

with the integral and series converging in the Hilbert-Schmidt norm.

PROOF. First we prove (2) including the statement about the convergence in the Hilbert-Schmidt norm. In what follows, $K \in S_2$, and $\gamma = \sup_{P \in \mathbb{P}} \|F(P)\|$. For $A \in \mathcal{L}(H)$ and any partition $\pi = \{P_0, P_1, \ldots, P_n\}$ of $\mathbb{P}$, define

$$S_A(\pi) = \sum_{j=1}^n \Delta P_j A F(P_j) K \Delta P_j, \qquad \widetilde{S}_A = \sum_\nu (P_\nu^+ - P_\nu^-) A F(P_\nu^+) K (P_\nu^+ - P_\nu^-).$$

If B is in $\mathcal{L}(H)$, then

$$
\|\widetilde{S}_A - \widetilde{S}_B\|_2^2 = \sum_\nu \|(P_\nu^+ - P_\nu^-)(A - B)F(P_\nu^+)K(P_\nu^+ - P_\nu^-)\|_2^2
$$
(3)
$$
\leq \gamma^2 \|A - B\|^2 \sum_\nu \|K(P_\nu^+ - P_\nu^-)\|_2^2 \leq \gamma^2 \|A - B\|^2 \|K\|_2^2.
$$

Similarly,

(4)
$$
\|S_A(\pi) - S_B(\pi)\|_2 \leq \gamma \|A - B\| \|K\|_2.
$$

Since every compact operator is the limit in norm of a sequence of linear combinations of operators of the form $A_\phi x = \langle x, \phi \rangle \phi$, $\|\phi\| = 1$, it suffices, in light of (3) and (4), to assume that $T = A_\phi$. Let $\varepsilon > 0$ be given. Choose M so that

(5)
$$
\left\| \sum_{\nu > M} (P_\nu^+ - P_\nu^-)\phi \right\| < \varepsilon/6,
$$

and let $\psi = \phi - \sum_{\nu > M}(P_\nu^+ - P_\nu^-)\phi$. Then $\|\psi\| \leq 1$ and

(6)
$$
\|A_\phi - A_\psi\| \leq 2\|\phi - \psi\| < \varepsilon/3.
$$

Clearly, $(P_\nu^+ - P_\nu^-)\psi = 0$, $\nu > M$. Therefore,

$$
(P_\nu^+ - P_\nu^-)A_\psi F(P_\nu^+)K(P_\nu^+ - P_\nu^-) = 0, \qquad \nu > M.
$$

Hence

$$
\widetilde{S}_{A_\psi} = \sum_{\nu=1}^{M}(P_\nu^+ - P_\nu^-)A_\psi F(P_\nu^+)K(P_\nu^+ - P_\nu^-).
$$

By taking $F(P) = PA_\psi$ in Lemma 3.2 it follows that there exists a partition $\pi_\varepsilon = \{\widetilde{P}_0, \widetilde{P}_1, \ldots, \widetilde{P}_m\}$ of $\mathbb{P}$ which contains all projections $P_1^-, P_1^+, \ldots, P_M^-, P_M^+$ and has the property that $\|(\Delta\widetilde{P}_j)\psi\| < \varepsilon/3$ whenever $(\widetilde{P}_{j-1}, \widetilde{P}_j)$ is not one of the jumps (P_ν^-, P_ν^+), $1 \leq \nu \leq M$. Note that $(P_\nu^+ - P_\nu^-)\psi = 0$ for $\nu > M$. Let $\pi = \{P_0, P_1, \ldots, P_n\}$ be a refinement of π_ε. If Λ is the set of indices j for which (P_{j-1}, P_j) does not coincide with any jump (P_ν^-, P_ν^+), $1 \leq \nu \leq M$, then

$$
\|\widetilde{S}_{A_\psi} - S_{A_\psi}(\pi)\|_2^2 \leq \sum_{j \in \Lambda} \|\Delta P_j A_\psi F(P_j)K\Delta P_j\|_2^2
$$
(7)
$$
\leq \gamma^2 \sum_{j \in \Lambda} \|(\Delta P_j)\psi\|^2 \|K\Delta P_j\|_2^2
$$
$$
\leq (\varepsilon/3)^2 \gamma^2 \|K\|_2^2.
$$

From (3)–(7) we get

$$
\|\widetilde{S}_{A_\phi} - S_{A_\phi}(\pi)\|_2 \leq \|\widetilde{S}_{A_\phi} - \widetilde{S}_{A_\psi}\|_2 + \|\widetilde{S}_{A_\psi} - S_{A_\psi}(\pi)\|_2 + \|S_{A_\psi}(\pi) - S_{A_\phi}(\pi)\|_2 \leq \varepsilon\gamma\|K\|_2,
$$

which completes the proof of (2) (including the statement about the convergence in Hilbert-Schmidt norm). The proof of (1) follows along the same lines by replacing K by the identity operator and the Hilbert-Schmidt norm $\| \cdot \|_2$ by the usual operator norm $\| \cdot \|$. $\square$

THEOREM 3.4. *Let K be a compact operator on H, and let $I - K$ be invertible. Then $I - K$ admits an LU-factorization with respect to a maximal chain $\mathbb{P}$ if and only if the following two conditions hold:*

(a) *$I - PKP$ is invertible for each $P \in \mathbb{P}$,*

(b) *at least one of the integrals*

$$X_- = \int_{[\mathbb{P}} (dP)K(I - PKP)^{-1}P, \qquad X_+ = \int_{[\mathbb{P}} (I - PKP)^{-1}PK\,dP$$

converges in $\mathcal{L}(H)$. In this case, both integrals converge and

$$I - K = (I + Y_-)D(I + Y_+),$$

where $Y_\pm$ are Volterra operators in $A_\pm(\mathbb{P})$, $D \in A_0(\mathbb{P})$,

$$(I + Y_\pm)^{-1} = I + X_\pm,$$

$$
\begin{aligned}
(8) \qquad D^{-1} &= \int_{\mathbb{P}]} (dP)(I - PKP)^{-1}\,dP \\
&= I + \sum_\nu (P_\nu^+ - P_\nu^-)K(I - P_\nu^+ K P_\nu^+)^{-1}(P_\nu^+ - P_\nu^-)
\end{aligned}
$$

where (P_ν^-, P_ν^+), $\nu = 1, 2, \ldots$, are the jumps of the maximal chain $\mathbb{P}$.

PROOF. Since a Hilbert space chain is a uniform chain, we can apply Theorem 2.1 and the remark following its proof. So we need only to show that the integral in (8) converges and has the desired value. Now for any $A \in \mathcal{L}(H)$ we have

$$(I - A)^{-1} = I + A(I - A)^{-1} = I + (I - A)^{-1}A,$$

provided $I - A$ is invertible. Hence, if $\{P_0, P_1, \ldots, P_n\}$ is a partition of $\mathbb{P}$, then

$$
\begin{aligned}
\sum_{j=1}^{n} (\Delta P_j)(I - P_j K P_j)^{-1}\Delta P_j &= \sum_{j=1}^{n} (\Delta P_j)\{I + P_j K P_j(I - P_j K P_j)^{-1}\}\Delta P_j \\
&= I + \sum_{j=1}^{n} (\Delta P_j)K(I - P_j K P_j)^{-1}\Delta P_j.
\end{aligned}
$$

Here we used that $P(I - PKP)^{-1} = (I - PKP)^{-1}P$ and $(\Delta P_j)P_j = \Delta P_j$. Next we apply (1) in Theorem 3.3 with K in place of T and with $F(P) = (I - PKP)^{-1}$. By

Lemma 3.1(b) the map F is continuous on $\mathbb{P}$, and hence F is bounded. Note that our chain $\mathbb{P}$ is closed (by Proposition XX.4.1). Thus Theorem 3.3 can be used to show that the second equality in (8) holds true. $\square$

COROLLARY 3.5. *If the chain $\mathbb{P}$ in the previous theorem is continuous and (a) and (b) hold, then D in (8) is the identity operator and*

$$I - K = (I + Y_-)(I + Y_+).$$

XXII.4 *LU*-FACTORIZATIONS OF HILBERT-SCHMIDT PERTURBATIONS OF THE IDENTITY

Our aim in this section is to show that if the operator K in Theorem 3.4 is Hilbert-Schmidt, then $I - K$ admits an LU-factorization if and only if condition (a) holds. We assume that H is a separable Hilbert space and the chains are chains of orthogonal projections.

PROPOSITION 4.1. *Let $\mathbb{P}$ be a closed chain on H, and let $F : \mathbb{P} \to \mathcal{L}(H)$ be continuous. If K is a Hilbert-Schmidt operator on H, then the integral*

$$(1) \qquad \int_{[\mathbb{P}} F(P) K \, dP$$

converges in the Hilbert-Schmidt norm. If, in addition, the chain $\mathbb{P}$ is continuous, then

$$(2) \qquad \int_{\mathbb{P}} F(P) K \, dP$$

converges in the Hilbert-Schmidt norm.

PROOF. Given $\varepsilon > 0$ there exists, by Lemma 3.2, a partition $\pi_\varepsilon = \{P_0, P_1, \ldots, P_m\}$ of $\mathbb{P}$ such that for any partition $\pi' = \{P_0', P_1', \ldots, P_r'\}$ of $\mathbb{P}$ finer than π_ε

$$\|F(P_k') - F(P_{k-1}')\| \le \varepsilon$$

provided (P_{k-1}', P_k') is not a jump which lies in π_ε.

Let $\pi = \{Q_0, Q_1, \ldots, Q_n\}$ be a partition of $\mathbb{P}$ finer than π_ε. Define

$$S(\pi) = \sum_{k=1}^{n} F(Q_{k-1}) K \Delta Q_k, \qquad S(\pi_\varepsilon) = \sum_{j=1}^{m} F(P_{j-1}) K \Delta P_j.$$

Since π is finer than π_ε, we have

$$S(\pi_\varepsilon) - S(\pi) = \sum_{k=1}^{n} (\Delta F_k) K \Delta Q_k,$$

where $\Delta F_k = F(Q_{k-1}) - F(\widetilde{P}_{k-1})$ and each $\widetilde{P}_{k-1}$ coincides with P_{j-1} whenever $P_{j-1} \leq Q_{k-1} < Q_k \leq P_j$.

The property of π_ε mentioned in the first paragraph of the proof implies that

$$(3) \qquad \|\Delta F_k\| \leq \varepsilon, \qquad k = 1, \ldots, n.$$

To see this, fix $1 \leq k \leq n$, and choose j such that $P_{j-1} \leq Q_{k-1} < Q_k \leq P_j$. If $Q_{k-1} = P_{j-1}$, then $\Delta F_k = 0$ by definition. Therefore, assume that $P_{j-1} < Q_{k-1}$, and consider the partition $\widetilde{\pi} = \pi_\varepsilon \cup \{Q_{k-1}\}$. Then $\widetilde{\pi}$ is finer than π_ε and (P_{j-1}, Q_{k-1}) does not lie in π_ε. Thus by the result mentioned in the first paragraph of the proof (applied to $\widetilde{\pi}$ in place of π') we have

$$\|F(Q_{k-1}) - F(P_{j-1})\| \leq \varepsilon,$$

which proves (3).

From (3) it follows that

$$\|S(\pi) - S(\pi_\varepsilon)\|_2^2 = \mathrm{tr}\big(S(\pi) - S(\pi_\varepsilon)\big)\big(S(\pi) - S(\pi_\varepsilon)\big)^*$$

$$= \mathrm{tr} \sum_{k=1}^n (\Delta F_k) K (\Delta Q_k) K^* (\Delta F_k)^*$$

$$= \sum_{k=1}^n \mathrm{tr}(\Delta F_k) K (\Delta Q_k) K^* (\Delta F_k)^*$$

$$= \sum_{k=1}^n \|\Delta F_k K \Delta Q_k\|_2^2$$

$$\leq \Big(\max_k \|\Delta F_k\|\Big)^2 \sum_{k=1}^n \|K \Delta Q_k\|_2^2$$

$$\leq \varepsilon^2 \|K\|_2^2.$$

Here we use that

$$\|K\|_2^2 = \mathrm{tr}\, K^* K = \mathrm{tr} \sum_{k=1}^n K^* K \Delta Q_k$$

$$= \sum_{k=1}^n \mathrm{tr}(\Delta Q_k) K^* K \Delta Q_k = \sum_{k=1}^n \|K \Delta Q_k\|^2.$$

Now the Cauchy criterion for convergence guarantees the existence of the integral (1) in the Hilbert-Schmidt norm.

If $\mathbb{P}$ is continuous, then $\mathbb{P}$ has no jumps and we may repeat the above reasoning with $S(\pi)$ and $S(\pi_\varepsilon)$ replaced by

$$R(\pi) = \sum_{k=1}^n F(Q''_{k-1}) K \Delta Q_k, \qquad R(\pi_\varepsilon) = \sum_{j=1}^m F(P''_{j-1}) K \Delta P_j,$$

where $Q_{k-1} \leq Q''_{k-1} \leq Q_k$ and $P_{j-1} \leq P''_{j-1} \leq P_j$. The same arguments prove now the convergence of (2) in S_2. $\square$

THEOREM 4.2. *Let K be a Hilbert-Schmidt operator on H, and let $I - K$ be invertible. Then $I - K$ admits an LU-factorization with respect to a maximal chain $\mathbb{P}$ if and only if $I - PKP$ is invertible for each $P \in \mathbb{P}$. In this case,*

$$I - K = (I + Y_-)D(I + Y_+),$$

where $Y_\pm$ are Hilbert-Schmidt operators in $A_\pm(\mathbb{P})$, $D \in A_0(\mathbb{P})$,

$$(4) \qquad (I + Y_-)^{-1} = I + \int_{[\mathbb{P}} (dP)KP(I - PKP)^{-1}$$

$$(5) \qquad (I + Y_+)^{-1} = I + \int_{[\mathbb{P}} (I - PKP)^{-1}PKdP$$

$$(6) \quad D^{-1} = \int_{\mathbb{P}]} (dP)(I - PKP)^{-1}dP = I + \sum_\nu (P_\nu^+ - P_\nu^-)K(I - P_\nu^+ K P_\nu^+)^{-1}(P_\nu^+ - P_\nu^-),$$

where (P_ν^-, P_ν^+), $\nu = 1, 2, \ldots$ are the jumps of the chain $\mathbb{P}$. The integrals converge in the Hilbert-Schmidt norm.

PROOF. If $I - K$ admits an LU-factorization, then $I - PKP$ is invertible for each $P \in \mathbb{P}$ by Theorem 3.4. So we have to prove the reverse implication.

In what follows we assume that $I - PKP$ is invertible for every $P \in \mathbb{P}$. Put $F(P) = (I - PKP)^{-1}P$. Note that our chain $\mathbb{P}$ is closed by Proposition XX.4.1. Thus we can apply Lemma 3.1 to show that the map F is continuous. But then Proposition 4.1 shows that the integral

$$X_+ = \int_{[\mathbb{P}} (I - PKP)^{-1}PKdP$$

converges in the Hilbert-Schmidt norm. In particular, X_+ exists in $\mathcal{L}(H)$, and we may apply Theorem 3.4 to show that $I - K$ has the desired LU-factorization with the factors given by (4), (5) and (6).

It remains to prove that the integrals in (4) and (6) converge in S_2. Note that

$$(7) \qquad \int_{[\mathbb{P}} (dP)KP(I - PKP)^{-1} = \left(\int_{[\mathbb{P}} (I - PK^*P)^{-1}PK^*dP \right)^*.$$

The last integral converges in S_2 by what has been proved so far applied to K^*. Thus, by duality, one obtains that the same holds true for the integral in (4). Next, if $\pi = \{P_0, P_1, \ldots, P_n\}$ is a partition of $\mathbb{P}$, then

$$(\Delta P_j)(I - P_j K P_j)^{-1}(\Delta P_j) = \Delta P_j + (\Delta P_j)(I - P_j K P_j)^{-1}P_j K P_j(\Delta P_j)$$

$$= \Delta P_j + (\Delta P_j)(I - P_j K P_j)^{-1}P_j K(\Delta P_j).$$

Thus in order to prove the convergence in S_2 of the integral in (6), it suffices to show that

$$\int_{\mathbb{P}]} (dP)F(P)K(dP)$$

converges in S_2 with $F(P) = (I - PKP)^{-1}P$. But the latter follows from Proposition 4.1 because the map F is continuous and the chain $\mathbb{P}$ is closed. $\quad\square$

COROLLARY 4.3. *Suppose that K is a Hilbert-Schmidt operator on H and $I - K$ is strictly positive. Then $I - K$ admits an LU-factorization*

$$(8) \qquad\qquad I - K = (I + Y_-)D(I + Y_-^*)$$

with respect to any maximal chain $\mathbb{P}$ on H.

PROOF. Recall that $I - K$ is strictly positive (see Section I.6) if there exists $\delta > 0$ such that

$$\langle (I - K)y, y \rangle \geq \delta \langle y, y \rangle, \qquad y \in H.$$

Without loss of generality we may assume that $0 < \delta < 1$. Hence, if $\|x\| = 1$ and $P \in \mathbb{P}$, then

$$\|(I - PKP)x\| \geq \langle (I - PKP)x, x \rangle = 1 - \langle KPx, Px \rangle$$
$$\geq 1 - (1 - \delta)\langle Px, Px \rangle \geq 1 - (1 - \delta) = \delta > 0.$$

Thus $I - PKP$ is injective and has closed range. Since PKP is selfadjoint, it follows that $I - PKP$ is invertible. Thus $I - K$ admits an LU-factorization by Theorem 4.2. Formula (7) and $K = K^*$ imply that $Y_-^* = Y_+$. $\quad\square$

The remark made in the last paragraph of the introduction to this chapter implies that the operator D in (8) is selfadjoint.

The factors $I + Y_+$ and $I + Y_-$ in Theorem 4.2 may also be represented by the multiplicative integrals appearing in Theorem 2.1. Moreover, under the conditions of Theorem 4.2, these multiplicative integrals converge in the Hilbert-Schmidt norm. To see this, one first notes that

$$D = \int_{[\mathbb{P}} (dP)[I - K - KP(I - PKP)^{-1}K]\,dP,$$

with convergence in the Hilbert-Schmidt norm, and next one uses identities like formula (8) in Section XXII.2. Since the left hand side of this identity has a limit in the Hilbert-Schmidt norm, the same holds for the right hand side.

XXII.5 *LU*-FACTORIZATIONS OF INTEGRAL OPERATORS

The aim of this section is to apply Theorem 4.2 to integral operators with Hilbert-Schmidt kernel functions. Throughout this section $H = L_2^m([a, b])$, the Hilbert space of all square integrable $\mathbb{C}^m$-valued functions on $[a, b]$, and K is an integral operator on $L_2^m([a, b])$ with an $m \times m$ Hilbert-Schmidt matrix kernel function k. Thus

$$(Kf)(t) = \int_a^b k(t, s)f(s)ds, \qquad a \leq t \leq b,$$

where k is an $m \times m$ matrix function whose entries are square integrable on $[a, b] \times [a, b]$. In particular, K is a Hilbert-Schmidt operator.

Let $\mathbb{P} = \{P_\tau \mid a \leq \tau \leq b\}$ be the chain on $L_2^n([a, b])$ defined by

$$(P_\tau f)(t) = \begin{cases} f(t) & \text{for} \quad a \leq t \leq \tau, \\ 0 & \text{for} \quad \tau < t \leq b. \end{cases}$$

Then $\mathbb{P}$ is a maximal Hilbert space chain and $\mathbb{P}$ has no jumps. Thus, by Theorem 4.2, the operator $I - K$ has an LU-factorization

(1) $$I - K = (I + Y_-)(I + Y_+)$$

if and only if $I - P_\tau K P_\tau$ is invertible for each $\tau \in [a, b]$. Since

$$(P_\tau K P_\tau f)(t) = \begin{cases} \displaystyle\int_a^\tau k(t, s)f(s)ds, & a \leq t \leq \tau, \\ \\ 0 & , \quad \tau < t \leq b, \end{cases}$$

and $P_\tau K P_\tau$ is compact, it follows that $I - P_\tau K P_\tau$ is invertible if and only if the equation

$$f(t) - \int_a^\tau k(t, s)f(s)ds = 0, \qquad a \leq t \leq \tau,$$

has only the trivial solution in $L_2^m([a, b])$. The operators $Y_\pm$ in (1) are Hilbert-Schmidt operators on $L_2^n([a, b])$, and hence $Y_\pm$ are integral operators with kernel functions $y_\pm$, say. Since $\mathbb{P}$ is invariant under Y_+ and $\mathbb{P}^c$ is invariant under Y_-, we conclude that $y_+(t, s) = 0$ for $s < t$ and $y_-(t, s) = 0$ for $s > t$. Thus we have obtained the following result.

THEOREM 5.1. *Let K be an integral operator on $L_2^m([a, b])$ with Hilbert-Schmidt kernel function $k(t, s)$. The operator $I - K$ admits the LU-factorization*

$$I - K = (I + Y_-)(I + Y_+)$$

where $Y_\pm$ are Hilbert-Schmidt integral operators with kernel functions $y_\pm(t, s)$ satisfying

$$y_+(t, s) = 0 \quad (a \leq s < t \leq b), \qquad y_-(t, s) = 0 \quad (a \leq t < s \leq b)$$

if and only if for every $\tau \in [a,b]$ the equation

$$(2) \qquad f(t) - \int_a^\tau k(t,s)f(s)ds = 0, \qquad a \leq t \leq \tau,$$

has only the trivial solution in $L_2^m([a,b])$.

In the particular case that the kernel function k is continuous on $[a,b] \times [a,b]$ we have the following result.

THEOREM 5.2. *Let the kernel function k of the integral operator K be continuous on $[a,b] \times [a,b]$. For τ in $(a,b]$ define K_τ on $L_2^m([a,\tau])$ by*

$$(K_\tau f)(t) = \int_a^\tau k(t,s)f(s)ds, \qquad a \leq t \leq \tau.$$

Suppose $I - K_\tau$ is invertible for each $\tau \in (a,b]$ and $\gamma_\tau(t,s)$ is the kernel function corresponding to $\Gamma_\tau = (I - K_\tau)^{-1} - I$. Then

$$I - K = (I + Y_-)(I + Y_+),$$

where $(I + Y_\pm)^{-1} - I$ have kernel functions $\gamma_\pm$ given by

$$(3a) \qquad\qquad \gamma_+(t,s) = \gamma_s(t,s), \qquad a \leq t \leq s \leq b,$$
$$(3b) \qquad\qquad \gamma_-(t,s) = \gamma_t(t,s), \qquad a \leq s \leq t \leq b,$$

and zero otherwise. Furthermore, the functions $\gamma_\pm$ are continuous on their respective triangles.

PROOF. Since $I - K_\tau$ is invertible, equation (2) has the trivial solution only. Thus, by Theorem 5.1, $I - K$ has the desired factorization. It remains to show that the kernel functions $\gamma_\pm$ of $X_\pm := (I + Y_\pm)^{-1} - I$ have the desired representation.

From $(I - K)(I + X_+) = I + Y_-$ or $-K + X_+ - KX_+ = Y_-$ we get

$$(4) \qquad -k(t,s) + \gamma_+(t,s) - \int_a^b k(t,\eta)\gamma_+(\eta,s)d\eta = y_-(t,s), \qquad a \leq s,t \leq b,$$

where y_- is the kernel function corresponding to Y_-. Since $\gamma_+(t,s) = 0$ for $s < t$, equation (4) yields

$$(5) \qquad \gamma_+(t,s) - \int_a^s k(t,\eta)\gamma_+(\eta,s)d\eta = k(t,s), \qquad a \leq t \leq s \leq b.$$

Since $(I - K_\tau)(I + \Gamma_\tau) = I$, it follows that

$$(6) \qquad \gamma_\tau(t,s) - \int_a^\tau k(t,\eta)\gamma_\tau(\eta,s)d\eta = k(t,s), \qquad a \leq s,t \leq \tau.$$

In general, (6) is an identity in the L_2-sense and the equality holds almost everywhere. However, since the kernel function k is continuous on $[a, b] \times [a, b]$, an inspection of Fredholm's formula representing γ_τ as the quotient of a Fredholm minor and the Fredholm determinant (see Mikhlin [1], Section I.9; also formula (23) in Section VII.7) establishes that γ_τ will be jointly continuous with respect to the variables τ, t, s as well as continuously differentiable with respect to τ. In particular, it follows that the identity (6) holds pointwise. If we put $\tau = s$ in (6), we see from (5) that both $\gamma_+(\cdot, s)$ and $\gamma_s(\cdot, s)$ satisfy the equation

$$(I - K_s)y = k(\cdot, s)$$

on $[a, s]$. Hence $\gamma_+(t, s) = \gamma_s(t, s)$ for $a \leq t \leq s \leq b$ and γ_+ has the desired continuity property. A similar argument shows that $\gamma_-(t, s) = \gamma_t(t, s)$ for $a \leq s \leq t \leq b$ and the continuity of γ_- on this triangle. $\square$

Next we show that if K has finite rank, then the formulas for $\gamma_\pm$ in (3a), (3b) are still valid even though k is not necessarily continuous. So, suppose that

$$(7) \qquad Kf = \sum_{i=1}^{n} \langle f, \varphi_i \rangle \psi_i,$$

where φ_i and ψ_i are in $L_2^m([a, b])$. Let $\Phi(t)$ and $\Psi(t)$ be the $m \times n$ matrices with j-th column equal to $\varphi_j(t)$ and $\psi_j(t)$, respectively. Then the kernel function of K in (7) is given by

$$(8) \qquad k(t, s) = \Psi(t)\Phi(s)^*.$$

In what follows we retain the notations used in Theorem 5.2 and its proof.

First we determine $\gamma_\tau(t, s)$. Fix $a < \tau \leq b$. Note that $K_\tau = BA$, where A and B are the operators defined by

$$A : L_2^m([a, \tau]) \to \mathbb{C}^n, \qquad Af = \int_a^\tau \Phi(s)^* f(s)\,ds,$$

$$B : \mathbb{C}^n \to L_2^m([a, \tau]), \qquad Bx = \Psi(\cdot)x.$$

It follows (cf., the last paragraph of Section III.4) that $I - K_\tau$ is invertible if and only if $I - BA$ is invertible, and in that case

$$(9) \qquad (I - K_\tau)^{-1} = I + B(I - AB)^{-1}A.$$

The matrix of $I - AB$ relative to the standard basis of $\mathbb{C}^n$ is equal to

$$(10) \qquad M(\tau) := I - \int_a^\tau \Phi(s)^* \Psi(s)\,ds.$$

Thus (9) may be rewritten as

$$((I - K_\tau)^{-1}f)(t) = f(t) + \Psi(t)M(\tau)^{-1} \int_a^\tau \Phi(s)^* f(s)\,ds, \qquad a \leq t \leq \tau,$$

and we see that

$$(11) \qquad \gamma_\tau(t,s) = \Psi(t)M(\tau)^{-1}\Phi(s)^*, \qquad a \leq s,t \leq \tau.$$

Our final step is to prove the validity of formulas (3a), (3b) for $\gamma_\pm$. Note that γ_b is the kernel function of $(I-K)^{-1} - I$. Since $(I-K)^{-1} = (I+Y_+)^{-1}(I+Y_-)^{-1}$, we have

$$\gamma_b(t,s) = \gamma_+(t,s) + \gamma_-(t,s) + \int_a^b \gamma_+(t,\eta)\gamma_-(\eta,s)d\eta,$$

and thus

$$(12) \qquad \gamma_b(t,s) = \begin{cases} \gamma_+(t,s) + \displaystyle\int_s^b \gamma_+(t,\eta)\gamma_-(\eta,s)d\eta, & a \leq t < s \leq b, \\[2em] \gamma_-(t,s) + \displaystyle\int_t^b \gamma_+(t,\eta)\gamma_-(\eta,s)d\eta, & a \leq s < t \leq b. \end{cases}$$

Recall that the *LU*-factorization is unique. Thus to prove that $\gamma_\pm$ have the form (3a), (3b), it suffices to show that (12) holds with $\gamma_+(t,s)$ replaced by $\gamma_s(t,s) = \Psi(t)M(s)^{-1}\Phi(s)^*$ and $\gamma_-(t,s)$ replaced by $\gamma_t(t,s) = \Psi(t)M(t)^{-1}\Phi(s)^*$. Take $a \leq t \leq s \leq b$. Then

$$F(t,s) := \gamma_s(t,s) + \int_s^b \gamma_s(t,\eta)\gamma_t(\eta,s)d\eta$$

$$= \Psi(t)M(s)^{-1}\Phi(s)^* + \int_s^b \Psi(t)M(\eta)^{-1}\Phi(\eta)^*\Psi(\eta)M(\eta)^{-1}\Phi(s)^*d\eta$$

$$= \Psi(t)\left[M(s)^{-1} + \int_s^b M(\eta)^{-1}\Phi(\eta)^*\Psi(\eta)M(\eta)^{-1}d\eta\right]\Phi(s)^*.$$

Note that $M(\cdot)$ is absolutely continuous and $\frac{d}{d\tau}M(\tau) = \Phi(\tau)^*\Psi(\tau)$ almost everywhere. It follows that $M(\cdot)^{-1}$ is absolutely continuous and

$$\frac{d}{d\tau}M(\tau)^{-1} = -M(\tau)^{-1}\left(\frac{d}{d\tau}M(\tau)\right)M(\tau)^{-1}$$

$$= M(\tau)^{-1}\Phi(\tau)^*\Psi(\tau)M(\tau)^{-1}, \qquad \text{a.e. on } a \leq \tau \leq b.$$

Thus

$$\int_s^b M(\eta)^{-1}\Phi(\eta)^*\Psi(\eta)M(\eta)^{-1}d\eta = M(b)^{-1} - M(s)^{-1},$$

and hence

$$F(t,s) = \Psi(t)M(b)^{-1}\Phi(s)^* = \gamma_b(t,s), \qquad a \leq t < s \leq b.$$

This shows that the first of the formulas in (12) holds with $\gamma_+(t,s) = \gamma_s(t,s)$ and $\gamma_-(t,s) = \gamma_t(t,s)$. A similar argument proves the analogous result for the second of the formulas in (12), and we are done. We summarize the results in the following theorem.

THEOREM 5.3. *Let K be the finite rank integral operator on $L_2^m([a,b])$ with kernel function $\Psi(t)\Phi(s)^*$, where Ψ and Φ are $m \times n$ matrix functions whose entries are square integrable on $[a,b]$. Put*

$$M(\tau) = I - \int_a^\tau \Phi(s)^*\Psi(s)ds, \qquad a \leq t \leq b.$$

Then $I - K$ admits an LU-factorization if and only if $\det M(\tau) \neq 0$ for each $a \leq \tau \leq b$. In this case

$$(13) \qquad I - K = (I + Y_-)(I + Y_+),$$

where $(I + Y_\pm)^{-1} - I$ have kernel functions $\gamma_\pm$ given by

$$\gamma_+(t,s) = \Psi(t)M(s)^{-1}\Phi(s)^*, \qquad a \leq t \leq s \leq b,$$
$$\gamma_-(t,s) = \Psi(t)M(t)^{-1}\Phi(s)^*, \qquad a \leq s \leq t \leq b,$$

and zero otherwise.

We conclude this section with an example. Let K be the operator defined on $L_2([0,\alpha])$ by

$$(Kf)(t) = \int_0^\alpha f(s)ds, \qquad 0 \leq t \leq \alpha.$$

We shall show that $I - K$ admits an LU-factorization

$$(14) \qquad I - K = (I + Y_-)(I + Y_-^*)$$

if and only if $0 < \alpha < 1$. We then exhibit the kernel functions of Y_- and Y_-^*. Here the factorization is relative to the chain

$$(P_\tau f)(t) = \begin{cases} f(t) & \text{for} \;\; 0 \leq t \leq \tau, \\ 0 & \text{for} \;\; \tau < t \leq \alpha. \end{cases}$$

First note that K is a self-adjoint operator of rank one. In fact, $K = \langle \cdot, \varphi \rangle \varphi$, with $\varphi(t) = 1$ for $0 \leq t \leq \alpha$. It follows from Theorem 5.3 that $I - K$ admits an LU-factorization if and only if $M(\tau) = 1 - \tau \neq 0$ for all $0 \leq \tau \leq \alpha$. The latter happens if and only if $0 < \alpha < 1$.

Assume $0 < \alpha < 1$. Then, by Theorem 5.3, the operator $I - K$ admits an LU-factorization of the form (13). Since $I - K$ is selfadjoint, the remark made in the last paragraph of the introduction to this chapter implies that $Y_+ = Y_-^*$ and we have (14). Furthermore, by Theorem 5.3, the operator $X_- := (I + Y_-)^{-1} - I$ and its adjoint are given by

$$(X_- f)(t) = \frac{1}{1-t} \int_0^t f(s)ds, \qquad 0 \le t \le \alpha$$

$$(X_-^* f)(t) = \int_t^\alpha \frac{1}{1-s} f(s)ds, \qquad 0 \le t \le \alpha.$$

Since $I + Y_- = (I - K)(I + X_-^*)$, we see that

$$(Y_- f)(t) = - \int_0^\alpha f(s)ds + \int_t^\alpha \frac{1}{1-s} f(s)ds - \int_0^\alpha \int_t^\alpha \frac{1}{1-s} f(s)ds\,dt.$$

But

$$\int_0^\alpha \int_t^\alpha \frac{1}{1-s} f(s)ds\,dt = \int_0^\alpha \int_0^s \frac{1}{1-s} f(s)dt\,ds = \int_0^\alpha \frac{s}{1-s} f(s)ds.$$

Hence

$$(Y_- f)(t) = - \int_0^\alpha \frac{1}{1-s} f(s)ds + \int_t^\alpha \frac{1}{1-s} f(s)ds = - \int_0^t \frac{1}{1-s} f(s)ds,$$

and

$$(Y_-^* f)(t) = - \frac{1}{1-t} \int_t^\alpha f(s)ds.$$

Note that $I + Y_-$ is almost unitary since (14) holds and K has rank one.

XXII.6 TRIANGULAR REPRESENTATIONS OF OPERATORS CLOSE TO UNITARY

Throughout this section H is a separable Hilbert space and the chains are Hilbert space chains. Let $A \in \mathcal{L}(H)$, and let $\mathbb{P}$ be an A-invariant chain on H. We are interested in triangular representations of A. The triangular representations which appear in Section XXI.1 are in terms of the chain $\mathbb{P}$ and the imaginary part of A. Furthermore, they are based on additive LU-decompositions of A and apply to operators which are in a certain sense close to selfadjoint operators. In this section we show that LU-factorizations give rise to another type of triangular representation, namely in terms of the chain and the operator $I - A^*A$. Note that the latter operator is a measure for the distance of A to the unitary operators.

THEOREM 6.1. *Let $A: H \to H$ be an invertible operator, and let $\mathbb{P}$ be an invariant chain for A. Assume the diagonal of A with respect to $\mathbb{P}$ exists and is an invertible operator. Put $K = I - A^*A$. Then*

$$(1) \qquad A = E \int_{[\mathbb{P}}^{\frown} I - (I - PKP)^{-1} PK \, dP,$$

$$(2) \qquad A^{-1} = \left(I + \int_{[\mathbb{P}} (I - PKP)^{-1} PK \, dP \right) E^{-1},$$

where E is the diagonal of A. If, in addition, the chain $\mathbb{P}$ is continuous and K is compact, then the operator E is unitary.

PROOF. Put $A_+ = A - E$. Then A_+ leaves $\mathbb{P}$ invariant and has a zero diagonal. So $A_+ \in A_+(\mathbb{P})$. Let us prove that also $E^{-1}A_+$ is in $A_+(\mathbb{P})$. To see this, we first note that $E^{-1}A_+$ leaves the chain $\mathbb{P}$ invariant. Next, if $\pi = \{P_0, P_1, \ldots, P_n\}$ is a partition of $\mathbb{P}$, then the operators A_+ and E^{-1} are block upper triangular matrices with respect to the decomposition

$$H = \operatorname{Im} \Delta P_1 \oplus \operatorname{Im} \Delta P_2 \oplus \cdots \oplus \operatorname{Im} \Delta P_n.$$

It follows that $\operatorname{diag}(E^{-1}A_+; \pi) = \operatorname{diag}(E^{-1}; \pi) \operatorname{diag}(A_+; \pi)$, and by taking limits we see that the diagonal of $E^{-1}A_+$ exists and is equal to zero. So $E^{-1}A_+ \in A_+(\mathbb{P})$.

Now, note that

$$(3) \qquad I - K = A^*A = \left(I + (E^{-1}A_+)^* \right) E^* E (I + E^{-1}A_+),$$

and the right hand side of (3) is an LU-factorization of $I - K$. Since a Hilbert space chain is uniform, we can apply Theorem 2.1 to show that

$$I + E^{-1}A_+ = \int_{[\mathbb{P}}^{\frown} I - (I - PKP)^{-1} PK \, dP,$$

$$(I + E^{-1}A_+)^{-1} = I + \int_{[\mathbb{P}} (I - PKP)^{-1} PK \, dP.$$

But $A = E(I + E^{-1}A_+)$. So (1) and (2) are proved.

Next, assume additionally that $\mathbb{P}$ is continuous and K is compact. Then, by virtue of Corollary 3.5 and the uniqueness of the LU-factorization. the operator E^*E

appearing in the right hand side of (3) is the identity operator on H. Since E is invertible, this implies that E is unitary. $\square$

The next theorem concerns the converse of Theorem 6.1.

THEOREM 6.2. *Let* $\mathbb{P}$ *be a maximal chain on* H, *and let* $K\colon H \to H$ *be a Hilbert-Schmidt operator such that* $I - K$ *is strictly positive. Put*

$$(4) \qquad A = U\sqrt{D}\left(\overset{\curvearrowright}{\underset{[\mathbb{P}}{\int}} I - (I - PKP)^{-1}KP\,dP\right),$$

where U *is an arbitrary unitary operator which is reduced by* $\mathbb{P}$, *and*

$$(5) \qquad D = \underset{[\mathbb{P}}{\int}(dP)\big[I - K - KP(I - PKP)^{-1}K\big]\,dP.$$

Then $A\colon H \to H$ *is invertible,* $\mathbb{P}$ *is an invariant chain for* A, *the diagonal of* A *with respect to* $\mathbb{P}$ *exists and is an invertible operator, and* $I - A^*A = K$.

PROOF. By Corollary 4.3 and Theorem 2.1, the operator $I - K$ admits an *LU*-factorization

$$(6) \qquad I - K = (I + Y_+^*)D(I + Y_+),$$

with respect to $\mathbb{P}$, where D is given by (5) and $I + Y_+$ by the multiplicative integral in the right hand side of (4). Since $I - K$ is strictly positive, the same holds for D, and hence $\sqrt{D}$ is well-defined.

Now, let A be defined by (4). Thus $A = U\sqrt{D}(I + Y_+)$, and $I - A^*A = K$ by (6) and the fact that U is unitary. Since $\sqrt{D}$ appears as the limit in the operator norm of polynomials in D (see Section V.6), the operator $\sqrt{D}$ is reduced by $\mathbb{P}$. By our hypotheses, the same holds for U. So $F := U\sqrt{D} \in A_0(\mathbb{P})$. By using an argument similar to the one used in the first paragraph of the proof of Theorem 6.1, one sees that $FY_+ \in A_+(\mathbb{P})$. It follows that $A = F + FY_+$ leaves the chain $\mathbb{P}$ invariant, and the diagonal of A exists and is equal to F. Obviously, A and F are invertible, and hence A has the desired properties. $\square$

XXII.7 *LU*-FACTORIZATION OF SEMI-SEPARABLE INTEGRAL OPERATORS

In this section we develop further Theorems 5.1 and 5.2 for an integral operator K with a semi-separable kernel function. Thus (cf., Section IX.1) in what follows K is an integral operator on $L_2^m([a, b])$ and its kernel function k has a semi-separable representation, i.e.,

$$(1) \qquad k(t, s) = \begin{cases} F_1(t)G_1(s), & a \le s \le t \le b, \\ F_2(t)G_2(s), & a \le t < s \le b. \end{cases}$$

Here $F_\nu(t)$ and $G_\nu(t)$ ($\nu = 1, 2$) are matrices of sizes $m \times n_\nu$ and $n_\nu \times m$, respectively, and as functions of t the entries of $F_\nu(t)$ and $G_\nu(t)$ are square integrable on $[a, b]$. The integer $n = n_1 + n_2$ is called the *order* of the representation (1). From Proposition IX.1.1 we know that k in (1) is a Hilbert-Schmidt kernel function. With (1) we associate the following matrix function:

$$(2) \qquad A(t) = \begin{bmatrix} G_1(t)F_1(t) & G_1(t)F_2(t) \\ -G_2(t)F_1(t) & -G_2(t)F_2(t) \end{bmatrix}, \qquad a \le t \le b.$$

As in Section XXII.5 we consider LU-factorizations of $I - K$ relative to the chain $\{P_\tau \mid a \le \tau \le b\}$, where

$$(P_\tau f)(t) = \begin{cases} f(t) & \text{for} \quad a \le t \le \tau, \\ 0 & \text{for} \quad \tau < t \le b. \end{cases}$$

We shall prove the following theorem.

THEOREM 7.1. *Let K be an integral operator with a semi-separable kernel function k, and let (1) be a semi-separable representation of k of order n, say. Let $U(t)$ be the unique absolutely continuous $n \times n$ matrix function such that*

$$(3) \qquad U(t) = I_n + \int_a^t A(s)U(s)\,ds, \qquad a \le t \le b,$$

where $A(t)$ is given by (2) and I_n is the $n \times n$ identity matrix. Partition $U(t)$ as a 2×2 block matrix according to the partitioning of the right hand side of (2):

$$(4) \qquad U(t) = \begin{bmatrix} U_{11}(t) & U_{12}(t) \\ U_{21}(t) & U_{22}(t) \end{bmatrix}, \qquad a \le t \le b.$$

Then $I - K$ admits an LU-factorization if and only if $\det U_{22}(\tau) \ne 0$ for each $a \le \tau \le b$. In this case

$$(5) \qquad I - K = (I + Y_-)(I + Y_+),$$

where $Y_\pm$ have kernel functions $y_\pm$ given by

$$(6a) \qquad y_+(t, s) = -F_1(t)U_{12}(t)U_{22}(t)^{-1}G_2(s) - F_2(t)G_2(s), \qquad a \le t \le s \le b,$$

$$(6b) \qquad y_-(t, s) = -F_1(t)G_1(s) - F_1(t)U_{12}(s)U_{22}(s)^{-1}G_2(s), \qquad a \le s \le t \le b,$$

and zero otherwise.

PROOF. Fix $a < \tau \le b$. Note that K_τ is also an integral operator with a semi-separable kernel function. In fact, the kernel function k_τ of K_τ is given by

$$k_\tau(t, s) = \begin{cases} F_1(t)G_1(s), & a \le s \le t \le \tau, \\ F_2(t)G_2(s), & a \le t < s \le \tau, \end{cases}$$

where F_ν and G_ν ($\nu = 1, 2$) are as in (1). Thus we can apply Theorem IX.2.1 to show that $I - K_\tau$ is invertible if and only if $U_{22}(\tau)$ is non-singular. But then it follows from Theorem 5.1 (and the compactness of K_τ) that for $I - K$ to admit an LU-factorization it is necessary and sufficient that $\det U_{22}(\tau) \neq 0$ for each $a < \tau \leq b$. Since $U(a) = I$, the latter condition is automatically fulfilled for $\tau = a$.

To derive the formulas (6a) and (6b) it will be convenient to introduce the following notation. Put

$$C(t) = \begin{bmatrix} F_1(t) & F_2(t) \end{bmatrix}, \qquad B(t) = \begin{bmatrix} G_1(t) \\ -G_2(t) \end{bmatrix}, \qquad P = \begin{bmatrix} 0 & 0 \\ 0 & I \end{bmatrix},$$

and

$$R(t) = \begin{bmatrix} 0 & -U_{12}(t)U_{22}(t)^{-1} \\ 0 & 0 \end{bmatrix} = (I - P)\{I - P + U(t)P\}^{-1}P.$$

Then the right hand sides of (6a) and (6b) may be rewritten as:

$$(7a) \qquad a_+(t, s) = C(t)\{P - R(t)\}B(s), \qquad a \leq t \leq s \leq b,$$
$$(7b) \qquad a_-(t, s) = -C(t)\{I - P + R(s)\}B(s), \qquad a \leq s \leq t \leq b.$$

Set $a_+(t, s) = 0$ on $a \leq s < t \leq b$ and $a_-(t, s) = 0$ on $a \leq t < s \leq b$. Furthermore, let $A_\pm$ be the integral operator on $L_2^m([a, b])$ with kernel function $a_\pm$. We shall prove that

$$(8) \qquad I - K = (I + A_-)(I + A_+).$$

Obviously, (8) is an LU-factorization. Since $I - K$ is invertible, such a factorization is unique, and hence $A_+ = Y_+$ and $A_- = Y_-$, which proves (6a) and (6b).

Thus it remains to establish (8). From (3) it follows that the function $R(\cdot)$ is absolutely continuous. In particular, the derivative $R'(t)$ exists almost everywhere on $a \leq t \leq b$. To compute $R'(t)$ we use formula (10) in Section I.1. We have

$$R'(t) = -(I - P)\{I - P + U(t)P\}^{-1}A(t)U(t)P\{I - P + U(t)P\}^{-1}P.$$

Now insert $I - P + P$ before the term $A(t)$, replace the term $U(t)P$ directly after $A(t)$ by

$$\{I - P + U(t)P\} - (I - P),$$

and use that $\{I - P + U(t)P\}^{-1}(I - P) = I - P$. In this way we obtain

$$(9) \qquad R'(t) = -\{I - P + R(t)\}A(t)\{P - R(t)\}$$

almost everywhere on $a \leq t \leq b$.

Next, let a be the kernel function of A_-A_+. For $a \le t \le s \le b$ we have

$$a(t,s) = \int_a^b a_-(t,\eta)a_+(\eta,s)d\eta$$

$$= -\int_a^t C(t)\{I - P + R(\eta)\}B(\eta)C(\eta)\{P - R(\eta)\}B(s)d\eta$$

$$= -C(t)\left(\int_a^t \{I - P + R(\eta)\}A(\eta)\{P - R(\eta)\}d\eta\right)B(s)$$

$$= C(t)\{R(\eta)|_a^t\}B(s)$$

$$= C(t)R(t)B(s), \qquad a \le t \le s \le b.$$

Here we used (9), the identity $A(t) = B(t)C(t)$ and the fact that $R(a) = 0$. Thus for $a \le t \le s \le b$ we have

$$a_-(t,s) + a_+(t,s) + a(t,s) = C(t)\{P - R(t)\}B(s) + C(t)R(t)B(s)$$

$$= C(t)PB(s) = -F_2(t)G_2(s)$$

$$= -k(t,s).$$

In a similar way one shows that

$$(10) \qquad a_-(t,s) + a_+(t,s) + a(t,s) = -k(t,s)$$

for $a \le s \le t \le b$. Thus (10) holds for all (t,s), and therefore we have (8). $\square$

The proof of formulas (6a), (6b) does not explain how one comes to such formulas. For this purpose we use Theorem 5.2. Let us assume that the kernel function k in (1) is also continuous. In what follows we continue to use the notations introduced in the above proof, and we assume that $\det U_{22}(\tau) \ne 0$ for each $a \le \tau \le b$. Since the kernel function of $I - K_\tau$ is semi-separable, we can apply Theorem IX.2.1 to show that the kernel function corresponding to $(I - K_\tau)^{-1} - I$ is given by

$$\gamma_\tau(t,s) = \begin{cases} C(t)U(t)(I - P(\tau))U(s)^{-1}B(s), & a \le s \le t \le \tau, \\ -C(t)U(t)P(\tau)U(s)^{-1}B(s), & a \le t < s \le \tau, \end{cases}$$

where $P(\tau)$ is the $n \times n$ matrix

$$P(\tau) = \begin{bmatrix} 0 & 0 \\ U_{22}(\tau)^{-1}U_{21}(\tau) & I \end{bmatrix}.$$

It follows from Theorem 5.2 that $X_\pm := (I + Y_\pm)^{-1} - I$ have kernel functions $\gamma_\pm$ given by

$$(11a) \qquad \gamma_+(t,s) = -C(t)U(t)P(s)U(s)^{-1}B(s), \qquad a \le t \le s \le b,$$

$$(11b) \qquad \gamma_-(t,s) = C(t)U(t)(I - P(t))U(s)^{-1}B(s), \qquad a \le s \le t \le b,$$

and zero otherwise. Note that the kernel functions of $X_\pm$ are semi-separable. So to obtain the kernel functions of $Y_\pm = (I + X_\pm)^{-1} - I$ we may apply again Theorem IX.2.1. Let us carry out the inversion procedure of Theorem IX.2.1 for X_-.

First we simplify the expression for γ_-. Introduce the following auxiliary functions:

$$\Delta(t) = U_{11}(t) - U_{21}(t)U_{22}(t)^{-1}U_{21}(t),$$
$$\rho(t) = U_{12}(t)U_{22}(t)^{-1},$$
$$S(t) = I + R(t),$$
$$V(t) = I - P(t) + P.$$

For each $a \leq t \leq b$ the matrices $S(t)$ and $V(t)$ are invertible,

$$S(t)^{-1} = I - R(t), \qquad V(t)^{-1} = I + P(t) - P.$$

Furthermore, we have the following identities:

$$S(t)U(t)V(t) = \begin{bmatrix} \Delta(t) & 0 \\ 0 & U_{22}(t) \end{bmatrix}$$

$$I - P(t) = V(t)(I - P), \qquad (I - P)V(t) = I - P,$$

which hold for each $t \in [a, b]$. It follows that

$$
\begin{aligned}
C(t)U(t)\big(I - P(t)\big) &= C(t)U(t)V(t)(I - P) \\
&= C(t)\begin{bmatrix} \Delta(t) & U_{12}(t) \\ 0 & U_{22}(t) \end{bmatrix}\begin{bmatrix} I & 0 \\ 0 & 0 \end{bmatrix} \\
&= [F_1(t)\Delta(t) \quad 0], \qquad a \leq t \leq b,
\end{aligned}
$$

and

$$
\begin{aligned}
(I - P)U(s)^{-1}B(s) &= (I - P)V(s)^{-1}U(s)^{-1}B(s) \\
&= (I - P)\begin{bmatrix} \Delta(s)^{-1} & 0 \\ 0 & U_{22}(s)^{-1} \end{bmatrix} S(s)B(s) \\
&= \begin{bmatrix} \Delta(s)^{-1} & 0 \\ 0 & 0 \end{bmatrix}\begin{bmatrix} I & -\rho(s) \\ 0 & I \end{bmatrix}\begin{bmatrix} G_1(s) \\ -G_2(s) \end{bmatrix} \\
&= \Delta(s)^{-1}\big(G_1(s) + \rho(s)G_2(s)\big), \qquad a \leq s \leq b.
\end{aligned}
$$

By combining these identities we obtain:

$$(12) \qquad \gamma_-(t, s) = F_1(t)\Delta(t)\Delta(s)^{-1}\big(G_1(s) + \rho(s)G_2(s)\big), \qquad a \leq s \leq t \leq b.$$

Now, put

$$\widetilde{A}_-(t) = -\Delta(t)^{-1}\big(G_1(t) + \rho(t)G_2(t)\big)F_1(t)\Delta(t), \qquad a \leq t \leq b.$$

To invert $I + X_-$ according to the procedure of Theorem IX.2.1 we have to find the fundamental matrix (normalized to I_n at $t = 0$) of the differential equation

$$(13) \qquad x'(t) = \widetilde{A}_-(t)x(t), \qquad a \leq t \leq b.$$

Note that

$$\begin{bmatrix} \Delta(t) & 0 \\ 0 & 0 \end{bmatrix} = (I - P)S(t)U(t)(I - P), \qquad a \le t \le b.$$

Furthermore, $S'(t) = R'(t)$ almost everywhere. Thus we can use (3) and (9) to show that $\Delta(t)$ is absolutely continuous on $[a, b]$ and

$$(14) \qquad \Delta'(t) = \big(G_1(t) + \rho(t)G_2(t)\big)F_1(t)\Delta(t)$$

almost everywhere on $a \le t \le b$. But then we may conclude (cf., Lemma IX.2.2) that

$$\frac{d}{dt}\big(\Delta(t)^{-1}\big) = -\Delta(t)^{-1}\big(G_1(t) + \rho(t)G_2(t)\big)F_1(t)$$
$$= \widetilde{A}_-(t)\Delta(t)^{-1}$$

almost everywhere on $a \le t \le b$. Since $\Delta(t) = I$ at $t = a$, we see that $\Delta(\cdot)^{-1}$ is the fundamental matrix of (13). A straightforward application of Theorem IX.2.1 now shows that the kernel function $x_-^{\times}$ of $(I + X_-)^{-1} - I$ is given by

$$x_-^{\times}(t, s) = \begin{cases} -F_1(t)\big(G_1(s) + \rho(s)G_2(s)\big), & a \le s \le t \le b, \\ 0 & , \quad a \le t < s \le b. \end{cases}$$

In other words, $x_-^{\times}$ coincides with the kernel function y_- introduced in Theorem 7.1. In a similar way, by inverting $I + X_+$, one sees that the choice of y_+ in (6a) is a natural one.

The next corollary shows that formula (12) for γ_- also holds when k is not continuous.

COROLLARY 7.2. *Let K be the semi-separable integral operator with kernel function k given by* (1), *and assume that $I - K$ admits the LU-factorization*

$$I - K = (I + Y_-)(I + Y_+).$$

Then the kernel functions $\gamma_\pm$ of $(I + Y_\pm)^{-1} - I$ are given by

$$(15a) \qquad \gamma_+(t, s) = \big(F_1(t)\rho(t) + F_2(t)\big)U_{22}(t)U_{22}(s)^{-1}G_2(s), \qquad a \le t \le s \le b,$$

$$(15b) \qquad \gamma_-(t, s) = F_1(t)\Delta(t)\Delta(s)^{-1}\big(G_1(s) + \rho(s)G_2(s)\big), \qquad a \le s \le t \le b,$$

and zero otherwise. Here

$$\Delta(t) = U_{11}(t) - U_{12}(t)U_{22}(t)^{-1}U_{21}(t), \qquad a \le t \le b,$$
$$\rho(t) = U_{12}(t)U_{22}(t)^{-1}, \qquad a \le t \le b,$$

with

$$U(t) = \begin{bmatrix} U_{11}(t) & U_{12}(t) \\ U_{21}(t) & U_{22}(t) \end{bmatrix}, \qquad a \le t \le b,$$

as in (3) and (4).

PROOF. From (6a) we see that the kernel function y_+ is semi-separable. So to find $(I + Y_+)^{-1}$ we may apply Theorem IX.2.1 with $k(t,s) = -y_+(t,s)$. Note that for this case the corresponding functions $F_1(\cdot)$ and $G_1(\cdot)$ are identically zero, and thus we have to determine the fundamental matrix of the differential equation

$$(16) \qquad x'(t) = -G_2(t)\{F_1(t)\rho(t) + F_2(t)\}, \qquad a \le t \le b.$$

Since $U(t)$ is given by (3) and (4) the matrix $U_{22}(0)$ is an identity matrix and

$$\begin{aligned} U'_{22}(t) &= \begin{bmatrix} -G_2(t)F_1(t) & -G_2(t)F_2(t) \end{bmatrix} \begin{bmatrix} U_{12}(t) \\ U_{22}(t) \end{bmatrix} \\ &= -G_2(t)F_1(t)U_{12}(t) - G_2(t)F_2(t)U_{22}(t) \\ &= -G_2(t)\{F_1(t)\rho(t) + F_2(t)\}U_{22}(t), \end{aligned}$$

almost everywhere on $a \le t \le b$. Thus U_{22} is the fundamental matrix corresponding to (16). A straightforward application of Theorem IX.2.1 now yields the desired expression for γ_+.

In a similar way one derives formula (15b). The identity (14) shows that for this case the associated fundamental matrix is the matrix function Δ. $\square$

XXII.8 GENERALIZED WIENER-HOPF EQUATIONS

In this section we present other necessary and sufficient conditions for an operator of the form $I - K$, K Hilbert-Schmidt, to admit an LU-factorization. First, let us introduce some notation. Let $\mathbb{P}$ be a chain of orthogonal projections on a separable Hilbert space H, and let $A^{(2)}$ denote the set of all operators A on H of the form $A = \lambda I + K$ with $\lambda \in \mathbb{C}$ and K Hilbert-Schmidt. Put

$$A_\alpha^{(2)}(\mathbb{P}) = A^{(2)} \cap A_\alpha(\mathbb{P}), \qquad \alpha = -, 0, +.$$

Since $I \in A_0(\mathbb{P})$ and each Hilbert-Schmidt operator on H admits an additive LU-decomposition relative to $\mathbb{P}$ of which the factors are again Hilbert-Schmidt operators (by Corollary XX.8.2), we have

$$(1) \qquad A^{(2)} = A_-^{(2)}(\mathbb{P}) \oplus A_0^{(2)}(\mathbb{P}) \oplus A_+^{(2)}(\mathbb{P}).$$

THEOREM 8.1. *Let K be a Hilbert-Schmidt operator on a separable Hilbert space H. Assume $I - K$ is invertible. Then $I - K$ admits a factorization with respect to a chain $\mathbb{P}$ on H if and only if there exist $X_\pm \in A_0^{(2)}(\mathbb{P}) \oplus A_\pm^{(2)}(\mathbb{P})$ satisfying the following two equations:*

$$(2a) \qquad X_+ - Q_+(KX_+) = Q_+(K),$$

$$(2b) \qquad X_- - Q_-(X_-K) = Q_-(K),$$

where $Q_\pm$ is the projection of $A^{(2)}$ onto $A_0^{(2)} \oplus A_\pm^{(2)}(\mathbb{P})$ along $A_\mp^{(2)}(\mathbb{P})$. In this case

$$I - K = (I + Y_-)D(I + Y_+)$$

with

$$Y_- = X_+ - KX_+ - K \in A_-^{(2)}(\mathbb{P}),$$

$$Y_+ = X_- - X_-K - K \in A_+^{(2)}(\mathbb{P}),$$

$$D = (I + Y_-)^{-1}(I + X_-)^{-1} = (I + X_+)^{-1}(I + Y_+)^{-1} \in A_0^{(2)}(\mathbb{P}).$$

For a continuous chain $\mathbb{P}$ equations (2a) and (2b) can also be written in the form

$$X_+ - \int_{\mathbb{P}} PKX_+ dP = \int_{\mathbb{P}} PK\,dP,$$

$$X_- - \int_{\mathbb{P}} (dP)X_-KP = \int_{\mathbb{P}} (dP)KP.$$

Equations (2a), (2b) are called *generalized Wiener-Hopf equations*, because they may be viewed as analogues of the equations

$$x_+(t) - \int_0^\infty k(t - s)x_+(s)ds = k(t), \qquad 0 \le t < \infty,$$

$$x_-(t) - \int_{-\infty}^0 x_-(s)k(t - s)ds = k(t), \qquad -\infty < t \le 0,$$

which arise naturally in problems of Wiener-Hopf factorization.

To prove Theorem 8.1 it will be convenient first to consider a more abstract version. Let A be an algebra with unit e which is a direct sum

$$A = A_- \oplus A_0 \oplus A_+$$

of subalgebras $A_\pm$ and A_0 with the following properties:

(i) If $a_\pm \in A_\pm$, then $e + a_\pm$ is invertible and $(e + a_\pm)^{-1} - e \in A_\pm$.

(ii) The unit e belongs to A_0, and if $a_0 \in A_0$ is invertible, then $a_0^{-1} \in A_0$.

(iii) If $a_0 \in A_0$ and $a_\pm \in A_\pm$, then $a_0 a_\pm$ and $a_\pm a_0$ are in $A_\pm$.

We shall call A with the above properties a *strongly decomposable* algebra.

An element $e - a$ in A is said to admit *an LU-factorization* if it can be represented in the form

$$(3) \qquad\qquad e - a = (e + a_-)a_0(e + a_+), \qquad a_\pm \in A_\pm,\ a_0 \in A_0.$$

An *LU*-factorization, if it exists, is unique, as may be seen from the proof of the uniqueness statement of the last paragraph of the introduction to the present chapter.

THEOREM 8.2. *Let $e-a$ be an invertible element in a strongly decomposable algebra* A. *A necessary and sufficient condition that $e - a$ admits an LU-factorization is that there exist $x_\pm \in$ A$_0 \oplus$ A$_\pm$ satisfying the following two equations*:

$$(4a) \qquad\qquad x_+ - Q_+(ax_+) = Q_+(a),$$

$$(4b) \qquad\qquad x_- - Q_-(x_-a) = Q_-(a),$$

where $Q_\pm$ is the projection of A *onto* A$_0 \oplus$ A$_\pm$ *along* A$_\mp$. *In this case*

$$e - a = (e + a_-)a_0(e + a_+),$$

where

$$a_- = x_+ - ax_+ - a \in \text{A}_-,$$
$$a_+ = x_- - x_-a - a \in \text{A}_+,$$
$$a_0 = (e + a_-)^{-1}(e + x_-)^{-1} = (e + x_+)^{-1}(e + a_+)^{-1} \in \text{A}_0.$$

PROOF. Suppose that $e - a$ admits the *LU*-factorization given in (3). Since $e - a$ is invertible, condition (i) implies that a_0 is invertible and (use conditions (i), (ii) and (iii))

$$e + x_+ := (e + a_+)^{-1}a_0^{-1} \in \text{A}_0 + \text{A}_+.$$

Then $(e - a)(e + x_+) = e + a_-$, and hence

$$(5) \qquad\qquad x_+ - ax_+ = a + a_-.$$

Applying Q_+ to both sides of (5) yields (4a). Similarly, put

$$e + x_- := a_0^{-1}(e + a_-)^{-1} \in \text{A}_0 \oplus \text{A}_-.$$

Then $(e + x_-)(e - a) = e + a_+$, and hence

$$(6) \qquad\qquad x_- - x_-a = a + a_+.$$

Applying Q_- to both sides of (6) yields (4b).

Conversely, suppose there exist $x_\pm \in$ A$_0 \oplus$ A$_\pm$ which satisfy equations (4a) and (4b). Then for some $a_\pm \in$ A$_\pm$,

$$(7) \qquad\qquad e + x_+ - ax_+ = e + a + a_-,$$

$$(8) \qquad\qquad e + x_- - x_-a = e + a + a_+.$$

Thus

$$(9) \qquad\qquad (e - a)(e + x_+) = e + a_-,$$

$$(10) \qquad\qquad (e + x_-)(e - a) = e + a_+.$$

Multiplying both sides of (9) on the left by $e + x_-$ and both sides of (10) on the right by $e + x_+$ yields

$$(11) \qquad\qquad (e + x_-)(e + a_-) = (e + a_+)(e + x_+) = d$$

and

$$d \in (A_0 \oplus A_-) \cap (A_0 \oplus A_+) = A_0.$$

Condition (i) and (9) imply that $e + x_+$ is invertible, and so we have that d is invertible,

$$(e + x_+)^{-1} = d^{-1}(e + a_+),$$

and

$$e - a = (e + a_-)d^{-1}(e + a_+).$$

The formulas for $a_\pm$ and $a_0 = d^{-1}$ follow immediately from (7), (8) and (11). $\quad\square$

Theorem 8.1 follows by applying Theorem 8.2 to the algebra $A^{(2)}$. Since for a Hilbert-Schmidt operator K the operator $(I - K)^{-1} - I$ is again Hilbert-Schmidt, we see from (1) and Corollary XX.8.2 that $A^{(2)}$ with the decomposition (1) is a strongly decomposable algebra, and thus Theorem 8.2 applies.

To illustrate the above results we now give an example of an LU-factorization in a reproducing kernel Hilbert space. In the remaining part of this section, H is the RKHS, considered in Section XX.11 with the chain $(P_\lambda f)(t) = f(\lambda \wedge t)$, $0 \le \lambda \le 1$. Let A be the operator on H with defining kernel $a(t, s) = ts$. Thus

$$(Af)(t) = \langle f, a(t, \cdot) \rangle = t \int_0^1 f'(s)ds = f(1)t.$$

Since A has rank 1, it is Hilbert-Schmidt. Also,

$$\langle Af, f \rangle = f(1) \int_0^1 f'(s)ds = f(1)^2,$$

whence $I + A \ge I$. Therefore $I + A$ has an LU-factorization of the form

$$I + A = (I + Y_-)(I + Y_-^*),$$

by virtue of Corollary XXII.4.3 and the fact that the chain is continuous. Put $I + X_- = (I + Y_-)^{-1}$. We shall compute the defining kernel for X_- by solving equation (2b) in the present setting, i.e.,

$$(12) \qquad\qquad X_- + (X_- A)_- + A_- = 0.$$

From formula (4) in Section XX.11 we know that the defining kernel $a_-(\cdot, \cdot)$ for A_- is

$$a_-(t, s) = \int_0^t (\lambda \wedge s) d\lambda.$$

If $\gamma(\cdot, \cdot)$ is the defining kernel for X_-, then, by Proposition XX.10.2, the defining kernel for $X_- A$ is given by

$$(13) \qquad (X_- A)(t, s) = \langle \gamma(t, \cdot), \overline{a(\cdot, s)} \rangle = s \int_0^1 \frac{\partial}{\partial \eta} \gamma(t, \eta) d\eta = s\gamma(t, 1),$$

and therefore

$$(X_- A)_-(t, s) = \int_0^t (\lambda \wedge s)\gamma_1(\lambda, 1) d\lambda, \qquad \gamma_1(t, s) = \frac{\partial}{\partial t} \gamma(t, s).$$

Here we used (4) in Section XX.11 with γ in place of a. Hence in terms of defining kernels equation (12) takes the form

$$(14) \qquad \gamma(t, s) + \int_0^t (\lambda \wedge s)\gamma_1(\lambda, 1) d\lambda + \int_0^t (\lambda \wedge s) d\lambda = 0,$$

which yields

$$(15) \qquad\qquad \gamma_1(t, s) + (t \wedge s)\gamma_1(t, 1) + (t \wedge s) = 0.$$

Take $s = 1$. Then (15) reduces to $\gamma_1(t, 1) + t\gamma_1(t, 1) + t = 0$, and thus

$$\gamma_1(t, 1) = \frac{-t}{1 + t}, \qquad 0 \le t \le 1.$$

So from (14) we get

$$(16) \qquad \gamma(t, s) = - \int_0^t (\lambda \wedge s)\left(\frac{1}{1 + \lambda}\right) d\lambda, \qquad 0 \le t \le 1, \ 0 \le s \le 1.$$

Note that γ and $\frac{\partial}{\partial t}\gamma$ are continuous on $[0, 1] \times [0, 1]$, and hence formula (4) in Section XX.11 also applies to γ. We conclude that the defining kernel γ of X_- is given by (16).

According to Theorem 8.1 we have $Y_-^* = X_- + X_- A + A$. Thus the defining kernel of Y_-^* is given by

$$y_-^*(t, s) = \gamma(t, s) + s\gamma(t, 1) + ts,$$

because of (13). The defining kernel of Y_- is equal to $\overline{y_-^*(s, t)}$, by virtue of Proposition XX.10.2.

XXII.9 GENERALIZED *LU*-FACTORIZATION RELATIVE TO DISCRETE CHAINS

The *LU*-factorizations we considered this far were of invertible operators which have a diagonal. In this section we remove this restriction and introduce a weaker type factorization relative to discrete chains. The factors now are strong limits rather than limits in the operator norm.

Throughout this section

$$\mathbb{P} := \{0 = P_0 < P_1 < P_2 < \cdots < P_j < \cdots < I\}$$

is an infinite discrete chain of orthogonal projections on the Hilbert space H. The chain $\mathbb{P}$ is assumed to be closed. In this case the latter means that $P_j x \to x$ for each $x \in H$ when $j \to \infty$. With $\mathbb{P}$ we associate the following sets

$$\widetilde{A}_+(\mathbb{P}) = \big\{A \in \mathcal{L}(H) \mid AP_j = P_{j-1}AP_j,\ j = 1,2,\ldots\big\},$$
$$\widetilde{A}_-(\mathbb{P}) = \big\{A \in \mathcal{L}(H) \mid P_jA = P_jAP_{j-1},\ j = 1,2,\ldots\big\},$$
$$\widetilde{A}_0(\mathbb{P}) = \big\{A \in \mathcal{L}(H) \mid P_jA = AP_j,\ j = 1,2,\ldots\big\}.$$

Note that $\widetilde{A}_0(\mathbb{P})$ is just equal to the set $A_0(\mathbb{P})$ introduced in Section XX.5, but the sets $\widetilde{A}_+(\mathbb{P})$ and $\widetilde{A}_-(\mathbb{P})$ are, in general, much larger than their counterparts in Section XX.5. For example, if $H = \ell_2$ and P_j is defined by

$$P_j(x_1, x_2, \ldots) = (x_1, \ldots, x_j, 0, 0, \ldots), \qquad j = 1,2,\ldots,$$

then the forward shift on ℓ_2 is in $\widetilde{A}_-(\mathbb{P})$, but not in $A_-(\mathbb{P})$. Similarly, the backward shift on ℓ_2 is in $\widetilde{A}_+(\mathbb{P})$, but not in $A_+(\mathbb{P})$. The main reason for this difference is that the operators in $A_\pm(\mathbb{P})$ are required to have a diagonal and those in $\widetilde{A}_\pm(\mathbb{P})$ not.

The sets $\widetilde{A}_+(\mathbb{P})$ and $\widetilde{A}_-(\mathbb{P})$ may also be described as follows:

$$\widetilde{A}_+(\mathbb{P}) = \big\{A \in \mathcal{L}(H) \mid AP_j = P_jAP_j,\ (\Delta P_j)A\Delta P_j = 0,\ j = 1,2,\ldots\big\},$$
$$\widetilde{A}_-(\mathbb{P}) = \big\{A \in \mathcal{L}(H) \mid P_jA = P_jAP_j,\ (\Delta P_j)A\Delta P_j = 0,\ j = 1,2,\ldots\big\}.$$

It is easy to see that $\widetilde{A}_\pm(\mathbb{P})$ and $\widetilde{A}_0(\mathbb{P})$ are linear manifolds in $\mathcal{L}(H)$ which are closed in the strong operator topology. Also,

$$\widetilde{A}_\pm(\mathbb{P}) \cap \widetilde{A}_0(\mathbb{P}) = \{0\}, \qquad \widetilde{A}_+(\mathbb{P}) \cap \widetilde{A}_-(\mathbb{P}) = \{0\}.$$

If S is the forward or backward shift on ℓ_2, then $I + S$ is not invertible. Thus $A_\pm \in \widetilde{A}_\pm(\mathbb{P})$ does not imply that $I + A_\pm$ is invertible. On the other hand, we have the following proposition.

PROPOSITION 9.1. *If* $A_\pm \in \widetilde{A}_\pm(\mathbb{P})$ *and* $I + A_\pm$ *is invertible, then*

$$(I + A_\pm)^{-1} - I \in \widetilde{A}_\pm(\mathbb{P}).$$

PROOF. We prove the proposition for $A = A_- \in \widetilde{A}_-(\mathbb{P})$ (the analogous result for A_+ is proved by a duality argument). Fix a positive integer N, and write $I + A$ as an $(N + 1) \times (N + 1)$ operator matrix relative to the decomposition

$$H = \mathrm{Im}(P_1 - P_0) \oplus \cdots \oplus \mathrm{Im}(P_N - P_{N-1}) \oplus \mathrm{Im}(I - P_N).$$

Then $I + A$ is block lower triangular and the first N diagonal elements are identity operators. Since $I + A$ is invertible, it follows that the $(N + 1)$-th diagonal element is invertible, and hence $(I + A)^{-1}$ is a block lower triangular operator matrix whose first N diagonal elements are identity operators. So for $A^\times = (I + A)^{-1} - I$ we have

$$P_j A^\times = P_j A^\times P_j, \quad (\Delta P_j) A^\times \Delta P_j = 0, \qquad j = 1, \ldots, N.$$

Since N is arbitrary, we conclude that $A^\times \in \widetilde{A}_-(\mathbb{P})$. $\square$

We say that an invertible operator $T \in \mathcal{L}(H)$ admits a *generalized LU-factorization* with respect to the discrete closed chain $\mathbb{P} = \{P_0, P_1, \ldots, I\}$ if there exist $Y_\pm \in \widetilde{A}_\pm(\mathbb{P})$ and $D \in \widetilde{A}_0(\mathbb{P})$ such that $I + Y_\pm$ is invertible and

$$T = (I + Y_-)D(I + Y_+).$$

An operator has at most one generalized LU-factorization. This follows from an argument which is essentially the same as the one used in the next to last paragraph of the introduction to the present chapter.

THEOREM 9.2. *Let $I - K$ be an invertible operator on a Hilbert space H, and let $\mathbb{P} = \{P_0, P_1, \ldots, I\}$ be a discrete closed chain. A necessary and sufficient condition for $I - K$ to admit a generalized LU-factorization with respect to $\mathbb{P}$ is that the following two conditions hold.*

(a) *For each j the operator $I - P_j K P_j$ is invertible and*
$$\sup_j \|(I - P_j K P_j)^{-1} P_j\| < \infty;$$

(b) *Each of the series*

$$X_- = \sum_{j=1}^{\infty} \Delta P_j K P_{j-1}(I - P_{j-1} K P_{j-1})^{-1},$$

$$X_+ = \sum_{j=1}^{\infty} (I - P_{j-1} K P_{j-1})^{-1} P_{j-1} K \Delta P_j,$$

converges strongly on H. If (a) *and* (b) *hold, then*

(1) $$I - K = (I + Y_-)D(I + Y_+),$$

where

(2) $$I + Y_- = \prod_{j=1}^{\infty} {}^{\curvearrowleft} \left[I - \Delta P_j K P_{j-1}(I - P_{j-1} K P_{j-1})^{-1} \right],$$

(3) $$I + Y_+ = \prod_{j=1}^{\infty} {}^{\curvearrowright} \left[I - (I - P_{j-1} K P_{j-1})^{-1} P_{j-1} K \Delta P_j \right],$$

$$(4) \qquad D = \sum_{j=1}^{\infty} \Delta P_j \left[I - K - K P_{j-1}(I - P_{j-1} K P_{j-1})^{-1} K \right] \Delta P_j.$$

The inverses of these operators are given by

$$(5) \qquad (I + Y_{\pm})^{-1} = I + X_{\pm}$$

$$(6) \qquad D^{-1} = \sum_{j=1}^{\infty} \Delta P_j (I - P_j K P_j)^{-1} \Delta P_j.$$

Each series and product converges strongly.

PROOF. We split the proof into three parts. In the first part we only assume that condition (a) is satisfied.

Part (α). Assume condition (a) holds. We prove that the series in (4) and (6) converge strongly. To do this we first note that $(I - P_n K P_n)^{-1} P_n$ converges strongly to $(I - K)^{-1}$. This follows from the inequality

$$\|(I - P_n K P_n)^{-1} P_n x - (I - K)^{-1} x\| \leq \|(I - P_n K P_n)^{-1} P_n\| \|P_n x - (I - P_n K P_n)(I - K)^{-1} x\|,$$

where $x \in H$, and the convergence $P_n x \to x$ if $n \to \infty$. By the principle of uniform boundedness, the strong convergence of the sequence $\big((I - P_j K P_j)^{-1} P_j\big)$ implies that this sequence is bounded in the operator norm. Now, in general, if (B_n) is a bounded sequence in $\mathcal{L}(H)$, then

$$\sum_{j=1}^{\infty} \|(\Delta P_j) B_j (\Delta P_j) x\|^2 = \sum_{j=1}^{\infty} \langle B_j (\Delta P_j) x, (\Delta P_j) x \rangle$$

$$\leq \left(\sup_j \|B_j\| \right) \sum_{j=1}^{\infty} \|(\Delta P_j) x\|^2$$

$$= \left(\sup_j \|B_j\| \right) \|x\|^2,$$

and hence the series $\sum_{j=1}^{\infty} (\Delta P_j) B_j (\Delta P_j) x$ converges to each x. By applying this result to $B_j = (I - P_j K P_j)^{-1} P_j$ we see that the series in (6) converges strongly. With

$$B_j = I - K - K P_{j-1}(I - P_{j-1} K P_{j-1})^{-1} K$$

we get the strong convergence of the sequence in (4).

Part (β). Suppose that the operator $I - K$ admits the generalized LU-factorization (1), and let us prove that (a) and (b) hold. From (1) we have for $j = 1, 2, \ldots$

$$(7) \qquad P_j (I - K) P_j = \big(P_j (I + Y_-) P_j \big) \big(P_j D P_j \big) \big(P_j (I + Y_+) P_j \big).$$

Since $Y_\pm$ and $(I + Y_\pm)^{-1} - I$ are both in $\widetilde{A}_\pm(\mathbb{P})$, we see that on $\operatorname{Im} P_j$ the operator $P_j(I + Y_\pm)P_j$ is invertible and its inverse is given by $P_j(I + Y_\pm)^{-1}P_j$. Similarly, $P_j D P_j$ is invertible on $\operatorname{Im} P_j$, and its inverse is given by $P_j D^{-1} P_j$. So (7) implies that $I - P_j K P_j$ is invertible and

$$(I - P_j K P_j)^{-1} = [P_j(I + Y_+)^{-1}P_j](P_j D^{-1} P_j)[P_j(I + Y_-)^{-1}P_j] + I - P_j.$$

Thus (a) holds. Let $H_n = P_n H$ and $\mathbb{P}_n = \{P_0, \ldots, P_{n-1}, I_n\}$, where I_n is the identity operator on H_n. By Theorem 1.1 applied to $I_n - P_n K P_n$ and the chain $\mathbb{P}_n$, we have

$$(8) \qquad P_n(I + Y_-)^{-1}P_n = P_n + \sum_{j=1}^{n} (\Delta P_j) K P_{j-1}(I - P_{j-1} K P_{j-1})^{-1},$$

$$(9) \qquad P_n(I + Y_+)^{-1}P_n = P_n + \sum_{j=1}^{n} (I - P_{j-1} K P_{j-1})^{-1} P_{j-1} K \Delta P_j,$$

$$(10) \qquad P_n D^{-1} P_n = \sum_{j=1}^{n} (\Delta P_j)(I - P_j K P_j)^{-1} \Delta P_j,$$

$$(11) \qquad P_n(I + Y_-)P_n = P_n \overset{\curvearrowleft}{\prod_{j=1}^{n}} \left[I - (\Delta P_j) K P_{j-1}(I - P_{j-1} K P_{j-1})^{-1} \right],$$

$$(12) \qquad P_n(I + Y_n)P_n = P_n \overset{\curvearrowright}{\prod_{j=1}^{n}} \left[I - (I - P_{j-1} K P_{j-1})^{-1} P_{j-1} K (\Delta P_j) \right],$$

$$(13) \qquad P_n D P_n = \sum_{j=1}^{n} \Delta P_j \left[I - K - K P_{j-1}(I - P_{j-1} K P_{j-1})^{-1} K \right] \Delta P_j.$$

Since $P_n B P_n$ converges strongly to B for each $B \in \mathcal{L}(H)$ whenever $n \to \infty$, we know that

$$(14) \qquad \lim_{n \to \infty} P_n(I + Y_n)^{-1} P_n x$$

exists for each $x \in H$. In particular, the series in the right hand side of (9) converges strongly. It follows that the second series in (b) converges strongly and $(I + Y_+)^{-1} = I + X_+$. In a similar way one shows that the first series in (b) converges strongly and $(I + Y_-)^{-1} = I + X_-$. Thus (b) holds. Moreover, we have also proved (5).

In the left hand sides of (10)–(13) we may take the limit for $n \to \infty$ in the strong operator topology. It follows that the series and products appearing in the right hand sides of (10)–(13) converge strongly. In this way we obtain (2) from (11), (3) from (12), (4) from (13) and (6) from (10).

Part (γ). Suppose now that (a) and (b) hold, and let us derive the generalized LU-factorization (1). It is easy to verify that $X_{\pm} \in \widetilde{A}_{\pm}(\mathbb{P})$. By Theorem 1.1, the operator $P_n(I - K)P_n$ admits an LU-factorization

$$(15) \qquad P_n(I - K)P_n = (I_n + Y_{n-})D_n(I_n + Y_{n+})$$

with respect to the finite chain $\mathbb{P}_n$ defined above. Here $(I_n + Y_{n-})^{-1}$, $(I_n + Y_{n+})^{-1}$, D_n^{-1}, $I_n + Y_{n-}$, $I_n + Y_{n+}$ and D_n are equal to the right hand sides of (8)–(13), respectively. From Part (α) of the proof we know that the sequences $(P_n D_n P_n)$ and $(P_n D_n^{-1} P_n)$ converge strongly, say to D and D^{-1}, respectively. Clearly, D and D^{-1} are in $\widetilde{A}_0(\mathbb{P})$. Since

$$I + Y_{n+} = D_n^{-1}(I_n + Y_{n-})^{-1}\big(P_n(I - K)P_n\big),$$

it follows from (b) that for each $x \in H$

$$D^{-1}(I + X_-)(I - K)x = \lim_{n\to\infty}(I + Y_{n+})P_n x = (I + Y_+)x,$$

where $I + Y_+$ is defined by (3). In a similar way, one shows that

$$(I - K)(I + X_+)D^{-1}x = \lim_{n\to\infty}(I + Y_{n-})P_n x = (I + Y_-)x,$$

where $I + Y_-$ is defined by (2). Finally, passing to the strong limit in (15) we get

$$I - K = (I + Y_-)D(I + Y_+),$$

which is the desired generalized LU-factorization. $\square$

Arguments analogous to the ones used in the preceding proof show that Theorem 9.2 remains true if the convergence of the two series in (b) is replaced by strong convergence of the products in (2) and (3), or the product in (2) and the series representing X_- in (b), or the product in (3) and the series representing X_+ in (b).

THEOREM 9.3. *A strictly positive operator on a Hilbert space H admits a generalized LU-factorization with respect to any discrete closed chain $\mathbb{P} = \{P_0, P_1, \ldots, I\}$ on H.*

PROOF. Let $I - K$ be a positive definite operator on H. So there exists a positive number c such that

$$\langle (I - K)x, x \rangle \geq c\|x\|^2, \qquad x \in H.$$

It follows that

$$\langle P_j(I - K)P_j x, x \rangle \geq c\|x\|^2, \qquad x \in \operatorname{Im} P_j,$$

which implies that $I - P_j K P_j$ is invertible and $\|(I - P_j K P_j)^{-1}\| \leq \max\{c^{-1}, 1\}$. Thus (a) in Theorem 9.2 holds.

Next, we prove that condition (b) in Theorem 9.2 also holds. By Theorem 1.1 the operator $P_n(I - K)P_n$ admits the *LU*-factorization

$$(16) \qquad P_n(I - K)P_n = (I_n + Y_{n-})D_n(I_n + Y_{n+})$$

with respect to the finite chain $\mathbb{P}_n = \{P_0, \ldots, P_n = I_n\}$ on the Hilbert space $H_n = \operatorname{Im} P_n$. The representations of the factors in the right hand side of (16) and their inverses are given by the right hand sides of the corresponding equations (8)–(13). From the result proved in Part (α) of the proof of Theorem 9.1 we know that the sequences (D_n) and (D_n^{-1}) converge strongly on H. Since K and P_n are selfadjoint,

$$P_n(I - K)P_n = (I + Y_{n+}^*)D_n^*(I_n + Y_{n-}^*)$$

is also an *LU*-factorization of $P_n(I - K)P_n$ with respect to the same finite chain $\mathbb{P}_n$. Hence

$$Y_{n+}^* = Y_{n-}, \qquad D_n^* = D_n = [(I + Y_{n+})^{-1}]^* P_n(I - K)P_n(I + Y_{n+})^{-1},$$

which shows that D_n is positive definite. By equation (15),

$$P_n(I - K)P_n = \widetilde{Y}_{n+}^* \widetilde{Y}_{n+}, \qquad \widetilde{Y}_{n+} := D_n^{1/2}(I + Y_{n+}).$$

Thus

$$\|P_n(I - K)P_n\| = \|\widetilde{Y}_{n+}\|^2 \geq \|D_n^{-1/2}\|^{-2}\|I + Y_{n+}\|^2.$$

Hence

$$\|I + Y_{n-}\| = \|I + Y_{n+}\| \leq \|D_n^{-1/2}\| \|P_n(I - K)P_n\|^{1/2}.$$

Since $\sup_n \|D_n^{-1/2}\|^2 = \sup_n \|D_n^{-1}\| < \infty$, because (D_n^{-1}) converges strongly, it follows that $\sup_n \|I_n + Y_{n\pm}\| < \infty$. From (15) and the uniform boundedness of $P_n(I - K)^{-1}P_n$ we have that

$$(I_n + Y_{n+})^{-1} = P_n(I - K)^{-1}P_n(I + Y_{n-})D_n$$

is uniformly bounded. Also, from the series given in (9), we see that

$$(I_n + Y_{n+})^{-1} = P_n(I + Y_{n+})^{-1}P_n$$

converges strongly on $\bigcup_{k=1}^{\infty} \operatorname{Im} P_k$, which is dense in H. Since $P_n(I + Y_{n+})^{-1}P_n$ is uniformly bounded, it follows that the series defining X_+ in (b) of Theorem 9.2 converges strongly on H. Also, the representation of $I_n + Y_{n+}$ given by the right hand side of (12) converges on $\bigcup_{k=1}^{\infty} \operatorname{Im} P_k$. Since $I_n + Y_{n+}$ is uniformly bounded, we have that $P_n(I_n + Y_{n+})P_n$ converges strongly on H. Thus

$$(I_n + Y_{n-})^{-1} = P_n(I - K)^{-1}P_n D_n(I + Y_{n+})$$

also converges strongly on H which proves that (b) in Theorem 9.2 is valid. Hence $I - K$ admits a generalized LU-factorization. $\square$

Generalized LU-factorization may also be defined for continuous chains. In fact, the factorization of a Wiener-Hopf integral operator T on $L_2^m([0,\infty])$ associated with a canonical factorization of its symbol (see Sections XIII.2 and XIII.3) can be understood as a generalized LU-factorization with respect to the chain $\{P_\tau\}_{0\leq\tau\leq\infty}$, where $P_\infty = I$ and

$$(P_\tau f)(t) = \begin{cases} f(t), & 0 \leq t \leq \tau < \infty, \\ 0, & t > \tau. \end{cases}$$

COMMENTS ON PART V

Part V presents an introduction to the theory of additive and multiplicative lower-upper decompositions of operators. The main source for the material in this part is the book Gohberg-Kreĭn [4]. Also later developments from the papers Barkar-Gohberg [1] and V.M. Brodskii-Gohberg-Kreĭn [1] have been taken into account. This allowed us to give a more complete and systematic presentation. In particular, the special role played by diagonals is emphasized. Sections XX.10 and XX.11 about chains in reproducing kernel Hilbert spaces are based on the papers Kailath [3] and Kailath-Duttweiler [1]. Section XXII.7 is taken from the paper Gohberg-Kaashoek [2]; related material may be found in Gohberg-Goldberg [1]–[3], Kailath [2] and Schumitzky [1]. Livšic's original result (see Livšic [1], [2]) describing the unitary equivalence class of the operator of integration appears in Section XXI.3. Theorem XXI.1.5 about triangular representations of Volterra operators is due to M.S. Brodskii [1], and the triangular integral appearing in the representation is called the *Brodskii triangular integral*. The results in Sections XXII.2–XXII.4 appeared for the first time in Gohberg-Kreĭn [5]; the derivation of the Hilbert-Schmidt factorization theorem in Section XXII.4 is different from the original one. The factorization theorems for integral operators in Section XII.6 are due to M.G. Kreĭn [1], [2]; for their relevance in numerical applications see Sobolev [1], Bellman [1] and the more recent paper Gohberg-Koltracht [1]. Let us mention that this part of the book focusses mainly on the general theory of chains and the triangular representation theorems for Hilbert-Schmidt operators. For the theory for other classes of operators as well as for other applications see Gohberg-Kreĭn [4]. In recent years much attention has been given to the problem of classification up to unitary equivalence of chains and nest algebras. For these topics, which are not treated here, we refer to the recent book Davidson [1]. In the later book it is proved that, in general, for uncountable chains the analogue of Theorem XXII.9.3 does not hold true (a result which is from Larson [1]). For applications of chains and triangular decompositions to mathematical system theory one may consult the book Feintuch-Saeks [1].

EXERCISES TO PART V

1. Let $H = L_2^2([0,1])$. Define the chain $\mathbb{P} = \{Q(\tau)\}_{0 \leq \tau \leq 1}$ on H by

$$Q(\tau) \begin{pmatrix} f_1 \\ f_2 \end{pmatrix} = \begin{pmatrix} P(\tau)f_1 \\ P(\tau)f_2 \end{pmatrix},$$

where $P(\tau)$ is the usual truncation operator, i.e.,

$$P(\tau)f = \begin{cases} f & \text{on} & [0,\tau], \\ 0 & \text{on} & (\tau,1]. \end{cases}$$

(a) Show that $\mathbb{P}$ is a continuous chain on H.

(b) Given the matrix function $k(t,s) = \big(k_{ij}(t,s)\big)_{i,j=1}^{2}$, where k_{ij} is square integrable on $[0,1] \times [0,1]$, define $K \in \mathcal{L}(H)$ by

$$(Kf)(t) = \int_0^1 k(t,s)f(s)ds.$$

Find the kernel functions of the operators $K_{\pm}$ in the additive LU-decomposition of K with respect to $\mathbb{P}$.

2. Let $H = L_2^2([0,1])$. Define the chain $\mathbb{P} = \{Q(\tau)\}_{0 \leq \tau \leq 1}$ on H by

$$Q(\tau) \begin{pmatrix} f_1 \\ f_2 \end{pmatrix} = \begin{cases} \begin{pmatrix} P(2\tau)f_1 \\ 0 \end{pmatrix}, & 0 \leq \tau \leq \tfrac{1}{2}, \\[2mm] \begin{pmatrix} f_1 \\ P(2\tau - 1)f_2 \end{pmatrix}, & \tfrac{1}{2} < \tau \leq 1, \end{cases}$$

where $P(\eta)$ is the usual truncation operator defined in Exercise 1.

(a) Show that $\mathbb{P}$ is a continuous chain on H.

(b) Given the operator K in Exercise 1(b), find the kernel functions of the operators $K_{\pm}$ in the additive LU-decomposition of K with respect to $\mathbb{P}$.

3. Given a chain $\mathbb{P}$ on a separable Hilbert space H, the *rank* of $\mathbb{P}$ is the smallest number of vectors $v_1, v_2, \ldots$ in H such that $H = \overline{\text{span}}\{Pv_j \mid P \in \mathbb{P},\ j = 1, 2, \ldots\}$. Show that the chain in Exercise 1 has rank 2.

4. Show that the chain in Exercise 2 has rank 1.

5. Two Hilbert space chains $\mathbb{P}$ and $\mathbb{P}'$ are called *unitarily equivalent* if there exists a unitary operator U such that

$$\mathbb{P}' = \{U^*PU \mid P \in \mathbb{P}\}.$$

Are the chains in Exercises 1 and 2 unitarily equivalent? (Hint: consider the ranks of equivalent chains.)

6. Let $H = L_2([0,1])$. Define the chain $\mathbb{P} = \{P(\tau)\}_{0 \leq \tau \leq 1}$ on H by

$$(P(\tau)f)(t) = \begin{cases} f(t), & t \in [0, \frac{\tau}{2}) \cup [\frac{1}{2}, \frac{1}{2} + \frac{\tau}{2}], \\ 0, & \text{elsewhere.} \end{cases}$$

    (a) Show that $\mathbb{P}$ is a continuous chain.

    (b) Compute the rank of $\mathbb{P}$.

7. Let H and $\mathbb{P}$ be as in Exercise 6. Define $K \in \mathcal{L}(H)$ by

$$(Kf)(t) = \int_0^1 k(t,s)f(s)ds,$$

where $k \in L_2([0,1] \times [0,1])$. Find the kernel functions of the operators $K_\pm$ in the additive LU-decomposition of K with respect to $\mathbb{P}$.

8. Let $H = L_2([0,1])$. Define the chain $\mathbb{P} = \{P(\tau)\}_{0 \leq \tau \leq 1}$ on H by

$$(P(\tau)f)(t) = \begin{cases} f(t), & t \in [0, \frac{\tau}{3}] \cup (\frac{1}{3}, \frac{1}{3} + \frac{\tau}{3}] \cup (\frac{2}{3}, \frac{2}{3} + \frac{\tau}{3}], \\ 0, & \text{elsewhere.} \end{cases}$$

Compute the rank of $\mathbb{P}$.

9. Let K and $\mathbb{P}$ be as in Exercises 7 and 8, respectively. Find the kernel functions of the operators $K_\pm$ in the additive LU-decomposition of K with respect to $\mathbb{P}$.

10. Given a positive integer n, find a chain of rank n. Next, give an example of a chain of infinite rank.

11. Let H be the space of absolutely continuous complex-valued functions defined on $[0,1]$, with derivative (which exists a.e.) in $L_2([0,1])$. Define an inner product on H by

$$\langle f, g \rangle = \int_0^1 f'(s)\overline{g'(s)}ds + f(0)\overline{g(0)}.$$

    (a) Prove that $R(t,s) = 1 + \min(t,s)$, $0 \leq s,t \leq 1$ is the reproducing kernel for H.

    (b) Describe the chain associated with H and $R(\cdot,\cdot)$.

12. Let H be the space in Exercise 11 with inner product

$$\langle f, g \rangle = \int_0^1 (f(s) + f'(s))(\overline{g(s)} + \overline{g'(s)})ds + 2f(0)\overline{g(0)}.$$

(a) Prove that H is a Hilbert space.

(b) Show that the function $R(t,s) = \frac{1}{2}\exp\{-|t - s|\}$, $0 \le s, t \le 1$, is the reproducing kernel for H.

(c) Describe the chain associated with H and $R(\cdot,\cdot)$.

13. Let H be the space in Exercise 11 with inner product

$$\langle f, g \rangle = \frac{1}{2}\left(f(0) + f(1)\right)\left(\overline{g(0)} + \overline{g(1)}\right) + \frac{1}{2}\int\limits_0^1 f'(s)\overline{g'(s)}ds.$$

(a) Show that $R(t,s) = 1 - |t - s|$ is the reproducing kernel for H.

(b) Prove that the chain associated with H and $R(\cdot,\cdot)$ is given by 0 and

$$(P_\lambda f)(t) = \begin{cases} f(t) & , \quad 0 \le t \le \lambda, \\ f(\lambda) - \frac{t-\lambda}{2-\lambda}\left(f(0) + f(\lambda)\right), & \lambda < t \le 1. \end{cases}$$

14. Let H be the RKHS in Exercise 11. Define $\mathbb{P}$ to be the chain $\{0\} \cup \{P_\lambda \mid \lambda \in [0,1]\}$, where

$$(P_\lambda f)(t) = f(t \wedge \lambda), \qquad t \wedge \lambda = \min(t, \lambda).$$

Suppose A is a Hilbert-Schmidt operator on H with corresponding kernel function

$$a(t,s) = \left[A^* R(t,\cdot)\right](s), \qquad 0 \le s, t \le 1.$$

Assume that $a(t,s)$ and $a_1(t,s) = \frac{\partial a}{\partial t}(t,s)$ are continuous on $[0,1] \times [0,1]$.

(a) Show that the kernel function $(a_-)_m(t,s)$ corresponding to $(A_-)_m = \int_{[\mathbb{P}} (dP)AP$ is given by

$$(a_-)_m(t,s) = \int\limits_0^t a_1(\lambda, s \wedge \lambda)d\lambda, \qquad s \wedge \lambda = \min(s, \lambda).$$

(Hint: see the example in Section XX.11.)

(b) Let $(a_-)_M(t,s)$ be the kernel function corresponding to

$$(A_-)_M = \int\limits_{\mathbb{P}]} (dP)AP.$$

Show that

$$(A_-)_M = (A_-)_m + P_0 A P_0,$$

$$(a_-)_M(t,s) = \int\limits_0^t a_1(\lambda, s \wedge \lambda)d\lambda + a(0,0).$$

15. In Exercise 14 take $a(t, s)$ to be the kernel function $-\frac{1}{2}(1 + t + s)$. Show that $I + A$ has an LU-factorization and find the factors.

16. Let H be the RKHS defined in Exercise 13(a), and let $K \in \mathcal{L}(H)$ be the operator with defining kernel $a(t, s) = ts$. Does $I + K$ have an LU-factorization? If so, find the defining kernels for the factors.

17. Let $e_1, e_2, \ldots$ be an orthonormal basis in the Hilbert space H, and let $\mathbb{P} = \{P_j\}_{j=0}^{\infty}$ be the discrete chain on H with $P_0 = 0$ and

$$P_j x = \sum_{\nu=1}^{j} \langle x, e_\nu \rangle e_\nu, \qquad j = 1, 2, \ldots, \infty.$$

Let $K \in \mathcal{L}(H)$ be given by $K = \alpha \langle \cdot, \varphi \rangle \varphi$, where $\varphi \in H$, $\|\varphi\| = 1$ and $\alpha \in \mathbb{C}$.

(a) Show that for $|\alpha| < 1$ the operator $I - K$ has an LU-factorization with respect to $\mathbb{P}$.

(b) Determine the α's for which $I - K$ has an LU-factorization with respect to $\mathbb{P}$.

18. Let H and $\mathbb{P}$ be as in the previous exercise, and let $K \in \mathcal{L}(H)$ be the finite rank operator

$$K = \sum_{j=1}^{n} \langle \cdot, \varphi_j \rangle \psi_j,$$

where $\varphi_1, \ldots, \varphi_n$ and $\psi_1, \ldots, \psi_n$ are vectors in H. State and prove the analogue of Theorem XXII.5.3 for the case considered here.

19. Let $\ldots, e_{-2}, e_{-1}, e_0, e_1, e_2, \ldots$ be an orthonormal basis in the Hilbert space H, and let $\mathbb{P} = \{P_j\}_{j=-\infty}^{\infty}$ be the double infinite discrete chain on H with $P_{-\infty} = 0$ and

$$P_j = \sum_{\nu=-\infty}^{j} \langle x, e_\nu \rangle e_\nu, \qquad j = \ldots, -1, 0, 1, \ldots, \infty.$$

Let $K \in \mathcal{L}(H)$ be given by $K = \alpha \langle \cdot, \varphi \rangle \varphi$, where $\varphi \in H$, $\|\varphi\| = 1$ and $\alpha \in \mathbb{C}$. Do problems (a) and (b) in Exercise 17 for the chain $\mathbb{P}$ introduced here.

20. Let H and $\mathbb{P}$ be as in the previous exercise, and let K be as in Exercise 18. State and prove the analogue of Theorem XXII.5.3 for the case considered here.

21. Let H and $\mathbb{P}$ be as in Exercise 17, and let K be the operator on H whose matrix $(k_{ij})_{i,j=1}^{\infty}$ with respect to the orthonormal basis $e_1, e_2, \ldots$ is given by

$$k_{ij} = \begin{cases} \langle K_1 e_j, e_i \rangle, & j < i, \\ \langle K_2 e_j, e_i \rangle, & j \geq i, \end{cases}$$

where $K_1, K_2 \in \mathcal{L}(H)$ are given finite rank operators. Find necessary and sufficient conditions in terms of K_1, K_2 in order that $I - K$ admits an LU-factorization with respect to $\mathbb{P}$, and determine in this case the factors in such a factorization. (Hint: generalize Theorem XXII.7.1 to the case considered here. This first requires one to develop the analogue of Theorem IX.2.1.)

22. Let K be the operator on $L_2([0,1])$ defined by

$$(Kf)(t) = \alpha \int_0^1 f(s)ds, \qquad 0 \le t \le 1.$$

Consider the chain $\mathbb{P}$ defined in Exercise 6.

   (a) Determine conditions on α which guarantee that $I - K$ admits an LU-factorization with respect to $\mathbb{P}$.

   (b) Find the kernel functions of the lower and upper triangular factors $Y_\pm$ in the LU-factorization.

23. Do problems (a) and (b) in the previous exercise for the same K and with $\mathbb{P}$ the chain given in Exercise 8.

24. In equality (6) of Section XIII.1 a factorization is given of a Wiener-Hopf operator. Can this be considered as a factorization along a chain? If so, find the chain.

25. Let $\varphi : [a, b] \to \mathcal{L}(H)$ be continuous, where H is a separable Hilbert space. Suppose $P(t)$, $0 \le t \le 1$, is a continuous chain on H and $A \in \mathcal{L}(H)$ has the property that

$$\sup \sum_{j=1}^n \|A\Delta P(t_j)A^*\| < \infty, \qquad \Delta P(t_j) = P(t_j) - P(t_{j-1})$$

where the supremum is taken over all partitions $0 = t_0 < t_1 < \cdots < t_n = 1$ of $[0,1]$. Prove that $\int_0^1 \varphi(t)AdP(t)$ converges.

PART VI

CLASSES OF TOEPLITZ OPERATORS

This part deals with equations of the following type:

$$\sum_{j=1}^{\infty} A_{i-j} x_j = y_i, \qquad i = 1, 2, \ldots .$$

Here $y_1, y_2, \ldots$ are given vectors in $\mathbb{C}^m$ and A_j $(j \in \mathbb{Z})$ is a linear transformation on $\mathbb{C}^m$. To treat the above equation it is natural to consider the operator

$$(*) \qquad T = \begin{bmatrix} A_0 & A_{-1} & A_{-2} & \cdots \\ A_1 & A_0 & A_{-1} & \cdots \\ A_2 & A_1 & A_0 & \cdots \\ \vdots & \vdots & \vdots & \ddots \end{bmatrix},$$

which has to be considered on an appropriate sequence space. An operator of the form $(*)$ is called a *(block) Toeplitz operator*.

In this part we restrict our attention to Toeplitz operators acting on Hilbert spaces which appear in many applications. They form an important class of concrete operators. As the discrete analogues of Wiener-Hopf integral operators these operators provide another illustration of the Fredholm theory with topological interpretations of the Fredholm characteristics. Also, Toeplitz operators form one of the few classes of operators for which there exist algorithms for inversion and explicit inversion formulas.

This part contains a concise introduction to the theory of block Toeplitz operators on Hilbert spaces. Special attention is paid to two subclasses, namely to block Toeplitz operators defined by rational matrix functions and to block Toeplitz operators defined by piecewise continuous functions. As in the Wiener-Hopf case, for the first subclass explicit formulas for solutions and Fredholm properties are obtained.

CHAPTER XXIII
BLOCK TOEPLITZ OPERATORS

The first three sections of this chapter have an introductory character. Section 2 contains a short introduction to Laurent operators. In Section 3 the first properties of block Toeplitz operators are derived. Sections 4 and 5 develop the Fredholm theory of block Toeplitz operators defined by continuous functions.

XXIII.1 PRELIMINARIES

Throughout this chapter $\mathbb{T}$ denotes the unit circle in $\mathbb{C}$. By definition, $L_2(\mathbb{T})$ is the space of all functions $f: \mathbb{T} \to \mathbb{C}$ such that

$$t \mapsto f(e^{it})$$

is Lebesgue measurable and square integrable on $[-\pi, \pi]$. As usual we identify two functions $f_1, f_2: \mathbb{T} \to \mathbb{C}$ whenever $\{t \mid f_1(e^{it}) \neq f_2(e^{it})\}$ has measure zero. The space $L_2(\mathbb{T})$ is a Hilbert space with inner product and norm given by

$$\langle f, g \rangle = \frac{1}{2\pi} \int_{-\pi}^{\pi} f(e^{it})\overline{g(e^{it})}dt,$$

$$\|f\| = \left(\frac{1}{2\pi} \int_{-\pi}^{\pi} |f(e^{it})|^2 dt \right)^{1/2}.$$

The functions ζ^n, $\zeta = e^{it}$, $n \in \mathbb{Z}$, form an orthonormal basis for $L_2(\mathbb{T})$. The numbers

$$c_n = \frac{1}{2\pi} \int_{-\pi}^{\pi} f(e^{it})e^{-int}dt, \qquad n = 0, \pm 1, \pm 2, \ldots$$

are called the *Fourier coefficients* of f. We denote by $H_2(\mathbb{T})$ the subspace of $L_2(\mathbb{T})$ consisting of all functions f for which the Fourier coefficients $c_{-1}, c_{-2}, \ldots$ are zero, i.e.,

$$H_2(\mathbb{T}) = \{f \in L_2(\mathbb{T}) \mid \langle f, e^{int} \rangle = 0, n = -1, -2, \ldots\}.$$

The space $H_2(\mathbb{T})$ is called the *Hardy space* of square integrable functions on the circle.

The symbol ℓ_2 stands for the usual Hilbert space of all square summable sequences of complex numbers. By $\ell_2(\mathbb{Z})$ we denote the Hilbert space of all square summable double infinite sequences of complex numbers. We shall identify ℓ_2 with its canonical image in $\ell_2(\mathbb{Z})$, i.e.,

$$\ell_2 = \{(w_j)_{j \in \mathbb{Z}} \in \ell_2(\mathbb{Z}) \mid w_j = 0 \quad (j < 0)\}.$$

The map U which assigns to a function $f \in L_2(\mathbb{T})$ its sequence of Fourier coefficients, i.e.,

$$(1) \qquad Uf = (c_n)_{n\in\mathbb{Z}}, \qquad c_n = \langle f, e^{int}\rangle,$$

is a unitary operator from $L_2(\mathbb{T})$ onto $\ell_2(\mathbb{Z})$, which carries $H_2(\mathbb{T})$ over into ℓ_2.

Given a Hilbert space H, we denote by H^m the Cartesian product of m copies of H. An element of H^m is an m-tuple of elements from H written as a column. Thus $x = \mathrm{col}(x_i)_{i=1}^m$ with $x_1, \ldots, x_m$ in H. The space H^m is a Hilbert space with inner product and norm given by

$$\langle x, y\rangle = \sum_{j=1}^m \langle x_j, y_j\rangle,$$

$$\|x\| = \left(\sum_{j=1}^m \|x_j\|^2\right)^{1/2}.$$

If H is a space of functions (or a space of sequences), then each element of H^m is a function with values in $\mathbb{C}^m$ (resp. a sequence with elements in $\mathbb{C}^m$). For example, given $f = \mathrm{col}(f_i)_{i=1}^m \in L_2^m(\mathbb{T}) = L_2(\mathbb{T})^m$, then

$$f(t) = \mathrm{col}\big(f_i(t)\big)_{i=1}^m \in \mathbb{C}^m.$$

Thus an element f of $L_2^m(\mathbb{T})$ is a function $f\colon \mathbb{T} \to \mathbb{C}^m$ whose component functions $f_1, \ldots, f_m$ are in $L_2(\mathbb{T})$. In a similar way we shall view the elements of ℓ_2^m and $\ell_2^m(\mathbb{Z}) = \ell_2(\mathbb{Z})^m$ as sequences with entries from $\mathbb{C}^m$. Indeed, let $x = \mathrm{col}(x_i)_{i=1}^m \in \ell_2^m$. Then $x_i = (w_{1i}, w_{2i}, \ldots) \in \ell_2$ for $i = 1, \ldots, m$, and thus x is a sequence of vectors in $\mathbb{C}^m$, $x = (\eta_1, \eta_2, \ldots)$, where

$$\eta_k = \mathrm{col}(w_{ki})_{i=1}^m \in \mathbb{C}^m.$$

The unitary map $U\colon L_2(\mathbb{T}) \to \ell_2(\mathbb{Z})$ defined by (1) extends in a natural way to a unitary operator, also denoted by U, from $L_2^m(\mathbb{T})$ onto $\ell_2^m(\mathbb{Z})$, namely

$$Uf = U\,\mathrm{col}(f_i)_{i=1}^m = \mathrm{col}(Uf_i)_{i=1}^m \in \ell_2^m(\mathbb{Z}).$$

We call U the *Fourier transformation* on $L_2^m(\mathbb{T})$ and Uf is called the *Fourier transform* of f. Note that

$$(2) \qquad f = \sum_{n=-\infty}^{\infty} e^{int} a_n$$

whenever $Uf = (a_n)_{n\in\mathbb{Z}}$. The series in the right hand side of (2), which converges in the norm of $L_2^m(\mathbb{T})$, is called the *Fourier series* of f. From (2) it is clear that the elements of the Hardy space $H_2^m(\mathbb{T})$ may be identified as those functions $f \in L_2^m(\mathbb{T})$ that have an extension to an analytic $\mathbb{C}^m$-valued function inside the unit circle.

We shall use the symbol $L_2^{m\times m}(\mathbb{T})$ to denote the set of all $m\times m$ matrices with entries in $L_2(\mathbb{T})$. If $\Phi\in L_2^{m\times m}(\mathbb{T})$, then

$$(3)\qquad A_k = \frac{1}{2\pi}\int_{-\pi}^{\pi}\Phi(e^{it})e^{-ikt}\,dt$$

is the $m\times m$ matrix whose (i,j)-th entry is equal to the k-th Fourier coefficient of the (i,j)-th entry of Φ. We shall refer to the left hand side of (3) as the k-th *Fourier coefficient* of the matrix function Φ.

XXIII.2 BLOCK LAURENT OPERATORS

A bounded linear operator $L:\ell_2^m(\mathbb{Z})\to\ell_2^m(\mathbb{Z})$ may be represented by a double infinite matrix whose entries are operators acting on $\mathbb{C}^m$:

$$(1)\qquad L = [A_{ij}]_{i,j=-\infty}^{\infty}.$$

In fact, $A_{ij}=\pi_i L\tau_j$, where

$$\tau_j:\mathbb{C}^m\to\ell_2^m(\mathbb{Z}),\qquad \tau_j x=(\delta_{nj}x)_{n\in\mathbf{Z}};$$
$$\pi_i:\ell_2^m(\mathbb{Z})\to\mathbb{C}^m,\qquad \pi_i(x_n)_{n\in\mathbf{Z}}=x_i.$$

Here δ_{nj} stands for the Kronecker delta. The representation (1) means that the action of L is given by

$$L\big((x_n)_{n\in\mathbf{Z}}\big)=(y_n)_{n\in\mathbf{Z}},\quad y_i=\sum_{j=-\infty}^{\infty}A_{ij}x_j\qquad(i\in\mathbb{Z}).$$

We call L a *block Laurent operator* if its matrix elements A_{ij} depend only on the difference $i-j$. The word "block" refers to the fact that the matrix entries are operators and not scalars; in the sequel we shall often omit the word block. Note that L is a Laurent operator on $\ell_2^m(\mathbb{Z})$ if and only if its double infinite matrix has the following form:

$$(2)\qquad \begin{bmatrix} \ddots & & & & \\ & A_0 & A_{-1} & A_{-2} & \\ & A_1 & \boxed{A_0} & A_{-1} & \\ & A_2 & A_1 & A_0 & \\ & & & & \ddots \end{bmatrix}.$$

Here $\boxed{A_0}$ denotes the $(0,0)$ entry which acts on the 0-th coordinate space.

A Laurent operator may also be characterized as a bounded linear operator acting on $\ell_2^m(\mathbb{Z})$ which commutes with the forward shift on $\ell_2^m(\mathbb{Z})$. Indeed, if S is the forward shift on $\ell_2^m(\mathbb{Z})$, then $LS=SL$ is equivalent to the statement that

$$(3)\qquad A_{ij}=A_{i+1,j+1}\qquad(i,j\in\mathbb{Z}).$$

To see this, note that

$$S(\ldots, x_{-2}, x_{-1}, \boxed{x_0}, x_1, x_2, \ldots) = (\ldots, x_{-3}, x_{-2}, \boxed{x_{-1}}, x_0, x_1, \ldots).$$

Thus $S\tau_j = \tau_{j+1}$ and $\pi_{i+1}S = \pi_i$. It follows that

$$A_{ij} = \pi_{i+1}SL\tau_j, \qquad A_{i+1,j+1} = \pi_{i+1}LS\tau_j,$$

and hence (3) holds if and only if $SL = LS$.

Since $\ell_2^m(\mathbb{Z})$ is a direct sum of m copies of $\ell_2(\mathbb{Z})$, an operator L on $\ell_2^m(\mathbb{Z})$ may also be represented by an $m \times m$ matrix whose entries are operators acting on $\ell_2(\mathbb{Z})$. Thus

$$(4) \qquad L = \begin{bmatrix} L_{11} & \cdots & L_{1m} \\ \vdots & & \vdots \\ L_{m1} & \cdots & L_{mm} \end{bmatrix} : \ell_2^m(\mathbb{Z}) \to \ell_2^m(\mathbb{Z}).$$

In this representation the forward shift S on $\ell_2^m(\mathbb{Z})$ is an $m \times m$ diagonal matrix whose diagonal elements are equal to the forward shift S_0 on $\ell_2(\mathbb{Z})$. It follows that $SL = LS$ if and only if

$$S_0 L_{rs} = L_{rs} S_0, \qquad r, s = 1, \ldots, m.$$

Thus L is a Laurent operator on $\ell_2^m(\mathbb{Z})$ if and only if all entries L_{rs} in the matrix representation (4) are Laurent operators on $\ell_2(\mathbb{Z})$.

Let L be a Laurent operator on $\ell_2^m(\mathbb{Z})$ with matrix representations (2) and (4). Then

$$L_{rs} = [A_{i-j}^{rs}]_{i,j=-\infty}^{\infty},$$

where A_n^{rs} is the (r, s)-th entry of the matrix of A_n with respect to the standard basis of $\mathbb{C}^m$. Note that

$$L_{rs}(\ldots, 0, 0, \boxed{1}, 0, 0, \ldots) = (A_n^{rs})_{n \in \mathbb{Z}} \in \ell_2(\mathbb{Z}).$$

Put

$$\Phi(e^{it}) = \begin{bmatrix} \varphi_{11}(e^{it}) & \cdots & \varphi_{1m}(e^{it}) \\ \vdots & & \vdots \\ \varphi_{m1}(e^{it}) & \cdots & \varphi_{mm}(e^{it}) \end{bmatrix},$$

where φ_{rs} is the function in $L_2(\mathbb{T})$ whose n-th Fourier coefficient is equal to A_n^{rs}. Thus

$$\varphi_{rs}(e^{it}) = \sum_{n=-\infty}^{\infty} e^{int} A_n^{rs}.$$

The $m \times m$ matrix function Φ is called the *defining function* of the Laurent operator L. We shall also say that L is *defined by* Φ. Note that $\Phi \in L_2^{m \times m}(\mathbb{T})$ and

$$A_k = \frac{1}{2\pi} \int_{-\pi}^{\pi} \Phi(e^{it}) e^{-ikt} dt.$$

Thus the defining function of L is the matrix function $\Phi \in L_2^{m \times m}(\mathbb{T})$ whose k-th Fourier coefficient is equal to A_k. In other words

$$\Phi(\zeta) = \sum_{n=-\infty}^{\infty} \zeta^n A_n, \qquad |\zeta| = 1.$$

The next theorem shows that the defining function of L is essentially bounded and that L is unitarily equivalent with the operator of multiplication by Φ on $L_2^m(\mathbb{T})$.

       THEOREM 2.1. *The defining function Φ of the block Laurent operator L is a matrix whose entries are measurable, essentially bounded functions. For $f \in L_2^m(\mathbb{T})$ we have*

$$\left((U^{-1}LU)f\right)(e^{it}) = \Phi(e^{it})f(e^{it}) \text{ a.e. },$$

where U is the Fourier transformation on $L_2^m(\mathbb{T})$.

       PROOF. It suffices to prove the theorem for $m = 1$. So assume that L is given by

$$(L\eta)_i = \sum_{j=-\infty}^{\infty} \alpha_{i-j}\eta_j \qquad (i \in \mathbb{Z}),$$

and let φ be its defining function. Let $U : L_2(\mathbb{T}) \to \ell_2(\mathbb{Z})$ be the Fourier transformation. Consider the functions

$$\varepsilon_k(e^{it}) = e^{ikt} \qquad (k \in \mathbb{Z}).$$

We have

$$(U^{-1}LU)\varepsilon_k = U^{-1}L\tau_k(1) = U^{-1}\left((\alpha_{n-k})_{n \in \mathbb{Z}}\right).$$

It follows that

(5) $$\left((U^{-1}LU)\varepsilon_k\right)(e^{it}) = \sum_{n=-\infty}^{\infty} \alpha_{n-k}e^{int} = \varphi(e^{it})\varepsilon_k(e^{it}), \text{ a.e. }.$$

       Let f be an arbitrary element of $L_2(\mathbb{T})$ with Fourier coefficients $(c_n)_{n \in \mathbb{Z}}$. Write

$$f_r(e^{it}) = \sum_{n=-r}^{r} c_n e^{int}.$$

By taking finite linear combinations in (5) we see that

(6) $$\left((U^{-1}LU)f_r\right)(e^{it}) = \varphi(e^{it})f_r(e^{it}), \text{ a.e. }.$$

Note that $f_r \to f$ in $L_2(\mathbb{T})$. Since $U^{-1}LU$ is a bounded linear operator, we have $U^{-1}LUf_r \to U^{-1}LUf$ in $L_2(\mathbb{T})$. By passing to subsequences if necessary we may assume that

$$f_r(e^{it}) \to f(e^{it}), \text{ a.e. },$$

$$((U^{-1}LU)f_r)(e^{it}) \to ((U^{-1}LU)f)(e^{it}), \quad \text{a.e. .}$$

But then (6) implies that

$$((U^{-1}LU)f)(e^{it}) = \varphi(e^{it})f(e^{it}), \quad \text{a.e. .}$$

We already know that the defining function φ is measurable. It remains to show that φ is essentially bounded. Put

$$E_q = \left\{ t \in [-\pi, \pi] \mid |\varphi(e^{it})| \geq q \right\},$$

and consider the function

$$k_q(e^{it}) = \chi_{E_q}(t), \qquad -\pi \leq t \leq \pi.$$

Here q is an arbitrary positive number, and the symbol χ_{E_q} denotes the characteristic function of the set E_q. One easily computes that

$$\|k_q\|^2 = \frac{1}{2\pi}m(E_q), \qquad \|(U^{-1}LU)k_q\|^2 \geq q^2\left(\frac{1}{2\pi}m(E_q)\right),$$

where $m(E_q)$ is the Lebesque measure of the set E_q. It follows that

$$\|(U^{-1}LU)k_q\| \geq q\|k_q\|, \qquad q > 0.$$

Since $U^{-1}LU$ is bounded, this implies that $\|k_q\| = 0$ whenever $q > \|U^{-1}LU\|$. Thus $m(E_q) = 0$ if $q > \|U^{-1}LU\|$, which shows that

$$(7) \qquad \operatorname*{ess\,sup}_{t} |\varphi(e^{it})| \leq \|U^{-1}LU\| = \|L\|,$$

and hence φ is essentially bounded. $\square$

The converse of Theorem 2.1 is also true. To see this, let Φ be an $m \times m$ matrix whose entries are measurable, essentially bounded functions on $\mathbb{T}$. Define M to be the operator of multiplication by Φ on $L_2^m(\mathbb{T})$, i.e.,

$$(8) \qquad M: L_2^m(\mathbb{T}) \to L_2^m(\mathbb{T}), \qquad (Mf)(e^{it}) = \Phi(e^{it})f(e^{it}), \quad \text{a.e. .}$$

Then M is a bounded operator on $L_2^m(\mathbb{T})$, and one easily checks that $L = UMU^{-1}$ is a Laurent operator on $\ell_2^m(\mathbb{Z})$ defined by the function Φ.

COROLLARY 2.2. *Let L be the block Laurent operator with defining function Φ. Then*

$$(9) \qquad \|L\| = \operatorname*{ess\,sup}_{t} \|\Phi(e^{it})\|.$$

PROOF. We know that $L = UMU^{-1}$, where M is the operator of multiplication by Φ on $L_2^m(\mathbb{T})$ (see (8)) and U is unitary. So it suffices to prove (9) for M in place of L. From

$$\|Mf\|^2 = \frac{1}{2\pi}\int\limits_{-\pi}^{\pi} \|\Phi(e^{it})f(e^{it})\|^2 dt,$$

it is clear that $\|M\| \leq \|\Phi\|_\infty$. Here $\|\Phi\|_\infty$ stands for the right hand side of (9). Put

$$\mu = \sup\left\{\operatorname*{ess\,sup}_t \|\Phi(e^{it})x\| \mid x \in \mathbb{C}^m,\ \|x\| = 1\right\}.$$

First, we show that $\mu \leq \|M\|$. Take $\varepsilon > 0$, and for $x \in \mathbb{C}^m$, $\|x\| = 1$, consider the set

$$\sigma = \sigma(\varepsilon, x) = \left\{t \mid \|\Phi(e^{it})x\| \geq \|M\| + \varepsilon\right\}.$$

Let χ_σ be the characteristic function for σ, and put $f(e^{it}) = x$ for each t. Then

$$\|M\|^2 \|\chi_\sigma f\|^2 \geq \|M(\chi_\sigma f)\|^2 = \|\chi_\sigma(Mf)\|^2$$

$$= \frac{1}{2\pi} \int_{-\pi}^{\pi} \|\chi_\sigma(t)\Phi(e^{it})x\|^2 dt$$

$$\geq (\|M\| + \varepsilon)^2 \left(\frac{1}{2\pi} \int_{-\pi}^{\pi} \|\chi_\sigma(t)x\|^2 dt\right)$$

$$= (\|M\| + \varepsilon)^2 \|\chi_\sigma f\|^2.$$

It follows that $\chi_\sigma f = 0$, and hence σ has measure zero. The latter fact implies that

$$\operatorname*{ess\,sup}_t \|\Phi(e^{it})x\| \leq \|M\| + \varepsilon.$$

This holds for each $\varepsilon > 0$ and $x \in \mathbb{C}^m$ with $\|x\| = 1$. Thus $\mu \leq \|M\|$.

To complete the proof it suffices to show that $\|\Phi\|_\infty \leq \mu$. Take $\varepsilon > 0$. Without loss of generality (change Φ on a set of measure zero if necessary) we may assume that for each $x \in \mathbb{C}^n$ with rational (complex) coordinates

$$\sup_t \|\Phi(e^{it})x\| = \operatorname*{ess\,sup}_t \|\Phi(e^{it})x\|.$$

Now, take $\varepsilon > 0$. Then the set

$$\tau = \left\{t \mid \|\Phi(e^{it})\| > \|\Phi\|_\infty - \varepsilon\right\}$$

has positive measure. Fix $t_0 \in \tau$. There exists $x_0 \in \mathbb{C}^m$, $\|x_0\| = 1$, with rational coordinates such that $\|\Phi(e^{it_0})x_0\| \geq \|\Phi\|_\infty - \varepsilon$. It follows that

$$\operatorname*{ess\,sup}_t \|\Phi(e^{it})x_0\| = \sup_t \|\Phi(e^{it})x_0\| \geq \|\Phi(e^{it_0})x_0\| \geq \|\Phi\|_\infty - \varepsilon.$$

Hence $\mu \geq \|\Phi\|_\infty - \varepsilon$, and this inequality holds for each $\varepsilon > 0$. Thus $\|\Phi\|_\infty \leq \mu$. $\quad\square$

COROLLARY 2.3. *Laurent operators on $\ell_2(\mathbb{Z})$ commute with one another.*

PROOF. The statement follows from the fact that operators of multiplication by scalar functions on $L_2(\mathbb{T})$ commute with one another. $\quad\square$

Now let us return to the matrix representation (4) for a block Laurent operator L. We know that the operators L_{rs} are Laurent operators on $\ell_2(\mathbb{Z})$. So, by the previous corollary, the entries of the matrix in (4) commute with one another. Hence

$$det\, L = \sum_\sigma (\mathrm{sgn}\,\sigma) L_{1\sigma_1} \cdots L_{m\sigma_m}$$

is a well-defined operator on $\ell_2(\mathbb{Z})$ (cf. Section XI.7). Note that $L_{1\sigma_1} \cdots L_{m\sigma_m}$ is a Laurent operator with defining function $\prod_j \varphi_{j\sigma_j}$, where φ_{rs} is the defining function of the Laurent operator L_{rs}. It follows that $det\, L$ is the Laurent operator defined by the function $\det \Phi(\cdot)$.

THEOREM 2.4. *Let L be the block Laurent operator defined by Φ. Then L is invertible if and only if there exists $\gamma > 0$ such that*

$$(10) \qquad \left\{ t \mid |\det \Phi(e^{it})| < \gamma \right\}$$

has measure zero, and in this case L^{-1} is the Laurent operator defined by $\Phi(\cdot)^{-1}$.

PROOF. Assume L is invertible. Then $det\, L$ is invertible (see Proposition XI.7.2) and $det\, L$ is the Laurent operator with defining function $\det \Phi(e^{it})$. So in order to prove that the set (10) has measure zero, it suffices to consider the case when $m = 1$. Write φ for the defining function of the (scalar) Laurent operator L, and assume L is invertible. Put

$$(11) \qquad E_n = \left\{ t \mid |\varphi(e^{it})| \leq \frac{1}{n} \right\},$$

and consider the function

$$(12) \qquad k_n(e^{it}) = \chi_{E_n}(t), \qquad -\pi \leq t \leq \pi.$$

As in the proof of Theorem 2.1 one easily computes that $\|(U^{-1}LU)k_n\| \leq n^{-1}\|k_n\|$. Since $U^{-1}LU$ has a bounded inverse, there exists $\delta > 0$ such that $\delta\|k_n\| \leq \|(U^{-1}LU)k_n\|$. It follows that $\|k_n\| = 0$ for $1/n < \delta$. But then we may conclude that E_n has measure zero for $1/n < \delta$.

To prove the converse statement, assume that the set (10) has measure zero for some $\gamma > 0$. Then outside a set of measure zero the inverse matrix $\Phi(e^{it})^{-1}$ is well-defined. By Cramer's rule $\Phi(\cdot)^{-1}$ is an $m \times m$ matrix whose entries are measurable, essentially bounded functions. Here we used that the entries φ_{rs} of Φ are measurable and essentially bounded. Let N be the operator of multiplication by $\Phi(\cdot)^{-1}$ on $L_2^m(\mathbb{T})$, and put $K = UNU^{-1}$. Then K is the Laurent operator defined by $\Phi(\cdot)^{-1}$ and $K = L^{-1}$. □

COROLLARY 2.5. *A block Laurent operator L is a Fredholm operator if and only if L is invertible.*

PROOF. We know that L is invertible if and only if $det\, L$ is invertible, and L is Fredholm if and only if $det\, L$ is Fredholm (see Section XI.7). Thus it suffices to consider the case $m = 1$.

Write φ for the defining function of the (scalar) Laurent operator L, and assume that L is a Fredholm operator. Consider the set

$$(13) \qquad E = \{t \in [-\pi, \pi] \mid \varphi(e^{it}) = 0\},$$

and put $\widetilde{E} = \{e^{it} \mid t \in E\}$. We have

$$(14) \qquad \{Uf\chi_{\widetilde{E}} \mid f \in L_2(\mathbb{T})\} \subset \operatorname{Ker} L.$$

Indeed, $U^{-1}LUf\chi_{\widetilde{E}} = \varphi f\chi_{\widetilde{E}} = (\varphi\chi_{\widetilde{E}})f = 0$, and thus $Uf\chi_{\widetilde{E}} \in \operatorname{Ker} L$. If E is a set of positive measure, then the left hand side of (14) is a linear space of infinite dimension. But $\dim \operatorname{Ker} L < \infty$. So E must have measure zero.

The fact that E has measure zero implies that L is injective. To see this, assume that $LUf = 0$. Then

$$0 = \big((U^{-1}LU)f\big)(e^{it}) = \varphi(e^{it})f(e^{it}) \text{ a.e. } .$$

It follows that $f(e^{it}) = 0$ a.e., and thus $f = 0$. So L is injective. Further, since L has a closed range, the same is true for $U^{-1}LU$, and hence there exists $\delta > 0$ such that $\delta\|f\| \le \|(U^{-1}LU)f\|$ for each $f \in L_2(\mathbb{T})$ (cf. Theorem XI.2.1). Now, let E_n be defined by (11), and take k_n as in (12). We have $\|(U^{-1}LU)k_n\| \le \frac{1}{n}\|k_n\|$. So as in the proof of Theorem 2.4 we may conclude that $\|k_n\| = 0$ for $1/n < \delta$. Hence E_n has measure zero for $1/n < \delta$. According to Theorem 2.4 this implies that L is invertible. The converse statement is trivial. $\square$

XXIII.3   BLOCK TOEPLITZ OPERATORS

A bounded linear operator $T: \ell_2^m \to \ell_2^m$ may be represented by an infinite matrix whose entries are operators acting on $\mathbb{C}^m$:

$$(1) \qquad T = [A_{ij}]_{i,j=0}^{\infty}.$$

As in the previous section, $A_{ij} = \pi_i T \tau_j$, where now

$$\begin{aligned} \tau_j &: \mathbb{C}^m \to \ell_2^m, & \tau_j x &= (\delta_{nj}x)_{n=0}^{\infty}, \\ \pi_i &: \ell_2^m \to \mathbb{C}^m, & \pi_i(x_n)_{n=0}^{\infty} &= x_i. \end{aligned}$$

We call T a *block Toeplitz operator* if its matrix elements A_{ij} depend only on the difference $i - j$. So T is a block Toeplitz operator if and only if its infinite matrix has the following form:

$$(2) \qquad \begin{bmatrix} A_0 & A_{-1} & A_{-2} & \cdots \\ A_1 & A_0 & A_{-1} & \cdots \\ A_2 & A_1 & A_0 & \\ \vdots & \vdots & & \ddots \end{bmatrix}.$$

The word "block" refers to the fact that in (2) the entries are not scalar but linear transformations on $\mathbb{C}^m$. In the sequel the word "block" will often be omitted.

Let S be the forward shift on ℓ_2^m. Thus

$$S(x_0, x_1, x_2, \ldots) = (0, x_0, x_1, \ldots).$$

An operator T on ℓ_2^m is a Toeplitz operator if and only if $T = S^*TS$. To see this, note that

$$\pi_i S^* = \pi_{i+1}, \quad S\tau_j = \tau_{j+1} \qquad (i, j = 0, 1, 2, \ldots).$$

Thus $\pi_i(S^*TS)\tau_j = A_{i+1,j+1}$. It follows that $T = S^*TS$ if and only if $A_{ij} = A_{i+1,j+1}$ for $i, j = 0, 1, \ldots$.

Since ℓ_2^m is a direct sum of m copies of ℓ_2, an operator T on ℓ_2^m may also be represented by an $m \times m$ matrix whose entries are operators acting on ℓ_2. Thus

$$(3) \qquad T = \begin{bmatrix} T_{11} & \cdots & T_{1m} \\ \vdots & & \vdots \\ T_{m1} & \cdots & T_{mm} \end{bmatrix} : \ell_2^m \to \ell_2^m.$$

From the characterization of Toeplitz operators in terms of the shift operator, it is clear that T is a Toeplitz operator on ℓ_2^m if and only if all entries T_{rs} in the matrix representation (3) are Toeplitz operators on ℓ_2.

THEOREM 3.1. *If* $T = [A_{i-j}]_{i,j=0}^{\infty}$ *is a Toeplitz operator on* ℓ_2^m, *then* $L = [A_{i-j}]_{i,j=-\infty}^{\infty}$ *is a well-defined Laurent operator on* $\ell_2^m(\mathbb{Z})$ *and*

$$(4) \qquad \|T\| = \|L\| < \infty.$$

PROOF. For $k = 0, 1, 2, \ldots$ let $T_k : \ell_2^m(\mathbb{Z}) \to \ell_2^m(\mathbb{Z})$ be defined by

$$\pi_i T_k \tau_j = \begin{cases} A_{i-j} & \text{for } i, j \geq -k, \\ 0 & \text{otherwise.} \end{cases}$$

Thus the operator T_k has the following representation as a double infinite block matrix:

$$T_k = \left[\begin{array}{c|cccccccc} 0 & & & & & 0 & & & \\ \hline & A_0 & A_{-1} & \cdots & A_{-k} & A_{-k-1} & A_{-k-2} & \cdots \\ & A_1 & A_0 & \cdots & A_{-k+1} & A_{-k} & A_{-k-1} & \cdots \\ & \vdots & \vdots & & \vdots & \vdots & \vdots & \\ 0 & A_k & A_{k-1} & \cdots & \boxed{A_0} & A_{-1} & A_{-2} & \cdots \\ & A_{k+1} & A_k & \cdots & A_1 & A_0 & A_{-1} & \cdots \\ & A_{k+2} & A_{k+1} & \cdots & A_2 & A_1 & A_0 & \cdots \\ & \vdots & \vdots & & \vdots & \vdots & \vdots & \end{array} \right].$$

From this matrix representation for T_k it is clear that $\|T_k\| = \|T_0\| = \|T\|$. We shall see that the sequence $T_0, T_1, \ldots$ converges in a weak sense to L.

Let $\mathcal{M}$ be the subspace of $\ell_2^m(\mathbb{Z})$ consisting of all double infinite sequences with a finite number of non-zero elements. Note that $\mathcal{M}$ is dense in $\ell_2^m(\mathbb{Z})$. Further, for $\widetilde{x}_0 \in \mathcal{M}$ and fixed $i \in \mathbb{Z}$ the i-th element in the sequence $T_k \widetilde{x}_0$ is independent of k for k sufficiently large. It follows that for $\widetilde{x}_0$ and $\widetilde{y}_0$ in $\mathcal{M}$ the inner product $\langle T_k \widetilde{x}_0, \widetilde{y}_0 \rangle$ is independent of k for k sufficiently large. In particular, $\lim_{k \to \infty} \langle T_k \widetilde{x}_0, \widetilde{y}_0 \rangle$ exists for each $\widetilde{x}_0$ and $\widetilde{y}_0$ in $\mathcal{M}$. Now take arbitrary elements $\widetilde{x}$ and $\widetilde{y}$ in $\ell_2^m(\mathbb{Z})$. We claim that the sequence $\left(\langle T_k \widetilde{x}, \widetilde{y} \rangle \right)_k$ is a Cauchy sequence. To see this, take $\varepsilon > 0$ and choose $\widetilde{x}_0$ and $\widetilde{y}_0 \in \mathcal{M}$ such that

$$\|T\|\left(\|\widetilde{x} - \widetilde{x}_0\|\|\widetilde{y}\| + \|\widetilde{x} - \widetilde{x}_0\|\|\widetilde{y} - \widetilde{y}_0\| + \|\widetilde{x}\|\|\widetilde{y} - \widetilde{y}_0\| \right) < \frac{1}{3}\varepsilon.$$

Then $|\langle T_k \widetilde{x}, \widetilde{y} \rangle - \langle T_k \widetilde{x}_0, \widetilde{y}_0 \rangle| < \frac{1}{3}\varepsilon$ for each k, and hence

$$|\langle T_k \widetilde{x}, \widetilde{y} \rangle - \langle T_\ell \widetilde{x}, \widetilde{y} \rangle| < \frac{2}{3}\varepsilon + |\langle T_k \widetilde{x}_0, \widetilde{y}_0 \rangle - \langle T_\ell \widetilde{x}_0, \widetilde{y}_0 \rangle|.$$

Choose n_0 such that $|\langle T_k \widetilde{x}_0, \widetilde{y}_0 \rangle - \langle T_\ell \widetilde{x}_0, \widetilde{y}_0 \rangle| < \frac{1}{3}\varepsilon$ for $k, \ell \geq n_0$. Then

$$|\langle T_k \widetilde{x}, \widetilde{y} \rangle - \langle T_\ell \widetilde{x}, \widetilde{y} \rangle| < \varepsilon \qquad (k, \ell \geq n_0).$$

Thus $\left(\langle T_k \widetilde{x}, \widetilde{y} \rangle \right)_k$ is a Cauchy sequence.

Put

$$(5) \qquad B(\widetilde{x}, \widetilde{y}) = \lim_{k \to \infty} \langle T_k \widetilde{x}, \widetilde{y} \rangle.$$

Note that $B(\widetilde{x}, \widetilde{y})$ is an indefinite inner product on $\ell_2^m(\mathbb{Z})$. From (5) it is clear that

$$(6) \qquad |B(\widetilde{x}, \widetilde{y})| \leq \|T\|\|\widetilde{x}\|\|\widetilde{y}\|.$$

For fixed $\widetilde{x}$ the function $g(\cdot) = \overline{B(\widetilde{x}, \cdot)}$ is a continuous linear functional on $\ell_2^m(\mathbb{Z})$. By the Riesz representation theorem there exists $\widetilde{u} \in \ell_2^m(\mathbb{Z})$ such that $g(\widetilde{y}) = \langle \widetilde{y}, \widetilde{u} \rangle$. Note that $\widetilde{u}$ is a function of $\widetilde{x}$. Write $\widetilde{u} = K(\widetilde{x})$. Then

$$(7) \qquad \langle K(\widetilde{x}), \widetilde{y} \rangle = B(\widetilde{x}, \widetilde{y}).$$

From (7) it is easy to deduce that K is a linear transformation on $\ell_2^m(\mathbb{Z})$. From (6) it is clear that K is a bounded operator and $\|K\| \leq \|T\|$. According to (5)

$$\langle K\widetilde{x}, \widetilde{y} \rangle = \lim_{k \to \infty} \langle T_k \widetilde{x}, \widetilde{y} \rangle,$$

which implies that

$$\langle \pi_i K \tau_j x, y \rangle = \lim_{k \to \infty} \langle \pi_i T_k \tau_j x, y \rangle$$

for each $x, y \in \mathbb{C}^m$. But $\pi_i T_k \tau_j = A_{i-j}$ for k sufficiently large. Thus K is the Laurent operator with double infinite matrix $[A_{i-j}]_{i,j=-\infty}^{\infty}$. In particular, $\|T\| \leq \|K\|$, and thus $\|K\| = \|T\|$. $\square$

Let $T = [A_{i-j}]_{i,j=0}^{\infty}$ be a block Toeplitz operator. According to the previous theorem (and Theorem 2.1) the entries of the $m \times m$ matrix function Φ,

$$\text{(8)} \qquad \Phi(\zeta) = \sum_{j=-\infty}^{\infty} \zeta^j A_j, \qquad |\zeta| = 1,$$

are measurable and essentially bounded on $\mathbb{T}$. We shall refer to Φ as the *defining function* of T. If (8) holds, then we shall also say that T is *defined by* Φ. From the remark preceding Corollary 2.2 it is clear that any $m \times m$ matrix function Φ on $\mathbb{T}$ with measurable and essentially bounded entries is the defining function of some block Toeplitz operator T. Indeed, if A_n is the n-th Fourier coefficient of Φ ($n \in \mathbb{Z}$), then $T = [A_{i-j}]_{i,j=0}^{\infty}$ is a block Toeplitz operator and its defining function is Φ.

COROLLARY 3.2. *Let T be the block Toeplitz operator with defining function Φ. Then*

$$\text{(9)} \qquad \|T\| = \operatorname{ess\,sup}_{t} \|\Phi(e^{it})\|.$$

PROOF. Apply Theorem 3.1 and Corollary 2.2. $\square$

Theorems 3.1 and 2.1 together imply that a block Toeplitz operator with defining function Φ is unitarily equivalent to the compression to the Hardy space $H_2^m(\mathbb{T})$ of the operator of multiplication by Φ on $L_2^m(\mathbb{T})$. More precisely the following corollary holds true.

COROLLARY 3.3. *Let T be the block Toeplitz operator with defining function Φ. Then*

$$U^{-1} T U f = \mathbb{P} M_\Phi f, \qquad f \in H_2^m(\mathbb{T}),$$

where U is the Fourier transformation on $H_2^m(\mathbb{T})$, the operator M_Φ is the operator of multiplication by Φ on $L_2^m(\mathbb{T})$ and $\mathbb{P}$ is the orthogonal projection of $L_2^m(\mathbb{T})$ onto $H_2^m(\mathbb{T})$.

XXIII.4 BLOCK TOEPLITZ OPERATORS DEFINED BY CONTINUOUS FUNCTIONS

We begin with a general remark. Let $T = [A_{i-j}]_{i,j=0}^{\infty}$ be a block Toeplitz operator, and let $L = [A_{i-j}]_{i,j=-\infty}^{\infty}$ be the corresponding Laurent operator. By J we denote the operator which identifies $\ell_2^m(\mathbb{Z})$ with two copies of ℓ_2^m. More precisely, $J\colon \ell_2^m(\mathbb{Z}) \to \ell_2^m \oplus \ell_2^m$ is defined by

$$J(\ldots, x_{-2}, x_{-1}, \boxed{x_0}, x_1, x_2, \ldots) = ((x_{-1}, x_{-2}, \ldots), (x_0, x_1, x_2, \ldots)).$$

Then JLJ^{-1} acts on the direct sum $\ell_2^m \oplus \ell_2^m$. Write JLJ^{-1} as a 2×2 operator matrix with respect to the direct sum $\ell_2^m \oplus \ell_2^m$:

$$JLJ^{-1} = \begin{bmatrix} R_{11} & R_{12} \\ R_{21} & R_{22} \end{bmatrix} \colon \ell_2^m \oplus \ell_2^m \to \ell_2^m \oplus \ell_2^m.$$

The operators R_{11}, R_{12}, R_{21} and R_{22} are bounded linear operators on ℓ_2^m given by the following infinite matrices:

$$R_{11} = \begin{bmatrix} A_0 & A_1 & A_2 & \cdots \\ A_{-1} & A_0 & A_1 & \cdots \\ A_{-2} & A_{-1} & A_0 & \\ \vdots & \vdots & & \ddots \end{bmatrix}, \qquad R_{12} = \begin{bmatrix} A_{-1} & A_{-2} & A_{-3} & \cdots \\ A_{-2} & A_{-3} & A_{-4} & \cdots \\ A_{-3} & A_{-4} & A_{-5} & \cdots \\ \vdots & \vdots & \vdots & \end{bmatrix},$$

$$R_{21} = \begin{bmatrix} A_1 & A_2 & A_3 & \cdots \\ A_2 & A_3 & A_4 & \cdots \\ A_3 & A_4 & A_5 & \cdots \\ \vdots & \vdots & \vdots & \end{bmatrix}, \qquad R_{22} = \begin{bmatrix} A_0 & A_{-1} & A_{-2} & \cdots \\ A_1 & A_0 & A_{-1} & \cdots \\ A_2 & A_1 & A_0 & \\ \vdots & \vdots & & \ddots \end{bmatrix}.$$

Thus R_{22} is just equal to the original Toeplitz operator T. The operator R_{11} is also a Toeplitz operator; its defining function is the matrix function $\Phi(e^{-it})$, where Φ is the defining function of T. The entries of the matrices for R_{12} and R_{21} are constant on lines perpendicular to the main diagonal. An operator H on ℓ_2^m with this property, i.e.,

$$\pi_i H \tau_{j+1} = \pi_{i+1} H \tau_j \qquad (i, j = 0, 1, 2, \ldots),$$

is called a *(block) Hankel operator*. Note that $\|R_{pq}\| \le \|JLJ^{-1}\| = \|L\|$, and hence, by virtue of Corollary 2.2, we have the following estimate for the norm of the operator R_{pq}:

$$(1) \qquad \|R_{pq}\| \le \operatorname*{ess\,sup}_{t} \|\Phi(e^{it})\|, \qquad p, q = 1, 2.$$

LEMMA 4.1. *If the defining function Φ of the block Toeplitz operator $T = [A_{i-j}]_{i,j=0}^{\infty}$ is continuous, then the block Hankel operators*

$$H_1 = [A_{i+j+1}]_{i,j=0}^{\infty}, \qquad H_2 = [A_{-i-j-1}]_{i,j=0}^{\infty}$$

are compact operators.

PROOF. Since Φ is continuous, there exists a sequence of trigonometric polynomials,

$$P_n(e^{it}) = \sum_{k=-r_n}^{r_n} e^{itk} C_k^{(n)}, \qquad n = 1, 2, \ldots,$$

where $C_k^{(n)}$ are operators on $\mathbb{C}^m$, such that

$$(2) \qquad \max_{-\pi \le t \le \pi} \|\Phi(e^{it}) - P_n(e^{it})\| \to 0 \qquad (n \to \infty).$$

Let L_n be the Laurent operator defined by the function P_n. Consider

$$JL_n J^{-1} = \begin{bmatrix} R_{11}^{(n)} & R_{12}^{(n)} \\ R_{21}^{(n)} & R_{22}^{(n)} \end{bmatrix}.$$

Note that P_n has only a finite number of non-zero Fourier coefficients. It follows that the operators $R_{12}^{(n)}$ and $R_{21}^{(n)}$ are operators of finite rank. From (2) we may conclude that the sequence $L_1, L_2, \ldots$ converges in the operator norm to the Laurent operator defined by Φ. But then

$$R_{12}^{(n)} \to H_2, \quad R_{21}^{(n)} \to H_1 \qquad (n \to \infty)$$

in the operator norm. Thus H_1 and H_2 are compact. $\square$

Let us write T_Φ for the block Toeplitz operator defined by Φ and L_Φ for the corresponding Laurent operator. Note that

$$(3) \qquad\qquad L_{\Phi_1} L_{\Phi_2} = L_{\Phi_1 \Phi_2},$$

where $(\Phi_1 \Phi_2)(e^{it})$ is the product of the matrices $\Phi_1(e^{it})$ and $\Phi_2(e^{it})$. The analogue of (3) for Toeplitz operators does not hold in general. To see this, consider

$$J L_{\Phi_j} J^{-1} = \begin{bmatrix} R_{11}^{(j)} & R_{12}^{(j)} \\ R_{21}^{(j)} & T_{\Phi_j} \end{bmatrix}, \qquad j = 1, 2.$$

According to (3) we have $J L_{\Phi_1 \Phi_2} J^{-1}$ is the product of $J L_{\Phi_1} J^{-1}$ and $J L_{\Phi_2} J^{-1}$. But then it is clear that

$$(4) \qquad\qquad T_{\Phi_1 \Phi_2} = T_{\Phi_1} T_{\Phi_2} + R_{21}^{(1)} R_{12}^{(2)}.$$

Formula (4) is a useful identity. From Lemma 4.1 and its proof it is clear that the following corollary holds true.

COROLLARY 4.2. *If all the entries in one of the defining matrix functions Φ_1 and Φ_2 are continuous (resp., trigonometric polynomials), then the operator*

$$(5) \qquad\qquad T_{\Phi_1 \Phi_2} - T_{\Phi_1} T_{\Phi_2}$$

is compact (resp., has finite rank).

THEOREM 4.3. *Assume that the defining function Φ of the block Toeplitz operator T is continuous. Then T is Fredholm if and only if*

$$(6) \qquad\qquad \det \Phi(\zeta) \neq 0, \qquad |\zeta| = 1.$$

PROOF. Assume (6) holds true. Let Φ^{-1} be the function defined by $\Phi^{-1}(\zeta) = \Phi(\zeta)^{-1}$, $\zeta \in \mathbb{T}$. Note that $\Phi \Phi^{-1} = \Phi^{-1} \Phi = E$, where E is the matrix function on $\mathbb{T}$ which is identically equal to the $m \times m$ identity matrix. Of course T_E is the identity operator on ℓ_2^m. So Corollary 4.2 implies that $I - T_\Phi T_{\Phi^{-1}}$ and $I - T_{\Phi^{-1}} T_\Phi$ are compact operators. Thus in the Calkin algebra $\mathcal{L}(\ell_2^m)/\mathcal{K}(\ell_2^m)$ the coset $[T_\Phi]$ is invertible and

$$[T_\Phi]^{-1} = [T_{\Phi^{-1}}].$$

In particular (see Theorem XI.5.2), the operator $T = T_\Phi$ is Fredholm.

To prove the converse, assume T is Fredholm. First take $m = 1$. Let L be the Laurent operator defined by φ, and write

$$(7) \qquad JLJ^{-1} = \begin{bmatrix} R_{11} & R_{12} \\ R_{21} & T \end{bmatrix}.$$

We shall prove that R_{11} is similar to the Banach dual T' of T. Recall that $T' : \ell_2' \to \ell_2'$ is defined by

$$(T'F)(x) = F(Tx), \qquad F \in \ell_2', x \in \ell_2.$$

Define $V : \ell_2 \to \ell_2'$ by setting $(Vx)(y) = \sum_{j=0}^{\infty} x_j y_j$. Here $x = (x_0, x_1, x_2, \dots)$ and $y = (y_0, y_1, y_2, \dots)$ are in ℓ_2. The operator V is invertible, and we shall see that

$$(8) \qquad R_{11} = V^{-1} T' V.$$

To prove (8), put $e_i = (\delta_{in})_{n=0}^{\infty}$, where δ_{ij} is the Kronecker delta. For $x = (x_0, x_1, \dots)$ in ℓ_2 we have

$$(T'Vx)(e_i) = (Vx)(Te_i) = \sum_{j=0}^{\infty} A_{j-i} x_j.$$

On the other hand $(VR_{11}x)(e_i)$ is equal to the i-th element of the sequence $R_{11}x$. Thus

$$(VR_{11}x)(e_i) = \sum_{j=0}^{\infty} A_{j-i} x_j.$$

This shows that the continuous linear functionals $T'Vx$ and $VR_{11}x$ coincide on the vectors e_i, $i = 0, 1, 2, \dots$. This implies $T'V = VR_{11}$, and (8) is proved.

Since T is Fredholm, the same is true for T'. Indeed, there exists an operator S such that $TS - I$ and $ST - I$ are finite rank operators. This implies that $S'T' - I$ and $T'S' - I$ are also finite rank operators, and therefore T' is Fredholm (cf. Theorem XI.5.1). By similarity (see formula (8)) the operator R_{11} is also Fredholm, and hence the operator

$$(9) \qquad \begin{bmatrix} R_{11} & 0 \\ 0 & T \end{bmatrix}$$

is Fredholm. From Lemma 4.1 we know that the operators R_{12} and R_{21} are compact. Thus the operator JLJ^{-1} is the sum of a Fredholm operator and a compact operator, which implies that JLJ^{-1} is Fredholm. Thus L is Fredholm. But then we can use Corollary 2.5 to show that L is invertible, which, by Theorem 2.4, implies that for some $\gamma > 0$ the set

$$\{t \mid |\varphi(e^{it})| < \gamma\}$$

has measure zero. Since φ is continuous, this yields (6).

Next, take $m \neq 1$. Let φ_{rs} be the (r,s)-th entry of Φ, and let T_{rs} be the Toeplitz operator on ℓ_2 defined by φ_{rs}. Recall that φ_{rs} is continuous. We know that

$$(10) \qquad T = \begin{bmatrix} T_{11} & \cdots & T_{1m} \\ \vdots & & \vdots \\ T_{m1} & \cdots & T_{mm} \end{bmatrix}.$$

Since the product of scalar functions is commutative, we may conclude from Corollary 4.2 that Toeplitz operators on ℓ_2 defined by continuous functions commute modulo the compact operators. In particular, the entries T_{rs} of the matrix in (10) commute with one another modulo the compact operators. So, since T is Fredholm, the same is true for $det\ T$ (see Section XI.7).

Let us compare $det\ T$ with $T_{\det \Phi}$, the Toeplitz operator defined by $\det \Phi$. Recall that

$$det\ T = \sum_{\sigma} (\operatorname{sgn} \sigma) T_{1\sigma_1} T_{2\sigma_2} \cdots T_{m\sigma_m}.$$

By repeatedly applying Corollary 4.2 we see that modulo the compact operators

$$(\operatorname{sgn} \sigma) T_{1\sigma_1} T_{2\sigma_2} \cdots T_{m\sigma_m}$$

is equal to the Toeplitz operator with symbol $(\operatorname{sgn} \sigma) \prod_{j=1}^{m} \varphi_{j\sigma_j}$. It follows that $det\ T - T_{\det \Phi}$ is compact. Since $det\ T$ is Fredholm, we see that $T_{\det \Phi}$ is Fredholm. But then we can apply the result for $m = 1$ and the proof is complete. $\sqsubset$

Note that the method used to prove Theorem 4.3 does not yield any information about the Fredholm index of T.

We conclude this section with a corollary about the essential spectrum for Toeplitz operators defined by scalar functions.

COROLLARY 4.4. *Let T be a Toeplitz operator on ℓ_2 defined by the scalar continuous function φ. Then*

$$(11) \qquad \sigma_{\mathrm{ess}}(T) = \{\varphi(\zeta) \mid \zeta \in \mathbb{T}\}.$$

PROOF. Recall (see Section XI.5) that $\lambda \in \sigma_{\mathrm{ess}}(T)$ if and only if $\lambda I - T$ is not a Fredholm operator. Now $\lambda I - T$ is the Toeplitz operator defined by the scalar function $\psi(\zeta) = \lambda - \varphi(\zeta)$, $\zeta \in \mathbb{T}$. So, according to Theorem 4.3, we have $\lambda \in \sigma_{\mathrm{ess}}(T)$ if and only if $\lambda - \varphi(\zeta) = 0$ for some $\zeta \in \mathbb{T}$, which proves (11). $\square$

XXIII.5 THE FREDHOLM INDEX OF A BLOCK TOEPLITZ OPERATOR DEFINED BY A CONTINUOUS FUNCTION

THEOREM 5.1. *Let T be the block Toeplitz operator defined by the continuous function Φ. Assume that $\det \Phi(\zeta) \neq 0$ for all $|\zeta| = 1$. Then T is a Fredholm operator and the index of T is equal to the negative of the winding number relative to the origin of the curve parametrized by the function*

$$(1) \qquad \rho : [-\pi, \pi] \to \mathbb{C}, \qquad \rho(t) = \det \Phi(e^{it}).$$

PROOF. We already know that the conditions on Φ imply that T is Fredholm (see Theorem 4.3). So we have to prove the statement about the index. To do this we first show that we may assume without loss of generality that the entries of Φ are trigonometric polynomials.

Since Φ is continuous, there exists a sequence of trigonometric polynomials

$$P_n(e^{it}) = \sum_{k=-r_n}^{r_n} e^{itk} C_k^{(n)}, \qquad n = 1, 2, \ldots,$$

where $C_k^{(n)}$ are operators on $\mathbb{C}^m$, such that

$$(2) \qquad \max_{|\zeta|=1} \|\Phi(\zeta) - P_n(\zeta)\| \to 0 \qquad (n \to \infty).$$

Let T_n be the Toeplitz operator defined by P_n. From (2) and Corollary 3.2 we may conclude that $T_n \to T$ in the operator norm. Since T is Fredholm, there exists a positive integer n_0 such that for $n \geq n_0$ the operator T_n is Fredholm and $\operatorname{ind} T_n = \operatorname{ind} T$.

From (2) it also follows that $\det P_n(\zeta) \to \det \Phi(\zeta)$ uniformly in ζ. So there exists $n_1 \geq n_0$ such that

$$(3) \qquad |\det P_n(\zeta) - \det \Phi(\zeta)| < |\det \Phi(\zeta)|, \qquad |\zeta| = 1,$$

for each $n \geq n_1$. This implies that for $n \geq n_1$ the curve parametrized by

$$\rho_n\colon [-\pi, \pi] \to \mathbb{C}, \qquad \rho_n(t) = \det P_n(e^{it})$$

does not pass through the origin and the winding number relative to the origin of this curve is equal to the winding number relative to the origin of the curve parameterized by (1) (see [C], Theorem V.3.8). So it remains to show that for $n \geq n_1$ the index of T_n is equal to the negative of the winding number relative to the origin of the curve parametrized by ρ_n. In other words it suffices to prove the theorem for the case when the defining function is a trigonometric matrix polynomial.

Assume that the entries of Φ are trigonometric polynomials, and denote by φ_{rs} the (r, s)-th entry of Φ. Let T_{rs} be the Toeplitz operator with symbol φ_{rs}. We know that

$$(4) \qquad T = \begin{bmatrix} T_{11} & \cdots & T_{1m} \\ \vdots & & \vdots \\ T_{m1} & \cdots & T_{mm} \end{bmatrix}.$$

Since the product of (scalar) trigonometric polynomials is commutative, we may conclude from Corollary 4.2 that the entries T_{rs} of the operator matrix (4) commute with one another modulo the operators of finite rank. It follows (see Theorem XI.7.6) that $\det T$ is a Fredholm operator and $\operatorname{ind}(\det T) = \operatorname{ind} T$.

Let $T_{\det \Phi}$ be the Toeplitz operator on ℓ_2 defined by $\det \Phi$. Note that $\det \Phi$ is a trigonometric polynomial. We already know (see the last paragraph of the proof of Theorem 4.3) that $det\, T - T_{\det \Phi}$ is a compact operator. Thus $T_{\det \Phi}$ is a Fredholm operator and

$$\text{ind}\, T_{\det \Phi} = \text{ind}(det\, T) = \text{ind}\, T.$$

Hence it suffices to prove the theorem for the case when $m = 1$ and the defining function is a (scalar) trigonometric polynomial.

So let T be a Toeplitz operator on ℓ_2, and let

$$\varphi(e^{it}) = \sum_{n=-r}^{s} \alpha_n e^{int}$$

be its symbol. We assume that $\varphi(\zeta) \neq 0$ for each $\zeta \in \mathbb{T}$. Note that $\lambda^r \varphi(\lambda)$ is a polynomial which has no zeros on $|\lambda| = 1$. So $\varphi(\lambda)$ admits a factorization of the following form:

$$(5) \qquad \varphi(\lambda) = c\lambda^{-r} \prod_{j=1}^{k} (\lambda - t_j^+) \prod_{j=1}^{\ell} (\lambda - t_j^-).$$

Here c is a non-zero constant, $t_1^+,\ldots,t_k^+$ are the zeros of $\lambda^r \varphi(\lambda)$ inside the unit circle and $t_1^-,\ldots,t_\ell^-$ are the zeros of $\lambda^r \varphi(\lambda)$ outside the unit circle. Let κ be the winding number relative to the origin of the curve parametrized by $t \mapsto \varphi(e^{it})$. Since φ is a rational function, the number κ is just equal to the number of zeros of φ inside the unit circle minus the number of poles of φ inside the unit circle. In other words, $\kappa = k - r$.

Let $T_{\omega(\lambda)}$ denote the Toeplitz operator on ℓ_2 defined by ω. By repeatedly applying Corollary 4.2 we see from the factorization (5) that

$$(6) \qquad T - c(T_{\lambda^{-1}})^r \prod_{j=1}^{k} T_{\lambda - t_j^+} \prod_{j=1}^{\ell} T_{\lambda - t_j^-}$$

is a compact operator. Let us investigate each of the factors in the second term of (6). Obviously, $T_{\lambda^{-1}} = S^*$, where S is the forward shift on ℓ_2. Thus $T_{\lambda^{-1}}$ is a Fredholm operator of index 1, and hence $(T_{\lambda^{-1}})^r$ is a Fredholm operator of index r. Next, observe that $T_{\lambda - \alpha} = S - \alpha I$. Thus for $|\alpha| > 1$ the operator $T_{\lambda - \alpha}$ is invertible. In particular, the operators

$$T_{\lambda - t_j^-}, \qquad j = 1,\ldots,\ell$$

are invertible. For $|\alpha| < 1$ the operator $S - \alpha I$ is a Fredholm operator, $n(S - \alpha I) = 0$ and $d(S - \alpha I) = 1$. In fact, for $|\alpha| < 1$ the Toeplitz operator

$$\begin{bmatrix} 0 & 1 & \alpha & \alpha^2 & \alpha^3 & \cdots \\ 0 & 0 & 1 & \alpha & \alpha^2 & \cdots \\ 0 & 0 & 0 & 1 & \alpha & \cdots \\ \vdots & \vdots & \vdots & \vdots & & \end{bmatrix}$$

is a left inverse of $S - \alpha I$ and its kernel consists of all sequences $(\eta, 0, 0, \ldots)$ with $\eta \in \mathbb{C}$. Thus the operators

$$T_{\lambda - t_j^+}, \qquad j = 1, \ldots, k$$

are Fredholm operators of index -1. By applying the product rule for Fredholm operators (see Theorem XI.3.2) we see that the second term in (6) is a Fredholm operator with index $r - k$. Since the operator (6) is compact, it follows that $\operatorname{ind} T = r - k = -\kappa$. $\square$

Let Φ be an $m \times m$ matrix function of which the entries are trigonometric polynomials, and assume that $\det \Phi(\zeta) \neq 0$ for each $\zeta \in \mathbb{T}$. In the above proof of Theorem 5.1 we have used Theorem XI.7.6 (which requires knowledge of trace class operators) to show that

$$(7) \qquad\qquad \operatorname{ind} T_\Phi = \operatorname{ind}(\det T_\Phi).$$

In the remaining part of this section we give an alternative proof for (7), based on Corollary XI.7.5, which does not employ the trace class theory.

Since the entries of Φ are trigonometric polynomials, we can choose a non-negative integer q such that $\Psi(\lambda) = \lambda^q \Phi(\lambda)$ is a matrix polynomial, i.e., each entry of Ψ is a scalar polynomial. We first show that it suffices to prove (7) for Ψ instead of Φ. Obviously, $\det \Psi(\zeta) \neq 0$ for each $\zeta \in \mathbb{T}$. Thus the block Toeplitz operator T_Ψ defined by Ψ is a Fredholm operator. Put $\Theta(\lambda) = \lambda^{-q} I$, and let T_Θ be the block Toeplitz operator defined by Θ. Note that $T_\Theta = (S^*)^q$, where S is the forward shift on ℓ_2^m. Thus T_Θ is a Fredholm operator and $\operatorname{ind} T_\Theta = qm$. By Corollary 4.2 the difference $T_\Phi - T_\Theta T_\Psi$ is a compact operator, and hence (use Theorems XI.4.2 and XI.3.2)

$$(8) \qquad\qquad \operatorname{ind} T_\Phi = qm + \operatorname{ind} T_\Psi.$$

Let T_θ be the Toeplitz operator on ℓ_2 defined by the function $\theta(\lambda) = \lambda^{-mq}$. The operator T_θ is Fredholm and $\operatorname{ind} T_\theta = mq$. Since the (i,j)-th entry of Φ is λ^{-q} times the (i,j)-th entry of Ψ, we can use Corollary 4.2 to prove that the operator $\det T_\Phi - T_\theta(\det T_\Psi)$ is compact. It follows that

$$(9) \qquad\qquad \operatorname{ind}(\det T_\Phi) = mq + \operatorname{ind}(\det T_\Psi).$$

Formulas (8) and (9) show that it suffices to prove (7) for Ψ instead of Φ.

So we may assume that Φ is a matrix polynomial. Then (see the next paragraph) there exists a sequence of matrix polynomials,

$$(10) \qquad \Phi^{(\nu)}(\lambda) = \begin{bmatrix} \varphi_{11}^{(\nu)}(\lambda) & \cdots & \varphi_{1m}^{(\nu)}(\lambda) \\ \vdots & & \vdots \\ \varphi_{m1}^{(\nu)}(\lambda) & \cdots & \varphi_{mm}^{(\nu)}(\lambda) \end{bmatrix}, \qquad \nu = 1, 2, \ldots,$$

such that

$$(11) \qquad \max_{|\zeta| = 1} \|\Phi(\zeta) - \Phi^{(\nu)}(\zeta)\| \to 0 \qquad (\nu \to \infty).$$

and for each ν

$$(12) \qquad \det \begin{bmatrix} \varphi_{11}^{(\nu)}(\zeta) & \cdots & \varphi_{1k}^{(\nu)}(\zeta) \\ \vdots & & \vdots \\ \varphi_{k1}^{(\nu)}(\zeta) & \cdots & \varphi_{kk}^{(\nu)}(\zeta) \end{bmatrix} \neq 0, \qquad \zeta \in \mathbb{T},\ k = 1,\ldots,m.$$

Let $T_{ij}^{(\nu)}$ be the Toeplitz operator on ℓ_2 defined by $\varphi_{ij}^{(\nu)}$, and put

$$T^{(\nu)} = \begin{bmatrix} T_{11}^{(\nu)} & \cdots & T_{1m}^{(\nu)} \\ \vdots & & \vdots \\ T_{m1}^{(\nu)} & \cdots & T_{mm}^{(\nu)} \end{bmatrix}.$$

Of course, $T^{(\nu)}$ is the block Toeplitz operator on ℓ_2^m defined by $\Phi^{(\nu)}$. From (11) and Corollary 3.2 it follows that $T^{(\nu)} \to T_\Phi$ in the operator norm. Formula (12) implies (use Theorem 4.3) that for $k = 1,\ldots,m$ and each ν the operator

$$\begin{bmatrix} T_{11}^{(\nu)} & \cdots & T_{1k}^{(\nu)} \\ \vdots & & \vdots \\ T_{k1}^{(\nu)} & \cdots & T_{kk}^{(\nu)} \end{bmatrix}$$

is Fredholm. It follows that T_Φ can be approximated in the operator norm as precise as we wish by an operator satisfying the conditions of Theorem XI.7.4. But then Corollary XI.7.5 yields (7).

So to prove (7) it remains to establish the existence of a sequence (10) with the properties (11) and (12). To do this put

$$M_k(\zeta) = \begin{bmatrix} \varphi_{11}(\zeta) & \cdots & \varphi_{1k}(\zeta) \\ \vdots & & \vdots \\ \varphi_{k1}(\zeta) & \cdots & \varphi_{kk}(\zeta) \end{bmatrix}, \qquad k = 1,\ldots,m,$$

where $\varphi_{ij}(\zeta)$ is the (i,j)-th entry of $\Phi(\zeta)$. Take $1 \le \ell \le m$, and let us assume that for $k = 1,\ldots,\ell-1$

$$\det M_k(\zeta) \neq 0, \qquad \zeta \in \mathbb{T},$$

while $\det M_\ell(\zeta)$ has a zero on $\mathbb{T}$. Since $M_\ell(\zeta)$ is a matrix polynomial, we can use the Smith canonical form (see Gohberg-Lancaster-Rodman [2], Appendix A.1) to write

$$(13) \qquad M_\ell(\lambda) = F_0(\lambda) \begin{bmatrix} p_1(\lambda) & & \\ & \ddots & \\ & & p_\ell(\lambda) \end{bmatrix} E_0(\lambda), \qquad \lambda \in \mathbb{C},$$

where $E_0(\lambda)$ and $F_0(\lambda)$ are $\ell \times \ell$ matrix polynomials such that $\det E_0(\lambda)$ and $\det F_0(\lambda)$ are non-zero for each $\lambda \in \mathbb{C}$ and the diagonal elements $p_1(\lambda),\ldots,p_\ell(\lambda)$ are certain scalar polynomials. Since $\det M_\ell(\zeta)$ is zero for some $\zeta \in \mathbb{T}$, some of the polynomials $p_1, \ldots, p_\ell$ must also have zeros on $\mathbb{T}$. We can remove these zeros from $\mathbb{T}$ by small perturbations. Next, make a partitioning of $\Phi(\zeta)$ according to the decomposition $\mathbb{C}^m = \mathbb{C}^\ell \oplus \mathbb{C}^{m-\ell}$ as follows:

$$\Phi(\zeta) = \begin{bmatrix} M_\ell(\zeta) & C(\zeta) \\ B(\zeta) & A(\zeta) \end{bmatrix},$$

and put

$$\widetilde{\Phi}(\zeta) = \begin{bmatrix} F_0(\zeta) & 0 \\ 0 & I_{m-\ell} \end{bmatrix} \begin{bmatrix} \widetilde{D}(\zeta) & \widetilde{C}(\zeta) \\ \widetilde{B}(\zeta) & A(\zeta) \end{bmatrix} \begin{bmatrix} E_0(\zeta) & 0 \\ 0 & I_{m-\ell} \end{bmatrix},$$

where

$$\widetilde{B}(\zeta) = B(\zeta)E_0(\zeta)^{-1}, \qquad \widetilde{C}(\zeta) = F_0(\zeta)^{-1}C(\zeta),$$

and $\widetilde{D}(\zeta)$ is an $\ell \times \ell$ diagonal matrix of which the diagonal elements $\widetilde{p}_1(\zeta), \ldots, \widetilde{p}_\ell(\zeta)$ are polynomials with no zeros on $\mathbb{T}$ such that $\widetilde{p}_j(\zeta) = p_j(\zeta)$ if $p_j(\zeta)$ has no zeros on $\mathbb{T}$ and the coefficients of $\widetilde{p}_j(\zeta)$ are small perturbations of the coefficients of $p_j(\zeta)$ otherwise. Let $\widetilde{M_k}(\zeta)$ be the $k \times k$ matrix in the left upper corner of $\widetilde{\Phi}(\zeta)$. Of course our construction now implies that $\det \widetilde{M_\ell}(\zeta)$ has no zeros on $\mathbb{T}$. Recall that $\det M_k(\zeta) \neq 0$ on $\mathbb{T}$ for each $1 \leq k \leq \ell-1$. This property is preserved under small perturbations. So, if the coefficients of $\widetilde{p}_j(\zeta)$ are chosen to be sufficiently close to those of p_j, then

$$(14) \qquad\qquad \det \widetilde{M_k}(\zeta) \neq 0, \qquad \zeta \in \mathbb{T}, k = 1, \ldots, \ell.$$

Note that in this construction the number

$$(15) \qquad\qquad \max_{\zeta \in \mathbb{T}} \|\Phi(\zeta) - \widetilde{\Phi}(\zeta)\|$$

can be made arbitrarily small, keeping the property (14). Now repeat the construction with ℓ replaced by $\ell + 1$. We obtain in a finite number of steps matrix polynomials $\widetilde{\Phi}$, which are close to Φ with respect to the distance measured by (15) and which have the extra property that for $1 \leq k \leq m$ the determinant of the $k \times k$ matrix in the left upper corner of $\widetilde{\Phi}$ has no zeros on $\mathbb{T}$. The construction of the desired sequence $(\Phi^{(\nu)})$ is now clear.

CHAPTER XXIV
TOEPLITZ OPERATORS DEFINED BY
RATIONAL MATRIX FUNCTIONS

In this chapter we study in more detail block Toeplitz operators defined by rational functions. The technique of Wiener-Hopf factorization is employed. The fact that the defining functions are rational allows us to represent the corresponding Toeplitz operators in a special way. We use this representation to construct explicitly the factors in a canonical Wiener-Hopf factorization. This yields explicit formulas for the inverse and the Fredholm characteristics. The results obtained here may be viewed as discrete analogues of the inversion and Fredholm theorems for Wiener-Hopf integral operators in Chapter XIII.

XXIV.1 PRELIMINARIES

Let T_Φ be the block Toeplitz operator on ℓ_2^m defined by the matrix function Φ. We call Φ a *plus-function* if all its Fourier coefficients with negative index are zero, i.e.,

$$A_n = \frac{1}{2\pi} \int_{-\pi}^{\pi} e^{-int}\Phi(e^{it})dt = 0, \qquad n = -1, -2, \dots .$$

Thus Φ is a plus-function if and only if T_Φ is lower triangular, that is,

$$T_\Phi = \begin{bmatrix} A_0 & 0 & 0 & \cdots \\ A_1 & A_0 & 0 & \cdots \\ A_2 & A_1 & A_0 & \cdots \\ \vdots & \vdots & \vdots & \ddots \end{bmatrix}.$$

If Φ is a rational matrix function (i.e., each entry of Φ is the quotient of two polynomials), then Φ is a plus-function if and only if Φ has no poles on the closed unit disc $|\zeta| \leq 1$ (which means that each entry of Φ has no poles on $|\zeta| \leq 1$). To see this note, that Φ has no poles on $|\zeta| = 1$ because Φ is a defining function, and hence for some $\varepsilon > 0$

$$(1) \qquad \Phi(\zeta) = \sum_{n=-\infty}^{\infty} \zeta^n A_n, \qquad 1 - \varepsilon < |\zeta| < 1 + \varepsilon,$$

where A_n is the n-th Fourier coefficient of Φ. It follows that Φ is analytic on $|\zeta| < 1 + \varepsilon$ if and only if Φ is a plus-function.

We say that Φ is a *minus-function* if all the Fourier coefficients with a (strictly) positive index are zero. In other words, Φ is a minus-function if and only if T_Φ is upper triangular. If Φ is a rational matrix function, then Φ is a minus-function if and only if Φ has no poles on $|\zeta| \geq 1$ (the point infinity included). In the sequel we write Φ^{-1} for the matrix function $\Phi(\cdot)^{-1}$ whenever for some $\gamma > 0$ the set

$$\{t \mid |\det \Phi(e^{it})| < \gamma\}$$

has measure zero (cf. Theorem XXIII.2.4).

THEOREM 1.1. *If Φ^{-1} exists and both Φ and Φ^{-1} are plus-functions or both are minus-functions, then T_Φ is invertible and*

$$(2) \qquad\qquad (T_\Phi)^{-1} = T_{\Phi^{-1}}.$$

The above theorem is an immediate corollary of the following lemma, which is an addition to Corollary XXIII.4.2.

LEMMA 1.2. *Consider the block Toeplitz operators T_{Φ_1} and T_{Φ_2}. If Φ_1 is a minus-function or Φ_2 is a plus-function, then*

$$(3) \qquad\qquad T_{\Phi_1 \Phi_2} = T_{\Phi_1} T_{\Phi_2}.$$

PROOF. We use formula (4) in Section XXIII.4. If Φ_1 is a minus-function, then the operator $R_{21}^{(1)}$ is the zero operator, and if Φ_2 is a plus-function, then $R_{12}^{(2)}$ is zero. Thus under the hypotheses of the present lemma the product $R_{21}^{(1)} R_{12}^{(2)}$ is always zero, and hence (3) holds true. $\square$

Let $\mathbb{P}$ be the orthogonal projection of $L_2^m(\mathbb{T})$ onto $H_2^m(\mathbb{T})$. If Φ is a plus-function, then

$$(4) \qquad\qquad \mathbb{P}M_\Phi g = M_\Phi g, \qquad g \in H_2^m(\mathbb{T}).$$

Here M_Φ is the operator of multiplication by Φ on $L_2^m(\mathbb{T})$. For later purposes we mention the following simple fact. If $g \in L_2^m(\mathbb{T})$ and g is rational (i.e., each component of g is the quotient of two polynomials), then $\mathbb{P}g$ is a rational function without poles on $|\zeta| \leq 1$ and

$$(5) \qquad\qquad (\mathbb{P}g)(\zeta) = \frac{1}{2\pi i} \int_{\mathbb{T}} \frac{g(\mu)}{\mu - \zeta} d\mu, \qquad |\zeta| < 1.$$

To prove (5), note that our hypotheses on g imply that g has no poles on $|\zeta| = 1$, and hence for some $\varepsilon > 0$

$$g(\zeta) = \sum_{n=-\infty}^{\infty} \zeta^n g_n, \qquad 1 - \varepsilon < |\zeta| < 1 + \varepsilon,$$

where g_n is the n-th Fourier coefficient of g. It follows that $(\mathbb{P}g)(\zeta) = \sum_{n=0}^{\infty} \zeta^n g_n$ is analytic on $|\zeta| < 1 + \varepsilon$. For $|\zeta| < 1$ the latter formula also yields (5).

XXIV.2 INVERTIBILITY AND FREDHOLM INDEX (SCALAR CASE)

For $|\alpha| < 1$ the scalar functions $1 - \zeta^{-1}\alpha$ and $(1 - \zeta^{-1}\alpha)^{-1}$ are minus-functions and for $|\alpha| > 1$ the functions $\zeta - \alpha$ and $(\zeta - \alpha)^{-1}$ are plus-functions. Since the product of minus-functions is again a minus-function, we see that the functions φ_- and φ_-^{-1}, where

$$(1) \qquad \varphi_-(\zeta) = c \frac{\prod_{j=1}^{k^+}(1 - \zeta^{-1}t_j^+)}{\prod_{j=1}^{\ell^+}(1 - \zeta^{-1}\tau_j^+)},$$

are minus-functions whenever $|t_j^+| < 1$ and $|\tau_j^+| < 1$ for all j. Similarly, the product of plus-functions is again a plus-function, and thus, if

$$(2) \qquad \varphi_+(\zeta) = d \frac{\prod_{j=1}^{k^-}(\zeta - t_j^-)}{\prod_{j=1}^{\ell^-}(\zeta - \tau_j^-)},$$

where $|t_j^-| > 1$ and $|\tau_j^-| > 1$ for all j, then φ_+ and φ_+^{-1} are plus-functions. It follows that any (scalar) rational function, which has no poles and zeros on $\mathbb{T}$, can be written in the form

$$(3) \qquad \varphi(\zeta) = \varphi_-(\zeta)\zeta^\kappa \varphi_+(\zeta), \qquad \zeta \in \mathbb{T},$$

where φ_- and φ_-^{-1} are minus-functions and φ_+ and φ_+^{-1} are plus-functions.

To derive the factorization (3), write $\varphi(\zeta)$ as a quotient $q_1(\zeta)/q_2(\zeta)$ of two polynomials which have no common zeros. Since φ has no poles and zeros on $\mathbb{T}$, the polynomials q_1 and q_2 have no zeros on $\mathbb{T}$. Thus we may write

$$q_1(\lambda) = c_1 \prod_{j=1}^{k^+}(\lambda - t_j^+) \prod_{j=1}^{k^-}(\lambda - t_j^-),$$

$$q_2(\lambda) = c_2 \prod_{j=1}^{\ell^+}(\lambda - \tau_j^+) \prod_{j=1}^{\ell^-}(\lambda - \tau_j^-),$$

where t_j^+ and τ_j^+ are inside the unit circle and the points t_j^- and τ_j^- are outside $\mathbb{T}$. It follows that (3) holds with $\kappa = k^+ - \ell^+$, the function φ_- is of the form (1) and φ_+ is of the form (2). Note that κ is equal to the winding number relative to the origin of the curve ρ_φ parametrized by the function $t \mapsto \varphi(e^{it})$. We shall refer to (3) as the *Wiener-Hopf factorization* of φ relative to $\mathbb{T}$.

THEOREM 2.1. *Let T be the Toeplitz operator on ℓ_2 defined by the (scalar) rational function φ, and assume that $\varphi(\zeta) \neq 0$ for each $\zeta \in \mathbb{T}$. Let*

$$\varphi(\zeta) = \varphi_-(\zeta)\zeta^\kappa \varphi_+(\zeta), \qquad \zeta \in \mathbb{T},$$

be the Wiener-Hopf factorization of φ relative to $\mathbb{T}$. Then T is invertible if and only if $\kappa = 0$, and in this case

$$T^{-1} = T_{\varphi_+^{-1}} \circ T_{\varphi_-^{-1}}.$$

Furthermore, if $\kappa > 0$, then T is left invertible, $d(T) = \kappa$ and a left inverse of T is given by

$$T^+ = T_{\varphi_+^{-1}}(S^*)^\kappa T_{\varphi_-^{-1}}.$$

If $\kappa < 0$, then T is right invertible, $n(T) = -\kappa$ and a right inverse of T is given by

$$T^+ = T_{\varphi_+^{-1}} S^{-\kappa} T_{\varphi_-^{-1}}.$$

Here S is the forward shift on ℓ_2 and S^* is its adjoint.

PROOF. Because of Lemma 1.2 we may write

$$T = T_{\varphi_-} T_{\zeta^\kappa} T_{\varphi_+}.$$

The factors T_{φ_-} and T_{φ_+} are invertible with inverses $T_{\varphi_-^{-1}}$ and $T_{\varphi_+^{-1}}$, respectively. To finish the proof it remains to observe that

$$T_{\zeta^\kappa} = \begin{cases} S^\kappa & \text{for} \quad \kappa \geq 0, \\ (S^*)^{-\kappa} & \text{for} \quad \kappa < 0. \end{cases} \qquad \square$$

Note that Theorem 2.1 gives an effective method to find the inverse of a Toeplitz operator. To illustrate this, let us consider the operator

$$T = \begin{bmatrix} -5 & 2 & 0 & 0 & \cdots \\ 2 & -5 & 2 & 0 & \cdots \\ 0 & 2 & -5 & 2 & \cdots \\ 0 & 0 & 2 & -5 & \cdots \\ \vdots & \vdots & \vdots & \vdots & \ddots \end{bmatrix}.$$

The operator T is a Toeplitz operator defined by $\varphi(\zeta) = 2\zeta^{-1} - 5 + 2\zeta$. Note that

$$\varphi(\zeta) = \zeta^{-1}(2 - 5\zeta + 2\zeta^2) = 2\left(1 - \zeta^{-1}\frac{1}{2}\right)(\zeta - 2),$$

and the latter factorization is the Wiener-Hopf factorization of φ relative to $\mathbb{T}$. It follows that the winding number of the curve $\varphi(e^{it})$, $-\pi \leq t \leq \pi$, relative to the origin is zero. Thus T is invertible and

$$T^{-1} = \frac{1}{2}T_{(\zeta-2)^{-1}} \circ T_{(1-\zeta^{-1}/2)^{-1}}.$$

Now $(e^{it} - 2)^{-1} = -\frac{1}{2}\sum_{n=0}^{\infty}\left(\frac{1}{2}\right)^n e^{int}$ and $\left(1 - \frac{1}{2}e^{-it}\right)^{-1} = \sum_{n=0}^{\infty}\left(\frac{1}{2}\right)^n e^{-int}$. Thus

$$T^{-1} = -\frac{1}{4}\begin{bmatrix} 1 & 0 & 0 & \cdots \\ \frac{1}{2} & 1 & 0 & \cdots \\ \left(\frac{1}{2}\right)^2 & \frac{1}{2} & 1 & \cdots \\ \vdots & \vdots & \vdots & \ddots \end{bmatrix}\begin{bmatrix} 1 & \frac{1}{2} & \left(\frac{1}{2}\right)^2 & \cdots \\ 0 & 1 & \frac{1}{2} & \cdots \\ 0 & 0 & 1 & \cdots \\ \vdots & \vdots & \vdots & \ddots \end{bmatrix}$$

$$= -\frac{1}{4}[\alpha_{ij}]_{i,j=0}^{\infty},$$

where

$$\alpha_{ij} = \sum_{\nu=0}^{\min(i,j)} \left(\frac{1}{2}\right)^{i-\nu} \left(\frac{1}{2}\right)^{j-\nu} = \frac{4}{3}\left(\frac{1}{2}\right)^{|i-j|} - \frac{1}{3}\left(\frac{1}{2}\right)^{i+j}.$$

The following corollary describes the spectrum of a Toeplitz operator defined by a (scalar) rational function.

COROLLARY 2.2. *Let T be a Toeplitz operator on ℓ_2 with a rational symbol φ, and let ρ_φ be the curve parametrized by $\varphi(e^{it})$, $-\pi \leq t \leq \pi$. Then the spectrum of T consists of ρ_φ and all points $\lambda \notin \rho_\varphi$ such that the winding number of the curve ρ_φ relative to λ is different from zero.*

PROOF. Note that $\lambda I - T$ is a Toeplitz operator defined by $\varphi_\lambda(e^{it}) = \lambda - \varphi(e^{it})$. We know (see Theorem XXIII.4.3) that $\lambda I - T$ is a Fredholm operator if and only if $\varphi_\lambda(e^{it}) \neq 0$ for all t. Thus $\lambda \in \rho_\varphi$ implies that $\lambda I - T$ is not a Fredholm operator. In particular, $\rho_\varphi \subset \sigma(T)$. If $\lambda \notin \rho_\varphi$, then by Theorem 2.1 the operator $\lambda I - T$ is invertible if and only if the winding number with respect to the origin of the curve parametrized by $\varphi_\lambda(e^{it})$, $-\pi \leq t \leq \pi$, is zero. Thus $\lambda \in \sigma(T)\backslash\rho_\varphi$ if and only if the winding number of ρ_φ relative to λ is different from zero. $\square$

XXIV.3 WIENER-HOPF FACTORIZATION

The following theorem gives the analogue of formula (3) in the previous section for rational matrix functions.

THEOREM 3.1. *Let Φ be a rational $m \times m$ matrix function with no poles on $\mathbb{T}$, and assume that $\det \Phi(\zeta) \neq 0$ for all $\zeta \in \mathbb{T}$. Then there exist integers $\kappa_1 \leq \kappa_2 \leq \cdots \leq \kappa_m$ and rational $m \times m$ matrix functions Φ_- and Φ_+, which have no poles on $\mathbb{T}$, such that*

$$(1) \qquad \Phi(\zeta) = \Phi_-(\zeta) \begin{bmatrix} \zeta^{\kappa_1} & & & 0 \\ & \zeta^{\kappa_2} & & \\ & & \ddots & \\ 0 & & & \zeta^{\kappa_m} \end{bmatrix} \Phi_+(\zeta), \qquad \zeta \in \mathbb{T},$$

and

(i) *Φ_+ has no poles on $|\zeta| \leq 1$,*

(ii) *$\det \Phi_+(\zeta) \neq 0$ for $|\zeta| \leq 1$,*

(iii) *Φ_- has no poles on $|\zeta| \geq 1$ (the point ∞ included),*

(iv) *$\det \Phi_-(\zeta) \neq 0$ for $|\zeta| \geq 1$ (the point ∞ included).*

In particular, Φ_-^{-1} and Φ_+^{-1} exist, the functions Φ_- and Φ_-^{-1} are minus-functions and Φ_+ and Φ_+^{-1} are plus-functions.

Theorem 3.1 is proved in Section XIII.2 (see the proof of Theorem XIII.2.1). The factorization (1) is called a *right Wiener-Hopf factorization* of Φ relative to $\mathbb{T}$. One obtains a *left Wiener-Hopf factorization* in (1) if the positions of the functions Φ_- and

Φ_+ are interchanged. In general we shall omit the work "right". We shall see in the next section that the integers $\kappa_1, \ldots, \kappa_m$ are uniquely determined by Φ; they are called the (*right*) *factorization indices*. If in (1) all indices $\kappa_1, \ldots, \kappa_m$ are equal to zero, then (1) is said to be a (*right*) *canonical factorization*. The factors Φ_- and Φ_+ in (1) are not uniquely determined by Φ. In contrast to the scalar case the standard construction of the Wiener-Hopf factorization (which is given in the proof of Theorem XIII.2.1) does not yield explicit formulas for the factors Φ_- and Φ_+ nor for the indices, but it provides an algorithm to obtain Φ_-, Φ_+ and the indices. In Section XXIV.6 we shall employ the notion of a realization to obtain explicit formulas for the factors in a canonical factorization.

XXIV.4 INVERTIBILITY AND FREDHOLM INDEX (MATRIX CASE)

THEOREM 4.1. *Let T be a block Toeplitz operator on ℓ_2^m defined by a rational matrix function Φ. Then T is invertible if and only if*

(i) $\det \Phi(\zeta) \neq 0$ *for each* $\zeta \in \mathbb{T}$,

(ii) Φ *admits a (right) canonical factorization relative to* $\mathbb{T}$.

In this case the inverse of T is obtained in the following way. Construct a canonical factorization $\Phi(\zeta) = \Phi_-(\zeta)\Phi_+(\zeta)$, $\zeta \in \mathbb{T}$, *and write the Fourier series*

$$\Phi_-(\zeta)^{-1} = \sum_{j=-\infty}^{0} \zeta^j \gamma_j^-, \qquad \Phi_+(\zeta)^{-1} = \sum_{j=0}^{\infty} \zeta^j \gamma_j^+.$$

Then $T^{-1} = [\Gamma_{ij}]_{i,j=0}^{\infty}$, *where*

$$\tag{1} \Gamma_{ij} = \begin{cases} \sum_{r=0}^{j} \gamma_{i-r}^+ \gamma_{r-j}^-, & i \geq j, \\ \sum_{r=0}^{i} \gamma_{i-r}^+ \gamma_{r-j}^-, & i \leq j. \end{cases}$$

First, it will be convenient to prove the theorem that gives the Fredholm properties of T.

THEOREM 4.2. *Let T be a block Toeplitz operator on ℓ_2^m defined by a rational matrix function Φ. Assume that $\det \Phi(\zeta) \neq 0$ for all $\zeta \in \mathbb{T}$, and let*

$$\tag{2} \Phi(\zeta) = \Phi_-(\zeta)\big([\zeta^{\kappa_j}\delta_{ij}]_{i,j=1}^{m}\big)\Phi_+(\zeta), \qquad \zeta \in \mathbb{T},$$

be a Wiener-Hopf factorization of Φ relative to $\mathbb{T}$. Then T is a Fredholm operator,

$$\tag{3} n(T) = \sum_{\kappa_j \leq 0} -\kappa_j, \qquad d(T) = \sum_{\kappa_j \geq 0} \kappa_j,$$

and a generalized inverse of T is given by the operator

$$(4) \qquad T^+ = T_{\Phi_+^{-1}} \begin{bmatrix} S^{-\kappa_1} & & & & & & & & 0 \\ & \ddots & & & & & & & \\ & & S^{-\kappa_r} & & & & & & \\ & & & I & & & & & \\ & & & & \ddots & & & & \\ & & & & & I & & & \\ & & & & & & (S^*)^{\kappa_{s+1}} & & \\ & & & & & & & \ddots & \\ 0 & & & & & & & & (S^*)^{\kappa_m} \end{bmatrix} T_{\Phi_-^{-1}},$$

where S is the forward shift on ℓ_2, the operator S^* is its adjoint, $\kappa_1, \ldots, \kappa_r$ are the negative factorization indices, and $\kappa_{s+1}, \ldots, \kappa_m$ are the positive factorization indices.

PROOF. From Theorem 1.1 we know that T_{Φ_-} and T_{Φ_+} are invertible operators with inverses $T_{\Phi_-^{-1}}$ and $T_{\Phi_+^{-1}}$, respectively. Furthermore, by Lemma 1.2,

$$(5) \qquad T = T_{\Phi_-} T_D T_{\Phi_+},$$

where T_D is the block Toeplitz operator on ℓ_2^m defined by $D(\lambda) = [\lambda^{\kappa_j} \delta_{ij}]_{i,j=1}^m$. Thus

$$T_D = \begin{bmatrix} (S^*)^{-\kappa_1} & & & & & & & & 0 \\ & \ddots & & & & & & & \\ & & (S^*)^{-\kappa_r} & & & & & & \\ & & & I & & & & & \\ & & & & \ddots & & & & \\ & & & & & I & & & \\ & & & & & & S^{\kappa_{s+1}} & & \\ & & & & & & & \ddots & \\ 0 & & & & & & & & S^{\kappa_m} \end{bmatrix}$$

where S is the forward shift on ℓ_2. From the special form of the operator T_D it is clear that the two identities in (3) hold true for T replaced by T_D. From (5) and the invertibility of the operators T_{Φ_-} and T_{Φ_+} one sees that $n(T) = n(T_D)$ and $d(T) = d(T_D)$. Thus (3) is proved.

Since S^* is a left inverse of S, it is readily checked that the middle term in the right hand side of (4) is a generalized inverse of T_D. But then it follows from (5) that T^+ is a generalized inverse of T. $\square$

Note that in the previous theorem the operator T is left invertible when all indices are non-negative and T is right invertible when all indices are non-positive. In particular, we see that Theorem 4.2 contains Theorem 2.1 as a special case.

PROOF OF THEOREM 4.1. Assume that T is invertible. Then T is a Fredholm operator, and thus $\det \Phi(\zeta) \neq 0$ for each $\zeta \in \mathbb{T}$ by Theorem XXIII.4.3. But then we can apply Theorem 4.2. According to formula (3) the invertibility of T implies that all factorization indices of Φ are equal to zero, and hence Φ admits a canonical factorization. So (i) and (ii) are proved.

Next, assume (i) and (ii) hold true, and let $\Phi(\zeta) = \Phi_-(\zeta)\Phi_+(\zeta)$, $\zeta \in \mathbb{T}$, be a canonical factorization. Theorem 4.2 implies that T is invertible and

$$(6) \qquad T^{-1} = T_{\Phi_+^{-1}} T_{\Phi_-^{-1}}.$$

Now use that

$$
T_{\Phi_+^{-1}} = \begin{bmatrix} \gamma_0^+ & 0 & 0 & \cdots \\ \gamma_1^+ & \gamma_0^+ & 0 & \cdots \\ \gamma_2^+ & \gamma_1^+ & \gamma_0^+ & \cdots \\ \vdots & \vdots & \vdots & \ddots \end{bmatrix}, \qquad
T_{\Phi_-^{-1}} = \begin{bmatrix} \gamma_0^- & \gamma_{-1}^- & \gamma_{-2}^- & \cdots \\ 0 & \gamma_0^- & \gamma_{-1}^- & \cdots \\ 0 & 0 & \gamma_0^- & \cdots \\ \vdots & \vdots & \vdots & \ddots \end{bmatrix},
$$

and formula (6) yields the description of the (i,j)-th entry in the block matrix representation of T^{-1} as given in (1). $\square$

Formula (3) can be used to prove that the factorization indices are uniquely determined by Φ. Let

$$\Phi(\zeta) = \Phi_-(\zeta)\big([\zeta^{\kappa_j}\delta_{ij}]_{i,j=1}^m\big)\Phi_+(\zeta), \qquad \zeta \in \mathbb{T},$$

be a Wiener-Hopf factorization of Φ relative to $\mathbb{T}$. Put $\Phi_\kappa(\zeta) = \zeta^{-\kappa}\Phi(\zeta)$. Note that $\widehat{\kappa}_j = \kappa_j - k$, $j = 1,\ldots,m$, are the factorization indices for Φ_k. Now apply the second identity in (3) to both Φ_k and Φ_{k+1}. One obtains that

$$
\begin{aligned}
d(T_{\Phi_k}) - d(T_{\Phi_{k+1}}) &= \sum_{\kappa_j - k \geq 0} (\kappa_j - k) - \sum_{\kappa_j - k - 1 \geq 0} (\kappa_j - k - 1) \\
&= \sum_{\kappa_j - k \geq 1} (\kappa_j - k) - \sum_{\kappa_j - k \geq 1} (\kappa_j - k - 1) \\
&= \sum_{\kappa_j - k \geq 1} 1 = \#\{j \mid \kappa_j \geq k + 1\}.
\end{aligned}
$$

It follows that

$$\#\{j \mid \kappa_j = k\} = d(T_{\Phi_{k-1}}) - 2d(T_{\Phi_k}) + d(T_{\Phi_{k+1}}),$$

which shows that the factorization indices are uniquely determined by Φ.

XXIV.5 INTERMEZZO ABOUT REALIZATION

In this section we prove that a block Toeplitz operator defined by a rational matrix function admits a special representation. As usual we identify a $p \times q$ matrix with the linear transformation from $\mathbb{C}^q$ into $\mathbb{C}^p$ defined by the canonical action of the matrix relative to the standard bases of $\mathbb{C}^q$ and $\mathbb{C}^p$. In the sequel I denotes a square identity matrix; we shall specify its size only when necessary.

THEOREM 5.1. *Assume that the block Toeplitz operator* $T = [\Phi_{i-j}]_{i,j=0}^{\infty}$ *is defined by a rational matrix function. Then its entries admit the following representation:*

(1)
$$\Phi_k = \begin{cases} -C\Omega^k(I - P)B, & k = 1, 2, \ldots, \\ I - C(I - P)B, & k = 0, \\ C\Omega^{-k-1}PB, & k = -1, -2, \ldots. \end{cases}$$

Here Ω *is a square matrix of order* n, *say,* Ω *has all its eigenvalues in the open unit disc* $\mathbb{D}$, *the* $n \times n$ *matrix* P *is a projection which commutes with* Ω, *and the matrices* B *and* C *have sizes* $n \times m$ *and* $m \times n$, *respectively.*

PROOF. Let T be defined by Φ. By assumption Φ is a rational matrix function. Since Φ is the defining function for T, Φ has no poles on $\mathbb{T}$. Let $\lambda_1, \ldots, \lambda_p$ be the poles of Φ in the open unit disc $\mathbb{D}$. Fix $1 \leq j \leq p$, and consider the Laurent series expansion of Φ in a punctured neighbourhood of λ_j:

$$\Phi(\lambda) = \sum_{\nu=-q_j}^{\infty} (\lambda - \lambda_j)^{\nu} A_{j,\nu}.$$

Introduce the following block matrices:

$$N_j = \begin{bmatrix} \lambda_j I & I & & & \\ & \lambda_j I & I & & \\ & & \ddots & \ddots & \\ & & & \lambda_j I & I \\ & & & & \lambda_j I \end{bmatrix}, \qquad Q_j = \begin{bmatrix} A_{j,-1} \\ A_{j,-2} \\ \vdots \\ A_{j,-q_j} \end{bmatrix},$$

$$R_j = [I \quad 0 \quad \cdots \quad 0].$$

Here I denotes the $m \times m$ identity matrix, the blanks in N_j stand for zero entries, and N_j has the size $q_j \times q_j$. The matrix $\lambda - N_j$ is invertible for $\lambda \neq \lambda_j$, and the first row in the block matrix representation of $(\lambda - N_j)^{-1}$ is given by

$$[(\lambda - \lambda_j)^{-1}I \quad (\lambda - \lambda_j)^{-2}I \quad \cdots \quad (\lambda - \lambda_j)^{-q_j}I].$$

It follows that $\Phi(\lambda) - R_j(\lambda - N_j)^{-1}Q_j$ is analytic in λ_j. We carry out this construction for each j, and define

$$A_2 = \begin{bmatrix} N_1 & & & \\ & N_2 & & \\ & & \ddots & \\ & & & N_p \end{bmatrix}, \qquad B_2 = \begin{bmatrix} Q_1 \\ Q_2 \\ \vdots \\ Q_p \end{bmatrix},$$

$$C_2 = [R_1 \quad R_2 \cdots R_p].$$

Note that A_2 is a block diagonal matrix with diagonal elements $N_1, N_2, \ldots, N_p$. So the eigenvalues of A_2 are precisely the poles $\lambda_1, \ldots, \lambda_p$. In particular, the eigenvalues of A_2 are in $\mathbb{D}$. Observe that

$$C_2(\lambda - A_2)^{-1} B_2 = \sum_{j=1}^{p} R_j(\lambda - N_j)^{-1} Q_j,$$

and hence $\Phi_+(\lambda) := \Phi(\lambda) - C_2(\lambda - A_2)^{-1} B_2$ has no poles in $\mathbb{D}$. Obviously, Φ_+ also has no poles on $|\lambda| = 1$. Therefore the function

$$(2) \qquad \Psi(\lambda) = \frac{1}{\lambda}\left[I - \Phi_+\left(\frac{1}{\lambda}\right) \right]$$

is analytic for each $|\lambda| \geq 1$ and

$$(3) \qquad \lim_{\lambda \to \infty} \Psi(\lambda) = 0.$$

Now repeat the above construction for Ψ instead of Φ. So there exist matrices A_1, B_1 and C_1 such that A_1 is a square matrix with all its eigenvalues in $\mathbb{D}$ and

$$(4) \qquad \Psi(\lambda) - C_1(\lambda - A_1)^{-1} B_1$$

is analytic on $\mathbb{D}$. Since both terms in (4) are analytic for $|\lambda| \geq 1$, we conclude that (4) defines an entire matrix function, which tends to 0 if $\lambda \to \infty$, because of (3). It follows that the function defined by (4) is identically zero. Thus $\Psi(\lambda) = C_1(\lambda - A_1)^{-1} B_1$. Together with (2) this yields:

$$(5) \qquad \Phi(\lambda) = C_1(\lambda A_1 - I_1)^{-1} B_1 + I + C_2(\lambda I_2 - A_2)^{-1} B_2,$$

where I_ν is the identity matrix of the same order as A_ν ($\nu = 1, 2$). Now put

$$(6) \qquad \Omega = \begin{bmatrix} A_1 & 0 \\ 0 & A_2 \end{bmatrix}, \qquad P = \begin{bmatrix} 0 & 0 \\ 0 & I_2 \end{bmatrix},$$

$$(7) \qquad C = [C_1 \quad C_2], \qquad B = \begin{bmatrix} B_1 \\ B_2 \end{bmatrix}.$$

Then Ω is a square matrix which has all its eigenvalues in $\mathbb{D}$ and P is a projection which commutes with Ω. Furthermore, the k-th Fourier coefficient Φ_k of $\Phi(\cdot)$ is given by

$$\begin{aligned}
\Phi_k &= -C_1 A_1^k B_1 = -C\Omega^k (I - P) B, & k > 0, \\
\Phi_0 &= I - C_1 B_1 = I - C(I - P) B, & \\
\Phi_{-k} &= C_2 A_2^{k-1} B_2 = C\Omega^{k-1} P B, & k > 0,
\end{aligned}$$

and hence (1) is proved. $\square$

THEOREM 5.2. *A rational $m \times m$ matrix function Φ without poles on $\mathbb{T}$ admits the following representation:*

$$(8) \qquad \Phi(\zeta) = I + C(\zeta G - A)^{-1}B, \qquad \zeta \in \mathbb{T}.$$

Here G and A are square matrices of the same size $n \times n$, say, $\det(\zeta G - A) \neq 0$ for each $\zeta \in \mathbb{T}$, and B and C are matrices of sizes $n \times m$ and $m \times n$, respectively.

PROOF. We have already shown that Φ can be written as in (5). Put

$$(9) \qquad G = \begin{bmatrix} A_1 & 0 \\ 0 & I_2 \end{bmatrix}, \qquad A = \begin{bmatrix} I_1 & 0 \\ 0 & A_2 \end{bmatrix},$$

and let B and C be as in (7). Then G and A have the desired properties and (8) holds. $\square$

If Φ is as in (8), then we shall say that Φ is in *realized form*, and we shall call the right hand side of (8) a *realization* of Φ. This terminology comes from mathematical system theory and refers to the fact that the right hand side of (8) is the transfer function of a (possibly singular) system.

The representations (1) and (8) are equivalent in the following sense. If the entries Φ_k of T are given by (1), then the defining function Φ of T admits the realization (8) with

$$(10) \qquad G = \Omega(I - P) + P, \qquad A = I - P + \Omega P.$$

Conversely, if the defining function Φ of T has the realized form (8), then the entries Φ_k of T are given by

$$(11) \qquad \Phi_k = \begin{cases} -CE\Omega^k(I - P)B, & k = 1, 2, \ldots \\ I - CE(I - P)B, & k = 0, \\ CE\Omega^{-k-1}PB, & k = -1, -2, \ldots \end{cases}$$

where

$$(12a) \qquad P = \frac{1}{2\pi i} \int_{\mathbb{T}} G(\zeta G - A)^{-1}d\zeta,$$

$$(12b) \qquad E = \frac{1}{2\pi i} \int_{\mathbb{T}} (1 - \zeta^{-1})(\zeta G - A)^{-1}d\zeta,$$

$$(12c) \qquad \Omega = \frac{1}{2\pi i} \int_{\mathbb{T}} (\zeta - \zeta^{-1})G(\zeta G - A)^{-1}d\zeta.$$

To prove (11) we use the spectral theory for operator pencils developed in Section IV.1. From Corollary IV.1.2 we know that P is a projection which commutes with Ω and

$$(13) \qquad \sigma(\Omega) \subset \mathbb{D},$$

where $\mathbb{D}$ is the open unit disc. Furthermore, E is invertible and

$$(14) \qquad (\zeta G - A)E = (\zeta\Omega - I)(I - P) + (\zeta I - \Omega)P.$$

It follows that

$$\Phi(\zeta) = I + CE(\zeta\Omega - I)^{-1}(I - P)B + CE(\zeta I - \Omega)^{-1}PB.$$

Since the spectrum of Ω is in $\mathbb{D}$, the latter identity yields (11).

Keeping the terminology introduced in Section IV.1 we refer to the operator E in (12b) as the *right equivalence operator* associated with the pencil $\zeta G - A$ and the curve $\mathbb{T}$. From (14) it follows that Ω in (12c) is also given by

$$\Omega = GE(I - P) + AEP.$$

The next lemma summarizes the spectral results of Section IV.1 in a form which will be useful in Sections XXIV.7 and 9.

LEMMA 5.3. *Consider the pencil $\zeta G - A$, where G and A are square matrices of order n, say. Assume $\det(\zeta G - A) \neq 0$ for all $\zeta \in \mathbb{T}$, and put*

$$P = \frac{1}{2\pi i}\int_{\mathbb{T}} G(\zeta G - A)^{-1}d\zeta, \qquad Q = \frac{1}{2\pi i}\int_{\mathbb{T}} (\zeta G - A)^{-1}Gd\zeta.$$

Then P and Q are projections, and relative to the decompositions $\mathbb{C}^n = \operatorname{Ker}Q \oplus \operatorname{Im}Q$ and $\mathbb{C}^n = \operatorname{Ker}P \oplus \operatorname{Im}P$ the pencil $\zeta G - A$ admits the following partitioning:

$$\zeta G - A = \begin{bmatrix} \zeta G_1 - A_1 & 0 \\ 0 & \zeta G_2 - A_2 \end{bmatrix} : \operatorname{Ker}Q \oplus \operatorname{Im}Q \to \operatorname{Ker}P \oplus \operatorname{Im}P,$$

where $A_1 : \operatorname{Ker}Q \to \operatorname{Ker}P$ and $G_2 : \operatorname{Im}Q \to \operatorname{Im}P$ are invertible. Furthermore for the functions

$$(15) \qquad F_1(\zeta) = \left[(\zeta G - A)|\operatorname{Ker}Q\right]^{-1} : \operatorname{Ker}P \to \operatorname{Ker}Q, \qquad \zeta \in \mathbb{T},$$

$$(16) \qquad F_2(\zeta) = \left[(\zeta G - A)|\operatorname{Im}Q\right]^{-1} : \operatorname{Im}P \to \operatorname{Im}Q, \qquad \zeta \in \mathbb{T},$$

we have that $F_1(\cdot)$ extends to an analytic function on the open unit disc $\mathbb{D}$ and $F_2(\cdot)$ has an analytic extension on $\mathbb{C}\backslash\overline{\mathbb{D}}$ which vanishes at infinity.

PROOF. According to Theorem IV.1.1 (with P and Q interchanged) the matrices P and Q are projections and the desired decompositions hold. Furthermore, the pencils $\zeta G_1 - A_1$ and $\zeta G_2 - A_2$ are $\mathbb{T}$-regular and for their spectra we have

$$(17). \qquad\qquad \sigma(G_1, A_1) \subset \mathbb{C}_\infty\backslash\overline{\mathbb{D}}, \qquad \sigma(G_2, A_2) \subset \mathbb{D}$$

These inclusions imply that A_1 and G_2 are invertible and the operators $G_1 A_1^{-1}$ and $A_2 G_2^{-1}$ have their spectra in $\mathbb{D}$. Since

$$F_1(\zeta) = A_1(\zeta G_1 A_1^{-1} - I)^{-1}, \qquad F_2(\zeta) = G_2(\zeta I - A_2 G_2^{-1})^{-1},$$

the analyticity properties of $F_1(\cdot)$ and $F_2(\cdot)$ are obvious. $\square$

XXIV.6 INVERSION OF A BLOCK LAURENT OPERATOR

In this section $L = [\Phi_{i-j}]_{i,j=-\infty}^{\infty}$ is a block Laurent operator on $\ell_2^m(\mathbb{Z})$ defined by a rational $m \times m$ matrix function. Theorem 5.1 implies that we may assume that the entries Φ_k are given by

$$
(1) \qquad \Phi_k = \begin{cases} -C\Omega^k(I-P)B, & k = 1,2,\ldots, \\ I - C(I-P)B\,, & k = 0, \\ C\Omega^{-k-1}PB & , & k = -1,-2,\ldots\,. \end{cases}
$$

Here Ω is a square matrix of order n, say, Ω has all its eigenvalues in the open unit disc $\mathbb{D}$ and P is a projection of $\mathbb{C}^n$ which commutes with Ω.

THEOREM 6.1. *Let the block Laurent operator* $L = [\Phi_{i-j}]_{i,j=-\infty}^{\infty}$ *be given by* (1). *Put*

$$
(2) \qquad G = \Omega(I-P) + P, \qquad A^{\times} = I - P + \Omega P - BC.
$$

Then L *is invertible if and only if* $\det(\zeta G - A^{\times}) \neq 0$ *for each* $\zeta \in \mathbb{T}$. *In this case the inverse of* L *is obtained in the following way. Put*

$$
(3) \qquad P^{\times} = \frac{1}{2\pi i} \int_{\mathbb{T}} G(\zeta G - A^{\times})^{-1} d\zeta,
$$

and let $E^{\times}$ *be the right equivalence operator associated with* $\zeta G - A^{\times}$ *and* $\mathbb{T}$. *Then* $L^{-1} = [\Phi_{i-j}^{\times}]_{i,j=-\infty}^{\infty}$ *with*

$$
(4) \qquad \Phi_k^{\times} = \begin{cases} C^{\times}(\Omega^{\times})^k(I-P^{\times})B, & k = 1,2,\ldots, \\ I + C^{\times}(I-P^{\times})B & , & k = 0, \\ -C^{\times}(\Omega^{\times})^{-k-1}P^{\times}B, & k = -1,-2,\ldots\,, \end{cases}
$$

where $\Omega^{\times} = GE^{\times}(I-P^{\times}) + A^{\times}E^{\times}P^{\times}$ *and* $C^{\times} = CE^{\times}$.

For the proof of the above theorem we need the following lemma.

LEMMA 6.2. *Let* $\Phi(\zeta) = I + C(\zeta G - A)^{-1}B$, $\zeta \in \mathbb{T}$, *be a given realization with* $\det(\zeta G - A) \neq 0$ *for* $\zeta \in \mathbb{T}$. *Then* $\det \Phi(\zeta) \neq 0$ *for each* $\zeta \in \mathbb{T}$ *if and only if* $\det(\zeta G - A + BC) \neq 0$ *for each* $\zeta \in \mathbb{T}$, *and in this case*

$$
(5) \qquad \Phi(\zeta)^{-1} = I - C\big[\zeta G - (A - BC)\big]^{-1}B, \qquad \zeta \in \mathbb{T}.
$$

PROOF. We use a variation of formula (2) in Section III.2. Let n be the order of G and A, and write I_n (resp. I_m) for the $n \times n$ (resp., $m \times m$) identity matrix. Take $\zeta \in \mathbb{T}$. Then the following formula holds true:

$$
(6) \qquad \begin{bmatrix} \Phi(\zeta) & 0 \\ 0 & I_n \end{bmatrix} = F(\zeta) \begin{bmatrix} \zeta G - A + BC & 0 \\ 0 & I_m \end{bmatrix} E(\zeta),
$$

where

$$E(\zeta) = \begin{bmatrix} -(\zeta G - A)^{-1}B & (\zeta G - A)^{-1} \\ \Phi(\zeta) & -C(\zeta G - A)^{-1} \end{bmatrix},$$

$$F(\zeta) = \begin{bmatrix} C(\zeta G - A)^{-1} & \Phi(\zeta) \\ I_n & B \end{bmatrix}.$$

Moreover, the factors $E(\zeta)$ and $F(\zeta)$ are invertible and

$$E(\zeta)^{-1} = \begin{bmatrix} C & I_m \\ \zeta G - A + BC & B \end{bmatrix},$$

$$F(\zeta)^{-1} = \begin{bmatrix} -B & I_n + BC(\zeta G - A)^{-1} \\ I_m & -C(\zeta G - A)^{-1} \end{bmatrix}.$$

The identity (6) implies that $\Phi(\zeta)$ is invertible if and only if $\zeta G - A + BC$ is invertible, and in this case

$$\Phi(\zeta)^{-1} = [I \quad 0]E(\zeta)^{-1}\begin{bmatrix} (\zeta G - A + BC)^{-1} & 0 \\ 0 & I \end{bmatrix} F(\zeta)^{-1}\begin{bmatrix} I \\ 0 \end{bmatrix}$$

$$= [C \quad I]\begin{bmatrix} (\zeta G - A + BC)^{-1} & 0 \\ 0 & I \end{bmatrix}\begin{bmatrix} -B \\ I \end{bmatrix}$$

$$= I - C(\zeta G - A + BC)^{-1}B. \quad \square$$

PROOF OF THEOREM 6.1. Put $A = I - P + \Omega P$. Then the function $\Phi(\cdot)$ defining L is given by

$$\Phi(\zeta) = I + C(\zeta G - A)^{-1}B, \qquad \zeta \in \mathbb{T}.$$

Since Φ has no poles on $\mathbb{T}$, the function Φ is continuous on $\mathbb{T}$, and hence, by Theorem XXIII.2.4, the operator L is invertible if and only if $\det \Phi(\zeta) \neq 0$ for each $\zeta \in \mathbb{T}$. Note that $A^{\times} = A - BC$. Thus we can apply Lemma 6.2 to show that L is invertible if and only if $\det(\zeta G - A^{\times}) \neq 0$ for each $\zeta \in \mathbb{T}$. This proves the first part of the theorem.

From Theorem XXIII.2.4 it also follows that $L^{-1} = [\Phi^{\times}_{i-j}]^{\infty}_{i,j=-\infty}$, where $\Phi^{\times}_{k}$ is the k-th Fourier coefficient of the function $\Phi(\cdot)^{-1}$. Now use formula (5), and apply formula (14) in the previous section to the pencil $\zeta G - A^{\times}$. We obtain

$$\Phi(\zeta)^{-1} = I - C(\zeta G - A^{\times})^{-1}B$$

$$= I - C^{\times}(\zeta \Omega^{\times} - I)^{-1}(I - P^{\times})B - C^{\times}(\zeta - \Omega^{\times})^{-1}P^{\times}B.$$

Here we used that $\Omega^{\times}$ commutes with $P^{\times}$. Since all the eigenvalues of $\Omega^{\times}$ are in $\mathbb{D}$, it follows that $\Phi^{\times}_{k}$ is given by (4). $\quad \square$

The following lemma will be used in the next sections.

LEMMA 6.3. *Let* $\Phi(\zeta) = I + C(\zeta G - A)^{-1}B$, $\zeta \in \mathbb{T}$, *be a given realization, and assume that* $\det \Phi(\zeta) \neq 0$ *for each* $\zeta \in \mathbb{T}$. *Put* $A^{\times} = A - BC$. *Then for* $\zeta \in \mathbb{T}$

$$C(\zeta G - A^{\times})^{-1} = \Phi(\zeta)^{-1}C(\zeta G - A)^{-1},$$

$$(\zeta G - A^{\times})^{-1}B = (\zeta G - A)^{-1}B\Phi(\zeta)^{-1},$$

$$(\zeta G - A^{\times})^{-1} = (\zeta G - A)^{-1} - (\zeta G - A)^{-1}B\Phi(\zeta)^{-1}C(\zeta G - A)^{-1}.$$

PROOF. A direct computation, using (5) and the fact that

$$BC = A - A^\times = (\zeta G - A^\times) - (\zeta G - A),$$

gives the desired formulas. $\square$

XXIV.7 EXPLICIT CANONICAL FACTORIZATION

Let Φ be an $m \times m$ rational matrix function without poles on $\mathbb{T}$ given in realized form:

$$(1) \qquad \Phi(\zeta) = I + C(\zeta G - A)^{-1}B, \qquad \zeta \in \mathbb{T}.$$

Put $A^\times = A - BC$. In this section we use (1) to construct explicitly a (right) canonical factorization of Φ. According to the definition, for such a factorization to exist it is necessary (but, in general, not sufficient) that $\det \Phi(\zeta) \neq 0$ for each $\zeta \in \mathbb{T}$. So Lemma 6.2 implies that we have to assume that $\det(\zeta G - A^\times) \neq 0$ for each $\zeta \in \mathbb{T}$.

THEOREM 7.1. *Let* $\Phi(\zeta) = I + C(\zeta G - A)^{-1}B$, $\zeta \in \mathbb{T}$, *be a given realization, and let n be the order of G and A. Put $A^\times = A - BC$, and assume that* $\det(\zeta G - A^\times) \neq 0$ *for all $\zeta \in \mathbb{T}$. Introduce the projections*

$$Q = \frac{1}{2\pi i}\int_{\mathbb{T}}(\zeta G - A)^{-1}Gd\zeta, \qquad Q^\times = \frac{1}{2\pi i}\int_{\mathbb{T}}(\zeta G - A^\times)^{-1}Gd\zeta,$$

$$P = \frac{1}{2\pi i}\int_{\mathbb{T}}G(\zeta G - A)^{-1}d\zeta, \qquad P^\times = \frac{1}{2\pi i}\int_{\mathbb{T}}G(\zeta G - A^\times)^{-1}d\zeta.$$

Then Φ admits a right canonical factorization relative to the circle $\mathbb{T}$ if and only if

$$(2) \qquad \mathbb{C}^n = \operatorname{Im} Q \oplus \operatorname{Ker} Q^\times, \qquad \mathbb{C}^n = \operatorname{Im} P \oplus \operatorname{Ker} P^\times.$$

In this case $\Phi(\zeta) = \Phi_-(\zeta)\Phi_+(\zeta)$, $\zeta \in \mathbb{T}$, is a right canonical factorization relative to $\mathbb{T}$ with

$$(3) \qquad \Phi_-(\zeta) = I + C(\zeta G - A)^{-1}(I - \rho)B, \qquad \zeta \in \mathbb{T},$$
$$(4) \qquad \Phi_+(\zeta) = I + C_\tau(\zeta G - A)^{-1}B, \qquad \zeta \in \mathbb{T},$$

$$(5) \qquad \Phi_-(\zeta)^{-1} = I - C(I - \tau)(\zeta G - A^\times)^{-1}B, \qquad \zeta \in \mathbb{T},$$
$$(6) \qquad \Phi_+(\zeta)^{-1} = I - C(\zeta G - A^\times)^{-1}\rho B, \qquad \zeta \in \mathbb{T}.$$

Here τ is the projection of $\mathbb{C}^n$ along $\operatorname{Im} Q$ onto $\operatorname{Ker} Q^\times$ and ρ is the projection of $\mathbb{C}^n$ along $\operatorname{Im} P$ onto $\operatorname{Ker} P^\times$. Furthermore, the first direct sum decomposition in (2) implies the second and conversely.

PROOF. Assume that the direct sum decompositions in (2) hold true. Write A, G, B, C and $A^\times = A - BC$ as block matrices relative to these decompositions:

$$(7) \qquad A = \begin{bmatrix} A_{11} & A_{12} \\ 0 & A_{22} \end{bmatrix} : \operatorname{Im} Q \oplus \operatorname{Ker} Q^\times \to \operatorname{Im} P \oplus \operatorname{Ker} P^\times,$$

$$(8) \qquad G = \begin{bmatrix} G_1 & 0 \\ 0 & G_2 \end{bmatrix} : \operatorname{Im} Q \oplus \operatorname{Ker} Q^\times \to \operatorname{Im} P \oplus \operatorname{Ker} P^\times,$$

$$(9) \qquad B = \begin{bmatrix} B_1 \\ B_2 \end{bmatrix} : \mathbb{C}^m \to \operatorname{Im} P \oplus \operatorname{Ker} P^\times,$$

$$(10) \qquad C = [C_1 \quad C_2] : \operatorname{Im} Q \oplus \operatorname{Ker} Q^\times \to \mathbb{C}^m,$$

$$(11) \qquad A^\times = \begin{bmatrix} A_{11}^\times & 0 \\ A_{21}^\times & A_{22}^\times \end{bmatrix} : \operatorname{Im} Q \oplus \operatorname{Ker} Q^\times \to \operatorname{Im} P \oplus \operatorname{Ker} P^\times.$$

From Lemma 5.3 (applied to $\zeta G - A$ as well as to $\zeta G - A^\times$) we know that

$$(12) \qquad AQ = PA, \qquad A^\times Q^\times = P^\times A^\times,$$
$$(13) \qquad GQ = PG, \qquad GQ^\times = P^\times G.$$

The first identity in (12) implies that A maps $\operatorname{Im} Q$ into $\operatorname{Im} P$. This explains the zero entry in the left lower corner of the block matrix for A. From (13) we conclude that G has the desired block diagonal form. From the second identity in (12) it follows that $A^\times$ maps $\operatorname{Ker} Q^\times$ into $\operatorname{Ker} P^\times$, which justifies the zero in the right upper corner of the block matrix for $A^\times$. The fact that $A^\times = A - BC$ implies

$$(14) \qquad A_{12} = B_1 C_2, \qquad A_{21}^\times = -B_2 C_1,$$
$$(15) \qquad A_{11}^\times = A_{11} - B_1 C_1, \qquad A_{22}^\times = A_{22} - B_2 C_2.$$

Define the matrix functions Φ_- and Φ_+ by (3) and (4), respectively. Using the block matrix representation (7)–(10) we may rewrite Φ_- and Φ_+ in the following form:

$$(16) \qquad \Phi_-(\zeta) = I + C_1(\zeta G_1 - A_{11})^{-1} B_1, \qquad \zeta \in \mathbb{T},$$
$$(17) \qquad \Phi_+(\zeta) = I + C_2(\zeta G_2 - A_{22})^{-1} B_2, \qquad \zeta \in \mathbb{T}.$$

From (7) and the first identity in (14) we see that

$$\Phi_-(\zeta)\Phi_+(\zeta) = I + [C_1 \quad C_2] \begin{bmatrix} \zeta G_1 - A_{11} & -B_1 C_2 \\ 0 & \zeta G_2 - A_{22} \end{bmatrix}^{-1} \begin{bmatrix} B_1 \\ B_2 \end{bmatrix}$$
$$= I + C(\zeta G - A)^{-1} B = \Phi(\zeta), \qquad \zeta \in \mathbb{T},$$

which gives us the desired factorization. Next, we check the analytical properties of the factors. Obviously, Φ_- and Φ_+ have no poles on $\mathbb{T}$. Note that

$$\zeta G_1 - A_{11} = (\zeta G - A)|\operatorname{Im} Q : \operatorname{Im} Q \to \operatorname{Im} P.$$

Thus we know from the analyticity properties of the function $F_2(\cdot)$ in formula (16) of Lemma 5.3 that $(\zeta G_1 - A_{11})^{-1}$ exists for $|\zeta| \geq 1$ and vanishes at infinity. So Φ_- has no poles on $|\zeta| > 1$ (including the point ∞). To see that the same is true for Φ_+ on $|\zeta| < 1$, we first note that the maps

$$J = (I - Q)|\operatorname{Ker} Q^\times : \operatorname{Ker} Q^\times \to \operatorname{Ker} Q,$$
$$H = (I - P)|\operatorname{Ker} P^\times : \operatorname{Ker} P^\times \to \operatorname{Ker} P$$

are invertible. In fact,

$$J^{-1} = \tau|\operatorname{Ker} Q, \qquad H^{-1} = \rho|\operatorname{Ker} P,$$

where τ is the projection along $\operatorname{Im} Q$ onto $\operatorname{Ker} Q^{\times}$ and ρ is the projection along $\operatorname{Im} P$ onto $\operatorname{Ker} P^{\times}$. Next, take $x \in \operatorname{Ker} Q^{\times}$. Then

$$\begin{aligned}
(\zeta G_2 - A_{22})x &= \rho(\zeta G - A)x \\
&= \rho(\zeta G - A)(I - Q)x \\
&= \rho(\zeta G - A)Jx,
\end{aligned}$$

which shows that

$$(18) \qquad H(\zeta G_2 - A_{22}) = \left[(\zeta G - A)|\operatorname{Ker} Q\right] J.$$

But then we can use the analyticity properties of the function $F_1(\cdot)$ in formula (15) of Lemma 5.3 to show that $(\zeta G_2 - A_{22})^{-1}$ is analytic on $|\zeta| < 1$. Hence Φ_+ has no poles on $|\zeta| < 1$.

From the factorization $\Phi(\zeta) = \Phi_-(\zeta)\Phi_+(\zeta)$ for $\zeta \in \mathbb{T}$ it follows that $\det \Phi_-(\zeta)$ and $\det \Phi_+(\zeta)$ are both non-zero for each $\zeta \in \mathbb{T}$. So we can apply Lemma 6.2 to show that

$$(19) \qquad \Phi_-(\zeta)^{-1} = I - C_1(\zeta G_1 - A_{11}^{\times})^{-1} B_1, \qquad \zeta \in \mathbb{T},$$
$$(20) \qquad \Phi_+(\zeta)^{-1} = I - C_2(\zeta G_2 - A_{22}^{\times})^{-1} B_2, \qquad \zeta \in \mathbb{T}.$$

Here we used the two identities in (15). Using the partitionings in (8)–(11), it is clear that (19) and (20) yield the formulas (5) and (6), respectively.

Finally, we check the analytical properties of Φ_-^{-1} and Φ_+^{-1}. First note that

$$\zeta G_2 - A_{22}^{\times} = (\zeta G - A^{\times})|\operatorname{Ker} Q^{\times} : \operatorname{Ker} Q^{\times} \to \operatorname{Ker} P^{\times}.$$

Now apply Lemma 5.3 to $\zeta G - A^{\times}$ in place of $\zeta G - A$. It follows that $(\zeta G_2 - A_{22}^{\times})^{-1}$ is analytic on $\mathbb{D}$, and hence $\Phi_+(\cdot)^{-1}$ has no poles in $|\zeta| < 1$. To prove the analogous result for $\Phi_-(\cdot)^{-1}$ on $|\zeta| > 1$ we use that

$$(21) \qquad H^{\times}(\zeta G_1 - A_{11}^{\times}) = \left[(\zeta G - A^{\times})|\operatorname{Im} Q^{\times}\right] J^{\times},$$

where

$$J^{\times} = Q^{\times}|\operatorname{Im} Q : \operatorname{Im} Q \to \operatorname{Im} Q^{\times}, \qquad H^{\times} = P^{\times}|\operatorname{Im} P : \operatorname{Im} P \to \operatorname{Im} P^{\times}$$

are invertible linear transformations of which the inverses are given by

$$(J^{\times})^{-1} = (I - \tau)|\operatorname{Im} Q^{\times}, \qquad (H^{\times})^{-1} = (I - \rho)|\operatorname{Im} P^{\times}.$$

Since $\left[(\zeta G - A^\times)|\operatorname{Im}Q^\times\right]^{-1}$ is analytic on $|\zeta| > 1$ (the point ∞ included) by virtue of Lemma 5.3 (applied to $\zeta G - A^\times$), formula (21) implies that $\Phi_-(\cdot)^{-1}$ has no poles on $|\zeta| > 1$ (the point ∞ included). We have proved that $\Phi(\zeta) = \Phi_-(\zeta)\Phi_+(\zeta)$ is a right canonical factorization.

Next we prove the necessity of the equalities in (2). So in what follows we assume that Φ admits a right canonical factorization relative to the circle:

$$(22) \qquad \Phi(\zeta) = \Phi_-(\zeta)\Phi_+(\zeta), \qquad \zeta \in \mathbb{T}.$$

Take $x \in \operatorname{Im}P \cap \operatorname{Ker}P^\times$, and put

$$\varphi_-(\zeta) = C(\zeta G - A)^{-1}x, \qquad \varphi_+(\zeta) = C(\zeta G - A^\times)^{-1}x \qquad (\zeta \in \mathbb{T}).$$

Since $x \in \operatorname{Im}P$, the first identity in (12) allows us to rewrite φ_- as follows

$$\varphi_-(\zeta) = (C|\operatorname{Im}Q)\left[(\zeta G - A)|\operatorname{Im}Q\right]^{-1}x, \qquad \zeta \in \mathbb{T},$$

and hence the results of Lemma 5.3 imply that φ_- has an analytic continuation to $|\zeta| > 1$ and $\varphi_-(\zeta) \to 0$ if $\zeta \to \infty$. Similarly, since

$$\varphi_+(\zeta) = (C|\operatorname{Ker}Q^\times)\left[(\zeta G - A^\times)|\operatorname{Ker}Q^\times\right]^{-1}x, \qquad \zeta \in \mathbb{T},$$

we conclude that φ_+ has an analytic continuation to $|\zeta| < 1$. Note that $\Phi(\zeta)^{-1}\varphi_-(\zeta) = \varphi_+(\zeta)$, $\zeta \in \mathbb{T}$, because of Lemma 6.3. It follows that

$$(23) \qquad \Phi_-(\zeta)^{-1}\varphi_-(\zeta) = \Phi_+(\zeta)\varphi_+(\zeta), \qquad \zeta \in \mathbb{T}.$$

Now use the properties of the factors Φ_- and Φ_+. We conclude that $\Phi_-(\zeta)^{-1}\varphi_-(\zeta)$ has an analytic continuation to $|\zeta| > 1$ and $\Phi_-(\zeta)^{-1}\varphi_-(\zeta) \to 0$ if $\zeta \to \infty$. On the other hand $\Phi_+(\zeta)\varphi_+(\zeta)$ has an analytic continuation on $|\zeta| < 1$. Liouville's theorem implies that both terms in (23) are identically zero. It follows that $\varphi_-(\zeta) = 0$ for each $\zeta \in \mathbb{T}$. But then we can apply the third identity in Lemma 6.3 to conclude that

$$(24) \qquad (\zeta G - A^\times)^{-1}x = (\zeta G - A)^{-1}x, \qquad \zeta \in \mathbb{T}.$$

Now, repeat part of the above reasoning. Note that $(\zeta G - A)^{-1}x$ has an analytic continuation on $|\zeta| > 1$ and $(\zeta G - A)^{-1}x \to 0$ if $\zeta \to \infty$, while $(\zeta G - A^\times)^{-1}x$ has an analytic continuation on $|\zeta| < 1$. Thus both terms in (24) are equal to zero on $\mathbb{T}$, which implies that $x = 0$. We proved that $\operatorname{Im}P \cap \operatorname{Ker}P^\times = \{0\}$. From Lemma 5.3 we know that G maps $\operatorname{Im}Q \cap \operatorname{Ker}Q^\times$ in a one-one manner into $\operatorname{Im}P \cap \operatorname{Ker}P^\times$, and thus $\operatorname{Im}Q \cap \operatorname{Ker}Q^\times = \{0\}$.

We proceed by showing that $\operatorname{Im}Q + \operatorname{Ker}Q^\times = \mathbb{C}^n$. Take $y \in \mathbb{C}^n$ such that $y \perp (\operatorname{Im}Q + \operatorname{Ker}Q^\times)$. Let y^* be the row vector of which the j-th entry is equal to the complex conjugate of the j-th entry of y $(j = 1, \ldots, n)$. Put

$$\psi_-(\zeta) = y^*(\zeta G - A^\times)^{-1}B, \quad \psi_+(\zeta) = y^*(\zeta G - A)^{-1}B \qquad (\zeta \in \mathbb{T}).$$

Since $y^*(I - Q^\times) = 0$, the function ψ_- has an analytic continuation on $|\zeta| > 1$ and $\psi_-(\zeta) \to 0$ if $\zeta \to \infty$ by virtue of Lemma 5.3 applied to $\zeta G - A^\times$. From $y^*Q = 0$, it follows that ψ_+ has an analytic continuation on $|\zeta| < 1$ (again apply Lemma 5.3). The second identity in Lemma 6.3 shows that

$$(25) \qquad \psi_+(\zeta)\Phi_+(\zeta)^{-1} = \psi_-(\zeta)\Phi_-(\zeta), \qquad \zeta \in \mathbb{T},$$

and Liouville's theorem implies that both terms in (25) are zero. It follows that $\psi_+(\zeta) = 0$ for each $\zeta \in \mathbb{T}$, and we can use the third identity in Lemma 6.3 to show that

$$(26) \qquad y^*(\zeta G - A^\times)^{-1} = y^*(\zeta G - A)^{-1}, \qquad \zeta \in \mathbb{T}.$$

By repeating part of the above reasoning, we obtain that both terms in (26) are zero, and hence $y = 0$. Thus $\mathbb{C}^n = \operatorname{Im} Q + \operatorname{Ker} Q^\times$. It remains to prove that $\mathbb{C}^n = \operatorname{Im} P + \operatorname{Ker} P^\times$. To do this first note that

$$\operatorname{Im} G = G[\operatorname{Im} Q + \operatorname{Ker} Q^\times] \subset \operatorname{Im} P + \operatorname{Ker} P^\times.$$

So

$$\mathbb{C}^n = \operatorname{Im} P^\times + \operatorname{Ker} P^\times$$
$$= G(\operatorname{Im} Q^\times) + \operatorname{Ker} P^\times \subset \operatorname{Im} G + \operatorname{Ker} P^\times \subset \operatorname{Im} P + \operatorname{Ker} P^\times \subset \mathbb{C}^n.$$

Here we used that G maps $\operatorname{Im} Q^\times$ onto $\operatorname{Im} P^\times$, which follows by applying Lemma 5.3 to the pencil $\zeta G - A^\times$.

It remains to establish the connections between the two direct sums in (2). Consider the operators

$$(27) \qquad Q^\times|\operatorname{Im} Q: \operatorname{Im} Q \to \operatorname{Im} Q^\times, \qquad P^\times|\operatorname{Im} P: \operatorname{Im} P \to \operatorname{Im} P^\times.$$

Note that $\operatorname{Ker} Q^\times \cap \operatorname{Im} Q = \operatorname{Ker}(Q^\times|\operatorname{Im} Q)$. Furthermore, we have

$$Q^\times x \in Q^\times(\operatorname{Im} Q) \Leftrightarrow x \in \operatorname{Ker} Q^\times + \operatorname{Im} Q.$$

Thus $\operatorname{Im}(Q^\times|\operatorname{Im} Q) = \operatorname{Im} Q^\times$ if and only if $\operatorname{Ker} Q^\times + \operatorname{Im} Q = \mathbb{C}^n$. From these observations it follows that the first equality in (2) is equivalent to the invertibility of the operator $Q^\times|\operatorname{Im} Q$. In a similar way one shows that the second equality in (2) is equivalent to the invertibility of $P^\times|\operatorname{Im} P$. Now recall (see Lemma 5.3) that G maps $\operatorname{Im} Q$ (resp., $\operatorname{Im} Q^\times$) in a one-one manner onto $\operatorname{Im} P$ (resp., $\operatorname{Im} P^\times$). Thus the operators

$$(28) \qquad F = G|\operatorname{Im} Q: \operatorname{Im} Q \to \operatorname{Im} P, \qquad F^\times = G|\operatorname{Im} Q^\times: \operatorname{Im} Q^\times - \operatorname{Im} P^\times$$

are invertible. Furthermore, from (13) we see that

$$F^\times(Q^\times|\operatorname{Im} Q) = (P^\times|\operatorname{Im} P)F.$$

So the operators in (27) are equivalent, and hence the first operator in (27) is invertible if and only if the same is true for the second operator in (27). It follows that the first equality in (2) implies the second and conversely. $\quad\square$

XXIV.8 EXPLICIT INVERSION FORMULAS

In this section $T = [\Phi_{i-j}]_{i,j=0}^{\infty}$ is a block Toeplitz operator on ℓ_2^m defined by a rational $m \times m$ matrix function. By Theorem 5.1 we may assume that its entries are given by

$$(1) \qquad \Phi_k = \begin{cases} -C\Omega^k(I-P)B, & k = 1, 2, \ldots, \\ I - C(I-P)B, & k = 0, \\ C\Omega^{-k-1}PB, & k = -1, -2, \ldots. \end{cases}$$

Here Ω is a square matrix of order n, say, Ω has all its eigenvalues in the open unit disc $\mathbb{D}$, and P is a projection of $\mathbb{C}^n$ which commutes with Ω.

THEOREM 8.1. *Let the block Toeplitz operator $T = [\Phi_{i-j}]_{i,j=0}^{\infty}$ be given by* (1). *Put*

$$(2) \qquad G = \Omega(I-P) + P, \qquad A^{\times} = I - P + \Omega P - BC.$$

Then T is invertible if and only if the following two conditions hold:

    (i) $\det(\zeta G - A^{\times}) \neq 0$ *for each* $\zeta \in \mathbb{T}$,

    (ii) $\mathbb{C}^n = \operatorname{Im} P \oplus \operatorname{Ker} P^{\times}$,

where

$$P^{\times} = \frac{1}{2\pi i} \int_{\mathbb{T}} G(\zeta G - A^{\times})^{-1} d\zeta.$$

In this case the inverse of T is obtained in the following way. Let $E^{\times}$ be the right equivalence operator associated with $\zeta G - A^{\times}$ and $\mathbb{T}$, and put

$$(3) \qquad \Omega^{\times} = GE^{\times}(I - P^{\times}) + A^{\times}E^{\times}P^{\times}, \qquad C^{\times} = CE^{\times}.$$

Then the entries of the inverse $T^{-1} = [\Gamma_{ij}]_{i,j=0}^{\infty}$ are given by

$$\Gamma_{ij} = \Phi_{i-j}^{\times} + K_{ij}, \qquad i, j = 0, 1, 2, \ldots,$$

$$\Phi_k^{\times} = \begin{cases} C^{\times}(\Omega^{\times})^k(I - P^{\times})B, & k = 1, 2, \ldots, \\ I + C^{\times}(I - P^{\times})B, & k = 0, \\ -C^{\times}(\Omega^{\times})^{-k-1}P^{\times}B, & k = -1, -2, \ldots, \end{cases}$$

$$K_{ij} = C^{\times}(\Omega^{\times})^i(I - P^{\times})\rho P^{\times}(\Omega^{\times})^j B,$$

where ρ is the projection of $\mathbb{C}^n$ along $\operatorname{Im} P$ onto $\operatorname{Ker} P^{\times}$.

     PROOF. Put $A = I - P + \Omega P$. Then the function Φ defining T is given by

$$(4) \qquad \Phi(\zeta) = I + C(\zeta G - A)^{-1}B, \qquad \zeta \in \mathbb{T}.$$

Note that $A^\times = A - BC$. According to Theorem 4.1, the operator T is invertible if and only if $\det \Phi(\zeta) \neq 0$ for each $\zeta \in \mathbb{T}$ and Φ admits a right canonical factorization relative to $\mathbb{T}$. Thus by Lemma 6.2 and Theorem 7.1 the conditions (i) and (ii) are necessary and sufficient for T to be invertible.

In the remaining part of the proof we assume that (i) and (ii) hold. Let $\Omega^\times$ be as in (3). From Lemma 5.3 applied to the pencil $\zeta G - A^\times$ (see also the two paragraphs after the proof of Theorem 5.2) we know that $\Omega^\times$ has all its eigenvalues inside the open unit disc and $\Omega^\times$ commutes with $P^\times$. To find the entries Γ_{ij} of T^{-1} we apply Theorem 4.1. Write $\Phi(\zeta) = \Phi_-(\zeta)\Phi_+(\zeta)$, $\zeta \in \mathbb{T}$, where Φ_- and Φ_+ are given by formulas (3) and (4) in the previous section. We need to compute the Fourier coefficients of the functions Φ_-^{-1} and Φ_+^{-1}. The latter functions are given by the formulas (5) and (6) in the previous section. We use the matrix $\Omega^\times$ to rewrite these formulas. Formula (14) in Section XXIV.5 (applied to $\zeta G - A^\times$) yields

$$(\zeta G - A^\times)E^\times = (\zeta\Omega^\times - I)(I - P^\times) + (\zeta - \Omega^\times)P^\times.$$

In the sequel τ denotes the projection of $\mathbb{C}^n$ along $\operatorname{Im} Q$ onto $\operatorname{Ker} Q^\times$, where Q and $Q^\times$ are as in Theorem 7.1. Since $(I - \tau)E^\times(I - P^\times) = 0$ and $P^\times\rho = 0$, we get

$$\Phi_-(\zeta)^{-1} = I - C(I - \tau)E^\times(\zeta - \Omega^\times)^{-1}P^\times B, \qquad \zeta \in \mathbb{T},$$
$$\Phi_+(\zeta)^{-1} = I - CE^\times(\zeta\Omega^\times - I)^{-1}(I - P^\times)\rho B, \qquad \zeta \in \mathbb{T},$$

and hence

$$\Phi_-(\zeta)^{-1} = \sum_{j=-\infty}^{0} \zeta^j \gamma_j^-, \qquad \Phi_+(\zeta)^{-1} = \sum_{j=0}^{\infty} \zeta^j \gamma_j^+,$$

where

$$\gamma_j^- = -C(I - \tau)E^\times(\Omega^\times)^{-j-1}P^\times B, \qquad j = -1, -2, \ldots,$$
$$\gamma_0^- = I, \qquad \gamma_0^+ = I + C^\times(I - P^\times)\rho B,$$
$$\gamma_j^+ = C^\times(\Omega^\times)^j(I - P^\times)\rho B, \qquad j = 1, 2, \ldots .$$

Now recall (see Theorem 4.1) that the (i,j)-th entry Γ_{ij} of T^{-1} is given by

$$(5) \qquad \Gamma_{ij} = \sum_{r=0}^{\min(i,j)} \gamma_{i-r}^+ \gamma_{r-j}^-.$$

To compute the products in (5) we first show that

$$(6) \qquad \Omega^\times(I - P^\times)\rho BC(I - \tau)E^\times P^\times = (I - P^\times)[\rho - \Omega^\times\rho\Omega^\times]P^\times.$$

Recall that $BC = A - A^\times$. From the definitions of τ and ρ it follows that $\rho A(I - \tau) = 0$ and $\rho A^\times\tau = A^\times\tau$. Thus

$$(7) \qquad \rho BC(I - \tau) = A^\times\tau - \rho A^\times.$$

Next, observe that the following identities hold:

$$\Omega^{\times} A^{\times}\tau = G\tau = \rho G, \tag{8}$$

$$A^{\times} E^{\times} P^{\times} = \Omega^{\times} P^{\times}, \qquad G E^{\times} P^{\times} = P^{\times}. \tag{9}$$

From these identities and the fact that $\Omega^{\times} P^{\times} = P^{\times}\Omega^{\times}$ it follows that

$$\begin{aligned}
\Omega^{\times}(I - P^{\times})\rho B C(I - \tau)E^{\times} P^{\times} &= (I - P^{\times})\Omega^{\times} A^{\times}\tau E^{\times} P^{\times} - (I - P^{\times})\Omega^{\times}\rho A^{\times} E^{\times} P^{\times} \\
&= (I - P^{\times})\rho G E^{\times} P^{\times} - (I - P^{\times})\Omega^{\times}\rho\Omega^{\times} P^{\times} \\
&= (I - P^{\times})\rho P^{\times} - (I - P^{\times})\Omega^{\times}\rho\Omega^{\times} P^{\times} \\
&= (I - P^{\times})[\rho - \Omega^{\times}\rho\Omega^{\times}]P^{\times},
\end{aligned}$$

and (6) is proved. From (6) and $\Omega^{\times} P^{\times} = P^{\times}\Omega^{\times}$ one derives that for $p > 0$ and $q < 0$

$$\begin{aligned}
\gamma_p^{+}\gamma_q^{-} = {}& C^{\times}(\Omega^{\times})^p(I - P^{\times})\rho P^{\times}(\Omega^{\times})^{-q} B \\
&- C^{\times}(\Omega^{\times})^{p-1}(I - P^{\times})\rho P^{\times}(\Omega^{\times})^{-q-1} B.
\end{aligned}$$

We are now ready to compute the (i, j)-th entry Γ_{ij}. First, assume that $i > j$. Then

$$\begin{aligned}
\Gamma_{ij} &= \sum_{r=0}^{j} \gamma_{i-r}^{+}\gamma_{r-j}^{-} \\
&= \gamma_{i-j}^{+} + \sum_{r=0}^{j-1} \gamma_{i-r}^{+}\gamma_{r-j}^{-} \\
&= C^{\times}(\Omega^{\times})^{i-j}(I - P^{\times})\rho B \\
&\quad + \sum_{r=0}^{j-1} C^{\times}(\Omega^{\times})^{i-r}(I - P^{\times})\rho P^{\times}(\Omega^{\times})^{j-r} B \\
&\quad - \sum_{r=0}^{j-1} C^{\times}(\Omega^{\times})^{i-r-1}(I - P^{\times})\rho P^{\times}(\Omega^{\times})^{j-r-1} B \\
&= C^{\times}(\Omega^{\times})^{i-j}(I - P^{\times})\rho B \\
&\quad + C^{\times}(\Omega^{\times})^{i}(I - P^{\times})\rho P^{\times}(\Omega^{\times})^{j} B \\
&\quad - C^{\times}(\Omega^{\times})^{i-j}(I - P^{\times})\rho P^{\times} B \\
&= C^{\times}(\Omega^{\times})^{i-j}(I - P^{\times})\rho(I - P^{\times})B + K_{ij}.
\end{aligned}$$

Since $\rho(I - P^{\times}) = I - P^{\times}$, we proved that

$$\Gamma_{ij} = \Phi_{i-j}^{\times} + K_{ij} \tag{10}$$

for $i > j$. Next, take $i = j$. Then

$$\Gamma_{ii} = \gamma_0^+ + \sum_{r=0}^{i-1} \gamma_{i-r}^+ \gamma_{r-i}^-$$

$$= I + C^\times (I - P^\times)\rho B$$

$$+ \sum_{r=0}^{i-1} C^\times (\Omega^\times)^{i-r}(I - P^\times)\rho P^\times (\Omega^\times)^{i-r} B$$

$$- \sum_{r=0}^{i-1} C^\times (\Omega^\times)^{i-r-1}(I - P^\times)\rho P^\times (\Omega^\times)^{i-r-1} B$$

$$= I + C^\times (I - P^\times)\rho B$$

$$+ C^\times (\Omega^\times)^i (I - P^\times)\rho P^\times (\Omega^\times)^i B$$

$$- C^\times (I - P^\times)\rho P^\times B$$

$$= I + C^\times (I - P^\times)B + K_{ii},$$

and (10) is proved for $i = j$. Finally, take $i < j$. Then

$$\Gamma_{ij} = \gamma_0^+ \gamma_{i-j}^- + \sum_{r=0}^{i-1} \gamma_{i-r}^+ \gamma_{r-j}^-$$

$$= \gamma_0^+ \gamma_{i-j}^- + \sum_{r=0}^{i-1} C^\times (\Omega^\times)^{i-r}(I - P^\times)\rho P^\times (\Omega^\times)^{j-r} B$$

$$- \sum_{r=0}^{i-1} C^\times (\Omega^\times)^{i-r-1}(I - P^\times)\rho P^\times (\Omega^\times)^{j-r-1} B$$

$$= -C(I - \tau)E^\times (\Omega^\times)^{j-i-1} P^\times B$$

$$- C^\times (I - P^\times)\rho BC(I - \tau)E^\times (\Omega^\times)^{j-i-1} P^\times B$$

$$+ C^\times (\Omega^\times)^i (I - P^\times)\rho P^\times (\Omega^\times)^j B$$

$$- C^\times (I - P^\times)\rho P^\times (\Omega^\times)^{j-i} B.$$

To simplify the preceding expression further we use (7), (9) and $E^\times A^\times \tau = \tau$ to obtain

$$E^\times (I - P^\times)\rho BC(I - \tau)E^\times P^\times$$

$$= -E^\times (I - P^\times)\rho A^\times E^\times P^\times + E^\times (I - P^\times)\rho A^\times \tau E^\times P^\times$$

$$= -E^\times (I - P^\times)\rho \Omega^\times P^\times + E^\times A^\times \tau E^\times P^\times$$

$$= -E^\times (I - P^\times)\rho \Omega^\times P^\times + \tau E^\times P^\times.$$

Since $C^\times = CE^\times$ and $\Omega^\times P^\times = P^\times \Omega^\times$, we see that

$$
\begin{aligned}
\Gamma_{ij} &= -C(I - \tau)E^\times(\Omega^\times)^{j-i-1}P^\times B \\
&\quad + CE^\times(I - P^\times)\rho P^\times(\Omega^\times)^{j-i}B \\
&\quad - C\tau E^\times(\Omega^\times)^{j-i-1}P^\times B \\
&\quad + K_{ij} - CE^\times(I - P^\times)\rho P^\times(\Omega^\times)^{j-i}B \\
&= -CE^\times(\Omega^\times)^{j-i-1}P^\times B + K_{ij},
\end{aligned}
$$

which proves (10) for $i < j$. $\square$

We conclude with a few remarks about the case when the matrix G in (2) is the $n \times n$ identity matrix. Then the defining function Φ admits the following realization

$$
(11) \qquad\qquad \Phi(\zeta) = I + C(\zeta - A)^{-1}B, \qquad \zeta \in \mathbb{T},
$$

where A has no eigenvalues on $\mathbb{T}$. Note that now P is simply the Riesz projection of A corresponding to the eigenvalues inside the open unit disc $\mathbb{D}$, and Theorem 8.1 assumes the following form. The block Toeplitz operator T defined by (11) is invertible if and only if

(i)$'$ $A^\times = A - BC$ has no eigenvalues on $\mathbb{T}$,

(ii)$'$ $\mathbb{C}^n = \operatorname{Im} P \oplus \operatorname{Ker} P^\times$,

where $P^\times$ is the Riesz projection of $A^\times$ corresponding to the eigenvalues in $\mathbb{D}$. Furthermore, in this case $T^{-1} = [\Gamma_{ij}]_{i,j=0}^\infty$ with

$$
(12) \qquad\qquad \Gamma_{ij} = \begin{cases}
C(A^\times)^{-i-1}\rho(A^\times)^j B & , \ i > j, \\
I + C(A^\times)^{-i-1}\rho(A^\times)^j B & , \ i = j, \\
-C(A^\times)^{-i-1}(I - \rho)(A^\times)^j B, & i < j,
\end{cases}
$$

where ρ is the projection of $\mathbb{C}^n$ along $\operatorname{Im} P$ onto $\operatorname{Ker} P^\times$. The expressions for Γ_{ij} in the right hand side of (12) have to be handled with a bit of care. They make sense if $A^\times$ is invertible. If $A^\times$ is not invertible, the inverses in (12) should be understood as follows:

$$
(A^\times)^{-i-1}\rho = (A^\times \,|\, \operatorname{Ker} P^\times)^{-i-1}\rho
$$

$$
(A^\times)^{-i-1}(I - \rho)(A^\times)^j = (A^\times)^{j-i-1}P^\times - (A^\times \,|\, \operatorname{Ker} P^\times)^{-i-1}(I - P^\times)\rho P^\times (A^\times)^i.
$$

Note that $A^\times \,|\, \operatorname{Ker} P^\times$ has all its eigenvalues in $|\zeta| > 1$. Thus $A^\times \,|\, \operatorname{Ker} P^\times$ is invertible.

XXIV.9 EXPLICIT FORMULAS FOR FREDHOLM CHARACTERISTICS

This section concerns the Fredholm characteristics of a block Toeplitz operator $T = [\Phi_{i-j}]_{i,j=0}^\infty$ defined by a rational matrix function. So we may assume that the entries are given by

$$(1) \qquad \Phi_k = \begin{cases} -C\Omega^k(I-P)B, & k=1,2,\ldots, \\ I - C(I-P)B\,, & k=0, \\ C\Omega^{-k-1}PB & , \quad k=-1,-2,\ldots, \end{cases}$$

where Ω and P are as in Theorem 5.1.

THEOREM 9.1. *Let the block Toeplitz operator $T = [\Phi_{i-j}]_{i,j=0}^{\infty}$ be given by* (1). *Put*

$$(2) \qquad G = \Omega(I-P) + P, \qquad A^{\times} = I - P + \Omega P - BC.$$

Then T is a Fredholm operator if and only if

$$(3) \qquad \det(\zeta G - A^{\times}) \neq 0, \qquad \zeta \in \mathbb{T}.$$

Assume that (3) *holds. Let $E^{\times}$ be the right equivalence operator associated with the pencil $\zeta G - A^{\times}$ and $\mathbb{T}$, and let $P^{\times}$ be the projection*

$$P^{\times} = \frac{1}{2\pi i} \int_{\mathbb{T}} G(\zeta G - A^{\times})^{-1} d\zeta.$$

Put $\Omega^{\times} = GE^{\times}(I - P^{\times}) + A^{\times}E^{\times}P^{\times}$ and $C^{\times} = CE^{\times}$. Then

$$(4) \qquad \operatorname{Ker} T = \big\{ (C^{\times}(\Omega^{\times})^j(I - P^{\times})x)_{j=0}^{\infty} \mid x \in \operatorname{Im} P \cap \operatorname{Ker} P^{\times} \big\},$$

$$(5) \qquad \operatorname{Im} T = \left\{ (\varphi_j)_{j=0}^{\infty} \in \ell_2^m \;\Big|\; \sum_{j=0}^{\infty} P^{\times}(\Omega^{\times})^j B\varphi_j \in \operatorname{Im} P + \operatorname{Ker} P^{\times} \right\},$$

$$(6) \qquad n(T) = \dim(\operatorname{Im} P \cap \operatorname{Ker} P^{\times}), \qquad d(T) = \dim \frac{\mathbb{C}^n}{\operatorname{Im} P + \operatorname{Ker} P^{\times}},$$

$$(7) \qquad \operatorname{ind}(T) = \operatorname{rank} P - \operatorname{rank} P^{\times},$$

and a generalized inverse of T is given by $T^{+} = [\Gamma_{ij}^{+}]_{i,j=0}^{\infty}$ with

$$(8) \qquad \Gamma_{ij}^{+} = \Phi_{i-j}^{\times} + K_{ij}^{+},$$

$$(9) \qquad \Phi_k^{\times} = \begin{cases} C^{\times}(\Omega^{\times})^k(I - P^{\times})B, & k=1,2,\ldots, \\ I + C^{\times}(I - P^{\times})B & , \quad k=0, \\ -C^{\times}(\Omega^{\times})^{-k-1}P^{\times}B, & k=-1,-2,\ldots, \end{cases}$$

$$K_{ij}^{+} = -C^{\times}(\Omega^{\times})^{i}(I - P^{\times})(J^{\times})^{+}P^{\times}(\Omega^{\times})^{j}B,$$

where $(J^{\times})^{+}$ is a generalized inverse of the operator

$$(10) \qquad J^{\times} = P^{\times}|\operatorname{Im}P\colon \operatorname{Im}P \to \operatorname{Im}P^{\times}.$$

PROOF. Put $A = I - P + \Omega P$. Then the function Φ defining T is given by

$$\Phi(\zeta) = I + C(\zeta G - A)^{-1}B, \qquad \zeta \in \mathbb{T}.$$

Note that $A^{\times} = A - BC$. From Theorem XXIII.4.3 we know that T is a Fredholm operator if and only if $\det \Phi(\zeta) \neq 0$ for each $\zeta \in \mathbb{T}$. By Lemma 6.2 the latter condition is equivalent to (3). So the first part of the theorem is proved.

In the following we assume that (3) holds. To establish the Fredholm properties of T we use matricial coupling (see Section III.4). Introduce the following operators:

$$\begin{bmatrix} T & U \\ R & J \end{bmatrix} : \ell_2^m \oplus \operatorname{Im}P^{\times} \to \ell_2^m \oplus \operatorname{Im}P,$$

$$\begin{bmatrix} T^{\times} & U^{\times} \\ R^{\times} & J^{\times} \end{bmatrix} : \ell_2^m \oplus \operatorname{Im}P \to \ell_2^m \oplus \operatorname{Im}P^{\times},$$

$$(Ux)_j = -C\Omega^j(I - P)x, \qquad x \in \operatorname{Im}P^{\times},$$

$$(U^{\times}x)_j = -C^{\times}(\Omega^{\times})^j(I - P^{\times})x, \qquad x \in \operatorname{Im}P,$$

$$R\eta = \sum_{j=0}^{\infty} P\Omega^j B\varphi_j, \qquad \eta = (\varphi_0, \varphi_1, \ldots) \in \ell_2^m,$$

$$R^{\times}\eta = -\sum_{j=0}^{\infty} P^{\times}(\Omega^{\times})^j B\varphi_j, \qquad \eta = (\varphi_0, \varphi_1, \ldots) \in \ell_2^m,$$

$$Jx = Px \quad (x \in \operatorname{Im}P^{\times}), \qquad J^{\times}x = P^{\times}x \quad (x \in \operatorname{Im}P).$$

Note that $J^{\times}$ is the operator defined by (10). The operator $T^{\times}$ is the block Toeplitz operator defined by $\Phi(\cdot)^{-1}$. We shall prove that

$$(11) \qquad \begin{bmatrix} T & U \\ R & J \end{bmatrix}^{-1} = \begin{bmatrix} T^{\times} & U^{\times} \\ R^{\times} & J^{\times} \end{bmatrix}.$$

Proving (11) boils down to verifying eight identities. Here we shall establish four of them, namely

$$(12) \qquad TT^{\times} + UR^{\times} = I_{\ell_2^m},$$

$$(13) \qquad RT^\times + JR^\times = 0,$$

$$(14) \qquad TU^\times + UJ^\times = 0,$$

$$(15) \qquad RU^\times + JJ^\times = I_{\operatorname{Im}P}.$$

The other four identities can be obtained by interchanging the roles of Φ and Φ^{-1}. It will be convenient to use the Fourier transform

$$F \colon H_2^m(\mathbb{T}) \to \ell_2^m, \qquad F\varphi = (c_j)_{j=0}^\infty,$$

where c_j is the j-th Fourier coefficient of φ. Put

$$S_\Phi = F^{-1}TF, \qquad V = F^{-1}U, \qquad N = RF,$$

$$S_{\Phi^{-1}} = F^{-1}T^\times F, \qquad V^\times = F^{-1}U^\times, \qquad N^\times = R^\times F.$$

Then

$$(Vx)(\zeta) = C(\zeta G - A)^{-1}(I - P)x, \qquad x \in \operatorname{Im}P^\times, \ \zeta \in \mathbb{T},$$

$$(V^\times x)(\zeta) = C(\zeta G - A^\times)^{-1}(I - P^\times)x, \qquad x \in \operatorname{Im}P, \ \zeta \in \mathbb{T},$$

$$N\varphi = \frac{1}{2\pi i} \int_\mathbb{T} PG(\zeta G - A)^{-1}B\varphi(\zeta)d\zeta, \qquad \varphi \in H_2^m(\mathbb{T}),$$

$$N^\times \varphi = \frac{-1}{2\pi i} \int_\mathbb{T} P^\times G(\zeta G - A^\times)^{-1}B\varphi(\zeta)d\zeta, \qquad \dot{\varphi} \in H_2^m(\mathbb{T}),$$

$$S_\Phi = \mathbb{P}M_\Phi, \qquad S_{\Phi^{-1}} = \mathbb{P}M_{\Phi^{-1}},$$

where $\mathbb{P}$ is the orthogonal projection of $L_2^m(\mathbb{T})$ onto $H_2^m(\mathbb{T})$ and M_Φ (resp., $M_{\Phi^{-1}}$) is the operator of multiplication by Φ (resp., Φ^{-1}) on $L_2^m(\mathbb{T})$. It suffices to prove the following identities:

$$(16) \qquad S_\Phi S_{\Phi^{-1}} + VN^\times = I_{H_2^m(\mathbb{T})},$$

$$(17) \qquad NS_{\Phi^{-1}} + JN^\times = 0,$$

$$(18) \qquad S_\Phi V^\times + VJ^\times = 0,$$

$$(19) \qquad NV^\times + JJ^\times = I_{\operatorname{Im}P}.$$

First we compute $S_\Phi S_{\Phi^{-1}}$. Note that

$$(20) \qquad BC = (\mu G - A^\times) - (\zeta G - A) - (\mu - \zeta)G.$$

Thus

$$
\begin{aligned}
\Phi(\zeta)\Phi(\mu)^{-1} &= \left[I + C(\zeta G - A)^{-1}B \right]\left[I - C(\mu G - A^\times)^{-1}B \right] \\
&= I - C(\mu G - A^\times)^{-1}B + C(\zeta G - A)^{-1}B \\
&\quad - C(\zeta G - A)^{-1}BC(\mu G - A^\times)^{-1}B \\
&= I + (\mu - \zeta)C(G - A)^{-1}G(\mu G - A^\times)^{-1}B.
\end{aligned}
$$

Take $g \in H_2^m(\mathbb{T})$, and assume that g is a polynomial. Then $M_{\Phi^{-1}}g$ is rational, and so by formula (5) in Section XXIV.1:

$$(S_{\Phi^{-1}}g)(\zeta) = \frac{1}{2\pi i} \int_{\mathbb{T}} \frac{\Phi(\mu)^{-1}g(\mu)}{\mu - \zeta}d\mu, \qquad |\zeta| < 1.$$

It follows that for $|\zeta| < 1$,

$$
\begin{aligned}
(M_\Phi S_{\Phi^{-1}}g)(\zeta) &= \frac{1}{2\pi i} \int_{\mathbb{T}} \frac{\Phi(\zeta)\Phi(\mu)^{-1}g(\mu)}{\mu - \zeta}d\mu \\
&= g(\zeta) + C(\zeta G - A)^{-1}\left(\frac{1}{2\pi i} \int_{\mathbb{T}} G(\mu G - A^\times)^{-1}Bg(\mu)d\mu \right).
\end{aligned}
$$

Lemma 5.3 applied to $\zeta G - A^\times$ shows that $(I - P^\times)G(\zeta G - A^\times)^{-1}$ extends to a function which is analytic on $\mathbb{D}$. Since $g \in H_2^m(\mathbb{T})$, we may conclude that

$$(21) \qquad \frac{1}{2\pi i} \int_{\mathbb{T}} (I - P^\times)G(\mu G - A^\times)^{-1}Bg(\mu)d\mu = 0.$$

Thus

$$(M_\Phi S_{\Phi^{-1}}g)(\zeta) = g(\zeta) - C(\zeta G - A)^{-1}N^\times g, \qquad |\zeta| < 1.$$

From Lemma 5.3 we also know that $C(\zeta G - A)^{-1}P$ has an analytic extension on $\mathbb{C}_\infty \setminus \overline{\mathbb{D}}$ which is zero at infinity and the function $C(\zeta G - A)^{-1}(I - P)$ has an analytic extension on $\mathbb{D}$. Note that all functions involved are rational. It follows that $S_\Phi S_{\Phi^{-1}}g = g - VN^\times g$ for each polynomial g in $H_2^m(\mathbb{T})$. But the polynomials are dense in $H_2^m(\mathbb{T})$. So the identity (16) is proved.

Again let $g \in H_2^m(\mathbb{T})$ be a polynomial. Put $h = (I - \mathbb{P})M_{\Phi^{-1}}g$. Then h is a minus-function and $h(\lambda) \to 0$ if $\lambda \to \infty$. Thus (using the results of Lemma 5.3) the function $PG(\lambda G - A)^{-1}Bh(\lambda)$ is a minus-function which has a zero of order 2 at infinity. It follows that

$$\frac{1}{2\pi i} \int_{\mathbb{T}} PG(\zeta G - A)^{-1}Bh(\zeta)d\zeta = 0.$$

Note that $S_{\Phi^{-1}}g = \mathbb{P}M_{\Phi^{-1}}g = M_{\Phi^{-1}}g - h$. Hence

$$NS_{\Phi^{-1}}g = \frac{1}{2\pi i}\int_{\mathbb{T}} PG(\zeta G - A)^{-1}B(M_{\Phi^{-1}}g - h)(\zeta)d\zeta$$

$$= \frac{1}{2\pi i}\int_{\mathbb{T}} PG(\zeta G - A)^{-1}B\Phi(\zeta)^{-1}g(\zeta)d\zeta.$$

Now apply Lemma 6.3 and formula (21). We get

$$NS_{\Phi^{-1}}g = \frac{1}{2\pi i}\int_{\mathbb{T}} PG(\zeta G - A^{\times})^{-1}Bg(\zeta)d\zeta$$

$$= P\left(\frac{1}{2\pi i}\int_{\mathbb{T}} P^{\times}G(\zeta G - A^{\times})^{-1}Bg(\zeta)d\zeta\right)$$

$$= -JN^{\times}g.$$

Since the polynomials are dense in $H_2^m(\mathbb{T})$, formula (17) is proved.

Next, we take $x \in \operatorname{Im} P$. Note that $(I - P)(I - P^{\times})x = -(I - P)P^{\times}x$. Thus, using Lemma 6.3,

$$(M_{\Phi}V^{\times}x)(\zeta) = \Phi(\zeta)C(\zeta G - A^{\times})^{-1}(I - P^{\times})x$$

$$= C(\zeta G - A)^{-1}(I - P^{\times})x$$

$$= C(\zeta G - A)^{-1}P(I - P^{\times})x - (VJ^{\times}x)(\zeta).$$

Now use that $(\zeta G - A)^{-1}P$ has an analytic extension on $\mathbb{C}_{\infty}\backslash\overline{\mathbb{D}}$ which is zero at infinity. It follows that $S_{\Phi}V^{\times} = -VJ^{\times}$, and (18) is proved.

Formula (20) (with $\mu = \zeta$) implies that

$$(\zeta G - A)^{-1}BC(\zeta G - A^{\times})^{-1} = (\zeta G - A)^{-1} - (\zeta G - A^{\times})^{-1}.$$

Thus for $x \in \operatorname{Im} P$,

$$NV^{\times}x = \frac{1}{2\pi i}\int_{\mathbb{T}} PG\big[(\zeta G - A)^{-1} - (\zeta G - A^{\times})^{-1}\big](I - P^{\times})xd\zeta$$

$$= P(I - P^{\times})x - PP^{\times}(I - P^{\times})x$$

$$= x - JJ^{\times}x,$$

which proves (19).

We have now shown that the operator T is matricially coupled to the operator $J^{\times}$ via the identity (11). But then we can apply Theorem III.4.1 and Corollary III.4.3 to show that

$$(22)\qquad\qquad \operatorname{Ker} T = \{U^{\times}x \mid x \in \operatorname{Ker} J^{\times}\},$$

$$(23) \qquad \operatorname{Im} T = \left\{ \eta = (\varphi_j)_{j=0}^{\infty} \in \ell_2^m \mid R^{\times} \eta \in \operatorname{Im} J^{\times} \right\},$$

$$(24) \qquad n(T) = \dim \operatorname{Ker} J^{\times}, \qquad d(T) = \dim[\operatorname{Im} P^{\times} / \operatorname{Im} J^{\times}].$$

Note that

$$(25) \qquad \operatorname{Ker} J^{\times} = \operatorname{Im} P \cap \operatorname{Ker} P^{\times},$$

$$(26) \qquad P^{\times} z \in \operatorname{Im} J^{\times} \Leftrightarrow z \in \operatorname{Im} P + \operatorname{Ker} P^{\times}.$$

From (22) and (25) we get the desired description of $\operatorname{Ker} T$. Formulas (23) and (26) yield (5). Note that (26) also implies

$$(27) \qquad \dim[\operatorname{Im} P^{\times} / \operatorname{Im} J^{\times}] = \dim \frac{\mathbb{C}^n}{\operatorname{Im} P + \operatorname{Ker} P^{\times}}.$$

From (18), (19) and (21) our formulas for $n(T)$ and $d(T)$ in (6) are clear. According to (24)

$$n(T) = \operatorname{rank} P - \operatorname{rank} J^{\times}, \qquad d(T) = \operatorname{rank} P^{\times} - \operatorname{rank} J^{\times},$$

which proves (7). Finally, if $(J^{\times})^+$ is a generalized inverse of $J^{\times}$, then

$$(28) \qquad T^+ = T^{\times} - U^{\times}(J^{\times})^+ R^{\times}$$

is a generalized inverse of T. The operator $T^{\times}$ is the block Toeplitz operator defined by $\Phi(\cdot)^{-1}$. Thus $T^{\times} = [\Phi_{i-j}^{\times}]_{i,j=0}^{\infty}$ with $\Phi_k^{\times}$ given by (9) (see Theorem 6.1). But then $T^+ = [\Gamma_{ij}^+]_{i,j=0}^{\infty}$ with Γ_{ij}^+ as in (8). $\square$

Note that the operator $J^{\times}$ in (10) is invertible if and only if

$$\mathbb{C}^n = \operatorname{Im} P \oplus \operatorname{Ker} P^{\times},$$

and in that case $(J^{\times})^{-1} = (I - \rho)|\operatorname{Im} P^{\times}$, where ρ is the projection of $\mathbb{C}^n$ along $\operatorname{Im} P$ onto $\operatorname{Ker} P^{\times}$. It follows that Theorem 9.1 provides an alternative proof for Theorem 8.1.

XXIV.10 AN EXAMPLE

In this section we illustrate the results of this chapter with an example. Let $T = [\Phi_{i-j}]_{i,j=0}^{\infty}$ be the block Toeplitz operator defined by

$$(1) \qquad \Phi(\zeta) = \begin{bmatrix} 1 - \zeta^{-1} & (2\zeta)^{-1} \\ -3\zeta & 1 + \zeta \end{bmatrix}.$$

Note that T is tridiagonal and the off diagonal entries are singular 2×2 matrices. We shall compute the inverse of the block Laurent operator L defined by Φ, construct a right canonical factorization of Φ, and compute the inverse of T by applying, respectively, the recipes given by Theorems 6.1, 7.1 and 8.1.

The first step is to find a representation of the entries Φ_k as in formula (1) of Section XXIV.5. Such a representation may be constructed by using the procedure outlined in the proof of Theorem 5.1. The following choices of Ω, P, B and C will do:

$$\Omega = \begin{bmatrix} 0 & 1 & 0 \\ 0 & 0 & 0 \\ 0 & 0 & 0 \end{bmatrix}, \qquad P = \begin{bmatrix} 0 & 0 & 0 \\ 0 & 0 & 0 \\ 0 & 0 & 1 \end{bmatrix},$$

$$B = \begin{bmatrix} 0 & 0 \\ 3 & -1 \\ -1 & -\frac{1}{2} \end{bmatrix}, \qquad C = \begin{bmatrix} 0 & 0 & 1 \\ 1 & 0 & 0 \end{bmatrix}.$$

Now let us investigate the invertibility of the Laurent operator L defined by Φ. This requires one to analyse the pencil $\zeta G - A^{\times}$, where

$$G = \Omega(I - P) + P = \begin{bmatrix} 0 & 1 & 0 \\ 0 & 0 & 0 \\ 0 & 0 & 1 \end{bmatrix},$$

$$A^{\times} = I - P + \Omega P - BC = \begin{bmatrix} 1 & 0 & 0 \\ 1 & 1 & -3 \\ -\frac{1}{2} & 0 & 1 \end{bmatrix}.$$

One computes that $\det(\zeta G - A^{\times}) = (\zeta + 2)(\zeta - \frac{1}{2})$, and thus $\det(\zeta G - A^{\times}) \neq 0$ for $\zeta \in \mathbb{T}$. Hence, by Theorem 6.1, the Laurent operator L is invertible. To find its inverse, (by the method of Theorem 6.1) we have to compute the projection

$$(2) \qquad P^{\times} = \frac{1}{2\pi i} \int_{\mathbb{T}} G(\zeta G - A^{\times})^{-1} d\zeta,$$

the right equivalence operator $E^{\times}$ associated with the pencil $\zeta G - A^{\times}$ and the curve $\mathbb{T}$, and the operator

$$\Omega^{\times} = GE^{\times}(I - P^{\times}) + A^{\times} E^{\times} P^{\times}.$$

The first step is the inversion of $\zeta G - A^{\times}$:

$$(3) \qquad (\zeta G - A^{\times})^{-1} = (\zeta + 2)^{-1}\left(\zeta - \frac{1}{2}\right)^{-1} \begin{bmatrix} 1 - \zeta & \zeta(1 - \zeta) & 3\zeta \\ \zeta + \frac{1}{2} & 1 - \zeta & 3 \\ \frac{1}{2} & \frac{1}{2}\zeta & \zeta + 1 \end{bmatrix}.$$

From Corollary IV.1.2 applied to $\zeta G - A^{\times}$ we know that $E^{\times}$ is given by

$$(4) \qquad E^{\times} = \frac{1}{2\pi i} \int_{\mathbb{T}} (1 - \zeta^{-1})(\zeta G - A^{\times})^{-1} d\zeta.$$

By using (3) in (2) and (4) one finds that

$$P^\times = \frac{1}{5}\begin{bmatrix} 2 & 1 & 6 \\ 0 & 0 & 0 \\ 1 & \frac{1}{2} & 3 \end{bmatrix}, \qquad E^\times = \frac{1}{5}\begin{bmatrix} 4 & -\frac{1}{2} & -3 \\ \frac{1}{2} & 4 & 9 \\ \frac{3}{2} & -\frac{1}{2} & 2 \end{bmatrix},$$

and hence

$$\Omega^\times = \frac{1}{5}\begin{bmatrix} -\frac{1}{2} & \frac{7}{2} & 6 \\ 0 & 0 & 0 \\ 1 & -\frac{3}{4} & \frac{1}{2} \end{bmatrix}.$$

We have now all ingredients to compute the entries $\Phi^\times_{i-j}$ of L^{-1} (cf., formula (4) in Section XXIV.6). To simplify the computations note that $\Omega^\times$ has three different eigenvalues, namely $\frac{1}{2}$, $-\frac{1}{2}$ and 0. This allows us to diagonalize $\Omega^\times$. Put

$$D = \begin{bmatrix} \frac{1}{2} & 0 & 0 \\ 0 & -\frac{1}{2} & 0 \\ 0 & 0 & 0 \end{bmatrix}, \qquad S = \begin{bmatrix} -2 & -3 & 2 \\ 0 & 0 & 2 \\ -1 & 1 & -1 \end{bmatrix}.$$

Then S is invertible,

$$S^{-1} = \frac{1}{5}\begin{bmatrix} -1 & -\frac{1}{2} & -3 \\ -1 & 2 & 2 \\ 0 & \frac{5}{2} & 0 \end{bmatrix}, \qquad \Omega^\times = SDS^{-1}.$$

Furthermore,

$$E^\times S = \begin{bmatrix} -1 & -3 & 2 \\ -2 & \frac{3}{2} & 0 \\ -1 & -\frac{1}{2} & 0 \end{bmatrix}, \qquad S^{-1}P^\times S = \begin{bmatrix} 1 & 0 & 0 \\ 0 & 0 & 0 \\ 0 & 0 & 0 \end{bmatrix}.$$

It follows that

$$CE^\times S = \begin{bmatrix} -1 & -\frac{1}{2} & 0 \\ -1 & -3 & 2 \end{bmatrix},$$

$$S^{-1}P^\times B = (S^{-1}P^\times S)S^{-1}B = \frac{1}{10}\begin{bmatrix} 3 & -2 \\ 0 & 0 \\ 0 & 0 \end{bmatrix},$$

$$S^{-1}(I - P^\times)B = \frac{1}{10}\begin{bmatrix} 0 & 0 \\ 8 & -2 \\ 15 & -5 \end{bmatrix}.$$

Using these expressions one finds that

$$(5) \qquad \Phi^\times_k = \begin{cases} \frac{1}{10}\left(-\frac{1}{2}\right)^k \begin{bmatrix} -4 & 1 \\ -24 & 6 \end{bmatrix}, & k = 1, 2, \ldots, \\[2ex] \frac{1}{10}\begin{bmatrix} 6 & 1 \\ 6 & 6 \end{bmatrix}, & k = 0, \\[2ex] -\frac{1}{10}\left(\frac{1}{2}\right)^{-k-1}\begin{bmatrix} -3 & 2 \\ -3 & 2 \end{bmatrix}, & k = -1, -2, \ldots. \end{cases}$$

Next, we compute a canonical factorization of Φ. From the discussion in the second paragraph following the proof of Theorem 5.2 we know that Φ admits the following realization:

$$\Phi(\zeta) = I + C(\zeta G - A)^{-1}B.$$

Here B, C and G are as above and

$$A = I - P + \Omega P = \begin{bmatrix} 1 & 0 & 0 \\ 0 & 1 & 0 \\ 0 & 0 & 0 \end{bmatrix}.$$

Note that $A - BC$ is precisely the operator $A^\times$ introduced above. One computes that

$$(\zeta G - A)^{-1} = \begin{bmatrix} -1 & -\zeta & 0 \\ 0 & -1 & 0 \\ 0 & 0 & \zeta^{-1} \end{bmatrix},$$

and hence

$$G(\zeta G - A)^{-1} = (\zeta G - A)^{-1}G = \begin{bmatrix} 0 & -1 & 0 \\ 0 & 0 & 0 \\ 0 & 0 & \zeta^{-1} \end{bmatrix}.$$

It follows that in this case the projections P and Q appearing in Theorem 7.1 are given by

$$P = Q = \begin{bmatrix} 0 & 0 & 0 \\ 0 & 0 & 0 \\ 0 & 0 & 1 \end{bmatrix}.$$

We already computed the projection $P^\times$. We also need

$$Q^\times = \frac{1}{2\pi i}\int_{\mathbb{T}} (\zeta G - A^\times)^{-1}Gd\zeta = \frac{1}{5}\begin{bmatrix} 0 & 1 & 3 \\ 0 & 2 & 6 \\ 0 & 1 & 3 \end{bmatrix}.$$

One checks that

$$\operatorname{Im} P = \operatorname{span}\left(\left\{\begin{bmatrix} 0 \\ 0 \\ 1 \end{bmatrix}\right\}\right), \quad \operatorname{Ker} P^\times = \operatorname{span}\left\{\begin{bmatrix} -3 \\ 0 \\ 1 \end{bmatrix}, \begin{bmatrix} 2 \\ 2 \\ -1 \end{bmatrix}\right\},$$

$$\operatorname{Im} Q = \operatorname{span}\left\{\begin{bmatrix} 0 \\ 0 \\ 1 \end{bmatrix}\right\}, \quad \operatorname{Ker} Q^\times = \operatorname{span}\left\{\begin{bmatrix} 1 \\ 0 \\ 0 \end{bmatrix}, \begin{bmatrix} 0 \\ -3 \\ 1 \end{bmatrix}\right\}.$$

Hence

$$(6) \qquad \mathbb{C}^3 = \operatorname{Im} P \oplus \operatorname{Ker} P^\times, \qquad \mathbb{C}^3 = \operatorname{Im} Q \oplus \operatorname{Ker} Q^\times,$$

and Theorem 7.1 implies that Φ admits a right canonical factorization relative to $\mathbb{T}$. To compute the factors we have to determine the projections corresponding to the decompositions in (6). This is easily done. Put

$$\rho = \begin{bmatrix} 1 & 0 & 0 \\ 0 & 1 & 0 \\ -\frac{1}{3} & -\frac{1}{6} & 0 \end{bmatrix}, \qquad \tau = \begin{bmatrix} 1 & 0 & 0 \\ 0 & 1 & 0 \\ 0 & -\frac{1}{3} & 0 \end{bmatrix}.$$

Then ρ is the projection of $\mathbb{C}^3$ along $\operatorname{Im} P$ onto $\operatorname{Ker} P^\times$, and τ is the projection along $\operatorname{Im} Q$ onto $\operatorname{Ker} Q^\times$. Now use formulas (3) and (4) in Section XXIV.7, and insert the present data. One computes that

$$\Phi_-(\zeta) = \begin{bmatrix} 1 - (2\zeta)^{-1} & (3\zeta)^{-1} \\ 0 & 1 \end{bmatrix}, \qquad \Phi_+(\zeta) = \begin{bmatrix} 2 & -\frac{1}{3} \\ -3\zeta & \zeta+1 \end{bmatrix},$$

are the factors in a right canonical factorization of Φ relative to $\mathbb{T}$.

We know now that the block Toeplitz operator T with symbol Φ is invertible. The next step is to compute its inverse. The recipe to find the entries of the inverse $T^{-1} = [\Gamma_{ij}]_{i,j=0}^\infty$ is given in Theorem 8.1. We already computed $\Phi_k^\times$. It remains to compute

$$K_{ij} = C^\times (\Omega^\times)^i (I - P^\times) \rho P^\times (\Omega^\times)^j B,$$

where $C^\times = CE^\times$. First we determine $S^{-1}(I - P^\times)\rho P^\times S$. Since

$$S^{-1}(I - P^\times)\rho P^\times S = S^{-1}(I - P^\times)S(S^{-1}\rho S)S^{-1}P^\times S,$$

it suffices to compute that $(2,1)$ and $(3,1)$ entries in $S^{-1}\rho S$. This is simple to do, and one finds that

$$(7) \qquad\qquad S^{-1}(I - P^\times)\rho P^\times S = \begin{bmatrix} 0 & 0 & 0 \\ \frac{2}{3} & 0 & 0 \\ 0 & 0 & 0 \end{bmatrix}.$$

Using (7) together with the formulas for $CE^\times S$ and $S^{-1}P^\times B$ which we derived earlier, one finds that

$$(8) \qquad\qquad K_{ij} = \frac{1}{10}(-1)^i \left(\frac{1}{2}\right)^{i+j} \begin{bmatrix} -1 & \frac{2}{3} \\ -6 & 4 \end{bmatrix}, \qquad i,j = 0,1,2,\dots .$$

So $T^{-1} = [\Phi_{i-j}^\times + K_{ij}]_{i,j=0}^\infty$, where $\Phi_k^\times$ is given by (5) and K_{ij} by (8).

XXIV.11 ASYMPTOTIC FORMULAS FOR DETERMINANTS OF BLOCK TOEPLITZ MATRICES

Let T be a block Toeplitz operator defined by a continuous $m \times m$ matrix function

$$\Phi(\zeta) = \sum_{\nu=-\infty}^{\infty} \zeta^\nu \Phi_\nu, \qquad \zeta \in \mathbb{T}.$$

By definition, the N-*th section* of T is the $N \times N$ block Toeplitz matrix

$$T_N = \begin{bmatrix} \Phi_0 & \Phi_{-1} & \cdots & \Phi_{-(N-1)} \\ \Phi_1 & \Phi_0 & \cdots & \Phi_{-(N-2)} \\ \vdots & \vdots & \ddots & \vdots \\ \Phi_{N-1} & \Phi_{N-2} & \cdots & \Phi_0 \end{bmatrix}.$$

Note that T_N is an $Nm \times Nm$ matrix, and hence its determinant $\det T_N$ is well-defined. The latter number will be denoted by $D_N(\Phi)$. Our aim is to analyse the asymptotic behaviour of the sequence $\big(D_N(\Phi)\big)_{N=1}^{\infty}$.

In the case when $\Phi(\zeta)$ is identically equal to Φ_0 on $\mathbb{T}$, we have $D_N(\Phi) = (\det \Phi_0)^N$, and this identity gives a hint of what one may expect in general. In fact, it turns out that under certain restrictions on Φ,

$$(1) \qquad \lim_{N \to \infty} \frac{D_N(\Phi)}{\Sigma_1^N} = \Sigma_2,$$

for some constants Σ_1 and Σ_2 depending on Φ.

Limit formulas of the type (1) have a long and interesting history which starts in the beginning of this century with two famous papers of G. Szegö (see Szegö [1], [2]). In these papers for positive scalar functions a limit of the type

$$\lim_{N \to \infty} D_N(\Phi)^{1/N} = \Sigma_1$$

was established, which is nowadays called the *first Szegö limit formula*. The limit (1), which is usually referred to as the *second Szegö limit formula*, appeared about 30 years later in Szegö [3]. Generalizations of these limit formulas for matrix-valued functions were obtained by H. Widom (see Widom [1], [2]). The literature on the subject is extensive (see Böttcher-Silbermann [1] for more details).

In the present section we restrict ourselves to the case when Φ is a rational matrix function. Furthermore, we shall assume that Φ is analytic at zero and its value at zero is the $m \times m$ identity matrix I_m. This allows us to represent Φ in the form

$$(2). \qquad \Phi(\zeta) = I_m + \zeta C(I - \zeta A)^{-1} B, \qquad \zeta \in \mathbb{T}.$$

Here A is a square matrix of order n, say, the matrix A has no eigenvalues on $\mathbb{T}$, and B and C are matrices of sizes $n \times m$ and $m \times n$, respectively. To obtain a representation (2) we apply the construction in the proof of Theorem XIII.4.1 to $W(\zeta) = \Phi(\zeta^{-1})$. Note that W is analytic at infinity and has the value I_m at infinity. Thus, by the construction in the proof of Theorem XIII.4.1, the rational matrix function W admits the following representation

$$(3) \qquad W(\zeta) = I + C(\zeta - A)^{-1} B,$$

where A, B and C are matrices of the desired sizes and the eigenvalues of A coincide with the poles of W. Note that W has no poles on $\mathbb{T}$ since Φ is continuous on $\mathbb{T}$. It follows that A has no eigenvalues on $\mathbb{T}$, and by replacing ζ by ζ^{-1} in (3) we obtain the representation (2).

To obtain a limit formula of the type (1) we shall assume that the following condition is fulfilled:

(F) $\det \Phi(\zeta) \neq 0$ for each $\zeta \in \mathbb{T}$.

In other words (cf., Theorem XXIII.4.3) we shall assume that the block Toeplitz operator defined by Φ is a Fredholm operator. In terms of the representation (2), condition (F) is equivalent to the requirement that $A^\times := A - BC$ has no eigenvalues on $\mathbb{T}$. To see this, put $W(\zeta) = \Phi(\zeta^{-1})$. Then $W(\zeta)$ is given by (3), and we may apply Lemma 6.2 to $W(\zeta)$ to obtain the desired result. Thus, if (F) is satisfied, then both A and $A^\times$ have no eigenvalues on $\mathbb{T}$, and hence the following Riesz projections are well-defined:

$$(4) \qquad P = I - \frac{1}{2\pi i} \int_{\mathbb{T}} (\zeta - A)^{-1} d\zeta,$$

$$(5) \qquad P^\times = I - \frac{1}{2\pi i} \int_{\mathbb{T}} (\zeta - A^\times)^{-1} d\zeta.$$

The following theorem is the main result of this section.

THEOREM 11.1. *Let T_N be the N-th section of the block Toeplitz operator defined by the rational $m \times m$ matrix function*

$$(6) \qquad \Phi(\zeta) = I_m + \zeta C (I - \zeta A)^{-1} B, \qquad \zeta \in \mathbb{T}.$$

Assume that the condition (F) *is fulfilled. Then*

$$(7) \qquad \lim_{N \to \infty} \frac{\det T_N}{\Sigma_1^N} = \Sigma_2$$

with

$$(8) \qquad \Sigma_1 = \frac{\det(I - P^\times + A^\times P^\times)}{\det(I - P + AP)},$$

$$(9) \qquad \Sigma_2 = \det\left[(I - P)(I - P^\times) + PP^\times\right].$$

Here $A^\times = A - BC$, and the matrices P and $P^\times$ are the Riesz projections defined by (4) *and* (5).

The proof of Theorem 11.1 will be based on the following lemma.

LEMMA 11.2. *Let T_N be the N-th section of the block Toeplitz operator defined by (2). Then*

$$\text{(10)} \qquad \det T_N = \det\{(I - P + PA)^{-N}(I - P + P(A^\times)^N)\},$$

where $A^\times = A - BC$ and P is the Riesz projection given by (4).

PROOF. In what follows PA^{-N} or $A^{-N}P$ stands for the operator $P(I - P + PA)^{-N}$. Note that (because of the definition of P) the operator

$$A|\operatorname{Im} P \colon \operatorname{Im} P \to \operatorname{Im} P$$

is invertible, and thus PA^{-N} and $A^{-N}P$ are well defined. Put

$$\text{(11)} \qquad \tau_k = \begin{cases} CA^{k-1}(I - P)B, & k > 0, \\ -CA^{k-1}PB, & k \le 0. \end{cases}$$

From (2) it follows that the k-th Fourier coefficient Φ_k of Φ is equal to τ_k for $k \ne 0$ and $\Phi_0 = I + \tau_0$. Thus $T_N = I + [\tau_{k-j}]_{k,j=0}^{N-1}$. Next define

$$\text{(12)} \qquad V_N = I - P + PA^{-N}(A^\times)^N.$$

We shall prove that

$$\text{(13)} \qquad V_N = I - S(I + H)^{-1}R,$$

where $R = \operatorname{col}[CA^j]_{j=0}^{N-1}$, $S = \operatorname{row}[PA^{-j-1}B]_{j=0}^{N-1}$ and $H = [H_{kj}]_{k,j=0}^{N-1}$ with

$$H_{kj} = \begin{cases} CA^{k-j-1}B, & 0 \le j < k \le N - 1, \\ 0, & 0 \le k \le j \le N - 1. \end{cases}$$

From the lower triangular block matrix representation of $I + H$ it is clear that $I + H$ is invertible. A direct computation shows that $(I + H)^{-1} = I + H^\times$ with $H^\times = [H_{kj}^\times]_{k,j=0}^{N-1}$ and

$$H_{kj}^\times = \begin{cases} -C(A^\times)^{k-j-1}B, & 0 \le j < k \le N - 1, \\ 0, & 0 \le k \le j \le N - 1. \end{cases}$$

It follows that $(I + H)^{-1}R = \operatorname{col}\big[C(A^\times)^j\big]_{j=0}^{N-1}$, and hence

$$S(I + H)^{-1}R = \sum_{j=0}^{N-1} PA^{-j-1}BC(A^\times)^j = \sum_{j=0}^{N-1} PA^{-j-1}(A - A^\times)(A^\times)^j$$

$$= P - PA^{-N}(A^\times)^N,$$

which proves (13). Note that $T_N = I + H - RS$. Furthermore, $\det(I + H) = 1$. So

$$\det V_N = \det\big[I - S(I + H)^{-1}R\big] = \det\big[I - (I + H)^{-1}RS\big]$$

$$= \det\big[(I + H)^{-1}T_N\big] = \det T_N. \quad \square$$

PROOF OF THEOREM 11.1. Let V_N be given by (12). We rewrite V_N in the form

$$V_N = (I - P + PA)^{-N} Q_N (I - P^\times + P^\times A^\times)^N,$$

where $Q_N = (I - P + P(A^\times)^N)(I - P^\times + P^\times A^\times)^{-N}$. According to Lemma 11.2 this implies that

$$\det T_N = \det Q_N \left[\frac{\det(I - P^\times + P^\times A^\times)}{\det(I - P + PA)} \right]^N.$$

Now note that

$$Q_N = (I - P)(I - P^\times + P^\times A^\times)^{-N} + P\big((I - P^\times)A^\times + P^\times\big)^N$$
$$\to (I - P)(I - P^\times) + PP^\times \qquad (N \to \infty).$$

The convergence follows because $P^\times$ is the Riesz projection for $A^\times$ corresponding to the eigenvalues outside $\mathbb{T}$ and thus $(A^\times)^N(I - P^\times)$ and $P^\times(A^\times)^{-N}$ tend to 0 if $N \to \infty$. Thus

$$\frac{\det T_N}{\Sigma_1^N} = \det Q_N \to \Sigma_2 \qquad (N \to \infty),$$

by the continuity of the determinant. $\square$

Let us analyse further the constants Σ_1 and Σ_2 appearing in Theorem 11.1. Put

$$V = (I - P)(I - P^\times) + PP^\times.$$

Thus $\Sigma_2 = \det V$. Note that $PV = VP^\times$. It is straightforward to check that $\det V \neq 0$ if and only if the following two identities hold:

$$\mathbb{C}^n = \operatorname{Ker} P \oplus \operatorname{Im} P^\times, \qquad \mathbb{C}^n = \operatorname{Ker} P^\times \oplus \operatorname{Im} P.$$

It follows (apply Theorem 7.1 to $W(\zeta) = \Phi(\zeta^{-1})$) that $\det V \neq 0$ is equivalent to the requirement that the rational matrix function Φ admits a left and a right canonical (Wiener-Hopf) factorization relative to $\mathbb{T}$. So we have the following corollary.

COROLLARY 11.3. *Let T_N be the N-th section of the block Toeplitz operator defined by*

$$\Phi(\zeta) = I_m + \zeta C(I - \zeta A)^{-1} B, \qquad \zeta \in \mathbb{T}.$$

Assume condition (F) *is fulfilled, and let Σ_1 be the constant defined by* (8). *Then*

$$\det T_N = o(\Sigma_1^N), \qquad N \to \infty,$$

if and only if Φ does not admit a left or does not admit a right canonical factorization.

Under an additional condition on Φ the constant Σ_1 defined by (8) can be expressed directly in terms of Φ. This is the content of the next proposition.

PROPOSITION 11.4. *Let the $m \times m$ rational matrix function Φ be given by (2). Assume that condition (F) is fulfilled and that the winding number relative to the origin of the curve parametrized by $t \mapsto \det \Phi(e^{it})$, $-\pi \leq t \leq \pi$, is zero. Then*

$$(14) \qquad \Sigma_1 = \frac{\det(I - P^\times + A^\times P^\times)}{\det(I - P + AP)} = \exp\left\{ \frac{1}{2\pi} \int_{-\pi}^{\pi} \log \det \Phi(e^{it}) dt \right\}.$$

PROOF. Since $\det(I + XY) = \det(I + YX)$ for matrices X and Y (cf., Corollary VII.6.2), we can use the representation (2) to obtain

$$\det \Phi(\zeta) = \det\left[(I - \zeta A)^{-1}(I - \zeta A^\times)\right],$$

where $A^\times = A - BC$. It follows that

$$(15) \qquad \det \Phi(\zeta) = \prod_{j=1}^{n} \frac{1 - \zeta \lambda_j(A)}{1 - \zeta \lambda_j(A^\times)}.$$

Here the numbers $\lambda_j(A)$ and $\lambda_j(A^\times)$ denote the eigenvalues of A and $A^\times$ (counted according to their multiplicities). From (15) and the winding number condition we see that the matrices A and $A^\times$ have the same number of eigenvalues outside $\mathbb{T}$. One computes that

$$(16) \qquad \frac{1}{2\pi} \int_{-\pi}^{\pi} \log \det \Phi(e^{i\theta}) d\theta = \log \prod_{i=1}^{k} \frac{\lambda_i(A^\times)}{\lambda_i(A)},$$

where $\lambda_1(A), \ldots, \lambda_k(A), \lambda_1(A^\times), \ldots, \lambda_k(A^\times)$ have absolute value larger than 1 and the remaining eigenvalues of A and $A^\times$ are inside $\mathbb{T}$. The right-hand side of (16) and $\log \Sigma_1$ (where Σ_1 is given by (8)) are equal, which proves (14). $\square$

Under the conditions mentioned in Proposition 11.4 it can also be shown that

$$(17) \qquad \Sigma_2 = \det\left\{ (I - P)(I - P^\times) + PP^\times \right\} = \det(TT^\times),$$

where T and $T^\times$ are the block Toeplitz operators defined by Φ and $\Phi(\cdot)^{-1}$, respectively. In fact, $\det(TT^\times)$ is Widom's formula (see Widom [1], [2], [3]) for the strong Szegö limit, and hence by Theorem 11.1 and Proposition 11.4 this quantity is equal to Σ_2, where Σ_2 is defined by (9).

There is a close connection between the limit formula (1) and the projection method which allows one to compute the inverse of a block Toeplitz operator T as a limit of the inverse T_N^{-1} of the N-th finite section of T for $N \to \infty$ (see Gohberg-Fel'dman [1]). For a block Toeplitz operator T defined by a continuous matrix function Φ this method works if and only if Φ admits a left and a right canonical factorization. In particular, if Φ is given by (2), then the projection method works if and only if

$$\det\left\{ (I - P)(I - P^\times) + PP^\times \right\} \neq 0,$$

and in this case for N sufficiently large the finite section T_N is invertible and T_N^{-1} can be explicitly computed by using the representation $T_N = I + H - RS$, where H, R and S are as in the proof of Lemma 11.2. Furthermore, in this case $T_N^{-1} \to T^{-1}$ follows from the arguments used in the proof of Theorem 11.1. A treatment of the projection method for block Toeplitz operators defined by rational matrix functions in terms of realizations, along the lines sketched above, may be found in Section 8 of Gohberg-Kaashoek [3]. For a systematic analysis of the projection method of (block) Toeplitz operators and its connections with Szegö type limit formulas we refer to the books Gohberg-Fel'dman [1] and Böttcher-Silbermann [1]. See Gohberg-Kaashoek [4] for recent results in this direction.

CHAPTER XXV

TOEPLITZ OPERATORS DEFINED BY
PIECEWISE CONTINUOUS MATRIX FUNCTIONS

In this chapter we introduce and study the symbol and Fredholm index of Toeplitz and block Toeplitz operators defined by piecewise continuous functions with a finite number of discontinuities. Sums and products of such operators are also considered. The chapter provides the necessary tools to develop the theory of Banach algebras generated by Toeplitz operators defined by piecewise continuous functions, which will be treated in Chapter XXXII.

XXV.1 PIECEWISE CONTINUOUS FUNCTIONS

This section has a preliminary character. We develop the properties of piecewise continuous functions and the associated symbols. By $PC(\mathbb{T})$ we denote the set of all complex valued functions φ on the unit circle $\mathbb{T}$ that are piecewise continuous on $\mathbb{T}$ and continuous from the left, that is, for each $\zeta = e^{it}$ on $\mathbb{T}$

$$\varphi(\zeta-) = \lim_{s \uparrow t} \varphi(e^{is}), \qquad \varphi(\zeta+) = \lim_{s \downarrow t} \varphi(e^{is})$$

exist and $\varphi(\zeta-) = \varphi(\zeta)$. In general, the number of discontinuities of φ can be infinite, but in this chapter we restrict our attention to functions for which this number is finite.

Let $\zeta_1, \ldots, \zeta_k$ be k different points on $\mathbb{T}$. By $PC(\mathbb{T}; \zeta_1, \ldots, \zeta_k)$ we denote the family of all $\varphi \in PC(\mathbb{T})$ such that φ has its discontinuities only in the set $\zeta_1, \ldots, \zeta_k$. Thus $\varphi \in PC(\mathbb{T}; \zeta_1, \ldots, \zeta_k)$ is continuous at each point $\zeta \neq \zeta_j$ $(j = 1, \ldots, k)$ and may be (but does not have to be) discontinuous in $\zeta_1, \ldots, \zeta_k$. With the supremum norm and the usual algebraic operations $PC(\mathbb{T}; \zeta_1, \ldots, \zeta_k)$ is a complex Banach space. Furthermore, the product of two functions in $PC(\mathbb{T}; \zeta_1, \ldots, \zeta_k)$ belongs again to this class.

Let $\varphi \in PC(\mathbb{T}; \zeta_1, \ldots, \zeta_k)$. With φ we associate a function $\widehat{\varphi}$ defined on the cylinder $\mathbb{T} \times [0, 1]$ as follows:

$$(1) \qquad \widehat{\varphi}(\zeta, \mu) = \mu \varphi(\zeta+) + (1 - \mu) \varphi(\zeta), \qquad \zeta \in \mathbf{T}, \ 0 \leq \mu \leq 1.$$

We shall refer to $\widehat{\varphi}$ as the *symbol* associated with (the Toeplitz operator T_φ defined by) the function φ.

If φ is continuous in ζ, then $\widehat{\varphi}(\zeta, \mu) = \varphi(\zeta)$ for all $0 \leq \mu \leq 1$, but if ζ is a point of discontinuity for φ, then

$$(2) \qquad \widehat{\varphi}(\zeta, 0) = \varphi(\zeta-) = \varphi(\zeta), \qquad \widehat{\varphi}(\zeta, 1) = \varphi(\zeta+).$$

Thus the set $\Gamma_{\widehat{\varphi}}$ of points $\widehat{\varphi}(\zeta, \mu)$ with $\zeta \in \mathbb{T}$ and $0 \leq \mu \leq 1$ consists of the curve γ parametrized by φ and the straight line elements which close the gaps in the curve γ. It

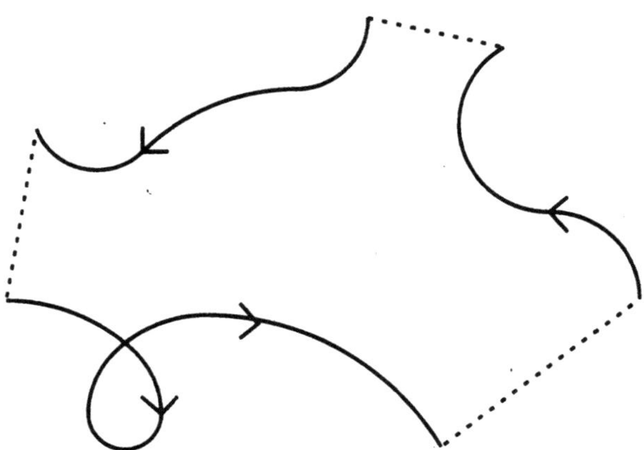

Figure 1

follows that $\Gamma_{\widehat{\varphi}}$ may be regarded as a closed oriented curve parametrized by a continuous function (see Figure 1).

To make the latter statement more precise we deform the unit circle in the following way. We split each point ζ_j of discontinuity in two points, namely ζ_j- and ζ_j+, and we connect these two points by a "handle" as indicated by Figure 2. We denote the deformed circle by $\widetilde{\mathbb{T}} = \widetilde{\mathbb{T}}(\zeta_1, \ldots, \zeta_k)$. The orientation on $\widetilde{\mathbb{T}}$ is the natural orientation which it inherits from $\mathbb{T}$ (see Figure 2).

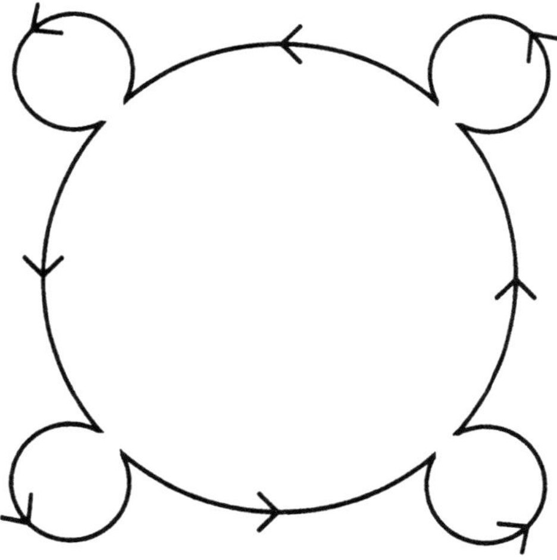

Figure 2

Given φ we define a function $\widetilde{\varphi}$ on $\widetilde{\mathbb{T}}$ by the following rule. In $\zeta \in \mathbb{T}$, $\zeta \neq \zeta_j$ $(j = 1, \ldots, k)$, the functions φ and $\widetilde{\varphi}$ coincide,

$$(3) \qquad \widetilde{\varphi}(\zeta_j -) = \varphi(\zeta_j), \quad \widetilde{\varphi}(\zeta_j +) = \varphi(\zeta_j +), \qquad j = 1, \ldots, k,$$

and on the "handles" $\widetilde{\varphi}$ is defined by linear interpolation (identifying each handle with the interval $[0, 1]$). Obviously, $\widetilde{\varphi}$ is continuous on the deformed circle $\widetilde{\mathbb{T}}$ and the (oriented) curve parametrized by $\widetilde{\varphi}$ is precisely equal to $\Gamma_{\widehat{\varphi}}$. Note that $\mathbb{T}$ and the deformed circle $\widetilde{\mathbb{T}}$ are topologically equivalent curves, and hence $\Gamma_{\widehat{\varphi}}$ is also parametrized by a continuous function on $\mathbb{T}$.

In the sequel we shall often meet the condition that $\widehat{\varphi}(\zeta, \mu) \neq 0$ for all $\zeta \in \mathbb{T}$ and $0 \leq \mu \leq 1$. This means that the closed curve $\Gamma_{\widehat{\varphi}}$ (which is parametrized by the continuous function $\widetilde{\varphi}$) does not go through the origin. In that case the winding number of $\Gamma_{\widehat{\varphi}}$ relative to the origin (see [C], Section IV.5) is well-defined. We shall denote this integer by $n(\widehat{\varphi}; 0)$, and we shall refer to it as the *winding number of $\widehat{\varphi}$ relative to zero*.

We shall also need winding numbers for more general functions. For example, assume

$$(4) \qquad \omega(\zeta, \mu) = \sum_{i=1}^{p} \widehat{\varphi}_{i1}(\zeta, \mu)\widehat{\varphi}_{i2}(\zeta, \mu) \cdots \widehat{\varphi}_{iq}(\zeta, \mu),$$

and let $\omega(\zeta, \mu) \neq 0$ for all $(\zeta, \mu) \in \mathbb{T} \times [0, 1]$. Here for each i and j the function $\widehat{\varphi}_{ij}$ is the symbol associated with a φ_{ij} in $PC(\mathbb{T}; \zeta_1, \ldots, \zeta_k)$. For each φ_{ij} we define a function $\widetilde{\varphi}_{ij}$ on the deformed circle as indicated above for φ. Then ω induces in a canonical way a continuous function $\widetilde{\omega}$ on $\widetilde{\mathbb{T}}$, namely

$$(5) \qquad \widetilde{\omega}(\lambda) = \sum_{i=1}^{p} \widetilde{\varphi}_{i1}(\lambda)\widetilde{\varphi}_{i2}(\lambda) \cdots \widetilde{\varphi}_{iq}(\lambda).$$

The closed oriented curve Γ parametrized by $\widetilde{\omega}$ does not go through zero. By definition $n(\omega; 0)$ is the winding number of the curve Γ relative to 0, and we refer to $n(\omega; 0)$ as the *winding number of ω relative to* 0. The definition of $n(\omega; 0)$ does not depend on the choice of the representation (4).

Let us denote by $\mathcal{A}$ the family of all sums of products of functions $\widehat{\varphi}$ with φ from $PC(\mathbb{T}; \zeta_1, \ldots, \zeta_k)$. In other words, $\omega \in \mathcal{A}$ if and only if ω admits a representation as in (4). Since each function $\omega \in \mathcal{A}$ can be identified in a canonical way with a continuous function on the deformed circle $\widetilde{\mathbb{T}}$ (as indicated above), the usual properties of the winding number extend to functions from $\mathcal{A}$. Thus, for $\omega_1, \omega_2 \in \mathcal{A}$ we have

$$(6) \qquad n(\omega_1 \omega_2; 0) = n(\omega_1; 0) + n(\omega_2; 0),$$

whenever ω_1 and ω_2 do not vanish on $\mathbb{T} \times [0, 1]$. Furthermore, from

$$(7) \qquad |\omega_2(\zeta, \mu) - \omega_1(\zeta, \mu)| < |\omega_1(\zeta, \mu)|$$

for all $(\zeta,\mu) \in \mathbb{T} \times [0,1]$, it follows that $n(\omega_1;0) = n(\omega_2;0)$. In the next section we shall need the following two lemmas.

LEMMA 1.1. *Let* $\varphi_1, \varphi_2 \in PC(\mathbb{T};\zeta_1,\ldots,\zeta_k)$, *and assume that the associated symbols do not vanish on* $\mathbb{T} \times [0,1]$. *If, in addition, the points of discontinuity of* φ_1 *and* φ_2 *are different, then the symbol* $\widehat{\varphi_1\varphi_2}$ *associated with* $\varphi_1\varphi_2$ *does not vanish on* $\mathbb{T} \times [0,1]$ *and*

$$(8) \qquad n(\widehat{\varphi_1\varphi_2};0) = n(\widehat{\varphi}_1;0) + n(\widehat{\varphi}_2;0).$$

PROOF. Put $\varphi = \varphi_1\varphi_2$. Then $\varphi \in PC(\mathbb{T};\zeta_1,\ldots,\zeta)$. Since the points of discontinuity of φ_1 differ from those of φ_2,

$$(9) \qquad \widehat{\varphi}(\zeta,\mu) = \widehat{\varphi}_1(\zeta,\mu)\widehat{\varphi}_2(\zeta,\mu), \qquad \zeta \in \mathbb{T},\; 0 \le \mu \le 1.$$

Indeed, take $\zeta \in \mathbb{T}$, and let us assume that φ_2 is continuous at ζ. Then $(\varphi_1\varphi_2)(\zeta+) = \varphi_1(\zeta+)\varphi_2(\zeta)$, and hence

$$\begin{aligned}
\widehat{\varphi_1\varphi_2}(\zeta,\mu) &= \mu(\varphi_1\varphi_2)(\zeta+) + (1-\mu)(\varphi_1\varphi_2)(\zeta) \\
&= \mu\varphi_1(\zeta+)\varphi_2(\zeta) + (1-\mu)\varphi_1(\zeta)\varphi_2(\zeta) \\
&= \widehat{\varphi}_1(\zeta,\mu)\varphi_2(\zeta).
\end{aligned}$$

But in this case $\widehat{\varphi}_2(\zeta,\mu) = \varphi_2(\zeta)$, and therefore (9) holds. It follows that $\widehat{\varphi}(\zeta,\mu) \neq 0$ for all $(\zeta,\mu) \in \mathbb{T} \times [0,1]$, and (8) is just a special case of (6). $\square$

In general, without the additional condition that φ_1 and φ_2 have different points of discontinuity, formula (8) fails to hold true. To see this, take

$$(10) \qquad \varphi_1(e^{it}) = \varphi_2(e^{it}) = \exp\left(\frac{1}{3}it\right), \qquad 0 < t \le 2\pi.$$

Then φ_1 and φ_2 are continuous on $\mathbb{T}$ except at $\zeta = 1$, where both φ_1 and φ_2 are discontinuous from the right. The origin lies in the outer domain of the curve parametrized by $\widehat{\varphi}_1 = \widehat{\varphi}_2$. Thus the symbols $\widehat{\varphi}_1$ and $\widehat{\varphi}_2$ do not vanish on $\mathbb{T} \times [0,1]$ and $n(\widehat{\varphi}_i;0) = 0$ for $i = 1,2$. Note that $(\varphi_1\varphi_2)(e^{it}) = \exp(\frac{2}{3}it)$, and hence $\widehat{\varphi_1\varphi_2}$ does not vanish on $\mathbb{T} \times [0,1]$. But $n(\widehat{\varphi_1\varphi_2};0) = 1$, and so (8) is not satisfied. The main obstacle is the fact that $\widehat{\varphi_1\varphi_2} \neq \widehat{\varphi}_1\widehat{\varphi}_2$ (i.e., formula (9) does not hold true). Note that by definition $\widehat{\varphi_1\varphi_2}$ is linear on $0 \le \mu \le 1$ at $\zeta = 1$, but $\widehat{\varphi}_1\widehat{\varphi}_2$ is not. By replacing the $1/3$ in (10) by $1/4$, one sees that it may happen that $\widehat{\varphi}_1(\zeta,\mu) \neq 0$, $\widehat{\varphi}_2(\zeta,\mu) \neq 0$ and $\widehat{\varphi_1\varphi_2}(\zeta,\mu) = 0$ for some (ζ,μ) (in fact, $(\zeta,\mu) = \left(1,\frac{1}{2}\right)$ will do).

LEMMA 1.2. *Let* $\varphi \in PC(\mathbb{T};\zeta_1,\ldots,\zeta_k)$, *and assume that the associated symbol* $\widehat{\varphi}$ *does not vanish on* $\mathbb{T} \times [0,1]$. *Then* φ *can be represented in the form* $\varphi = \varphi_0\varphi_1$, *where* φ_0 *is continuous on* $\mathbb{T}$, $\varphi_1 \in PC(\mathbb{T};\zeta_1,\ldots,\zeta_k)$ *and the values of* φ_1 *are in the right half plane* $\Re\lambda \ge \delta_0 > 0$.

PROOF. Put $\rho(\zeta) = |\varphi(\zeta)|$. From our hypotheses it follows that $\rho \in PC(\mathbb{T})$ and there exists $\delta > 0$ such that $\rho(\zeta) \ge \delta > 0$ for all $\zeta \in \mathbb{T}$. So we can put ρ into the factor φ_1, and therefore we may assume without loss of generality that $\varphi(\zeta) = \exp\big(i\theta(\zeta)\big)$.

Choose $\zeta_0 \in \mathbb{T}$, $\zeta_0 \neq \zeta_j$ ($j = 1, \ldots, k$), as the point from where we start counting the argument. The fact that

$$\widehat{\varphi}(\zeta, \mu) = \mu\varphi(\zeta+) + (1 - \mu)\varphi(\zeta) \neq 0 \tag{11}$$

for all $(\zeta, \mu) \in \mathbb{T} \times [0, 1]$ allows us to choose the real-valued function θ in such a way that θ is continuous at all points $\zeta \in \mathbb{T}$ different from ζ_j ($j = 0, 1, \ldots, k$), continuous from the left at $\zeta_0, \zeta_1, \ldots, \zeta_k$, and for some $\delta > 0$

$$|\theta(\zeta_j) - \theta(\zeta_j+)| < \pi - \delta \qquad (j = 1, \ldots, k),$$

while $\varphi(\zeta_0) - \varphi(\zeta_0+)$ is a multiple of 2π. Now define real-valued functions $b(\cdot)$ and $c(\cdot)$ in $PC(\mathbb{T})$ by setting

$$b(\zeta_j) = \theta(\zeta_j), \quad b(\zeta_j+) = \theta(\zeta_j+), \qquad j = 0, 1, \ldots, k,$$
$$c(\zeta_0) = \theta(\zeta_0), \quad c(\zeta_0+) = \theta(\zeta_0+),$$
$$c(\zeta_j) = c(\zeta_j+) = \frac{1}{2}\big(\theta(\zeta_j) + \theta(\zeta_j+)\big), \qquad j = 1, \ldots, k,$$

and on the remaining arcs $b(\cdot)$ and $c(\cdot)$ are defined by linear interpolation. Then

$$\sup_{|\zeta|=1} |b(\zeta) - c(\zeta)| < \frac{1}{2}(\pi - \delta). \tag{12}$$

The choice of $b(\cdot)$ and $c(\cdot)$ implies that the functions $\theta(\cdot) - b(\cdot)$ and $\exp\big(ic(\cdot)\big)$ are continuous on $\mathbb{T}$. Hence

$$f(\zeta) = \exp\big[i\big(\theta(\zeta) - b(\zeta) + c(\zeta)\big)\big] \tag{13}$$

is continuous on $\mathbb{T}$. Of course, $|f(\zeta)| = 1$ for all $\zeta \in \mathbb{T}$. So, by the second Weierstrass approximation theorem, there exists a trigonometric polynomial p such that $p(\zeta) \neq 0$ for all $\zeta \in \mathbb{T}$ and for $m = 1 - f/p$ the following holds true:

$$\sup_{|\zeta|=1} |m(\zeta)| < \frac{1}{2}, \tag{14}$$

$$-\frac{1}{4}\delta < \arg\big(1 - m(\zeta)\big) < \frac{1}{4}\delta \qquad (\zeta \in \mathbb{T}). \tag{15}$$

Note that (15) can always be obtained by taking the left hand side of (14) sufficiently small. Put $\varphi_0 = p$, and define φ_1 by

$$\varphi_1(\zeta) = \big(1 - m(\zeta)\big) \exp\big[i\big(b(\zeta) - c(\zeta)\big)\big], \qquad \zeta \in \mathbb{T}.$$

Then φ_0 is continuous on $\mathbb{T}$ and $\varphi_1 \in PC(\mathbb{T})$. Since the function f in (13) is equal to $p(1 - m)$, we conclude that

$$\varphi(\zeta) = \exp\big(i\theta(\zeta)\big) = f(\zeta)\exp\big[i\big(b(\zeta) - c(\zeta)\big)\big] = \varphi_0(\zeta)\varphi_1(\zeta), \qquad \zeta \in \mathbb{T}.$$

From (12) and (15) it follows that $|\arg \varphi_1(\zeta)| < \frac{1}{2}\pi - \frac{1}{4}\delta$, and (14) implies that $|\varphi_1(\zeta)| \geq \frac{1}{2}$. Thus the values of φ_1 are in some right half plane $\Re\varphi_1(\zeta) \geq \delta_0 > 0$. Note that (11) and $\varphi = \varphi_0\varphi_1$ imply that $\varphi_0(\zeta) \neq 0$ for all $\zeta \in \mathbb{T}$. Thus φ_1 has the same discontinuities as φ. In particular, φ_1 belongs to $PC(\mathbb{T}; \zeta_1, \ldots, \zeta_k)$. $\square$

The next lemma is not needed in the present chapter, but it will be used in Section XXXII.3. First some preparations. Let $\varphi \in PC(\mathbb{T}; \zeta_1, \ldots, \zeta_k)$, and let $g_1, \ldots, g_k$ be complex valued continuous functions on $[0, 1]$ such that

$$(16) \qquad g_j(0) = \varphi(\zeta_j), \quad g_j(1) = \varphi(\zeta_j+), \qquad j = 1, \ldots, k.$$

Put

$$(17) \qquad f(\zeta, \mu) = \begin{cases} \varphi(\zeta) & \text{if } \zeta \neq \zeta_j \quad (j = 1, \ldots, k), \\ g_j(\mu) & \text{if } \zeta = \zeta_j. \end{cases}$$

If $g_j(\mu) = \mu\varphi(\zeta_j+) + (1 - \mu)\varphi(\zeta_j)$ for $j = 1, \ldots, k$, then the function f defined by (17) is precisely the symbol $\widehat{\varphi}$ associated with φ. A sum of products of symbols $\widehat{\varphi}$ with φ from $PC(\mathbb{T}; \zeta_1, \ldots, \zeta_k)$ is also a function of the form (17). Indeed, if ω is as in (4), then ω is given by the right hand side of (17) with

$$(18a) \qquad \varphi = \sum_{i=1}^{p} \varphi_{i1}\varphi_{i2}\cdots\varphi_{iq},$$

$$(18b) \qquad g_j(\mu) = \sum_{i=1}^{p} \prod_{\nu=1}^{q} \left(\mu\varphi_{i\nu}(\zeta_j+) + (1 - \mu)\varphi_{i\nu}(\zeta_j)\right).$$

The set of all functions f of the form (17) will be denoted by $C[\mathbb{T}[\zeta_1, \ldots, \zeta_k]]$.

With the usual algebraic operations and endowed with the supremum norm $C[\mathbb{T}[\zeta_1, \ldots, \zeta_k]]$ is a complex Banach space. Of course, any function f in $C[\mathbb{T}[\zeta_1, \ldots, \zeta_k]]$ may be identified in a canonical way with a continuous function on the deformed circle $\widetilde{\mathbb{T}} = \widetilde{\mathbb{T}}(\zeta_1, \ldots, \zeta_k)$, and hence $C[\mathbb{T}[\zeta_1, \ldots, \zeta_k]]$ is isometrically isomorphic to the Banach space $C\left(\widetilde{\mathbb{T}}(\zeta_1, \ldots, \zeta_k)\right)$ of all continuous functions on the deformed circle $\widetilde{\mathbb{T}}$ endowed with the supremum norm.

LEMMA 1.3. *The set $\mathcal{A}$ of all sums of products of symbols $\widehat{\varphi}$ with φ from $PC(\mathbb{T}; \zeta_1, \ldots, \zeta_k)$ is dense in $C[\mathbb{T}[\zeta_1, \ldots, \zeta_k]]$.*

PROOF. Let f be as in (17), and take $\varepsilon > 0$. By the Weierstrass approximation theorem we can find complex polynomials $p_1, \ldots, p_k$ such that for each j

$$(19) \qquad |g_j(\mu) - p_j(\mu)| < \varepsilon, \qquad 0 \leq \mu \leq 1,$$

$$(20) \qquad p_j(0) \neq 0, \qquad p_j(1) \neq 0.$$

From (19) it follows that $|\varphi(\zeta_j) - p_j(0)|$ and $|\varphi(\zeta_j+) - p_j(1)|$ are both strictly less than ε for each j. This allows us to choose ψ in $PC(\mathbb{T}; \zeta_1, \ldots, \zeta_k)$ such that

$$(21a) \qquad \psi(\zeta_j) = p_j(0), \quad \psi(\zeta_j+) = p_j(1), \qquad j = 1, \ldots, k,$$

$$(21b) \qquad |\varphi(\zeta) - \psi(\zeta)| < \varepsilon, \qquad \zeta \in \mathbb{T}.$$

Put

$$h(\zeta, \mu) = \begin{cases} \psi(\zeta) & \text{if } \zeta \neq \zeta_j \quad (j = 1, \ldots, k), \\ p_j(\mu) & \text{if } \zeta = \zeta_j. \end{cases}$$

Then $h \in C\big[\mathbb{T}[\zeta_1, \ldots, \zeta_k]\big]$ and $\|f - h\| < \varepsilon$ by (19) and (21b). It suffices to prove that $h \in \mathcal{A}$.

Fix j, $1 \leq j \leq k$. Write $p_j(\mu) = c \prod_{\nu=1}^n (\mu - \alpha_\nu)$. Choose $\varphi_1, \ldots, \varphi_n$ in $PC(\mathbb{T}; \zeta_1, \ldots, \zeta_k)$ such that $\varphi_1, \ldots, \varphi_n$ are continuous at each point of $\mathbb{T}$ except perhaps at $\zeta = \zeta_j$, where

$$\varphi_\nu(\zeta_j) = -\alpha_\nu, \qquad \varphi_\nu(\zeta_j+) = 1 - \alpha_\nu.$$

From (20) it follows that $1 - \alpha_\nu$ and $-\alpha_\nu$ are both nonzero for each ν. Hence we can choose $\varphi_1, \ldots, \varphi_n$ such that they do not vanish at any point of $\mathbb{T}$. Put $\psi_j = c\varphi_1 \cdots \varphi_n$ and

$$h_j(\zeta, \mu) = c\widehat{\varphi}_1(\zeta, \mu)\widehat{\varphi}_2(\zeta, \mu) \cdots \widehat{\varphi}_n(\zeta, \mu).$$

Note that $\mu - \alpha_\nu = \mu(1 - \alpha_\nu) + (1 - \mu)(-\alpha_\nu)$. Thus

$$(22) \qquad h_j(\zeta, \mu) = \begin{cases} \psi_j(\zeta) & \text{if } \zeta \neq \zeta_j, \\ p_j(\mu) & \text{if } \zeta = \zeta_j. \end{cases}$$

Furthermore, $h_j \in \mathcal{A}$ and

$$(23) \qquad \psi_j(\zeta_j) = p_j(0), \qquad \psi_j(\zeta_j+) = p_j(1).$$

We carry out the above construction for each j. This yields $\psi_1, \ldots, \psi_k$ in $PC(\mathbb{T}; \zeta_1, \ldots, \zeta_k)$ and $h_1, \ldots, h_k \in \mathcal{A}$ such that (22) and (23) hold true. Moreover for each j and $\zeta \in \mathbb{T}$ the numbers $\psi_j(\zeta)$ and $\psi_j(\zeta+)$ are different from zero. Put

$$\psi_0(\zeta) = \frac{\psi(\zeta)}{\psi_1(\zeta) \cdots \psi_k(\zeta)}, \qquad \zeta \in \mathbb{T}.$$

Formulas (21a) and (23) imply that ψ_0 is continuous on $\mathbb{T}$. Put $h_0(\zeta, \mu) = \widehat{\psi}_0(\zeta, \mu)$. Then $h_0 \in \mathcal{A}$ and $h_0(\zeta, \mu) = \psi_0(\zeta)$ for all (ζ, μ) in $\mathbb{T} \times [0, 1]$. It is now straightforward to check that $h = h_0 h_1 \cdots h_k$, and $h \in \mathcal{A}$. $\square$

XXV.2 SYMBOL AND FREDHOLM INDEX (SCALAR CASE)

The next theorem gives the necessary and sufficient conditions in order that a Toeplitz operator on ℓ_2 defined by a piecewise continuous function φ with a finite number of discontinuities is a Fredholm operator. Also, the Fredholm index of the operator is identified. Recall that T_φ denotes the Toeplitz operator defined by φ.

THEOREM 2.1. *Let $\varphi \in PC(\mathbb{T})$, and assume that φ has a finite number of discontinuities. Then the Toeplitz operator T_φ is Fredholm if and only if*

$$(1) \qquad \widehat{\varphi}(\zeta,\mu) := \mu\varphi(\zeta+) + (1-\mu)\varphi(\zeta) \neq 0$$

for each $\zeta \in \mathbb{T}$ and $0 \leq \mu \leq 1$. In that case the index of T_φ is the negative of the winding number of $\widehat{\varphi}$ relative to zero, i.e.,

$$(2) \qquad \operatorname{ind} T_\varphi = -n(\widehat{\varphi}; 0).$$

We refer to the function $\widehat{\varphi}$ as the *symbol* of T_φ (see Section XXV.1). The above theorem justifies this terminology. If φ is continuous (and hence $\widehat{\varphi}(\cdot,\mu) = \varphi$ for $0 \leq \mu \leq 1$), then Theorem 2.1 is the scalar version of Theorems 4.3 and 5.1 in Chapter XXIII. For the proof of Theorem 2.1 we need the following lemma.

LEMMA 2.2. *Assume $\varphi = \varphi_0\varphi_1$, where φ_0 is continuous on $\mathbb{T}$ and φ_1 is a measurable essentially bounded function on $\mathbb{T}$ such that for some $\delta > 0$*

$$(3) \qquad \Re\varphi_1(\zeta) \geq \delta > 0 \qquad (\zeta \in \mathbb{T}).$$

If, in addition, $\varphi(\zeta) \neq 0$ for each $\zeta \in \mathbb{T}$, then T_φ is Fredholm and $\operatorname{ind} T_\varphi$ is equal to the negative of the winding number relative to zero of the curve parametrized by φ_0.

PROOF. Given a measurable essentially bounded function ψ on $\mathbb{T}$, let L_ψ be the Laurent operator on $\ell_2(\mathbb{Z})$ defined by ψ (see Section XXIII.2). Define P to be the orthogonal projection of $\ell_2(\mathbb{Z})$ onto ℓ_2. Thus

$$P(\ldots, \eta_{-1}, \eta_0, \eta_1, \ldots) = (\ldots, 0, 0, \eta_0, \eta_1, \ldots).$$

Take $x \in \ell_2$. Then

$$\begin{aligned}
T_\varphi x &= PL_\varphi x = PL_{\varphi_0}L_{\varphi_1}Px \\
&= PL_{\varphi_0}PL_{\varphi_1}Px + PL_{\varphi_0}(I-P)L_{\varphi_1}Px \\
&= T_{\varphi_0}T_{\varphi_1}x + Kx.
\end{aligned}$$

Since φ_0 is continuous, $PL_{\varphi_0}(I-P)$ is a compact operator (Lemma XXIII.4.1) and hence K is compact. Thus T_φ is Fredholm if and only if $T_{\varphi_0}T_{\varphi_1}$ is Fredholm and in that case $\operatorname{ind} T_\varphi = \operatorname{ind} T_{\varphi_0} + \operatorname{ind} T_{\varphi_1}$.

From our hypothesis on φ_1 it follows that the values $\varphi_1(\zeta)$ for almost all $\zeta \in \mathbb{T}$ lie in a truncated sector as indicated in Figure 1. It follows that we can choose $\varepsilon > 0$ such that for almost all $\zeta \in \mathbb{T}$ the value $\varepsilon\varphi_1(\zeta)$ is in the unit circle and

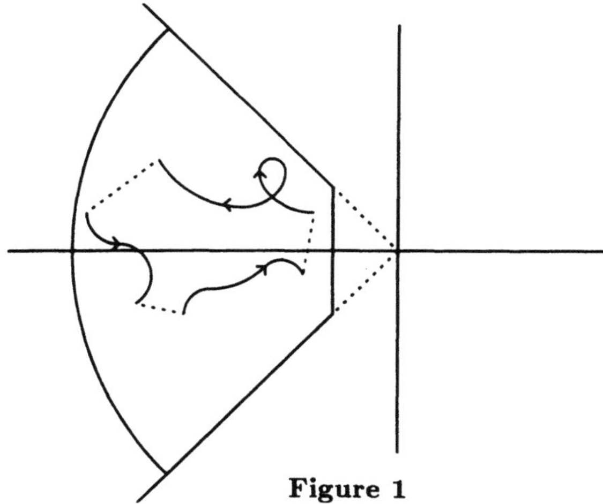

Figure 1

$$\|1 - \varepsilon\varphi_1\| = \operatorname*{ess\ sup}_{\zeta\in\mathbb{T}} |1 - \varepsilon\varphi_1(\zeta)| < 1.$$

In particular, $\|T_{1-\varepsilon\varphi_1}\| < 1$ (Corollary XXIII.3.2), and hence $I - T_{1-\varepsilon\varphi_1}$ is invertible. But then T_{φ_1} is invertible, because

$$T_{\varphi_1} = \frac{1}{\varepsilon}T_{\varepsilon\varphi_1} = \frac{1}{\varepsilon}(I - T_{1-\varepsilon\varphi_1}).$$

Now, assume that $\varphi(\zeta) \neq 0$ for all $\zeta \in \mathbb{T}$. Then $\varphi_0(\zeta) \neq 0$ for all $\zeta \in \mathbb{T}$, and hence T_{φ_0} is Fredholm (Theorem XXIII.4.3). Since T_{φ_1} is invertible, $T_{\varphi_0}T_{\varphi_1}$ is also Fredholm and $\operatorname{ind} T_{\varphi_0}T_{\varphi_1} = \operatorname{ind} T_{\varphi_0}$. We conclude that T_φ is Fredholm and $\operatorname{ind} T_\varphi = \operatorname{ind} T_{\varphi_0}$, which (by Theorem XXIII.5.1) gives the desired result. $\square$

PROOF OF THEOREM 2.1. Let $\varphi \in PC(\mathbb{T})$ have a finite number of discontinuities, and assume that (1) holds. Thus $\varphi \in PC(\mathbb{T};\zeta_1,\ldots,\zeta_k)$ for some $\zeta_1,\ldots,\zeta_k$ in $\mathbb{T}$ and the associated symbol $\widehat{\varphi}$ does not vanish on $\mathbb{T} \times [0,1]$. Thus φ can be represented in the form $\varphi = \varphi_0\varphi_1$ with φ_0 and φ_1 as in Lemma 1.2. According to Lemma 2.2 this implies that T_φ is a Fredholm operator and $\operatorname{ind} T_\varphi = -n_0$, where n_0 is the winding number relative to zero of the curve parametrized by φ_0. Since φ_0 is continuous, Lemma 1.1 implies that

$$n(\widehat{\varphi};0) = n(\widehat{\varphi}_0;0) + n(\widehat{\varphi}_1;0)$$

Now use that the values of φ_1 are in a closed right half plane $\Re\lambda \geq \delta_0 > 0$. It follows that $n(\widehat{\varphi}_1;0) = 0$. The continuity of φ_0 implies that $n(\widehat{\varphi}_0;0) = n_0$. Hence (2) is proved.

Next, assume that T_φ is a Fredholm operator. We have to prove that $\widehat{\varphi}$ does not vanish on $\mathbb{T} \times [0,1]$. Assume not. So there exists $(\zeta_0,\mu_0) \in \mathbb{T} \times [0,1]$ such that $\widehat{\varphi}(\zeta_0,\mu_0) = 0$. From the perturbation theory for Fredholm operators we know (see Theorem XI.4.1) that there exists a constant $\gamma > 0$ such that $T \in \mathcal{L}(\ell_2)$ is Fredholm whenever $\|T_\varphi - T\| < \gamma$. Let $\eta \in PC(\mathbb{T})$ have the same points of discontinuity as φ, and assume that $|\eta(\zeta)| \leq \frac{1}{2}\gamma$ for all $\zeta \in \mathbb{T}$. Then, by Corollary XXIII.3.2,

$$\|T_\varphi - T_{\varphi+\eta}\| = \sup_{\zeta \in \mathbb{T}} |\eta(\zeta)| < \gamma,$$

and hence $T_{\varphi+\eta}$ is also Fredholm. This remark allows us to modify φ.

Let $\zeta_1, \ldots, \zeta_k$ be the points of discontinuity of φ. Assume $\varphi(\zeta_j) = 0$ (resp., $\varphi(\zeta_j+) = 0$). Then we can choose the above η in such a way that η differs from 0 only on a small neighbourhood of ζ_j, the function $\varphi + \eta$ has a zero on the arc left (resp., right) of ζ_j, while both $(\varphi + \eta)(\zeta_j)$ and $(\varphi + \eta)(\zeta_j+)$ are nonzero. It follows that we may assume without loss of generality that

$$(4) \qquad \varphi(\zeta_j) \neq 0, \quad \varphi(\zeta_j+) \neq 0 \qquad (j = 1, \ldots, k).$$

Let $\omega \in PC(\mathbb{T})$ have the same points of discontinuity as φ, and assume that $|\omega(\zeta)| \leq \frac{1}{4}\delta$ for all $\zeta \in \mathbb{T}$, where $0 < \delta \leq \gamma$ and

$$\delta < \min\{|\varphi(\zeta_j)|, |\varphi(\zeta_j+)|\}, \qquad j = 1, \ldots, k.$$

Then also $|\widehat{\omega}(\zeta, \mu)| \leq \frac{1}{4}\delta$ for all (ζ, μ) in $\mathbb{T} \times [0, 1]$. Now take $\eta(\zeta) = \omega(\zeta) - \widehat{\omega}(\zeta_0, \mu_0)$. Then $|\eta(\zeta)| \leq \frac{1}{2}\delta \leq \frac{1}{2}\gamma$, and hence $T_{\varphi+\eta}$ is Fredholm. Furthermore

$$(5) \qquad \widehat{(\varphi + \eta)}(\zeta_0, \mu_0) = \widehat{\eta}(\zeta_0, \mu_0) = 0.$$

This allows us to assume that on each of the arcs determined by the points $\zeta_1, \ldots, \zeta_k$ the function φ is the restriction of a non-constant trigonometric polynomial. Making another small perturbation, if necessary, this implies that the curve $\Gamma_{\widehat{\varphi}}$ does not intersect itself at 0 and has at 0 a tangent line.

Now, let ε_1 and ε_2 be nonzero complex numbers, and put $\psi_j(\zeta) = \varepsilon_j + \varphi(\zeta)$ for $j = 1, 2$. From the smoothness condition on φ it follows that we can choose ε_1 and ε_2 in such a way that 0 is inside the curve $\Gamma_{\widehat{\psi_1}}$ and outside the curve $\Gamma_{\widehat{\psi_2}}$ (cf. Section XXV.1). According to the part of the theorem which has already been proved, this implies that T_{ψ_1} and T_{ψ_2} are Fredholm operators and $\operatorname{ind} T_{\psi_1} \neq 0 = \operatorname{ind} T_{\psi_2}$. On the other hand we may choose ε_1 and ε_2 as small as we want. So by the perturbation theorem for Fredholm operators (Section XI.4)

$$\operatorname{ind} T_{\psi_1} = \operatorname{ind} T_\varphi = \operatorname{ind} T_{\psi_2}.$$

Contradiction. So $\widehat{\varphi}$ does not vanish on $\mathbb{T} \times [0, 1]$. $\square$

Let us illustrate the results of this section with an example. Consider on ℓ_2 the following Toeplitz operator:

$$(6) \qquad T(\alpha) = \begin{bmatrix} \frac{1}{\alpha} & \frac{1}{\alpha+1} & \frac{1}{\alpha+2} & \cdots \\ \frac{1}{\alpha-1} & \frac{1}{\alpha} & \frac{1}{\alpha+1} & \cdots \\ \frac{1}{\alpha-2} & \frac{1}{\alpha-1} & \frac{1}{\alpha} & \cdots \\ \vdots & \vdots & \vdots & \ddots \end{bmatrix}.$$

Here α is assumed to be real, but $\alpha \notin \mathbb{Z}$. The operator $T(\alpha)$ is the Toeplitz operator defined by the function:

$$(7) \qquad \varphi_\alpha(e^{it}) = 2\pi i (e^{2\pi i \alpha} - 1)^{-1} e^{i\alpha t}, \qquad 0 < t \leq 2\pi.$$

Obviously, $\varphi_\alpha \in PC(\mathbb{T})$ with a discontinuity only at $\zeta = 1$. In fact,

$$(8) \qquad \varphi_\alpha(1) = \varphi_\alpha(1+) e^{2\pi i \alpha}.$$

First we prove that $T(\alpha)$ is invertible for $0 < |\alpha| < \frac{1}{2}$. Take such an α. Then the straight line ℓ through $\varphi_\alpha(1)$ and $\varphi_\alpha(1+)$ does not go through the origin. Moreover, from (7) and the fact that $0 < |\alpha| < \frac{1}{2}$ it follows that the origin and the curve parametrized by φ_α are on different sides of the line ℓ. So for a suitable $\gamma = e^{i\theta}$ the function $\psi = \gamma \varphi_\alpha$ has all its values in a right half plane $\Re \lambda \geq \delta > 0$. According to the proof of Lemma 2.2 this implies that T_ψ is invertible, and thus $T(\alpha) = \frac{1}{\gamma} T_\psi$ is also invertible.

We shall see that for all other values of α the operator $T(\alpha)$ is not invertible. First take $\alpha = k + \frac{1}{2}$, where k is an arbitrary integer. Then the straight line element that closes the gap between $\varphi_\alpha(1)$ and $\varphi_\alpha(1+)$ contains the point zero. Thus the symbol $\widehat{\varphi}_\alpha$ has a zero on $\mathbb{T} \times [0, 1]$. In fact, from (8) we see that

$$\widehat{\varphi}_{k+1/2}\left(1, \tfrac{1}{2}\right) = 0, \qquad k = 0, \pm 1, \pm 2, \ldots .$$

Thus, by Theorem 2.1, the operator $T\left(k + \frac{1}{2}\right)$ is not even a Fredholm operator for $k \in \mathbb{Z}$. Next, take $0 < |\alpha - k| < \frac{1}{2}$, where k is some nonzero integer. Then the symbol $\widehat{\varphi}_\alpha$ does not vanish on $\mathbb{T} \times [0, 1]$ and one computes that the winding number $n(\widehat{\varphi}_\alpha; 0) = k$. Theorem 2.1 implies that $T(\alpha)$ is Fredholm and

$$\operatorname{ind} T(\alpha) = -k, \qquad 0 < |\alpha - k| < \frac{1}{2}.$$

In particular, $T(\alpha)$ is not invertible for these values of α.

We conclude this section with a lemma that will be useful in proving the analogue of Theorem 2.1 for matrix-valued functions (in the next section).

LEMMA 2.3. *If $\varphi_1, \varphi_2 \in PC(\mathbb{T}; \zeta_1, \ldots, \zeta_k)$, then the corresponding Toeplitz operators T_{φ_1} and T_{φ_2} commute modulo the compact operators.*

PROOF. Choose $\psi \in PC(\mathbb{T}; \zeta_1, \ldots, \zeta_k)$ such that $\psi(\zeta_j) = 0$ and $\psi(\zeta_j +) = 1$ for $j = 1, \ldots, k$. Each $\varphi \in PC(\mathbb{T}; \zeta_1, \ldots, \zeta_k)$ may be represented as $\varphi = \alpha + \beta \psi$, where α and β are continuous functions on $\mathbb{T}$ (depending on φ). Indeed, first one chooses $\beta \in C(\mathbb{T})$ such that

$$\beta(\zeta_j) = \varphi(\zeta_j +) - \varphi(\zeta_j), \qquad j = 1, \ldots, k,$$

and next one sets $\alpha = \varphi - \beta\psi$, which is continuous on $\mathbb{T}$. It follows that we may write

$$(9) \qquad \varphi_\nu = \alpha_\nu + \beta_\nu \psi, \qquad \nu = 1, 2.$$

where α_1, α_2 and β_1, β_2 are continuous functions on $\mathbb{T}$.

Given $T, S \in \mathcal{L}(\ell_2)$, let us write $T \equiv S$ whenever $T - S$ is compact. Since the functions β_1 and β_2 in (9) are continuous on $\mathbb{T}$, we may apply Corollary XXIII.4.2 to show that

$$T_{\varphi_\nu} \equiv T_{\alpha_\nu} + T_{\beta_\nu} T_\psi, \qquad \nu = 1, 2.$$

Corollary XXIII.4.2 also shows that the five operators $T_{\alpha_1}, T_{\alpha_2}, T_{\beta_1}, T_{\beta_2}$ and T_ψ commute with one another modulo the compact operators, because the functions α_1, α_2, β_1, β_2 are continuous functions and, clearly, T_ψ commutes with itself. Therefore,

$$\begin{aligned}
T_{\varphi_1} T_{\varphi_2} &\equiv (T_{\alpha_1} + T_{\beta_1} T_\psi)(T_{\alpha_2} + T_{\beta_2} T_\psi) \\
&= T_{\alpha_1} T_{\alpha_2} + T_{\beta_1} T_\psi T_{\alpha_2} + T_{\alpha_1} T_{\beta_2} T_\psi + T_{\beta_1} T_\psi T_{\beta_2} T_\psi \\
&\equiv T_{\alpha_2} T_{\alpha_1} + T_{\alpha_2} T_{\beta_1} T_\psi + T_{\beta_2} T_\psi T_{\alpha_1} + T_{\beta_2} T_\psi T_{\beta_1} T_\psi \\
&= (T_{\alpha_2} + T_{\beta_2} T_\psi)(T_{\alpha_1} + T_{\beta_1} T_\psi) \\
&\equiv T_{\varphi_2} T_{\varphi_1},
\end{aligned}$$

and hence $T_{\varphi_1} T_{\varphi_2} - T_{\varphi_2} T_{\varphi_1}$ is compact. $\square$

Obviously, Lemma 2.3 remains true if $\varphi_1, \varphi_2 \in PC(\mathbb{T})$ and the functions φ_1, φ_2 have a finite number of discontinuities. More generally, we shall see later (cf., Theorem XXXII.4.2) that Lemma 2.3 holds for any pair of functions in $PC(\mathbb{T})$, without a restriction on the number of discontinuities.

XXV.3 SYMBOL AND FREDHOLM INDEX (MATRIX CASE)

By $PC^{m \times m}(\mathbb{T})$ we denote the set of all $m \times m$ matrix functions that are piecewise continuous on $\mathbb{T}$ and continuous from the left. Thus

$$(1) \qquad \Phi = \begin{bmatrix} \varphi_{11} & \cdots & \varphi_{1m} \\ \vdots & & \vdots \\ \varphi_{m1} & \cdots & \varphi_{mm} \end{bmatrix}$$

belongs to $PC^{m \times m}(\mathbb{T})$ if and only if each entry φ_{ij} in (1) belongs to $PC(\mathbb{T})$. In this section we assume that $\Phi \in PC^{m \times m}(\mathbb{T})$ and has a finite number of discontinuities. With Φ we associate that $m \times m$ matrix function $\widehat{\Phi}$ defined on $\mathbb{T} \times [0,1]$ by

$$(2) \qquad \widehat{\Phi}(\zeta, \mu) = \mu \Phi(\zeta+) + (1 - \mu)\Phi(\zeta), \qquad \zeta \in \mathbb{T}, \ 0 \leq \mu \leq 1.$$

If Φ is given by (1), then $\widehat{\Phi} = [\widehat{\varphi}_{ij}]_{i,j=1}^m$, where $\widehat{\varphi}_{ij}$ is defined by formula (1) in the first section of this chapter. It follows that $\det \widehat{\Phi}$ is a sum of products of (scalar) symbols. Now, assume that $\det \widehat{\Phi}(\zeta, \mu) \neq 0$ for all (ζ, μ) in $\mathbb{T} \times [0,1]$. Then $\omega = \det \widehat{\Phi}$ is a function of the type appearing in formula (4) of Section XXV.1, and hence the winding number $n(\det \widehat{\Phi}; 0)$ of $\det \widehat{\Phi}$ relative to zero is well-defined. The next theorem, which is the matrix analogue of Theorem 2.1, is the main theorem of this section.

THEOREM 3.1. *Let $\Phi \in PC^{m \times m}(\mathbb{T})$, and assume that Φ has a finite number of discontinuities. Put*

$$\text{(3)} \qquad \widehat{\Phi}(\zeta, \mu) := \mu \Phi(\zeta+) + (1 - \mu)\Phi(\zeta).$$

Then the block Toeplitz operator T_Φ is Fredholm if and only if $\det \widehat{\Phi}(\zeta, \mu) \neq 0$ for each $\zeta \in \mathbb{T}$ and $0 \leq \mu \leq 1$. In that case the index of T_Φ is the negative of the winding number of $\det \widehat{\Phi}$ relative to zero, i.e.,

$$\text{(4)} \qquad \operatorname{ind} T_\Phi = -n(\det \widehat{\Phi}; 0).$$

We shall refer to $\widehat{\Phi}$ as the *symbol* of the block Toeplitz operator T_Φ. Thus Theorem 3.1 states that T_Φ is Fredholm if and only if the determinant of its symbol does not vanish. For the proof of Theorem 3.1 we need an additional lemma. In what follows $C^{m \times m}(\mathbb{T})$ stands for the set of all $m \times m$ matrix functions that are continuous on $\mathbb{T}$.

LEMMA 3.2. *Let $\Phi \in PC^{m \times m}(\mathbb{T})$, and assume that Φ has a finite number of discontinuities. If $\det \Phi(\zeta) \neq 0$ and $\det \Phi(\zeta+) \neq 0$ for all $\zeta \in \mathbb{T}$, then there exists $M, R \in C^{m \times m}(\mathbb{T})$ and $X \in PC^{m \times m}(\mathbb{T})$ such that $X(\zeta)$ is upper triangular for each $\zeta \in \mathbb{T}$ and*

$$\text{(5)} \qquad \Phi(\zeta) = M(\zeta)X(\zeta)R(\zeta), \qquad \zeta \in \mathbb{T}.$$

PROOF. Assume that the factorization (5) has been constructed. Then all factors have nonzero determinants and

$$\text{(6)} \qquad R(\zeta)\left[\Phi(\zeta)^{-1}\Phi(\zeta + 0)\right] R(\zeta)^{-1} = X(\zeta)^{-1} X(\zeta + 0).$$

Note that the right hand side of (6) is upper triangular. If Φ is continuous at ζ, then the left hand side of (6) is just the $m \times m$ identity matrix I. If ζ is a point of discontinuity for Φ, then $R(\zeta)$ is a similarity matrix which brings $\Phi(\zeta)^{-1}\Phi(\zeta+0)$ into upper triangular form. This gives us a clue for the construction of R.

Let $\zeta_1, \ldots, \zeta_k$ be the points of discontinuity for Φ. Choose invertible matrices $R_1, \ldots, R_k$ such that $U_j := R_j\left[\Phi(\zeta_j)^{-1}\Phi(\zeta_j+0)\right] R_j^{-1}$ is in upper triangular form, and let $R(\cdot)$ be any continuous $m \times m$ matrix function on $\mathbb{T}$ such that $R(\zeta_j) = R_j$ for $j = 1, \ldots, k$ and $\det R(\zeta) \neq 0$ for all $\zeta \in \mathbb{T}$. Note that $\det U_j \neq 0$ for $j = 1, \ldots, k$. Next, we choose $X \in PC^{m \times m}(\mathbb{T})$ in such a way that $X(\zeta)$ is upper triangular for each $\zeta \in \mathbb{T}$, the function X has only discontinuities in $\zeta_1, \ldots, \zeta_k$,

$$X(\zeta_j) = I, \quad X(\zeta_j + 0) = U_j \qquad (j = 1, \ldots, k),$$

and $\det X(\zeta) \neq 0$ for each $\zeta \in \mathbb{T}$. Finally, we define M by

$$M(\zeta) = \Phi(\zeta)R(\zeta)^{-1}X(\zeta)^{-1}, \qquad \zeta \in \mathbb{T}.$$

It remains to prove that M is continuous on $\mathbb{T}$. Obviously, M can have only a discontinuity in one of the points $\zeta_1, \ldots, \zeta_k$. But

$$M(\zeta_j + 0) = \Phi(\zeta_j + 0)R(\zeta_j)^{-1}X(\zeta_j - 0)^{-1}$$
$$= \Phi(\zeta_j + 0)R_j^{-1}U_j^{-1}$$
$$= \Phi(\zeta_j)R_j^{-1}$$
$$= \Phi(\zeta_j)R(\zeta_j)^{-1}X(\zeta_j)^{-1} = M(\zeta_j).$$

Thus M is continuous on $\mathbb{T}$. $\quad\square$

PROOF OF THEOREM 3.1. Assume $\det \widehat{\Phi}(\zeta, \mu) \neq 0$ for all $(\zeta, \mu) \in \mathbb{T} \times [0, 1]$. By L_Φ we denote the block Laurent operator on $\ell_2^m(\mathbb{Z})$ defined by Φ. Let P be the orthogonal projection of $\ell_2^m(\mathbb{Z})$ onto ℓ_2^m, i.e.,

$$P(\ldots, \eta_{-2}, \eta_{-1}, \eta_0, \eta_1, \eta_2, \ldots) = (\ldots, 0, 0, \eta_0, \eta_1, \eta_2, \ldots).$$

From our hypotheses on $\widehat{\Phi}$ it follows that the conditions of Lemma 3.2 are fulfilled. So we may write Φ in the form $\Phi(\zeta) = M(\zeta)X(\zeta)R(\zeta)$, where X, R and M are as in Lemma 3.2. Since M and R are continuous, the operators $PL_M(I - P)$ and $(I - P)L_R P$ are compact (Lemma XXIII.4.1). Thus for each $x \in \ell_2^m$

$$
\begin{aligned}
T_\Phi x = PL_\Phi P x &= PL_M L_X L_R P x \\
&= PL_M PL_X PL_R P x + K x \\
&= T_M T_X T_R x + K x,
\end{aligned}
$$

where K is some compact operator on ℓ_2^m. Hence, by Theorem XI.4.2, it suffices to show that $T_M T_X T_R$ is Fredholm and to establish (4) with $T_M T_X T_R$ instead of T_Φ.

Recall that M and R are continuous on $\mathbb{T}$. From our hypotheses on Φ it follows that $\det M(\zeta) \neq 0$ and $\det R(\zeta) \neq 0$ for all $\zeta \in \mathbb{T}$. Hence we can apply Theorem XXIII.5.1 to show that T_M and T_R are Fredholm operators and $\operatorname{ind} T_M$ (resp., $\operatorname{ind} T_R$) is equal to the negative of the winding number relative to zero of the curve parametrized by $\det M$ (resp., $\det R$). The continuity of M and R also implies that $\widehat{M}(\zeta, \mu) = M(\zeta)$ and $\widehat{R}(\zeta, \mu) = R(\zeta)$ for all $(\zeta, \mu) \in \mathbb{T} \times [0, 1]$. So $\operatorname{ind} T_M = -n(\det \widehat{M}; 0)$ and $\operatorname{ind} T_R = -n(\det \widehat{R}; 0)$. Next, we consider T_X. For $\zeta \in \mathbb{T}$ the matrix $X(\zeta)$ is upper triangular. Therefore,

$$
(7) \qquad T_X = \begin{bmatrix} T_{x_{11}} & \cdots & T_{x_{1m}} \\ & \ddots & \vdots \\ 0 & & T_{x_{mm}} \end{bmatrix},
$$

where $x_{ij}(\zeta)$ are the entries of $X(\zeta)$. Since Φ has a finite number of discontinuities, the same holds for X. It follows that all x_{ij} belong to $PC(\mathbb{T}; \zeta_1, \ldots, \zeta_k)$ for a suitable choice of points $\zeta_1, \ldots, \zeta_k$ on $\mathbb{T}$. But then we can apply Lemma 2.3 to show that the operators $T_{x_{ij}}$, $1 \leq i, j \leq m$, commute modulo the compact operators. This allows us to apply Theorem XI.7.3 to the operator T_X in (7). We conclude that T_X is Fredholm if and only if the product $T_{x_{11}} T_{x_{22}} \cdots T_{x_{mm}}$ is Fredholm. The latter is equivalent to the requirement that $T_{x_{11}}, \ldots, T_{x_{mm}}$ are Fredholm, because these operators commute with one another modulo the compact operators. Therefore, T_X is Fredholm if and only if the operators $T_{x_{11}}, \ldots, T_{x_{mm}}$ on the diagonal are Fredholm, and in that case (by Theorem XI.7.4)

$$
(8) \qquad \operatorname{ind} T_X = \sum_{j=1}^m \operatorname{ind} T_{x_{jj}}.
$$

According to our hypothesis on Φ,

$$\prod_{j=1}^m \widehat{x}_{jj}(\zeta, \mu) = \det \widehat{X}(\zeta, \mu) \neq 0$$

for all $(\zeta, \mu) \in \mathbb{T} \times [0, 1]$. But then we can use Theorem 2.1 (the scalar version) to show that for each $j = 1, \ldots, m$ the operator $T_{x_{jj}}$ is Fredholm and

$$(9) \qquad \operatorname{ind} T_{x_{jj}} = -n(\widehat{x}_{jj}; 0).$$

It follows that T_X is Fredholm and, using (8) and (9), we obtain that

$$\operatorname{ind} T_X = \sum_{j=1}^{m} -n(\widehat{x}_{jj}; 0) = -n\left(\prod_{j=1}^{m} \widehat{x}_{jj}; 0\right) = -n(\det \widehat{X}; 0).$$

Thus in the product $T_M T_X T_R$ all three factors are Fredholm. We conclude (Theorem XI.3.2) that $T_M T_X T_R$ is Fredholm and

$$\begin{aligned}
\operatorname{ind} T_M T_X T_R &= \operatorname{ind} T_M + \operatorname{ind} T_X + \operatorname{ind} T_R \\
&= -n(\det \widehat{M}; 0) - n(\det \widehat{X}; 0) - n(\det \widehat{R}; 0) \\
&= -n\left(\det(\widehat{M}\widehat{X}\widehat{R}); 0\right).
\end{aligned}$$

By the continuity of M and R, the product $\widehat{M}\widehat{X}\widehat{R}$ is equal to $\widehat{MXR} = \widehat{\Phi}$ (cf., Lemma 1.1). Thus T_Φ is Fredholm and (4) holds.

Next, assume that T_Φ is Fredholm. We have to show that $\det \widehat{\Phi}$ does not vanish on $\mathbb{T} \times [0, 1]$. From Theorem XI.4.1 we know that there exists $\gamma > 0$ such that $T \in \mathcal{L}(\ell_2^m)$ is Fredholm whenever $\|T_\varphi - T\| < \gamma$. Let $A \in PC^{m \times m}(\mathbb{T})$ have the same points of discontinuity as Φ, and assume that $\|A(\zeta)\| \leq \frac{1}{2}\gamma$ for all $\zeta \in \mathbb{T}$. Then

$$(10) \qquad \|T_\Phi - T_{\Phi+A}\| \leq \sup_{\zeta \in \mathbb{T}} \|A(\zeta)\| < \gamma,$$

and hence $T_{\Phi+A}$ is also Fredholm. We shall use this remark to modify Φ.

Let $\zeta_1, \ldots, \zeta_k$ be the points of discontinuity of Φ ordered according increasing argument. Put $\zeta_{k+1} = \zeta_1$, and let Γ_j be the closed arc from ζ_j to ζ_{j+1} $(j = 1, \ldots, k)$. Take $\Omega \in PC^{m \times m}(\mathbb{T})$ such that Ω and Φ have the same points of discontinuity, and assume $\|\Omega(\zeta)\| \leq \frac{1}{4}\gamma$, where γ is as in the previous paragraph. Put $A(\zeta) = \Omega(\zeta) - \widehat{\Omega}(\zeta_0, \mu_0)$, where $(\zeta_0, \mu_0) \in \mathbb{T} \times [0, 1]$ and $\widehat{\Phi}(\zeta_0, \mu_0) = 0$. Then $\|A(\zeta)\| \leq \frac{1}{2}\gamma$, and hence $T_{\Phi+A}$ is Fredholm. Moreover,

$$(\widehat{\Phi + A})(\zeta_0, \mu_0) = \widehat{\Phi}(\zeta_0, \mu_0) + \widehat{\Omega}(\zeta_0, \mu_0) - \widehat{\Omega}(\zeta_0, \mu_0) = 0.$$

By the Weierstrass approximation theorem the function $t \mapsto \Phi(e^{it})$ may be approximated by matrix polynomials in t on each of the closed arcs $\Gamma_1, \ldots, \Gamma_k$. Hence by choosing Ω in a suitable way and replacing Φ by $\Phi + A$ (with A as in the present paragraph), we may assume without loss of generality that $\Phi(e^{it})$ is a polynomial $P_j(t)$ on the arc Γ_j.

Next we use the Smith form for a matrix polynomial (see Gohberg-Lancaster-Rodman [2], Theorem A.1.1). That is, we write

$$P_j(\lambda) = M_j(\lambda)D_j(\lambda)R_j(\lambda),$$

where M_j, D_j and R_j are matrix polynomials, $\det M_j(\lambda) \neq 0$ and $\det R_j(\lambda) \neq 0$ for each $\lambda \in \mathbb{C}$, and $D_j(\lambda)$ is a diagonal matrix of which the diagonal entries $p_{j1}(\lambda), \ldots, p_{jm}(\lambda)$ are polynomials in λ. Assume $\det \Phi(\zeta_j+) = 0$. Write $\zeta_j = e^{is}$. Then $\det P_j(s) = 0$, and hence s is zero for some of the polynomials $p_{j1}, \ldots, p_{jm}$. By a small change in the coefficients of these polynomials we can move the zero to a point to the right of s. If $\det \Phi(\zeta_{j+1}) \neq 0$, a small change in P_j does not change this property. If $\det \Phi(\zeta_{j+1}) = 0$, then we can choose the perturbation in such a way that also the perturbed function has a nonzero determinant in ζ_{j+1}. These arguments show that the perturbation A in formula (10) may be chosen such that $\det(\Phi + A)(\zeta_j)$ and $\det(\Phi + A)(\zeta_j+)$ are nonzero for all $j = 1, \ldots, k$.

So without loss of generality we may assume that the matrices $\Phi(\zeta_j)$ and $\Phi(\zeta_j+)$ are invertible for each $j = 1, \ldots, k$. Now, choose $\Psi \in PC^{m \times m}(\mathbb{T})$ so that

$$\Psi(\zeta_j) = \Phi(\zeta_j), \quad \Psi(\zeta_j+) = \Phi(\zeta_j+), \qquad j = 1, \ldots, k,$$

Ψ is continuous at all other points of $\mathbb{T}$ and $\det \Psi(\zeta) \neq 0$ for all $\zeta \in \mathbb{T}$. Put

$$B(\zeta) = \Phi(\zeta)\Psi(\zeta)^{-1}, \qquad C(\zeta) = \Psi(\zeta)^{-1}\Phi(\zeta).$$

Then B and C are continuous on $\mathbb{T}$, and we can apply Corollary XXIII.4.2 to show that the operators

$$T_\Phi - T_B T_\Psi, \qquad T_\Phi - T_\Psi T_C$$

are in $\mathcal{K}$, the set of all compact operators on ℓ_2^m. Denote by $[T]$ the coset $T + \mathcal{K}$. An application of Theorem XI.5.2 shows that $[T_B][T_\Psi]$ and $[T_\Psi][T_C]$ are invertible in the Calkin algebra $\mathcal{B} = \mathcal{L}(\ell_2^m)/\mathcal{K}$. This implies that $[T_\Psi]$ is invertible in $\mathcal{B}$ and hence, because of that, also $[T_B]$ and $[T_C]$ are invertible in $\mathcal{B}$. We conclude (from Theorem XI.5.2) that T_Ψ, T_B and T_C are Fredholm. But then $\det B(\zeta)$ and $\det C(\zeta)$ are nonzero for all $\zeta \in \mathbb{T}$ (by Theorem XXIII.4.3). Since $\widehat{\Phi}(\zeta,\mu) = B(\zeta)\widehat{\Psi}(\zeta,\mu)$, it remains to show that $\det \widehat{\Psi}(\zeta,\mu)$ does not vanish.

So, without loss of generality, we may assume that $\det \Phi(\zeta) \neq 0$ and $\det \Phi(\zeta+) \neq 0$ for all $\zeta \in \mathbb{T}$. But then we can apply Lemma 3.2 and write Φ as in (5). Since $\det \Phi(\zeta) \neq 0$, also $\det M(\zeta)$ and $\det R(\zeta)$ are different from zero. Theorem XXIII.4.3 implies that T_M and T_R are Fredholm operators. As in the first paragraph of this proof one shows that $T_\Phi - T_M T_X T_R$ is compact. So the product $T_M T_X T_R$ is Fredholm. By going to the Calkin algebra one sees that T_X is Fredholm. But then we can use (7), and deduce that the operators $T_{x_{11}}, \ldots, T_{x_{mm}}$ are Fredholm. Since the functions $x_{11}, \ldots, x_{mm}$ are scalar, we are allowed to use the "only if" part of Theorem 2.1. It follows that $\widehat{x}_{jj}(\zeta,\mu) \neq 0$ for all $(\zeta,\mu) \in \mathbb{T} \times [0,1]$ and each $j = 1, \ldots, k$. So $\det \widehat{X}(\zeta,\mu) \neq 0$ for all $(\zeta,\mu) \in \mathbb{T} \times [0,1]$, and hence

$$\det \widehat{\Phi}(\zeta,\mu) = \det(\widehat{MXR})(\zeta,\mu) = \big(\det M(\zeta)\big)\big(\det \widehat{X}(\zeta,\mu)\big)\big(\det R(\zeta)\big)$$

does not vanish on $\mathbb{T} \times [0,1]$. $\square$

We conclude this section with an example. Let A be an $m \times m$ matrix with real eigenvalues which are not in $\mathbb{Z}$. Consider on ℓ_1^m the following block Toeplitz operator:

$$T(A) = \begin{bmatrix} A^{-1} & (A+I)^{-1} & (A+2I)^{-1} & \cdots \\ (A-I)^{-1} & A^{-1} & (A+I)^{-1} & \cdots \\ (A-2I)^{-1} & (A-I)^{-1} & A^{-1} & \cdots \\ \vdots & \vdots & \vdots & \end{bmatrix}.$$

The operator $T(A)$ is the matrix analogue of the operator $T(\alpha)$ in formula (6) of the previous section. Note that $T(A)$ is defined by the $m \times m$ matrix function

$$\Phi_A(e^{it}) = 2\pi i (e^{2\pi i A} - I)^{-1} e^{itA}, \qquad 0 < t \leq 2\pi.$$

Since the eigenvalues of A are not in $\mathbb{Z}$, both Φ_A and $T(A)$ are well-defined. From

$$(11) \qquad \Phi_A(e^{it}) = \Phi_A(1+)e^{itA}, \qquad 0 < t \leq 2\pi,$$

it is clear that $\Phi_A \in PC^{m \times m}(\mathbb{T})$ with the only discontinuity at $\zeta = 1$.

We shall prove that $T(A)$ is Fredholm if and only if all eigenvalues of A are different from $\frac{1}{2} + k$, where k is an arbitrary integer. To do this we may assume without loss of generality that A is an upper triangular matrix with diagonal elements $\alpha_1, \ldots, \alpha_m$, say. Note that $\alpha_1, \ldots, \alpha_m$ are real but not in $\mathbb{Z}$. To determine when $T(A)$ is Fredholm we have to analyze the symbol

$$\widehat{\Phi}_A(\zeta, \mu) = \begin{cases} \Phi_A(\zeta) & , \quad \zeta \neq 1, \\ \mu \Phi_A(1+) + (1-\mu)\Phi_A(1), & \zeta = 1. \end{cases}$$

For $\zeta = e^{it}$, $0 < t < 2\pi$, we have

$$\det \widehat{\Phi}_A(\zeta, \mu) = \det \Phi_A(e^{it}) = \left(\prod_{j=1}^m e^{it\alpha_j} \right) \det \Phi_A(1+) \neq 0.$$

For $\zeta = 1$ the situation is different:

$$\det \widehat{\Phi}_A(1, \mu) = \left\{ \prod_{j=1}^m \left[\mu + (1-\mu)e^{2\pi i \alpha_j} \right] \right\} \det \Phi_A(1+).$$

It follows that $\det \widehat{\Phi}_A$ does not vanish on $\mathbb{T} \times [0,1]$ if and only if none of the α_j's is of the form $\frac{1}{2} + k$ for some integer k. Since $\alpha_1, \ldots, \alpha_m$ are precisely the eigenvalues of A (counted according to algebraic multiplicity), Theorem 3.1 implies that $T(A)$ is Fredholm if and only if the eigenvalues of A are not of the form $\frac{1}{2} + k$ for some integer k.

To compute the index of $T(A)$, let us assume that $\alpha_1, \ldots, \alpha_m$ are located as follows:

$$0 < |\alpha_j - k_j| < \frac{1}{2}, \qquad j = 1, \ldots, m,$$

where $k_1, \ldots, k_m$ are integers. Since

$$\det \widehat{\Phi}_A(\zeta, \mu) = \left\{ \prod_{j=1}^{m} \omega_j(\zeta, \mu) \right\} \det \Phi_A(1+),$$

with

$$\omega_j(\zeta, \mu) = \begin{cases} e^{it\alpha_j} & \text{for} \quad \zeta = e^{it}, \ 0 < t < 2\pi, \\ \mu + (1 - \mu)e^{2\pi i \alpha_j} & \text{for} \quad \zeta = 1, \ 0 \le \mu \le 1, \end{cases}$$

we conclude that

$$n(\det \widehat{\Phi}_A; 0) = \sum_{j=1}^{m} n(\omega_j, 0) = \sum_{j=1}^{m} k_j.$$

So Theorem 3.1 implies that the Fredholm index of A is given by

$$\operatorname{ind} T(A) = \sum_{k \in \mathbf{Z}} -m(k),$$

where $m(k)$ is the number of eigenvalues λ of A, counted according to algebraic multiplicity, such that $|\lambda - k| < \frac{1}{2}$.

XXV.4   SUMS OF PRODUCTS OF TOEPLITZ OPERATORS DEFINED BY PIECEWISE CONTINUOUS FUNCTIONS

In this section we study Fredholm properties of operators on ℓ_2 which can be represented as a sum of products of Toeplitz operators defined by functions in $PC(\mathbb{T}; \zeta_1, \ldots, \zeta_k)$. Let S be such an operator. Thus

$$(1) \qquad S = \sum_{i=1}^{p} T_{\varphi_{i1}} T_{\varphi_{i2}} \cdots T_{\varphi_{iq}},$$

where $\varphi_{ij} \in PC(\mathbb{T}; \zeta_1, \ldots, \zeta_k)$ for $i = 1, \ldots, p$ and $j = 1, \ldots, q$. Define

$$(2) \qquad \omega(\zeta, \mu) = \sum_{i=1}^{p} \widehat{\varphi}_{i1}(\zeta, \mu) \widehat{\varphi}_{i2}(\zeta, \mu) \cdots \widehat{\varphi}_{iq}(\zeta, \mu).$$

We shall refer to ω as the *symbol* of the operator S. At present it seems that the definition of the symbol depends on the representation (1), but later (Corollary 4.3 below) we shall see that this is not the case. From Section XXV.1 we know that $\omega \in C\big[\mathbb{T}[\zeta_1, \ldots, \zeta_k]\big]$. Furthermore, if $\omega(\zeta, \mu) \ne 0$ for each $\zeta \in \mathbb{T}$ and $0 \le \mu \le 1$, then the winding number $n(\omega; 0)$ is well-defined. The use of the word symbol for the function ω is justified by the next theorem.

THEOREM 4.1. *The operator S defined by (1) is Fredholm if and only if its symbol ω does not vanish on* $\mathbb{T} \times [0, 1]$. *In this case*

$$(3) \qquad \operatorname{ind} S = -n(\omega; 0).$$

If all the functions φ_{ij} appearing in (1) are continuous, then Theorem 4.1 is an easy consequence of the theory developed in Chapter XXIII. Indeed, in this case $\psi = \sum_{j=1}^{p} \prod_{i=1}^{q} \varphi_{ij}$ is continuous on $\mathbb{T}$,

$$\omega(\zeta, \mu) = \psi(\zeta), \qquad \zeta \in \mathbb{T},\ 0 \leq \mu \leq 1,$$

and $S - T_\psi$ is compact (by Corollary XXIII.4.2). Thus S is Fredholm if and only if T_ψ is Fredholm, and in this case $\operatorname{ind} S = \operatorname{ind} T_\psi$. It remains to apply Theorems 4.3 and 5.1 in Chapter XXIII to get the desired result.

For piecewise continuous φ_{ij} this argument does not work. The main obstacle is the fact that for φ and ψ in $PC(\mathbb{T}; \zeta_1, \ldots, \zeta_k)$ it may happen that $T_\varphi T_\psi - T_{\varphi\psi}$ is not compact.

To prove Theorem 4.1 we use the following linearization lemma, which will allow us to reduce Theorem 4.1 to the block Toeplitz case considered in Theorem 3.1. Since the lemma has to be applied in different contexts it is formulated for sums of products of elements from an arbitrary ring.

LEMMA 4.2. *Let $\mathcal{R}$ be a ring with unit e, and let a be a sum of products of elements in $\mathcal{R}$,*

$$(4) \qquad a = \sum_{i=1}^{p} a_{i1} a_{i2} \cdots a_{iq}.$$

Introduce matrices B and C of sizes $p \times 1$ and $1 \times p$, respectively, and $p \times p$ diagonal matrices $D_1, \ldots, D_q$ as follows:

$$B = \begin{bmatrix} e \\ e \\ \vdots \\ e \end{bmatrix}, \quad D_j = \begin{bmatrix} a_{1j} & & & \\ & a_{2j} & & \\ & & \ddots & \\ & & & a_{pj} \end{bmatrix}, \qquad j = 1, \ldots, q,$$

$$C = [e \quad e \quad \cdots \quad e],$$

and let E be the $p \times p$ diagonal matrix with e on the main diagonal. Then

$$a = C D_1 D_2 \cdots D_q B$$

and

$$(5) \qquad \begin{bmatrix} Z & X \\ Y & 0 \end{bmatrix} = \begin{bmatrix} I & 0 \\ W & e \end{bmatrix} \begin{bmatrix} I & 0 \\ 0 & a \end{bmatrix} \begin{bmatrix} Z & X \\ 0 & e \end{bmatrix},$$

where

$$Z = \begin{bmatrix} E & -D_1 & & & \\ & E & -D_2 & & \\ & & \ddots & \ddots & \\ & & & E & -D_q \\ & & & & E \end{bmatrix}, \quad X = \begin{bmatrix} 0 \\ 0 \\ \vdots \\ 0 \\ -B \end{bmatrix},$$

$$Y = [C \quad 0 \quad \cdots \quad 0],$$

$$W = [C \quad W_1 \quad \cdots \quad W_q], \qquad W_j = C D_1 D_2 \cdots D_j,$$

and I is the $(q+1) \times (q+1)$ block diagonal matrix with E on the main diagonal. The unspecified entries in the block matrix representation of Z are zero matrices.

PROOF. The representation $a = CD_1D_2 \cdots D_qB$ is trivial. To prove (5) one only has to check that

$$(6) \qquad\qquad WZ = Y, \qquad WX + a = 0.$$

The first identity in (6) follows from $W_1 = CD_1$ and $W_{j+1} = W_jD_{j+1}$ for $j = 1,\ldots,q-1$. Obviously, $WX = -W_qB = -CD_1D_2 \cdots D_qB = -a$, and hence (6) is proved. $\square$

If a is given by (4), we write $\Xi(a)$ for the 2×2 block matrix $\begin{bmatrix} Z & X \\ Y & 0 \end{bmatrix}$ in the left hand side of (5). We shall refer to $\Xi(a)$ as the *linearization* of a (relative to the representation (4)). To explain this term, assume that $\mathcal{R}$ is a ring of operators on a linear space X. Consider the equation $a(x) = y$. Since $a = CD_1D_2 \cdots D_qB$, we may introduce new unknowns $x_1,\ldots,x_{q+1}$ by setting

$$(7) \qquad\qquad x_1 = D_1x_2,\ldots,x_q = D_qx_{q+1}, \qquad x_{q+1} = Bx.$$

Obviously, (7) may be rewritten in the form

$$Z \begin{bmatrix} x_1 \\ \vdots \\ x_q \\ x_{q+1} \end{bmatrix} = \begin{bmatrix} 0 \\ \vdots \\ 0 \\ x_{q+1} \end{bmatrix} = -Xx.$$

Since $y = Cx_1$, it follows that $a(x) = y$ is equivalent to the linear equation

$$\begin{bmatrix} Z & X \\ Y & 0 \end{bmatrix} \begin{bmatrix} x_1 \\ \vdots \\ x_{q+1} \\ x \end{bmatrix} = \begin{bmatrix} 0 \\ \vdots \\ 0 \\ y \end{bmatrix}.$$

Formula (5) makes this equivalence explicit.

Note that Z in (5) is a $(q+1) \times (q+1)$ block matrix of which the entries are $p \times p$ matrices with entries from the set

$$(8) \qquad\qquad V = \{a_{ij} \mid i = 1,\ldots,p,\ j = 1,\ldots,q\} \cup \{0,e\}.$$

Similarly, X and Y are matrices with entries in V of sizes $(pq+p) \times 1$ and $1 \times (pq+p)$, respectively. It follows that $\Xi(a)$ is a square matrix of order $pq + p + 1$ and has entries in V.

Since $\Xi(A)$ is a square matrix over $\mathcal{R}$, it has a well-defined determinant (cf., formula (2) in Section XI.7) and $\det \Xi(a) \in \mathcal{R}$. In the case when $\mathcal{R}$ is commutative, formula (5) allows us to conclude that

$$(9) \qquad\qquad \det \Xi(a) = a.$$

PROOF OF THEOREM 4.1. Let S be given by (1), and consider $\Xi(S)$. Put $\ell = pq+p+1$. From the preceding remarks we know that $\Xi(S)$ is an $\ell \times \ell$ operator matrix of which the entries are Toeplitz operators defined by functions φ from $PC(\mathbb{T}; \zeta_1, \ldots, \zeta_k)$. It follows that $\Xi(S) = T_\Phi$, where $\Phi \in PC^{\ell \times \ell}(\mathbb{T})$ and Φ has a finite number of discontinuities. So we can apply Theorem 3.1 to show that $\Xi(S)$ is a Fredholm operator if and only if the determinant of

$$\widehat{\Phi}(\zeta, \mu) = \mu \Phi(\zeta) + (1 - \mu)\Phi(\zeta)$$

is different from zero for all $(\zeta, \mu) \in \mathbb{T} \times [0, 1]$.

Next, consider the linearization $\Xi(\omega)$ of ω relative to the representation (2). From the construction of $\Xi(S)$ and $\Xi(\omega)$ it is clear that $\widehat{\Phi}$ is precisely equal to $\Xi(\omega)$. Since the functions $\widehat{\varphi}_{ij}$ are scalar, they commute with one another, and hence we can apply formula (9) to show that $\det \widehat{\Phi} = \omega$.

To finish the proof we apply Lemma 4.2 to S and ω. Note that the first and third factor in the right hand side of (5) are invertible. This implies that $\Xi(S)$ is Fredholm if and only if S is Fredholm, and in that case

$$(10) \qquad \qquad \operatorname{ind} S = \operatorname{ind} \Xi(S).$$

The first part of Theorem 4.1 is now proved. It remains to establish (3). Assume that S is Fredholm. Then, according to Theorem 3.1 and formula (10), $\operatorname{ind} S = -n(\det \widehat{\Phi}; 0)$. But $\det \widehat{\Phi} = \omega$, and the proof is complete. $\square$

COROLLARY 4.3. *Let S be defined by (1). Then the symbol of S does not depend on the way S is represented as a sum of products.*

PROOF. Let ω be as in (2). We have to prove that ω does not depend on the special form of the representation of S in (1). To do this, it suffices to show that $S = 0$ implies that ω is identically zero on $\mathbb{T} \times [0, 1]$. Let λ be an arbitrary complex number, and let η be the function $\eta(\zeta) = \lambda$ for all $\zeta \in \mathbb{T}$. Then

$$\lambda - S = T_\eta - \sum_{i=1}^{p} T_{\varphi_{i1}} T_{\varphi_{i2}} \cdots T_{\varphi_{iq}},$$

and hence $\lambda - S$ is a sum of products of Toeplitz operators defined by functions from $PC(\mathbb{T}; \zeta_1, \ldots, \zeta_k)$. The corresponding symbol is the function $\alpha(\zeta, \mu) = \lambda - \omega(\zeta, \mu)$. It follows (by Theorem 4.1) that $\lambda - S$ is Fredholm if and only if $\omega(\zeta, \mu) \neq \lambda$ for all $(\zeta, \mu) \in \mathbb{T} \times [0, 1]$. Now assume that $S = 0$. Then $\lambda - S$ is Fredholm for each $\lambda \neq 0$. Hence $\omega(\zeta, \mu) = 0$ for all $(\zeta, \mu) \in \mathbb{T} \times [0, 1]$. $\square$

XXV.5 SUMS OF PRODUCTS OF BLOCK TOEPLITZ OPERATORS DEFINED BY PIECEWISE CONTINUOUS FUNCTIONS

In this section we prove the matrix version of Theorem 4.1. By $PC^{m \times m}(\mathbb{T}; \zeta_1, \ldots, \zeta_k)$ we denote the set of all $m \times m$ matrix functions Φ from $PC^{m \times m}(\mathbb{T})$

that are continuous at each $\zeta \neq \zeta_j$ $(j = 1, \ldots, k)$. Let S on ℓ_2^m be a sum of products of block Toeplitz operators:

$$(1) \qquad S = \sum_{i=1}^{p} T_{\Phi_{i1}} T_{\Phi_{i2}} \cdots T_{\Phi_{iq}},$$

with $\Phi_{ij} \in PC^{m \times m}(\mathbb{T}; \zeta_1, \ldots, \zeta_k)$ for each i and j. With S we associate the $m \times m$ matrix function

$$(2) \qquad \Omega(\zeta, \mu) = \sum_{i=1}^{p} \widehat{\Phi}_{i1}(\zeta, \mu) \widehat{\Phi}_{i2}(\zeta, \mu) \cdots \widehat{\Phi}_{iq}(\zeta, \mu).$$

For each i and j the function $\widehat{\Phi}_{ij}$ is the symbol of the block Toeplitz operator defined by Φ_{ij} (see Section XXV.3). Thus the entries of $\Omega(\zeta, \mu)$ are sums of products of symbols $\widehat{\varphi}$ with φ from $PC(\mathbb{T}; \zeta_1, \ldots, \zeta_k)$, and hence $\det \Omega$ is also a sum of products of functions of this type. It follows that the winding number $n(\det \Omega; 0)$ is well-defined if $\det \Omega(\zeta, \mu) \neq 0$ for all $(\zeta, \mu) \in \mathbb{T} \times [0, 1]$.

THEOREM 5.1. *Let S on ℓ_2^m be a sum of products of block Toeplitz operators as in (1), and let Ω be given by (2). Then S is a Fredholm operator if and only if $\det \Omega(\zeta, \mu) \neq 0$ for all $\zeta \in \mathbb{T}$ and $0 \leq \mu \leq 1$. In that case*

$$(3). \qquad \operatorname{ind} S = -n(\det \Omega; 0).$$

PROOF. We apply Lemma 4.2 to reduce the theorem to Theorem 3.1. Let S be given by (1), and let $\Xi(S)$ be the linearization of S relative to the representation (1). Put $\ell = pq + p + 1$. Then $\Xi(S)$ is an $\ell \times \ell$ operator matrix of which the entries are block Toeplitz operators T_Φ defined by Φ from $PC^{m \times m}(\mathbb{T}; \zeta_1, \ldots, \zeta_k)$. Such a Φ is of the form $\Phi = [\varphi_{ij}]_{i,j=1}^{m}$, where $\varphi_{ij} \in PC(\mathbb{T}; \zeta_1, \ldots, \zeta_k)$ for each i and j, and hence

$$T_\Phi = \begin{bmatrix} T_{\varphi_{11}} & \cdots & T_{\varphi_{1m}} \\ \vdots & & \vdots \\ T_{\varphi_{m1}} & \cdots & T_{\varphi_{mm}} \end{bmatrix}.$$

This allows us to consider $\Xi(S)$ as an $m\ell \times m\ell$ matrix of which the entries are Toeplitz operators defined by functions from $PC(\mathbb{T}; \zeta_1, \ldots, \zeta_k)$. Thus $\Xi(S) = T_\Psi$ with Ψ from $PC^{m\ell \times m\ell}(\mathbb{T}; \zeta_1, \ldots, \zeta_k)$. But then we can apply Theorem 3.1 to show that $\Xi(S)$ is a Fredholm operator if and only if the determinant of

$$\widehat{\Psi}(\zeta, \mu) = \mu \Psi(\zeta+) + (1 - \mu)\Psi(\zeta)$$

is different from zero for all $(\zeta, \mu) \in \mathbb{T} \times [0, 1]$.

Next, consider the linearization $\Xi(\Omega)$ of Ω relative to the representation (2). Thus $\Xi(\Omega)$ is an $\ell \times \ell$ matrix of which each entry belongs to $PC^{m \times m}(\mathbb{T}; \zeta_1, \ldots, \zeta_k)$. In particular, each entry of $\Xi(\Omega)$ is an $m \times m$ matrix. It follows that $\Xi(\Omega)$ may also be viewed as an $m\ell \times m\ell$ matrix with entries from $PC(\mathbb{T}; \zeta_1, \ldots, \zeta_k)$. From the construction of the $m\ell \times m\ell$ matrix functions $\Xi(S)$ and $\Xi(\Omega)$ it is clear that $\widehat{\Psi} = \Xi(\Omega)$.

Next, we use formula (5) in Section XXV.4 for $a = \Omega$. Note that each factor in this formula is an $\ell \times \ell$ matrix of which the entries are $m \times m$ matrix functions. So each factor may be considered as an $m\ell \times m\ell$ matrix with entries from $PC(\mathbb{T}; \zeta_1, \ldots, \zeta_k)$. The first and third factor in the right hand side of formula (5) in Section XXV.4 are lower and upper triangular, respectively, and their diagonal elements are all equal to the constant (scalar) function 1. It follows that

$$(4) \qquad \det \widehat{\Psi} = \det \Xi(\Omega) = \det \Omega.$$

To finish the proof we apply Lemma 4.2 to both S and Ω. Note that the first and third factor in the right hand side of formula (5) in Section XXV.4 are invertible. This implies that $\Xi(S)$ is Fredholm if and only if S is Fredholm, and in this case

$$(5) \qquad \operatorname{ind} S = \operatorname{ind} \Xi(S).$$

It is now clear that S is Fredholm if and only if $\det \Omega$ does not vanish on $\mathbb{T} \times [0,1]$. To get the formula for the index, assume that S is Fredholm. Then, according to formula (5) and Theorem 3.1, $\operatorname{ind} S = -n(\det \widehat{\Psi}; 0)$. But $\det \widehat{\Psi} = \det \Omega$ by (4), and the proof is complete. $\square$

Let S be as in (1) and Ω as in (2). We shall refer to Ω as the *symbol* of the operator S. The definition of the symbol does not depend on the particular representation of S as a sum of products. To see this, recall that S acts on ℓ_2^m. So we may write S as an $m \times m$ operator matrix of which the entries act on ℓ_2,

$$S = \begin{bmatrix} S_{11} & \cdots & S_{1m} \\ \vdots & & \vdots \\ S_{m1} & \cdots & S_{mm} \end{bmatrix}.$$

Fix i and j. The entry S_{ij} is a sum of products of Toeplitz operators defined by functions from $PC(\mathbb{T}; \zeta_1, \ldots, \zeta_k)$. Let ω_{ij} be the corresponding symbol (see Section XXV.4). From the way Ω and S are built it is clear that ω_{ij} is precisely the (i,j)-th entry of Ω. Note that S_{ij} is uniquely determined by S, the symbol ω_{ij} is uniquely determined by S_{ij} (Corollary 4.3), and hence the (i,j)-th entry of Ω is uniquely determined by S. It follows that Ω does not depend on the particular representation of S as a sum of products.

COMMENTS ON PART VI

Toeplitz operators have become a popular topic in mathematical analysis and its applications. These operators form a non-trivial and relatively easy to handle class of operators in which many operator theoretical operations can be performed explicitly.

For instance, as we have seen, for Toeplitz operators there is an algorithm to compute the inverse when it exists. Also, the class of Toeplitz operators belongs to one of the first classes of operators for which an explicit geometrical formula for the Fredholm index has been found. Furthermore, Toeplitz operators form one of the few classes of non-selfadjoint operators which allow an explicit description of the spectrum. On the other hand, these operators and their matrix analogues appear in different branches of mathematics and applications; good lists of references may be found in Böttcher-Silbermann [1] and Heinig-Rost [1].

The material presented in this part forms a concise introduction to the general theory of block Toeplitz operators on Hilbert spaces. The first chapter and the first four sections of the second chapter are based on Gohberg-Kreĭn [2] and use also some elements from Gohberg-Fel'dman [1]. Section XXIV.2 has its origin in M.G. Kreĭn [3]. The sections 5–10 of Chapter XXIV, which concern block Toeplitz operators defined by rational matrix functions, are taken from Gohberg-Kaashoek [3]. The last section of Chapter XXIV has its origin in Gohberg-Kaashoek-Van Schagen [1]. Chapter XXV is based on Gohberg-Krupnik [1] (see also Gohberg-Krupnik [2] and Chapter IV in Gohberg-Fel'dman [1]). Banach algebras generated by Toeplitz operators defined by piecewise continuous functions will be one of the topics in the Banach algebra part.

For additional reading on Toeplitz operators we recommend the books Böttcher-Silbermann [1], Douglas [1], Gohberg-Krupnik [4], [5], Nikolskii [1], Rosenblum-Rovnyak [1]. For Toeplitz operators defined by functions in several variables, the so-called multi-dimensional case, we refer to Böttcher-Silbermann [1], Boutet de Monvel-Guillemin [1], and the survey paper Guillemin [1].

EXERCISES TO PART VI

1. Let A on ℓ_2 be given by

$$
A = \begin{bmatrix}
-7 & 2 & 0 & 0 & \cdots \\
3 & -7 & 2 & 0 & \cdots \\
0 & 3 & -7 & 2 & \cdots \\
0 & 0 & 3 & 7 & \cdots \\
\vdots & \vdots & \vdots & \vdots & \ddots
\end{bmatrix} : \ell_2 \to \ell_2.
$$

(a) Show that A is invertible and construct its inverse.

(b) Solve in ℓ_2 the equation $Ax = y$ for

    (i) $y = (1, 0, 0, \ldots)$,

    (ii) $y = \left(1, \frac{1}{2}, \left(\frac{1}{2}\right)^2, \ldots\right)$,

    (iii) $y = (1, q, q^2, \ldots)$, where $|q| < 1$.

2. Let α be a complex parameter, and consider on ℓ_2 the operator B_α given by

$$
B_\alpha = \begin{bmatrix}
1 + 2\alpha^2 & \alpha & 0 & 0 & \cdots \\
2\alpha & 1 + 2\alpha^2 & \alpha & 0 & \cdots \\
0 & 2\alpha & 1 + 2\alpha^2 & \alpha & \cdots \\
0 & 0 & 2\alpha & 1 + 2\alpha^2 & \cdots \\
\vdots & \vdots & \vdots & \vdots & \ddots
\end{bmatrix}.
$$

Investigate for different values of α the invertibility of B_α (left, right, two-sided or non invertible), construct the inverse if it exists, describe the image and kernel of B_α.

3. Consider for $a = e^\lambda$, $\Re\lambda < 0$, the equation

$$
\sum_{k=0}^{\infty} a^{|j-k|} x_k = q^j, \qquad j = 0, 1, 2, \ldots,
$$

where $|q| < 1$. Show that the equation is uniquely solvable in ℓ_2, and construct its solution.

4. Let T be the Toeplitz operator on ℓ_2 defined by

$$
\varphi(\lambda) = \frac{2\lambda^2 + 5\lambda + 1}{2\lambda^2 - 5\lambda + 1}.
$$

    (a) Prove that T is invertible, and construct its inverse by using the method of Theorem XXIV.2.1.

    (b) Construct a realization of φ of the form

$$
\varphi(\lambda) = 1 + C(\lambda I - A)^{-1} B,
$$

and construct the inverse of T via the method of Theorem XXIV.8.1.

5. Let A be the operator on ℓ_2 given by

$$
A = \begin{bmatrix}
0 & -1 & -\frac{1}{2} & -\frac{1}{3} & \cdots \\
1 & 0 & -1 & -\frac{1}{2} & \cdots \\
\frac{1}{2} & 1 & 0 & -1 & \cdots \\
\frac{1}{3} & \frac{1}{2} & 1 & 0 & \cdots \\
\vdots & \vdots & \vdots & \vdots & \ddots
\end{bmatrix}.
$$

(a) Show that A is a well-defined bounded operator and compute $\|A\|$.

(b) Let A_+ (resp., A_-) be the infinite upper (resp., lower) triangular matrix which one obtains from A by replacing the entries of A below (resp., above) its main diagonal by zero. Show that $A_\pm$ do not define bounded linear operators on ℓ_2.

(c) Let $P_0, P_1, \ldots$ be the orthogonal projections on ℓ_2 defined by

$$
P_n(x_1, x_2, \ldots) = (x_1, \ldots, x_n, 0, 0, \ldots), \qquad n = 0, 1, 2, \ldots,
$$

and let $P_\infty = I$. Show that A does not admit an additive LU-decomposition (see Section XX.7) relative to the chain $\mathbb{P} = \{P_n\}_{n=0}^\infty$.

6. Let L be the block Laurent operator on $\ell_2^n(\mathbb{Z})$ defined by

$$
(Lx)_i = Ax_{i+1}, \qquad i \in \mathbb{Z},
$$

where A is an $n \times n$ matrix. Describe the spectrum of L in terms of A.

7. Let T be the block Toeplitz operator on ℓ_2^n defined by

$$
(Tx)_i = Ax_{i+1}, \qquad i = 0, 1, 2, \ldots .
$$

Describe (in terms of A) the spectrum and the essential spectrum of T. If $cI - T$ is Fredholm ($c \in \mathbb{C}$), compute its Fredholm index, the dimension of its kernel and the codimension of its image.

8. Let T be the block Toeplitz operator on ℓ_2^n defined by

$$
\varphi(\lambda) = \sum_{k=-\infty}^{\infty} \lambda A^{|k|},
$$

where A is an $n \times n$ matrix with all its eigenvalues in $\mathbb{D}$. Prove that T is invertible, and construct its inverse.

9. Let $L = [A_{i-j}]_{i,j=-\infty}^\infty$ be a block Laurent operator on $\ell_2^m(\mathbb{Z})$ which is lower triangular (i.e., $A_k = 0$ for $k < 0$), and let T be the corresponding Toeplitz operator on ℓ_2^m. Show that T is invertible if and only if L is invertible and L^{-1} is lower triangular.

10. Consider the matrix function

$$
\Phi(\lambda) = \begin{bmatrix}
\lambda^n & (\lambda - \frac{1}{2})(\lambda - 2)^{-1} \\
0 & 1
\end{bmatrix},
$$

where n is an integer. Construct a right and a left Wiener-Hopf factorization of Φ relative to $\mathbb{T}$.

11. Do the problem of the previous exercise for the matrix function

$$\Phi(\lambda) = \begin{bmatrix} \lambda^n & (\lambda - \tfrac{1}{2})(\lambda - 2)^{-1} \\ 0 & \lambda^{-n} \end{bmatrix},$$

where, as before, n is an integer.

12. Construct a right and a left Wiener-Hopf factorization relative to $\mathbb{T}$ for

$$\Phi(\lambda) = \begin{bmatrix} \lambda^n & (\lambda - \tfrac{1}{2})(\lambda - 2)^{-1} \\ 0 & \lambda^m \end{bmatrix}.$$

Here n and m are integers.

13. Consider the following input/output system:

$$\Delta \begin{cases} x_{n+1} = Ax_n + Bu_n, & n \in \mathbb{Z}, \\ y_n = Cx_n + u_n. \end{cases}$$

Here A, B and C are matrices of sizes $n \times n$, $n \times m$ and $m \times n$, respectively, and the matrix A has no eigenvalue on $\mathbb{T}$.

(a) Show that for each sequence $\vec{u} = (u_n)_{n \in \mathbb{Z}}$ in $\ell_2^m(\mathbb{Z})$ there exists a unique sequence $\vec{x} = (x_n)_{n \in \mathbb{Z}}$ in $\ell_2^n(\mathbb{Z})$ satisfying the first equation in Δ.

(b) Let $\vec{y} = (y_n)_{n \in \mathbb{Z}}$ be the sequence which one obtains by inserting the solution of the first equation into the second. The corresponding map $\vec{u} \mapsto \vec{y}$ is called the input-output map of Δ, and is denoted by L_Δ. Prove that L_Δ is a block-Laurent operator, and compute its defining function.

(c) Let $\Delta^\times$ be the system which one obtains if in Δ the role of input and output are interchanged, i.e.,

$$\Delta^\times \begin{cases} x_{n+1} = (A - BC)x_n + By_n, & n \in \mathbb{Z}, \\ u_n = -Cx_n + y_n. \end{cases}$$

Is $T_{\Delta^\times}$ the inverse of T_Δ? If so, prove this statement. If not, find necessary and sufficient conditions in order that the statement is true.

14. Consider the following input-output system:

$$\Sigma_0 \begin{cases} x_{n+1} = Ax_n + Bu_n, & 0, 1, 2, \ldots, \\ y_n = Cx_n + u_n, \\ x_0 \in \operatorname{Ker} P. \end{cases}$$

Here A, B and C are as in the previous exercise, and P is the Riesz projection of A corresponding to the eigenvalues of A in $\mathbb{D}$.

(a) Show that for each sequence $\vec{u} = (u_0, u_1, \ldots)$ in ℓ_2^m there exists a unique sequence $\vec{x} = (x_0, x_1, \ldots)$ in ℓ_2^n satisfying the first equation in Σ_0 and the initial condition.

(b) Let $\vec{y} = (y_0, y_1, \ldots)$ be the sequence which one obtains by inserting the solution of the first equation in Σ_0 into the second. The map $\vec{u} \mapsto \vec{y}$ is called the input-output map for Σ_0. Prove that the input-output of Σ_0 is a block Toeplitz operator and computes its defining function.

15. Consider the following input-output system:

$$\Sigma \begin{cases} x_{n+1} = Ax_n + Bu_n, & n = 0, 1, 2, \ldots, \\ y_n = Cx_n + u_n, \\ x_0 \in L. \end{cases}$$

Here A, B and C are as in Exercise 13, and L is a subspace of $\mathbb{C}^n$ such that

$$\mathbb{C}^n = L \oplus \operatorname{Im} P,$$

where P is the Riesz projection of A corresponding to the eigenvalues of A in $\mathbb{D}$.

(a) Do problem (a) in the previous exercise for Σ in place of Σ_0.

(b) As in (b) of the previous exercise, show that Σ has a well-defined input-output map T_Σ, which is a bounded linear operator on ℓ_2^m. Is T_Σ a block Toeplitz operator?

(c) Let $\Sigma^\times$ be the system which one obtains if in Σ the role of the input and output are interchanged, i.e.,

$$\Sigma^\times \begin{cases} x_{n+1} = (A - BC)x_n + By_n, & n = 0, 1, 2, \ldots, \\ u_n = -Cx_n + y_n, \\ x_0 \in L. \end{cases}$$

Assume that

(i) $A - BC$ has no eigenvalue on $\mathbb{T}$,

(ii) $\mathbb{C}^n = L \oplus \operatorname{Im} P^\times$,

where $P^\times$ is the Riesz projection of $A - BC$ corresponding to its eigenvalues in $\mathbb{D}$. Prove that T_Σ is invertible and $T_\Sigma^{-1} = T_{\Sigma^\times}$.

(d) Show that T_Σ is Fredholm if and only if condition (i) in (c) is fulfilled. Find a formula for the Fredholm index.

(e) Show that for T_Σ to be invertible it is necessary and sufficient that conditions (i) and (ii) in (c) are fulfilled.

(f) Specify the above for the system Σ_0 of the previous exercise, and compare your results with those of Sections XXIV.8 and XXIV.9.

16. Let $W(\lambda) = I + \lambda C(\lambda G - A)^{-1}B$, where G and A are square matrices of order n, say, $\det(\lambda G - A) \neq 0$ for $\lambda \in \mathbb{T}$, and the matrices B and C have sizes $n \times m$ and $m \times n$,

respectively. Find a necessary and sufficient condition in order that Φ admits a right canonical factorization relative to $\mathbb{T}$, and compute the factors in such a factorization. (Hint: use the transformation $\lambda \mapsto \lambda^{-1}$.)

17. Let T be the Toeplitz operator on ℓ_2 defined by the function $\varphi(e^{it}) = \exp\left(\frac{1}{2}it\right)$ for $0 \le t < 2\pi$. Compute the entries of the infinite matrix representing T. Find the spectrum and essential spectrum of T. If $cI - T$ is Fredholm $(c \in \mathbb{C})$, compute its Fredholm index.

18. Consider the scalar rational function

$$\varphi(\zeta) = \frac{2\zeta^2 + 5\zeta + 2}{2\zeta^2 - 5\zeta + 2}.$$

Show that φ admits a representation as in formula (2) of Section XXIV.11, with

$$A = \begin{bmatrix} \frac{1}{2} & 0 \\ 0 & 2 \end{bmatrix}, \qquad B = \begin{bmatrix} 1 \\ 1 \end{bmatrix}, \qquad C = \begin{bmatrix} -\frac{5}{3} & \frac{20}{3} \end{bmatrix}.$$

Let T_N be the N-th section of the Toeplitz operator defined by φ. Use Theorem XXIV.11.1 to show that

$$\lim_{N \to \infty} (-1)^N \det T_N = \frac{25}{9}.$$

19. Let T_1 and T_2 be block Toeplitz operators on ℓ_2^m defined by the rational matrix functions $C_1(\zeta - A_1)^{-1}B_1$ and $C_2(\zeta - A_2)^{-1}B_2$, respectively. Here A_1 and A_2 are square matrices which have no eigenvalue on $\mathbb{T}$. Assume $B_2 C_1 = 0$. Show that the operator $I - T_1 T_2$ is Fredholm. Analyse the invertibility of $I - T_1 T_2$ in terms of the given realizations.

20. Let T be the (block) Toeplitz operator on ℓ_2^m defined by

$$\Phi(e^{it}) = 2\pi i(e^{2\pi iA} - I)^{-1}e^{itA}, \qquad 0 < t \le 2\pi,$$

where A is an $m \times m$ matrix whose eigenvalues are real and not in $\mathbb{Z}$. Analyse invertibility and Fredholm properties of the operator $I - T^2$. First consider the case when $m = 1$.

PART VII

CONTRACTIVE OPERATORS AND
CHARACTERISTIC OPERATOR FUNCTIONS

One of the important directions of research in modern operator theory deals with unitary invariants, analysis and construction of invariant subspaces, and triangular representations of operators, with operator models and characteristic operator functions as the main tools. This part of the book presents a concise introduction to these topics. In order to give a good cross-section of the main ideas and results involved the presentation is restricted to the class of contractive operators, which are viewed here as "perturbations" of unitary operators with the defect operators providing the measure of deviation from unitary.

This part starts with the study of shift operators and their invariant subspaces (Chapter XXVI). Dilation to unitary operators and the commutant lifting theorem are the main topics in Chapter XXVII. The results are illustrated on applications to interpolation problems of Nevanlinna-Pick type. The main body of this part (Chapter XXVIII) is devoted to the theory of unitary systems and their transfer functions (characteristic operator functions), and to model theory. Connections with characteristic operator functions for other classes of operators are also considered.

CHAPTER XXVI
BLOCK SHIFT OPERATORS

This chapter, which has an introductory character, contains the main elements of the theory of block shift operators. These operators are among the simplest infinite dimensional operators, and they may serve as building blocks for more complicated operators. In the first section block forward shifts are identified as pure isometries. In the second section it is shown that block backward shifts provide universal models for arbitrary operators. The third section describes the invariant subspaces of $\mathbb{C}^m$-block forward shifts. Throughout this chapter all operators are Hilbert space operators.

XXVI.1 FORWARD SHIFTS AND ISOMETRIES

Throughout this section H is a (complex) Hilbert space. Recall (see [GG], Section VIII.3) that $A \in \mathcal{L}(H)$ is called an *isometry* if $\|Ax\| = \|x\|$ for each $x \in H$. We say that A is a *pure isometry* if, in addition,

$$(1) \qquad \bigcap \{ \operatorname{Im} A^k \mid k = 0, 1, 2, \ldots \} = \{0\}.$$

This terminology is justified by the following theorem.

THEOREM 1.1. (Von Neumann-Wold decomposition). *Let $A \in \mathcal{L}(H)$ be an isometry. Then there exists a unique subspace L of H such that the following conditions are satisfied:*

(i) *$AL \subset L$, $AL^\perp \subset L^\perp$,*

(ii) *$A|L$ is a pure isometry,*

(iii) *$A|L^\perp$ is a unitary operator.*

Furthermore, $L = \mathcal{R}^\perp$, where $\mathcal{R} = \bigcap \{ \operatorname{Im} A^k \mid k \geq 0 \}$.

PROOF. It is straightforward to check that the image of an isometry is closed (cf., Theorem XI.2.1). It follows that $\operatorname{Im} A^k$ is closed for each k. Therefore, $\mathcal{R}$ is a closed linear submanifold of H. From $A(\operatorname{Im} A^k) = \operatorname{Im} A^{k+1} \subset \operatorname{Im} A^k$ it follows that $\mathcal{R}$ is invariant under A. We shall prove that $A\mathcal{R} = \mathcal{R}$. Indeed, take $y \in \mathcal{R}$. Then $y \in \operatorname{Im} A$, and hence there exists $z \in H$ such that $Az = y$. Since A is an isometry, A is injective, and hence z is uniquely determined by y. Now, $y \in \operatorname{Im} A^{k+1}$ for any $k \geq 0$. Thus $y = A^{k+1} x_{k+1}$ for some x_{k+1} in H. By virtue of the uniqueness, $z = A^k x_{k+1}$, and thus $z \in \operatorname{Im} A^k$. This holds for each $k \geq 0$, and therefore $z \in \mathcal{R}$. From $A\mathcal{R} = \mathcal{R}$ and A is an isometry, we conclude that $A|\mathcal{R}$ is unitary.

Since A is an isometry, A preserves the inner product $\langle \cdot, \cdot \rangle$ on H. This follows from the so-called polar identity,

$$(2) \quad \langle x, y \rangle = \frac{1}{4} \{ \langle x + y, x + y \rangle - \langle x - y, x - y \rangle + i \langle x + iy, x + iy \rangle - i \langle x - iy, x - iy \rangle \},$$

which holds for each x and y in H. If in (2) the vectors x and y are replaced by Ax and Ay, then the right hand side of (2) does not change because A is an isometry. So

$$(3) \qquad \langle Ax, Ay \rangle = \langle x, y \rangle \qquad (x, y \in H).$$

Now, take $x \perp \mathcal{R}$, then $Ax \perp A\mathcal{R}$. But $A\mathcal{R} = \mathcal{R}$. Thus $L := \mathcal{R}^{\perp}$ is invariant under A.

We have now proved the statements (i) and (iii). Next, we prove (ii). Let

$$x \in \bigcap \{ \mathrm{Im}(A|L)^k \mid k \geq 0 \}.$$

Then, obviously, $x \in \mathrm{Im}\, A^k$ for all $k \geq 0$, and hence $x \in \mathcal{R}$. Thus $x \in L \cap \mathcal{R} = \{0\}$. It follows that (1) holds with $A|L$ in place of A. So $A|L$ is a pure isometry.

It remains to prove the uniqueness statement. Let N be a second subspace of H such that (i), (ii) and (iii) hold with N in place of L. Since $A|N^{\perp}$ is unitary, $A^k N^{\perp} = N^{\perp}$ for each $k \geq 0$, and hence $N^{\perp} \subset \mathcal{R}$. Put $H_0 = N \cap \mathcal{R}$. Then H_0 is invariant under A and A^*, and $A|H_0$ is a unitary operator, because $A|\mathcal{R}$ is unitary. It follows that $A^k H_0 = H_0$ for each $k \geq 0$, and thus

$$H_0 \subset \bigcap \{ \mathrm{Im}(A|N)^k \mid k = 0, 1, 2, \ldots \}.$$

But the latter space consists of the zero vector only, and hence $H_0 = \{0\}$. So $N^{\perp} = \mathcal{R}$, and therefore $N = \mathcal{R}^{\perp} = L$. $\square$

Examples of isometries are provided by block shift operators. To define the latter operators we need some notation. Let K be a Hilbert space. By $\ell_2(K)$ we denote the space of all infinite sequences $x = (x_0, x_1, \ldots)$ with entries $x_j \in K$ that are square summable in norm, i.e., $\sum_{j=0}^{\infty} \|x_j\|^2 < \infty$. The space $\ell_2(K)$ is a Hilbert space with inner product and norm being given by

$$(4a) \qquad \langle x, y \rangle = \sum_{j=0}^{\infty} \langle x_j, y_j \rangle,$$

$$(4b) \qquad \|x\| = \left(\sum_{j=0}^{\infty} \|x_j\|^2 \right)^{1/2}.$$

The inner product in the right hand side of (4a) is the inner product in K, and the norm in the right hand side of (4b) is the norm in K. Instead of $\ell_2(K)$ we also write $\overset{\infty}{\underset{0}{\boxplus}} K$. The *forward shift* on $\ell_2(K)$ is operator V on $\ell_2(K)$ defined by

$$(5) \qquad V(x_0, x_1, x_2, \ldots) = (0, x_0, x_1, \ldots).$$

Its adjoint, whose action is given by

$$(6) \qquad V^*(x_0, x_1, x_2, \ldots) = (x_1, x_2, x_3, \ldots),$$

is called the *backward shift* on $\ell_2(K)$. To distinguish V from the usual forward shift on ℓ_2 we refer to V as the *K-block forward shift*.

A block forward shift is a pure isometry. The next theorem shows that the converse is true up to unitary equivalence. Two Hilbert space operators, $A_1 \in \mathcal{L}(H_1)$ and $A_2 \in \mathcal{L}(H_2)$ are said to be *unitarily equivalent* if there exists a unitary operator $U\colon H_1 \to H_2$ such that $UA_1 = A_2U$.

THEOREM 1.2. *Let* $A \in \mathcal{L}(H)$ *be a pure isometry, and put* $M = \mathrm{Ker}\, A^*$. *Then* A *is unitarily equivalent to the forward shift on* $\ell_2(M)$.

PROOF. Put $M_j = A^j M$ for $j = 0, 1, 2, \dots$. In particular, $M_0 = M = \mathrm{Ker}\, A^*$. First we prove that

$$(7) \qquad\qquad M_i \perp M_j, \qquad i \neq j.$$

Assume $j > i \geq 0$. Take $x \in M_i$ and $y \in M_j$. Thus $x = A^i x_0$ and $y = A^j y_0$ with x_0 and y_0 in $\mathrm{Ker}\, A^*$. Since $j > i$, the vector $A^{j-i} y_0 \in \mathrm{Im}\, A$. But $\mathrm{Im}\, A \perp \mathrm{Ker}\, A^*$, and hence $\langle x_0, A^{j-i} y_0 \rangle = 0$. Now use that an isometry preserves the inner product. It follows that

$$0 = \langle x_0, A^{j-i} x_0 \rangle = \langle A^i x_0, A^j y_0 \rangle = \langle x, y \rangle,$$

which proves (7).

Next, consider the operator

$$(8) \qquad\qquad J\colon \ell_2(M) \to H, \qquad J(x_0, x_1, \dots) = \sum_{j=0}^{\infty} A^j x_j.$$

We have to prove that J is well-defined. Fix $x = (x_0, x_1, \dots)$ in $\ell_2(M)$, and put $s_n = \sum_{j=0}^{n} A^j x_j$ for $n \geq 0$. Take $m > n$. Since $A^i x_i$ is orthogonal to $A^j x_j$ for $i \neq j$ (by virtue of (7)), we have

$$\|s_m - s_n\|^2 = \Big\| \sum_{j=n+1}^{m} A^j x_j \Big\|^2 = \sum_{j=n+1}^{m} \|A^j x_j\|^2.$$

Now, recall that A is an isometry. So $\|A^j x_j\| = \|x_j\|$ for each $j \geq 0$. Furthermore, the series $\sum_{j=0}^{\infty} \|x_j\|^2$ is convergent, because $x \in \ell_2(M)$. It follows that

$$\|s_m - s_n\|^2 = \sum_{j=n+1}^{m} \|x_j\|^2 \leq \sum_{j=n+1}^{\infty} \|x_j\|^2 \to 0 \qquad (n \to \infty).$$

Thus (s_n) is a Cauchy sequence in H. We conclude that $\lim_{n \to \infty} s_n$ exists, and hence the series in (8) is convergent. Furthermore,

$$\|Jx\|^2 = \lim_{n \to \infty} \|s_n\|^2 = \lim_{n \to \infty} \Big\| \sum_{j=0}^{n} A^j x_j \Big\|^2$$

$$= \lim_{n \to \infty} \sum_{j=0}^{n} \|x_j\|^2 = \sum_{j=0}^{\infty} \|x_j\|^2 = \|x\|^2,$$

which shows that J is an isometry.

We want to show that J is unitary. Since J is an isometry, $\operatorname{Im} J$ is closed. So in order to prove that J is unitary it suffices to show that $y \perp \operatorname{Im} J$ implies $y = 0$. Since A is an isometry, $\operatorname{Im} A^j$ is closed for each j. Let us show for $j \geq 1$ that

$$(9) \qquad\qquad H = M_0 \oplus \cdots \oplus M_{j-1} \oplus \operatorname{Im} A^j,$$

and that the spaces in this decomposition are mutually orthogonal. Note that

$$M_0^\perp = (\operatorname{Ker} A^*)^\perp = \overline{\operatorname{Im} A} = \operatorname{Im} A,$$

and thus (9) holds for $j = 1$. We proceed by induction. Assume (9) holds for some $j \geq 1$. Apply A to both sides of (9), and recall that A preserves orthogonality. It follows that

$$\operatorname{Im} A = M_1 \oplus \cdots \oplus M_j \oplus \operatorname{Im} A^{j+1}$$

and the spaces in this decomposition are mutually orthogonal. Since $H = M_0 \oplus \operatorname{Im} A$, we conclude that (9) holds with j replaced by $j + 1$ and with the desired orthogonality of the subspaces. Now, take $y \perp \operatorname{Im} J$. From the definition of J in (8) we see that $M_j = A^j M \subset \operatorname{Im} J$ for each j. Thus

$$y \perp M_0 \oplus \cdots \oplus M_{j-1}, \qquad j \geq 1.$$

But then the orthogonality of the subspaces in (9) implies that $y \in \operatorname{Im} A^j$ for each $j \geq 0$. But A is a pure isometry, i.e., (1) holds. Hence $y = 0$, and therefore J is unitary.

Let V be the forward shift on $\ell_2(M)$, and take $x = (x_0, x_1, \ldots) \in \ell_2(M)$. Then

$$JVx = J(0, x_0, x_1, \ldots) = \sum_{j=1}^{\infty} A^j x_{j-1} = A\left(\sum_{j=1}^{\infty} A^{j-1} x_{j-1}\right) = AJx,$$

which shows that A is unitarily equivalent to V. $\square$

COROLLARY 1.3. *An isometry is unitarily equivalent to a Hilbert space direct sum of a block forward shift and a unitary operator.*

Before we prove the corollary let us first explain the terminology. Let $A_1 \in \mathcal{L}(H_1)$ and $A_2 \in \mathcal{L}(H_2)$ be Hilbert space operators. By $H_1 \boxplus H_2$ we denote the *Hilbert space direct sum* of H_1 and H_2, and $A_1 \boxplus A_2$ stands for the operator

$$A_1 \boxplus A_2 = \begin{bmatrix} A_1 & 0 \\ 0 & A_2 \end{bmatrix} : H_1 \boxplus H_2 \to H_1 \boxplus H_2.$$

Note that $H_1 \boxplus H_2$ consists of all pairs $\binom{x_1}{x_2}$ with $x_1 \in H_1$ and $x_2 \in H_2$, and the inner product on $H_1 \boxplus H_2$ is given by

$$\left\langle \begin{pmatrix} x_1 \\ x_2 \end{pmatrix}, \begin{pmatrix} y_1 \\ y_2 \end{pmatrix} \right\rangle = \langle x_1, y_1 \rangle + \langle x_2, y_2 \rangle.$$

The operator $A_1 \boxplus A_2$ assigns to the vector $\binom{x_1}{x_2}$ the vector $\binom{A_1 x_1}{A_2 x_2}$. We shall refer to $A_1 \boxplus A_2$ as the *Hilbert space direct sum* of A_1 and A_2. If L is a subspace of the Hilbert space H, then we identify the space $L \boxplus L^\perp$ with H by identifying $\binom{\ell}{k}$ with $\ell + k$ for $\ell \in L$ and $k \in L^\perp$. Similar notation and terminology will be used for any finite number of spaces and operators.

PROOF OF COROLLARY 1.3. Define L as in Theorem 1.1. Put $A_1 = A|L$ and $U = A|L^\perp$. Now A_1 is a pure isometry. So A_1 is unitarily equivalent to the forward shift V on $\ell_2(M)$, where $M = \operatorname{Ker} A_1^*$. According to Theorem 1.1 we have $A = A_1 \boxplus U$. It follows that A is unitarily equivalent to $V \boxplus U$. Since U is unitary, this proves the corollary. $\square$

Let $A \in \mathcal{L}(H)$ be an isometry. A subspace M of H is called *wandering* for A if $A^i M \perp A^j M$ for each $i \neq j$. Since A preserves the inner product the latter condition is equivalent to the requirement that M is orthogonal to each of the spaces $A^i M$ $(i \geq 1)$. The first paragraph of the proof of Theorem 1.2 shows that $\operatorname{Ker} A^*$ is a wandering subspace for A.

XXVI.2 PARTS OF BLOCK SHIFT OPERATORS

Let $A \in \mathcal{L}(H)$ and $T \in \mathcal{L}(K)$ be Hilbert space operators. We call A a *part* of T if H is a subspace of K which is invariant under T and $T|H = A$. Thus, if A is a part of T, then T has the following 2×2 operator matrix representation:

$$(1) \qquad T = \begin{bmatrix} A & * \\ 0 & * \end{bmatrix} : H \boxplus H^\perp \to H \boxplus H^\perp.$$

One way to study operators is to see them (if possible) as parts of simpler operators. For that reason one wants to know what kind of parts a (block) shift operator can have. If T is a (pure) isometry, then (1) shows that any part of T is again a (pure) isometry. Thus Theorem 1.2 tells us that a part of a block forward shift is (up to unitary equivalence) again a block forward shift. The next theorem shows that for block backward shifts the situation is very different.

THEOREM 2.1. *An operator $A \in \mathcal{L}(H)$ is unitarily equivalent to a part of a block backward shift if and only if the following two conditions are fulfilled:*

(i) $\|A\| \leq 1$,

(ii) $A^n x \to 0$ $(n \to \infty)$ *for each* $x \in H$.

PROOF. First we prove the necessity of the conditions (i) and (ii). Let V be the forward shift on $\ell_2(K)$, and let $M \subset \ell_2(K)$ be an invariant subspace of V^*. Assume A is unitarily equivalent to $V^*|M$. Note that $\|V^*\| = \|V\| = 1$, and hence $\|V^*|M\| \leq 1$. Since $A = J^{-1}(V^*|M)J$ with J unitary, we also have $\|A\| \leq 1$. Next, let $x = (x_0, x_1, x_2, \ldots) \in \ell_2(K)$. Then

$$(V^*)^n x = (x_n, x_{n+1}, \ldots),$$

and thus

$$(2) \qquad \|(V^*)^n x\| = \left(\sum_{j=n}^{\infty} \|x_j\|^2 \right)^{1/2} \to 0 \qquad (n \to \infty).$$

Now, take $y \in H$ and put $x = Jy$. Then $x \in M$, and

$$A^n y = \left\{ J^{-1}(V^*|M)J \right\}^n y = J^{-1}(V^*|M)^n Jy$$
$$= J^{-1}(V^*|M)^n x = J^{-1}(V^*)^n x \to 0 \qquad (n \to \infty),$$

by virtue of (2). Thus (ii) is also fulfilled.

To prove the convergence implication, assume $A \in \mathcal{L}(H)$ satisfies (i) and (ii). Since $\|A\| \leq 1$, we have

$$\langle (I - A^*A)x, x \rangle = \langle x, x \rangle - \langle A^*Ax, x \rangle$$
$$(3) \qquad\qquad = \langle x, x \rangle - \langle Ax, Ax \rangle$$
$$= \|x\|^2 - \|Ax\|^2 \geq 0$$

for each $x \in H$. Thus $I - A^*A$ is non-negative, and hence $D_A := (I - A^*A)^{1/2}$ is well-defined (see Section V.6). Put $K = \overline{\operatorname{Im} D_A}$. Then K is a Hilbert space in its own right. Define

$$(4) \qquad J : H \to \ell_2(K), \qquad Jx = (D_A x, D_A A x, D_A A^2 x, \ldots).$$

The operator J is well-defined. Indeed, if $x \in H$, then $D_A A^k x \in K$ and

$$\|D_A A^k x\|^2 = \langle D_A A^k x, D_A A^k x \rangle$$
$$= \langle (A^*)^k D_A D_A A^k x, x \rangle$$
$$= \langle (A^*)^k A^k x, x \rangle - \langle (A^*)^{k+1} A^{k+1} x, x \rangle$$
$$= \|A^k x\|^2 - \|A^{k+1} x\|^2.$$

Hence

$$\sum_{k=0}^{n} \|D_A A^k x\|^2 = \|x\|^2 - \|A^{n+1} x\|^2 \to \|x\|^2 \qquad (n \to \infty),$$

because of condition (ii). It follows that J is well-defined and $\|Jx\| = \|x\|$ for each $x \in H$.

Now, put $M = \operatorname{Im} J$. Since J is an isometry, $\operatorname{Im} J$ is closed. Let V be the forward shift on $\ell_2(K)$. Then

$$JAx = V^* Jx, \qquad x \in H.$$

It follows that $M = \operatorname{Im} J$ is invariant under V^* and $A = U^{-1}(V^*|M)U$, where $U : H \to M$ is defined by $Ux = Jx$. Thus A is unitarily equivalent to $V^*|M$, which is a part of V^*.
$\square$

Note that conditions (i) and (ii) in Theorem 2.1 are automatically fulfilled if $\|A\| < 1$. Thus up to a positive scalar factor and unitary equivalence any Hilbert space operator is a part of a block backward shift. In other words the latter operators provide a universal model for Hilbert space operators.

XXVI.3 INVARIANT SUBSPACES OF FORWARD SHIFT OPERATORS

In this section we determine the invariant subspaces of the forward shift S on $\ell_2^m = \ell_2(\mathbb{C}^m)$. To do this it will be more convenient to work with the operator V acting on the Hardy space $H_2^m(\mathbb{T})$ defined by

$$(1) \qquad\qquad (V\varphi)(z) = z\varphi(z), \qquad z \in \mathbb{T}.$$

The operators S and V are unitarily equivalent; in fact

$$(2) \qquad\qquad S = FVF^{-1},$$

where $F\colon H_2^m(\mathbb{T}) \to \ell_2^m$ is the Fourier transformation (see Section XXIII.1) for the definition. From (2) it follows that M is an invariant subspace for the $\mathbb{C}^m$-block shift S if and only if $F^{-1}(M)$ is an invariant subspace of V, and hence it suffices to determine the invariant subspaces of V.

For the latter purpose we need the following definition. An $m \times \ell$ matrix function Φ on $\mathbb{T}$ is called *inner* if its entries are measurable and essentially bounded functions on $\mathbb{T}$ and

(i) $\frac{1}{2\pi} \int_{-\pi}^{\pi} \Phi(e^{it}) e^{-ikt}\, dt = 0$, $k = -1, -2, \ldots$,

(ii) $\Phi(e^{it})^* \Phi(e^{it}) = I$ a.e. on $-\pi \leq t \leq \pi$.

Let Φ be such a function. Since its entries are in $L_\infty(\mathbb{T})$, the function Φ defines a bounded linear operator from $L_2^\ell(\mathbb{T})$ into $L_2^m(\mathbb{T})$, namely by multiplication as follows:

$$(3) \qquad\qquad (M_\Phi g)(z) = \Phi(z)g(z), \qquad z \in \mathbb{T}.$$

Condition (i) above implies that M_Φ maps $H_2^\ell(\mathbb{T})$ into $H_2^m(\mathbb{T})$ and from (ii) it follows that M_Φ is an isometry. (The converses of the statements in the previous sentence are also true.)

THEOREM 3.1. *A subspace $M \subset H_2^m(\mathbb{T})$ is invariant under the operator V defined by (1) if and only if there exists an $m \times \ell$ matrix function Φ such that Φ is inner and*

$$(4) \qquad\qquad M = \{ M_\Phi g \mid g \in H_2^\ell(\mathbb{T}) \}.$$

In this case $\ell = \dim \operatorname{Ker}(V|M)^*$ and Φ is uniquely determined up to a constant unitary $\ell \times \ell$ matrix on the right.

By $H_\infty(\mathbb{T})$ we shall denote the set of all $\varphi \in L_\infty(\mathbb{T})$ such that

$$(5) \qquad\qquad \frac{1}{2\pi} \int_{-\pi}^{\pi} \varphi(e^{it}) e^{-ikt}\, dt = 0, \qquad k = -1, -2, \ldots .$$

For the proof of Theorem 3.1 we need the following lemma, which may be viewed as an addition to the material in Section XXIII.3.

LEMMA 3.2. *Let T be an operator on ℓ_2 which commutes with the forward shift on ℓ_2. Then T is a Toeplitz operator defined by a function in $H_\infty(\mathbb{T})$.*

PROOF. Let S be the forward shift. Note that $S^*S = I$, and thus $ST = TS$ implies that $S^*TS = T$. But then we know (from Section XXIII.3) that T is a Toeplitz operator, $T = [a_{i-j}]_{i,j=0}^\infty$ say. Put $e_0 = (1,0,0,\ldots)$. From $ST = TS$, we see that $T^*S^* = S^*T^*$. Since S^* is the backward shift on ℓ_2, it follows that

$$\begin{aligned}
0 = T^*S^*e_0 &= S^*T^*e_0 \\
&= S^*(\overline{a}_0, \overline{a}_{-1}, \overline{a}_{-2}, \ldots) \\
&= (\overline{a}_{-1}, \overline{a}_{-2}, \ldots).
\end{aligned}$$

Thus, (5) holds for the defining function φ of T, i.e., $\varphi \in H_\infty(\mathbb{T})$. $\square$

PROOF OF THEOREM 3.1. We split the proof into four parts. The first part concerns the sufficiency.

Part (a). Let M be given by (4), where Φ is an inner $m \times \ell$ matrix function. We shall show that M is an invariant subspace for V. We know that M_Φ maps $H_2^\ell(\mathbb{T})$ into $H_2^m(\mathbb{T})$, and thus $M \subset H_2^m(\mathbb{T})$. Since M_Φ is an isometry, the restriction of M_Φ to $H_2^\ell(\mathbb{T})$ is an isometry, and hence M is closed in $H_2^m(\mathbb{T})$. It remains to prove that $VM \subset M$. Here as well as in the sequel we use the symbol V for the operator of multiplication by z both on $H_2^m(\mathbb{T})$ and $H_2^\ell(\mathbb{T})$. Thus, for $g \in H_2^\ell(\mathbb{T})$ we have

$$\begin{aligned}
(VM_\Phi g)(z) = z(M_\Phi g)(z) &= z\Phi(z)g(z) \\
&= \Phi(z)(zg(z)) \\
&= (M_\Phi Vg)(z),
\end{aligned}$$

(6)

which shows that V maps M into M. Hence M is a V-invariant subspace.

Part (b). Let $M \subset H_2^m(\mathbb{T})$ be a subspace invariant under V, and put $\ell = \dim \mathrm{Ker}(V|M)^*$. In this part we show that $\ell \leq m$ and that there exists an isometry $J: H_2^\ell(\mathbb{T}) \to H_2^m(\mathbb{T})$ such that $\mathrm{Im}\, J = M$ and

$$(7) \qquad\qquad VJg = JVg, \qquad g \in H_2^\ell(\mathbb{T}).$$

Recall (see (2)) that V is unitarily equivalent to the $\mathbb{C}^m$-block forward shift. Thus V is a pure isometry. This property is preserved by restriction to an invariant subspace. So, by Theorem 1.2, the operator $V|M$ is unitarily equivalent to the forward shift on $\ell_2(L)$, where $L = \mathrm{Ker}(V|M)^*$. Let us prove that

$$\ell = \dim L \leq \dim \mathrm{Ker}\, V^* = \dim \mathrm{Ker}\, S^* = m.$$

Choose an orthonormal basis $f_1, \ldots, f_m$ of $\mathrm{Ker}\, V^*$. It follows (from the proof of Theorem 1.2) that the functions

$$V^j f_k, \qquad k = 1, \ldots, m, \; j = 0, 1, 2, \ldots$$

form an orthonormal basis for $H_2^m(\mathbb{T})$. Let P denote the orthogonal projection of $H_2^m(\mathbb{T})$ onto M. Since $V^*V = I$, the operator VPV^* is the orthogonal projection of $H_2^m(\mathbb{T})$ onto VM, and hence $Q := P - VPV^*$ is the orthogonal projection of $H_2^m(\mathbb{T})$ onto $M \cap (VM)^\perp$. The latter space is precisely L. Now, let $\{g_\nu\}_{\nu \in \Lambda}$, where Λ is some index set, be an orthonormal basis of $L = \operatorname{Im} Q$. Since Q is an orthogonal projection, we can use Parseval's equality twice to show that

$$
\begin{aligned}
\ell = \dim L &= \sum_{\nu \in \Lambda} \|g_\nu\|^2 \\
&= \sum_{\nu \in \Lambda} \sum_{k=1}^{m} \sum_{j=0}^{\infty} \langle g_\nu, V^j f_k \rangle \\
&= \sum_{k=1}^{m} \sum_{j=0}^{\infty} \sum_{\nu \in \Lambda} \langle g_\nu, QV^j f_k \rangle \\
&= \sum_{k=1}^{m} \sum_{j=0}^{\infty} \|QV^j f_k\|^2 .
\end{aligned}
$$

On the other hand, since $V^*V = I$, we have

$$
\begin{aligned}
\|QV^j f_k\|^2 &= \langle QV^j f_k, V^j f_k \rangle \\
&= \langle (P - VPV^*)V^j f_k, V^j f_k \rangle \\
&= \langle PV^j f_k, V^j f_k \rangle - \langle PV^{j-1} f_k, V^{j-1} f_k \rangle \\
&= \|PV^j f_k\|^2 - \|PV^{j-1} f_k\|^2 ,
\end{aligned}
$$

whenever $j \geq 1$. Furthermore, $Qf_k = Pf_k$, because $V^* f_k = 0$. It follows that

$$
\begin{aligned}
\sum_{k=1}^{m} \sum_{j=0}^{\infty} \|QV^j f_k\|^2 &= \lim_{N \to \infty} \sum_{k=1}^{m} \sum_{j=0}^{N} \|QV^j f_k\|^2 \\
&= \lim_{N \to \infty} \sum_{k=1}^{m} \|PV^N f_k\|^2 \\
&\leq \sum_{k=1}^{m} \|f_k\|^2 = \dim \operatorname{Ker} V^* = m.
\end{aligned}
$$

Thus $\ell \leq m$, and without loss of generality (apply unitary equivalence if necessary) we may assume that $L = \mathbb{C}^\ell$. Since the forward shift on $\ell_2(\mathbb{C}^\ell)$ is unitarily equivalent to the operator V acting on $H_2^\ell(\mathbb{T})$, we have proved that there exists a unitary operator $J_0\colon H_2^\ell(\mathbb{T}) \to M$ such that $V|M = J_0 V J_0^{-1}$. Now, define $J\colon H_2^\ell(\mathbb{T}) \to H_2^m(\mathbb{T})$ by setting $Jg = J_0 g$ for each $g \in H_2^\ell(\mathbb{T})$. Then J is the desired isometry.

     Part (c). Let J be as in the previous part. We now show that $Jg = M_\Phi g$, where Φ is an inner $m \times \ell$ matrix function. Since $H_2^n(\mathbb{T})$ is the Hilbert space direct sum of n copies of $H_2(\mathbb{T})$, we may represent J by an $m \times \ell$ operator matrix,

$$J = \begin{bmatrix} J_{11} & \cdots & J_{1\ell} \\ \vdots & & \vdots \\ J_{m1} & \cdots & J_{m\ell} \end{bmatrix},$$

where each entry J_{ij} acts on $H_2(\mathbb{T})$. From (7) it follows that J_{ij} commutes with V, where V is defined by (1) and acts now on $H_2(\mathbb{T})$. Let F be the Fourier transformation on $H_2(\mathbb{T})$, and recall that FVF^{-1} is the forward shift S on ℓ_2 (cf., (2)). We have that $FJ_{ij}F^{-1}$ commutes with S, and hence, by Lemma 3.2, the operator $FJ_{ij}F^{-1}$ is a Toeplitz operator defined by a function $\varphi_{ij} \in H_\infty(\mathbb{T})$. Now apply Corollary XXIII.3.3, and use the analyticity of φ_{ij} to show that $\varphi_{ij}f \in H_2(\mathbb{T})$ whenever $f \in H_2(\mathbb{T})$. We see that $J_{ij}f = \varphi_{ij}f$, and thus $Jg = M_\Phi g$ with

$$\Phi = \begin{bmatrix} \varphi_{11} & \cdots & \varphi_{1\ell} \\ \vdots & & \vdots \\ \varphi_{m1} & \cdots & \varphi_{m\ell} \end{bmatrix}.$$

It remains to check property (ii) in the definition of an inner function. Note that $J^*g = \mathbb{P}M_{\Phi^*}g$, where $\mathbb{P}$ is the orthogonal projection of $L_2^\ell(\mathbb{T})$ onto $H_2^\ell(\mathbb{T})$ and

$$\Phi^*(e^{it}) = \Phi(e^{it})^*, \qquad -\pi \leq t \leq \pi.$$

Now J is an isometry. Thus $\langle Jg, Jh \rangle = \langle g, h \rangle$ for g and h in $H_2^\ell(\mathbb{T})$, and hence $J^*J = I$. On the other hand, $J^*Jg = \mathbb{P}M_{\Phi^*\Phi}g$. Thus, by Corollary XXIII.3.3, the Toeplitz operator defined by $\Phi^*\Phi$ is the identity operator. Since the defining function of a Toeplitz operator is unique up to changes on a set of measure zero, we obtain the desired property (ii).

Part (d). Let Φ be an $m \times \ell$ matrix function such that (4) holds, and assume that Φ is inner. If U is a constant unitary $\ell \times \ell$ matrix, then, clearly, $\Psi(\cdot) = \Phi(\cdot)U$ is inner and (4) holds with Ψ in place of Φ. To complete the proof we have to establish the reverse implication. So, let Ψ be an $m \times \ell$ matrix function such that Ψ is inner and

$$(8) \qquad M = \{M_\Psi g \mid g \in H_2^\ell(\mathbb{T})\}.$$

Put $U(e^{it}) = \Phi(e^{it})^*\Psi(e^{it})$. Since Φ is inner, we see from (4) that M_{Φ^*} maps M into $H_2^\ell(\mathbb{T})$. This observation, together with (8), shows that M_U leaves $H_2^\ell(\mathbb{T})$ invariant. But then we must have

$$(9) \qquad \frac{1}{2\pi} \int_{-\pi}^{\pi} U(e^{it})e^{-ikt}\,dt = 0, \qquad k = -1, -2, \ldots .$$

Indeed, let x be an arbitrary vector in $\mathbb{C}^\ell$, and let $g(e^{it}) = x$ for each $-\pi \leq t \leq \pi$. Then $g \in H_2^\ell(\mathbb{T})$, and thus $U(\cdot)x = M_U g$ belongs to $H_2^\ell(\mathbb{T})$. It follows that

$$\frac{1}{2\pi} \int_{-\pi}^{\pi} e^{-ikt}U(e^{it})x\,dt = 0, \qquad k = -1, -2, \ldots .$$

This holds for each $x \in \mathbb{C}^\ell$, and therefore we have (9). By applying the above arguments with the roles of Φ and Ψ interchanged, we see that (9) also holds with $U(\cdot)^*$ in place of $U(\cdot)$. It follows that

$$\frac{1}{2\pi} \int_{-\pi}^{\pi} U(e^{it}) e^{-ikt} dt = 0, \qquad k \neq 0,$$

and hence $U(\cdot) = U_0$, where U_0 is a constant $\ell \times \ell$ matrix.

Next, we show that $\Phi(\cdot)U_0 = \Psi(\cdot)$ and U_0 is unitary. Since $M \subset \operatorname{Im} M_\Phi$, the operator $M_{\Phi\Phi^*}$ acts as the identity operator on M. Thus $M_\Phi M_{\Phi^*} M_\Psi g = M_\Psi g$ for all $g \in H_2^\ell(\mathbb{T})$. In particular, the latter identity holds for $g(e^{it}) \equiv x$, where x is an arbitrary vector in $\mathbb{C}^\ell$. Hence

(10) $$\Phi(e^{it})U_0 = \Psi(e^{it}), \text{ a.e.}, \qquad -\pi \leq t \leq \pi.$$

By using property (i) in the definition of an inner function, we see from (10) that

$$U_0^* U_0 = \Psi(\cdot)^* \Phi(\cdot) U_0 = \Psi(\cdot)^* \Psi(\cdot) = I,$$

and hence U_0 has the desired properties. $\quad\square$

CHAPTER XXVII
DILATION THEORY

In this chapter a general theory for contractive operators is developed. Such operators may be viewed as compressions of isometric and unitary operators. The minimal isometric and minimal unitary dilations of a given contraction are to a large extent unique, which implies that those operators are useful instruments for the analysis of contractions. In this chapter we also prove the commutant lifting theorem and present some of its applications to interpolation problems.

XXVII.1 PRELIMINARIES ABOUT CONTRACTIONS

In this section we discuss some of the elementary properties of contractive operators.

A Hilbert space operator $A: H \to K$ is called a *contraction* (or a *contractive operator*) if $\|A\| \leq 1$. The identity

$$(1) \qquad \langle (I - A^*A)x, x \rangle = \|x\|^2 - \|Ax\|^2, \qquad x \in H,$$

implies that A is a contraction if and only if $I - A^*A$ is non-negative. Thus for a contraction A the operator $D_A := (I - A^*A)^{1/2}$ is a well-defined non-negative operator. One calls D_A the *defect operator* associated with A. Note that D_A acts on H, and D_A is invertible if and only if $\|A\| < 1$.

LEMMA 1.1. *Let* $A: H \to K$ *be a contraction, and let* D_A *be the associated defect operator. Then*

(i) $\|x\|^2 = \|Ax\|^2 + \|D_A x\|^2$ *for each* $x \in H$,

(ii) A^* *is a contraction and*

$$(2) \qquad\qquad\qquad A D_A = D_{A^*} A.$$

PROOF. Statement (i) is an immediate consequence of the identity (1). Since $\|A\| = \|A^*\|$, the first part of (ii) is obvious. It remains to prove the identity (2).

Choose $\alpha < 0$ and $\beta > \|A\| = \|A^*\|$, and put

$$g(t) = \begin{cases} 0 & \text{for} \quad \alpha \leq t \leq 0, \\ t^{1/2} & \text{for} \quad 0 \leq t \leq \beta. \end{cases}$$

By the Weierstrass approximation theorem there exists a sequence of polynomials $p_1, p_2, \ldots$ such that $p_n \to g$ $(n \to \infty)$ uniformly on $[\alpha, \beta]$. Put $S = I - A^*A$ and $S_* = I - AA^*$. From the theory of selfadjoint operators (see the proof of Theorem V.6.1) we know that in the operator norm

$$(3) \qquad p_n(S) \to D_A, \quad p_n(S_*) \to D_{A^*} \qquad (n \to \infty).$$

Now note that $AS = S_*A$. Thus

$$AD_A = \lim_{n\to\infty} Ap_n(S) = \lim_{n\to\infty} p_n(S_*)A = D_{A^*}A,$$

which proves (2). $\square$

Let $T: K_1 \to K_2$ and $A: H_1 \to H_2$ be Hilbert space operators. We call A a *compression* of T if H_1 and H_2 are subspaces of K_1 and K_2, respectively, and

$$(4) \qquad\qquad\qquad A = \pi_{H_2} T \tau_{H_1}.$$

Let us explain the notation in (4). If H is a subspace of the Hilbert space K, then $\tau_H: H \to K$ will denote the canonical embedding operator and $\pi_H: K \to H$ the orthogonal projection of K onto H. Thus

$$(5) \qquad\qquad \tau_H x = x \quad (x \in H), \qquad \pi_H = \tau_H^*.$$

Note that (4) implies that T has the following 2×2 operator matrix representation:

$$(6) \qquad\qquad T = \begin{bmatrix} A & * \\ * & * \end{bmatrix} : H_1 \boxplus H_1^{\perp} \to H_2 \boxplus H_2^{\perp}.$$

Conversely, if T is as in (6), then A is a compression of T. Since the operators τ_{H_1} and π_{H_2} are contractions, we see that compressions of contractions are again contractions.

The next proposition shows that all contractions may be obtained as compressions of unitary operators.

PROPOSITION 1.2. *If* $A: H_1 \to H_2$ *is a contraction, then the operator*

$$(7) \qquad\qquad U_A := \begin{bmatrix} A & D_{A^*} \\ D_A & -A^* \end{bmatrix} : H_1 \boxplus H_2 \to H_2 \boxplus H_1$$

is unitary and $U_A^{-1} = U_{A^*}$.

PROOF. Obviously, $(U_A)^* = U_{A^*}$. So it suffices to show that $U_A U_{A^*}$ is equal to the identity operator on $H_2 \boxplus H_1$ and $U_{A^*} U_A$ to the identity operator on $H_1 \boxplus H_2$. But these equalities follow by a direct computation, using (2) both for A and A^*. $\square$

Let $A: H_1 \to H_2$ be a contraction. Put $M = \overline{\operatorname{Im} D_A}$ and $M_* = \overline{\operatorname{Im} D_{A^*}}$. By applying (2) to A^* we see that $A^* D_{A^*} = D_A A^*$. It follows that the operator U_A in (7) maps the Hilbert space $H_1 \boxplus M_*$ into the Hilbert space $H_2 \boxplus M$. In the sequel we let R_A denote the corresponding restriction, i.e.,

$$(8) \qquad\qquad R_A = U_A|(H_1 \boxplus M_*): H_1 \boxplus M_* \to H_2 \boxplus M.$$

By interchanging the roles of A and A^* we see that U_{A^*} maps $H_2 \boxplus M$ into $H_1 \boxplus M_*$. Thus Proposition 1.2 implies that R_A is unitary and $R_A^{-1} = R_{A^*}$. One calls R_A the *rotation operator* associated with A.

Let us illustrate the above with a simple example. Take

$$
(9) \qquad A = \begin{bmatrix} 0 & 0 & \cdots & 0 & 0 \\ 1 & 0 & \cdots & 0 & 0 \\ 0 & 1 & \cdots & 0 & 0 \\ \vdots & \vdots & & \vdots & \vdots \\ 0 & 0 & \cdots & 1 & 0 \end{bmatrix} : \mathbb{C}^n \to \mathbb{C}^n.
$$

Obviously, A is a contraction. The defect operators of A and A^* are given by

$$
D_A \begin{bmatrix} x_1 \\ \vdots \\ x_n \end{bmatrix} = \begin{bmatrix} 0 \\ \vdots \\ 0 \\ x_n \end{bmatrix}, \qquad D_{A^*} \begin{bmatrix} x_1 \\ \vdots \\ x_n \end{bmatrix} = \begin{bmatrix} x_1 \\ 0 \\ \vdots \\ 0 \end{bmatrix},
$$

and hence

$$
M = \mathrm{span}\left\{ \begin{bmatrix} 0 \\ \vdots \\ 0 \\ 1 \end{bmatrix} \right\}, \qquad M_* = \mathrm{span}\left\{ \begin{bmatrix} 1 \\ 0 \\ \vdots \\ 0 \end{bmatrix} \right\}.
$$

Note that $D_{A^*}|M_*$ is the canonical embedding of M_* into $\mathbb{C}^n$, and $D_A : \mathbb{C}^n \to M$ is the orthogonal projection of $\mathbb{C}^n$ onto M. Furthermore, $-A^*|M_*$ is the zero operator. By identifying M and M_* with $\mathbb{C}$ one sees that the rotation operator R_A associated with (9) is unitarily equivalent to the operator

$$
\begin{bmatrix} 0 & 0 & \cdots & 0 & 0 & 1 \\ 1 & 0 & \cdots & 0 & 0 & 0 \\ 0 & 1 & \cdots & 0 & 0 & 0 \\ \vdots & \vdots & & \vdots & \vdots & \vdots \\ 0 & 0 & \cdots & 1 & 0 & 0 \\ 0 & 0 & \cdots & 0 & 1 & 0 \end{bmatrix} : \mathbb{C}^{n+1} \to \mathbb{C}^{n+1}.
$$

XXVII.2 PRELIMINARIES ABOUT DILATIONS

Let $T \in \mathcal{L}(K)$ and $A \in \mathcal{L}(H)$ be Hilbert space operators. The operator T is said to be a *dilation* of A if H is a subspace of K and

$$
(1) \qquad A^n = \pi_H T^n \tau_H, \qquad n = 1, 2, \ldots .
$$

Here $\tau_H : H \to K$ is the canonical embedding and $\pi_H : K \to H$ is the orthogonal projection of K onto H (see formula (5) in the previous section). In the terminology of the previous section, T is a dilation of A if and only if A^n is a compression of T^n for each n. Note that (1) is equivalent to the requirement that

$$
(2) \qquad \langle A^n x, y \rangle = \langle T^n x, y \rangle, \qquad x, y \in H, \ n = 1, 2, \ldots .
$$

If A is a part of the operator T (see Section XXVI.2 for the definition), then T is a dilation of A, but the converse statement is not true in general. To describe dilations in terms of invariant subspaces we need the following definition.

Let $T \in \mathcal{L}(K)$. A subspace H of K is called *semi-invariant* under T if there exist two T-invariant subspaces M and L such that $H \subset M$ and $L = M \cap H^\perp$. In this case T may be represented as an upper triangular 3×3 operator matrix:

$$
(3) \qquad T = \begin{bmatrix} T_{11} & T_{10} & T_{12} \\ 0 & T_{00} & T_{02} \\ 0 & 0 & T_{22} \end{bmatrix} : L \boxplus H \boxplus M^\perp \to L \boxplus H \boxplus M^\perp.
$$

PROPOSITION 2.1. *Let $A \in \mathcal{L}(H)$ be a compression of $T \in \mathcal{L}(K)$. Then T is a dilation of A if and only if H is semi-invariant under T.*

PROOF. Assume H is semi-invariant under T. So we may represent T in the form (3). This implies that

$$
(4) \qquad T^n = \begin{bmatrix} T_{11}^n & * & * \\ 0 & T_{00}^n & * \\ 0 & 0 & T_{22}^n \end{bmatrix}, \qquad n = 1, 2, \ldots,
$$

where the $*$'s denote entries which we shall not specify further. Since A is a compression of T, we have $T_{00} = A$, and we see from (4) that (1) holds. So T is a dilation of A.

To prove the converse implication, assume that T is a dilation of H. Set

$$
(5) \qquad M = \overline{\operatorname{span}\{T^n H \mid n = 0, 1, 2, \ldots\}}.
$$

Obviously, $H \subset M$ and $TM \subset M$. It remains to prove that the space $L := M \cap H^\perp$ is invariant under T. Take $h \in H$. Then

$$
\pi_H T(T^n h) = \pi_H T^{n+1} h = A^{n+1} h = A(\pi_H T^n h) = A \pi_H (T^n h).
$$

This holds for $n = 0, 1, \ldots$. Thus we see from the definition of M in (5) that

$$
(6) \qquad \pi_H T x = A \pi_H x, \qquad x \in M.
$$

Now, take $x \in L$. Then $x \in M$ and $x \perp H$. Thus the right hand side of (6) is zero. It follows that $Tx \perp H$. Also, $Tx \in M$, because $x \in M$ and M is T-invariant. So $Tx \in L$, and we have shown that L is T-invariant. $\square$

Let $A \in \mathcal{L}(H)$ be a contraction. In this chapter we shall be interested in dilations of A that are unitary. Proposition 1.2 tells us that A always appears as the compression of a unitary operator, but, in general, the construction in Proposition 1.2 does not yield a unitary dilation. To see this let us consider the operator

$$
(7) \qquad A = \begin{bmatrix} 0 & 0 & \cdots & 0 & 0 \\ 1 & 0 & \cdots & 0 & 0 \\ 0 & 1 & \cdots & 0 & 0 \\ \vdots & \vdots & & \vdots & \vdots \\ 0 & 0 & \cdots & 1 & 0 \end{bmatrix} : \mathbb{C}^n \to \mathbb{C}^n.
$$

In this case the unitary operator U_A constructed in Proposition 1.2 acts on the space $\mathbb{C}^{2n}$, but the operator A in (7) does not have a unitary dilation acting on a finite dimensional space. To see this, let $T \in \mathcal{L}(K)$ be a unitary dilation of A, and assume that $\dim K < \infty$. Write T in the form (3) with $H = \mathbb{C}^n$ and $T_{00} = A$. Then

$$T^{-1} = T^* = \begin{bmatrix} T_{11}^* & 0 & 0 \\ T_{10}^* & T_{00}^* & 0 \\ T_{12}^* & T_{02}^* & T_{22}^* \end{bmatrix}.$$

By computing the first column in the product T^*T we see that

$$(8) \qquad\qquad T_{11}^*T_{11} = I_L, \qquad T_{10}^*T_{11} = 0, \qquad T_{12}^*T_{11} = 0.$$

Since $\dim L < \infty$, the first identity in (8) shows that T_{11} is invertible. But then it follows from the other identities in (8) that $T_{10} = 0$ and $T_{12} = 0$. Next, compute the second column in T^*T. This yields $T_{00}^*T_{00} = I$ and $T_{02}^*T_{00} = 0$, and in the same way as before we conclude that $T_{02} = 0$. Thus

$$T = \begin{bmatrix} T_{11} & 0 & 0 \\ 0 & A & 0 \\ 0 & 0 & T_{22} \end{bmatrix}.$$

But then T is unitary implies that A is unitary which is false.

So the operator A in (7) does not have a unitary dilation acting on a finite dimensional space. On the other hand, the forward shift U on $\ell_2(\mathbb{Z})$, i.e.,

$$U\big((x_\nu)_{\nu \in \mathbb{Z}}\big) = (x_{\nu-1})_{\nu \in \mathbb{Z}}$$

is a unitary dilation of A in (7). To see this, consider the following subspaces:

$$H = \big\{(x_\nu)_{\nu \in \mathbb{Z}} \in \ell_2(\mathbb{Z}) \mid x_j = 0 \text{ for } j \notin \{1,\ldots,n\}\big\}.$$
$$M = \big\{(x_\nu)_{\nu \in \mathbb{Z}} \in \ell_2(\mathbb{Z}) \mid x_j = 0 \text{ for } j \leq 0\big\},$$
$$L = \big\{(x_\nu)_{\nu \in \mathbb{Z}} \in \ell_2(\mathbb{Z}) \mid x_j = 0 \text{ for } j \leq n\big\}.$$

Then $\ell_2(\mathbb{Z}) = L \boxplus H \boxplus M^\perp$, the spaces L and M are invariant under U, and by identifying H with $\mathbb{C}^n$ we see that U is a dilation of A in (7).

We conclude this section with a bit of notation. Let $W_1, W_2, \ldots$ be subsets of a Hilbert space K. By $\bigvee_{n=1}^{\infty} W_n$ we denote the smallest closed linear manifold in K containing the sets W_j ($j \geq 1$). In other words

$$(9) \qquad\qquad \bigvee_{n=1}^{\infty} W_n = \overline{\text{span}\{W_n \mid n \geq 1\}}.$$

In particular, the space in the right hand side of (5) is equal to $\bigvee_{n=0}^{\infty} T^n H$.

XXVII.3 ISOMETRIC DILATIONS

In this section we construct dilations that are isometric. Note that this requires the original operator to be a contraction.

THEOREM 3.1. *Let* $A \in \mathcal{L}(H)$ *be a contraction, and put* $M = \overline{\mathrm{Im}\, D_A}$. *Define* T *on* $K = H \boxplus \ell_2(M)$ *by setting*

$$(1) \qquad T = \begin{bmatrix} A & 0 \\ B & C \end{bmatrix} : H \boxplus \ell_2(M) \to H \boxplus \ell_2(M),$$

where

$$(2) \qquad B \colon H \to \ell_2(M), \qquad Bh = (D_A h, 0, 0, \ldots),$$

and C *is the forward shift on* $\ell_2(M)$. *Then*

(i) T *is a dilation of* A,

(ii) T *is an isometry*,

(iii) $K = \bigvee_{n=0}^{\infty} T^n H$.

PROOF. From the representation (1) it is clear that T is a dilation of A. Let us prove that T is an isometry. Let us write the elements of K as $(h, \underline{x})$, where $h \in H$ and $\underline{x} = (x_1, x_2, \ldots) \in \ell_2(M)$. Then

$$T(h, \underline{x}) = \big(Ah, (D_A h, x_1, x_2, \ldots)\big),$$

and thus

$$\|T(h, \underline{x})\|^2 = \|Ah\|^2 + \|D_A h\|^2 + \sum_{j=1}^{\infty} \|x_j\|^2$$
$$= \|h\|^2 + \|\underline{x}\|^2 = \|(h, \underline{x})\|^2,$$

which proves that T is an isometry.

To prove (iii), take $(h, \underline{x}) \perp T^n H$ for each $n \geq 0$. For $n = 0$ this yields $h = 0$. Thus $(0, \underline{x}) \perp T^n H$ for $n \geq 1$. For $n \geq 1$ we have

$$(3) \qquad T^n H = \big\{ \big(A^n h, (D_A A^{n-1} h, \ldots, D_A h, 0, 0, \ldots)\big) \mid h \in H \big\}.$$

It follows that $x_j = 0$ for $j = 1, 2, \ldots$. We prove the latter statement by induction. We have $(0, \underline{x}) \perp T H$. So (3) implies that $x_1 \perp D_A h$ for each $h \in H$. But $x_1 \in M = \overline{\mathrm{Im}\, D_A}$. So $x_1 = 0$. Next, assume that $x_j = 0$ for $1 \leq j \leq k$. Then $(0, \underline{x}) \perp T^{k+1} H$ yields that $x_{k+1} \perp \mathrm{Im}\, D_A$, and thus $x_{k+1} = 0$, because x_{k+1} is an element of $\overline{\mathrm{Im}\, D_A}$. So, by induction, $\underline{x} = 0$. We see that $(h, \underline{x}) \perp T^n H$ for $n \geq 0$ implies that h and $\underline{x}$ are zero. It follows that $\mathrm{span}\{ T^n H \mid n \geq 0 \}$ is dense in K, and therefore (iii) holds. $\square$

Let $A \in \mathcal{L}(H)$ be a contraction. We call $T \in \mathcal{L}(K)$ an *isometric dilation* of A if T has the properties (i) and (ii) in Theorem 3.1. If also (iii) in Theorem 3.1 is fulfilled, then T is called a *minimal isometric dilation*.

Let $T \in \mathcal{L}(K)$ be an isometric dilation of $A \in \mathcal{L}(H)$, and let $U: K \to K'$ be a unitary operator such that $Uh = h$ for each $h \in H$ (and hence, in particular, $H \subset K'$). Then UTU^{-1} is again a dilation of A. Furthermore, if T is minimal, then so is UTU^{-1}. The next theorem shows that the above unitary equivalence describes all the freedom one has in the choice of a minimal isometric dilation.

THEOREM 3.2. *Let $A \in \mathcal{L}(H)$ be a contraction, and let $T \in \mathcal{L}(K)$, $T' \in \mathcal{L}(K')$ be minimal isometric dilations of A. Then there exists a unique unitary operator $U: K \to K'$ such that*

(i) $Uh = h$ *for each* $h \in H$,

(ii) $T' = UTU^{-1}$.

PROOF. Let $h_0, \ldots, h_N$ be in H, and consider the finite sum

$$(4) \qquad s = \sum_{j=0}^{N} T^j h_j.$$

We claim that $\|s\|$ does not depend on the choice of the isometric dilation T, but only on A. To see this we use formula (2) in the previous section and the fact that T preserves the inner product. This yields

$$\|s\|^2 = \left\langle \sum_{j=0}^{N} T^j h_j, \sum_{k=0}^{N} T^k h_k \right\rangle = \sum_{j,k=0}^{N} \langle T^j h_j, T^k h_k \rangle$$

$$= \sum_{\substack{j,k=0 \\ j \geq k}}^{N} \langle T^{j-k} h_j, h_k \rangle + \sum_{\substack{j,k=0 \\ j < k}}^{N} \langle h_j, T^{k-j} h_k \rangle$$

$$= \sum_{\substack{j,k=0 \\ j \geq k}}^{N} \langle A^{j-k} h_j, h_k \rangle + \sum_{\substack{j,k=0 \\ j < k}}^{N} \langle h_j, A^{k-j} h_k \rangle.$$

Thus $\|s\|$ does not change if T is replaced by T'.

The latter fact implies that the operator

$$(5a) \qquad U_0: \operatorname{span}\{T^n H \mid n \geq 0\} \to \operatorname{span}\{(T')^n H \mid n \geq 0\},$$

$$(5b) \qquad U_0\left(\sum_{j=0}^{N} T^j h_j \right) = \sum_{j=0}^{N} (T')^j h_j,$$

is well-defined. Indeed, assume $\sum_{j=0}^{N} T^j h_j = \sum_{i=0}^{R} T^i h'_i$, where h_j and h'_i are in H for all i and j. Without loss of generality we may assume that $N = R$ (add zero vectors if necessary). Thus

$$\sum_{j=0}^{N} T^j (h_j - h'_j) = 0.$$

But then, by the result of the previous paragraph, we have

$$\sum_{j=0}^{N}(T')^j(h_j - h'_j) = 0,$$

which implies that $\sum_{j=0}^{N}(T')^j h_j = \sum_{j=0}^{N}(T')^j h'_j$, and hence U_0 is well-defined. The result of the previous paragraph also implies that U_0 is an isometry and U_0 is bijective.

Since $\mathrm{span}\{T^n H \mid n \geq 0\}$ is dense in K and U_0 is an isometry, we can extend U_0 by continuity to all of K. The resulting operator is denoted by U. From $\|Ux\| = \|x\|$ for $x \in \mathrm{span}\{T^n H \mid n \geq 0\}$, it follows by continuity that U is an isometry. Furthermore,

$$\overline{\mathrm{Im}\, U} = \mathrm{Im}\, U \supset \mathrm{Im}\, U_0 = \mathrm{span}\{(T')^n H \mid n \geq 0\}.$$

But the latter space is dense in K'. Thus $\mathrm{Im}\, U = K'$, and U is unitary.

Let us check the properties (i) and (ii). Take $h \in H$. By definition, $Uh = U_0 h = h$, and so (i) holds. Furthermore,

$$\begin{aligned}
UT(T^n h) = U(T^{n+1}h) &= U_0(T^{n+1}h) \\
&= (T')^{n+1}h = T'\{(T')^n h\} \\
&= T'U(T^n h),
\end{aligned}$$

which implies that UT and $T'U$ coincide on the space $\mathrm{span}\{T^n h \mid n \geq 0\}$. Since this space is dense in K, we conclude that $UT = T'U$.

Finally, let $\widetilde{U} \colon K \to K'$ be a second unitary operator such that (i) and (ii) hold with $\widetilde{U}$ in place of U. Take $h_0, \ldots, h_N$ in H. Then

$$\widetilde{U}\left(\sum_{j=0}^{N}T^j h_j\right) = \sum_{j=0}^{N}\widetilde{U}T^j h_j = \sum_{j=0}^{N}(T')^j\widetilde{U}h_j = \sum_{j=0}^{N}(T')^j h_j,$$

because of (i) and (ii) for $\widetilde{U}$. The above calculation shows that $\widetilde{U}$ and U coincide on $\mathrm{span}\{T^n H \mid n \geq 0\}$. Again use that the latter space is dense in K. So, by continuity, we have $\widetilde{U} = U$. $\square$

The previous theorem allows one to speak about *the* minimal isometric dilation T of the contraction A. From Theorem 3.1 we know that T has the following operator matrix representation

$$(6) \qquad T = \begin{bmatrix} A & 0 & 0 & 0 & \cdots \\ D_A & 0 & 0 & 0 & \cdots \\ 0 & I & 0 & 0 & \cdots \\ 0 & 0 & I & 0 & \\ \vdots & \vdots & \vdots & \ddots & \ddots \end{bmatrix} : H \boxplus \ell_2(M) \to H \boxplus \ell_2(M),$$

where $M = \overline{\mathrm{Im}\, D_A}$.

Let $A \in \mathcal{L}(H)$ be a contraction. It turns out that the minimal isometric dilation of A can be obtained by a step by step procedure in the following way. Put $M = \overline{\operatorname{Im} D_A}$, and consider the operator

$$A_{[1]} = \begin{bmatrix} A & 0 \\ D_A & 0 \end{bmatrix} : H \boxplus M \to H \boxplus M.$$

The operator $A_{[1]}$ is called the *one step dilation* of A. Note that $A_{[1]}$ is again a contraction. Indeed, for $\begin{bmatrix} h \\ m \end{bmatrix} \in H \boxplus M$ we have

$$\left\| A_{[1]} \begin{bmatrix} h \\ m \end{bmatrix} \right\|^2 = \left\| \begin{bmatrix} Ah \\ D_A h \end{bmatrix} \right\|^2$$
$$= \|Ah\|^2 + \|D_A h\|^2$$
$$= \|h\|^2 \le \left\| \begin{bmatrix} h \\ m \end{bmatrix} \right\|^2,$$

by virtue of Lemma 1.1(i). It is interesting to compute the one step dilation of $A_{[1]}$. To do this we have to determine $\operatorname{Im} D_{A_{[1]}}$. We have

$$I - (A_{[1]})^* A_{[1]} = \begin{bmatrix} I_H & 0 \\ 0 & I_M \end{bmatrix} - \begin{bmatrix} A^* & D_A \\ 0 & 0 \end{bmatrix} \begin{bmatrix} A & 0 \\ D_A & 0 \end{bmatrix}$$
$$= \begin{bmatrix} I_H & 0 \\ 0 & I_M \end{bmatrix} - \begin{bmatrix} I_H & 0 \\ 0 & 0 \end{bmatrix}$$
$$= \begin{bmatrix} 0 & 0 \\ 0 & I_M \end{bmatrix}.$$

Here I_H and I_M denote the identity operators on H and M, respectively. We see that

$$D_{A_{[1]}} = \begin{bmatrix} 0 & 0 \\ 0 & I_M \end{bmatrix} : H \boxplus M \to H \boxplus M.$$

Thus $\overline{\operatorname{Im} D_{A_{[1]}}} = M$, and according to the definition, the one step dilation of $A_{[1]}$ is the operator

$$A_{[2]} = (A_{[1]})_{[1]} = \begin{bmatrix} A & 0 & 0 \\ D_A & 0 & 0 \\ 0 & I & 0 \end{bmatrix} : H \boxplus M \boxplus M \to H \boxplus M \boxplus M.$$

Let us define the *N-step dilation* $A_{[N]}$ of A to be the one step dilation of $A_{[N-1]}$. Then, by induction, one finds that

$$(7) \qquad A_{[N]} = \begin{bmatrix} A & 0 & \cdots & 0 & 0 \\ D_A & 0 & \cdots & 0 & 0 \\ 0 & I & \cdots & 0 & 0 \\ \vdots & \vdots & & \vdots & \vdots \\ 0 & 0 & \cdots & I & 0 \end{bmatrix} : H \boxplus \underbrace{M \boxplus \cdots \boxplus M}_{N \times} \to H \boxplus \underbrace{M \boxplus \cdots \boxplus M}_{N \times}.$$

PROPOSITION 3.3. *Let $A \in \mathcal{L}(H)$ be a contraction, and for $N \geq 1$ let $A_{[N]}$ be the N-step dilation of A. Put $M = \overline{\operatorname{Im} D_A}$, and define operators*

$$\tau_N \colon H \boxplus M \boxplus \cdots \boxplus M \to H \boxplus \ell_2(M),$$
$$\pi_N \colon H \boxplus \ell_2(M) \to H \boxplus M \boxplus \cdots \boxplus M$$

by setting

$$\tau_N(h, x_0, \ldots, x_{N-1}) = (h, x_0, \ldots, x_{N-1}, 0, 0, \ldots),$$

$$\pi_N(h, x_0, x_1, \ldots) = (h, x_0, \ldots, x_{N-1}).$$

Then

(8a)
$$\lim_{N \to \infty} \tau_N A_{[N]} \pi_N y = T y \qquad (y \in K),$$

(8b)
$$A_{[N]} = \pi_N T \tau_N,$$

where $K = H \boxplus \ell_2(M)$ and $T \in \mathcal{L}(K)$ is the minimal isometric dilation of A.

PROOF. Let us write $y = (h, x_0, x_1, \ldots)$ for an arbitrary element of K. Here $h \in H$ and $(x_0, x_1, \ldots) \in \ell_2(M)$. It follows that

$$Ty = (Ah, D_A h, x_0, x_1, \ldots),$$

$$\tau_N A_{[N]} \pi_N y = (Ah, D_A h, x_0, \ldots, x_{N-2}, 0, 0, \ldots),$$

and thus

$$\|Ty - \tau_N A_{[N]} \pi_N y\|^2 = \sum_{j=N-1}^{\infty} \|x_j\|^2,$$

which converges to zero if $N \to \infty$. Thus (8a) holds. The identity (8b) follows directly from the definitions. $\square$

THEOREM 3.4. *Let $A \in \mathcal{L}(H)$ be a contraction. The minimal isometric dilation of A is a pure isometry if and only if for each $h \in H$ we have $(A^*)^n h \to 0$ if $n \to \infty$.*

PROOF. Assume that the minimal isometric dilation $T \in \mathcal{L}(K)$ is a pure isometry. By Theorem XXVI.1.2 this implies that T is unitarily equivalent to a block forward shift. Thus T^* is unitarily equivalent to a block backward shift. Hence, by virtue of Theorem XXVI.2.1, we have $(T^*)^n y \to 0$ $(n \to \infty)$ for each $y \in K$. From the construction of the minimal isometric dilation in Theorem 3.1 (see formula (1)) it follows that H is invariant under T^* and $A^* = T^*|H$. Thus for each $h \in H$ we obtain

$$(A^*)^n h = (T^*)^n h \to 0 \qquad (n \to \infty).$$

To prove the converse, assume that $(A^*)^n h \to 0$ if $n \to \infty$ for each $h \in H$. Consider the Von Neumann-Wold decomposition of T, i.e., write $K = K_1 \boxplus K_2$, where K_1 and K_2 are T-invariant subspaces of K such that $T|K_1$ is unitary and $T|K_2$ is a pure isometry. We have to prove that K_1 consists of the zero element only. Take $h \in H$, and write $h = k_1 + k_2$ with $k_1 \in K_1$ and $k_2 \in K_2$. We already proved that $A^* = T^*|H$. Thus

$$\begin{aligned} (A^*)^n h = (T^*)^n h &= (T^*)^n k_1 + (T^*)^n k_2 \\ &= \left((T|K_1)^*\right)^n k_1 + \left((T|K_2)^*\right)^n k_2. \end{aligned}$$

Now $T|K_2$ is a pure isometry. So $T|K_2$ is unitarily equivalent to a block forward shift, and hence $\left((T|K_2)^*\right)^n k_2 \to 0$ if $n \to \infty$. By assumption, $(A^*)^n h \to 0$ if $n \to 0$. It follows that $\left((T|K_1)^*\right)^n k_1 \to 0$ if $n \to \infty$. But $T|K_1$ is unitary, and hence we may conclude that $k_1 = 0$. Thus we have shown that $H \subset K_2$. Since K_2 is invariant under T, we see that $T^n H \subset K_2$ for each $n \geq 0$. Now use that $\operatorname{span}\{T^n H \mid n \geq 0\}$ is dense in K. So $K = K_2$, and $T = T|K_2$ is a pure isometry. $\square$

We conclude with an example. Consider the operator

$$(9) \qquad A = \begin{bmatrix} 0 & 0 & \cdots & 0 & 0 \\ 1 & 0 & \cdots & 0 & 0 \\ 0 & 1 & \cdots & 0 & 0 \\ \vdots & \vdots & & \vdots & \vdots \\ 0 & 0 & \cdots & 1 & 0 \end{bmatrix} : \mathbb{C}^n \to \mathbb{C}^n.$$

We already know (see Section XXVII.1) that

$$M := \operatorname{Im} D_A = \operatorname{span}\left\{ \begin{bmatrix} 0 \\ \vdots \\ 0 \\ 1 \end{bmatrix} \right\}.$$

Let us identify M with $\mathbb{C}$ by setting $\lambda = [0 \cdots 0 \ \lambda]^T$ (where superscript T denotes the transpose). Then the minimal isometric dilation of A in (9) is the operator

$$T = \begin{bmatrix} A & 0 \\ B & C \end{bmatrix} : \mathbb{C}^n \boxplus \ell_2 \to \mathbb{C}^n \boxplus \ell_2,$$

where

$$B : \mathbb{C}^n \to \ell_2, \qquad B = \begin{bmatrix} 0 & \cdots & 0 & 1 \\ 0 & \cdots & 0 & 0 \\ 0 & \cdots & 0 & 0 \\ \vdots & & \vdots & \vdots \end{bmatrix}$$

and C is the forward shift on ℓ_2. In other words, if $x = (x_1, \ldots, x_n, x_{n+1}, x_{n+2}, \ldots)$ denotes an arbitrary element of $\mathbb{C}^n \boxplus \ell_2$, then

$$Tx = (0, x_1, \ldots, x_{n-1}, x_n, x_{n+1}, \ldots),$$

and thus T is just a copy of the forward shift on ℓ_2.

XXVII.4 UNITARY DILATIONS

Let $A \in \mathcal{L}(H)$ be a contraction. We call $U \in \mathcal{L}(K)$ a *unitary dilation* of A if U is a unitary operator and U is a dilation of A. A *minimal unitary dilation* of A is a unitary dilation U of A with the additional property that

$$(1) \qquad \bigvee_{n=-\infty}^{\infty} U^n H = K.$$

An operator $V \colon K \to K$ is called a *co-isometry* if V^* is an isometry. Since $V^{**} = V$, co-isometries are adjoints of isometries. To construct unitary dilations we shall employ the next lemma.

LEMMA 4.1. *The minimal isometric dilation of a co-isometry is unitary.*

PROOF. Let $A \in \mathcal{L}(H)$ be a co-isometry, and let $T \in \mathcal{L}(K)$ be its minimal isometric dilation. We know (see Theorem 3.1) that $K = H \boxplus \ell_2(M)$, where $M = \overline{\operatorname{Im} D_A}$, and that

$$T = \begin{bmatrix} A & 0 \\ B & C \end{bmatrix} \colon H \boxplus \ell_2(M) \to H \boxplus \ell_2(M),$$

where C is the forward shift on $\ell_2(M)$ and

$$B \colon H \to \ell_2(M), \qquad Bh = (D_A h, 0, 0, \ldots).$$

Since T is an isometry, we already know that $T^*T = I$. We have to prove that $TT^* = I$.

Since A is a co-isometry, the operator A^* preserves the inner product, and thus $AA^* = I$. It follows that $D_{A^*} = 0$, and hence

$$(2) \qquad AD_A = D_{A^*}A = 0.$$

From $AA^* = I$ it also follows that A^*A is a projection. Indeed, $(A^*A)^2 = A^*(AA^*)A = A^*A$. Thus $I - A^*A$ is an orthogonal projection, and we may conclude that $D_A^2 = D_A$. In other words $D_A|M$ is the identity operator on M. Now note that $B^* \colon \ell_2(M) \to H$ is given by

$$(3) \qquad B^*(x_0, x_1, \ldots) = D_A x_0 = x_0.$$

The second equality in (3) stems from the fact that the entries x_j in (3) are in M. In particular, $x_0 \in M$ and hence $D_A x_0 = x_0$, by the remarks made above. From the first equality in (3) and (2) we see that $AB^* = 0$. The second equality in (3) implies that BB^* is the orthogonal projection of $\ell_2(M)$ onto the first coordinate space, and hence $BB^* + CC^*$ is the identity operator on $\ell_2(M)$. We have now all ingredients to check that $TT^* = I$. $\square$

THEOREM 4.2. *Let $A \in \mathcal{L}(H)$ be a contraction. Let $\widetilde{T}$ be the minimal isometric dilation of A, and let V be the minimal isometric dilation of $\widetilde{T}^*$. Then $U = V^*$ is a minimal unitary dilation of A.*

PROOF. By Lemma 4.1 the operator U is unitary. Let $\widetilde{K}$ be the space on which $\widetilde{T}$ acts. From the construction of V (cf., Theorem 3.1) we see that $\widetilde{K}$ is invariant under U and $U|\widetilde{K} = \widetilde{T}$. Now, take h_1, h_2 in H. Then

$$(4) \qquad \langle A^n h_1, h_2 \rangle = \langle \widetilde{T}^n h_1, h_2 \rangle = \langle U^n h_1, h_2 \rangle, \qquad n = 1, 2, \dots .$$

The first identity in (4) follows from the fact that $\widetilde{T}$ is a dilation of A, and for the second we use that $U|\widetilde{K} = \widetilde{T}$. From (4) we conclude that U is a dilation of A. To prove the minimality, note that

$$(5) \qquad \widetilde{K} = \bigvee_{n=0}^{\infty} \widetilde{T}^n H = \bigvee_{n=0}^{\infty} U^n H.$$

For the first equality in (5) we use the minimality of $\widetilde{T}$ and second follows from $U|\widetilde{K} = \widetilde{T}$. Now, (5) shows that $\widetilde{K} \subset \bigvee_{n=-\infty}^{\infty} U^n H$. But then also

$$V^r \widetilde{K} = U^{-r} \widetilde{K} \subset \bigvee_{n=-\infty}^{\infty} U^n H, \qquad r \geq 0,$$

and thus (because of the minimality of the dilation V) the identity (1) holds with K equal to the space on which U acts. $\square$

Let $A \in \mathcal{L}(H)$ be a contraction. It is instructive to compute in more detail the minimal unitary dilation U of A defined in Theorem 4.2. Let $\widetilde{T} \in \mathcal{L}(\widetilde{K})$ be as in Theorem 4.2. To find $U = V^*$ we apply Theorem 3.1 and take adjoints. Put $L = \overline{\operatorname{Im} D_{\widetilde{T}^*}}$. Then

$$U = \begin{bmatrix} \widetilde{C} & 0 \\ \widetilde{B} & \widetilde{T} \end{bmatrix} : \ell_2(L) \boxplus \widetilde{K} \to \ell_2(L) \boxplus \widetilde{K},$$

where $\widetilde{C}$ is the backward shift on $\ell_2(L)$ and

$$\widetilde{B} : \ell_2(L) \to \widetilde{K}, \qquad \widetilde{B}(y_0, y_1, \dots) = y_0.$$

Here we used (see the proof of Lemma 4.1) that $D_{\widetilde{T}^*}|L$ is the identity operator on L. To determine L we use that $\widetilde{K} = H \boxplus \ell_2(M)$, where $M = \overline{\operatorname{Im} D_A}$, and that

$$\widetilde{T} = \begin{bmatrix} A & 0 \\ B & C \end{bmatrix} : H \boxplus \ell_2(M) \to H \boxplus \ell_2(M),$$

with C the forward shift on $\ell_2(M)$ and

$$B : H \to \ell_2(M), \qquad Bh = (D_A h, 0, 0, \dots).$$

Let $\tau_0 : M \to \ell_2(M)$ be the operator $\tau_0(m) = (m, 0, 0, \dots)$. Then $B = \tau_0 D_A$. Since $\widetilde{T}$ is an isometry, we know (see the proof of Lemma 4.1) that

$$D_{\widetilde{T}^*} = I - \widetilde{T}\widetilde{T}^* = \begin{bmatrix} I - AA^* & -AB^* \\ -BA^* & I - BB^* - CC^* \end{bmatrix}.$$

From the definitions of B and C it follows that

$$I - BB^* - CC^* = \tau_0^* A^* A \tau_0.$$

Next use formula (2) in Section XXVII.1 to show that

$$AB^* = AD_A \tau_0^* = D_{A^*} A \tau_0^*.$$

It follows that $D_{\widetilde{T}_*}$ factorizes as follows:

$$(6) \qquad D_{\widetilde{T}_*} = \begin{bmatrix} D_{A^*} \\ -\tau_0 A^* \end{bmatrix} [D_{A^*} \quad -A\tau_0^*] \colon H \boxplus \ell_2(M) \to H \boxplus \ell_2(M).$$

Note that $A(\operatorname{Im} D_A) \subset \operatorname{Im} D_{A^*}$, and hence $AM \subset M_*$, where $M_* = \overline{\operatorname{Im} D_{A^*}}$. We conclude that $L = JM_*$, where

$$J \colon M_* \to L, \qquad Jz = \begin{bmatrix} D_{A^*} z \\ -\tau_0 A^* z \end{bmatrix}.$$

Note that $\|Jz\| = \|z\|$ by Lemma 1.1(i) applied to A^*. Hence J is unitary. Now define

$$\widetilde{J} \colon \ell_2(M_*) \boxplus H \boxplus \ell_2(M) \to \ell_2(L) \boxplus H \boxplus \ell_2(M),$$

$$\widetilde{J} \begin{pmatrix} (x_\nu)_{\nu=0}^\infty \\ h \\ (m_\nu)_{\nu=0}^\infty \end{pmatrix} = \begin{pmatrix} (Jx_\nu)_{\nu=0}^\infty \\ h \\ (m_\nu)_{\nu=0}^\infty \end{pmatrix}.$$

Then $\widetilde{J}$ is a unitary operator and

$$(7) \qquad \widetilde{J}^{-1} U \widetilde{J} = \begin{bmatrix} C_* & 0 & 0 \\ B_* & A & 0 \\ D & B & C \end{bmatrix},$$

where C_* is the backward shift on $\ell_2(M_*)$ and the operators B_* and D are given by

$$B_* \colon \ell_2(M_*) \to H, \qquad B_*(z_0, z_1, \ldots) = D_{A^*} z_0,$$

$$D \colon \ell_2(M_*) \to \ell_2(M), \qquad D(z_0, z_1, \ldots) = (-A^* z_0, 0, 0, \ldots).$$

Note that $\widetilde{J}^{-1} U \widetilde{J}$ acts on $\ell_2(M_*) \boxplus H \boxplus \ell_2(M)$. Let us write an element of the latter space as a double infinite column, as follows

$$(\ldots, z_2, z_1, z_0, h, m_0, m_1, m_2, \ldots)^T,$$

where the superscript T denotes transpose. Then $\widetilde{J}^{-1} U \widetilde{J}$ is represented by the following double infinite operator matrix:

$$(8) \qquad \begin{bmatrix} \ddots & & & & & & \\ & I & & & & & \\ & & I & 0 & 0 & 0 & \\ & & 0 & D_{A^*} & A & 0 & \\ & & 0 & -A^* & D_A & 0 & \\ & & 0 & 0 & 0 & I & \\ & & & & & & I \\ & & & & & & & \ddots \end{bmatrix},$$

with the blank spots denoting zero operator entries. Note that A acts from H into H, and hence A is the $(0,0)$-entry. By interchanging the columns containing A and $-A^*$ the 2×2 operator matrix in the center becomes the rotation operator associated with A. Since $\widetilde{J}h = h$ for each h, it is straightforward to check that operator (8) (which is equal to $\widetilde{J}^{-1}U\widetilde{J}$) is again a minimal unitary dilation of A.

THEOREM 4.3. *Let $A \in \mathcal{L}(H)$ be a contraction, and let $U \in \mathcal{L}(K)$, $U' \in \mathcal{L}(K')$ be minimal unitary dilations of A. Then there exists a unique unitary operator $J\colon K \to K'$ such that*

(i) *$Jh = h$ for each $h \in H$,*

(ii) *$U' = JUJ^{-1}$.*

The proof of Theorem 4.3 follows the same line of reasoning as that of Theorem 3.2. Instead of the finite sums appearing in formula (4) of Section XXVII.3 one uses finite sums of the following type

$$\sum_{j=-N}^{N} U^j h_j,$$

where $h_{-N}, \ldots, h_N$ are in H. We omit further details.

Theorem 4.3 shows that in essence the minimal unitary dilation is unique, and it allows us to speak about *the* minimal unitary dilation of a contraction.

As an illustration let us compute the minimal unitary dilation of A for the case when $A = S$ is the forward shift on ℓ_2. Since $S^*S = I$, we have $M = \overline{\operatorname{Im} D_S} = \{0\}$. Furthermore, $I - SS^*$ is the projection of ℓ_2 onto the first coordinate space. It follows that

$$M_* = \operatorname{span}\left\{ \begin{bmatrix} 1 \\ 0 \\ 0 \\ \vdots \end{bmatrix} \right\}.$$

Let us identify M_* with $\mathbb{C}$ by setting $\lambda = [\lambda, 0, 0, \ldots]^T$ for each $\lambda \in \mathbb{C}$. Then $\ell_2(M_*) = \ell_2$, and the minimal unitary dilation of S is the operator

$$\begin{bmatrix} S^* & 0 \\ D & S \end{bmatrix} \colon \ell_2 \boxplus \ell_2 \to \ell_2 \boxplus \ell_2,$$

where $D\colon \ell_2 \to \ell_2$ is given by $D(x_1, x_2, \ldots) = (x_1, 0, 0, \ldots)$. By identifying $\ell_2 \boxplus \ell_2$ with $\ell_2(\mathbb{Z})$, via

$$\begin{pmatrix} (x_\nu)_{\nu=1}^{\infty} \\ (y_\nu)_{\nu=1}^{\infty} \end{pmatrix} = (\ldots, x_2, x_1, y_1, y_2, \ldots),$$

we see that the bilateral forward shift on $\ell_2(\mathbb{Z})$ is the minimal unitary dilation of S.

XXVII.5 INTERMEZZO ABOUT 2×2 OPERATOR MATRIX COMPLETIONS

Consider the following 2×2 operator matrix:

$$(1) \qquad T = \begin{bmatrix} T_{11} & Z \\ T_{21} & T_{22} \end{bmatrix} : H_1 \boxplus H_2 \to K_1 \boxplus K_2.$$

Here T_{11}, T_{21} and T_{22} are given Hilbert space operators, and, as before, $\boxplus$ stands for the Hilbert space direct sum. This section presents the solution of the following problem. Determine $Z: H_2 \to K_1$ (if possible) such that the operator T in (1) is a contraction. If this problem has a solution, then the operators

$$(2) \qquad \begin{bmatrix} T_{11} \\ T_{21} \end{bmatrix} : H_1 \to K_1 \boxplus K_2, \qquad [T_{21} \quad T_{22}]: H_1 \boxplus H_2 \to K_2$$

must be contractions. The next theorem shows that these conditions are not only necessary but also sufficient. Moreover the theorem gives a full parametrization of all possible solutions Z. Note that if one of the operators in (2) is a contraction, then T_{21} is a contraction, and hence in that case the operators $D_{T_{21}}: H_1 \to H_1$ and $D_{T_{21}^*}: K_2 \to K_2$ are well-defined operators.

THEOREM 5.1. *Assume that the operators in (2) are contractions. Put* $D = D_{T_{21}}$ *and* $D_* = D_{T_{21}^*}$. *Then*

(i) *there exists a unique contraction* $A: H_1 \to K_1$ *such that*

$$T_{11} = AD, \qquad \operatorname{Ker} A \supset \operatorname{Ker} D;$$

(ii) *there exists a unique contraction* $B: H_2 \to K_2$ *such that*

$$T_{22} = D_* B, \qquad \operatorname{Im} B \subset \overline{\operatorname{Im} D_*};$$

(iii) *the operator* T *in* (1) *is a contraction if and only if*

$$(3) \qquad Z = D_{A^*} \Delta D_B - A T_{21}^* B,$$

where $\Delta: H_2 \to H_1$ *is an arbitrary contraction. Furthermore, in* (3) *one may choose* Δ *so that* $\operatorname{Ker} \Delta \supset \operatorname{Ker} D_B$ *and* $\operatorname{Im} \Delta \subset \overline{\operatorname{Im} D_{A^*}}$, *and in this case* Δ *is uniquely determined by* Z.

To prove the theorem we need the following auxiliary result (which is usually called the *Douglas factorization lemma*).

LEMMA 5.2. *Let* $A: H \to K_1$ *and* $B: H \to K_2$ *be given Hilbert space operators. Then* $B^* B \le A^* A$ *if and only if* $B = CA$ *with* $C: K_1 \to K_2$ *a contraction. Furthermore, in this case the contraction* C *can be chosen such that*

$$(4) \qquad \operatorname{Ker} C \supset \operatorname{Ker} A^*,$$

and with this additional condition C is uniquely determined.

PROOF. Assume $B = CA$ with C a contraction. Then for each $x \in H$ we have

$$\langle B^*Bx, x \rangle = \langle Bx, Bx \rangle = \langle CAx, CAx \rangle = \|CAx\|^2 \le \|Ax\|^2 = \langle A^*Ax, x \rangle,$$

and hence $B^*B \le A^*A$.

To prove the converse implication, assume that $B^*B \le A^*A$. It follows that

$$(5) \qquad \|Bx\| \le \|Ax\|, \qquad x \in H.$$

Define $C_0 \colon \operatorname{Im} A \to \operatorname{Im} B$ by setting $C_0(Ax) = Bx$. Then C_0 is well-defined. Indeed, if $Ax_1 = Ax_2$, then $A(x_1 - x_2) = 0$, and thus (5) implies that $B(x_1 - x_2) = 0$, which yields $Bx_1 = Bx_2$. From (5) and the definition of C_0, we obtain $\|C_0\| \le 1$. But then C_0 extends by continuity to a contraction, also denoted by C_0, from $\overline{\operatorname{Im} A}$ to $\overline{\operatorname{Im} B}$. Now, define $C \colon K_1 \to K_2$ by setting

$$Cx = \begin{cases} C_0 x, & \text{if } x \in \overline{\operatorname{Im} A}, \\ 0, & \text{if } x \perp \overline{\operatorname{Im} A}. \end{cases}$$

Then C is a contraction, $CA = B$ and (4) is fulfilled.

It remains to prove the uniqueness. Assume $\widetilde{C} \colon K_1 \to K_2$ is a second contraction such that

$$(6) \qquad B = \widetilde{C}A, \qquad \operatorname{Ker} \widetilde{C} \supset \operatorname{Ker} A^*.$$

From the first part of (6) it follows that $\widetilde{C}$ and C coincide on $\operatorname{Im} A$. By continuity we see that C and $\widetilde{C}$ coincide on $\overline{\operatorname{Im} A}$. Since $\operatorname{Ker} A^* = \overline{\operatorname{Im} A}^{\perp}$, we also know that C and $\widetilde{C}$ coincide on the orthogonal complement of $\overline{\operatorname{Im} A}$. Hence $C = \widetilde{C}$. $\square$

PROOF OF THEOREM 5.1. We split the proof into five parts. Parts (a) and (b) contain the proofs of statements (i) and (ii), the other parts concern (iii).

Part (a). We prove (i). Since the first operator in (2) is a contraction, we have

$$I \ge [T_{11}^* \ \ T_{21}^*] \begin{bmatrix} T_{11} \\ T_{21} \end{bmatrix} = T_{11}^*T_{11} + T_{21}^*T_{21}.$$

Thus $T_{11}^*T_{11} \le D^2$, where $D^* = D$. By the Douglas factorization lemma there exists a unique contraction $A \colon H_1 \to K_1$ such that $T_{11} = AD$ and $\operatorname{Ker} A \supset \operatorname{Ker} D$, and so (i) is proved.

Part (b). We prove (ii). Note that (ii) is a dual version of (i). Since $[T_{21} \ \ T_{22}]$ is a contraction, we have

$$I \ge [T_{21} \ \ T_{22}] \begin{bmatrix} T_{21}^* \\ T_{22}^* \end{bmatrix} = T_{21}T_{21}^* + T_{22}T_{22}^*,$$

and therefore $D_*^2 \geq T_{22}T_{22}^*$, where $D_* = D_*^*$. By the Douglas factorization lemma there exists a unique contraction $C: K_2 \to H_2$ such that $T_{22}^* = CD_*$ and $\operatorname{Ker} C \supset \operatorname{Ker} D_*$. Put $B = C^*$. Then $B: H_2 \to K_2$ is a contraction, $T_{22} = D_*B$ and

$$\overline{\operatorname{Im} B} = (\operatorname{Ker} C)^\perp \subset (\operatorname{Ker} D_*)^\perp = \overline{\operatorname{Im} D_*},$$

which proves (ii).

   Part (c). In this part we show that

$$(7) \qquad I - \begin{bmatrix} T_{21}^* \\ T_{22}^* \end{bmatrix} [T_{21} \quad T_{22}] = \begin{bmatrix} D & -T_{21}^*B \\ 0 & D_B \end{bmatrix}^* \begin{bmatrix} D & -T_{21}^*B \\ 0 & D_B \end{bmatrix},$$

where B is the contraction defined in (ii). Thus $T_{22} = D_*B$. Note that $T_{21}^*D_* = DT_{21}^*$, by Lemma 1.1. It follows that

$$T_{21}^*T_{22} = T_{21}^*D_*B = DT_{21}^*B.$$

Furthermore,

$$I - T_{22}^*T_{22} = I - B^*D_*^2B = I - B^*B + B^*T_{21}T_{21}^*B.$$

It follows that

$$\begin{aligned}
I - \begin{bmatrix} T_{21}^* \\ T_{22}^* \end{bmatrix} [T_{21} \quad T_{22}] &= \begin{bmatrix} I - T_{21}^*T_{21} & -T_{21}^*T_{22} \\ -T_{22}^*T_{21} & I - T_{22}^*T_{22} \end{bmatrix} \\
&= \begin{bmatrix} D^2 & -DT_{21}^*B \\ -B^*T_{21}D & I - B^*B + B^*T_{21}T_{21}^*B \end{bmatrix} \\
&= \begin{bmatrix} D & 0 \\ -B^*T_{21} & I \end{bmatrix} \begin{bmatrix} I & 0 \\ 0 & I - B^*B \end{bmatrix} \begin{bmatrix} D & -T_{21}^*B \\ 0 & I \end{bmatrix}.
\end{aligned}$$

Now factorize $I - B^*B$ as $D_B^*D_B$, and we obtain the factorization (7).

   Part (d). Let $Z: H_2 \to K_1$ be given by (3). We shall show that with this choice of Z the operator T in (1) is a contraction. From the definition of Z and (i) it follows that

$$(8) \qquad [T_{11} \quad Z] = [A \quad D_{A^*}\Delta] \begin{bmatrix} D & -T_{21}^*B \\ 0 & D_B \end{bmatrix}.$$

Now $[A \quad D_{A^*}\Delta]$ is a contraction. Indeed,

$$\left\| \begin{bmatrix} A^*x \\ \Delta^*D_{A^*}x \end{bmatrix} \right\|^2 = \|A^*x\|^2 + \|\Delta^*D_{A^*}x\|^2$$

$$\leq \|A^*x\|^2 + \|D_{A^*}x\|^2 = \|x\|^2.$$

Here we used that Δ and A^* are contractions. Since the first term in the right hand side of (8) is a contraction, we can apply the Douglas factorization lemma to show that

$$\begin{bmatrix} T_{11}^* \\ Z^* \end{bmatrix} [T_{11} \quad Z] \leq \begin{bmatrix} D & -T_{21}^* B \\ 0 & D_B \end{bmatrix}^* \begin{bmatrix} D & -T_{21}^* B \\ 0 & D_B \end{bmatrix}.$$

Now use (7). It follows that

$$I - T^*T = I - \begin{bmatrix} T_{11}^* \\ Z^* \end{bmatrix} [T_{11} \quad Z] - \begin{bmatrix} T_{21}^* \\ T_{22}^* \end{bmatrix} [T_{21} \quad T_{22}] \geq 0,$$

and hence T is a contraction.

Part (e). Assume $Z: H_2 \to K_1$ is such that T in (1) is a contraction. We have to prove that Z has the desired representation. Since T is a contraction, we have

$$\begin{bmatrix} T_{11}^* \\ Z^* \end{bmatrix} [T_{11} \quad Z] \leq I - \begin{bmatrix} T_{21}^* \\ T_{22}^* \end{bmatrix} [T_{21} \quad T_{22}]$$

$$= \begin{bmatrix} D & -T_{21}^* B \\ 0 & D_B \end{bmatrix}^* \begin{bmatrix} D & -T_{21}^* B \\ 0 & D_B \end{bmatrix}.$$

because of (7). So, by the Douglas factorization lemma, the operator $[T_{11} \quad Z]$ factorizes as

$$(9) \qquad [T_{11} \quad Z] = [X_1 \quad X_2] \begin{bmatrix} D & -T_{21}^* B \\ 0 & D_B \end{bmatrix},$$

where $[X_1 \quad X_2]$ is a contraction such that

$$(10) \qquad \mathrm{Ker}[X_1 \quad X_2] \supset \mathrm{Ker} \begin{bmatrix} D & 0 \\ -B^* T_{21} & D_B \end{bmatrix}.$$

The latter condition implies that $[X_1 \quad X_2]$ is uniquely determined by Z. From (9) we conclude that

$$(11) \qquad T_{11} = X_1 D, \qquad Z = X_2 D_B - X_1 T_{21}^* B.$$

Next we prove that $\mathrm{Ker}\, D \subset \mathrm{Ker}\, X_1$. Take x in $\mathrm{Ker}\, D$. Then, by Lemma 1.1 applied to T_{21},

$$D_* T_{21} x = T_{21} D x = 0,$$

and thus $T_{21} x \in \mathrm{Ker}\, D_*$. According to our choice of B,

$$\mathrm{Ker}\, B^* = (\overline{\mathrm{Im}\, B})^\perp \supset (\overline{\mathrm{Im}\, D_*})^\perp = \mathrm{Ker}\, D_*.$$

Thus $B^* T_{21} x = 0$. But then, by (10), the vector $\begin{bmatrix} x \\ 0 \end{bmatrix}$ is in $\mathrm{Ker}[X_1 \quad X_2]$, and hence $x \in \mathrm{Ker}\, X_1$. So we see that X_1 is a contraction such that

$$T_{11} = X_1 D, \qquad \mathrm{Ker}\, X_1 \supset \mathrm{Ker}\, D.$$

Now apply the uniqueness part of (i), and we obtain $X_1 = A$.

Since $[X_1 \quad X_2] = [A \quad X_2]$ is a contraction, we have

$$0 \leq I - [A \quad X_2] \begin{bmatrix} A^* \\ X_2^* \end{bmatrix} = I - AA^* - X_2 X_2^*,$$

and hence $X_2 X_2^* \leq I - AA^* = D_{A^*}^2$. The Douglas factorization lemma yields a unique contraction $\widetilde{\Delta} \colon H_2 \to K_1$ such that

$$(12) \qquad\qquad X_2 = D_{A^*} \widetilde{\Delta}, \qquad \mathrm{Im}\, \widetilde{\Delta} \subset \overline{\mathrm{Im}\, D_{A^*}}.$$

From (12) it follows that $\mathrm{Ker}\, X_2 = \mathrm{Ker}\, \widetilde{\Delta}$. The inclusion $\mathrm{Ker}\, \widetilde{\Delta} \subset \mathrm{Ker}\, X_2$ is obvious. To prove the reverse inclusion, take $x \in \mathrm{Ker}\, X_2$. Then, by the first part of (12), the vector $\widetilde{\Delta} x$ is in $\mathrm{Ker}\, D_{A^*}$. By the second part of (12) we have $\widetilde{\Delta} x \in (\mathrm{Ker}\, D_{A^*})^{\perp}$. So $\widetilde{\Delta} x = 0$, and $\mathrm{Ker}\, \widetilde{\Delta} = \mathrm{Ker}\, X_2$. Now apply (10) to show that $\mathrm{Ker}\, D_B \subset \mathrm{Ker}\, X_2$, and we see that $\mathrm{Ker}\, \widetilde{\Delta} \supset \mathrm{Ker}\, D_B$. The second part of (11), the first part of (12) and $X_1 = A$ now yield the desired representation for Z. $\quad\square$

In the following corollaries we specify Theorem 5.1 for a number of special cases.

COROLLARY 5.3. *The operator*

$$[T_1 \quad T_2] \colon H_1 \boxplus H_2 \to K$$

is a contraction if and only if T_1 is a contraction and

$$T_2 = D_{T_1^*} \Delta,$$

where $\Delta \colon H_2 \to K$ is a contraction. Furthermore, in this case one can choose Δ so that $\mathrm{Im}\, \Delta \subset \overline{\mathrm{Im}\, D_{T_1^*}}$, *and with this extra property Δ is unique.*

PROOF. Apply Theorem 5.1 with $K_1 = K$ and $K_2 = \{0\}$. $\quad\square$

COROLLARY 5.4. *The operator*

$$\begin{bmatrix} T_1 \\ T_2 \end{bmatrix} \colon H \to K_1 \boxplus K_2$$

is a contraction if and only if T_1 is a contraction and $T_2 = \Delta D_{T_1}$, where $\Delta \colon H \to K_2$ is a contraction. Furthermore, in this case one can choose Δ so that $\mathrm{Ker}\, \Delta \supset \mathrm{Ker}\, D_{T_1}$, and with this extra property Δ is unique.

PROOF. Apply Theorem 5.1 with K_1 and K_2 interchanged, $H_1 = \{0\}$, and $H_2 = H$. $\square$

COROLLARY 5.5. *The operator*

$$\begin{bmatrix} T_{11} & Z \\ 0 & T_{22} \end{bmatrix} : H_1 \boxplus H_2 \to K_1 \boxplus K_2$$

is a contraction if and only if T_{11} and T_{22} are contractions and

$$Z = D_{T_{11}^*} \Delta D_{T_{22}},$$

where $\Delta: H_2 \to K_1$ is an arbitrary contraction. Furthermore, one can choose Δ so that

$$\operatorname{Ker} \Delta \supset \operatorname{Ker} D_{T_{22}}, \qquad \operatorname{Im} \Delta \subset \overline{\operatorname{Im} D_{T_{11}}},$$

and with these extra conditions Δ is uniquely determined by Z.

PROOF. Apply Theorem 5.1 with $T_{21} = 0$. $\square$

XXVII.6 THE COMMUTANT LIFTING THEOREM

In this section we prove the following theorem (which is known as the *commutant lifting theorem*).

THEOREM 6.1. *Let $A \in \mathcal{L}(H)$ be a contraction, and let $T \in \mathcal{L}(K)$ be a minimal isometric dilation of A. Assume $B \in \mathcal{L}(H)$ commutes with A. Then there exists $S \in \mathcal{L}(K)$ such that*

$$(1) \qquad S = \begin{bmatrix} B & 0 \\ * & * \end{bmatrix} : H \boxplus H^\perp \to H \boxplus H^\perp,$$

the operator S commutes with T and $\|S\| = \|B\|$.

To prove the commutant lifting theorem one may, without loss of generality, assume that B is a contraction. From Proposition 3.3 we know that the minimal isometric dilation of A can be constructed by repeatedly applying one step dilations and taking limits in the strong operator topology. Therefore it is convenient to prove first the following one step lifting lemma. As in Section XXVII.3, we write $C_{[1]}$ for the one step dilation of a contraction C.

LEMMA 6.2. *Let $A \in \mathcal{L}(H)$ and $A' \in \mathcal{L}(H')$ be contractions, and let $B: H \to H'$ be a contraction such that $A'B = BA$. Put $M = \overline{\operatorname{Im} D_A}$ and $M' = \overline{\operatorname{Im} D_{A'}}$. Then there exists an operator*

$$(2) \qquad S = \begin{bmatrix} B & 0 \\ C & D \end{bmatrix} : H \boxplus M \to H' \boxplus M$$

such that S is a contraction and

$$(3) \qquad (A')_{[1]} S = S A_{[1]}.$$

PROOF. We begin with a few remarks. Let

$$(4) \qquad S = \begin{bmatrix} B & F \\ C & D \end{bmatrix} : H \boxplus M \to H' \boxplus M'$$

be an arbitrary contraction such that (3) holds. Recall that

$$A_{[1]} = \begin{bmatrix} A & 0 \\ D_A & 0 \end{bmatrix} : H \boxplus M \to H \boxplus M,$$

$$(A')_{[1]} = \begin{bmatrix} A' & 0 \\ D_{A'} & 0 \end{bmatrix} : H' \boxplus M' \to H' \boxplus M'.$$

Thus, because of the intertwining relation (3), the following identities must hold:

$$(5a) \qquad A'F = 0, \qquad D_{A'}F = 0,$$

$$(5b) \qquad \begin{bmatrix} C & D \end{bmatrix} \begin{bmatrix} A \\ D_A \end{bmatrix} = D_{A'}B.$$

From the second identity in (5a) we see that $F = (A')^* A'F$, and hence the first identity in (5a) shows that $F = 0$, which explains the choice of the $(1,2)$-entry of the operator matrix in (2). Next, we multiply the operator S from right by the rotation operator R_A associated with A and apply (5b). Recall (see Section XXVII.1) that

$$R_A = \begin{bmatrix} A & D_{A^*} \\ D_A & -A^* \end{bmatrix} : H \boxplus M_* \to H \boxplus M,$$

where $M_* = \overline{\operatorname{Im} D_{A^*}}$. It follows that

$$(6) \qquad SR_A = \begin{bmatrix} B & 0 \\ C & D \end{bmatrix} R_A = \begin{bmatrix} BA & BD_{A^*} \\ D_{A'}B & X \end{bmatrix},$$

with

$$X = \begin{bmatrix} C & D \end{bmatrix} \begin{bmatrix} D_{A^*} \\ -A^* \end{bmatrix} : M_* \to M'.$$

On the other hand, by multiplying (6) on the right by $R_A^{-1} = R_{A^*}$, we see that

$$(7) \qquad S = \begin{bmatrix} B & 0 \\ C & D \end{bmatrix} = \begin{bmatrix} BA & BD_{A^*} \\ D_{A'}B & X \end{bmatrix} R_{A^*}.$$

Note that SR_A is a contraction (because S is a contraction and R_A is unitary). Thus the 2×2 operator matrix in the right hand side of (6) is a contraction, and the element

X can be determined via Theorem 5.1. As soon as X has been found, we can construct the entries C and D in S by using (7). From the above remarks it is clear how to prove the lemma.

The fact that A, A' and B are contractions implies that the operators

$$\begin{bmatrix} BA \\ D_{A'}B \end{bmatrix} = \begin{bmatrix} A' \\ D_{A'} \end{bmatrix} B \colon H \to H' \boxplus M',$$

$$[BA \quad BD_{A^*}] = B[A \quad D_{A^*}] \colon H \boxplus M_* \to M',$$

are contractions. Here we used that $BA = A'B$. So, by Theorem 5.1, there exists an operator $X \colon M_* \to M'$ such that

$$(8) \qquad \begin{bmatrix} BA & BD_{A^*} \\ D_{A'}B & X \end{bmatrix} \colon H \boxplus M_* \to H' \boxminus M'$$

is a contraction. It follows that

$$S := \begin{bmatrix} BA & BD_{A^*} \\ D_{A'}B & X \end{bmatrix} R_{A^*} \colon H \boxplus M \to H' \boxplus M'$$

is a contraction. By a direct multiplication one checks (using $AD_A = D_{A^*}A$) that

$$S = \begin{bmatrix} B & 0 \\ C & D \end{bmatrix},$$

with

$$(9a) \qquad C = D_{A'}BA^* + XD_{A^*} \colon H \to M',$$

$$(9b) \qquad D = D_{A'}BD_A - XA \colon M \to M'.$$

Since SR_A is equal to the operator (8), we see that with C and D as in (9a), (9b) the identity (5b) holds, and hence S satisfies (3). $\quad \square$

Note that the proof of Lemma 6.2 contains a method to obtain the set of all contractions (2) satisfying (3).

PROOF OF THEOREM 6.1. The case $B = 0$ is trivial. Therefore without loss of generality we may assume that $\|B\| = 1$. Put $M = \overline{\operatorname{Im} D_A}$, and set $H_N = H \boxplus M \boxplus \cdots \boxplus M$, where N is equal to the number of copies of M in this direct sum. By $A_{[N]}$ we denote the N-step dilation of A which (see Section XXVII.3) acts on H_N according to the following operator matrix:

$$\begin{bmatrix} A & 0 & \cdots & 0 & 0 \\ D_A & 0 & \cdots & 0 & 0 \\ 0 & I & \cdots & 0 & 0 \\ \vdots & \vdots & & \vdots & \vdots \\ 0 & 0 & \cdots & I & 0 \end{bmatrix}.$$

By definition $A_{[N]}$ is the one step dilation of $A_{[N-1]}$. Thus, by repeatedly applying the one step lifting result of Lemma 6.2, we find a sequence of contractions $S_1, S_2, \ldots$ such that

$$(10) \qquad S_1 = \begin{bmatrix} B & 0 \\ * & * \end{bmatrix} : H \boxplus M \to H \boxplus M,$$

$$(11) \qquad S_{N+1} = \begin{bmatrix} S_N & 0 \\ * & * \end{bmatrix} : H_N \boxplus M \to H_N \boxplus M,$$

$$(12) \qquad A_{[N]} S_N = S_N A_{[N]}, \qquad N = 1, 2, \ldots .$$

Let $\tau_N : H_N \to H \boxplus \ell_2(M)$ be defined as in Proposition 3.3, i.e.,

$$\tau_N(h, x_0, \ldots, x_{N-1}) = (h, x_0, \ldots, x_{N-1}, 0, 0, \ldots).$$

Put $\pi_N = \tau_N^*$, and set $K_N = \tau_N H_N$. Note that

$$(13) \qquad H \subset K_1 \subset K_2 \subset \cdots, \qquad \bigvee_{N=1}^{\infty} K_N = K = H \boxplus \ell_2(M).$$

Define $V_N : K_N \to K_N$ by $V_N \tau_N = \tau_N S_N^*$. From (11) it follows that K_N is invariant under V_{N+1} and

$$V_{N+1} | K_N = V_N, \qquad N \geq 1.$$

Thus there exists an operator V on $\bigcup \{K_N \mid N \geq 1\}$ such that V coincides with V_N on K_N. Obviously, V is a contraction. So, by continuity, V extends to a contraction on all of $K = H \boxplus \ell_2(M)$. Now, put $S = V^*$. We claim that S has the desired properties.

From the definition of V we see that $V \tau_N = \tau_N S_N^*$, and thus

$$(14) \qquad \pi_N S = S_N \pi_N, \qquad N \geq 1.$$

So, by (10), the operator S has the desired form (1). Next, take $y \in K$. Fix a positive integer n, and let $N \geq n$. Note that $\tau_N \pi_N$ is the orthogonal projection of K onto K_N. Hence $\pi_n = \pi_n \tau_N \pi_N$ because $N \geq n$. So

$$\pi_n S \tau_N = \pi_n \tau_N \pi_N S \tau_N = \pi_n \tau_N S_N \pi_N \tau_N = \pi_n \tau_N S_N,$$

because $\pi_N \tau_N$ is the identity on H_N. According to Proposition 3.3,

$$\tau_N A_{[N]} \pi_N y \to T y \qquad (N \to \infty).$$

Thus, use also (12) and (14) to obtain,

$$\pi_n S(Ty) = \lim_{N \to \infty} \pi_n S(\tau_N A_{[N]} \pi_N y)$$

$$= \pi_n \left(\lim_{N \to \infty} \tau_N S_N A_{[N]} \pi_N y \right)$$

$$= \pi_n \left(\lim_{N \to \infty} \tau_N A_{[N]} S_N \pi_N y \right)$$

$$= \pi_n \left(\lim_{N \to \infty} \tau_N A_{[N]} \pi_N S y \right) = \pi_n T S y.$$

This holds for each n, and so $ST = TS$. By virtue of (1), we have $\|B\| \leq \|S\|$. Since S is a contraction and $\|B\| = 1$, we obtain $\|B\| = \|S\|$. $\square$

There is a slightly more general version of Theorem 6.1. For this purpose we need the following definition. An operator $T \in \mathcal{L}(K)$ is called a *lifting* of $A \in \mathcal{L}(H)$ if A^* is a part of T^* or, equivalently, if T admits the following 2×2 operator matrix representation

$$(15) \qquad T = \begin{bmatrix} A & 0 \\ * & * \end{bmatrix} : H \boxplus H^\perp \to H \boxplus H^\perp.$$

Note that the operator S in (1) is a lifting of B.

If T is the minimal isometric dilation of a contraction A, then we know from Theorem 3.1 that T is a lifting of A. It turns out that Theorem 6.1 remains true with "minimal isometric dilation" replaced by "isometric lifting".

THEOREM 6.3. *Let $A \in \mathcal{L}(H)$ be a contraction, and let $T \in \mathcal{L}(K)$ be an isometric lifting of A. Assume that $B \in \mathcal{L}(H)$ commutes with A. Then B has a lifting $S \in \mathcal{L}(K)$ such that S commutes with T and $\|S\| = \|B\|$.*

PROOF. Set

$$(16) \qquad \widetilde{K} = \bigvee_{n=0}^{\infty} T^n H.$$

The space $\widetilde{K}$ is invariant under T, and $H \subset \widetilde{K}$. Put $\widetilde{T} = T|\widetilde{K}$. For $h_1, h_2 \in H$ we have

$$\langle \widetilde{T}^n h_1, h_2 \rangle = \langle T^n h_1, h_2 \rangle = \langle h_1, (T^*)^n h_2 \rangle$$
$$= \langle h_1, (A^*)^n h_2 \rangle = \langle A^n h_1, h_2 \rangle,$$

and thus $\widetilde{T}$ is a dilation of A. From (16) and $\widetilde{T} = T|\widetilde{K}$ we conclude that $\widetilde{T}$ is a minimal isometric dilation of A. Let L be the orthogonal complement of H in $\widetilde{K}$. By Theorem 6.1 there exists $\widetilde{S} \in \mathcal{L}(\widetilde{K})$ such that

$$\widetilde{S} = \begin{bmatrix} B & 0 \\ * & * \end{bmatrix} : H \boxplus L \to H \boxplus L,$$

the operator $\widetilde{S}$ commutes with $\widetilde{T}$ and $\|\widetilde{S}\| = \|B\|$. In particular, $\widetilde{S}$ is a lifting of B.

Since T is an isometric lifting of A, we have $T^*H = A^*H \subset H$ and $T^*(T^n H) = T^{n-1}H$ for $n \geq 1$. Thus the space $\widetilde{K}$ in (16) is invariant both under T and T^*. But then T can be represented in the following form:

$$T = \begin{bmatrix} \widetilde{T} & 0 \\ 0 & \widehat{T} \end{bmatrix} : \widetilde{K} \boxplus \widetilde{K}^\perp \to \widetilde{K} \boxplus \widetilde{K}^\perp,$$

where $\widehat{T}$ is the restriction of T to $\widetilde{K}^\perp$, and hence

$$S = \begin{bmatrix} \widetilde{S} & 0 \\ 0 & 0 \end{bmatrix} : \widetilde{K} \boxplus \widetilde{K}^\perp \to \widetilde{K} \boxplus \widetilde{K}^\perp$$

has all the desired properties. $\square$

XXVII.7 APPLICATIONS TO INTERPOLATION PROBLEMS

In this section classical interpolation problems of Carathéodory-Schur and Nevanlinna-Pick are analyzed by means of the commutant lifting theorem.

We begin with some preliminary remarks about $H_\infty(\mathbb{T})$. Let $\varphi \in L_\infty(\mathbb{T})$, and let φ_k denote its k-th Fourier coefficient, i.e.,

$$\varphi_k = \frac{1}{2\pi} \int_{-\pi}^{\pi} \varphi(e^{it}) e^{-ikt} dt.$$

By definition (see Section XXVI.3) the function $\varphi \in H_\infty(\mathbb{T})$ if $\varphi_k = 0$ for $k = -1, -2, \ldots$. It follows that each $\varphi \in H_\infty(\mathbb{T})$ induces an analytic function on $\mathbb{D}$, which we also denote by φ, namely

$$(1) \qquad \varphi(z) = \sum_{k=0}^{\infty} \varphi_k z^k, \qquad z \in \mathbb{D}.$$

The series in (1) converges for each $z \in \mathbb{D}$ because

$$|\varphi_k| \leq \|\varphi\|_\infty := \operatorname*{ess\,sup}_{-\pi \leq t \leq \pi} |\varphi(e^{it})|.$$

LEMMA 7.1. *If $\varphi \in H_\infty(\mathbb{T})$, then the function $\varphi(\cdot)$ in (1) is bounded on $\mathbb{D}$ and*

$$(2) \qquad \sup_{z \in \mathbb{D}} |\varphi(z)| = \|\varphi\|_\infty.$$

Furthermore, any bounded analytic function on $\mathbb{D}$ is obtained in this way.

PROOF. Let $\varphi \in H_\infty(\mathbb{T})$, and let T_φ be the Toeplitz operator on ℓ_2 defined by φ. Take $z \in \mathbb{D}$, and consider $u = (1, \overline{z}, \overline{z}^2, \ldots) \in \ell_2$. Then

$$T_\varphi^* u = \begin{bmatrix} \overline{\varphi_0} & \overline{\varphi_1} & \overline{\varphi_2} & \cdots \\ 0 & \overline{\varphi_0} & \overline{\varphi_1} & \cdots \\ 0 & 0 & \overline{\varphi_0} & \cdots \\ \vdots & \vdots & \vdots & \ddots \end{bmatrix} \begin{bmatrix} 1 \\ \overline{z} \\ \overline{z}^2 \\ \vdots \end{bmatrix} = \begin{bmatrix} \overline{\varphi(z)} \\ \overline{z}\,\overline{\varphi(z)} \\ \overline{z}^2\,\overline{\varphi(z)} \\ \vdots \end{bmatrix}$$

$$= \overline{\varphi(z)} u.$$

Hence $\overline{\varphi(z)}$ is an eigenvalue of T_φ^*, which implies that

$$|\varphi(z)| \leq \|T_\varphi^*\| = \|T_\varphi\| = \|\varphi\|_\infty,$$

where for the latter identity we use Corollary XXIII.3.2.

Conversely, let $\varphi(\cdot)$ be a bounded analytic function on $\mathbf{D}$. Put

$$m = \sup_{z \in \mathbf{D}} |\varphi(z)|,$$

and for $0 < r < 1$ set

$$\psi_r(z) = \varphi(rz) = \sum_{n=0}^{\infty} r^n \varphi_n z^n, \qquad z \in \mathbf{D}.$$

Note that $\|\psi_r\|_\infty \leq m$ for each r. Let T_{ψ_r} denote the Toeplitz operator defined by ψ_r. Put $e = (1, 0, 0, \ldots) \in \ell_2$. Then $T_{\psi_r} e \in \ell_2$, and hence for $n = 1, 2, \ldots$

$$\sum_{\nu=0}^{n} r^{2\nu} |\varphi_\nu|^2 \leq \sum_{\nu=0}^{\infty} r^{2\nu} |\varphi_\nu|^2 = \|T_{\psi_r} e\|^2$$

$$\leq \|T_{\psi_r}\|^2 = \|\psi_r\|^2 \leq m, \qquad 0 < r < 1.$$

By taking limits, first for $r \uparrow 1$ and next for $n \to \infty$, we see that $(\varphi_0, \varphi_1, \varphi_2, \ldots) \in \ell_2$. Set

$$\eta_r := (\varphi_0, r\varphi_1, r^2\varphi_2, \ldots) \in \ell_2, \qquad 0 < r \leq 1.$$

For each $N \geq 1$ we have

$$\|\eta_1 - \eta_r\|^2 \leq \sum_{\nu=0}^{N} (1 - r^{2\nu})|\varphi_\nu|^2 + \sum_{\nu=N+1}^{\infty} 2|\varphi_\nu|^2,$$

and hence

$$\limsup_{r \to 1} \|\eta_r - \eta_1\|^2 \leq \sum_{\nu=N+1}^{\infty} 2|\varphi_\nu|^2 \to 0, \qquad N \to \infty.$$

It follows that $\eta_r \to \eta_1$ in ℓ_2-norm if $r \uparrow 1$. Now take $y \in \ell_2$. Then $T_{\psi_r}^* y = (\langle y, V^\nu \eta_r \rangle)_{\nu=0}^{\infty}$, where V is the forward shift on ℓ_2. Hence for $n = 0, 1, 2, \ldots$ we have

$$\sum_{\nu=0}^{n} |\langle y, V^\nu \eta_r \rangle|^2 \leq \sum_{\nu=0}^{\infty} |\langle y, V^\nu \eta_r \rangle|^2 = \|T_{\psi_r}^* y\|^2$$

$$\leq \|T_{\psi_r}^*\|^2 \|y\|^2$$

$$\leq m^2 \|y\|^2,$$

where $0 < r < 1$. By taking limits, first for $r \uparrow 1$ and next for $n \to \infty$, we see that $(\langle y, V^\nu \eta \rangle)_{\nu=0}^{\infty} \in \ell_2$ and its ℓ_2-norm is less than or equal to $m\|y\|$. It follows that the infinite matrix

$$\begin{bmatrix} \overline{\varphi}_0 & \overline{\varphi}_1 & \overline{\varphi}_2 & \cdots \\ 0 & \overline{\varphi}_0 & \overline{\varphi}_1 & \cdots \\ 0 & 0 & \overline{\varphi}_0 & \cdots \\ \vdots & \vdots & \vdots & \end{bmatrix}$$

defines a bounded linear operator on ℓ_2 with operator norm less than or equal to m. Thus

$$\varphi(e^{it}) = \sum_{n=0}^{\infty} \varphi_n e^{int}, \qquad -\pi \le t \le \pi,$$

is the defining function of a Toeplitz operator T on ℓ_2 with $\|T\| \le m$. But then $\varphi \in H_\infty(\mathbb{T})$ and $\|\varphi\|_\infty \le m$. From the first part of the proof we know that $m \le \|\varphi\|_\infty$, and so the equality (2) is also proved. $\square$

Consider now the following *Carathéodory-Schur interpolation problem*. Given $b_0, b_1, \ldots, b_{N-1}$ in $\mathbb{C}$, find an analytic function φ on $\mathbb{D}$ such that

(CS1) $\varphi_j = b_j$, $j = 0, \ldots, N-1$,

(CS2) $\sup_{|z|<1} |\varphi(z)| \le 1$.

Assume we have a solution φ. Then condition (CS2) implies that the Toeplitz operator T_φ defined by φ,

$$T_\varphi = \begin{bmatrix} \varphi_0 & 0 & 0 & \cdots \\ \varphi_1 & \varphi_0 & 0 & \cdots \\ \varphi_2 & \varphi_1 & \varphi_0 & \cdots \\ \vdots & \vdots & \vdots & \ddots \end{bmatrix},$$

has norm $\|T_\varphi\| \le 1$. It follows that the same must hold true for the compressions of T_φ. In particular, all matrices

$$(3) \qquad \begin{bmatrix} \varphi_0 & 0 & 0 & \cdots & 0 \\ \varphi_1 & \varphi_0 & 0 & \cdots & 0 \\ \vdots & \vdots & \vdots & & \vdots \\ \varphi_{n-1} & \varphi_{n-2} & \varphi_{n-3} & \cdots & \varphi_0 \end{bmatrix}$$

must have norm ≤ 1. (Here the norm of the matrix (3) is the norm of the induced operator on $\mathbb{C}^n$ or, equivalently, the norm of (3) is equal to largest singular value of the matrix (3).) Note that for $n = N$ the entries in (3) are entirely determined by the given data. Thus in order for the Carathéodory-Schur interpolation problem to be solvable we must have

$$(4) \qquad \left\| \begin{bmatrix} b_0 & 0 & 0 & \cdots & 0 \\ b_1 & b_0 & 0 & \cdots & 0 \\ \vdots & \vdots & \vdots & & \vdots \\ b_{N-1} & b_{N-2} & b_{N-3} & \cdots & b_0 \end{bmatrix} \right\| \le 1.$$

Now let us use the commutant lifting theorem to show that the condition (4) is also sufficient.

Let B be the operator on $\mathbb{C}^N$ defined by the matrix in (4), and assume

$\|B\| \leq 1$. Let

$$A = \begin{bmatrix} 0 & 0 & \cdots & 0 & 0 \\ 1 & 0 & \cdots & 0 & 0 \\ 0 & 1 & \cdots & 0 & 0 \\ \vdots & \vdots & & \vdots & \vdots \\ 0 & 0 & \cdots & 1 & 0 \end{bmatrix} : \mathbb{C}^N \to \mathbb{C}^N.$$

Then A and B are contractions, and the special structure of the matrix in (4) implies that A and B commute. Let us identify $\mathbb{C}^N$ with the subspace of ℓ_2 consisting of all sequences $(x_0, x_1, x_2, \ldots)$ with $x_j = 0$ for $j \geq N$ by setting

$$(x_0, \ldots, x_{N-1}) = (x_0, \ldots, x_{N-1}, 0, 0, \ldots).$$

Then the forward shift V on ℓ_2 is the minimal isometric dilation of A (see the end of Section XXVII.3). So, by the commutant lifting theorem, B can be lifted to a contraction S on ℓ_2 which commutes with the forward shift V. But then we can apply Lemma XXVI.3.2 to show that S is a Toeplitz operator defined by some $\Psi \in H_\infty(\mathbb{T})$. Since S is a lifting of B, we must have $\Psi_j = b_j$, $j = 0, \ldots, N - 1$, and since S is a contraction, $\|\Psi\|_\infty \leq 1$ (by virtue of Corollary XXIII.3.2). Thus Ψ satisfies conditions (CS1) and (CS2). Moreover, all solutions can be obtained in this way. Indeed, if φ satisfies (CS1) and (CS2), then the Toeplitz operator T_φ is a contraction, T_φ is a lifting of B and T_φ commutes with the forward shift on ℓ_2. So we have proved the following proposition.

PROPOSITION 7.2. *The Carathéodory-Schur interpolation problem* (CS1), (CS2) *is solvable if and only if the operator*

$$B = \begin{bmatrix} b_0 & 0 & 0 & \cdots & 0 \\ b_1 & b_0 & 0 & \cdots & 0 \\ \vdots & \vdots & \vdots & & \vdots \\ b_{N-1} & b_{N-2} & b_{N-3} & \cdots & b_0 \end{bmatrix} : \mathbb{C}^N \to \mathbb{C}^N$$

is a contraction, and in this case φ is a solution if and only if the Toeplitz operator defined by φ is a contractive lifting of B which commutes with the forward shift on ℓ_2.

Next, we consider the following *Nevanlinna-Pick interpolation problem*. Given $z_1, \ldots, z_N$ in $\mathbb{D}$, $z_i \neq z_j$ for $i \neq j$, and $\alpha_1, \ldots, \alpha_N$ in $\mathbb{C}$, find an analytic function φ on $\mathbb{D}$ such that

(NP1) $\varphi(z_j) = \alpha_j$, $j = 1, \ldots, N$,

(NP2) $\sup_{|z|<1} |\varphi(z)| \leq 1$.

Assume we have a solution φ, and let T_φ be the Toeplitz operator defined by φ. Consider the vector

$$u_k = (1, \overline{z}_k, \overline{z}_k^2, \ldots) \in \ell_2.$$

Then $T_\varphi^* u_k = \overline{\varphi(z_k)} u_k = \overline{\alpha}_k u_k$. Thus the space H,

$$H := \mathrm{span}\{u_1, \ldots, u_N\} \subset \ell_2,$$

is invariant under T_φ^* and the action of T_φ^* on H is entirely determined by the given data. The condition that the points $z_1, \ldots, z_N$ are all different implies that $u_1, \ldots, u_N$ is a basis for H. Define $B \colon H \to H$ by setting $B^* u_k = \overline{\alpha}_k u_k$ for $k = 1, \ldots, N$. Then B is well-defined, and $B^* = T_\varphi^* | H$. It follows that for the Nevanlinna-Pick interpolation problem to have a solution it is necessary that $\|B\| \leq 1$. Now, let us use the commutant lifting theorem to show that this condition is also sufficient.

Let $B \colon H \to H$ be defined by $B^* u_k = \overline{\alpha}_k u_k$, $k = 1, \ldots, N$, and assume $\|B\| \leq 1$. Let V denote the forward shift on ℓ_2. Note that H is invariant under the backward shift V^*, because $V^* u_k = \overline{z}_k u_k$ for $k = 1, \ldots, N$. Define $A \colon H \to H$ by setting $A = (V^* | H)^*$. Then

$$V = \begin{bmatrix} A & 0 \\ * & * \end{bmatrix} \colon H \boxplus H^\perp \to H \boxplus H^\perp.$$

Since V is an isometry, we conclude that V is an isometric lifting of A. Obviously, B^* and $A^* = V^* | H$ commute with each other, and hence $AB = BA$. Now apply Theorem 6.3 (the second version of the commutant lifting theorem which does not require minimality). It follows that there exists a lifting $S \in \mathcal{L}(\ell_2)$ of B such that $SV = VS$ and $\|S\| = \|B\| \leq 1$. From $SV = VS$ we conclude that S is a Toeplitz operator T_φ on ℓ_2 defined by some $\varphi \in H_\infty$. Moreover, $\|\varphi\|_\infty \leq 1$, because $\|T_\varphi\| = \|S\| \leq 1$. Since $S = T_\varphi$ is a lifting of B, we have

$$T_\varphi^* u_k = S^* u_k = B^* u_k = \overline{\alpha}_k u_k.$$

On the other hand (see the proof of Lemma 7.1),

$$T_\varphi^* u_k = \overline{\varphi(z_k)} u_k.$$

Thus $\varphi(z_k) = \alpha_k$, $k = 1, \ldots, N$. We have proved that φ is analytic on $\mathbb{D}$ and satisfies (NP1) and (NP2). Moreover, *all solutions of the Nevanlinna-Pick interpolation problem may be obtained in this way.* Indeed, if φ is a solution, then (as we saw in the previous paragraph) the Toeplitz operator T_φ defined by φ is a lifting of B which commutes with the forward shift V and $\|T_\varphi\| \leq 1$.

The condition $\|B\| \leq 1$ appearing in the two previous paragraphs is equivalent to the requirement that $I - BB^* \geq 0$. The latter condition can be restated as

$$(5) \qquad \left[\langle (I - BB^*) u_i, u_j \rangle \right]_{i,j=1}^N \geq 0.$$

Now $\langle u_i, u_j \rangle = (1 - \overline{z}_i z_j)^{-1}$ and $B^* u_i = \overline{\alpha}_i u_i$. It follows that the $N \times N$ matrix in (5) is equal to the classical Pick matrix

$$(6) \qquad \left[\frac{1 - \overline{\alpha}_i \alpha_j}{1 - \overline{z}_i z_j} \right]_{i,j=1}^N,$$

and the Nevanlinna-Pick interpolation problem is solvable if and only if the matrix in (6) is non-negative.

XXVII.8 DILATION OF CONTRACTION SEMIGROUPS

In this section we use Naimark's theorem concerning the connection between positive definite functions and unitary representations (see Theorem 8.1 below) to extend the unitary dilation theorem in Section XXVII.4 to contraction semigroups.

A map $T(\cdot)$ from a group G into $\mathcal{L}(K)$, where K is a complex Hilbert space, is said to be *positive definite* if for all finite sets $g_1,\ldots,g_n$ in G and $x_1,\ldots,x_n$ in K,

$$(1) \qquad \sum_{j,k=1}^{n} \langle T(g_j^{-1}g_k)x_k, x_j \rangle \geq 0.$$

We note that

$$(2) \qquad T(g^{-1}) = T(g)^*.$$

To see this, take $n = 1$ and $g_1 = e$ in (1) and get $\langle T(e)x, x \rangle \geq 0$, $x \in K$. Now take $n = 2$ with $g_1 = g$, $g_2 = e$, $x_i \in K$, $i = 1,2$. Then

$$\langle T(e)x_1, x_1 \rangle + \langle T(g^{-1})x_2, x_1 \rangle + \langle T(g)x_1, x_2 \rangle + \langle T(e)x_2, x_2 \rangle \geq 0.$$

Hence $\langle T(g^{-1})x_2, x_1 \rangle + \langle T(g)x_1, x_2 \rangle$ is real. Replacing x_2 by ix_2 we get that $i(\langle T(g^{-1})x_2, x_1 \rangle - \langle T(g)x_1, x_2 \rangle)$ is also real. But this can only happen if

$$\langle T(g^{-1})x_2, x_1 \rangle = \overline{\langle T(g)x_1, x_2 \rangle} = \langle x_2, T(g)x_1 \rangle,$$

which proves (2).

A map $U(\cdot)$ from G into $\mathcal{L}(K)$ is called a *unitary representation* if $U(\cdot)$ is a homomorphism, i.e.,

$$U(g_1 g_2) = U(g_1)U(g_2), \qquad g_1, g_2 \in G,$$

and the values of U are unitary operators on K. In this case $U(e) = I$, where e is the unit in G and I is the identity operator on K.

For example, if U is a unitary operator on K, then $U(n) = U^n$, $n = 0, \pm 1, \pm 2, \ldots$, is a unitary representation of the group $\mathbb{Z}$ of integers on K.

The following theorem is known as Naimark's theorem.

THEOREM 8.1. *If $U(\cdot)$ is a unitary representation of a group G on a Hilbert space K, and H is a subspace of K, then the function*

$$(3) \qquad T(g) = PU(g)|H, \qquad g \in G,$$

is positive definite on G, where P is the orthogonal projection of K on H.

Conversely, if $T(\cdot): G \to \mathcal{L}(H)$ is positive definite with $T(e) = I$, then there exists a unitary representation $U(\cdot)$ of G on a Hilbert space K and a linear isometry J from H into K such that

$$(4) \qquad T(g) = PU(g)J, \qquad g \in G.$$

Furthermore, the representation $U(\cdot)$ is unique up to an isometric isomorphism if

$$(5) \qquad\qquad K = \bigvee_{g \in G} U(g)JH.$$

The representation $U(\cdot)$ in (4) is called a *dilation* of $T(\cdot)$. If K is given by (5), we say that $U(\cdot)$ is a *minimal dilation* of $T(\cdot)$.

PROOF OF THEOREM 8.1. Suppose $U(\cdot)$ is a unitary representation of G on K. Let $T(\cdot)$ be defined by (3). Then for $g_1, \ldots, g_n$ in G and $x_1, \ldots, x_n$ in H,

$$
\begin{aligned}
\sum_{j,k=1}^{n} \langle T(g_j^{-1}g_k)x_k, x_j \rangle &= \sum_{j,k=1}^{n} \langle PU(g_j^{-1}g_k)x_k, x_j \rangle \\
&= \sum_{j,k=1}^{n} \langle U(g_j)^{-1}U(g_k)x_k, x_j \rangle \\
&= \sum_{j,k=1}^{n} \langle U(g_k)x_k, U(g_j)x_j \rangle \\
&= \Big\| \sum_{i=1}^{n} U(g_i)x_i \Big\|^2 \geq 0.
\end{aligned}
$$

Hence $T(\cdot)$ is positive definite.

Given a positive definite $\mathcal{L}(H)$-valued function $T(\cdot)$ defined on the group G, let $\mathcal{F}$ be the set of all functions $f: G \to H$ which have only a finite number of non-zero values. With the usual definitions of addition and scalar multiplication, $\mathcal{F}$ becomes a vector space. Define on $\mathcal{F} \times \mathcal{F}$,

$$\langle f_1, f_2 \rangle_0 = \sum_{g,h \in G} \langle T(g^{-1}h)f_1(h), f_2(g) \rangle.$$

It follows from (1) and (2) that $\langle \cdot, \cdot \rangle_0$ has all the properties of an inner product except that $\langle f, f \rangle_0 = 0$ need not imply $f = 0$. As usual, we take $N = \{ f \in \mathcal{F} \mid \langle f, f \rangle_0 = 0 \}$. Then N is a subspace of $\mathcal{F}$ and the quotient space $\mathcal{F}/N$ is an inner product space with $\langle [f_1], [f_2] \rangle = \langle f_1, f_2 \rangle_0$. Let K be the Hilbert space completion of $\mathcal{F}/N$. The space H is embedded into K as follows. Given $x \in H$, define $f_x \in \mathcal{F}$ by $f_x(e) = x$ and $f_x(g) = 0$, $g \neq e$. Let $Jx = [f_x]$. Clearly, $J: H \to K$ is linear and

$$\|Jx\|^2 = \langle f_x, f_x \rangle_0 = \langle T(e)f_x(e), f_x(e) \rangle = \langle x, x \rangle = \|x\|^2.$$

Thus J is an isometry. In order to define our representation $U(\cdot)$ on K we define for each $g \in G$ a linear map $V(g): \mathcal{F} \to \mathcal{F}$ by

$$(6) \qquad\qquad (V(g)f)(h) = f(g^{-1}h), \qquad h \in G.$$

Then

$$\langle V(g_0)f, V(g_0)f \rangle_0 = \sum_{g,h} \langle T(g^{-1}h)(V(g_0)f)(h), (V(g_0)f)(g) \rangle$$

$$= \sum_{g,h} \langle T(g^{-1}h)f(g_0^{-1}h), f(g_0^{-1}g) \rangle$$

$$= \sum_{g',h'} T((g')^{-1}h')f(h'), f(g') \rangle,$$

where $h' = g_0^{-1}h$ and $g' = g_0^{-1}g$ range over G as do h and g. Thus $V(g)$ may be considered as a unitary map on $\mathcal{F}/\mathcal{N}$ with $V(g)^{-1} = V(g^{-1})$ for each $g \in G$. Hence $V(g)$ has a unique unitary extension $U(g)$. Since $V(g_1g_2) = V(g_1)V(g_2)$, the map $U(\cdot)$ is a homomorphism on G, and it follows that $g \mapsto U(g)$ is a unitary representation of G on K.

To see that (4) holds, let x, y be arbitrary elements in H. It follows from the definitions of f_x and f_y that

$$\langle U(g_0)Jx, Jy \rangle = \sum_{g,h \in G} \langle T(g^{-1}h)(V(g_0))f_x(h), f_y(g) \rangle$$

$$= \sum_{g,h \in G} \langle T(g^{-1}h)f_x(g_0^{-1}h), f_y(g) \rangle$$

$$= \langle T(g_0)x, y \rangle, \qquad g_0 \in G,$$

which proves (4). It is easy to see that (5) follows from (6) and the definition of $U(\cdot)$.

Next, suppose that

$$K_1 = \bigvee_{g \in G} U_1(g)J_1H,$$

where $U_1(\cdot)$ is another unitary dilation of $T(\cdot)$ and $J_1 \colon H \to K_1$ is a linear isometry. Define $A \colon \mathcal{F} \to K_1$ by

$$Af = \sum_{g \in G} U_1(g)J_1f(g).$$

Recall that the latter sum is finite. Since $U_1(\cdot)$ is a homomorphism and $U_1(h)$ is unitary, we have

$$\|Af\|^2 = \sum_{g,h \in G} \langle U_1(g)J_1f(g), U_1(h)J_1f(h) \rangle$$

$$= \sum_{g,h \in G} \langle U_1(h^{-1}g)J_1f(g), J_1f(h) \rangle$$

$$= \sum_{g,h \in G} \langle T(h^{-1}g)f(g), f(h) \rangle = \|f\|^2.$$

Therefore A induces a linear isometry A_1 which maps K onto K_1. Also, for $f \in \mathcal{F}$ and

$g \in G$,

$$AV(g)f = \sum_{g' \in G} U_1(g')J_1\big((V(g)f)(g')\big)$$

$$= \sum_{g' \in G} U_1(g')J_1 f(g^{-1}g') = U_1(g)Af,$$

and hence $A_1 U(g) = U_1(g)A_1$ for each $g \in G$. $\quad\square$

The map A_1 defined in the last paragraph of the proof of Theorem 8.1 also satisfies the identity $A_1 J = J_1$. Indeed, for $x \in H$ we have

$$A_1 Jx = Af_x = \sum_{g \in G} U_1(g)J_1 f_x(g) = J_1 x.$$

Since contractions have unitary dilation (by Theorem 4.2), the first part of Theorem 8.1 yields the following corollary.

COROLLARY 8.2. *Let $A \in \mathcal{L}(H)$ be a contraction. Define $A(\cdot)$ on the group* $\mathbb{Z}$ *of integers by*

$$(7) \qquad\qquad A(n) = \begin{cases} A^n & , \quad n \geq 0, \\ (A^{-n})^*, & n < 0. \end{cases}$$

Then $A(\cdot)$ is positive definite on $\mathbb{Z}$.

PROOF. Let U be a (minimal) unitary dilation of A. Then for $x, y \in H$ and $n = 0, 1, 2, \ldots,$

$$\langle A^n x, y \rangle = \langle U^n x, y \rangle,$$

$$\langle (A^n)^* x, y \rangle = \langle x, A^n y \rangle = \langle x, U^n y \rangle = \langle U^{-n} x, y \rangle.$$

Thus

$$A(n) = PU^n | H, \qquad n \in \mathbb{Z},$$

and $A(\cdot)$ is positive definite by the first part of Theorem 8.1 (applied with $G = \mathbb{Z}$ and $U(n) = U^n$). $\quad\square$

We are now ready to extend Theorem 4.2 to contraction semigroups (see Section XIX.4 for the definition of these semigroups).

THEOREM 8.3. *Let $T(\cdot)$ be a contraction semigroup on a Hilbert space H. Then there exists a minimal unitary dilation of $T(\cdot)$ on $\mathbb{R}$ which is unique up to an isometric isomorphism.*

PROOF. For $t < 0$, define $T(t) = T(-t)^*$. The theorem follows from Theorem 8.1 once we show that $T(\cdot)$ is positive definite on $\mathbb{R}$. First assume $t_1, t_2, \ldots, t_k$ are rational numbers and $x_1, \ldots, x_k$ are in H. Choose a rational number $r \geq 0$ and integers $n_1, \ldots, n_k$ so that $t_j = rn_j$ for $j = 1, \ldots, k$. E.g., if $t_j = p_j/q_j$ with p_j and q_j integers, take

$$r = \left(\prod_{j=1}^{k} |q_j| \right)^{-1}, \qquad n_j = \pm p_j \left(\prod_{i \neq j} q_i \right).$$

Put $A = T(r)$. Note that A is a contraction. Define $A(n)$ as in (7). Then

$$A(n) = T(nr), \quad A(-n) = A(n)^* = T(-nr), \qquad n = 0, 1, 2, \ldots .$$

By Corollary 8.2, the map $A(\cdot)$ is positive definite on $\mathbb{Z}$. Therefore,

$$(8) \qquad \sum_{i,j=1}^{k} \langle T(t_i - t_j)x_i, x_j \rangle = \sum_{i,j=1}^{k} \langle A(n_i - n_j)x_i, x_j \rangle \geq 0.$$

Since each t_j is an arbitrary rational number and the map $t \to T(t)x$ is continuous for each x, it follows that (8) holds for any $t_1, \ldots, t_k$ in $\mathbb{R}$. $\square$

CHAPTER XXVIII
UNITARY SYSTEMS AND CHARACTERISTIC
OPERATOR FUNCTIONS

This chapter presents an introduction to the theory of unitary systems and their transfer functions. We already met transfer functions of the form $I + C(\lambda - A)^{-1}B$ in earlier chapters. Here we are concerned with the case when the system matrix

$$\begin{bmatrix} A & B \\ C & D \end{bmatrix}$$

is a unitary operator acting on an infinite dimensional Hilbert space. Characteristic operator functions are transfer functions of such systems. In mathematical system theory often the starting point is the input-output map or the associated transfer function. In the theory of characteristic operator functions the situation is different. Here the main operator A, which is a contraction, comes first and the characteristic operator function serves as a unitary invariant for A.

This chapter consists of 12 sections. The first section has a preliminary character and concerns completely non-unitary operators. Sections 2–4 describe the main properties of unitary systems and the corresponding characteristic operator functions. Section 5 contains the realization theorem and presents the functional model for a completely non-unitary contraction. Sections 6–8 deal with the connections between invariant subspaces of the main operator and factorizations of the characteristic operator function. In Sections 9–11 we use elements of Part V to derive triangular representations for certain contractions and multiplicative representations for their characteristic operator functions. As an illustration a certain contractive integral operator of the second kind is proved to be unicellular and its unitary equivalence class is described. In the last section we review the symmetric analogue of the theory, which concerns dissipative operators and their characteristic operator functions.

XXVIII.1 COMPLETELY NON-UNITARY OPERATORS

An operator $A \in \mathcal{L}(H)$ is said to be *completely non-unitary* if H does not have an orthogonal direct sum decomposition, $H = H_0 \boxplus H_1$, with the following properties:

(a) $H_0 \neq \{0\}$,

(b) $AH_0 \subset H_0$ and $AH_1 \subset H_1$,

(c) $A|H_0$ is unitary.

From the Von Neumann-Wold decomposition for isometries (see Theorem XXVI.1.1) it follows that an isometry A is completely non-unitary if and only if A is a pure isometry. The next theorem gives the analogue of the Von Neumann-Wold decomposition for contractions.

THEOREM 1.1. *Let $A \in \mathcal{L}(H)$ be a contraction. Then there exists a unique subspace L of H such that the following holds:*

(i) *$AL \subset L$ and $AL^\perp \subset L^\perp$,*

(ii) *$A|L$ is completely non-unitary,*

(iii) *$A|L^\perp$ is unitary.*

Furthermore,

$$(1) \qquad L = \overline{\left(\bigvee_{n=0}^{\infty} \operatorname{Im} A^n D_{A^*} \right) + \left(\bigvee_{n=0}^{\infty} \operatorname{Im}(A^*)^n D_A \right)},$$

$$(2) \qquad L^\perp = \left\{ x \in H \mid \|A^n x\| = \|x\| = \|(A^*)^n x\|,\ n = 0, 1, 2, \ldots \right\}.$$

Here D_A and D_{A^*} are the defect operators associated with A and A^*, respectively.

PROOF. Let L be the subspace defined in (1), and write M for the set in the right hand side of (2). Note that

$$\left(\bigvee_{n=0}^{\infty} \operatorname{Im} A^n D_{A^*} \right)^\perp = \bigcap_{n=0}^{\infty} \operatorname{Ker} D_{A^*}(A^*)^n,$$

$$\left(\bigvee_{n=0}^{\infty} \operatorname{Im}(A^*)^n D_A \right)^\perp = \bigcap_{n=0}^{\infty} \operatorname{Ker} D_A A^n.$$

Thus $h \perp L$ is equivalent to the requirement that $D_A A^n h = 0$ and $D_{A^*}(A^*)^n h = 0$ for each $n \geq 0$. Since A and A^* are contractions, the latter happens (use item (i) in Lemma XXVII.1.1) if and only if $h \in M$. Thus $M = L^\perp$.

Obviously, $L^\perp$ is invariant under A and A^*, and hence (i) holds. From (2) it also follows directly that $A|L^\perp$ is unitary. Next, let us prove that $A|L$ is completely non-unitary. Let H_0 be a subspace of L, invariant under A and A^*, such that $A|H_0$ is unitary. Then we see from (2) that $H_0 \subset L^\perp \cap L$, and hence H_0 consists of the zero vector only. So (ii) holds.

It remains to prove the uniqueness statement. Let N be a second subspace of H such that (i), (ii) and (iii) hold with N in place of L. Since $A|N^\perp$ is unitary, we have $N^\perp \subset L^\perp$, where $L^\perp$ is as in (2). Put $H_0 = N \cap L^\perp$ and $H_1 = N \cap H_0^\perp$. Note that H_0 is invariant under both A and A^*. Furthermore, $A|H_0$ is unitary because $H_0 \subset L^\perp$. So N has an orthogonal direct sum decomposition with the properties (b) and (c) mentioned in the first paragraph of this section. But $A|N$ is completely non-unitary. So $H_0 = \{0\}$, and it follows that $N = L$. $\square$

The next remark (which we state in the form of a lemma) shows that a completely non-unitary contraction cannot have (non-zero) compressions that are unitary.

LEMMA 1.2. *Let the operator*

$$(3) \qquad A = \begin{bmatrix} A_{11} & A_{12} \\ A_{21} & A_{22} \end{bmatrix} : H_1 \boxplus H_2 \to K_1 \boxplus K_2$$

be a contraction, and assume that A_{11} is unitary. Then A_{12} and A_{21} are both zero operators.

PROOF. According to Corollaries 5.3 and 5.4 in Section XXVII.5, the operators A_{12} and A_{21} have the following form:

$$(4) \qquad\qquad A_{12} = D_* \Delta_1, \qquad A_{21} = \Delta_2 D,$$

where D and D_* are the defect operators associated with A_{11} and A_{11}^*, respectively, and Δ_1 and Δ_2 are certain contractions. Since A_{11} is unitary, the operators D and D_* are zero, and hence the same holds for A_{12} and A_{21}. $\square$

Lemma 1.2 shows that in the definition of a completely non-unitary operator the second condition in (b) is superfluous. More precisely, if $AH_0 \subset H_0$ and $A|H_0$ is unitary, then always $AH_0^\perp \subset H_0^\perp$, because of Lemma 1.2.

Since the spectrum of a unitary operator lies on the unit circle, Theorem 1.1 shows that a *strict contraction* on a Hilbert space H (i.e., an operator A with norm strictly less than 1) is necessarily completely non-unitary. More generally, if $A \in \mathcal{L}(H)$ is a contraction with all its spectrum in the open unit disc, then A is completely non-unitary. The next proposition shows that in the finite dimensional case the converse statement is also true.

PROPOSITION 1.3. *Let $A \in \mathcal{L}(H)$ be a contraction on a finite dimensional Hilbert space. Then A is completely non-unitary if and only if the eigenvalues of A lie in the open unit disc $\mathbb{D}$.*

PROOF. It suffices to prove the "only if" part. Assume that A is a completely non-unitary contraction on H. Let λ be an eigenvalue of A with eigenvector $e \neq 0$. Since A is a contraction, we have $|\lambda| \leq 1$. By Lemma 1.2, the operator A does not have non-zero compressions that are unitary. In particular, the compression of A to $\text{span}\{e\}$ cannot be unitary. It follows that $|\lambda| \neq 1$. Thus $\lambda \in \mathbb{D}$. $\square$

Proposition 1.3 is a special case of the following more general proposition.

PROPOSITION 1.4. *Let $A \in \mathcal{L}(H)$ be a contraction, and assume $I - A$ is a compact operator. Then A is completely non-unitary if and only if A has no eigenvalues on $\mathbf{T}$.*

PROOF. The argument given in the proof of Proposition 1.3 shows that A cannot have eigenvalues on $\mathbf{T}$ if A is completely non-unitary. We have to prove the reverse implication.

Assume that A has no eigenvalues on $\mathbf{T}$. Let L be the subspace with the properties described in Theorem 1.1. We want to show that $H_0 = L^\perp$ consists of the zero vector only. Put $U = A|H_0$. Then U is unitary, and according to our assumptions U has no eigenvalue on $\mathbf{T}$ and $I - U$ is compact. Note that $I + U = 2I - (I - U)$. Hence 2 is not an eigenvalue for the compact operator $I - U$, and so $2I - (I - U)$ is invertible. Put

$$(5) \qquad\qquad T = i(I - U)(I + U)^{-1}.$$

Then T is a a compact selfadjoint operator which does not have a real eigenvalue. Hence, by the spectral theorem for compact selfadjoint operators, $T = 0$. But then $U = I$, and each non-zero vector of H_0 is an eigenvector of U corresponding to the eigenvalue 1. But U has no eigenvalues on $\mathbb{T}$. So $H_0 = \{0\}$. $\square$

Note that Proposition 1.3 implies that for contractions on a finite dimensional space complete non-unitarity is determined by the spectrum, and hence is a similarity invariant. For contractions on infinite dimensional spaces the latter statement does not hold. In fact, there exist completely non-unitarity contractions that are similar to unitary operators. A concrete example will be given in Section XXVIII.5 as a corollary to the general theory developed there (see Corollary 5.3).

The symmetric analogue of complete non-unitarity is the notion of complete non-selfadjointness which we met in Section XXI.2. The connections between these two notions will be employed in the last section of this chapter.

In the remaining part of this section we extend Theorem 1.1 to contraction semigroups (see Section XIX.4 for the definition of these semigroups).

A strongly continuous semigroup $T(\cdot)$ of operators on a Hilbert space H is said to be *completely non-unitary* if there does not exist a subspace $N \neq (0)$ of H with the properties that for each $t \geq 0$ the space N is $T(t)$-invariant and $T(t)|N$ is unitary.

THEOREM 1.5. *Let $T(\cdot)$ be a contraction semigroup on H. Then there exists a unique subspace M of H such that*

(i) $T(t)M \subset M$, $T(t)M^\perp \subset M^\perp$, $t \geq 0$,

(ii) $T(t)|M$ *is unitary,* $t \geq 0$,

(iii) $T(\cdot)|M^\perp$ *is a completely non-unitary semigroup.*

Furthermore,

$$(6) \qquad M = \bigcap_{t \geq 0} T(t)Z, \qquad Z = \{x \mid \|T(t)x\| = \|x\| \text{ for all } t \geq 0\}.$$

PROOF. The set Z is a subspace of H. Indeed, Z is closed and $\alpha z \in Z$, $\alpha \in \mathbb{C}$, $z \in Z$. To see that Z is closed under addition, let u, v be in Z. By applying the parallellogram identity twice, one sees that for all $t \geq 0$,

$$2(\|u\|^2 + \|v\|^2) = \|u + v\|^2 + \|u - v\|^2 \geq \|T(t)(u + v)\|^2 + \|T(t)(u - v)\|^2$$
$$= 2(\|T(t)u\|^2 + \|T(t)v\|^2) = 2(\|u\|^2 + \|v\|^2).$$

Thus

$$\|T(t)(u + v)\|^2 + \|T(t)(u - v)\|^2 = \|u + v\|^2 + \|u - v\|^2.$$

Since $\|T(t)(u \pm v)\|^2 \leq \|u \pm v\|^2$, it follows that $\|T(t)(u + v)\| = \|u + v\|$, i.e., $u + v \in Z$.

Let M be as in (6). Note that $T(t)$ acts as an isometry on Z. Thus $T(t)Z$ is a closed linear manifold of H, and hence the same holds true for M. Furthermore, since $T(\cdot)$ is a semigroup, we have

$$\left\|T(s)(T(t)x)\right\| = \|T(s + t)x\| = \|x\| = \|T(t)x\|, \qquad x \in Z,$$

for each $s \geq 0$ and $t \geq 0$. It follows that Z is $T(t)$-invariant for each $t \geq 0$. Fix $t' \geq 0$. Then, by the semigroup property,

$$T(t')M = \bigcap_{t \geq 0} T(t'+t)Z = \bigcap_{t \geq 0} T(t)T(t')Z \subset \bigcap_{t \geq 0} T(t)Z \subset M,$$

which proves the first inclusion in (i). Furthermore,

$$T(t')M = \bigcap_{t \geq t'} T(t)Z \supset \bigcap_{t \geq 0} T(t)Z \supset M.$$

Hence $T(t)M = M$ for each $t \geq 0$. Since $M \subset Z$, we see that (ii) holds, which, by Lemma 1.2, implies that the second inclusion in (i) is also satisfied.

Next, let us prove (iii). Let W be a subspace of $M^{\perp}$ such that W is invariant under $T(t)$ and $T(t)|W$ is unitary for each $t \geq 0$. Then $W \subset Z$. Hence $W \subset M$, for if $w \in W$, then there exists for each $t \geq 0$ an element $w_t \in W$ such that

$$w = T(t)w_t \in T(t)Z.$$

Thus $W \subset M \cap M^{\perp} = (0)$.

To prove uniqueness, suppose that M_1 is a subspace of H and (i)–(iii) hold with M_1 in place of M. Then $M_1 \subset M$ by the above argument (which showed that $W \subset M$). Take $v \in M_1^{\perp} \cap M$, and fix $t \geq 0$. There exists $m_t \in M$ such that $T(t)m_t = v$. In fact, $m_t = T(t)^*v$, because $T(t)|M$ is unitary. Since $M_1^{\perp}$ is $T(t)$-invariant, $T(t)^*v \in M_1^{\perp}$. Thus $m_t \in M_1^{\perp} \cap M$. It follows that for each $t \geq 0$ the space $N = M_1^{\perp} \cap M$ is $T(t)$-invariant and $T(t)|N$ is unitary. Therefore $N = M_1^{\perp} \cap M = (0)$ by the assumption that $T(\cdot)|M_1^{\perp}$ is completely non-unitary on $M_1^{\perp}$. Hence $M = M_1$. $\square$

XXVIII.2 UNITARY SYSTEMS AND THEIR TRANSFER FUNCTIONS

Consider the following input-output systems:

$$(1) \qquad \begin{cases} x_{n+1} = Ax_n + Bu_n, & n = 0, \pm 1, \pm 2, \ldots, \\ y_n = Cx_n + Du_n. \end{cases}$$

Here $A\colon H \to H$, $B\colon K \to H$, $C\colon H \to L$ and $D\colon K \to L$ are bounded linear operators acting between Hilbert spaces. The spaces H, K and L are called *state space*, *input space* and *output space*, respectively. The operator A is called the *state space operator* or *main operator*. We refer to B, C and D as the *input operator*, the *output operator* and the *feedthrough operator*, respectively. Instead of (1) it will be more convenient to write

$$(2) \qquad \Sigma = (A, B, C, D; H, K, L).$$

We shall omit the spaces H, K and L in (2) and simply write $\Sigma = (A, B, C, D)$ when it is clear between which spaces the operators act. The system (1) is called *finite dimensional* if input space, state space and output space are all finite dimensional.

In this chapter we are interested in systems Σ for which the operator

$$(3) \qquad \begin{bmatrix} A & B \\ C & D \end{bmatrix} : H \boxplus K \to H \boxplus L$$

is unitary. Systems with this additional property are called *unitary*. Thus $\Sigma = (A, B, C, D)$ is a unitary system if and only if the following six identities hold:

$$(4a) \qquad AA^* + BB^* = I, \qquad A^*A + C^*C = I,$$

$$(4b) \qquad AC^* + BD^* = 0, \qquad A^*B + C^*D = 0,$$

$$(4c) \qquad CC^* + DD^* = I, \qquad B^*B + D^*D = I.$$

The 2×2 operator matrix in (3) is called the *system matrix*.

Assume that the system (1) starts operating at time N (in particular, $x_N = 0$), and that the input sequence $(u_n)_{n=N}^{\infty}$, the state sequence $(x_n)_{n=N}^{\infty}$ and the output sequence $(y_n)_{n=N}^{\infty}$ are square summable in norm. Consider on $0 < |\lambda| < 1$ the functions

$$u(\lambda) = \sum_{n=N}^{\infty} \lambda^n u_n, \qquad x(\lambda) = \sum_{n=N}^{\infty} \lambda^n x_n, \qquad y(\lambda) = \sum_{n=N}^{\infty} \lambda^n y_n.$$

Then the equations in (1) imply that

$$(5) \qquad \begin{cases} \lambda^{-1} x(\lambda) = Ax(\lambda) + Bu(\lambda), & 0 < |\lambda| < 1, \\ y(\lambda) = Cx(\lambda) + Du(\lambda). \end{cases}$$

Since for λ sufficiently small $I - \lambda A$ is invertible, we can solve $x(\lambda)$ from the first equation in (5). By inserting the solution into the second equation of (5) we see that $y(\lambda) = \theta_\Sigma(\lambda) u(\lambda)$, where

$$(6) \qquad \theta_\Sigma(\lambda) = D + \lambda C(I - \lambda A)^{-1} B.$$

The operator $\theta_\Sigma(\cdot)$ defined by (6) is called the *transfer function* of the system Σ. If Σ is a unitary system, then, by (4a), its main operator is a contraction, and hence in this case $I - \lambda A$ is invertible for all λ in the open unit disc $\mathbb{D}$. It follows that the transfer function of a unitary system is an operator-valued function on $\mathbb{D}$.

THEOREM 2.1. *The transfer function θ_Σ of the unitary system (2) is an analytic function on $\mathbb{D}$ whose values are bounded linear operators acting from K into L, and*

$$(7) \qquad \sup_{\lambda \in \mathbb{D}} \|\theta_\Sigma(\lambda)\| \leq 1.$$

PROOF. Since $\|A\| \leq 1$, we may expand $(I - \lambda A)^{-1}$ into a power series, which yields

$$\theta_\Sigma(\lambda) = D + \sum_{n=1}^{\infty} \lambda^n C A^{n-1} B, \qquad |\lambda| < 1,$$

with the series converging in the operator norm. Thus $\theta_\Sigma(\cdot)$ is analytic on $\mathbb{D}$.

Fix $\lambda \in \mathbb{D}$. We have

$$\begin{aligned}
\theta_\Sigma(\lambda)^* \theta_\Sigma(\lambda) &= \left[D^* + \overline{\lambda} B^*(I - \overline{\lambda} A^*)^{-1} C^*\right]\left[D + \lambda C(I - \lambda A)^{-1} B\right] \\
&= D^* D + \overline{\lambda} B^*(I - \overline{\lambda} A^*)^{-1} C^* D + \lambda D^* C(I - \lambda A)^{-1} B \\
&\quad + \overline{\lambda}\lambda B^*(I - \overline{\lambda} A^*)^{-1} C^* C(I - \lambda A)^{-1} B.
\end{aligned}$$

Now use the second identity in (4b) to rewrite $\overline{\lambda} C^* D$ as

$$\overline{\lambda} C^* D = (I - \overline{\lambda} A^*) B - B.$$

It follows that

$$\overline{\lambda} B^*(I - \overline{\lambda} A^*)^{-1} C^* D = B^* B - B^*(I - \overline{\lambda} A^*)^{-1} B.$$

By taking adjoints, the latter identity yields:

$$\lambda D^* C(I - \lambda A)^{-1} B = B^* B - B^*(I - \lambda A)^{-1} B.$$

Next use the second identity in (4a) to replace $\overline{\lambda}\lambda C^* C$ by $\overline{\lambda}\lambda I - \overline{\lambda}\lambda A^* A$, and use that

$$\overline{\lambda}\lambda A^* A = I + (I - \overline{\lambda} A^*)(I - \lambda A) - (I - \overline{\lambda} A^*) - (I - \lambda A).$$

One obtains that

$$\begin{aligned}
\overline{\lambda}\lambda B^*(I - \overline{\lambda} A^*)^{-1} C^* C(I - \lambda A)^{-1} B &= \\
&= (\overline{\lambda}\lambda - 1) B^*(I - \overline{\lambda} A^*)^{-1}(I - \lambda A)^{-1} B - B^* B \\
&\quad + B^*(I - \lambda A)^{-1} B + B^*(I - \overline{\lambda} A^*)^{-1} B.
\end{aligned}$$

Thus

$$\theta_\Sigma(\lambda)^* \theta_\Sigma(\lambda) = D^* D + B^* B + (\overline{\lambda}\lambda - 1) B^*(I - \overline{\lambda} A^*)^{-1}(I - \lambda A)^{-1} B.$$

But then, using the second identity in (4c), we see that

$$(8) \qquad I - \theta_\Sigma(\lambda)^* \theta_\Sigma(\lambda) = (1 - |\lambda|^2) B^*(I - \overline{\lambda} A^*)^{-1}(I - \lambda A)^{-1} B \geq 0,$$

and hence $\|\theta_\Sigma(\lambda)\| \leq 1$. $\square$

Formula (8) is a special case of the first of the following two identities:

$$(9a) \qquad I - \theta_\Sigma(\mu)^* \theta_\Sigma(\lambda) = (1 - \overline{\mu}\lambda) B^*(I - \overline{\mu} A^*)^{-1}(I - \lambda A)^{-1} B,$$

$$(9b) \qquad I - \theta_\Sigma(\lambda)\theta_\Sigma(\mu)^* = (1 - \lambda\overline{\mu})C(I - \lambda A)^{-1}(I - \overline{\mu}A^*)^{-1}C^*.$$

Here $\Sigma = (A, B, C, D)$ is a unitary system, and the complex numbers λ and μ are in $\mathbb{D}$. To prove (9a) one may use the same arguments as the ones used to establish (8). One obtains (9b) from (9a) by applying (9a) to the unitary system $\Sigma^* = (A^*, C^*, B^*, D^*)$ and interchanging the roles of λ and $\overline{\mu}$. The identities (9a), (9b) are very useful and contain much information. For example, if the main operator A is a strict contraction, then $\theta_\Sigma(\zeta)$ is well-defined for each $\zeta \in \mathbb{T}$, and by taking limits in (9a) and (9b) we see that $\theta_\Sigma(\zeta)$ must be unitary.

COROLLARY 2.2. *Let Σ in (2) be a finite dimensional unitary system with $K = \mathbb{C}^k$ and $L = \mathbb{C}^\ell$, say. Then $k = \ell$ and θ_Σ is a $k \times k$ rational matrix function with all its poles outside $\overline{\mathbb{D}}$, and on $\mathbb{T}$ the values of θ_Σ are unitary.*

PROOF. Here, as well in the sequel, we identify an operator $A\colon \mathbb{C}^k \to \mathbb{C}^\ell$ with the $\ell \times k$ matrix of A relative to the standard bases of $\mathbb{C}^k$ and $\mathbb{C}^\ell$. Since the state space of Σ is finite dimensional, we may assume that $H = \mathbb{C}^n$. From the Jordan normal form for A it is clear that $(I - \lambda A)^{-1}$ is an $n \times n$ matrix function all whose entries are quotients of polynomials. Thus $(I - \lambda A)^{-1}$ is rational in λ. But then the same holds true for $\theta_\Sigma(\lambda)$.

From Theorem 2.1 we know that θ has no poles on $\mathbb{D}$. Since $\theta_\Sigma(\cdot)$ is bounded in norm on $\mathbb{D}$, each entry $\theta_{ij}(\cdot)$ of $\theta_\Sigma(\cdot)$ is a scalar rational function which is bounded on $\mathbb{D}$. It follows that the entries of $\theta_\Sigma(\cdot)$ do not have poles on $\mathbb{T} = \partial\mathbb{D}$, and therefore $\theta_\Sigma(\cdot)$ does not have poles on $\overline{\mathbb{D}}$.

From the remarks made in the paragraph preceding the present corollary we know that $\theta_\Sigma(\zeta)$ is unitary for each point $\zeta \in \mathbb{T}$ such that $\overline{\zeta}$ is not an eigenvalue of A. Since the number of eigenvalues of A is finite, it follows that $\theta_\Sigma(\zeta)$ is unitary for $\zeta \in \mathbb{T}$ with the possible exception of a finite number of points. But, by the result proved in the previous paragraph, the function $\theta_\Sigma(\cdot)$ is continuous on $\mathbb{T}$. So, by continuity, $\theta_\Sigma(\zeta)$ is unitary for each $\zeta \in \mathbb{T}$. In particular, $k = \ell$. $\square$

With a unitary system $\Sigma = (A, B, C, D; H, K, L)$ we associate the following two subspaces of H:

$$\mathcal{N}(\Sigma) = \left(\bigcap_{n=0}^{\infty} \operatorname{Ker} CA^n\right) \bigcap \left(\bigcap_{n=0}^{\infty} \operatorname{Ker} B^*(A^*)^n\right),$$

$$\mathcal{R}(\Sigma) = \mathcal{N}(\Sigma)^\perp.$$

We shall refer to $\mathcal{N}(\Sigma)$ as the *excessive subspace* and to $\mathcal{R}(\Sigma)$ as the *principal subspace*. The next lemma explains this terminology.

LEMMA 2.3. *Let $\Sigma = (A, B, C, D; H, K, L)$ be a unitary system. Then the operators A, B and C admit the following partitionings:*

$$(10a) \qquad A = \begin{bmatrix} A_{11} & 0 \\ 0 & A_{00} \end{bmatrix} : \mathcal{N}(\Sigma)\boxplus\mathcal{R}(\Sigma) \to \mathcal{N}(\Sigma)\boxplus\mathcal{R}(\Sigma),$$

$$(10b) \qquad B = \begin{bmatrix} 0 \\ B_0 \end{bmatrix} : K \to \mathcal{N}(\Sigma) \boxplus \mathcal{R}(\Sigma),$$

$$(10c) \qquad C = [0 \quad C_0] : \mathcal{N}(\Sigma) \boxplus \mathcal{R}(\Sigma) \to L.$$

Here A_{11} is unitary, and the system $\Sigma_0 = (A_{00}, B_0, C_0, D; \mathcal{R}(\Sigma), K, L)$ is a unitary system which has the same transfer function as Σ. Moreover, the excessive subspace of the system Σ_0 consists of the zero vector only.

PROOF. To prove (10a) we have to show that $\mathcal{N}(\Sigma)$ and $\mathcal{R}(\Sigma)$ are both invariant under A. Since $\mathcal{R}(\Sigma) = \mathcal{N}(\Sigma)^{\perp}$, it suffices to show that $\mathcal{N}(\Sigma)$ is invariant under both A and A^*. Put

$$(11) \qquad \mathcal{N} = \bigcap_{n=0}^{\infty} \operatorname{Ker} C A^n, \qquad \mathcal{N}_* = \bigcap_{n=0}^{\infty} \operatorname{Ker} B^*(A^*)^n.$$

Thus $\mathcal{N}(\Sigma) = \mathcal{N} \cap \mathcal{N}_*$. Take $x \in \mathcal{N}(\Sigma)$. Then $C A^n x = 0$ for each $n \geq 0$, and hence

$$(CA^n) A x = C A^{n+1} x = 0, \qquad n = 0, 1, 2, \ldots,$$

which proves that $A\mathcal{N}(\Sigma) \subset \mathcal{N}$. By induction we show that also $A^*\mathcal{N}(\Sigma) \subset \mathcal{N}$. Again, let $x \in \mathcal{N}(\Sigma)$. From the first identity in (4b) we see that

$$C(A^* x) = C A^* x = -D B^* x = 0,$$

because $x \in \mathcal{N}(\Sigma) \subset \mathcal{N}_* \subset \operatorname{Ker} B^*$. Now, assume that $C A^n (A^* x) = 0$ for some $n \geq 0$. Then, by the first identity in (4a),

$$\begin{aligned} C A^{n+1}(A^* x) &= C A^n (A A^* x) \\ &= C A^n (x - B B^* x) = C A^n x = 0. \end{aligned}$$

Thus $A^* x \in \mathcal{N}$. In a similar way, one shows that $A\mathcal{N}(\Sigma) \subset \mathcal{N}_*$ and $A^*\mathcal{N}(\Sigma) \subset \mathcal{N}_*$. Since $\mathcal{N}(\Sigma) = \mathcal{N} \cap \mathcal{N}_*$, we have proved that $\mathcal{N}(\Sigma)$ is invariant under both A and A^*, and it follows that (10a) holds.

Note that $\mathcal{N}(\Sigma) \subset \operatorname{Ker} B^*$, and thus

$$\overline{\operatorname{Im} B} = (\operatorname{Ker} B^*)^{\perp} \subset \mathcal{N}(\Sigma)^{\perp} = \mathcal{R}(\Sigma),$$

which proves the partitioning in (10b). The partitioning in (10c) follows from the fact that $\mathcal{N}(\Sigma) \subset \operatorname{Ker} C$.

From the partitionings in (10a), (10b), (10c) we see that

$$(12) \qquad \begin{bmatrix} A & B \\ C & D \end{bmatrix} = \begin{bmatrix} A_{11} & 0 & 0 \\ 0 & A_{00} & B_0 \\ 0 & C_0 & D \end{bmatrix} : \mathcal{N}(\Sigma) \boxplus \mathcal{R}(\Sigma) \boxplus K \to \mathcal{N}(\Sigma) \boxplus \mathcal{R}(\Sigma) \boxplus L.$$

The operator in the left hand side of (12) is unitary. It follows that the operators $A_{11}: \mathcal{N}(\Sigma) \to \mathcal{N}(\Sigma)$ and

$$\begin{bmatrix} A_{00} & B_0 \\ C_0 & D \end{bmatrix} : \mathcal{R}(\Sigma) \boxplus K \to \mathcal{R}(\Sigma) \boxplus L$$

are both unitary. In particular, the system Σ_0 is a unitary system.

From the partitionings in (10a), (10b) and (10c) it also follows that

$$CA^n B = [0 \quad C_0] \begin{bmatrix} A_{11}^n & 0 \\ 0 & A_{00}^n \end{bmatrix} \begin{bmatrix} 0 \\ B_0 \end{bmatrix}$$
$$= C_0 A_{00}^n B_0, \qquad n = 0, 1, 2, \dots .$$

Hence $C(I - \lambda A)^{-1} B = C_0 (I_0 - \lambda A_{00})^{-1} B_0$ for each $\lambda \in \mathbb{D}$, and therefore Σ and Σ_0 have the same transfer function.

Finally, let us prove that $\mathcal{N}(\Sigma_0) = \{0\}$. Take $x_0 \in \mathcal{N}(\Sigma_0)$. Then

$$CA^n \begin{bmatrix} 0 \\ x_0 \end{bmatrix} = [0 \quad C_0] \begin{bmatrix} A_{11}^n & 0 \\ 0 & A_{00}^n \end{bmatrix} \begin{bmatrix} 0 \\ x_0 \end{bmatrix} = C_0 A_{00}^n x_0 = 0, \qquad n = 0, 1, 2, \dots ,$$

$$B^*(A^*)^n \begin{bmatrix} 0 \\ x_0 \end{bmatrix} = [0 \quad B_0^*] \begin{bmatrix} (A_{11}^*)^n & 0 \\ 0 & (A_{00}^*)^n \end{bmatrix} \begin{bmatrix} 0 \\ x_0 \end{bmatrix} = B_0^*(A_{00}^*)^n = 0, \qquad n = 0, 1, 2, \dots .$$

Thus $\begin{bmatrix} 0 \\ x_0 \end{bmatrix} \in \mathcal{N}(\Sigma) \cap \mathcal{R}(\Sigma)$, and hence $x_0 = 0$. $\square$

Let $\Sigma = (A, B, C, D; H, K, L)$ be a unitary system. The system Σ_0 constructed in the previous lemma is called the *principal part* of Σ. Note that the state space $\mathcal{R}(\Sigma)$ of Σ_0 is precisely the smallest closed linear manifold of H containing all vectors of the form

(13a)
$$A^n B x, \qquad x \in H,\ n = 0, 1, 2, \dots$$

(13b)
$$(A^*)^k C^* y, \qquad y \in H,\ k = 0, 1, 2, \dots .$$

To see this, recall that $\mathcal{N}(\Sigma) = \mathcal{N} \cap \mathcal{N}_*$, where $\mathcal{N}$ and $\mathcal{N}_*$ are defined by (11). It follows that

$$\mathcal{R}(\Sigma) = \mathcal{N}(\Sigma)^\perp = \overline{\mathcal{N}^\perp + \mathcal{N}_*^\perp}.$$

Now

$$\mathcal{N}^\perp = \bigvee_{n=0}^{\infty} \operatorname{Im} A^n B, \qquad \mathcal{N}_*^\perp = \bigvee_{n=0}^{\infty} \operatorname{Im}(A^*)^n C^*,$$

and thus $\mathcal{R}(\Sigma)$ is the closed linear hull of the vectors in (13a) and (13b).

A unitary system Σ is called *pure* if the excessive subspace of Σ consists of the zero vector only. Lemma 2.3 shows how one can reduce a unitary system to a pure one without changing its transfer function.

THEOREM 2.4. *A unitary system is pure if and only if its main operator is completely non-unitary.*

PROOF. Let $\Sigma = (A, B, C, D; H, K, L)$ be a unitary system. Assume A is completely non-unitary. Let $\mathcal{N}(\Sigma)$ be the excessive subspace, and $\mathcal{R}(\Sigma)$ the principal subspace. From Lemma 2.3 we know that A has the following partitioning

$$A = \begin{bmatrix} A_{11} & 0 \\ 0 & A_{00} \end{bmatrix} : \mathcal{N}(\Sigma) \boxplus \mathcal{R}(\Sigma) \to \mathcal{N}(\Sigma) \boxplus \mathcal{R}(\Sigma),$$

where A_{11} is unitary. Since A is completely non-unitary, this can only happen when $A_{11} = 0$. Hence $\mathcal{N}(\Sigma) = \{0\}$.

Next, assume that A is not completely non-unitary. So there exists a non-zero subspace N of H such that the spaces N and $N^{\perp}$ are invariant under A and $A|N$ is unitary. Consider the following partitionings:

$$A = \begin{bmatrix} A_{11} & 0 \\ 0 & A_{00} \end{bmatrix} : N \boxplus N^{\perp} \to N \boxplus N^{\perp},$$

$$B = \begin{bmatrix} B_1 \\ B_0 \end{bmatrix} : K \to N \boxplus N^{\perp}, \qquad C = [C_1 \quad C_0] : N \boxplus N^{\perp} \to L,$$

where A_{11} is unitary. It follows that

$$(14) \qquad \begin{bmatrix} A & B \\ C & D \end{bmatrix} = \begin{bmatrix} A_{11} & 0 & B_1 \\ 0 & A_{00} & B_0 \\ C_1 & C_0 & D \end{bmatrix} : N \boxplus N^{\perp} \boxplus K \to N \boxplus N^{\perp} \boxplus L.$$

The operator in (14) is unitary, and hence a contraction. Since A_{11} is unitary, this allows us to apply Lemma 1.2. It follows that the operators C_1 and B_1 in (14) are zero operators, and thus

$$N \subset \operatorname{Ker} C A^n, \qquad N \subset \operatorname{Ker} B^*(A^*)^n$$

for each $n \geq 0$, and therefore $N \subset \mathcal{N}(\Sigma)$. Since N is non-zero, we have proved that Σ is not pure. $\square$

COROLLARY 2.5. *A finite dimensional unitary system is pure if and only if its main operator has all its eigenvalues in* $\mathbb{D}$.

PROOF. By Proposition 1.3, a contraction on a finite dimensional Hilbert space is completely non-unitary if and only if all the eigenvalues of A are inside the open unit disc. This remark together with Theorem 2.4 yields the corollary. $\square$

The notion of purity of unitary systems is closely related to the notions of observability and controllability appearing in mathematical systems theory. A system $\Sigma = (A, B, C, D; H, K, L)$ is called (*approximately*) *observable* if

$$(15) \qquad \operatorname{Ker}(C|A) := \bigcap_{j=0}^{\infty} \operatorname{Ker} C A^j = \{0\},$$

and Σ is (*approximately*) *controllable* if

$$(16) \qquad \operatorname{Im}(A|B) := \bigvee_{j=0}^{\infty} \operatorname{Im} A^j B = H.$$

If a unitary system is observable and controllable, then its excessive part consists of the zero vector only, and hence such a system is pure. For finite dimensional systems the converse is also true.

PROPOSITION 2.6. *A finite dimensional unitary system is observable and controllable if and only if the system is pure.*

PROOF. Let $\Sigma = (A, B, C, D)$ be a pure unitary system with a finite dimensional state space H. We have to show that Σ is observable and controllable. To do this, note that the space $\operatorname{Ker}(C|A)$ in (15) is invariant under A. Thus, if $\operatorname{Ker}(C|A) \neq \{0\}$, then we can find $0 \neq x \in \operatorname{Ker}(C|A)$ so that $Ax = \lambda x$ for some $\lambda \in \mathbb{C}$. Since $Cx = 0$, the second identity in (4a) implies that

$$x = (I - C^*C)x = A^*Ax = \lambda A^*x.$$

It follows that $\|x\| \leq |\lambda|\|x\|$, because $\|A^*\| = \|A\| \leq 1$. So $|\lambda| \geq 1$. On the other hand, since Σ is pure, we know from Corollary 2.5 that $|\lambda| < 1$. Contradiction. Thus $\operatorname{Ker}(C|A) = \{0\}$, and Σ is observable. In a similar way, using

$$\operatorname{Im}(A|B)^{\perp} = \bigcap_{j=0}^{\infty} \operatorname{Ker} B^*(A^*)^j,$$

one shows that $\operatorname{Im}(A|B) = H$, and Σ is controllable. $\square$

In general, for arbitrary infinite dimensional systems Proposition 2.6 does not hold. In fact, it may happen that a pure unitary system with an infinite dimensional state space is neither observable nor controllable. A concrete example will be given in Corollary 5.3 of Section XXVIII.5.

The following proposition will be needed in Section XXVIII.10.

PROPOSITION 2.7. *Let $\Sigma = (A, B, C, D)$ be a unitary system, and fix $\lambda \in \mathbb{D}$. Then $\theta_{\Sigma}(\lambda)$ is invertible if and only if $\overline{\lambda}$ is in the resolvent of A, and in this case*

$$(17) \qquad \theta_{\Sigma}(\lambda)^{-1} = D^* + B^*(\lambda - A^*)^{-1}C^*.$$

PROOF. Since $\lambda A - I$ is invertible, we have

$$(18) \qquad \begin{bmatrix} \lambda A - I & \lambda B \\ C & D \end{bmatrix} = \begin{bmatrix} \lambda A - I & 0 \\ C & I \end{bmatrix} \begin{bmatrix} I & 0 \\ 0 & \theta_{\Sigma}(\lambda) \end{bmatrix} \begin{bmatrix} I & \lambda(\lambda A - I)^{-1}B \\ 0 & I \end{bmatrix}.$$

On the other hand,

$$\begin{bmatrix} \lambda A - I & \lambda B \\ C & D \end{bmatrix} = \begin{bmatrix} \lambda & 0 \\ 0 & I \end{bmatrix} \begin{bmatrix} A & B \\ C & D \end{bmatrix} - \begin{bmatrix} I & 0 \\ 0 & 0 \end{bmatrix}$$

$$= \left\{ \begin{bmatrix} \lambda & 0 \\ 0 & I \end{bmatrix} - \begin{bmatrix} I & 0 \\ 0 & 0 \end{bmatrix} \begin{bmatrix} A^* & C^* \\ B^* & D^* \end{bmatrix} \right\} \begin{bmatrix} A & B \\ C & D \end{bmatrix},$$

because Σ is a unitary system. It follows that

$$\begin{bmatrix} \lambda A - I & \lambda B \\ C & D \end{bmatrix} = \begin{bmatrix} \lambda - A^* & -C^* \\ 0 & I \end{bmatrix} \begin{bmatrix} A & B \\ C & D \end{bmatrix}. \tag{19}$$

The first and the third term in the right hand side of (18) are invertible 2×2 operator matrices. Also, the second term in the right hand side of (19) is invertible. It follows, by equivalence, that $\theta_\Sigma(\lambda)$ is invertible if and only if the left hand side of (18) is invertible. But, by (19), the latter is equivalent to the invertibility of $\lambda - A^*$. So $\theta_\Sigma(\lambda)$ is invertible if and only if $\overline{\lambda} \in \rho(A)$, and in this case

$$\begin{aligned} \theta_\Sigma(\lambda)^{-1} &= [0 \quad I] \begin{bmatrix} \lambda A - I & \lambda B \\ C & D \end{bmatrix}^{-1} \begin{bmatrix} 0 \\ I \end{bmatrix} \\ &= [0 \quad I] \begin{bmatrix} A^* & C^* \\ B^* & D^* \end{bmatrix} \begin{bmatrix} (\lambda - A^*)^{-1} & (\lambda - A^*)^{-1}C^* \\ 0 & I \end{bmatrix} \begin{bmatrix} 0 \\ I \end{bmatrix} \\ &= [B^* \quad D^*] \begin{bmatrix} (\lambda - A^*)^{-1}C^* \\ I \end{bmatrix} \\ &= D^* + B^*(\lambda - A^*)^{-1}C^*. \quad \square \end{aligned}$$

The arguments used to prove Proposition 2.7 show that the H-extension of $\theta_\Sigma(\cdot)$ is globally equivalent (see Section III.2 for the terminology) to the L-extension of $\lambda - A^*$. Here H is the state space of Σ and L is the output space. In fact, we have the following equivalence relation:

$$\begin{bmatrix} \theta_\Sigma(\lambda) & 0 \\ 0 & I_H \end{bmatrix} = E(\lambda) \begin{bmatrix} \lambda - A^* & 0 \\ 0 & I_L \end{bmatrix} F(\lambda), \qquad \lambda \in \mathbb{D}, \tag{20}$$

where

$$E(\lambda) = \begin{bmatrix} -C(\lambda A - I)^{-1} & I \\ (\lambda A - I)^{-1} & 0 \end{bmatrix} \begin{bmatrix} I & -C^* \\ 0 & I \end{bmatrix}, \tag{21a}$$

$$F(\lambda) = \begin{bmatrix} A & B \\ C & D \end{bmatrix} \begin{bmatrix} -\lambda(\lambda A - I)^{-1}B & 0 \\ I & I \end{bmatrix}, \tag{21b}$$

and the operators $E(\lambda)$ and $F(\lambda)$ are invertible for each $\lambda \in \mathbb{D}$.

XXVIII.3 UNITARY EQUIVALENCE

Two unitary systems Σ_1 and Σ_2,

$$\Sigma_\nu = (A_\nu, B_\nu, C_\nu, D_\nu; H_\nu, K_\nu, L_\nu), \qquad \nu = 1, 2, \tag{1}$$

are said to be *unitarily equivalent* if their feedthrough operators coincide (that is, $K_1 = K_2$, $L_1 = L_2$ and $D_1 = D_2$) and there exists a unitary operator $J: H_1 \to H_2$ such that

$$A_2 = JA_1J^{-1} \qquad B_2 = JB_1, \qquad C_2 = C_1J^{-1}. \tag{2}$$

Unitary equivalence is a reflexive, symmetric and transitive relation.

Main properties of unitary systems are preserved under unitary equivalence. For example, from (2) it follows that

$$(3) \qquad J\mathcal{N}(\Sigma_1) = \mathcal{N}(\Sigma_2),$$

where $\mathcal{N}(\Sigma_1)$ and $\mathcal{N}(\Sigma_2)$ are the excessive subspaces of Σ_1 and Σ_2, respectively. Thus, if Σ_1 and Σ_2 are unitarily equivalent and Σ_1 is pure, then so is Σ_2. From (2) it also follows that

$$(4) \qquad C_2 A_2^n B_2 = C_1 A_1^n B_1, \qquad n = 0, 1, 2, \ldots,$$

and thus $C_2(I - \lambda A_2)^{-1} B_2 = C_1(I - \lambda A_1)^{-1} B_1$ for each $\lambda \in \mathbb{D}$. Since unitarily equivalent systems have identical feedthrough operators, we conclude that the transfer function does not change when a unitary system is replaced by a unitarily equivalent one. The next theorem shows that for pure unitary systems the converse is also true.

THEOREM 3.1. *Two pure unitary systems have the same transfer function if and only if these systems are unitarily equivalent, and in this case the unitary operator establishing the unitary equivalence is unique.*

Theorem 3.1 tells us that for pure unitary systems the transfer function is a *complete unitary invariant*.

PROOF OF THEOREM 3.1. Let Σ_1 and Σ_2 be pure unitary systems, and assume that $\theta_{\Sigma_1}(\cdot) = \theta_{\Sigma_2}(\cdot)$. We have to prove that Σ_1 and Σ_2 are unitarily equivalent. Write Σ_1 and Σ_2 as in (1). Since $\theta_{\Sigma_1}(\cdot) = \theta_{\Sigma_2}(\cdot)$, these functions have the same value at 0, and therefore $D_1 = D_2$. In particular, $K_1 = K_2$ and $L_1 = L_2$. But then we can use formulas (6), (9a) and (9b) in the previous section to conclude that

$$C_1(I - \lambda A_1)^{-1} B_1 = C_2(I - \lambda A_2)^{-1} B_2,$$

$$B_1^*(I - \overline{\mu} A_1^*)^{-1}(I - \lambda A_1)^{-1} B_1 = B_2^*(I - \overline{\mu} A_2^*)^{-1}(I - \lambda A_2)^{-1} B_2,$$

$$C_1(I - \lambda A_1)^{-1}(I - \overline{\mu} A_1^*)^{-1} C_1^* = C_2(I - \lambda A_2)^{-1}(I - \overline{\mu} A_2^*)^{-1} C_2^*,$$

where μ and λ run over $\mathbb{D}$. By expanding the resolvents into power series, we obtain that for $i, j = 0, 1, 2, \ldots$

$$(5a) \qquad C_1 A_1^i B_1 = C_2 A_2^i B_2,$$

$$(5b) \qquad B_1^*(A_1^*)^i A_1^j B_1 = B_2^*(A_2^*)^j A_2^j B_2,$$

$$(5c) \qquad C_1 A_1^i (A_1^*)^j C_1^* = C_2 A_2^i (A_2^*)^j C_2^*.$$

Now, consider vectors $x \in H_1$ that admit a representation of the form

$$(6) \qquad x = \sum_{i=0}^{n} \alpha_i A_1^i B_1 y_i + \sum_{j=0}^{k} \beta_j (A_1^*)^j C_1^* z_j,$$

where $y_0, \ldots, y_n$ and $z_0, \ldots, z_k$ are arbitrary vectors in K_1 and L_1, respectively, and $\alpha_0, \ldots, \alpha_n$ and $\beta_0, \ldots, \beta_k$ are arbitrary complex numbers. Since Σ_1 is a pure unitary system, the set of vectors (6) is dense in H_1. For x as in (6) we set

$$
(7) \qquad Jx = \sum_{i=0}^{n} \alpha_i A_2^i B_2 y_i + \sum_{j=0}^{k} \beta_j (A_2^*)^j C_2^* z_j.
$$

In general, a vector x in (6) may have many different representations of the form considered in (6), but the definition of Jx does not depend on the particular choice of the representation in (6). To prove this, it suffices to show that the right hand side of (7) is the zero vector whenever x in (6) is the zero vector. So, consider

$$
0 = \sum_{i=0}^{n} \alpha_i A_1^i B_1 y_i + \sum_{j=0}^{k} \beta_j (A_1^*)^j C_1^* z_j,
$$

$$
\widetilde{x} = \sum_{i=0}^{n} \alpha_i A_2^i B_2 y_i + \sum_{j=0}^{k} \beta_j (A_2^*)^j C_2^* z_j.
$$

From these identities and (5a), (5c) we obtain:

$$
\begin{aligned}
C_2 A_2^r \widetilde{x} &= \sum_{i=0}^{n} \alpha_i C_2 A_2^{r+i} B_2 y_i + \sum_{j=0}^{k} \beta_j C_2 A_2^r (A_2^*)^j C_2^* z_j \\
&= \sum_{i=0}^{n} \alpha_i C_1 A_1^{r+i} B_1 y_i + \sum_{j=0}^{k} \beta_j C_1 A_1^r (A_1^*)^j C_1^* z_j \\
&= C_1 A_1^r 0 = 0, \qquad r = 0, 1, 2, \ldots .
\end{aligned}
$$

Similarly, using (5b) and the dual version of (5a), we obtain:

$$
B_2^* (A_2^*)^r \widetilde{x} = B_1^* (A_1^*)^r 0 = 0, \qquad r = 0, 1, 2, \ldots .
$$

It follows that $\widetilde{x}$ belongs to the excessive subspace of Σ_2. But Σ_2 is pure, so the latter subspace consists of the zero vector only, and hence $\widetilde{x} = 0$. So Jx in (7) is well-defined.

Let us denote by M_1 (resp., M_2) the set of all vectors given by the right hand side of (6) (resp., (7)). Both M_1 and M_2 are linear manifolds, M_1 is dense in H_1 (because Σ_1 is pure) and M_2 is dense in H_2 (because Σ_2 is pure). Let x be as in (6). Then

$$
\langle x, x \rangle = \sum_{i=0}^{n} \sum_{j=0}^{n} \alpha_i \overline{\alpha}_j \langle B_1^* (A_1^*)^j A_1^i B_1 y_i, y_j \rangle
$$

$$+ \sum_{i=0}^{n} \sum_{j=0}^{k} \alpha_i \overline{\beta}_j \langle C_1 A_1^{j+i} B_1 y_i, z_j \rangle$$

(8)
$$+ \sum_{i=0}^{k} \sum_{j=0}^{n} \beta_i \overline{\alpha}_j \langle B_1^* (A_1^*)^{j+i} C_1^* z_i, y_j \rangle$$

$$+ \sum_{i=0}^{k} \sum_{j=0}^{k} \beta_i \overline{\beta}_j \langle C_1 A_1^{j} (A_1^*)^{i} C_1^* z_i, z_j \rangle.$$

By the identities in (5a), (5b) and (5c), the right hand side of (8) does not change if throughout the subscript 1 is replaced by the subscript 2. It follows that

$$\langle Jx, Jx \rangle = \langle x, x \rangle, \qquad x \in M_1.$$

Obviously, J is a linear transformation. Thus J is an isometry on M_1. So, by continuity, J extends to an isometry on $\overline{M}_1 = H_1$, which we also denote by J. Note that $M_2 \subset \operatorname{Im} J$, and hence $H_2 \subset \overline{\operatorname{Im} J}$. But the range of an isometry is closed (cf. Theorem XI.2.1). Therefore, $\operatorname{Im} J = H_2$, and we have proved that J is a unitary operator.

From the definition of J (by the formulas (6) and (7)) we see that $JB_1 = B_2$ and $JC_1^* = C_2^*$. The latter identity implies that $C_1 J^{-1} = C_2$ because $J^* = J^{-1}$. Next, note (again use the definition of J) that

(9)
$$JA_1^i B_1 = A_2^i B_2, \qquad J(A_1^*)^j C_1^* = (A_2^*)^j C_2^*$$

for each i and j. The first identity in (9) yields

(10)
$$(JA_1 - A_2 J)A_1^i B_1 = JA_1^{i+1} B_1 - A_2^{i+1} B_2 = 0, \qquad i = 0, 1, 2, \ldots .$$

We also have

(11)
$$(JA_1 - A_2 J)(A_1^*)^i C_1^* = 0, \qquad i = 0, 1, 2, \ldots .$$

To prove (11) we first take $i = 0$. Since Σ_1 and Σ_2 are unitary, $B_\nu D_\nu^* = -A_\nu C_\nu^*$ for $\nu = 1, 2$ (see (4b) in the previous section). Now use $D_1 = D_2$ and (9). It follows that

(12)
$$JA_1 C_1^* = -JB_1 D_1^* = -B_2 D_2^* = A_2 C_2^*$$
$$= A_2 J C_1^*.$$

For $i > 0$ we use $A_\nu A_\nu^* = I - B_\nu B_\nu^*$ for $\nu = 1, 2$ (see (4a) in the previous section) and apply the formulas (9) and (5a) to get:

$$JA_1 (A_1^*)^i C_1^* = J(A_1 A_1^*)(A_1^*)^{i-1} C_1^*$$
$$= J(A_1^*)^{i-1} C_1^* - JB_1 B_1^* (A_1^*)^{i-1} C_1^*$$
$$= (A_2^*)^{i-1} C_2^* - B_2 B_2^* (A_2^*)^{i-1} C_2^*$$
$$= A_2 (A_2^*)^i C_2^* = A_2 J (A_1^*)^i C_1^*.$$

From (10) and (11) we see that the operator $JA_1 - A_2J$ is zero on all vectors of the form (6). But the set of all vectors in (6) is dense in H_1. So, by continuity, $JA_1 - A_2J = 0$. We have now proved that

$$(13) \qquad A_2 = JA_1J^{-1}, \qquad B_2 = JB_1, \qquad C_2 = C_1J^{-1}$$

with J unitary, and hence the systems Σ_1 and Σ_2 are unitarily equivalent.

It remains to prove the uniqueness statement. Let $U\colon H_1 \to H_2$ be a second unitary operator so that (13) holds with U in place of J. The first two identities in (13) with U in place of J give

$$UA_1^i B_1 = A_2^i B_2, \qquad i = 0, 1, 2, \dots .$$

Since U is unitary, the first and third identity in (13) with U in place of J show that $A_2^* = UA_1^*U^{-1}$ and $UC_1^* = C_2^*$, and hence

$$U(A_1^*)^j C_1^* = (A_2^*)^j C_2^*, \qquad j = 0, 1, 2, \dots .$$

So we see that (9) holds with U in place of J. But then $U - J$ acts as the zero operator on the vectors in (6). Since the set of vectors of the form (6) is dense in H, we conclude, by continuity, that U and J coincide. $\square$

Theorem 3.1 and Lemma 2.3 together yield the following corollary.

COROLLARY 3.2. *Two unitary systems have the same transfer function if and only if their principal parts are unitarily equivalent.*

XXVIII.4 CHARACTERISTIC OPERATOR FUNCTIONS AND EMBEDDING THEOREMS

The main operator of a unitary system is a contraction. Conversely, any contraction $A \in \mathcal{L}(H)$ can be *embedded* into a unitary system, i.e., given a contraction $A \in \mathcal{L}(H)$ one can find operators B, C and D such that $\Sigma = (A, B, C, D)$ is a unitary system. One such system can be obtained as follows. Put

$$(1a) \qquad R_{12}\colon \overline{\operatorname{Im} D_{A^*}} \to H, \qquad R_{12}u = D_{A^*}u,$$

$$(1b) \qquad R_{21}\colon H \to \overline{\operatorname{Im} D_A}, \qquad R_{21}x = D_A x,$$

$$(1c) \qquad R_{22}\colon \overline{\operatorname{Im} D_{A^*}} \to \overline{\operatorname{Im} D_A}, \qquad R_{22}u = -A^*u.$$

Then

$$(2) \qquad \Sigma_A := (A, R_{12}, R_{21}, R_{22}; H, \overline{\operatorname{Im} D_{A^*}}, \overline{\operatorname{Im} D_A})$$

is a unitary system. In fact, the system matrix associated with Σ_A is precisely the rotation operator R_A defined by A. The transfer function of the system Σ_A is denoted by $W_A(\cdot)$ and is called the *characteristic operator function* of the contraction A. Note that

$$(3) \qquad W_A(\lambda) = -A^* + \lambda D_A(I - \lambda A)^{-1} D_{A^*} : \overline{\operatorname{Im} D_{A^*}} \to \overline{\operatorname{Im} D_A}.$$

If A is replaced by a contraction $\widetilde{A}$ which is unitarily equivalent to A, i.e., $\widetilde{A} = UAU^{-1}$ for some unitary operator U, then the characteristic operator functions of A and $\widetilde{A}$ are essentially the same. To make this statement more precise, let us say that $W_A(\cdot)$ *coincides* with the operator function $W(\cdot) : \mathbb{D} \to \mathcal{L}(K, L)$ if there exist unitary operators

$$(4) \qquad V : \overline{\operatorname{Im} D_A} \to L, \qquad V_* : \overline{\operatorname{Im} D_{A^*}} \to K$$

such that $W(\lambda) = VW_A(\lambda)V_*^{-1}$ for each $\lambda \in \mathbb{D}$. Now, assume $\widetilde{A} = UAU^{-1}$ with U unitary. Then $\widetilde{A}^* = UA^*U^{-1}$, and it follows that

$$(5) \qquad D_{\widetilde{A}} = UD_A U^{-1}, \qquad D_{\widetilde{A}^*} = UD_{A^*}U^{-1}.$$

These identities imply that U maps $\overline{\operatorname{Im} D_A}$ onto $\overline{\operatorname{Im} D_{\widetilde{A}}}$ and $\overline{\operatorname{Im} D_{A^*}}$ onto $\overline{\operatorname{Im} D_{\widetilde{A}^*}}$. Let

$$(6) \qquad V : \overline{\operatorname{Im} D_A} \to \overline{\operatorname{Im} D_{\widetilde{A}}}, \qquad V_* : \overline{\operatorname{Im} D_{A^*}} \to \overline{\operatorname{Im} D_{\widetilde{A}^*}}$$

be the induced operators. Then V and V_* are unitary operators, and (by using $\widetilde{A} = UAU^{-1}$ and the identities in (5)) one obtains

$$(7) \qquad W_{\widetilde{A}}(\lambda) = VW_A(\lambda)V_*^{-1}, \qquad \lambda \in \mathbb{D}.$$

Thus the characteristic operator functions of A and $\widetilde{A}$ coincide. By applying Theorem 3.1 one may show that for completely non-unitary contractions the converse statement is also true.

THEOREM 4.1. *Two completely non-unitary contractions are unitarily equivalent if and only if their characteristic operator functions coincide.*

PROOF. Let $A \in \mathcal{L}(H)$ and $\widetilde{A} \in \mathcal{L}(\widetilde{H})$ be completely non-unitary contractions, and assume that W_A and $W_{\widetilde{A}}$ coincide. We have to show that A and $\widetilde{A}$ are unitarily equivalent. Let V and V_* be unitary operators, acting as in (6), such that (7) holds. Let Σ_A be as in (2), and let $\Sigma_{\widetilde{A}}$ be the corresponding unitary system for $\widetilde{A}$. Put

$$\Sigma = (A, R_{12}V_*^{-1}, VR_{21}, VR_{22}V_*^{-1}).$$

By virtue of (7) the unitary systems Σ and $\Sigma_{\widetilde{A}}$ have the same transfer function. Since A and $\widetilde{A}$ are completely non-unitary, the systems Σ and $\Sigma_{\widetilde{A}}$ are pure (by Theorem 2.4), and thus we can apply Theorem 3.1 to show that Σ and $\Sigma_{\widetilde{A}}$ are unitarily equivalent. In particular, A and $\widetilde{A}$ are unitarily equivalent. $\square$

There are many different ways to embed a contraction into a unitary system. The next theorem gives a complete description of the freedom one has.

THEOREM 4.2. *Let $A \in \mathcal{L}(H)$ be a contraction. Then the input operator B, the output operator C and the feedthrough operator D of an arbitrary unitary system Σ with A as its main operator are given by:*

$$(8a) \qquad B = [R_{12}V_*^{-1} \quad 0]: K_0 \boxplus K_1 \to H,$$

$$(8b) \qquad C = \begin{bmatrix} VR_{21} \\ 0 \end{bmatrix}: H \to L_0 \boxplus L_1,$$

$$(8c) \qquad D = \begin{bmatrix} VR_{22}V_*^{-1} & 0 \\ 0 & E \end{bmatrix}: K_0 \boxplus K_1 \to L_0 \boxplus L_1,$$

where R_{12}, R_{21} and R_{22} are the operators defined by (1a), (1b) and (1c), respectively, and

$$(9) \qquad V: \overline{\operatorname{Im} D_A} \to L_0, \qquad V_*: \overline{\operatorname{Im} D_{A^*}} \to K_0,$$

$$(10) \qquad E: K_1 \to L_1,$$

are arbitrary unitary operators. Moreover, there is a one-one correspondence between B, C and D and the operators V, V_* and E.

Theorem 4.2 has the following corollary; it describes the transfer functions of all possible unitary systems with a given state operator A.

COROLLARY 4.3. *Let $A \in \mathcal{L}(H)$ be a contraction. Then the general form of the transfer function $\theta_\Sigma(\cdot)$ of a unitary system Σ with A as its main operator is given by*

$$(11) \qquad \theta_\Sigma(\lambda) = \begin{bmatrix} W(\lambda) & 0 \\ 0 & E \end{bmatrix}: K_0 \boxplus K_1 \to L_0 \boxplus L_1,$$

where $W(\cdot)$ is an operator function which coincides with the characteristic operator function of A and E is an arbitrary unitary operator.

PROOF OF THEOREM 4.2. Let B, C and D be given by (8a), (8b) and (8c), respectively. Note

$$(12) \qquad \begin{bmatrix} A & R_{12}V_*^{-1} \\ VR_{21} & VR_{22}V_*^{-1} \end{bmatrix} = \begin{bmatrix} I & 0 \\ 0 & V \end{bmatrix} R_A \begin{bmatrix} I & 0 \\ 0 & V_*^{-1} \end{bmatrix},$$

where R_A is the rotation operator associated with A. Since V and V_* are unitary, it follows that the operator in the left hand side of (12) is unitary. Also E is unitary. Therefore, the operator

$$(13) \qquad \begin{bmatrix} A & R_{12}V_*^{-1} & 0 \\ VR_{21} & VR_{22}V_*^{-1} & 0 \\ 0 & 0 & E \end{bmatrix}: H \boxplus K_0 \boxplus K_1 \to H \boxplus L_0 \boxplus L_1$$

is unitary. But the operator in (13) is precisely the system matrix of $\Sigma = (A, B, C, D)$, and thus Σ is a unitary system.

To prove the converse, assume that $\Sigma = (A, B, C, D; H, K, L)$ is a unitary system. Put

$$K_0 = \overline{\operatorname{Im} B^*}, \qquad K_1 = \operatorname{Ker} B.$$

$$L_0 = \overline{\operatorname{Im} C}, \qquad L_1 = \operatorname{Ker} C^*.$$

Partition B, C and D as follows:

$$B = [B_0 \quad 0] : K_0 \boxplus K_1 \to H,$$

$$C = \begin{bmatrix} C_0 \\ 0 \end{bmatrix} : H \to L_0 \boxplus L_1,$$

$$D = \begin{bmatrix} D_{00} & D_{01} \\ D_{10} & D_{11} \end{bmatrix} : K_0 \boxplus K_1 \to L_0 \boxplus L_1.$$

Since $I - AA^* = BB^*$, we have $D_{A^*}^2 = BB^*$, and one can show (by using the same arguments as in the second paragraph of the proof of Lemma XXVII.5.2) that there exists a unitary operator $V_*: \overline{\operatorname{Im} D_{A^*}} \to \overline{\operatorname{Im} B^*}$ such that $V_* D_{A^*} x = B^* x$ for each $x \in H$. Let R_{12} be as in (1a). Then

$$B_0 V_*(D_{A^*} x) = B_0 B^* x = BB^* x$$
$$= D_{A^*}(D_{A^*} x) = R_{12}(D_{A^*} x), \qquad x \in H.$$

It follows that $B_0 V_* = R_{12}$. In a similar way one shows that there exists a unitary operator $V: \overline{\operatorname{Im} D_A} \to L_0$ such that $V R_{21} = C_0$, where R_{21} is defined by (1b).

Since $\Sigma = (A, B, C, D)$ is unitary, we have $AC^* = -BD^*$, and thus

$$[AC_0^* \quad 0] = AC^* = -BD^*$$
$$= -[B_0 \quad 0] \begin{bmatrix} D_{00}^* & D_{10}^* \\ D_{01}^* & D_{11}^* \end{bmatrix}$$
$$= -[B_0 D_{00}^* \quad B_0 D_{10}^*].$$

This implies that $B_0 D_{10}^* = 0$. But B_0 is 1-1, by definition, and thus $D_{10} = 0$. In a similar way, by using $A^* B = -C^* D$, one proves that $D_{01} = 0$. Thus the system matrix of $\Sigma = (A, B, C, D)$ partitions as follows:

$$(14) \qquad \begin{bmatrix} A & B_0 & 0 \\ C_0 & D_{00} & 0 \\ 0 & 0 & D_{11} \end{bmatrix} : H \boxplus K_0 \boxplus K_1 \to H \boxplus L_0 \boxplus L_1,$$

which implies that D_{11} is unitary.

To complete the proof it remains to show that $D_{00} = V R_{22} V_*^{-1}$. Put $\widetilde{D}_{00} = V R_{22} V_*^{-1}$, and consider the operators

$$(15) \qquad \begin{bmatrix} A & B_0 \\ C_0 & D_{00} \end{bmatrix}, \qquad \begin{bmatrix} A & B_0 \\ C_0 & \widetilde{D}_{00} \end{bmatrix}.$$

Since the operator in (14) is unitary, the same holds true for the first operator in (15). For the second operator in (15) the identity (12) holds true, and hence this operator is also unitary. Let K be the orthogonal complement of $\operatorname{Im}\begin{bmatrix} A \\ C_0 \end{bmatrix}$ in $H \boxplus L_0$. Then the operators

$$J_0 = \begin{bmatrix} B_0 \\ D_{00} \end{bmatrix} : K_0 \to K, \qquad \widetilde{J}_0 = \begin{bmatrix} B_0 \\ \widetilde{D}_{00} \end{bmatrix} : K_0 \to K$$

are both unitary. It follows that

$$\begin{bmatrix} B_0 \\ D_{00} \end{bmatrix} = \begin{bmatrix} B_0 \\ \widetilde{D}_{00} \end{bmatrix} J$$

for some unitary operator $J : K_0 \to K_0$. Since B_0 is 1–1, we conclude that J is the identity operator on K_0, and hence $D_{00} = \widetilde{D}_{00}$.

Note that V_* is uniquely determined by $B_0 V_* = R_{12}$ and the fact that B_0 is 1–1. Similarly, V is uniquely determined by $R_{21} = V^{-1} C_0$ and $\overline{\operatorname{Im} C_0} = L_0$. The uniqueness of E is obvious. $\square$

PROOF OF COROLLARY 4.3. Let $\theta(\cdot)$ be given by the right hand side of (11) with $W(\cdot)$ and E having the desired properties. Since $W(\cdot)$ and $W_A(\cdot)$ coincide, there exist unitary operators V and V_* as in (9). Now define B, C and D by (8a), (8b) and (8c), respectively. Then, by Theorem 4.2, the system $\Sigma = (A, B, C, D)$ is unitary, and its transfer function is precisely $\theta(\cdot)$. The converse statement follows directly from Theorem 4.2. $\square$

We conclude this section by computing the characteristic operator function of the operator $A : L_2([0,1]) \to L_2([0,1])$ defined by

$$(16) \qquad (Af)(t) = f(t) - 4 \int_t^1 e^{2(t-s)} f(s)\,ds, \qquad 0 \le t \le 1.$$

First, let us prove that A is a contraction. It will be convenient to write $A = I - 2K$, where

$$(Kf)(t) = 2 \int_t^1 e^{2(t-s)} f(s)\,ds, \qquad 0 \le t \le 1,$$

and thus

$$(K^* f)(t) = 2 \int_0^t e^{2(s-t)} f(s)\,ds, \qquad 0 \le t \le 1.$$

It follows that K^*K is the integral operator on $L_2([0,1])$ with kernel function $\ell(t,s)$ given by

$$\ell(t,s) = \int_0^{\min(t,s)} 4e^{-2t}e^{2\alpha}e^{2\alpha}e^{-2s}\,d\alpha$$

$$= \begin{cases} e^{2(t-s)} - e^{-2(t+s)}, & s \le t, \\ e^{2(s-t)} - e^{-2(t+s)}, & s \ge t. \end{cases}$$

Hence

$$I - A^*A = 2(K + K^* - 2K^*K) = \langle \cdot, u\rangle u,$$

where $u(t) = 2e^{-2t}$ for $0 \le t \le 1$. It follows that A is a contraction and

$$(17) \qquad D_A = \langle \cdot, u\rangle g, \qquad g := \frac{1}{\|u\|}u.$$

In a similar way, one computes that

$$I - AA^* = 2(K + K^* - 2KK^*) = \langle \cdot, u_*\rangle u_*,$$

where $u_*(t) = 2e^{2(t-1)}$, and thus

$$(18) \qquad D_{A^*} = \langle \cdot, u_*\rangle g_*, \qquad g_* := \frac{1}{\|u_*\|}u_*.$$

In the following we compute the function

$$(19) \qquad W(\lambda) := \langle W_A(\lambda)g_*, -g\rangle.$$

Since both defect operators have rank one, $W(\cdot)$ actually coincides with the characteristic operator function $W_A(\cdot)$.

Note that $I - A$ is a Volterra operator. So $I - \lambda A$ is invertible for $\lambda \ne 1$, and

$$(20) \qquad (I - \lambda A)^{-1}f = \frac{1}{1-\lambda}\left(I - \frac{2\lambda}{\lambda-1}K\right)^{-1}f, \qquad \lambda \ne 1.$$

To compute the operator in the right hand side of (20), we apply Theorem IX.2.1 to

$$k(t,s) = \begin{cases} 0, & 0 \le s \le t \le 1, \\ f(t)g(s), & 0 \le t \le s \le 1, \end{cases}$$

where $f(t) = 4\lambda(\lambda - 1)^{-1}e^{2t}$ and $g(t) = e^{-2t}$. The final result is:

$$(21a) \qquad \left(\left(I - \frac{2\lambda}{\lambda-1}K\right)^{-1}f\right)(t) = f(t) + \int_t^1 \gamma(t,s;\lambda)f(s)\,ds,$$

with

(21b)
$$\gamma(t,s;\lambda) = \frac{4\lambda}{\lambda-1}\left\{\exp\left(2t\left(\frac{1+\lambda}{1-\lambda}\right)\right)\right\}\left\{\exp\left(-2s\left(\frac{1+\lambda}{1-\lambda}\right)\right)\right\}.$$

Now

$$D_{A^*}g_* = \langle g_*, u_*\rangle g_* = u_*,$$

and therefore

$$D_A(I-\lambda A)^{-1}D_{A^*}g_* = \langle (I-\lambda A)^{-1}u_*, u\rangle g.$$

By using (21a) and (21b), we compute that

$$\left(\left(I-\frac{2\lambda}{\lambda-1}K\right)^{-1}u_*\right)(t) = 2\exp\left\{2(t-1)\left(\frac{1+\lambda}{1-\lambda}\right)\right\},$$

and thus

$$\langle (I-\lambda A)^{-1}u_*, u\rangle = \int_0^1 \frac{4}{1-\lambda}\left\{\exp\left(2(t-1)\left(\frac{1+\lambda}{1-\lambda}\right)\right)\right\}e^{-2t}\,dt$$

$$= \frac{4}{1-\lambda}\left\{\exp\left(-2\left(\frac{1+\lambda}{1-\lambda}\right)\right)\right\}\int_0^1 \exp\left(\frac{4\lambda t}{1-\lambda}\right)dt$$

$$= \frac{1}{\lambda}e^{-2} - \frac{1}{\lambda}\exp\left\{-2\left(\frac{1+\lambda}{1-\lambda}\right)\right\}.$$

Hence

(22)
$$\lambda D_A(I-\lambda A)^{-1}D_{A^*}g = \left(e^{-2} - \exp\left\{-2\left(\frac{1+\lambda}{1-\lambda}\right)\right\}\right)g.$$

Next, we compute that $-A^*g_* = -\|u_*\|^{-1}A^*u_*$. We have

$$(A^*u_*)(t) = ((I-2K^*)u_*)(t)$$

$$= u_*(t) - 4\int_0^t e^{2(s-t)}2e^{2(s-1)}\,ds$$

$$= u_*(t) - 2e^{-2t}e^{-2}\left(\int_0^t 4e^{4s}\,ds\right)$$

$$= 2e^{-2t}e^{-2} = e^{-2}u(t).$$

One computes that $\|u\| = \|u_*\| = (1-e^{-4})^{1/2}$, and thus

$$-A^*g_* = -e^{-2}\frac{\|u\|}{\|u_*\|}g = -e^{-2}g.$$

We conclude that

$$(23) \qquad W_A(\lambda)g_* = -\exp\left\{2\left(\frac{\lambda+1}{\lambda-1}\right)\right\}g,$$

and hence the characteristic operator function of the operator (16) coincides with the scalar function

$$(24) \qquad W(\lambda) = \exp\left\{2\left(\frac{\lambda+1}{\lambda-1}\right)\right\}.$$

XXVIII.5   REALIZATION THEOREM AND FUNCTIONAL MODEL

Let $\theta(\cdot)$ be an operator-valued function which is analytic on $\mathbb{D}$ and whose values are bounded linear operators acting from the Hilbert space K into the Hilbert space L. We say that $\theta(\cdot)$ has a *realization in the class of unitary systems* if there exists a unitary system Σ such that the transfer function of Σ coincides with $\theta(\cdot)$, and in this case we call Σ a *unitary realization* of $\theta(\cdot)$. For θ to have a unitary realization it is necessary (see Theorem 2.1) that

$$(1) \qquad \sup_{\lambda \in \mathbb{D}} \|\theta(\lambda)\| \le 1.$$

Condition (1) is also sufficient. In this section we shall prove the latter statement for the case when $K = \mathbb{C}^k$ and $L = \mathbb{C}^\ell$. (The restriction to finite dimensional Hilbert spaces is not essential and is made here only to avoid some technical difficulties connected with H_∞-functions with values in Hilbert spaces.)

We shall need a vector-valued version of Lemma XXVII.7.1. In the sequel we identify an operator A from $\mathbb{C}^k$ into $\mathbb{C}^\ell$ with the $\ell \times k$ matrix of A relative to the standard bases of $\mathbb{C}^k$ and $\mathbb{C}^\ell$, and, conversely, each $\ell \times k$ matrix is identified in this way with an operator from $\mathbb{C}^k$ into $\mathbb{C}^\ell$. By $H_\infty^{\ell \times k}(\mathbb{T})$ we denote the space of $\ell \times k$ matrices with entries in $H_\infty(\mathbb{T})$. Let $\widetilde\varphi \in H_\infty^{k \times \ell}(\mathbb{T})$, and let $\widetilde\varphi_{pq}$ be its (p,q)-th entry. From the second paragraph of Section XXVII.7 we know that each $\widetilde\varphi_{pq}$ induces a bounded analytic function φ_{pq} on $\mathbb{D}$ via the formulas

$$\varphi_{pq}(\lambda) = \sum_{n=0}^{\infty} (\varphi_{pq})_n \lambda^n, \quad \lambda \in \mathbb{D}, \qquad (\varphi_{pq})_n = \frac{1}{2\pi} \int_{-\pi}^{\pi} \widetilde\varphi_{pq}(e^{it}) e^{-int}\,dt.$$

Put

$$(2) \qquad \varphi(\lambda) = \begin{bmatrix} \varphi_{11}(\lambda) & \cdots & \varphi_{1k}(\lambda) \\ \vdots & & \vdots \\ \varphi_{\ell 1}(\lambda) & \cdots & \varphi_{\ell k}(\lambda) \end{bmatrix}, \qquad \lambda \in \mathbb{D}.$$

Then φ is analytic on $\mathbb{D}$ and $\|\varphi(\cdot)\|$ is bounded. (Here the norm of an $\ell \times k$ matrix A is the norm of A as an operator from $\mathbb{C}^k$ into $\mathbb{C}^\ell$.) Conversely, by Lemma XXVII.7.1, if

$$\varphi: \mathbb{D} \to \mathcal{L}(\mathbb{C}^k, \mathbb{C}^\ell) \tag{3}$$

is a bounded analytic function, then there exists a unique $\widetilde{\varphi} \in H_\infty^{\ell \times k}(\mathbb{T})$ such that φ and $\widetilde{\varphi}$ are related in the above way. In this case we refer to $\widetilde{\varphi}$ as the *boundary function* associated with φ.

   LEMMA 5.1. *Let φ in (2) be a bounded analytic function on $\mathbb{D}$, and let $\widetilde{\varphi}$ be its associated boundary function. Then*

$$\sup_{\lambda \in \mathbb{D}} \|\varphi(\lambda)\| = \operatorname*{ess\,sup}_{-\pi \le t \le \pi} \|\widetilde{\varphi}(e^{it})\|. \tag{4}$$

   PROOF. In the following we write $\|\widetilde{\varphi}\|_\infty$ for the right hand side of (4). Let M be the operator of multiplication by $\widetilde{\varphi}(\cdot)^*$ acting from $H_2^\ell(\mathbb{T})$ into $L_2^k(\mathbb{T})$. Thus

$$(Mg)(e^{it}) = \widetilde{\varphi}(e^{it})^* g(e^{it}), \qquad -\pi \le t \le \pi,$$

and hence

$$\|Mg\|^2 = \frac{1}{2\pi} \int_{-\pi}^{\pi} \|\widetilde{\varphi}(e^{it})^* g(e^{it})\|^2 dt \le \left(\operatorname*{ess\,sup}_{-\pi \le t \le \pi} \|\widetilde{\varphi}(e^{it})^*\| \right)^2 \|g\|^2.$$

Since $\|A^*\| = \|A\|$, it follows that $\|M\| \le \|\widetilde{\varphi}\|_\infty$.

   Next, fix $z \in \mathbb{D}$, and let x be an arbitrary vector in $\mathbb{C}^\ell$. Consider the function:

$$\widetilde{u}(\lambda) = (1 - \overline{z}\lambda)^{-1} x = \sum_{n=0}^{\infty} \overline{z}^n \lambda^n x, \qquad \lambda \in \mathbb{T}.$$

Obviously, $\widetilde{u}(\cdot) \in H_\infty^\ell(\mathbb{T})$ and

$$\|\widetilde{u}(\cdot)\|^2 = \frac{1}{1 - |z|^2} \|x\|^2. \tag{5}$$

Since

$$(1 - ze^{-it})^{-1} = \sum_{n=0}^{\infty} z^n e^{-int},$$

we see that

$$x^* \varphi(z) = \frac{1}{2\pi} \int_{-\pi}^{\pi} \widetilde{u}(e^{it})^* \widetilde{\varphi}(e^{it}) dt. \tag{6}$$

This identity holds for any $\widetilde{\varphi} \in H_\infty^{\ell \times k}(\mathbf{T})$. Thus in (6) we may replace $\widetilde{\varphi}(e^{it})$ by $e^{int}\widetilde{\varphi}(e^{it})$ with $n \geq 0$ which yields:

$$(7) \qquad z^n x^* \varphi(z) = \frac{1}{2\pi} \int_{-\pi}^{\pi} \widetilde{u}(e^{it})^* \widetilde{\varphi}(e^{it}) e^{int}\, dt, \qquad n = 0, 1, 2, \ldots .$$

By taking adjoints we see that for $n = 0, 1, 2, \ldots$ the Fourier coefficients of the functions $M\widetilde{u}(\cdot)$ and $\varphi(z)^*\widetilde{u}(\cdot)$ coincide. Note that the function $\varphi(z)^*\widetilde{u}(\cdot)$ is a vector function of the same type as $\widetilde{u}(\cdot)$. In fact, one obtains $\varphi(z)^*\widetilde{u}(\cdot)$ from the definition of $\widetilde{u}$ by replacing x by $\varphi(z)^*x$. Then $\varphi(z)^*\widetilde{u}(\cdot) \in H_\infty^k(\mathbf{T})$ and

$$(8) \qquad \|\varphi(z)^*\widetilde{u}(\cdot)\|^2 = \frac{1}{1 - |z|^2}\|\varphi(z)^*x\|^2.$$

Let $\mathbb{P}$ denote the orthogonal projection of $L_2^k(\mathbf{T})$ onto $H_2^k(\mathbf{T})$. Then the previous remarks imply that

$$\mathbb{P}M\widetilde{u}(\cdot) = \mathbb{P}\big(\varphi(z)^*\widetilde{u}(\cdot)\big) = \varphi(z)^*\widetilde{u}(\cdot),$$

and so

$$\|\varphi(z)^*\widetilde{u}(\cdot)\| = \|\mathbb{P}M\widetilde{u}(\cdot)\| \leq \|M\widetilde{u}(\cdot)\| \leq \|\widetilde{\varphi}\|_\infty \|\widetilde{u}(\cdot)\|.$$

Now use (5) and (8) to conclude that $\|\varphi(z)^*x\| \leq \|\widetilde{\varphi}\|_\infty \|x\|$ for each $x \in \mathbf{C}^\ell$, and thus

$$\|\varphi(z)\| = \|\varphi(z)^*\| \leq \|\widetilde{\varphi}\|_\infty, \qquad z \in \mathbb{D}.$$

To prove that this upper bound is sharp, put $\varphi_r(e^{it}) = \varphi(re^{it})$ for $0 < r < 1$. From the proof of Lemma XXVII.7.1 we know that for each p and q the (p, q)-th entry of φ_r converges in the norm of $L_2(\mathbf{T})$ to the (p, q)-th entry of $\widetilde{\varphi}$. So there exist a set $\Gamma \subset [-\pi, \pi]$ of measure zero and a sequence $r_1, r_2, \ldots, r_n \uparrow 1$, such that for each $t \in [-\pi, \pi]\backslash\Gamma$ and each p and q the (p, q)-th entry of $\varphi_{r_n}(e^{it})$ converges to the (p, q)-th entry of $\widetilde{\varphi}(e^{it})$ when $n \to \infty$. It follows that

$$\lim_{n \to \infty} \|\varphi_{r_n}(e^{it})\| = \|\widetilde{\varphi}(e^{it})\|$$

for all $t \in [-\pi, \pi]\backslash\Gamma$. Since

$$(9) \qquad \sup_{\lambda \in \mathbb{D}} \|\varphi(\lambda)\| \geq \|\varphi_{r_n}(e^{it})\|$$

for all t and n, we see that the left hand side of (9) is larger than or equal to $\|\widetilde{\varphi}\|_\infty$, which completes the proof. $\square$

Now assume that $\theta : \mathbb{D} \to \mathcal{L}(\mathbf{C}^k, \mathbf{C}^\ell)$ is an analytic operator-valued function such that (1) holds. Then θ has a well-defined boundary function, which we also denote by θ, and

$$(10) \qquad \operatorname*{ess\,sup}_{-\pi \leq t \leq \pi} \|\theta(e^{it})\| \leq 1.$$

In the sequel M_θ will denote the operator of multiplication by θ acting from $L_2^k(\mathbb{T})$ into $L_2^\ell(\mathbb{T})$. Condition (10) implies that M_θ is a contraction. We write Δ_θ for the associated defect operator. Put

$$(11) \qquad J_\theta = \begin{bmatrix} M_\theta \tau \\ \Delta_\theta \tau \end{bmatrix} : H_2^k(\mathbb{T}) \to H_2^\ell(\mathbb{T}) \boxplus L_2^k(\mathbb{T}).$$

Here τ is the canonical embedding of $H_2^k(\mathbb{T})$ into $L_2^k(\mathbb{T})$, that is,

$$\tau: H_2^k(\mathbb{T}) \to L_2^k(\mathbb{T}), \qquad \tau f = f.$$

Since the entries of θ are in $H_\infty(\mathbb{T})$, the operator M_θ maps $H_2^k(\mathbb{T})$ into $H_2^\ell(\mathbb{T})$, and thus J_θ is well-defined. Note that $\tau^*\tau$ is the identity operator on $H_2^k(\mathbb{T})$, and $\mathbb{P} := \tau\tau^*$ is the orthogonal projection of $L_2^\ell(\mathbb{T})$ onto $H_2^\ell(\mathbb{T})$.

THEOREM 5.2. *Let $\theta: \mathbb{D} \to \mathcal{L}(\mathbb{C}^k, \mathbb{C}^\ell)$ be an analytic operator-valued function such that (1) holds. Let M_θ be the operator of multiplication by θ acting from $L_2^k(\mathbb{T})$ into $L_2^\ell(\mathbb{T})$, let Δ_θ be the associated defect operator, and let J_θ be defined by (11). Put*

$$(12) \qquad H = \left\{ \begin{bmatrix} f \\ g \end{bmatrix} \in H_2^\ell(\mathbb{T}) \boxplus \overline{\mathrm{Im}\,\Delta_\theta} \; \Big| \; \begin{bmatrix} f \\ g \end{bmatrix} \perp \mathrm{Im}\,J_\theta \right\}$$

and consider the following operators

$$(13a) \qquad A: H \to H, \qquad A = \begin{bmatrix} S^* & 0 \\ 0 & V^* \end{bmatrix} \tau_H,$$

$$(13b) \qquad B: \mathbb{C}^k \to H, \qquad B = \begin{bmatrix} S^* M_\theta \tau_0 \\ V^* \Delta_\theta \tau_0 \end{bmatrix},$$

$$(13c) \qquad C: H \to \mathbb{C}^\ell, \qquad C = [\pi_0 \quad 0]\tau_H,$$

$$(13d) \qquad D: \mathbb{C}^k \to \mathbb{C}^\ell, \qquad D = \theta(0),$$

where τ_H is the canonical embedding of H into $H_2^\ell(\mathbb{T}) \boxplus L_2^k(\mathbb{T})$ and

$$(14) \qquad S: H_2^\ell(\mathbb{T}) \to H_2^\ell(\mathbb{T}), \qquad (Sf)(e^{it}) = e^{it} f(e^{it}),$$

$$(15) \qquad V: L_2^k(\mathbb{T}) \to L_2^k(\mathbb{T}), \qquad (Vg)(e^{it}) = e^{it} g(e^{it}),$$

$$(16) \qquad \tau_0: \mathbb{C}^k \to L_2^k(\mathbb{T}), \qquad (\tau_0 x)(e^{it}) = x,$$

$$(17) \qquad \pi_0 \colon H_2^\ell(\mathbb{T}) \to \mathbb{C}^\ell, \qquad \pi_0(f) = \frac{1}{2\pi} \int_{-\pi}^{\pi} f(e^{it})dt.$$

Then $\Sigma_\theta := (A, B, C, D; H, \mathbb{C}^k, \mathbb{C}^\ell)$ is a unitary realization of θ and Σ_θ is pure.

PROOF. We split the proof into six parts. In the first part we examine further the operator J_θ; the second part concerns the definitions of the operators A and B.

Part (a). In this part we show that J_θ is an isometry, and Im J_θ is invariant under the operator

$$(18) \qquad R = \begin{bmatrix} S & 0 \\ 0 & V \end{bmatrix} \colon H_2^\ell(\mathbb{T}) \boxplus L_2^k(\mathbb{T}) \to H_2^\ell(\mathbb{T}) \boxplus L_2^k(\mathbb{T}).$$

It will be convenient to use the operator

$$(19) \qquad \pi \colon L_2^\ell(\mathbb{T}) \to H_2^\ell(\mathbb{T}), \qquad g - \pi g \perp H_2^\ell(\mathbb{T}).$$

Note that π^* is the canonical embedding of $H_2^\ell(\mathbb{T})$ into $L_2^\ell(\mathbb{T})$, and thus $\pi^*\pi$ is the orthogonal projection of $L_2^\ell(\mathbb{T})$ onto $H_2^\ell(\mathbb{T})$. The operator J_θ in (11) can now be written more precisely as

$$J_\theta = \begin{bmatrix} \pi M_\theta \tau \\ \Delta_\theta \tau \end{bmatrix}.$$

It follows that

$$J_\theta^* J_\theta = [\tau^* M_\theta^* \pi^* \quad \tau^* \Delta_\theta] \begin{bmatrix} \pi M_\theta \tau \\ \Delta_\theta \tau \end{bmatrix}$$

$$= \tau^* M_\theta^* \pi^* \pi M_\theta \tau + \tau^* \Delta_\theta^2 \tau.$$

Recall that $M_\theta \tau$ maps into $H_2^\ell(\mathbb{T})$ and $\pi^*\pi$ acts as the identity operator on $H_2^\ell(\mathbb{T})$. Thus

$$J_\theta^* J_\theta = \tau^* M_\theta^* M_\theta \tau + \tau^* \tau - \tau^* M_\theta^* M_\theta \tau = \tau^* \tau = I,$$

where I is the identity operator on $H_2^k(\mathbb{T})$. We proved that J is an isometry.

To see that Im J_θ is invariant under the operator R, let $\widetilde{S}$ and $\widetilde{V}$ be the operators S and V in (14) and (15) with the roles of k and ℓ interchanged. Thus $\widetilde{S}$ and $\widetilde{V}$ are the operators of multiplication by e^{it} acting on $H_2^k(\mathbb{T})$ and $L_2^\ell(\mathbb{T})$, respectively. Then

$$(20) \qquad \widetilde{V} M_\theta = M_\theta V, \qquad V \Delta_\theta = \Delta_\theta V,$$

$$(21) \qquad V\tau = \tau \widetilde{S}, \qquad Sg = \pi \widetilde{V} g \qquad (g \in H_2^\ell(\mathbb{T})).$$

It follows that

$$(22) \qquad R J_\theta = \begin{bmatrix} S \pi M_\theta \tau \\ V \Delta_\theta \tau \end{bmatrix} = J_\theta \widetilde{S},$$

and hence $\operatorname{Im} J_\theta$ is invariant under R.

Part (b). In this part we show that the operators A and B are well-defined. Obviously, H is a closed linear manifold in the subspace $H_2^\ell(\mathbb{T})\boxplus\overline{\operatorname{Im}\Delta_\theta}$. Hence H is a Hilbert space in its own right. Note that $\operatorname{Im} J_\theta$ is also contained in $H_2^\ell\boxplus\overline{\operatorname{Im}\Delta_\theta}$. Since J_θ is an isometry, $\operatorname{Im} J_\theta$ is closed, and so

$$(23) \qquad H_2^\ell(\mathbb{T})\boxplus\overline{\operatorname{Im}\Delta_\theta} = H\boxplus\operatorname{Im} J_\theta.$$

The second identity in (20) implies that $V^*\Delta_\theta = \Delta_\theta V^*$, and thus $H_2^\ell(\mathbb{T})\boxplus\overline{\operatorname{Im}\Delta_\theta}$ is invariant under R^*. From Part (a) we know that the same is true for $(\operatorname{Im} J_\theta)^\perp$. It follows that H is also invariant under R^*, and thus $A = R^*\tau_H$ is well-defined.

Next, we consider B. Since τ_0 maps into $H_2^k(\mathbb{T})$, we have $\tau\tau^*\tau_0 = \tau_0$, and thus B may be rewritten as

$$(24) \qquad B = R^* J_\theta \tau^* \tau_0.$$

Now $\operatorname{Im} J_\theta$ is contained in $H_2^\ell(\mathbb{T})\boxplus\overline{\operatorname{Im}\Delta_\theta}$, and the latter space is invariant under R^*. So $\operatorname{Im} B \subset H_2^\ell(\mathbb{T})\boxplus\overline{\operatorname{Im}\Delta_\theta}$. Take x in $\mathbb{C}^k$ and $f \in H_2^k(\mathbb{T})$. Then we see from (22) that

$$\begin{aligned}
\langle Bx, J_\theta f\rangle &= \langle R^* J_\theta \tau^* \tau_0 x, J_\theta f\rangle\\
&= \langle J_\theta \tau^* \tau_0 x, J_\theta \widetilde{S} f\rangle\\
&= \langle \tau^* \tau_0 x, \widetilde{S} f\rangle\\
&= \langle \widetilde{S}^* \tau^* \tau_0 x, f\rangle = 0.
\end{aligned}$$

Here we used that J_θ is an isometry (and hence J_θ preserves the inner product) and that $\widetilde{S}^*$ acts on a constant function as the zero operator. The above calculation shows that $\operatorname{Im} B\perp\operatorname{Im} J_\theta$, and hence B maps into H. So B is well-defined.

Part (c). In this part we compute the adjoints of A, B and C. We start with B. Put $\pi_H = \tau_H^*$. Since B maps into H, we see from (24) that

$$B = \pi_H B = \pi_H R^* J_\theta \tau^* \tau_0,$$

and hence

$$(25) \qquad B^* = \tau_0^* \tau J_\theta^* R\tau_H.$$

Next, consider A. Since A maps into H, we have $A = R^*\tau_H = \pi_H R^*\tau_H$, and thus $A^* = \pi_H R\tau_H$. Note that $R\tau_H$ maps into $H_2^\ell(\mathbb{T})\boxplus\overline{\operatorname{Im}\Delta_\theta}$, because the latter space is invariant under R. Now, by (23), the projection π_H coincides on $H_2^\ell(\mathbb{T})\boxplus\overline{\operatorname{Im}\Delta_\theta}$ with the orthogonal projection on $(\operatorname{Im} J_\theta)^\perp$ restricted to $H_2^\ell(\mathbb{T})\boxplus\overline{\operatorname{Im}\Delta_\theta}$. Since J_θ is an isometry, it follows that the orthogonal projection of $H_2^\ell(\mathbb{T})\boxplus L_2^k(\mathbb{T})$ onto $(\operatorname{Im} J_\theta)^\perp$ is given by $I - J_\theta J_\theta^*$. Thus

$$\begin{aligned}
A^* &= (I - J_\theta J_\theta^*)R\tau_H\\
&= R\tau_H - J_\theta\{I - \widetilde{S}\widetilde{S}^* + \widetilde{S}\widetilde{S}^*\}J_\theta^* R\tau_H\\
&= R\tau_H - J_\theta(I - \widetilde{S}\widetilde{S}^*)J_\theta^* R\tau_H - J_\theta \widetilde{S}\widetilde{S}^* J_\theta^* R\tau_H.
\end{aligned}$$

According to (22) we have

$$\widetilde{S}^* J_\theta^* R\tau_H = J_\theta^* R^* R\tau_H = J_\theta^* \tau_H = 0.$$

Here we used that $R^* R$ is the identity operator on $H_2^\ell(\mathbb{T}) \boxplus L_2^k(\mathbb{T})$, and that $J_\theta^* \tau_H = 0$, which follows from $H \subset (\operatorname{Im} J_\theta)^\perp = \operatorname{Ker} J_\theta^*$. Next, note that

$$(I - \widetilde{S}\widetilde{S}^*)f = \tau^*(\tau_0 \tau_0^*)\tau f, \qquad f \in H_2^k(\mathbb{T}).$$

We conclude (use (25)) that

(26) $$A^* = R\tau_H - J_\theta \tau^*(\tau_0 \tau_0^*)\tau J_\theta^* R\tau_H = R\tau_H - J_\theta \tau^* \tau_0 B^*.$$

Since $\tau_H^* = \pi_H$, the adjoint of C is obtained in the following way:

$$C^* = \pi_H \begin{bmatrix} \pi_0^* \\ 0 \end{bmatrix}$$

$$= (I - J_\theta J_\theta^*) \begin{bmatrix} \pi_0^* \\ 0 \end{bmatrix}$$

$$= \begin{bmatrix} \pi_0^* \\ 0 \end{bmatrix} - J_\theta \tau^* M_\theta^* \pi^* \pi_0^*.$$

From the definition of π_0 we see that

(27) $$\pi_0^* \colon \mathbb{C}^\ell \to H_2^\ell(\mathbb{T}), \qquad (\pi_0^* x)(e^{it}) = x.$$

Hence $\pi^* \pi_0^*$ is the same map but now viewed as an operator from $\mathbb{C}^\ell$ into $L_2^\ell(\mathbb{T})$. Thus

$$(M_\theta^* \pi^* \pi_0^* x)(e^{it}) = \theta(e^{it})^* x.$$

Since the entries of $\theta(\cdot)$ are in $H_\infty(\mathbb{T})$, the function $\tau^* \theta(\cdot)^* x$ is a constant, namely $\theta(0)^* x$. It follows that $\tau^* M_\theta^* \pi^* \pi_0^* x = \tau^* \tau_0 D^* x$, and hence

(28) $$C^* = \begin{bmatrix} \pi_0^* \\ 0 \end{bmatrix} - J_\theta \tau^* \tau_0 D^*.$$

Part (d). In this part we show that Σ_θ is a unitary system. This requires one to check the six identities in formulas (4a), (4b) and (4c) in Section XXVIII.2. By definition, the operators A and R^* coincide on H. Thus, in virtue of the second equality in (26), we have

$$AA^* = R^* A^* = R^* R\tau_H - R^* J_\theta \tau^* \tau_0 B^*.$$

Now $R^* R\tau_H$ is the identity operator on H, and hence (use (24)), we see that $AA^* + BB^* = I_H$. Since A and R^* coincide on H, we also have

$$AC^* = R^* \begin{bmatrix} \pi_0^* \\ 0 \end{bmatrix} - R^* J_\theta \tau^* \tau_0 D^*$$

$$= \begin{bmatrix} S^* \pi_0^* \\ 0 \end{bmatrix} - BD^* = -BD^*.$$

Here we used that S^* maps constant functions into zero, and so $S^*\pi_0^* = 0$. The range of C^* is in H. Thus τ_H acts as the identity operator on $\operatorname{Im} C^*$, and therefore

$$CC^* = [\pi_0 \quad 0]\left\{\begin{bmatrix} \pi_0^* \\ 0 \end{bmatrix} - J_\theta \tau^* \tau_0 D^*\right\}.$$

Now use that $\pi_0\pi_0^*$ is the identity operator on $\mathbb{C}^\ell$, and that

$$[\pi_0 \quad 0]J_\theta\tau^*\tau_0 = \pi_0 M_\theta \tau\tau^*\tau_0 = \pi_0 M_\theta\tau_0 = D,$$

and one obtains that $CC^* + DD^* = I_{\mathbb{C}^\ell}$.

We have now proved the first identities in formulas (4a), (4b) and (4c) of Section XXVIII.2. Let us proceed with the second set of identities. First, we compute B^*A. Since τ_H acts as the identity operator on H and $J_\theta^*\tau_H = 0$, we have

$$\begin{aligned}
B^*A &= \tau_0^*\tau J_\theta^* RR^*\tau_H = \tau_0^*\tau J_\theta^*(RR^* - I)\tau_H \\
&= -\tau_0^*\tau J_\theta^* \begin{bmatrix} I - SS^* & 0 \\ 0 & 0 \end{bmatrix}\tau_H \\
&= -[\tau_0^*\tau\tau^* M_\theta^*\pi^*(I - SS^*) \quad 0]\tau_H \\
&= -[\tau_0^* M_\theta^*\pi_0^*\pi_0 \quad 0]\tau_H \\
&= -[D^*\pi_0 \quad 0]\tau_H = -D^*C.
\end{aligned}$$

It follows (use (28)) that

$$\begin{aligned}
A^*A &= R\tau_H A - J_\theta\tau^*\tau_0 B^*A \\
&= RR^*\tau_H + J_\theta\tau^*\tau_0 D^*C \\
&= RR^*\tau_H + \begin{bmatrix} \pi_0^* \\ 0 \end{bmatrix}[\pi_0 \quad 0]\tau_H - C^*C \\
&= \begin{bmatrix} SS^* + \pi_0^*\pi_0 & 0 \\ 0 & I \end{bmatrix}\tau_H - C^*C \\
&= I_H - C^*C,
\end{aligned}$$

because $SS^* + \pi_0^*\pi_0$ is the identity operator on $H_2^\ell(\mathbb{T})$. Finally,

$$\begin{aligned}
B^*B &= \tau_0^*\tau J_\theta^* R\tau_H B = \tau^*\tau J_\theta^* RB \\
&= \tau_0^*\tau J_\theta^* RR^* J_\theta\tau^*\tau_0 \\
&= \tau_0^*\tau J_\theta^* J_\theta\tau^*\tau_0 - \tau_0^*\tau J_\theta^* \begin{bmatrix} I - SS^* & 0 \\ 0 & 0 \end{bmatrix} J_\theta\tau^*\tau_0 \\
&= I_{\mathbb{C}^k} - \tau_0^*\tau\tau^* M_\theta^*\pi^*(I - SS^*)\pi M_\theta\tau\tau^*\tau_0 \\
&= I_{\mathbb{C}^k} - \tau_0^* M_\theta^*\pi^*\pi_0^*\pi_0\pi M_\theta\tau\tau^*\tau_0 \\
&= I_{\mathbb{C}^k} - D^*D.
\end{aligned}$$

We have now proved that Σ_θ is a unitary system.

Part (e). In this part we show that Σ_θ is a realization of θ. So we have to compute the transfer function of Σ_θ. Let K be the orthogonal complement of H in the space $H_2^\ell(\mathbb{T}) \boxplus L_2^k(\mathbb{T})$. We know that H is invariant under R^*. Thus

$$(29) \qquad I - \lambda R^* = \begin{bmatrix} I - \lambda A & * \\ 0 & * \end{bmatrix} : H \boxplus K \to H \boxplus K.$$

Since A is the main operator of a unitary system, the operator A is a contraction. Thus $I - \lambda A$ is invertible for $|\lambda| < 1$. The same holds true for $I - \lambda R^*$. It follows that

$$\tau_H(I - \lambda A)^{-1} = (I - \lambda R^*)^{-1}\tau_H, \qquad \lambda \in \mathbb{D}.$$

But then

$$
\begin{aligned}
C(I - \lambda A)^{-1} B &= \begin{bmatrix} \pi_0 & 0 \end{bmatrix} \begin{bmatrix} (I - \lambda S^*)^{-1} & 0 \\ 0 & (I - \lambda V)^{-1} \end{bmatrix} B \\
&= \pi_0(I - \lambda S^*)^{-1} S^* M_\theta \tau_0 \\
&= \sum_{\nu=0}^{\infty} \lambda^\nu \pi_0 (S^*)^{\nu+1} M_\theta \tau_0.
\end{aligned}
$$

Now

$$\pi_0(S^*)^{\nu+1} M_\theta \tau_0 = \frac{1}{2\pi} \int_{-\pi}^{\pi} e^{-i(\nu+1)t} \theta(e^{it})\,dt,$$

which shows that $C(I - \lambda A)^{-1} B = \lambda^{-1}\big(\theta(\lambda) - \theta(0)\big)$. Hence the transfer function of Σ_θ is precisely θ.

Part (f). In this final part we show that the system Σ_θ is pure. Assume

$$(30) \qquad \begin{bmatrix} f \\ g \end{bmatrix} \in \Big(\bigcap_{n=0}^{\infty} \operatorname{Ker} C A^n \Big) \bigcap \Big(\bigcap_{n=0}^{\infty} \operatorname{Ker} B^*(A^*)^n \Big).$$

From (29) we see that $\tau_H A^n = (R^*)^n \tau_H$. It follows that

$$
\begin{aligned}
0 = C A^n \begin{bmatrix} f \\ g \end{bmatrix} &= \begin{bmatrix} \pi_0 & 0 \end{bmatrix} (R^*)^n \begin{bmatrix} f \\ g \end{bmatrix} \\
&= \begin{bmatrix} \pi_0 & 0 \end{bmatrix} \begin{bmatrix} (S^*)^n & 0 \\ 0 & (V^*)^n \end{bmatrix} \begin{bmatrix} f \\ g \end{bmatrix} \\
&= \pi_0(S^*)^n f.
\end{aligned}
$$

So $\pi_0(S^*)^n f = 0$ for $n = 0, 1, 2, \dots$. In other words,

$$\frac{1}{2\pi} \int_{-\pi}^{\pi} e^{-int} f(e^{it})\,dt = 0, \qquad n = 0, 1, 2, \dots.$$

Since $f \in H_2^\ell(\mathbb{T})$, we may conclude that $f = 0$.

To prove that $g = 0$ we use that

$$(31) \qquad B^*(A^*)^n \begin{bmatrix} 0 \\ g \end{bmatrix} = 0, \qquad n = 0, 1, 2, \dots .$$

By induction, using the second equality in (26), one shows that (31) implies

$$(A^*)^n \begin{bmatrix} 0 \\ g \end{bmatrix} = \begin{bmatrix} 0 \\ V^n g \end{bmatrix}, \qquad n = 0, 1, 2, \dots .$$

But then we have

$$\begin{aligned}
0 = B^*(A^*)^n \begin{bmatrix} 0 \\ g \end{bmatrix} &= B^* \begin{bmatrix} 0 \\ V^n g \end{bmatrix} \\
&= \tau_0^* \tau J_\theta^* R \begin{bmatrix} 0 \\ V^n g \end{bmatrix} \\
&= \tau_0^* \tau \tau^* \Delta_\theta V^{n+1} g.
\end{aligned}$$

Put $\Xi(e^{it}) = \left(I - \theta(e^{it})^*\theta(e^{it})\right)^{1/2}$. Then the preceding calculations yield

$$(32) \qquad \frac{1}{2\pi} \int_{-\pi}^{\pi} e^{i(n+1)t} \Xi(e^{it}) g(e^{it}) dt = 0, \qquad n = 0, 1, 2, \dots .$$

Now recall that the space in the right hand side of (30) is invariant under A. So the conclusion we reached so far remains valid for

$$A^k \begin{bmatrix} f \\ g \end{bmatrix} = \begin{bmatrix} (S^*)^k f \\ (V^*)^k g \end{bmatrix}, \qquad k = 1, 2, \dots,$$

in place of $\begin{bmatrix} f \\ g \end{bmatrix}$. In particular, in (32) we may replace g by $(V^*)^k g$ with $k = 1, 2, \dots$. It follows that

$$\frac{1}{2\pi} \int_{-\pi}^{\pi} e^{i(n+1)t} \Xi(e^{it}) g(e^{it}) dt = 0, \qquad n \in \mathbb{Z},$$

and hence $\Xi(\cdot)g(\cdot) = 0$ a.e. So $\Delta_\theta g = 0$. On the other hand, $g \in \overline{\operatorname{Im} \Delta_\theta}$. However, because of the selfadjointness of Δ_θ, the operator Δ_θ is injective on $\overline{\operatorname{Im} \Delta_\theta}$, and thus $g = 0$. So the space in the right hand side of (30) consists of the zero element only, and therefore Σ_θ is pure. $\square$

The pure unitary system Σ_θ constructed in Theorem 5.2 will be called the *canonical realization* of θ. Let $T \in \mathcal{L}(K)$ be a completely non-unitary contraction, and assume that the defect operators D_T and D_{T^*} have finite rank. Then the characteristic operator function $W_T(\cdot)$ of T coincides with a matrix function of the type considered in

Theorem 5.2, and hence the latter theorem applies to $W_T(\cdot)$. The state operator of the canonical realization of $W_T(\cdot)$ is called the *functional model* of T. By Theorem 3.1, the completely non-unitary contraction T and its functional model are unitary equivalent.

In the following we specify Theorem 5.2 for certain special cases, which leads to a few corollaries. The first answers questions which were left open in the first two sections of this chapter.

COROLLARY 5.3. *Let θ be the scalar function which is constantly equal to $\frac{1}{2}\sqrt{2}$. Then the canonical realization Σ_θ of θ is a pure unitary system which is neither approximately observable nor approximately controllable. Moreover, the state operator of Σ_θ is a completely non-unitary operator which is similar to a unitary operator.*

PROOF. We use the notation introduced in Theorem 5.2. In this case $k = \ell = 1$. Furthermore,

$$M_\theta g = \frac{1}{2}\sqrt{2}g, \quad \Delta_\theta g = \frac{1}{2}\sqrt{2}g \qquad (g \in L_2(\mathbb{T})),$$

and hence

$$J_\theta h = \begin{bmatrix} \frac{1}{2}\sqrt{2}h \\ \frac{1}{2}\sqrt{2}h \end{bmatrix}, \qquad h \in H_2(\mathbb{T}).$$

It follows that the state space of Σ_θ is the space

$$(33) \qquad H = \left\{ \begin{bmatrix} -\pi g \\ g \end{bmatrix} \mid g \in L_2(\mathbb{T}) \right\}.$$

Here π is the orthogonal projection of $L_2(\mathbb{T})$ onto $H_2(\mathbb{T})$. From the description of the state space in (33) and Theorem 5.2 we conclude that the state operator A, the input operator B, and the output operator C of Σ_θ are given by:

$$(34a) \qquad A\colon H \to H, \qquad A\begin{bmatrix} -\pi g \\ g \end{bmatrix} = \begin{bmatrix} -\pi V^* g \\ V^* g \end{bmatrix},$$

$$(34b) \qquad B\colon \mathbb{C} \to H, \qquad B1 = \begin{bmatrix} 0 \\ \frac{1}{2}\sqrt{2}V^* e_0 \end{bmatrix},$$

$$(34c) \qquad C\colon H \to \mathbb{C}, \qquad C\begin{bmatrix} -\pi g \\ g \end{bmatrix} = \frac{1}{2\pi}\int_{-\pi}^{\pi} g(e^{it})dt.$$

Here V is the operator of multiplication by e^{it} on $L_2(\mathbb{T})$ and $e_0(e^{it}) = 1$ for $-\pi \leq t \leq \pi$.

From Theorems 5.2 and 2.4 we know that A is a completely non-unitary contraction. We claim that A is similar to the unitary operator V^*. Indeed, define $E\colon H \to L_2(\mathbb{T})$ by setting

$$E\begin{bmatrix} -\pi g \\ g \end{bmatrix} = g \qquad (g \in L_2(\mathbb{T})).$$

Obviously, E is a bounded operator which maps the Hilbert space H in a one-one manner onto $L_2(\mathbb{T})$. In particular, E is invertible. Furthermore,

$$
V^* E \begin{bmatrix} -\pi g \\ g \end{bmatrix} = V^* g = E \begin{bmatrix} -\pi V^* g \\ V^* g \end{bmatrix} = E A \begin{bmatrix} -\pi g \\ g \end{bmatrix},
$$

and therefore $A = E^{-1} V^* E$.

From the definitions of A, B and C in (34a), (34b) and (34c) one sees that

$$
(35a) \qquad \bigcap_{j=0}^{\infty} \operatorname{Ker} C A^j = \left\{ \begin{bmatrix} 0 \\ g \end{bmatrix} \,\Big|\, g \perp H_2(\mathbb{T}) \right\} \neq \{0\},
$$

$$
(35b) \qquad \bigvee_{j=0}^{\infty} \operatorname{Im} A^j B = \left\{ \begin{bmatrix} 0 \\ g \end{bmatrix} \,\Big|\, g \perp H_2(\mathbb{T}) \right\} \neq H.
$$

Thus Σ_θ is neither approximately observable nor approximately controllable. $\square$

The last part of Corollary 5.3 is a special case of a more general theorem (see Section IX.1 in Sz.Nagy-Foias [3]) which states that a completely non-unitary contraction T is similar to a unitary operator if and only if the values of its characteristic operator function $W_T(\cdot)$ are invertible at each $\lambda \in \mathbb{D}$ and

$$
\sup_{\lambda \in \mathbb{D}} \|W_T(\lambda)^{-1}\| < \infty.
$$

Corollary 2.2 states that the transfer function of a finite dimensional unitary system is a square rational matrix function without poles on $\overline{\mathbb{D}}$ such that its values on $\mathbb{T}$ are unitary matrices. The next corollary shows that, conversely, any matrix function of the latter type may be realized by a finite dimensional unitary system.

COROLLARY 5.4. *Let θ be a $k \times k$ rational matrix function without poles on $\overline{\mathbb{D}}$, and assume that $\theta(\lambda)$ is unitary for each $\lambda \in \mathbb{T}$. Then the canonical realization of θ is finite dimensional.*

PROOF. We use the notation introduced in Theorem 5.2. In this case $k = \ell$. Since $\theta(\lambda)$ is unitary for each $\lambda \in \mathbb{T}$, the operator Δ_θ is the zero operator. This implies that the state space of the canonical realization of θ is the space

$$
(36) \qquad H = \{ f \in H_2^k(\mathbb{T}) \mid f \perp M_\theta(H_2^k(\mathbb{T})) \}.
$$

Let $T_\theta \colon H_2^k(\mathbb{T}) \to H_2^k(\mathbb{T})$ be defined by $T_\theta f = M_\theta f$. Since θ is analytic in $\mathbb{D}$, the operator T_θ is well-defined and T_θ is unitarily equivalent to the Toeplitz operator on ℓ_2^k defined by θ. According to our hypothesis, θ is continuous on $\mathbb{T}$ and $\det \theta(\zeta) \neq 0$ for each $\zeta \in \mathbb{T}$. But then we can apply Theorem XXIII.4.3 to show that T_θ is Fredholm. In particular, $H = (\operatorname{Im} T_\theta)^\perp$ is finite dimensional. $\square$

COROLLARY 5.5. *Let $\theta\colon \mathbb{D} \to \mathcal{L}(\mathbb{C}^k, \mathbb{C}^\ell)$ be an analytic operator-valued function such that (1) holds, and let $A \in \mathcal{L}(H)$ be the main operator of the canonical realization of θ. Then θ is inner if and only if*

$$(37) \qquad (A^*)^n x \to 0 \qquad (x \in H).$$

PROOF. Assume that θ is inner. Then the operator Δ_θ appearing in Theorem 5.2 is equal to zero. It follows that H is a subspace of $H_2^\ell(\mathbb{T})$ which is invariant under S^* (where S is defined by (14)) and A is the restriction of S^* to H. In other words, A is unitarily equivalent to a part of a block backward shift, and hence (37) holds.

Conversely, assume that (37) holds. Since A is the main operator of Σ_θ, we have

$$m := \dim \operatorname{Im} D_A = \dim \operatorname{Im}(I - A^*A) = \dim \operatorname{Im} B^*B \le k.$$

Here B is the input operator of Σ_θ. It follows from the proof of Theorem XXVI.2.1 that A is unitarily equivalent to a part of the backward shift on $\ell_2(\mathbb{C}^m)$. Let S be the operator of multiplication by e^{it} on $H_2^m(\mathbb{T})$. Then we can find a subspace $L \subset H_2^m(\mathbb{T})$, invariant under S^*, such that A is unitarily equivalent to $\widetilde{A} = S^*|L$. Note that $L^\perp$ is invariant under S. So, by the Beurling-Lax theorem (Theorem XXVI.3.1) there exists an $m \times p$ matrix function Φ so that Φ is inner and

$$L^\perp = \{ M_\Phi g \mid g \in H_2^p(\mathbb{T}) \}.$$

Now apply Theorem 5.2 to Φ. It follows that $\widetilde{A}$ is precisely the main operator of the canonical realization of Φ. Since A and $\widetilde{A}$ are unitarily equivalent, we conclude that Φ and θ are transfer functions of unitary systems which have A as their main operator. But then we can apply Corollary 4.3 to show that θ is inner (because Φ is inner). $\square$

We conclude this section with a few remarks about minimal realizations. Let θ be a $k \times k$ rational matrix function with no poles on $\mathbb{D}$ such that (1) holds. Such a function always can be represented in the form

$$(38) \qquad \theta(\lambda) = D + \lambda C(I - \lambda A)^{-1} B, \qquad \lambda \in \mathbb{D},$$

where $\Sigma = (A, B, C, D)$ is a finite dimensional system and A is a contraction. However, in general, the system Σ is not unitary. In fact, as we know from Corollary 2.2, the condition that Σ is unitary forces the boundary function of θ to have unitary values which we do not require here.

To get the representation (38), let θ_n be the n-th Fourier coefficient of $\theta(\cdot)$, and let $K\colon \ell_2^k \to \ell_2^k$ be the block Hankel operator given by

$$(39) \qquad K = [\theta_{i+j-1}]_{i,j=1}^\infty.$$

Since θ is rational and has no poles on $\mathbb{D}$, condition (1) implies that $\theta(\cdot)$ has no poles on $\mathbb{T}$, and hence K is a bounded operator on ℓ_2^k. From θ is rational it also follows that K has finite rank (cf., the proof of Lemma XXIII.4.1). Let S be the forward shift on ℓ_2^k.

The matrix representation of K in (39) implies that $S^*K = KS$, and hence $H_0 = \operatorname{Im} K$ is invariant under S^*. Now, define

$$(40a) \qquad A_0\colon H_0 \to H_0, \qquad A_0 = S^*|H_0,$$

$$(40b) \qquad B_0\colon \mathbb{C}^k \to H_0, \qquad B_0 u = \begin{bmatrix} \theta_1 u \\ \theta_2 u \\ \vdots \end{bmatrix},$$

$$(40c) \qquad C_0\colon H_0 \to \mathbb{C}^k, \qquad C_0 \begin{bmatrix} x_0 \\ x_1 \\ \vdots \end{bmatrix} = x_0,$$

and set $D_0 = \theta_0$. Then $\Sigma_0 = (A_0, B_0, C_0, D_0)$ is a finite dimensional system, A_0 is a contraction, and the transfer function of Σ_0 is $\theta(\cdot)$. So (38) holds with (A_0, B_0, C_0, D_0) in place of (A, B, C, D). The system Σ_0 is called the *restricted shift realization* of $\theta(\cdot)$.

The restricted shift realization Σ_0 of $\theta(\cdot)$ has the smallest possible state space dimension among all possible realizations of $\theta(\cdot)$. To see this, assume (38) holds. Then the n-th Fourier coefficient of θ is given by $CA^{n-1}B$ for $n \geq 1$. It follows that for each p and q

$$[\theta_{i+j-1}]_{i=1,j=1}^{p,\ q} = \begin{bmatrix} C \\ CA \\ \vdots \\ CA^{p-1} \end{bmatrix} [B \quad AB \cdots A^{q-1}B],$$

which implies that the rank of Hankel operator (39) is less than or equal to the dimension of the state space H of Σ. Since $H_0 = \operatorname{Im} K$, we obtain $\dim H_0 \leq \dim H$.

Realizations Σ of $\theta(\cdot)$ with the smallest possible state space dimension are called *minimal* realizations.

COROLLARY 5.6. *Let θ be a $k \times k$ rational matrix function without poles on $\overline{\mathbb{D}}$, and assume that $\theta(\lambda)$ is unitary for each $\lambda \in \mathbb{T}$. Then the canonical realization of θ is unitarily equivalent to the restricted shift realization of θ, and hence is a minimal realization.*

PROOF. We use the notation introduced in Theorem 5.2. Consider the following partitioning of M_θ:

$$M_\theta = \begin{bmatrix} * & 0 \\ K_\theta & T_\theta \end{bmatrix} \colon H_2^k(\mathbb{T})^\perp \boxplus H_2^k(\mathbb{T}) \to H_2^k(\mathbb{T})^\perp \boxplus H_2^k(\mathbb{T}).$$

The zero in the upper right corner stems from the fact that θ has no poles on $\overline{\mathbb{D}}$. Since the values of θ on $\mathbb{T}$ are unitary matrices, the operator M_θ is unitary, and hence

$$(41) \qquad K_\theta K_\theta^* + T_\theta T_\theta^* = I, \qquad T_\theta^* K_\theta = 0.$$

We already know (see the proof of Corollary 5.4) that the state space H of the canonical realization of θ is equal to $\operatorname{Ker} T_\theta^*$. Thus the first identity in (41) implies that $H \subset \operatorname{Im} K_\theta$, and from the second we see that $\operatorname{Im} K_\theta \subset H$. So $H = \operatorname{Im} K_\theta$. Using the Fourier transformation $F: L_2^k(\mathbb{T}) \to \ell_2^k(\mathbb{Z})$, one sees that $F(\operatorname{Im} K_\theta) = \operatorname{Im} K$, where K is the block Hankel operator associated with θ (as in (39)). Since $\Delta_\theta = 0$, it is now straightforward to check that the canonical realization of θ is unitarily equivalent to the restricted shift realization of θ with the unitary equivalence being provided by the Fourier transformation. $\square$

XXVIII.6 CASCADE CONNECTIONS

Consider two input-output systems:

$$\Sigma_1 \begin{cases} x_{n+1}^{(1)} = A_1 x_n^{(1)} + B_1 u_n^{(1)}, & n = 0, \pm 1, \pm 2, \ldots, \\ y_n^{(1)} = C_1 x_n^{(1)} + D_1 u_n^{(1)}, \end{cases}$$

$$\Sigma_2 \begin{cases} x_{n+1}^{(2)} = A_2 x_n^{(2)} + B_2 u_n^{(2)}, & n = 0, \pm 1, \pm 2, \ldots, \\ y_n^{(2)} = C_2 x_n^{(2)} + D_2 u_n^{(2)}. \end{cases}$$

Assume that the output space L_2 of Σ_2 is equal to the input space K_1 of Σ_1. Then we may take as the input sequence for Σ_1 the output sequence of Σ_2 (see Figure 1). For the resulting system the connection between input and output is given by:

$$(1) \quad \begin{cases} \begin{bmatrix} x_{n+1}^{(1)} \\ x_{n+1}^{(2)} \end{bmatrix} = \begin{bmatrix} A_1 & B_1 C_2 \\ 0 & A_2 \end{bmatrix} \begin{bmatrix} x_n^{(1)} \\ x_n^{(2)} \end{bmatrix} + \begin{bmatrix} B_1 D_2 \\ B_2 \end{bmatrix} u_n, & n = 0, \pm 1, \pm 2, \ldots, \\ y_n = \begin{bmatrix} C_1 & D_1 C_2 \end{bmatrix} \begin{bmatrix} x_n^{(1)} \\ x_n^{(2)} \end{bmatrix} + D_1 D_2 u_n. \end{cases}$$

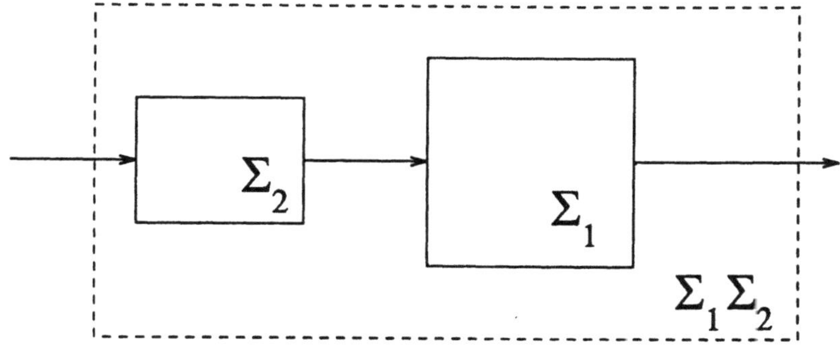

Figure 1

The system (1), which will be denoted by $\Sigma_1 \Sigma_2$, is called the *cascade connection* or *product* of Σ_1 and Σ_2. Note that the input space of $\Sigma_1 \Sigma_2$ is the input space

K_2 of Σ_2 and the outspace of $\Sigma_1\Sigma_2$ is the outspace L_1 of Σ_1. The state space of $\Sigma_1\Sigma_2$ is $H_1\boxplus H_2$, where H_1 and H_2 are the state spaces of Σ_1 and Σ_2, respectively. More formally, if

$$(2) \qquad \Sigma_\nu = (A_\nu, B_\nu, C_\nu, D_\nu; H_\nu, K_\nu, L_\nu), \qquad \nu = 1, 2,$$

and $L_2 = K_1$, then

$$(3) \qquad \Sigma_1\Sigma_2 = \left(\begin{bmatrix} A_1 & B_1 C_2 \\ 0 & A_2 \end{bmatrix}, \begin{bmatrix} B_1 D_2 \\ B_2 \end{bmatrix}, [C_1 \quad D_1 C_2], D_1 D_2; H_1\boxplus H_2, K_2, L_1 \right).$$

THEOREM 6.1. *If Σ_1 and Σ_2 are unitary systems, then the cascade connection $\Sigma_1\Sigma_2$ is a unitary system and*

$$(4) \qquad \theta_{\Sigma_1\Sigma_2}(\lambda) = \theta_{\Sigma_1}(\lambda)\theta_{\Sigma_2}(\lambda), \qquad \lambda \in \mathbb{D}.$$

PROOF. We have to prove that the operator U,

$$(5) \qquad U = \begin{bmatrix} A_1 & B_1 C_2 & B_1 D_2 \\ 0 & A_2 & B_2 \\ C_1 & D_1 C_2 & D_1 D_2 \end{bmatrix} : H_1\boxplus H_2\boxplus K_2 \to H_1\boxplus H_2\boxplus L_1,$$

is unitary. By interchanging in (5) the last two (operator) rows and the last two (operator) columns, we see that U is unitarily equivalent to the product

$$(6) \qquad \begin{bmatrix} A_1 & B_1 & 0 \\ C_1 & D_1 & 0 \\ 0 & 0 & I \end{bmatrix} \begin{bmatrix} I & 0 & 0 \\ 0 & D_2 & C_2 \\ 0 & B_2 & A_2 \end{bmatrix}.$$

In (6) each of the factors is unitary, because Σ_1 and Σ_2 are unitary systems. It follows that U is unitary.

To compute the transfer function of $\Sigma_1\Sigma_2$ note that

$$\begin{bmatrix} I - \lambda A_1 & -\lambda B_1 C_2 \\ 0 & I - \lambda A_2 \end{bmatrix}^{-1} = \begin{bmatrix} (I - \lambda A_1)^{-1} & \lambda(I - \lambda A_1)^{-1} B_1 C_2 (I - \lambda A_2)^{-1} \\ 0 & (I - \lambda A_2)^{-1} \end{bmatrix}.$$

Hence for $\lambda \in \mathbb{D}$ we have

$$\theta_{\Sigma_1\Sigma_2}(\lambda) = D_1 D_2 + \lambda[C_1 \quad D_1 C_2] \begin{bmatrix} I - \lambda A_1 & -\lambda B_1 C_2 \\ 0 & I - \lambda A_2 \end{bmatrix}^{-1} \begin{bmatrix} B_1 D_2 \\ B_2 \end{bmatrix}$$

$$= D_1 D_2 + \lambda C_1 (I - \lambda A_1)^{-1} B_1 D_2$$

$$\qquad + \lambda^2 C_1 (I - \lambda A_1)^{-1} B_1 C_2 (I - \lambda A_2)^{-1} B_2 + \lambda D_1 C_2 (I - \lambda A_2)^{-1} B_2$$

$$= \left[D_1 + \lambda C_1 (I - \lambda A_1)^{-1} B_1 \right] \left[D_2 + \lambda C_2 (I - \lambda A_2)^{-1} B_2 \right]$$

$$= \theta_{\Sigma_1}(\lambda)\theta_{\Sigma_2}(\lambda). \quad \square$$

THEOREM 6.2. *If the cascade connection $\Sigma_1\Sigma_2$ of two unitary systems Σ_1 and Σ_2 is pure, then Σ_1 and Σ_2 are both pure.*

PROOF. Let Σ_1 and Σ_2 be as in (2). Take x in the excessive space of Σ_1. So $C_1 A_1^n x$ and $B_1^*(A_1^*)^n x$ are zero for each n. This implies that

$$[C_1 \quad D_1 C_2]\begin{bmatrix} A_1 & B_1 C_2 \\ 0 & A_2 \end{bmatrix}^n \begin{bmatrix} x \\ 0 \end{bmatrix} = [C_1 \quad D_1 C_2]\begin{bmatrix} A_1^n x \\ 0 \end{bmatrix}$$
$$= C_1 A_1^n x = 0, \qquad n = 0, 1, 2, \dots .$$

Furthermore,

$$\begin{bmatrix} A_1^* & 0 \\ C_2^* B_1^* & A_2^* \end{bmatrix}\begin{bmatrix} x \\ 0 \end{bmatrix} = \begin{bmatrix} A_1^* x \\ 0 \end{bmatrix}.$$

Thus, by induction,

$$\begin{bmatrix} A_1^* & 0 \\ C_2^* B_1^* & A_2^* \end{bmatrix}^n \begin{bmatrix} x \\ 0 \end{bmatrix} = \begin{bmatrix} (A_1^*)^n x \\ 0 \end{bmatrix}, \qquad n = 0, 1, 2, \dots,$$

and we may conclude that

$$[D_2^* B_1^* \quad B_2^*]\begin{bmatrix} A_1^* & 0 \\ C_2^* B_1^* & A_2^* \end{bmatrix}^n \begin{bmatrix} x \\ 0 \end{bmatrix} = D_2^* B_1^*(A_1^*)^n x = 0, \qquad n = 0, 1, 2, \dots .$$

So we have proved that the vector $\begin{bmatrix} x \\ 0 \end{bmatrix}$ is in the excessive space of $\Sigma_1\Sigma_2$. Since $\Sigma_1\Sigma_2$ is pure, the latter space consists of the zero vector only, and hence $x = 0$. It follows that $\mathcal{N}(\Sigma_1) = \{0\}$, and therefore Σ_1 is pure.

Next, take $y \in \mathcal{N}(\Sigma_2)$. Thus $C_2 A_2^n y = 0$ and $B_2^*(A_2^*)^n y = 0$ for each n. This implies that

$$[D_2^* B_1^* \quad B_2^*]\begin{bmatrix} A_1^* & 0 \\ C_2^* B_1^* & A_2^* \end{bmatrix}^n \begin{bmatrix} 0 \\ y \end{bmatrix} = [D_2^* B_1^* \quad B_2^*]\begin{bmatrix} 0 \\ (A_2^*)^n y \end{bmatrix}$$
$$= B_2^*(A_2^*)^n y = 0, \qquad n = 0, 1, 2, \dots .$$

Furthermore,

$$\begin{bmatrix} A_1 & B_1 C_2 \\ 0 & A_2 \end{bmatrix}\begin{bmatrix} 0 \\ y \end{bmatrix} = \begin{bmatrix} 0 \\ A_2 y \end{bmatrix}.$$

Thus, by induction,

$$\begin{bmatrix} A_1 & B_1 C_2 \\ 0 & A_2 \end{bmatrix}^n \begin{bmatrix} 0 \\ y \end{bmatrix} = \begin{bmatrix} 0 \\ A_2^n y \end{bmatrix}, \qquad n = 0, 1, 2, \dots,$$

and therefore

$$[C_1 \quad D_1 C_2]\begin{bmatrix} A_1 & B_1 C_2 \\ 0 & A_2 \end{bmatrix}^n \begin{bmatrix} 0 \\ y \end{bmatrix} = D_1 C_2 A_2^n y = 0, \qquad n = 0, 1, 2, \dots .$$

Thus $y \in \mathcal{N}(\Sigma_1 \Sigma_2) = \{0\}$, and hence $\mathcal{N}(\Sigma_2) = \{0\}$. So we have proved that Σ_2 is also pure. $\square$

The converse of Theorem 6.2 is not true. To see this, let $\Sigma = (A, B, C, D)$ be the canonical realization of the scalar function $\theta(\lambda) \equiv \frac{1}{2}\sqrt{2}$. Then Σ is a pure unitary system (see Theorem 5.2), and we claim that the product $\widetilde{\Sigma} = \Sigma\Sigma$ is not pure. By definition, $\widetilde{\Sigma} = (\widetilde{A}, \widetilde{B}, \widetilde{C}, \widetilde{D})$, where $\widetilde{D} = D^2$ and

$$\widetilde{A} = \begin{bmatrix} A & BC \\ 0 & A \end{bmatrix}, \qquad \widetilde{B} = \begin{bmatrix} BD \\ B \end{bmatrix}, \qquad \widetilde{C} = [C \quad DC].$$

From the construction of A, B and C in the proof of Corollary 5.3 we know (see formulas (35a), (35b) in the previous section) that

$$N := \bigcap_{j=0}^{\infty} \operatorname{Ker} C A^j = \bigvee_{j=0}^{\infty} \operatorname{Im} A^j B.$$

In particular, $C A^n B = 0$ for $n = 0, 1, 2, \dots$. The latter fact implies that

$$\widetilde{C}(\widetilde{A})^j = [C A^j \quad DC A^j], \qquad j = 0, 1, 2, \dots,$$

$$(\widetilde{A})^j \widetilde{B} = \begin{bmatrix} A^j BD \\ A^j B \end{bmatrix}, \qquad j = 0, 1, 2, \dots .$$

Since $D: \mathbb{C} \to \mathbb{C}$ is just multiplication by $\frac{1}{2}\sqrt{2}$, we conclude that the excessive subspace of $\widetilde{\Sigma}$ is given by

$$\mathcal{N}(\widetilde{\Sigma}) = \left\{ \begin{bmatrix} 2h - \sqrt{2}k \\ 2k - \sqrt{2}h \end{bmatrix} \mid h \in N, \ k \in N^\perp \right\},$$

and thus $\widetilde{\Sigma}$ is not pure.

For finite dimensional unitary systems the converse of Theorem 6.2 is true.

PROPOSITION 6.3. *The cascade connection $\Sigma_1 \Sigma_2$ of two finite dimensional pure unitary systems Σ_1 and Σ_2 is again a pure unitary system.*

PROOF. Let A_1 and A_2 be the state operators of Σ_1 and Σ_2, respectively. Then the state operator A of $\Sigma_1 \Sigma_2$ is of the form

$$A = \begin{bmatrix} A_1 & * \\ 0 & A_2 \end{bmatrix}.$$

Since Σ_1 and Σ_2 are pure, we know from Corollary 2.5 that the eigenvalues of A_1 and A_2 are in the open unit disc. But then the above representation of A shows that the same holds true for A. So, again apply Corollary 2.5, the system $\Sigma_1 \Sigma_2$ is pure. $\square$

The next proposition will be useful later (in the next two sections).

PROPOSITION 6.4. *Consider the cascade connections* $\Sigma = \Sigma_1\Sigma_2$ *and* $\Sigma' = \Sigma_1'\Sigma_2'$, *where*

$$\Sigma_\nu = (A_\nu, B_\nu, C_\nu, D_\nu; H_\nu, K_\nu, L_\nu), \qquad \nu = 1, 2,$$

$$\Sigma_\nu' = (A_\nu', B_\nu', C_\nu', D_\nu'; H_\nu', K_\nu', L_\nu'), \qquad \nu = 1, 2.$$

For $\nu = 1, 2$ *assume that* $J_\nu \colon H_\nu \to H_\nu'$ *is a unitary operator which establishes a unitary equivalence between* Σ_ν *and* Σ_ν'. *Then* Σ *and* Σ' *are unitarily equivalent and the unitary operator* $J_1 \boxplus J_2$ *provides the unitary equivalence.*

PROOF. For $\nu = 1, 2$ we have $D_\nu = D_\nu'$ and

$$A_\nu' = J_\nu A_\nu J_\nu^{-1}, \qquad B_\nu' = J_\nu B_\nu, \qquad C_\nu' = C_\nu J_\nu^{-1}.$$

It follows that $D_1 D_2 = D_1' D_2'$ and

$$\begin{bmatrix} J_1 & 0 \\ 0 & J_2 \end{bmatrix} \begin{bmatrix} A_1 & B_1 C_2 \\ 0 & A_2 \end{bmatrix} = \begin{bmatrix} A_1' & B_1' C_2' \\ 0 & A_2' \end{bmatrix} \begin{bmatrix} J_1 & 0 \\ 0 & J_2 \end{bmatrix},$$

$$\begin{bmatrix} J_1 & 0 \\ 0 & J_2 \end{bmatrix} \begin{bmatrix} B_1 D_2 \\ B_2 \end{bmatrix} = \begin{bmatrix} B_1' D_2' \\ B_2' \end{bmatrix},$$

$$[C_1 \quad D_1 C_2] \begin{bmatrix} J_1 & 0 \\ 0 & J_2 \end{bmatrix}^{-1} = [C_1' \quad D_1' C_2'].$$

With these identities the proposition is proved. $\square$

XXVIII.7 FACTORIZATION AND INVARIANT SUBSPACES

Let $\Sigma = \Sigma_1\Sigma_2$ be a product of two unitary systems. Then the state operator A of Σ admits the following partitioning

$$(1) \qquad A = \begin{bmatrix} A_1 & * \\ 0 & A_2 \end{bmatrix} \colon H_1 \boxplus H_2 \to H_1 \boxplus H_2,$$

where $A_1 \colon H_1 \to H_1$ and $A_2 \colon H_2 \to H_2$ are the state operators of Σ_1 and Σ_2, respectively. It follows that the space H_1 (which is identified with $H_1 \boxplus \{0\}$ by setting x equal to $\begin{bmatrix} x \\ 0 \end{bmatrix}$) is an invariant subspace of A. In this section we shall see that, conversely, any invariant subspace of A yields a factorization of Σ into a product of two unitary systems. It is convenient first to prove the following lemma.

LEMMA 7.1. *Let* $A \colon H \to H$ *and* $C \colon H \to L$ *be Hilbert space operators. In order that* A *is the state operator and* C *is the output operator of a unitary system it is necessary and sufficient that* $A^*A + C^*C = I$. *Assume that the latter condition is fulfilled and put*

$$(2a) \qquad K_0 = \left\{ \begin{bmatrix} x \\ y \end{bmatrix} \in H \boxplus L \mid A^*x + C^*y = 0 \right\},$$

$$(2b) \qquad B_0 \colon K_0 \to H, \qquad B_0 \begin{bmatrix} x \\ y \end{bmatrix} = x,$$

$$(2c) \qquad D_0 \colon K_0 \to L, \qquad D_0 \begin{bmatrix} x \\ y \end{bmatrix} = y.$$

Then the input operator B and the feedthrough operator D of a unitary system with state operator A and output operator C are given by $B = B_0 U$ and $D = D_0 U$, where $U \colon K \to K_0$ is an arbitrary unitary operator. Moreover, there is a one-one correspondence between the pair B, D and the unitary operator U.

PROOF. From formula (4a) in Section XXVIII.2 we know that the condition $A^* A + C^* C = I$ is necessary. To prove sufficiency, assume that $A^* A + C^* C = I$. Then the map J defined by

$$(3) \qquad J = \begin{bmatrix} A \\ C \end{bmatrix} \colon H \to H \boxplus L$$

is an isometry, and hence Im J is closed. Note that K_0 is the orthogonal complement of Im J in $H \boxplus L$. Let B_0 and D_0 be defined by (2a) and (2b), respectively, and consider

$$V_0 = \begin{bmatrix} A & B_0 \\ C & D_0 \end{bmatrix} \colon H \boxplus K_0 \to H \boxplus L.$$

On H the operator V_0 acts as J. It follows that

$$\widetilde{J} = V_0 | H \colon H \to \operatorname{Im} J$$

is unitary. Furthermore, V_0 maps K_0 into K_0, and on K_0 the operator V_0 acts as the identity operator. It follows that V_0 admits the following partitioning:

$$V_0 = \begin{bmatrix} \widetilde{J} & 0 \\ 0 & I_{K_0} \end{bmatrix} \colon H \boxplus K_0 \to \operatorname{Im} J \boxplus K_0,$$

where $\widetilde{J}$ is unitary. So V_0 is unitary, and hence $\Sigma_0 = (A, B_0, C, D_0)$ is a unitary system.

Next, let $B = B_0 U$ and $D = D_0 U$, where $U \colon K \to K_0$ is unitary. Then

$$(4) \qquad \begin{bmatrix} A & B \\ C & D \end{bmatrix} = \begin{bmatrix} A & B_0 \\ C & D_0 \end{bmatrix} \begin{bmatrix} I & 0 \\ 0 & U \end{bmatrix}.$$

The two operator matrices in the right hand side of (4) are unitary. So the left hand side of (4) is unitary, and therefore $\Sigma = (A, B, C, D)$ is a unitary system. Conversely, if $\Sigma = (A, B, C, D)$ is a unitary system with input space K, then the product

$$\begin{bmatrix} A^* & C^* \\ B^* & D^* \end{bmatrix} \begin{bmatrix} A & B_0 \\ C & D_0 \end{bmatrix}$$

is a 2×2 operator matrix, which is unitary, and the $(1,1)$-entry in this operator matrix is equal to the identity operator. But then we can apply Lemma 1.2 to show that (4) holds for a unique unitary operator $U\colon K \to K_0$. In particular, $B = B_0 U$ and $D = D_0 U$.
$\square$

The unitary system $\Sigma_0 = (A, B_0, C, D_0)$ constructed in the preceding lemma will be called the *standard unitary system* associated with A and C.

Now, we turn to factorization. Let

$$(5) \qquad\qquad \Sigma = (A, B, C, D; H, K, L)$$

be a unitary system, and let H_1 be an A-invariant subspace of H. Put

$$(6) \qquad\qquad H_2 = H_1^{\perp}, \qquad K_2 = K, \qquad L_1 = L,$$

and consider the following partitionings:

$$A = \begin{bmatrix} A_{11} & A_{12} \\ 0 & A_{22} \end{bmatrix} \colon H_1 \boxplus H_2 \to H_1 \boxplus H_2,$$

$$B = \begin{bmatrix} B_{12} \\ B_{22} \end{bmatrix} \colon K_2 \to H_1 \boxplus H_2$$

$$C = [C_{11} \quad C_{12}] \colon H_1 \boxplus H_2 \to L_1.$$

Thus

$$(7) \qquad\qquad \begin{bmatrix} A & B \\ C & D \end{bmatrix} = \begin{bmatrix} A_{11} & A_{12} & B_{12} \\ 0 & A_{22} & B_{22} \\ C_{11} & C_{12} & D \end{bmatrix}.$$

Since the operator in the left hand side of (7) is unitary, the same is true for the operator in the right hand side of (7), and thus the pair

$$A_{11}\colon H_1 \to H_1, \qquad C_{11}\colon H_1 \to L_1$$

satisfies the necessary and sufficient condition in Lemma 7.1. The associated standard unitary system Σ_1,

$$(8) \qquad\qquad \Sigma_1 = (A_{11}, B_1, C_{11}, D_1; H, K_1, L_1),$$

is referred to as the *left projection* of Σ associated with the invariant subspace H_1.

Next, consider the product

$$(9) \qquad \begin{bmatrix} A_{11}^* & C_{11}^* & 0 \\ B_1^* & D_1^* & 0 \\ 0 & 0 & I_{H_2} \end{bmatrix} \begin{bmatrix} A_{11} & B_{12} & A_{12} \\ C_{11} & D & C_{12} \\ 0 & B_{22} & A_{22} \end{bmatrix} \colon H_1 \boxplus K_2 \boxplus H_2 \to H_1 \boxplus K_1 \boxplus H_2.$$

Since the operator in the right hand side of (7) is unitary, the $(1,1)$-entry in the 3×3 operator matrix defined by the product in (9) is equal to I_{H_1}. On the other hand, the product in (9) defines a unitary operator, because both its factors are unitary. Hence we may apply Lemma 1.2 to show that the product in (9) is of the form

$$(10) \qquad \begin{bmatrix} I_{H_1} & 0 & 0 \\ 0 & D_2 & C_2 \\ 0 & B_{22} & A_{22} \end{bmatrix}$$

for certain operators $C_2 \colon H_2 \to K_1$ and $D_2 \colon K_2 \to K_1$. Now put $L_2 = K_1$. Then the operators A_{22}, B_{22}, C_2 and D_2 form a unitary system Σ_2,

$$(11) \qquad \Sigma_2 = (A_{22}, B_{22}, C_2, D_2; H_2, K_2, L_2),$$

which we call the *right projection* of Σ associated with the invariant subspace H_1.

THEOREM 7.2. *Let Σ be a unitary system, and let M be an invariant subspace for the state operator of Σ. Then the left projection Σ_1 and the right projection Σ_2 of Σ associated with M are unitary systems, and $\Sigma = \Sigma_1 \Sigma_2$.*

PROOF. Put $H_1 = M$, and let us continue to use the notation introduced in the two paragraphs preceding the present theorem. We already know that Σ_1 and Σ_2 are unitary systems. From (9) and the fact that Σ_1 is unitary it follows that

$$(12) \qquad \begin{bmatrix} A_{11} & B_{12} & A_{12} \\ C_{11} & D & C_{12} \\ 0 & B_{22} & A_{22} \end{bmatrix} = \begin{bmatrix} A_{11} & B_1 & 0 \\ C_{11} & D_1 & 0 \\ 0 & 0 & I \end{bmatrix} \begin{bmatrix} I & 0 & 0 \\ 0 & D_2 & C_2 \\ 0 & B_{22} & A_{22} \end{bmatrix}.$$

But (as we saw in the proof of Theorem 6.1) the identity (12) is equivalent to the statement that $\Sigma = \Sigma_1 \Sigma_2$. $\square$

Let Σ_1 and Σ_2 be unitary systems. We say that the factorization $\Sigma = \Sigma_1 \Sigma_2$ is *supported* by the space M if M is the state space of Σ_1. This terminology is justified by the following theorem.

THEOREM 7.3. *If the factorization $\Sigma = \Sigma_1 \Sigma_2$ is supported by the space M, then M is an invariant subspace for the state operator of Σ and there exists a unique unitary operator U such that*

$$(13a) \qquad \Sigma_1 = (A_1, B_1 U^{-1}, C_1, D_1 U^{-1}),$$

$$(13b) \qquad \Sigma_2 = (A_2, B_2, U C_2, U D_2),$$

where (A_1, B_1, C_1, D_1) is the left projection and (A_2, B_2, C_2, D_2) is the right projection of Σ associated with M.

PROOF. Assume that

$$\Sigma_\nu = (\widetilde{A}_\nu, \widetilde{B}_\nu, \widetilde{C}_\nu, \widetilde{D}_\nu; H_\nu, \widetilde{K}_\nu, \widetilde{L}_\nu), \qquad \nu = 1, 2.$$

Then $M = H_1$, and the state operator A of Σ can be partitioned as in (1). Thus M is an invariant subspace for A. Let the left and right projections of Σ associated with M be given by

$$\Sigma_{\text{left}} = (A_1, B_1, C_1, D_1; H_1, K_1, L_1),$$

$$\Sigma_{\text{right}} = (A_2, B_2, C_2, D_2; H_2, K_2, L_2),$$

respectively. Note that $L_2 = K_1$ (by definition) and $\widetilde{L}_2 = \widetilde{K}_1$ (because the product $\Sigma_1 \Sigma_2$ is defined). Consider the following operators:

$$(14a) \qquad \begin{bmatrix} A_1 & B_1 & 0 \\ C_1 & D_1 & 0 \\ 0 & 0 & I \end{bmatrix} \begin{bmatrix} I & 0 & 0 \\ 0 & D_2 & C_2 \\ 0 & B_2 & A_2 \end{bmatrix} : H_1 \boxplus K_2 \boxplus H_2 \to H_1 \boxplus L_1 \boxplus H_2,$$

$$(14b) \qquad \begin{bmatrix} \widetilde{A}_1 & \widetilde{B}_1 & 0 \\ \widetilde{C}_1 & \widetilde{D}_1 & 0 \\ 0 & 0 & I \end{bmatrix} \begin{bmatrix} I & 0 & 0 \\ 0 & \widetilde{D}_2 & \widetilde{C}_2 \\ 0 & \widetilde{B}_2 & \widetilde{A}_2 \end{bmatrix} : H_1 \boxplus \widetilde{K}_2 \boxplus H_2 \to H_1 \boxplus \widetilde{L}_1 \boxplus H_2.$$

Since $\Sigma = \Sigma_1 \Sigma_2 = \Sigma_{\text{left}} \Sigma_{\text{right}}$, we have $\widetilde{K}_2 = K_2$, $\widetilde{L}_1 = L_1$, and the operators defined by the products in (14a) and (14b) are equal. The systems involved are all unitary, and hence the factors in (14a) and (14b) are unitary. It follows that the operator V,

$$(15) \qquad \begin{aligned} V :&= \begin{bmatrix} \widetilde{A}_1 & \widetilde{B}_1 & 0 \\ \widetilde{C}_1 & \widetilde{D}_1 & 0 \\ 0 & 0 & I \end{bmatrix}^{-1} \begin{bmatrix} A_1 & B_1 & 0 \\ C_1 & D_1 & 0 \\ 0 & 0 & I \end{bmatrix} \\[2mm] &= \begin{bmatrix} I & 0 & 0 \\ 0 & \widetilde{D}_2 & \widetilde{C}_2 \\ 0 & \widetilde{B}_2 & \widetilde{A}_2 \end{bmatrix} \begin{bmatrix} I & 0 & 0 \\ 0 & D_2 & C_2 \\ 0 & B_2 & A_2 \end{bmatrix}^{-1} : H_1 \boxplus K_1 \boxplus H_2 \to H_1 \boxplus \widetilde{K}_1 \boxplus H_2, \end{aligned}$$

is unitary. The two equalities in (15) imply that

$$V = \begin{bmatrix} I_{H_1} & 0 & 0 \\ 0 & U & 0 \\ 0 & 0 & I_{H_2} \end{bmatrix},$$

where $U : K_1 \to \widetilde{K}_1$ is unitary. From the first equality in (15) we see that

$$(16) \qquad \begin{bmatrix} \widetilde{A}_1 & \widetilde{B}_1 & 0 \\ \widetilde{C}_1 & \widetilde{D}_1 & 0 \\ 0 & 0 & I \end{bmatrix} \begin{bmatrix} I & 0 & 0 \\ 0 & U & 0 \\ 0 & 0 & I \end{bmatrix} = \begin{bmatrix} A_1 & B_1 & 0 \\ C_1 & D_1 & 0 \\ 0 & 0 & I \end{bmatrix},$$

which proves (13a). In a similar way one can use the second equality in (15) to derive (13b).

It remains to prove the uniqueness statement. Let $\widetilde{U} : K_1 \to \widetilde{K}_1$ be a second unitary operator so that (13a) and (13b) hold with $\widetilde{U}$ in place of U. Then (16) holds with $\widetilde{U}$ in place of U. But the first factor in the left hand side of (16) is invertible, and hence U in (16) is uniquely determined. Therefore, $\widetilde{U} = U$. $\quad\square$

Let $\Sigma = \Sigma_1\Sigma_2$ be a factorization into unitary systems supported by the space M, and let U be the unique unitary operator such that (13a) and (13b) hold. In the sequel we shall refer to the factorization $\Sigma = \Sigma_1\Sigma_2$ as the *factorization determined by the supporting subspace M and the unitary operator U.*

As an illustration of Lemma 7.1 let us construct a unitary system with state operator $A = \alpha \colon \mathbb{C} \to \mathbb{C}$ and with output operator

$$C = \Gamma \colon \mathbb{C} \to L.$$

By Lemma 7.1 for such a system to exist it is necessary and sufficient that

(17) $$|\alpha|^2 + \Gamma^*\Gamma = 1.$$

Assume (17) is fulfilled, and let $\Sigma_0 = (\alpha, B_0, \Gamma, D_0; \mathbb{C}, K_0, L)$ be the standard unitary system associated with α and Γ. Thus

$$K_0 = \left\{ \begin{bmatrix} x \\ y \end{bmatrix} \in \mathbb{C} \boxplus L \mid \bar{\alpha}x + \Gamma^*y = 0 \right\},$$

$$B_0 \colon K_0 \to \mathbb{C}, \qquad B_0 \begin{bmatrix} x \\ y \end{bmatrix} = x,$$

$$D_0 \colon K_0 \to L, \qquad D_0 \begin{bmatrix} x \\ y \end{bmatrix} = y.$$

To make the construction more transparent we have to understand better the input space K_0. If $\Gamma = 0$, then $\alpha \neq 0$, by virtue of (17), and K_0 is just a copy of L. In this case

(18) $$\Sigma = (\alpha, 0, 0, I; \mathbb{C}, L, L)$$

is a unitary system with state operator $A = \alpha$ and output operator $C = \Gamma = 0$. Note that the transfer function of the system in (18) is identically equal to the identity operator on L.

Next, consider the case when $\alpha = 0$. Then Γ is an isometry, and hence $P = \Gamma\Gamma^*$ is the orthogonal projection of L onto $\operatorname{Im}\Gamma$ along $\operatorname{Ker}\Gamma^*$. In particular, $\operatorname{rank} P = 1$. In this case

$$K_0 = \left\{ \begin{bmatrix} x_0 \\ y_0 \end{bmatrix} \mid x_0 \in \mathbb{C},\ y_0 \in \operatorname{Ker}\Gamma^* = \operatorname{Ker} P \right\}.$$

Hence

$$Jy := \begin{bmatrix} \Gamma^*y \\ (I - P)y \end{bmatrix} \in K_0, \qquad y \in L.$$

Furthermore, for $y \in L$ we have

$$\begin{aligned}
\|Jy\|^2 &= \|(I - P)y\|^2 + \|\Gamma^*y\|^2 \\
&= \|(I - P)y\|^2 + \langle \Gamma\Gamma^*y, y \rangle \\
&= \|(I - P)y\|^2 + \|Py\|^2 = \|y\|^2.
\end{aligned}$$

Also, $JL = K_0$. Indeed, if $x_0 \in \mathbb{C}$ and $y_0 \in \operatorname{Ker} P$, then we may choose $z_0 \in \operatorname{Im} P$ so that $\Gamma^* z_0 = x_0$, and it follows that

$$J(z_0 + y_0) = \begin{bmatrix} x_0 \\ y_0 \end{bmatrix}.$$

So $J: L \to K_0$ is a unitary operator. Note that $B_0 J = \Gamma^*$ and $D_0 J = I - P$. Thus the system

$$(19) \qquad \Sigma = (0, \Gamma^*, \Gamma, I - \Gamma\Gamma^*; \mathbb{C}, L, L)$$

is a unitary system with $A = 0$ as state operator and $C = \Gamma$ as output operator. Note that the transfer function of the system in (19) is equal to $I - P + \lambda P$. A function of the latter type, where P is an orthogonal projection of rank one, will be called an *elementary pencil*.

Finally, let us consider the more interesting case when $0 < |\alpha| < 1$. Introduce the operator

$$(20) \qquad P = \frac{1}{1 - |\alpha|^2} \Gamma\Gamma^* : L \to L.$$

Since $\Gamma^* \Gamma = 1 - |\alpha|^2$, because of (17), we have $\Gamma^* P = \Gamma^*$, and hence P is the orthogonal projection of L onto $\operatorname{Im} \Gamma$. In particular, rank P is one. Note that

$$(21) \qquad \begin{bmatrix} x_0 \\ y_0 \end{bmatrix} \in K_0 \Leftrightarrow x_0 = -\Gamma^*(\overline{\alpha}^{-1} y_0).$$

Hence, it follows from $\Gamma^* P = \Gamma^*$ that

$$Jy := \begin{bmatrix} -\frac{\alpha}{|\alpha|} \Gamma^* y \\ |\alpha| P y + (I - P)y \end{bmatrix} \in K_0, \qquad y \in L.$$

Furthermore, for $y \in L$ we have

$$\begin{aligned}
\|Jy\|^2 &= \||\alpha|Py + (I-P)y\|^2 + \| -\frac{\alpha}{|\alpha|}\Gamma^* y\|^2 \\
&= |\alpha|^2 \|Py\|^2 + \|(I-P)y\|^2 + \langle \Gamma\Gamma^* y, y \rangle \\
&= |\alpha|^2 \|Py\|^2 + \|(I-P)y\|^2 + (1 - |\alpha|^2)\langle Py, y \rangle \\
&= \|Py\|^2 + \|(I-P)y\|^2 = \|y\|^2.
\end{aligned}$$

Thus $J: L \to K_0$ is an isometry. Also $JL = K_0$. Indeed, if $\begin{bmatrix} x_0 \\ y_0 \end{bmatrix} \in K_0$, then $Jy = \begin{bmatrix} x_0 \\ y_0 \end{bmatrix}$ with

$$y = |\alpha|^{-1} P y_0 + (I - P)y_0,$$

by virtue of (21) and $\Gamma^* = \Gamma^* P$. So J is unitary. Note that $B_0 J = -\alpha |\alpha|^{-1} \Gamma^*$ and $D_0 J = |\alpha| P + I - P$. Thus the system

$$(22) \qquad \Sigma = \left(\alpha, -\frac{\alpha}{|\alpha|}\Gamma^*, \Gamma, I + \frac{|\alpha| - 1}{1 - |\alpha|^2}\Gamma\Gamma^*; \mathbb{C}, L, L \right)$$

is a unitary system with $A = \alpha$ as state operator $(0 < |\alpha| < 1)$ and with $C = \Gamma$ as output operator. The transfer function of the system in (22) is given by

$$(23) \qquad I - P + \frac{\overline{\alpha} - \lambda}{1 - \lambda\alpha}\frac{|\alpha|}{\overline{\alpha}}P,$$

where P is the orthogonal projection defined by (20). We shall refer to (23) as an *elementary Blaschke factor*. Note that P in (23) is an orthogonal projection of rank one.

The unitary systems in (19) and (22) are pure, because their state operators are strict contractions (cf., Theorem 2.4 and the paragraph preceding Proposition 1.3). Any system unitarily equivalent to one of the systems in (19) or (22) will be called an *elementary rotation* over L. Note that the transfer function of an elementary rotation is either an elementary pencil or an elementary Blaschke factor.

COROLLARY 7.4. *Let Σ be a pure unitary system with a finite dimensional state space. Then, up to a constant unitary operator on the right, Σ is equal to a product of n elementary rotations, where n is the state space dimension. Furthermore, the transfer function of Σ factorizes as*

$$\theta_\Sigma(\lambda) = \theta_1(\lambda)\theta_2(\lambda)U, \qquad \lambda \in \mathbb{D},$$

where U is a unitary operator, and

$$\theta_1(\lambda) = \overset{\curvearrowright}{\prod_{j=1}^{m}} (I - Q_j + \lambda Q_j),$$

$$\theta_2(\lambda) = \overset{\curvearrowright}{\prod_{j=1}^{n-m}} \left(I - P_j + \frac{\overline{\lambda}_j - \lambda}{1 - \lambda\lambda_j}\frac{|\lambda_j|}{\overline{\lambda}_j}P_j\right).$$

Here $Q_1, \ldots, Q_m$ and $P_1, \ldots, P_{n-m}$ are orthogonal projections of rank one, m is the algebraic multiplicity of 0 as an eigenvalue of the state operator A and $\lambda_1, \ldots, \lambda_{n-m}$ are the non-zero eigenvalues of A repeated according to multiplicity.

PROOF. We prove the corollary by induction on n. Take $n = 1$, and let

$$\Sigma = (\alpha, B, C, D; \mathbb{C}, K, L).$$

Since Σ is pure, we have $|\alpha| < 1$ (by Theorem 2.4 and Proposition 1.4). Put $\Gamma = C$, and let $\widetilde{\Sigma}$ be the elementary rotation defined by (19) if $\alpha = 0$ or by (22) if $\alpha \neq 0$. Then we know from Lemma 7.1 that there exists a unitary operator $U \colon K \to L$ such that $\widetilde{\Sigma} = (\alpha, BU^{-1}, C, DU^{-1})$. Note that

$$\theta_\Sigma(\lambda) = \theta_{\widetilde{\Sigma}}(\lambda)U, \qquad \lambda \in \mathbb{D}.$$

Since $\theta_{\widetilde{\Sigma}}(\lambda)$ is either an elementary pencil or an elementary Blaschke factor, the corollary is proved for $n = 1$.

Next, assume that the corollary has been proved for $n = k \geq 1$. Take $n = k + 1$. Let α be an eigenvalue of A with $|\alpha|$ minimal. (In particular, $\alpha = 0$ if A is not invertible.) Let $e \neq 0$ be a corresponding eigenvector. Since Σ is pure, we know from Theorem 2.4 and Proposition 1.4 that $|\alpha| < 1$. Put $H_1 = \mathrm{span}\{e\}$. Then H_1 is invariant under A. Let

$$\Sigma_1 = (A_1, B_1, C_1, D_1), \qquad \Sigma_2 = (A_2, B_2, C_2, D_2)$$

be the left projection and the right projection, respectively, associated with A and the invariant subspace H_1. Then $\Sigma = \Sigma_1 \Sigma_2$. By replacing Σ_1 and Σ_2 by

$$\widetilde{\Sigma}_1 = (A_1, B_1 U^{-1}, C_1, D_1 U^{-1}), \qquad \widetilde{\Sigma}_2 = (A_2, B_2, U C_2, U D_2),$$

where U is some unitary operator, we see from the result of the first paragraph of the proof that without loss of generality we may assume that Σ_1 is an elementary rotation. By Theorem 6.2 the system Σ_2 is pure. From the construction of the factorization it is clear that Σ_2 has state space dimension $n - 1 = k$. So Σ_2 has the desired factorization by our induction hypotheses. But then Σ also factors in the desired way. Furthermore, by Theorem 6.1,

$$\theta_{\Sigma}(\lambda) = \theta_{\Sigma_1}(\lambda)\theta_{\Sigma_2}(\lambda), \qquad \lambda \in \mathbf{D},$$

where θ_{Σ_1} is an elementary pencil if $\alpha = 0$ or an elementary Blaschke factor if $\alpha \neq 0$. By our induction hypotheses, θ_{Σ_2} has a factorization as described in the theorem with $n - 1$ in place of n and with the state operator A of Σ replaced by A_2. Now A partitions as

$$A = \begin{bmatrix} A_1 & * \\ 0 & A_2 \end{bmatrix},$$

and hence the eigenvalues of A are equal to $\alpha, \lambda_2, \ldots, \lambda_n$, where $\lambda_2, \ldots, \lambda_n$ are the eigenvalues of A_2 (multiplicities taken into account). So, by induction, θ_{Σ} has a factorization as stated in the theorem. $\square$

For later purposes (see the next section) we mention the following proposition.

PROPOSITION 7.5. *Let Σ and $\widetilde{\Sigma}$ be unitary systems with state spaces H and $\widetilde{H}$, respectively. Assume that Σ and $\widetilde{\Sigma}$ are unitarily equivalent, and let the unitary equivalence being given by the unitary operator $V : H \to \widetilde{H}$. Let $\widetilde{\Sigma} = \widetilde{\Sigma}_1 \widetilde{\Sigma}_2$ be a factorization of $\widetilde{\Sigma}$ supported by the space $\widetilde{M}$, and put $M = V^{-1}\widetilde{M}$. Then Σ has a factorization $\Sigma = \Sigma_1 \Sigma_2$ supported by the space M such that Σ_1 and $\widetilde{\Sigma}_1$ are unitarily equivalent, and Σ_2 and $\widetilde{\Sigma}_2$ are unitarily equivalent.*

PROOF. Since V is unitary, the identity $VM = \widetilde{M}$ implies that $VM^{\perp} = \widetilde{M}^{\perp}$. Put

$$V_1 = V|M : M \to \widetilde{M}, \qquad V_2 = V|M^{\perp} : M^{\perp} \to \widetilde{M}^{\perp}.$$

Note that the two operators V_1 and V are unitary. Let

$$\widetilde{\Sigma}_\nu = (\widetilde{A}_\nu, \widetilde{B}_\nu, \widetilde{C}_\nu, \widetilde{D}_\nu), \qquad \nu = 1, 2.$$

The state space of $\widetilde{\Sigma}_1$ is $\widetilde{M}$ and that of $\widetilde{\Sigma}_2$ is $\widetilde{M}^{\perp}$. So we may define

$$\Sigma_\nu = (V_\nu^{-1}\widetilde{A}_\nu V_\nu, V_\nu^{-1}\widetilde{B}_\nu, \widetilde{C}_\nu V_\nu, \widetilde{D}_\nu), \qquad \nu = 1, 2.$$

By virtue of Proposition 6.4, the product $\Sigma_1\Sigma_2$ is unitarily equivalent to $\widetilde{\Sigma}_1\widetilde{\Sigma}_2$ and the unitary equivalence is given by the operator $V_1 \boxplus V_2$. Now, $\widetilde{\Sigma}_1\widetilde{\Sigma}_2 = \widetilde{\Sigma}$ and $V = V_1 \boxplus V_2$. It follows that $\Sigma_1\Sigma_2 = \Sigma$, and the factors Σ_1 and Σ_2 have the desired properties. $\square$

XXVIII.8 REGULAR FACTORIZATION OF ANALYTIC MATRIX FUNCTIONS

By $\mathcal{S}^{\ell\times k}(\mathbb{D})$ we denote the set of all operator-valued functions $\theta\colon \mathbb{D} \to \mathcal{L}(\mathbb{C}^k, \mathbb{C}^\ell)$ that are analytic on $\mathbb{D}$ and such that

$$(1) \qquad \sup_{\lambda\in\mathbb{D}} \|\theta(\lambda)\| \leq 1.$$

The $\mathcal{S}$ stands for Schur. From Theorems 2.1 and 5.2 we see that $\mathcal{S}^{\ell\times k}(\mathbb{D})$ is precisely the set of all transfer functions of unitary systems which have $\mathbb{C}^k$ as input space and $\mathbb{C}^\ell$ as output space. If $\theta_1 \in \mathcal{S}^{\ell+r}(\mathbb{D})$ and $\theta_2 \in \mathcal{S}^{r\times k}(\mathbb{D})$, then the function θ defined by

$$(2) \qquad \theta(\lambda) = \theta_1(\lambda)\theta_2(\lambda), \qquad \lambda \in \mathbb{D},$$

belongs to $\mathcal{S}^{\ell\times k}(\mathbb{D})$. Assume we have a factorization as in (2), and let Σ_1 and Σ_2 be pure unitary realizations of θ_1 and θ_2, respectively. Such realizations may be constructed by employing Theorem 5.2. From Theorem 6.1 we know that $\Sigma_1\Sigma_2$ is a unitary realization of θ. The factorization (2) is said to be a *regular factorization* if $\Sigma_1\Sigma_2$ is pure. This definition does not depend on the choice of the unitary systems Σ_1 and Σ_2. Indeed, let $\widetilde{\Sigma}_1$ and $\widetilde{\Sigma}_2$ be other pure unitary realizations of θ_1 and θ_2, respectively. Since Σ_1 and $\widetilde{\Sigma}_1$ have the same characteristic operator function and both are pure, Theorem 3.1 implies that Σ_1 and $\widetilde{\Sigma}_1$ are unitarily equivalent. In the same way one shows that Σ_2 and $\widetilde{\Sigma}_2$ are unitarily equivalent. But then we can use Proposition 6.4 to show that $\Sigma_1\Sigma_2$ and $\widetilde{\Sigma}_1\widetilde{\Sigma}_2$ are unitarily equivalent. Since purity of systems is preserved under unitary equivalence, the system $\Sigma_1\Sigma_2$ is pure if and only if $\widetilde{\Sigma}_1\widetilde{\Sigma}_2$ is pure. Thus the definition of a regular factorization is independent of the choice of the realizations Σ_1 and Σ_2.

Factorizations do not have to be regular. This follows from the fact that the product of two pure unitary systems does not have to be pure. For example, if θ and θ_0 are the scalar functions which are identically equal on $\mathbb{D}$ to $\frac{1}{2}$ and $\frac{1}{2}\sqrt{2}$, respectively, then $\theta = \theta_0\theta_0$. But the latter factorization is not regular. To see this, let Σ_0 be the canonical realization of θ_0. The system Σ_0 is a pure unitary system, but the product $\Sigma_0\Sigma_0$ is not (as we have seen in Section XXVIII.6).

The next theorem tells us how to construct all regular factorizations of a given θ.

THEOREM 8.1. *Let $\theta \in \mathcal{S}^{\ell\times k}(\mathbb{D})$, and let Σ be a pure unitary realization of θ with state space operator A.*

(i) *Let M be an A-invariant subspace, let Σ_1 be the left projection and Σ_2 the right projection of Σ associated with M, and let θ_1 and θ_2 be the transfer functions of Σ_1 and Σ_2, respectively. Then $\theta(\cdot) = \theta_1(\cdot)\theta_2(\cdot)$, and this factorization is regular.*

(ii) *Let $\theta(\cdot) = \theta_1(\cdot)\theta_2(\cdot)$ be a regular factorization. Then there exists a unique A-invariant subspace M and a unique unitary operator U such that $\theta_1(\cdot)U^{-1}$ and $U\theta_2(\cdot)$*

are the transfer functions of the left projection and of the right projection of Σ associated with M, respectively.

PROOF. (i). Let Σ_1 and Σ_2 be as in (i). Then $\Sigma = \Sigma_1\Sigma_2$, by Theorem 7.2, and we can apply Theorem 6.1 to show that $\theta(\cdot) = \theta_1(\cdot)\theta_2(\cdot)$. According to our hypothesés $\Sigma_1\Sigma_2$ $(= \Sigma)$ is pure. So the factorization is regular.

(ii). Let $\theta(\cdot) = \theta_1(\cdot)\theta_2(\cdot)$ be a regular factorization. Choose pure unitary realizations $\widetilde{\Sigma}_1$ and $\widetilde{\Sigma}_2$ of θ_1 and θ_2, respectively. Put $\widetilde{\Sigma} = \widetilde{\Sigma}_1\widetilde{\Sigma}_2$. Then $\widetilde{\Sigma}$ is a unitary realization of θ (by Theorem 6.1), and $\widetilde{\Sigma}$ is pure because the factorization is regular. Now Σ and $\widetilde{\Sigma}$ are pure unitary systems with the same transfer function. So Σ and $\widetilde{\Sigma}$ are unitarily equivalent by virtue of Theorem 3.1. Let H and $\widetilde{H}$ be the state spaces of Σ and $\widetilde{\Sigma}$, respectively, and let $W: H \to \widetilde{H}$ be the (unique) unitary operator that establishes the unitary equivalence. Let the factorization $\widetilde{\Sigma} = \widetilde{\Sigma}_1\widetilde{\Sigma}_2$ be supported by the subspace $\widetilde{M}$ of $\widetilde{H}$. Put $M = W^{-1}\widetilde{M}$. Then Σ has a factorization $\Sigma = \Sigma_1\Sigma_2$ supported by M such that Σ_1 is unitarily equivalent to $\widetilde{\Sigma}_1$ and Σ_2 is unitarily equivalent to $\widetilde{\Sigma}_2$ (see Proposition 7.5). In particular,

$$\theta_\nu(\lambda) = \theta_{\widetilde{\Sigma}_\nu}(\lambda) = \theta_{\Sigma_\nu}(\lambda), \quad \lambda \in \mathbb{D} \quad (\nu = 1,2).$$

Now apply Theorem 7.3 to the factorization $\Sigma = \Sigma_1\Sigma_2$. Let Σ_ℓ be the left projection of Σ associated with M, and let Σ_r be the right projection of Σ associated with M. By Theorem 7.3 there exists a unitary operator V such that

$$\theta_{\Sigma_1}(\lambda) = \theta_{\Sigma_\ell}(\lambda)V^{-1}, \quad \theta_{\Sigma_2}(\lambda) = V\theta_{\Sigma_r}(\lambda) \quad (\lambda \in \mathbb{D}).$$

Hence with $U = V^{-1}$ we get

$$(3) \qquad \theta_1(\lambda)U^{-1} = \theta_{\Sigma_\ell}(\lambda), \quad U\theta_2(\lambda) = \theta_{\Sigma_r}(\lambda) \quad (\lambda \in \mathbb{D}),$$

and thus $\theta_1(\cdot)$ and $\theta_2(\cdot)$ have the desired form.

It remains to prove the uniqueness statement. Let N be an A-invariant subspace, and let V be a unitary operator such that

$$(4) \qquad \theta_1(\cdot)V^{-1} = \theta_{\Sigma'_\ell}(\cdot), \quad V\theta_2(\cdot) = \theta_{\Sigma'_r}(\cdot),$$

where Σ'_ℓ is the left projection and Σ'_r the right projection of Σ associated with N. Note that the state space of H admits the following decompositions:

$$(5) \qquad H = M \boxplus M^\perp, \quad H = N \boxplus N^\perp.$$

Let us write down in more detail the systems appearing in (3) and (4), as follows:

$$(6a) \qquad \Sigma_\nu = (A_\nu, B_\nu, C_\nu, D_\nu), \quad \nu = \ell, r,$$

$$(6b) \qquad \Sigma'_\nu = (A'_\nu, B'_\nu, C'_\nu, D'_\nu), \quad \nu = \ell, r.$$

With U as in (3) and V as in (4) put

$$(7a) \qquad \Delta_1 = (A_1, B_1 U, C_1, D_1 U),$$

$$(7b) \qquad \Delta_2 = (A_2, B_2, U^{-1} C_2, U^{-1} D_2),$$

$$(7c) \qquad \Delta_1' = (A_1', B_1' V, C_1', D_1' V),$$

$$(7d) \qquad \Delta_2' = (A_2', B_2', V^{-1} C_2', V^{-1} D_2').$$

Note that $\Delta_1 \Delta_2$ coincides with the product $\Sigma_\ell \Sigma_r$, and hence $\Sigma = \Delta_1 \Delta_2$. Similarly, $\Delta_1' \Delta_2'$ coincides with $\Sigma_\ell' \Sigma_r'$, and thus $\Sigma = \Delta_1' \Delta_2'$. Since Σ is pure, we can apply Theorem 6.2 to show that each of the systems in (6a), (6b) and (7a)–(7d) is pure. Now, note the first parts of (3) and (4) imply that Δ_1 and Δ_1' have the same transfer function, and hence, by Theorem 3.1, the systems Δ_1 and Δ_1' are unitarily equivalent. Let $J_1: M \to N$ provide the unitary equivalence between Δ_1 and Δ_1'. In a similar way, using the second parts of (3) and (4), one shows that there exists a unitary operator $J_2: M^\perp \to N^\perp$ that establishes a unitary equivalence between Δ_2 and Δ_2'. But then we can use Proposition 6.4 to show that $J_1 \boxplus J_2$ establishes a unitary equivalence between $\Delta_1 \Delta_2$ and $\Delta_1' \Delta_2'$. But $\Delta_1 \Delta_2$ and $\Delta_1' \Delta_2'$ are both equal to Σ, and Σ is pure. So, by the uniqueness statement in Theorem 3.1, we obtain that $J_1 \boxplus J_2$ is the identity operator on the state space H of Σ. It follows, by virtue of (5), that $M = N$ and $\Delta_\nu = \Delta_\nu'$ for $\nu = 1, 2$. Note that $M = N$ implies that $\Sigma_\nu = \Sigma_\nu'$ for $\nu = \ell, r$. But then, $\Delta_\nu = \Delta_\nu'$ for $\nu = 1, 2$ and the uniqueness statement in Theorem 7.3 yield $U = V$. $\square$

If θ_1 and θ_2 are transfer functions of pure finite dimensional unitary systems, then, by virtue of Proposition 6.3, the factorization $\theta = \theta_1 \theta_2$ is regular. More generally, the following can be proved. If at almost all points t of $[0, 2\pi]$ at least one of the identities $\theta_1(e^{it})^* \theta_1(e^{it}) = I$ and $\theta_2(e^{it})^* \theta_2(e^{it}) = I$ holds, then the factorization $\theta = \theta_1 \theta_2$ is regular (see Brodskii [2]).

XXVIII.9 INTERMEZZO ABOUT TRIANGULAR REPRESENTATIONS OF CONTRACTIONS

In this section $A \in \mathcal{L}(H)$ is a contraction with a compact defect operator D_A, and we assume that there exists a point $z_0 \in \mathbb{D}$ such that $z_0 - A$ is invertible. In this case, the operator

$$(1) \qquad B := (z_0 I - A)(I - \bar{z}_0 A)^{-1}$$

is again a contraction, the defect operator of B is compact and the operator B is invertible. For this reason in the following the point z_0 is assumed to be zero. Our aim is to obtain a triangular representation for A.

LEMMA 9.1. *Let $A \in \mathcal{L}(H)$ be an invertible contraction such that D_A is compact. Then $\sigma(A) \cap \mathbb{D}$ consists of eigenvalues of finite type only. Furthermore, if $\widetilde{E}_A$*

is the smallest closed linear manifold of H containing all the eigenvectors and generalized eigenvectors of A corresponding to eigenvalues of A in $\mathbb{D}$, then A has a triangular 2×2 operator matrix representation

$$(2) \qquad A = \begin{bmatrix} A_{11} & A_{12} \\ 0 & A_{22} \end{bmatrix} : \widetilde{E}_A \boxplus \widetilde{E}_A^{\perp} \to \widetilde{E}_A \boxplus \widetilde{E}_A^{\perp},$$

where A_{11} is an invertible contraction with a complete system of eigenvectors and generalized eigenvectors and A_{22} is an invertible contraction with $\sigma(A_{22}) \subset \mathbb{T}$.

PROOF. We split the proof into three parts. In the first part we show that A can be written in the form

$$(3) \qquad A = U(I - V),$$

where U is unitary and V is compact.

Part (a). We prove (3). Let $A = UR$ be the polar decomposition of A. Thus (see Theorem V.6.3) R is the square root of A^*A and U is a partial isometry with initial space $\overline{\operatorname{Im} R}$. Since A is invertible, the same holds for R, and hence U is unitary. Put $K = I - A^*A$. According to our hypotheses, K is a non-negative compact operator, and hence, by the spectral theorem,

$$Kx = \sum_j \lambda_j(K) \langle x, \varphi_j \rangle \varphi_j, \qquad x \in H,$$

where $\varphi_1, \varphi_2, \ldots$ is some orthonormal set. Note that $0 < \lambda_j(K) \leq 1$, because K and $I - K$ are both non-negative. It follows that

$$(4) \qquad (A^*A)^{1/2} x = x - \sum_j \left\{ 1 - \left(1 - \lambda_j(K) \right)^{1/2} \right\} \langle x, \varphi_j \rangle \varphi_j, \qquad x \in H,$$

and thus $R = (A^*A)^{1/2} = I - V$ with V compact.

Part (b). Here we prove the statement about $\sigma(A) \cap \mathbb{D}$. Take $\lambda \in \mathbb{D}$. Since U is unitary, $\lambda - U$ is invertible (by Lemma V.7.2). The difference $(\lambda - A) - (\lambda - U)$ is compact by (3). It follows from Theorem XI.4.1 that $\lambda - A$ is a Fredholm operator. Now, apply Corollary XI.8.4 to $W(\lambda) = \lambda - A$ and $\Omega = \mathbb{D}$. We see that $\sigma(A) \cap \mathbb{D}$ is at most countable and has no accumulation point in $\mathbb{D}$. Furthermore, for $\lambda_0 \in \sigma(A) \cap \mathbb{D}$ and λ sufficiently close to λ_0 we have

$$(\lambda - A)^{-1} = \sum_{n=-q}^{\infty} (\lambda - \lambda_0)^n A_n,$$

where A_{-1} is an operator of finite rank. Now, $A_{-1} = P_{\{\lambda_0\}}$, where $P_{\{\lambda_0\}}$ is the Riesz projection corresponding λ_0. Thus rank $P_{\{\lambda_0\}}$ is finite, and hence (see Section II.1) the point λ_0 is an eigenvalue of finite type.

Part (c). In this part we derive the properties of A_{11} and A_{22}. Let $\lambda_0 \in \sigma(A) \cap \mathbb{D}$, and put $M_0 = \operatorname{Im} P_{\{\lambda_0\}}$. Then $AM_0 \subset M_0$. Since M_0 is finite dimensional and A is invertible, we see that the smallest linear manifold M containing all eigenvectors and generalized eigenvectors of A corresponding to eigenvalues in $\mathbb{D}$ is invariant under both A and A^{-1}. By definition, $\widetilde{E}_A = \overline{M}$, and thus, by continuity, the space $\widetilde{E}_A$ is also invariant under both A and A^{-1}. The latter justifies the zero in the left lower corner of the 2×2 matrix in (2) and it shows that A_{11} is invertible. The other properties of A_{11} are obvious. Since A and A_{11} are both invertible, the triangular form of the matrix in (2) implies that A_{22} is invertible. Clearly, A_{22} is a contraction. It remains to prove the statement about $\sigma(A_{22})$.

By computing the $(2,2)$-entry in the product AA^* we see that $I - A_{22}A_{22}^*$ is compact, and hence by the spectral theorem, the defect operator of A_{22}^* is compact. In particular, A_{22}^* is an operator of the same type as A. Now, let $\mu \in \mathbb{D} \cap \sigma(A_{22})$. Then $\overline{\mu} \in \mathbb{D} \cap \sigma(A_{22}^*)$, and by Part (b) of the proof $\overline{\mu}$ is an eigenvalue of A_{22}^* of finite type. Let $x_0 \neq 0$ in $\widetilde{E}_A^{\perp}$ be a corresponding eigenvector. From the triangular representation (2) it follows that A^* leaves $\widetilde{E}_A^{\perp}$ invariant and on $\widetilde{E}_A^{\perp}$ the operators A^* and A_{22}^* coincide. Thus $A^* x_0 = \overline{\mu} x_0$. From

$$(5) \qquad D_{A^*} = A D_A A^{-1}$$

we see that A^* has the same properties as A. In particular, D_{A^*} is compact. Thus $\overline{\mu}$ is an eigenvalue of finite type, and hence

$$(6) \qquad 0 \neq x_0 \in \widetilde{E}_A^{\perp} \cap \operatorname{Im} P_{\{\overline{\mu}\}}(A^*).$$

On the other hand $\operatorname{Im} P_{\{\mu\}}(A) \subset \widetilde{E}_A$. So we can apply Proposition I.2.5 to show that

$$\widetilde{E}_A^{\perp} \subset (\operatorname{Im} P_{\{\mu\}}(A))^{\perp} = \operatorname{Ker} P_{\{\mu\}}(A)^* = \operatorname{Ker} P_{\{\overline{\mu}\}}(A^*),$$

which contradicts (6). So $\sigma(A_{22}) \cap \mathbb{D}$ is empty, and thus $\sigma(A_{22}) \subset \mathbb{T}$. $\square$

THEOREM 9.2. *Let $A \in \mathcal{L}(H)$ be a completely non-unitary invertible contraction, and assume that D_A is a Hilbert-Schmidt operator. Then*

$$(7) \qquad \det(A^* A) \leqq \prod_j |\lambda_j(A)|^2,$$

where $\lambda_1(A), \lambda_2(A), \ldots$ are the (non-zero) eigenvalues of A repeated according to algebraic multiplicity. Furthermore, equality holds in (7) if and only if A has a complete system of eigenvectors and generalized eigenvectors.

PROOF. We split the proof into three parts. However, first note that $\det(A^* A)$ is well-defined, because $I - A^* A = D_A^2$ is a trace class operator.

Part (a). By Lemma 9.1 the operator A has the representation (2). In this part we show that

$$(8) \qquad \det(A^* A) = \det(A_{11}^* A_{11}) \det(A_{22}^* A_{22}).$$

From (2) we see that

$$(9a) \qquad A^*A = \begin{bmatrix} A_{11}^* A_{11} & A_{11}^* A_{12} \\ A_{12}^* A_{11} & A_{12}^* A_{12} + A_{22}^* A_{22} \end{bmatrix},$$

$$(9b) \qquad AA^* = \begin{bmatrix} A_{11} A_{11}^* + A_{12} A_{12}^* & A_{12} A_{22}^* \\ A_{22} A_{12}^* & A_{22} A_{22}^* \end{bmatrix}.$$

By (5) the defect operator D_{A^*} is also Hilbert-Schmidt. It follows that both $A^*A - I$ and $AA^* - I$ are in S_1, the set of trace class operators. But then (9a) and (9b) imply that $A_{11}^* A_{11} - I_{11}$ and $A_{22} A_{22}^* - I_{22}$ are in S_1. Here I_{11} and I_{22} are the identity operators on $\widetilde{E}_A$ and $\widetilde{E}_A^{\perp}$, respectively. Since A_{22} is invertible, we also have

$$A_{22}^* A_{22} - I_{22} = A_{22}^{-1}(A_{22} A_{22}^* - I_{22})A_{22} \in S_1.$$

Thus the determinants appearing in the right hand side of (8) are well-defined.

The representation (2) implies that A factorizes as $A = M_2 M M_1$, where

$$M_1 = \begin{bmatrix} A_{11} & 0 \\ 0 & I_{22} \end{bmatrix}, \qquad M = \begin{bmatrix} I_{11} & A_{12} \\ 0 & I_{22} \end{bmatrix}, \qquad M_2 = \begin{bmatrix} I_{11} & 0 \\ 0 & A_{22} \end{bmatrix}.$$

Note that the three operators M_1, M and M_2 are invertible. From the result proved in the previous paragraph we see that $M_1^* M_1 - I$ and $M_2^* M_2 - I$ are trace class operators and

$$(10) \qquad \det(M_\nu^* M_\nu) = \det(A_{\nu\nu}^* A_{\nu\nu}), \qquad \nu = 1, 2.$$

Since $A^*A - I$ is a trace class operator, (9a) implies that $A_{11}^* A_{12} \in S_1$. But then $A_{12} = (A_{11}^*)^{-1}(A_{11}^* A_{12})$ is trace class, and hence $M - I \in S_1$. Now, $(M - I)^2 = 0$, and so $M - I$ has no non-zero eigenvalue, and therefore $\det M = 1$ (by Theorem VII.6.1(ii)). Note that

$$A^*A = M_1^* M^* M_2^* M_2 M M_1$$
$$= M_1^* M_1 \{ M_1^{-1} M^{-1}(MM^* M_2^* M_2)M M_1 \},$$

and therefore

$$\det(A^*A) = \det(M_1^* M_1) \det(MM^* M_2^* M_2)$$
$$= \det(M_1^* M_1) \det(M_2^* M_2),$$

which, by (10), yields the desired equality (8).

Part (b). In this part we prove the inequality (7). To do this we first compute $\det(A_{11}^* A_{11})$. The proof of Lemma II.3.3 shows that we may choose an orthonormal basis $\varphi_1, \varphi_2, \ldots$ in $\widetilde{E}_A$ such that for $j = 1, 2, \ldots$

$$(11) \qquad A\varphi_j = a_{1j}\varphi_1 + \cdots + a_{jj}\varphi_j, \qquad a_{jj} = \lambda_j(A).$$

Let P_n be the orthogonal projection of $\widetilde{E}_A$ onto span$\{\varphi_1, \ldots, \varphi_n\}$. Then

$$P_n A_{11}^* A_{11} P_n = P_n A_{11}^* P_n P_n A_{11} P_n,$$

and thus, by Theorem VII.3.2,

$$\det A_{11}^* A_{11} = \lim_{n \to \infty} \det(P_n A_{11}^* P_n P_n A_{11} P_n)$$
$$= \lim_{n \to \infty} \det(P_n A_{11}^* P_n) \det(P_n A_{11} P_n)$$
$$= \lim_{n \to \infty} \prod_{j=1}^{n} |\lambda_j(A)|^2 = \prod_{j} |\lambda_j(A)|^2.$$

Next, let $K = I - A_{22}^* A_{22}$. We know that $K \in S_1$ and $K \geq 0$ because $\|A_{22}\| \leq 1$. Furthermore, $I - K = A_{22}^* A_{22}$ is strictly positive by the invertibility of A_{22}. So the non-zero eigenvalues of K are in the open interval $(0, 1)$. Hence, by Theorem VII.6.1(ii),

$$(12) \qquad \det(A_{22}^* A_{22}) = \prod_{j}(1 - \lambda_j(K)) \leq 1.$$

We have now proved the inequality (7).

Part (c). This part concerns the statement about completeness. From the result proved in the previous paragraph we see that we have equality in (7) if and only if $\det(A_{22}^* A_{22}) = 1$. Since the non-zero eigenvalues of $K = I - A_{22}^* A_{22}$ are in $(0, 1)$, we see that $\det(A_{22}^* A_{22}) = 1$ if and only if $K = 0$. But $K = 0$ is equivalent to A_{22} being unitary (because of the invertibility of A_{22}). From Lemma 1.2 we know that a completely non-unitary contraction cannot have a non-zero compression which is unitary. So, since A is completely non-unitary, we have equality in (7) if and only if $\widetilde{E}_A = H$. The fact that A is completely non-unitary also implies that A has no eigenvalues on $\mathbb{T}$ (again apply Lemma 1.2), and hence $\widetilde{E}_A$ coincides with the smallest closed linear manifold in H containing all eigenvectors and generalized eigenvectors of A. With these remarks the theorem is proved. $\square$

THEOREM 9.3. *Let $A \in \mathcal{L}(H)$ be a completely non-unitary invertible contraction such that $I - A$ compact. Then A has a triangular 2×2 operator matrix representation*

$$(13) \qquad A = \begin{bmatrix} A_{11} & A_{12} \\ 0 & A_{22} \end{bmatrix} : H_1 \boxplus H_2 \to H_1 \boxplus H_2$$

where A_{11} is a completely non-unitary invertible contraction with a complete system of eigenvectors and generalized eigenvectors, and A_{22} is a completely non-unitary invertible contraction such that $I - A_{22}$ is Volterra. Furthermore, A_{22} has an invariant chain $\mathbb{P}$ of orthogonal projections, which is maximal and continuous, such that A_{22} and A_{22}^{-1} have the following triangular representations relative to $\mathbb{P}$:

$$(14) \qquad A_{22} = \int_{\mathbb{P}}^{\curvearrowright} (I - (I - PD^2P)^{-1} PD^2 dP),$$

$$(15) \qquad A_{22}^{-1} = I + \int_{\mathbb{P}} (I - PD^2P)^{-1} PD^2 \, dP,$$

where D is the defect operator of A_{22}.

PROOF. Put $B = I - A$. Then

$$(16) \qquad I - A^*A = I - (I - B^*)(I - B) = B^* + B - B^*B,$$

and hence $I - A^*A$ is a compact non-negative selfadjoint operator. It follows that D_A is compact, and we can apply Lemma 9.1. Take $H_1 = \widetilde{E}_A$ and $H_2 = \widetilde{E}_A^{\perp}$, where $\widetilde{E}_A$ is as in Lemma 9.1. Then we have the partitioning (13), and the operators A_{11} and A_{22} have the properties described in the lemma.

Let us prove that A_{11} is completely non-unitary. Let M be a subspace of H_1, invariant under A_{11} and A_{11}^*, on which A_{11} acts as a unitary operator. Then Lemma 1.2 implies that M is invariant under A and A^* and that A acts as a unitary operator on M. But then $M = \{0\}$, because A is completely non-unitary. By a similar argument, starting with A^* in place of A, one proves that A_{22} completely non-unitary.

Since $I - A$ is compact, the same is true for the operator $V := I - A_{22}$. From Lemma 9.1 we know that $\sigma(A_{22}) \subset \mathbb{T}$. On the other hand, since A_{22} is completely non-unitary, A_{22} has no eigenvalues on $\mathbb{T}$, by virtue of Proposition 1.4. It follows that $\sigma(A_{22}) = \{1\}$, and hence $\sigma(V) = \{0\}$. So $I - A_{22}$ is Volterra.

To get the triangular representation for A_{22} and A_{22}^{-1} we first apply Theorem XX.4.3 and Corollary XX.4.6 to the Volterra operator $V = I - A_{22}$. It follows that there exists a maximal chain $\mathbb{P}$ of orthogonal projections such that $\mathbb{P}$ is invariant under A, the diagonal E of A with respect to $\mathbb{P}$ exists, and E is equal to the identity operator on H. Let (P_ν^-, P_ν^+), $\nu = 1, 2, \ldots$, be the jumps in $\mathbb{P}$. The statement about the diagonal of A implies (cf., Theorem XX.4.4) that

$$(17) \qquad (P_\nu^+ - P_\nu^-)A(P_\nu^+ - P_\nu^-) = P_\nu^+ - P_\nu^-, \qquad \nu = 1, 2, \ldots .$$

Since A is completely non-unitary, Lemma 1.2 shows that A cannot have a non-zero compression which is unitary. But, by (17), the compression of A to $\mathrm{Im}(P_\nu^+ - P_\nu^-)$ is the identity operator. It follows that $\mathbb{P}$ does not have jumps, and hence $\mathbb{P}$ is a continuous chain. We have now proved that the chain $\mathbb{P}$ has the desired properties. The formulas (14) and (15) follow directly from Theorem XXII.6.1. $\square$

With minor modifications Theorem 9.3 also holds for non-invertible contractions. To see this, assume that A is a non-invertible contraction and $I - A$ is compact. Then 0 is an eigenvalue of A of finite type. Put $M = \mathrm{Im}\, P_0$, where P_0 is the Riesz projection of A corresponding to $\{0\}$. Then A partitions as

$$A = \begin{bmatrix} A_0 & * \\ 0 & \widetilde{A} \end{bmatrix} : M \boxplus M^{\perp} \to M \boxplus M^{\perp},$$

where $\dim M < \infty$ and $\widetilde{A}$ is an invertible contraction such that $I - \widetilde{A}$ is compact. Now apply Theorem 9.3 to $\widetilde{A}$, and one sees that the original operator A partitions as in (13), where, except for the invertibility of A_{11}, the operators A_{11} and A_{22} have the properties stated in Theorem 9.3.

XXVIII.10 MULTIPLICATIVE REPRESENTATIONS OF TRANSFER FUNCTIONS

In this section

$$(1) \qquad \Sigma = (A, B, C, D; H, K, L)$$

is a pure unitary system of which the main operator A is assumed to be of the form $A = I + V$ with V compact. The latter implies that $\sigma(A) \cap \mathbb{T}$ consists of a finite number of points, and hence on $\mathbb{T}$, with the exception of a finite number of points, the values of the transfer function θ_Σ are unitary operators. In particular, θ_Σ is inner.

Our aim is to obtain multiplicative representations of θ_Σ. The simplest result in this direction, when Σ is a finite dimensional system, is already given in Corollary 7.4. The next theorems concern the infinite dimensional case.

THEOREM 10.1. *Let Σ in (1) be a pure unitary system such that $I - A$ is compact, and assume that A has a complete system of eigenvectors and generalized eigenvectors. Then*

$$(2) \qquad \theta_\Sigma(\lambda) = \theta_1(\lambda)\theta_2(\lambda), \qquad \lambda \in \mathbb{D},$$

with

$$(3a) \qquad \theta_1(\lambda) = \overset{\curvearrowright}{\underset{j=0}{\overset{m}{\prod}}} (I - Q_j + \lambda Q_j),$$

$$(3b) \qquad \theta_2(\lambda) = \lim_{n \to \infty} \left\{ \overset{\curvearrowright}{\underset{j=1}{\overset{n}{\prod}}} \left(I - P_j + \frac{\overline{\lambda}_j - \lambda}{1 - \lambda\lambda_j} \frac{|\lambda_j|}{\overline{\lambda}_j} P_j \right) \right\} U_n.$$

Here $Q_1, \ldots, Q_m$ and $P_1, P_2, \ldots$ are orthogonal projections of L of rank one, m is the algebraic multiplicity of 0 as an eigenvector of A, $\lambda_1, \lambda_2, \ldots$ are the non-zero eigenvalues of A repeated according to algebraic multiplicity and $U_1, U_2, \ldots$ is a sequence of operators in $\mathcal{L}(K, L)$ such that U_n is unitary for n sufficiently large. Furthermore, the convergence in (3b) is uniform on compact subsets of $\mathbb{D}$.

THEOREM 10.2. *Let Σ in (1) be a pure unitary system such that $I - A$ is Volterra, and let $\mathbb{P}$ be a maximal continuous chain which is invariant under A. Then*

$$(4) \qquad \theta_\Sigma(\lambda) = \left\{ \overset{\curvearrowright}{\underset{\mathbb{P}}{\int}} \left(I - \frac{1 + \lambda}{1 - \lambda} d(FPF^*) \right) \right\} \theta_\Sigma(-1),$$

where

$$F = \sqrt{2}\, C(I + A)^{-1} : H \to L.$$

Furthermore, if the map $P \mapsto FPF^*$ is of bounded variation, that is, if

$$(5) \qquad \sup\left\{ \sum_{j=1}^{n} \|F(\Delta P_j)F^*\| \mid \pi = \{P_j\}_{j=0}^{n} \text{ is a partition of } \mathbb{P}\right\} < \infty,$$

then

$$(6) \qquad \theta_\Sigma(\lambda) = \left\{ \int_{\mathbb{P}}^{\curvearrowright} \exp\left(-\frac{1+\lambda}{1-\lambda}d(FPF^*)\right)\right\}\theta_\Sigma(-1).$$

THEOREM 10.3. *Let Σ be as in (1) be a pure unitary system such that $I - A$ is compact. Then*

$$(7) \qquad \theta(\lambda) = \theta_1(\lambda)\theta_2(\lambda)\theta_3(\lambda), \qquad \lambda \in \mathbb{D},$$

with

$$(8a) \qquad \theta_1(\lambda) = \prod_{j=1}^{m}{}^{\curvearrowright} (I - Q_j + \lambda Q_j),$$

$$(8b) \qquad \theta_2(\lambda) = \lim_{n\to\infty}\left\{ \prod_{j=1}^{n}{}^{\curvearrowright} \left(I - P_j + \frac{\overline{\lambda}_j - \lambda}{1 - \lambda\lambda_j}\frac{|\lambda_j|}{\overline{\lambda}_j}P_j\right)\right\}U_n,$$

$$(8c) \qquad \theta_3(\lambda) = \left\{ \int_{\mathbb{P}}^{\curvearrowright} \left(I - \frac{1+\lambda}{1-\lambda}d(FPF^*)\right)\right\}U.$$

Here $Q_1, \ldots, Q_m$ and $P_1, P_2, \ldots$ are orthogonal projections of L of rank one, m is the algebraic multiplicity of 0 as an eigenvalue of A, $\lambda_1, \lambda_2, \ldots$ are the non-zero eigenvalues of A repeated according to algebraic multiplicity, $U_1, U_2, \ldots$ is a sequence of operators on L such that U_n is unitary for n sufficiently large, $U: K \to L$ is unitary, $\mathbb{P}$ is a maximal continuous chain in $E_A^{\perp}$ and

$$(9) \qquad F: E_A^{\perp} \to L$$

is a compact operator. Here E_A is the smallest closed linear manifold of H containing all eigenvectors and generalized eigenvectors of A. Furthermore, the chain $\mathbb{P}$ is invariant under the operator $\tau^* A\tau$, where $\tau: E_A^{\perp} \to H$ is the canonical embedding, and the convergence in (8b) is uniform on compact subsets of $\mathbb{D}$.

It is convenient first to prove the following lemma.

LEMMA 10.4. *Let Σ in (1) be a unitary system such that $I - A$ is compact. Let $\widetilde{P}_1, \widetilde{P}_2, \ldots$ be a sequence of orthogonal projections of H such that $\widetilde{P}_n x \to x$ for each $x \in H$ when $n \to \infty$, and assume that*

$$(10) \qquad A\widetilde{P}_n = \widetilde{P}_n A\widetilde{P}_n, \qquad n = 1, 2, \ldots .$$

Let Σ_n be the left projection of Σ associated with the invariant subspace $\operatorname{Im}\widetilde{P}_n$ for $n = 1, 2, \ldots$. Then there exists a sequence of operators $U_1, U_2, \ldots$ such that U_n is unitary for n sufficiently large and

$$(11) \qquad \lim_{n \to \infty} \theta_{\Sigma_n}(\lambda)U_n = \theta_\Sigma(\lambda), \qquad \lambda \in \mathbb{D},$$

with uniform convergence on compact subsets of $\mathbb{D}$.

PROOF. We split the proof into five parts. The first part concerns a general fact about unitary systems of which the main operator is a compact perturbation of the identity.

Part (a). In this part we show that the operators B and C are compact, and that D is a compact perturbation of a unitary operator. Write A as $I + V$ with V compact. Then $I - A^*A = -V^* - V - V^*V$, and hence $I - A^*A$ is a non-negative compact operator. From the spectral theorem for compact selfadjoint operators it follows that D_A is compact. But then we can apply Theorem 4.2 to show that the same holds for C. Since $A^* = I + V^*$, the operator D_{A^*} is compact too, and Theorem 4.2 also shows that B is compact. By Theorem 4.2 the operator D is of the form

$$(12) \qquad D = \begin{bmatrix} E_1 D_0 E_2 & 0 \\ 0 & E \end{bmatrix},$$

where E_1, E_2 and E are unitary operators, and

$$D_0 : \overline{\operatorname{Im} D_{A^*}} \to \overline{\operatorname{Im} D_A}, \qquad D_0 u = -A^* u.$$

Since θ_Σ is inner, there exists a unitary operator $U : \overline{\operatorname{Im} D_A} \to \overline{\operatorname{Im} D_{A^*}}$. It follows that

$$E_1 D_0 E_2 = E_1 U^{-1}(-UV^* - U)E_2.$$

Therefore $E_1 D_0 E_2$ is a compact perturbation of a unitary operator, and, by (12), the same holds for D.

Part (b). In order to prove (11) we may without loss of generality assume that Σ is the standard unitary system associated with A and C. This means (see Lemma 7.1) that the input space K and the operators B and D are defined as follows. Set

$$J : H \to H \boxplus L, \qquad J = \begin{bmatrix} A \\ C \end{bmatrix}.$$

Then J is an isometry, and

$$(13a) \qquad K = (H \boxplus L) \cap (\operatorname{Im} J)^{\perp},$$

$$(13b) \qquad \begin{bmatrix} B \\ D \end{bmatrix} = I_{H \boxplus L} | K \colon K \to H \boxplus L.$$

Let Π be the orthogonal projection of $H \boxplus L$ onto K. In this part we show that the operator

$$(14) \qquad \begin{bmatrix} I_H & 0 \\ 0 & 0 \end{bmatrix} \Pi \colon H \boxplus L \to H \boxplus L$$

is compact. Indeed, let $\gamma \colon K \to H \boxplus L$ be the canonical embedding. Then $\gamma\gamma^* = \Pi$, and, by virtue of (13b),

$$\begin{bmatrix} I_H & 0 \\ 0 & 0 \end{bmatrix} \Pi = \begin{bmatrix} I_H & 0 \\ 0 & 0 \end{bmatrix} \gamma\gamma^* = \begin{bmatrix} B\gamma^* \\ 0 \end{bmatrix}.$$

By the result of Part (a), the operator $B\gamma^*$ is compact, and hence the operator (14) is also compact.

Part (c). Put $H_n = \operatorname{Im} \widetilde{P}_n$, and let $\Sigma_n = (A_n, B_n, C_n, D_n; H_n, K_n, L)$ be the left projection of Σ associated with the invariant subspace H_n. Thus A_n is the restriction of A to H_n viewed as an operator from H_n into H_n, the operator C_n is the restriction of C to H_n viewed as an operator from H_n into L, and

$$B_n \colon K_n \to H_n, \qquad D_n \colon K_n \to L,$$

are defined as follows. Let

$$J_n \colon H_n \to H_n \boxplus L, \qquad J_n = \begin{bmatrix} A_n \\ C_n \end{bmatrix}.$$

Then J_n is an isometry, and

$$(15a) \qquad K_n = (H_n \boxplus L) \cap (\operatorname{Im} J_n)^\perp,$$

$$(15b) \qquad \begin{bmatrix} B_n \\ D_n \end{bmatrix} = I_{H_n \boxplus L} | K_n \colon K_n \to H_n \boxplus L.$$

To relate Σ_n to Σ we need the canonical embedding $\tau_n \colon H_n \to H$ and the canonical embedding $\rho_n \colon H_n \boxplus L \to H \boxplus L$. Note that $\widetilde{P}_n = \tau_n \tau_n^*$, and we set

$$\widetilde{Q}_n := \rho_n \rho_n^* = \begin{bmatrix} \widetilde{P}_n & 0 \\ 0 & 0 \end{bmatrix} \colon H \boxplus L \to H \boxplus L.$$

Let Π_n be the orthogonal projection of $H_n \boxplus L$ onto the space K_n, and let Π be the orthogonal projection of $H \boxplus L$ onto K. In this part we show that

$$(16) \qquad \| \rho_n \Pi_n \rho_n^* - \Pi \| \to 0 \qquad (n \to \infty).$$

Since J_n is an isometry, $\Pi_n = I_n - J_n J_n^*$, where I_n is the identity operator on H_n. Similarly, $\Pi = I - JJ^*$. Now, $J_n = \rho_n^* J \tau_n$, and so

$$\rho_n \Pi_n \rho_n^* = \rho_n \rho_n^* - \rho_n \rho_n^* J \tau \tau_n^* J^* \rho_n \rho_n^*$$
$$= \widetilde{Q}_n - \widetilde{Q}_n J \widetilde{P}_n J^* \widetilde{Q}_n$$
$$= \widetilde{Q}_n \Pi \widetilde{Q}_n - \widetilde{Q}_n J (I - \widetilde{P}_n) J^* \widetilde{Q}_n.$$

Write A as $I + V$, where V is compact. Since $\widetilde{P}_n \to I$ $(n \to \infty)$ strongly, the compactness of V and C implies that

$$\begin{bmatrix} V \\ C \end{bmatrix} (I - \widetilde{P}_n) \to 0 \qquad (n \to \infty)$$

in the operator norm. So we may conclude that

$$J(I - \widetilde{P}_n)J^* = \begin{bmatrix} I - \widetilde{P}_n & 0 \\ 0 & 0 \end{bmatrix} + G_n,$$

where $\|G_n\| \to 0$ if $n \to \infty$. It follows that $\widetilde{Q}_n J(I - \widetilde{P}_n)J^* \widetilde{Q}_n$ converges to zero in the operator norm if $n \to \infty$, because

$$\widetilde{Q}_n \begin{bmatrix} I - \widetilde{P}_n & 0 \\ 0 & 0 \end{bmatrix} \widetilde{Q}_n = 0, \qquad n = 1, 2, \ldots .$$

Note that

$$(I - \widetilde{Q}_n)\Pi = \begin{bmatrix} I - \widetilde{P}_n & 0 \\ 0 & 0 \end{bmatrix} \Pi = \begin{bmatrix} I - \widetilde{P}_n & 0 \\ 0 & 0 \end{bmatrix} \begin{bmatrix} I_H & 0 \\ 0 & 0 \end{bmatrix} \Pi.$$

Since the operator (14) is compact and $I - \widetilde{P}_n \to 0$ $(n \to \infty)$ in the strong operator topology, we conclude that $(I - \widetilde{Q}_n)\Pi \to 0$ $(n \to \infty)$ in the operator norm. By taking adjoints we see that also $\Pi(I - \widetilde{Q}_n) \to 0$ $(n \to \infty)$ in the operator norm. But then $\widetilde{Q}_n \Pi \widetilde{Q}_n \to \Pi$ $(n \to \infty)$ in the operator norm, and (16) is proved.

Part (d). Put $\widehat{K}_n = \rho_n K_n$. Thus $\widehat{K}_n$ is the space K_n viewed as a subspace of $H \boxplus L$. The orthogonal projection $\widehat{\Pi}_n$ of $H \boxplus L$ onto $\widehat{K}_n$ is precisely equal to $\rho_n \Pi_n \rho_n^*$, and according to (16) we have $\widehat{\Pi}_n \to \Pi$ $(n \to \infty)$ in the operator norm. In this part we construct a sequence of operators $\widehat{U}_1, \widehat{U}_2, \ldots$ on $H \boxplus L$ such that $\widehat{U}_n \to I$ $(n \to \infty)$ in the operator norm and there exists a positive integer N so that $\widehat{U}_n$ is unitary for $n \geq N$ and

$$(17) \qquad \widehat{\Pi}_n = \widehat{U}_n \Pi \widehat{U}_n^{-1}, \qquad n \geq N.$$

Put

$$(18) \qquad S_n = \widehat{\Pi}_n \Pi + (I - \widehat{\Pi}_n)(I - \Pi), \qquad n = 1, 2, \ldots .$$

Note that S_n acts on $H \boxplus L$ and $S_n \Pi = \widehat{\Pi} S_n$ for $n = 1, 2, \ldots$. Since $S_n \to I$ $(n \to \infty)$ in the operator norm, S_n is invertible for n sufficiently large. Let $\widehat{U}_n \widehat{R}_n$ be the polar decomposition of S_n (see Section V.6). From the invertibility of S_n for n large, it follows that $\widehat{U}_n$ is a unitary operator for $n \geq N$, say. The projections $\widehat{\Pi}_n$ and Π appearing in (18) are orthogonal. Therefore,

$$(19) \qquad \Pi S_n^* S_n = S_n^* \widehat{\Pi}_n S_n = S_n^* S_n \Pi, \qquad n = 1, 2, \ldots .$$

Now use that $\widehat{R}_n = (S_n^* S_n)^{1/2}$ is the limit in the operator norm of a sequence of polynomials in $S_n^* S_n$. This together with (19) shows that $\Pi \widehat{R}_n = \widehat{R}_n \Pi$ for $n = 1, 2, \ldots$. But then we may conclude that (17) holds. Since $S_n \to I$ $(n \to \infty)$ in the operator norm, it follows that $S_n^* S_n \to I$ $(n \to \infty)$ in the operator norm. Hence, by the spectral theorem, $\widehat{R}_n \to I$ $(n \to \infty)$ in the operator norm, and thus $\|\widehat{U}_n - I\| \to 0$ $(n \to \infty)$.

Part (e). In this part we prove (11). By virtue of (17), we have $\widehat{U}_n K = \widehat{K}_n$ for $n \geq N$. So, for $n \geq N$, the operator

$$U_n \colon K \to K_n, \quad U_n x = \rho_n^* \widehat{U}_n x \qquad (x \in K)$$

is unitary. Put

$$(20) \qquad \begin{bmatrix} \widehat{B}_n \\ \widehat{D}_n \end{bmatrix} := \begin{bmatrix} \tau_n B_n \\ D_n \end{bmatrix} U_n = \rho_n \begin{bmatrix} B_n \\ D_n \end{bmatrix} \rho_n^* \widehat{U}_n \gamma$$
$$= \widetilde{Q}_n \widehat{U}_n \gamma,$$

where, as before, $\gamma \colon K \to H \boxplus L$ is the canonical embedding. We shall compare the left hand side of (20) to

$$(21) \qquad \begin{bmatrix} B \\ D \end{bmatrix} = I_{H \boxplus L}|K = \begin{bmatrix} I & 0 \\ 0 & 0 \end{bmatrix} \gamma + \begin{bmatrix} 0 & 0 \\ 0 & I \end{bmatrix} \gamma.$$

From $\gamma = \gamma \gamma^* \gamma = \Pi \gamma$ we see that

$$(22) \qquad \widetilde{Q}_n \widehat{U}_n \gamma = \begin{bmatrix} \widetilde{P}_n & 0 \\ 0 & 0 \end{bmatrix} \widehat{\Pi}_n \widehat{U}_n \gamma + \begin{bmatrix} 0 & 0 \\ 0 & I \end{bmatrix} \widehat{U}_n \gamma.$$

The second term in the right hand side of (22) converges for $n \to \infty$ in the operator norm to the second term in the right hand side of (21). To see what happens to the first term, we write

$$(23) \qquad \begin{bmatrix} \widetilde{P}_n & 0 \\ 0 & 0 \end{bmatrix} \widehat{\Pi}_n \widehat{U}_n \gamma = \begin{bmatrix} \widetilde{P}_n & 0 \\ 0 & 0 \end{bmatrix} \Pi \widehat{U}_n \gamma + \begin{bmatrix} \widetilde{P}_n & 0 \\ 0 & 0 \end{bmatrix} (\widehat{\Pi}_n - \Pi) \widehat{U}_n \gamma.$$

By the compactness of the operator in (14), the first term in the right hand side of (23) converges if $n \to \infty$ in the operator norm to the first term in the right hand side of (21). The second term in the right hand side of (23) converges if $n \to \infty$ in the operator norm to zero. Thus

$$\begin{bmatrix} \widehat{B}_n \\ \widehat{D}_n \end{bmatrix} \to \begin{bmatrix} B \\ D \end{bmatrix} \qquad (n \to \infty)$$

in the operator norm.

Take $n \geq N$, and put

$$\widehat{A}_n = \begin{bmatrix} A_n & 0 \\ 0 & I_{H_n^\perp} \end{bmatrix} \colon H_n \boxplus H_n^\perp \to H_n \boxplus H_n^\perp,$$

$$\widehat{C}_n = [C_n \;\; 0] \colon H_n \boxplus H_n^\perp \to L.$$

Then $\widehat{\Sigma}_n = (\widehat{A}_n, \widehat{B}_n, \widehat{C}_n, \widehat{D}_n; H, K, L)$ is a unitary system, and

$$
\begin{aligned}
\theta_{\widehat{\Sigma}_n}(\lambda) &= \widehat{D}_n + \lambda \widehat{C}_n (I - \lambda \widehat{A}_n)^{-1} \widehat{B}_n \\
&= \widehat{D}_n + \lambda C (I - \lambda A) \widehat{B}_n \\
&\to D + \lambda C (I - \lambda A)^{-1} B = \theta_\Sigma(\lambda).
\end{aligned}
$$

Here we used that $\widehat{B}_n \to B$ and $\widehat{D}_n \to D$ if $n \to \infty$ (which is proved in the previous paragraph), and the fact that $\operatorname{Im} \widehat{B}_n \subset H_n$, $n \geq N$. Finally, note that

$$
\begin{aligned}
\theta_{\Sigma_n}(\lambda) U_n &= \big(D_n + \lambda C_n (I - \lambda A_n)^{-1} B_n\big) U_n \\
&= \big(D_n + \lambda C (I - \lambda A)^{-1} \tau_n B_n\big) U_n \\
&= \widehat{D}_n + \lambda C (I - \lambda A)^{-1} \widehat{B}_n = \theta_{\widehat{\Sigma}_n}(\lambda),
\end{aligned}
$$

and the proof is complete. $\square$

PROOF OF THEOREM 10.1. Since the linear manifold spanned by all eigenvectors and generalized eigenvectors of A is dense in H, we can find an orthonormal basis $\varphi_1, \varphi_2, \ldots$ in H such that

$$
A\varphi_j = a_{j1}\varphi_1 + a_{j2}\varphi_2 + \cdots + a_{jj}\varphi_j, \qquad j = 1, 2, \ldots,
$$

with

$$
a_{jj} = \begin{cases} 0 &, \quad j = 1, \ldots, m, \\ \lambda_{j-m}, & j \geq m + 1. \end{cases}
$$

Let M be the space spanned by $\varphi_1, \ldots, \varphi_m$. Then A has the following partitioning:

$$
(24) \qquad A = \begin{bmatrix} A_1 & * \\ 0 & A_2 \end{bmatrix} : M \boxplus M^\perp \to M \boxplus M^\perp.
$$

The space M is invariant under A. So we may consider a factorization $\Sigma = \Sigma_1 \Sigma_2$ supported by M. Note that $\Sigma_1 = (A_1, B_1, C_1, D_1)$ has a finite dimensional state space and Σ_1 is pure by Theorem 6.2. Moreover, all the eigenvalues of A_1 are zero. So we may apply Corollary 7.4 to show that $\theta_{\Sigma_1}(\cdot) = \theta_1(\cdot)U$ for some unitary operator U, where θ_1 is as in (3a). By replacing Σ_1 by $\widetilde{\Sigma}_1 = (A_1, B_1 U^{-1}, C_1, D_1 U^{-1})$ we see that without loss of generality we may assume that $\theta_{\Sigma_1} = \theta_1$.

By Theorem 6.2 the system Σ_2 is also pure. Write $\widetilde{H}$ for $M^\perp$, and let $\widetilde{P}_n$ be the orthogonal projection of $\widetilde{H}$ onto the space spanned by the vectors $\varphi_{m+1}, \ldots, \varphi_{m+n}$. Then $\widetilde{P}_n x \to x$ $(n \to \infty)$ for each $x \in \widetilde{H}$. Furthermore, $\operatorname{Im} \widetilde{P}_n$ is invariant under A_2, the state operator of Σ_2. So we may apply Lemma 10.4. Let $\Sigma_{2,n}$ be the left projection of Σ_2 associated with the invariant subspace $\operatorname{Im} \widetilde{P}_n$. From Corollary 7.4 and its proof we know that there exist orthogonal projections $P_1, P_2, \ldots$ of rank one acting on L and unitary operators $V_1, V_2, \ldots$ such that

$$
\Theta_{\Sigma_{2,n}}(\lambda) = \Theta_{2,n}(\lambda) V_n, \qquad n = 1, 2, \ldots,
$$

where

$$\theta_{2,n}(\lambda) = \overset{\curvearrowright}{\prod_{j=m+1}^{m+n}} \left(I - P_j + \frac{\overline{\lambda}_j - \lambda}{1 - \lambda_j \lambda} \frac{|\lambda_j|}{\overline{\lambda}_j} P_j \right).$$

So, by Lemma 10.4, we can find a sequence of operators $U_1, U_2, \ldots$ such that U_n is unitary for n sufficiently large and

$$\lim_{n \to \infty} \theta_{2,n}(\lambda) U_n = \theta_{\Sigma_2}(\lambda), \qquad \lambda \in \mathbb{D},$$

with uniform convergence on compact subsets of $\mathbb{D}$. Since $\Sigma = \Sigma_1 \Sigma_2$, we have $\theta_\Sigma = \theta_{\Sigma_1} \theta_{\Sigma_2}$, by Theorem 6.1, and the theorem is proved. $\square$

The following proposition shows that under certain additional conditions the factor θ_2 in (2) may be written as an infinite product.

PROPOSITION 10.5. *If the sequence $\lambda_1, \lambda_2, \ldots$ in Theorem 10.1 satisfies*

$$(25) \qquad \sum_j 1 - |\lambda_j| < \infty,$$

then the factor θ_2 in (2) may be rewritten as

$$(26) \qquad \theta_2(\lambda) = \left\{ \overset{\curvearrowright}{\prod_{j=1}^{\infty}} \left(I - P_j + \frac{\overline{\lambda}_j - \lambda}{1 - \lambda \lambda_j} \frac{|\lambda_j|}{\overline{\lambda}_j} P_j \right) \right\} U,$$

where $U \colon K \to L$ is a unitary operator and the infinite product in (26) converges uniformly in the operator norm on compact subsets of $\mathbb{D}$.

PROOF. Put

$$(27) \qquad b_j(\lambda) = \frac{\overline{\lambda}_j - \lambda}{1 - \lambda \lambda_j} \frac{|\lambda_j|}{\overline{\lambda}_j}, \qquad j = 1, 2, \ldots .$$

Since Σ is pure, the state operator A of Σ is completely non-unitary, and hence A has no eigenvalues on $\mathbb{T}$. Thus $|\lambda_j| < 1$ for each j, and (25) implies that $\prod_{j=1}^{\infty} |b_j(\lambda)|$ converges uniformly on compact subsets of $\mathbb{D}$. Note that

$$\prod_{j=p}^{q} \|I - P_j + b_j(\lambda) P_j\| \le \prod_{j=p}^{q} |b_j(\lambda)|, \qquad \lambda \in \mathbb{D}.$$

It follows that the infinite product in the right hand side of (26) has the right convergence property. To prove the equality (11), set

$$\theta_{2,n}(\lambda) = \overset{\curvearrowright}{\prod_{j=1}^{n}} \left(I - P_j + b_j(\lambda) P_j \right), \qquad n = 1, 2, \ldots,$$

and choose $0 \neq z \in \mathbf{D}$ such that $\bar{z}$ is in the resolvent set of A. Such a z exists, because $\sigma(A)$ is at most countable. By Proposition 2.7, the operator $\theta_\Sigma(z)$ is invertible, and hence, by Theorem 10.1,

$$(28) \qquad \lim_{n \to \infty} \theta_\Sigma(z)^{-1} \theta_1(z) \theta_{2n}(z) U_n = I.$$

Since $\bar{z} \neq \lambda_j$ for each j, we have $\prod_{j=1}^\infty b_j(z) \neq 0$ (see [R], Theorem 15.21), and thus $\prod_{j=1}^\infty |b_j(z)^{-1}|$ converges. The latter implies that

$$\overset{\curvearrowleft}{\underset{j=1}{\overset{\infty}{\prod}}} \left(I - P_j + b_j(z)^{-1} P_j \right)$$

converges, and hence the infinite product in (26) defines an invertible operator. So

$$V = \lim_{n \to \infty} \theta_\Sigma(z)^{-1} \theta_1(z) \theta_{2n}(z)$$

is invertible, and (28) shows that $U_n \to V^{-1}$ if $n \to \infty$. Thus (26) holds with $U = V^{-1}$. $\square$

If the system Σ in Theorem 10.1 has a finite dimensional output space (or, equivalently, a finite dimensional input space), then (25) is fulfilled. Indeed, if $\dim L < \infty$, then $I - A^* A$ has finite rank, and hence D_A is a Hilbert-Schmidt operator. But then we can use Theorem 9.2 to show that $\prod_{j=1}^\infty |\lambda_j|^2$ converges. Since the square root is a continuous function, it follows that $\prod_{j=1}^\infty |\lambda_j|$ converges. The latter is equivalent to (10), because $|\lambda_j| < 1$ for each j.

Condition (25) is also fulfilled if $V = I - A$ is a trace class operator. Indeed, in this case

$$\left| 1 - |\lambda_j| \right| \leq |1 - \lambda_j| \leq |\lambda_j(V)|, \qquad j = 1, 2, \ldots,$$

and the sequence $\sum_{j=1}^\infty \lambda_j(V)$ converges absolutely.

Next, we turn to Theorem 10.2. Before we prove this theorem let us remark that the existence of a chain $\mathbf{P}$ of the type appearing in Theorem 10.2 is guaranteed by Theorem 9.3. Indeed, in this case, since $I - A$ is Volterra, the space H_1 in Theorem 9.3 consists of the zero vector only, and thus $A = A_{22}$.

PROOF OF THEOREM 10.2. We split the proof into four parts.

Part (a). In this part we show that

$$(29) \qquad \theta_\Sigma(\lambda) \theta_\Sigma(-1)^* = I + \frac{\lambda+1}{\lambda-1} F \left(I - \frac{\lambda+1}{\lambda-1} G \right)^{-1} F^*, \qquad \lambda \in \mathbf{D},$$

where

$$(30) \qquad G = (I - A)(I + A)^{-1}, \qquad F = \sqrt{2} C (I + A)^{-1}.$$

The fact that $V := I - A$ is Volterra implies that $I + A = 2I + V$ is invertible. So the operators G and F in (30) are well-defined. From $-1 \in \rho(A)$, it also follows that

$$\theta_\Sigma(-1) = \lim_{\lambda \to -1, \lambda \in \mathbb{D}} \theta_\Sigma(\lambda)$$

is well-defined. The identities (9a), (9b) in Section XXVIII.2 imply that $\theta_\Sigma(-1)$ is unitary. Now, write $\lambda = (i + z)(i - z)$ with $\Im z < 0$. Then

$$\theta_\Sigma\left(\frac{i+z}{i-z}\right) = D + \frac{i+z}{i-z}C\left(I - \frac{i+z}{i-z}A\right)^{-1}B$$
$$= D + (i+z)C(i - z - iA - zA)^{-1}B$$
$$= D + (i+z)C\big(i(I - A) - z(I + A)\big)^{-1}B$$
$$= D + (i+z)C(I + A)^{-1}(iG - z)^{-1}B$$

where G is given by the first identity in (30). Next, use that

$$(31) \qquad I + G = I + (I - A)(I + A)^{-1} = 2(I + A)^{-1},$$

and replace $(i + z)B$ by $\{i(I + G) - (iG - z)\}B$. We obtain that

$$\theta_\Sigma\left(\frac{i+z}{i-z}\right) = \theta_\Sigma(-1) - 2iC(I + A)^{-1}(z - iG)^{-1}(I + A)^{-1}B.$$

Let us compute $(I + A)^{-1}B\theta_\Sigma(-1)^*$. To do this we use the first identities in formulas (4a), (4b) of Section XXVIII.2. We have

$$(I + A)^{-1}B\theta_\Sigma(-1)^* = (I + A)^{-1}B\big[D^* - B^*(I + A^*)^{-1}C^*\big]$$
$$= (I + A)^{-1}BD^* - (I + A)^{-1}BB^*(I + A^*)^{-1}C^*$$
$$= -(I + A)^{-1}AC^* - (I + A)^{-1}(I - AA^*)(I + A^*)^{-1}C^*$$
$$= -(I + A)^{-1}\{A + (I - AA^*)(I + A^*)^{-1}\}C^*$$
$$= -(I + A)^{-1}\{A + (I + A^*) - A(I + A^* - I)(I + A^*)^{-1}\}C^*$$
$$= -(I + A)^{-1}\{(I + A^*)^{-1} + A(I - A^*)^{-1}\}C^*$$
$$= -(I + A^*)^{-1}C^*.$$

Let F be defined by the second identity in (30). Then

$$\theta_\Sigma\left(\frac{i+z}{i-z}\right)\theta_\Sigma(-1)^* = I + iF(z - iG)^{-1}F^*.$$

Recall that $\lambda = (i + z)(i - z)^{-1}$, and hence $z = i(\lambda - 1)(\lambda + 1)^{-1}$, which yields (29).

Part (b). In this part we show that

$$(32) \qquad F^*F = G + G^*,$$

where F and G are as in (30). Indeed,

$$\begin{aligned}
F^*F &= 2(I + A^*)^{-1}C^*C(I + A)^{-1} \\
&= 2(I + A^*)^{-1}(I - A^*A)(I + A)^{-1} \\
&= 2(I + A^*)^{-1}(I + A)^{-1} - 2(I + A^*)^{-1}A^*A(I + A)^{-1}.
\end{aligned}$$

Now, use that $A(I + A)^{-1} = I - (I + A)^{-1}$, and we get

$$\begin{aligned}
F^*F &= 2(I + A)^{-1} + 2(I + A^*)^{-1} - 2I \\
&= I + G + I + G^* - 2I = G + G^*,
\end{aligned}$$

by virtue of (31).

Part (c). Let $\mathbb{P}$ be a maximal continuous chain which is invariant under A. Then $\mathbb{P}$ is also invariant under $V = I - A$, and the diagonal of V with respect $\mathbb{P}$ exists and is equal to the zero operator (cf., Corollaries XX.4.5 and XX.4.6). Thus $V \in A_+(\mathbb{P})$ (see Section XX.5 for the notation). It follows that also

$$G = (I - A)(I + A)^{-1} = V(2I + V)^{-1} \in A_+(\mathbb{P}).$$

Note that G is Volterra. So, by Theorem XXI.1.5 applied to iG, we see from (32) that

$$(33) \qquad G = \int_{\mathbb{P}} PF^*F\,dP.$$

In this part we use (29) and (33) to prove (4).

Put

$$(34) \qquad \alpha = \frac{\lambda + 1}{\lambda - 1}G, \qquad \beta = F^*, \qquad \gamma = \frac{\lambda + 1}{\lambda - 2}F.$$

It suffices to show that

$$(35) \qquad I + \gamma(I - \alpha)^{-1}\beta = \int_{\mathbb{P}}^{\frown} (I + \gamma(dP)\beta).$$

From (33) we know that

$$\alpha = \int_{\mathbb{P}} P\beta\gamma\,dP.$$

Let $\pi = \{P_0, P_1, \ldots, P_n\}$ be a partition of $\mathbb{P}$. Note that the operator $S_\pi = \sum_{j=1}^n P_{j-1}\beta\gamma\Delta P_j$ is strictly upper triangular relative to the decomposition

$$H = \operatorname{Im}\Delta P_1 \boxplus \operatorname{Im}\Delta P_2 \boxplus \cdots \boxplus \operatorname{Im}\Delta P_n.$$

It follows that

$$I + \gamma\left(I - \sum_{j=1}^{n} P_{j-1}\beta\gamma\Delta P_j\right)\beta = I + \gamma\beta + \gamma S_\pi\beta + \cdots + \gamma S_\pi^{n-1}\beta$$

$$\begin{aligned}
(36) \qquad &= I + \sum_{j=1}^{m} \gamma(\Delta P_j)\beta + \sum_{1\leq i<j\leq m} \gamma(\Delta P_i)\beta\gamma(\Delta P_j)\beta \\
&\quad + \cdots + \gamma(\Delta P_1)\beta\gamma(\Delta P_2)\beta\cdots\gamma(\Delta P_n)\beta \\
&= \prod_{j=1}^{\overset{\frown}{n}} (I + \gamma(\Delta P_j)\beta).
\end{aligned}$$

By taking the limit in (36), we obtain

$$I + \gamma\left(I - \int_{\mathbb{P}} P\beta\gamma dP\right)^{-1}\beta = \int_{\mathbb{P}}^{\overset{\frown}{}} (I + \gamma(dP)\beta),$$

which is (35).

Part (d). In this part we prove (6). Let $\pi = \{P_0, P_1, \ldots, P_n\}$ be a partition of $\mathbb{P}$. Put

$$S_j = I + \gamma(\Delta P_j)\beta, \qquad j = 1,\ldots,n, \ S_0 = I,$$

$$E_j = e^{\gamma(\Delta P_j)\beta}, \qquad j = 1,\ldots,n, \ E_{n+1} = I,$$

where β and γ are as in (34). By induction one shows that

$$(37) \qquad \prod_{j=1}^{\overset{\frown}{n}} E_j - \prod_{j=1}^{\overset{\frown}{n}} S_j = \sum_{i=1}^{n} S_0 S_1 \cdots S_{i-1}(E_i - S_i)E_{i+1}\cdots E_n E_{n+1}.$$

Put

$$m(\pi) = \sum_{j=1}^{n} \|F(\Delta P_j)F^*\|, \qquad \alpha(\pi) = \sup_j \|F(\Delta P_j)F^*\|, \qquad c(\lambda) = \left|\frac{1+\lambda}{1-\lambda}\right|.$$

Then

$$(38a) \qquad \left\|\prod_{\nu=0}^{\overset{\frown}{i-1}} S_\nu\right\| \leq \prod_{\nu=0}^{i-1}(1 + \|\gamma(\Delta P_\nu)\beta\|) \leq \prod_{\nu=0}^{i-1} e^{\|\gamma(\Delta P_\nu)\beta\|} \leq e^{c(\lambda)m(\pi)},$$

$$(38b) \qquad \left\|\prod_{\nu=i+1}^{\overset{\frown}{n}} E_\nu\right\| \leq \prod_{\nu=i+1}^{n} e^{\|\gamma(\Delta P_\nu)\beta\|} \leq e^{c(\lambda)m(\pi)},$$

$$\|E_i - S_i\| = \| \sum_{\nu=2}^{\infty} \frac{1}{\nu!} (\gamma(\Delta P_i)\beta)^{\nu} \|$$

(38c)
$$\leq \|\gamma(\Delta P_i)\beta\|^2 \sum_{\nu=0}^{\infty} \frac{1}{(\nu+2)!} \|\gamma(\Delta P_i)\beta\|^{\nu}$$

$$\leq c(\lambda)^2 \alpha(\pi)^2 e^{\|\gamma(\Delta P_i)\beta\|}$$

$$\leq c(\lambda)^2 \alpha(\pi)^2 e^{c(\lambda)m(\pi)}.$$

From (37) and (38a)–(38c) it follows that

(39)
$$\| \prod_{j=1}^{\overset{\curvearrowright}{n}} E_j - \prod_{j=1}^{\overset{\curvearrowright}{n}} S_j \| \leq c(\lambda)^2 \alpha(\pi)^2 e^{3c(\lambda)m},$$

where m is the number defined by the left hand side of (5). Note that the map $P \mapsto FPF^*$ from $\mathbb{P}$ into $\mathcal{L}(H)$ is continuous, and recall that our chain $\mathbb{P}$ has no jumps. So, by Lemma XXII.3.2 (which also holds for non-separable Hilbert spaces), given $\varepsilon > 0$ there exists a partition π_0 such that for any partition $\pi = \{P_0, P_1, \ldots, P_n\}$ finer than π_0 we have

$$\alpha(\pi) < \sqrt{\varepsilon} \left| \frac{\lambda-1}{\lambda+1} \right| \exp\left(-\frac{3}{2} c(\lambda)m\right).$$

So we see from (39) that

$$\| \prod_{j=1}^{\overset{\curvearrowright}{n}} E_j - \prod_{j=1}^{\overset{\curvearrowright}{n}} S_j \| < \varepsilon, \qquad \pi \supset \pi_0.$$

Since the integral in (4) converges, we conclude that the integral in (6) converges, and both integrals have the same value. $\square$

PROOF OF THEOREM 10.3. Since Σ is pure, the state operator A of Σ is completely non-unitary. It follows that A has no eigenvalues on $\mathbb{T}$ (by virtue of Proposition 1.4). Let H_1 be the smallest closed linear manifold in H containing all eigenvectors and generalized eigenvectors of A. Put $H_2 = H_1^{\perp}$. Since H_1 is A-invariant, we may partition A as

$$A = \begin{bmatrix} A_1 & * \\ 0 & A_2 \end{bmatrix} : H_1 \boxplus H_2 \to H_1 \boxplus H_2.$$

Let $\Sigma = \Sigma_1 \Sigma_2$ be a factorization supported by H_1. Without loss of generality we may assume that the input space and the output space of Σ_1 are equal to L. By Theorem 6.2, both systems Σ_1 and Σ_2 are pure unitary systems. The state operator of Σ_1 is A_1, and thus Σ_1 satisfies the conditions of Theorem 10.1. It follows that

$$\theta_{\Sigma_1}(\lambda) = \theta_1(\lambda)\theta_2(\lambda), \qquad \lambda \in \mathbb{D},$$

where θ_1 and θ_2 are given by (8a) and (8b), respectively. Note that A_2 is the state operator of Σ_2. From Theorem 9.3 one may deduce (see also the remark after the proof of Theorem 9.3) that Theorem 10.2 applies to Σ_2. Note that the existence of the chain $\mathbb{P}$ is guaranteed by Theorem 9.3. It follows that $\theta_{\Sigma_2} = \theta_3$, where θ_3 is as in (8c). The compactness of the operator F in (9) follows from the formula for F in Theorem 10.2 and Part (a) of the proof of Lemma 10.4. By Theorem 6.1, we have $\theta_{\Sigma} = \theta_{\Sigma_1}\theta_{\Sigma_2}$, and thus the theorem is proved. $\square$

XXVIII.11 UNICELLULARITY

Let $\widetilde{A}\colon L_2([0,1]) \to L_2([0,1])$ be the operator defined by

$$(1) \qquad (\widetilde{A}f)(t) = f(t) - 4\int_t^1 e^{2(t-s)} f(s)\,ds, \qquad 0 \le t \le 1.$$

In this section we use the theory developed in the preceding sections to show that the subspaces

$$(2) \qquad M_\tau := \{ f \in L_2([0,1]) \mid f(s) = 0,\ (\tau < s \le 1) \}, \qquad 0 \le \tau \le 1,$$

are the only invariant subspaces of $\widetilde{A}$. In particular, using the terminology introduced in Section XXI.4, we shall prove that $\widetilde{A}$ is unicellular.

In Section XXVIII.4 (after the proof of Corollary 4.3) we have seen that $\widetilde{A} = I - V$, where V is the integral operator

$$(Vf)(t) = 4\int_t^1 e^{2(t-s)} f(s)\,ds, \qquad 0 \le t \le 1.$$

The latter operator is a Volterra operator and 0 is not eigenvalue of V. It follows that $\widetilde{A} = I - V$ has no eigenvalues, and so Proposition 1.4 shows that $\widetilde{A}$ is completely non-unitary. Formula (17) in Section XXVIII.4 shows that the defect operator of $\widetilde{A}$ has rank one.

THEOREM 11.1. *Let $A \in \mathcal{L}(H)$ be a completely non-unitary contraction such that $I - A$ is Volterra, and assume that the defect operator of A has rank one. Then the characteristic operator function of A coincides with a scalar function of the form*

$$(3) \qquad \exp\!\left\{ c\!\left(\frac{\lambda+1}{\lambda-1} \right) \right\},$$

where $c > 0$ is a constant uniquely determined by A.

Note that Theorem 11.1 agrees with the computation of the characteristic operator function for the operator $\widetilde{A}$ given in Section XXVIII.4. In fact, in Section XXVIII.4 we have shown that the characteristic operator function of $\widetilde{A}$ coincides with the scalar function

$$(4) \qquad \exp\!\left\{ 2\!\left(\frac{\lambda+1}{\lambda-1} \right) \right\}.$$

In the following we use Theorem 10.2 to derive (3).

PROOF OF THEOREM 11.1. Since $AD_A = D_{A^*}A$ and A is invertible, D_{A^*} also has rank one. Recall that

$$(5) \qquad W_A(\lambda) = -A^* + \lambda D_A (I - \lambda A)^{-1} D_{A^*}\colon \overline{\operatorname{Im} D_{A^*}} \to \overline{\operatorname{Im} D_A}.$$

According to the definition of the term "coincide" we may multiply W_A on the left and on the right by unitary operators. It follows that W_A may be replaced by the transfer function of a unitary system Σ which has $\mathbb{C}$ as input and as output space, i.e.,

$$(6) \qquad \Sigma = (A, B, C, D; H, \mathbb{C}, \mathbb{C}).$$

Since Σ is completely non-unitary, we know from Theorem 2.4 that Σ is pure. Let us use Theorem 10.2 to compute the transfer function of Σ.

Put

$$(7) \qquad F = \sqrt{2}\,C(I + A)^{-1} \colon H \to \mathbb{C},$$

and choose a maximal continuous chain $\mathbb{P}$ which is invariant under A. Let $e = F^*1$. Then $Fx = \langle x, e \rangle$ for each $x \in H$, and hence

$$FPF^*1 = FPe = \langle Pe, e \rangle.$$

The latter identities allow us to show that the map $P \mapsto FPF^*$ is of bounded variation. Indeed, let $\pi = \{P_0, P_1, \ldots, P_n\}$ be a partition of $\mathbb{P}$. Then $\|F(\Delta P_j)F^*\| = \langle (\Delta P_j)e, e \rangle$, and

$$\sum_{j=1}^{n} \|F(\Delta P_j)F^*\| = \sum_{j=1}^{n} \langle (\Delta P_j)e, e \rangle = \|e\|^2.$$

Thus, in this case, the number defined by the left hand side of formula (5) in Theorem 10.2 is equal to $\|e\|^2$, and hence $P \mapsto FPF^*$ is of bounded variation. So θ_Σ is given by formula (6) in Theorem 10.2. Since input and output spaces are equal to $\mathbb{C}$, the value of θ_Σ at -1 is a complex number d of modulus one. Furthermore, FPF^* is a scalar. Let us identify in the usual way operators on $\mathbb{C}$ with scalars. Then

$$\theta_\Sigma(\lambda) = d \int_{\mathbb{P}}^{\frown} \exp\left\{ \left(\frac{\lambda + 1}{\lambda - 1} \right) d(FPF^*) \right\}$$

$$= d \exp\left\{ \left(\frac{\lambda + 1}{\lambda - 1} \right) \int_{\mathbb{P}} d(FPF)^* \right\}$$

$$= d \exp\left\{ c \left(\frac{\lambda + 1}{\lambda - 1} \right) \right\},$$

where $c = FF^* = \|e\|^2$. Since $|d| = 1$, we conclude that W_A coincides with the function given by (3).

Note that $\|FF^*\| = \|F^*\|^2 = \|F\|^2 = \|F^*F\|$. Since $C^*C = I - A^*A$, formula (7) yields

$$(8) \qquad c = 2\|(I + A^*)^{-1}(I - A^*A)(I + A)^{-1}\|,$$

and hence c is uniquely determined by A. $\quad\square$

We shall refer to c in (3) as the *defect characteristic* of A.

COROLLARY 11.2. *For $\nu = 1,2$ let A_ν be a completely non-unitary contraction such that $I - A_\nu$ is a Volterra operator and such that the defect operator of A_ν has rank one. Then A_1 and A_2 are unitarily equivalent if and only if they have the same defect characteristic.*

PROOF. By Theorem 11.1 the operators A_1 and A_2 have the same defect characteristic if and only if their characteristic operator functions coincide. Hence, the corollary follows by applying Theorem 4.1. □

THEOREM 11.3. *The invariant subspaces of the operator $\widetilde{A}$ defined in (1) are precisely the spaces M_τ, $0 \leq \tau \leq 1$, defined by (2).*

PROOF. Note that the operator $\widetilde{A}$ is precisely the operator A defined by formula (16) in Section XXVIII.4. This allows us to use the computations made in Section XXVIII.4 following after the proof of Corollary 4.3. We split the proof into three parts.

Part (a). It is convenient to embed $\widetilde{A}$ into a unitary system. Introduce the following operators:

$$(9a) \qquad \widetilde{B}\colon \mathbb{C} \to L_2([0,1]), \qquad \widetilde{B}1 = D_{\widetilde{A}_*}g_*,$$

$$(9b) \qquad \widetilde{C}\colon L_2([0,1]) \to \mathbb{C}, \qquad \widetilde{C}f = \langle D_{\widetilde{A}}f, -g\rangle,$$

$$(9c) \qquad \widetilde{D}\colon \mathbb{C} \to \mathbb{C}, \qquad \widetilde{D}1 = d = e^{-2}.$$

Here g and g_* are the functions defined by formulas (17) and (18) in Section XXVIII.4, respectively. From the computations made in Section XXVIII.4 we know that $\Sigma = (\widetilde{A}, \widetilde{B}, \widetilde{C}, \widetilde{D})$ is a unitary system and its transfer function is given by

$$(10) \qquad \theta_\Sigma(\lambda) = \exp\left\{2\left(\frac{\lambda+1}{\lambda-1}\right)\right\}.$$

Thus the defect characteristic of $\widetilde{A}$ is 2.

Part (b). Let M be an invariant subspace of A, and let $\Sigma = \Sigma_1\Sigma_2$ be a factorization of Σ supported by M. We assume that $M \neq \{0\}$. In this part we compute θ_{Σ_1} and θ_{Σ_2}. Put $H_1 = M$, and let

$$A_1 = \widetilde{A}|H_1\colon H_1 \to H_1, \qquad C_1 = \widetilde{C}|H_1\colon H_1 \to \mathbb{C}.$$

The operators A_1 and C_1 are, respectively, the state operator and the output operator of Σ_1. Since $\widetilde{A}$ is completely non-unitary, Σ is pure, and hence, by Theorem 6.2, the same holds for Σ_1 and Σ_2. In particular, A_1 is completely non-unitary. Furthermore, $A_1 - I_1$ is Volterra, where I_1 is the identity operator on H_1. Let B_1 be the input operator of Σ_1. Since (see Part (a) of the proof of Theorem 10.2)

$$(11) \qquad (I_1 + A_1)^{-1}B_1 = -(I_1 + A_1^*)^{-1}C_1^*, \qquad \cdot$$

$C_1 = 0$ implies that $B_1 = 0$, and hence A_1 is unitary (by virtue of the two identities in (4b) of Section XXVIII.2). But A_1 is completely non-unitary, and hence $C_1 \neq 0$. It follows that the defect operator of A_1 has rank one. To compute the defect characteristic $c(A_1)$ of A_1, let $\rho\colon H_1 \to L_2([0,1])$ be the canonical embedding. Then

$$
\begin{aligned}
c(A_1) &= 2\|(I + A_1^*)^{-1} C_1^* C_1 (I + A_1)^{-1}\| \\
&= 2\|\rho^*(I + \widetilde{A})^{-1} \widetilde{C}^* \widetilde{C}(I + \widetilde{A})^{-1}\rho\| \\
&\leq 2\|(I + \widetilde{A})^{-1} \widetilde{C}^* \widetilde{C}(I + \widetilde{A})\| = 2.
\end{aligned}
$$

So there exists $0 < \tau_M \leq 1$ so that $c(A_1) = 2\tau_M$. Recall that Σ_1 is uniquely determined up to a unitary factor on the left. So without loss of generality we may assume that the input space of Σ_1 is $\mathbb{C}$ and $\theta_{\Sigma_1}(-1) = 1$. It follows that

$$(12a) \qquad\qquad \theta_{\Sigma_1}(\lambda) = \exp\left\{ 2\tau_M\left(\frac{\lambda+1}{\lambda-1}\right)\right\},$$

$$(12b) \qquad\qquad \theta_{\Sigma_2}(\lambda) = \exp\left\{ 2(1 - \tau_M)\left(\frac{\lambda+1}{\lambda-1}\right)\right\}.$$

Note that the formula for θ_{Σ_2} follows from formulas (10) and (12a) and the fact (use Theorem 6.1) that

$$(13) \qquad\qquad \theta_{\Sigma}(\lambda) = \theta_{\Sigma_1}(\lambda)\theta_{\Sigma_2}(\lambda), \qquad \lambda \in \mathbf{D}.$$

Part (c). Fix $0 < \tau \leq 1$, and let M_τ be the subspace defined by (2). Let

$$A_{1,\tau} = \widetilde{A}|M_\tau\colon M_\tau \to M_\tau, \qquad C_{1,\tau} = C|M_\tau\colon M_\tau \to \mathbb{C}.$$

From the results proved under Part (b) we know that $A_{1,\tau}$ is completely non-unitary, $A_{1,\tau} - I_{1,\tau}$ is Volterra, where $I_{1,\tau}$ is the identity operator on M_τ, and the defect operator of $A_{1,\tau}$ has rank one. In this part we prove that the defect characteristic $c(A_{1,\tau})$ of $A_{1,\tau}$ is equal to 2τ.

Let $\rho_\tau\colon M_\tau \to L_2([0,1])$ be the canonical embedding. Put

$$F_\tau = \sqrt{2}C_{1,\tau}(I + A_{1,\tau})^{-1} = \sqrt{2}\widetilde{C}(I + \widetilde{A})^{-1}\rho_\tau.$$

Let $e_\tau = F_\tau^*1$. Then $c(A_{1,\tau}) = \|e_\tau\|^2$. From Part (a) of the proof of Theorem 10.2 we know that

$$(I + \widetilde{A})^{-1}\widetilde{B} = -(I + \widetilde{A}^*)^{-1}\widetilde{C}^*,$$

and thus

$$
\begin{aligned}
e_\tau &= \sqrt{2}\rho_\tau^*(I + \widetilde{A}^*)^{-1}\widetilde{C}^* \\
&= -\sqrt{2}\rho_\tau^*(I + \widetilde{A})^{-1}\widetilde{B}1 \\
&= -\sqrt{2}\rho_\tau^*(I + \widetilde{A})^{-1}u_*,
\end{aligned}
$$

where $u_*(t) = 2e^{2(t-1)}$ (see Section XXVIII.4). By using formulas (20), (21a) and (21b) in Section XXVIII.4 one computes that $((I + \widetilde{A})^{-1}u_*)(t) = 1$ for $0 \le t \le 1$. Now, ρ_τ^* is the orthogonal projection of $L_2([0,1])$ onto M_τ, and thus

$$e_\tau(t) = \begin{cases} -\sqrt{2}, & 0 \le t \le \tau, \\ 0 & , & \tau < t \le 1. \end{cases}$$

It follows that $c(A_{1,\tau}) = \|e_\tau\|^2 = 2\tau$.

Part (d). In this part we prove the theorem. Let $M \ne \{0\}$ be an $\widetilde{A}$-invariant subspace. By the result of Part (b) we know that the defect characteristic of $\widetilde{A}|M$ is equal to 2τ for some $0 < \tau \le 1$. We claim that $M = M_\tau$. To see this we apply Theorem 8.1. Since Σ is pure, factorizations of Σ yield regular factorizations of θ_Σ. In particular, the factorization (13) is regular. By applying the result of Part (b) to M_τ, we see that Σ has a factorization $\Sigma = \Sigma_1' \Sigma_2'$ supported by M_τ so that $\theta_{\Sigma_1'} = \theta_{\Sigma_1}$ and $\theta_{\Sigma_2'} = \theta_{\Sigma_2}$. But then the uniqueness statement in Theorem 8.1 implies that $M_\tau = M$. $\square$

COROLLARY 11.4. *Let A be a completely non-unitary contraction such that $I - A$ is a Volterra operator, and assume that A has a defect operator of rank one. Then A is unicellular.*

PROOF. By replacing A by

$$A_t = (t + A)(I + tA)^{-1}, \qquad -1 < t < 1,$$

one obtains a completely non-unitary contraction of the same type as A. Indeed,

$$I - A_t^* A_t = (1 - t^2)(I + tA^*)^{-1}(I - A^*A)(I + tA)^{-1},$$

$$I - A_t = (1 - t)(I + tA)^{-1}(I - A),$$

and thus A_t has the same properties as A. Moreover, the defect characteristic $c(A_t)$ of A_t is related to the defect characteristic $c(A)$ of A in the following way:

$$c(A_t) = \frac{1-t}{1+t}c(A), \qquad -1 < t < 1.$$

Thus for a suitable choice of $t \in (-1,1)$ we have $c(A_t) = 2$. But for this t the operator A_t is unitarily equivalent to $\widetilde{A}$ in (1), and hence, by Theorem 11.3, the operator A_t is unicellular. Since A_t and A have the same invariant subspaces, the corollary is proved. $\square$

XXVIII.12 DISSIPATIVE OPERATORS AND THEIR CHARACTERISTIC OPERATOR FUNCTIONS

The theory of unitary systems and the corresponding characteristic operator functions has an analogue in which the contraction operators are replaced by i-dissipative operators and the role of the open unit disc $\mathbb{D}$ is taken over by the open lower half plane $\mathbb{C}_-$. In this version the corresponding systems are continuous time symmetric systems on the real line rather than discrete time unitary systems as in the previous sections. In this section we review the main results for the continuous time symmetric systems by relating them, via an appropriate Möbius transformation, to the analogous results for unitary systems.

Consider the following continuous time input-output system:

$$(1) \qquad \begin{cases} x'(t) = Tx(t) + Qu(t), & t \in \mathbb{R}, \\ y(t) = Rx(t) + Fu(t). \end{cases}$$

Here $T: H \to H$, $R: K \to H$, $Q: H \to L$ and $F: K \to L$ are bounded linear operators acting between Hilbert spaces. We shall employ the same terminology as introduced in Section XXVIII.2. So, for example, the operators T, R and Q are called the *state* (or *main*) *operator*, the *input operator* and the *output operator*, respectively, and we refer to H as the *state space* of the system. Assuming the system to be at rest at time $t = 0$ and applying the Laplace transform,

$$\widehat{f}(\lambda) = \int_0^\infty e^{-\lambda t} f(t)\, dt,$$

yields the so-called frequency domain version of (1), namely:

$$(2) \qquad \begin{cases} \lambda \widehat{x}(\lambda) = T\widehat{x}(\lambda) + Q\widehat{u}(\lambda), \\ \widehat{y}(\lambda) = R\widehat{x}(\lambda) + F\widehat{u}(\lambda), \end{cases}$$

which is well-defined for $\Im\lambda$ sufficiently large. By solving $\widehat{x}(\lambda)$ for the first equation in (2) and inserting the solution into the second equation one sees that in the frequency domain the connection between input and output is given by $\widehat{y}(\lambda) = \Xi(\lambda)\widehat{u}(\lambda)$ with

$$(3) \qquad \Xi(\lambda) = F + R(\lambda - T)^{-1}Q.$$

The operator function in (3), which is defined on the resolvent set of T, is called the *transfer function* of the system (1).

The system (1) is called *symmetric* if the feedthrough operator F is the identity operator on K (and hence $L = K$) and the state operator, the input operator and the output operator are related in the following way

$$(4) \qquad T - T^* = QR, \qquad R = 2iQ^*.$$

In this case we write the system (1) in a concise form as follows:

$$(5) \qquad \Delta = (T, Q, R; H, K).$$

Its transfer function will be denoted by Ξ_Δ.

THEOREM 12.1. *The transfer function* $\Xi_\Delta(\cdot)$ *of the symmetric system* (5) *is an operator-valued function which is analytic on the open lower half plane* $\mathbb{C}_-$ *and at infinity, its values are bounded linear operators acting on* K, *and* $\Xi_\Delta(\infty) = I$. *Furthermore,*

$$(6) \qquad \sup_{\lambda \in \mathbb{C}_-} \|\Xi_\Delta(\lambda)\| \leq 1.$$

We shall prove Theorem 12.1 later. First, we introduce some further terminology. A symmetric system $\Delta = (T, Q, R; H, K)$ is said to be *pure* if

$$(7) \qquad \mathrm{Ker}(R|T) = \bigcap_{j=0}^{\infty} \mathrm{Ker}\, RT^j = \{0\}.$$

Since $T - T^* = 2iQQ^*$ and $R = 2iQ^*$, one may prove by induction on n that $RT^j x = 0$ for $j = 0, \ldots, n$ if and only if $Q^*(T^*)^j x = 0$ for $j = 0, \ldots, n$. It follows that (7) holds if and only if

$$(8) \qquad \mathrm{Im}(T|Q) = \bigvee_{j=0}^{\infty} \mathrm{Im}\, T^j Q = H,$$

and hence the symmetric system Δ is pure if and only if Δ is approximately observable and approximately controllable.

The space $\mathrm{Ker}(R|T)$ in (7) is invariant under both T and T^*, and on $\mathrm{Ker}(R|T)$ the operators T and T^* coincide. Thus, if the main operator T of a symmetric system Δ is completely non-selfadjoint (see Section XXI.2 for the definition), then Δ must be pure. The converse statement is also true, for if H_0 is a subspace, invariant under both T and T^*, on which T and T^* coincide, then the identities in (4) imply that $H_0 \subset \mathrm{Ker}\, RT^j$ for each $j \geq 0$, and in this case (7) yields $H_0 = \{0\}$. So we conclude that a symmetric system is pure if and only if its main operator is completely non-selfadjoint.

Two symmetric systems Δ_1 and Δ_2,

$$(9) \qquad \Delta_\nu = (T_\nu, Q_\nu, R_\nu; H_\nu, K_\nu), \qquad \nu = 1, 2,$$

are said to be *unitarily equivalent* if $K_1 = K_2$ and there exists a unitary operator $J: H_1 \to H_2$ such that

$$T_2 = JT_1 J^{-1}, \qquad Q_2 = JQ_1, \qquad R_2 = R_1 J^{-1}.$$

THEOREM 12.2. *Two pure symmetric systems have the same transfer function if and only if these systems are unitarily equivalent, and in this case the unitary operator establishing the unitary equivalence is unique.*

Let Δ_1 and Δ_2 be as in (9). The *cascade connection* or *product* of Δ_1 and Δ_2 is the system $\Delta_1\Delta_2$ which one obtains by taking the output of Δ_2 as the input for Δ_1. This implies $K_1 = K_2$, and as for the unitary systems one computes that

$$(10) \qquad \Delta_1\Delta_2 = \left(\begin{bmatrix} T_1 & Q_1 R_2 \\ 0 & T_2 \end{bmatrix}, \begin{bmatrix} Q_1 \\ Q_2 \end{bmatrix}, [R_1 \quad R_2]; H_1 \boxplus H_2, K \right).$$

THEOREM 12.3. *If Δ_1 and Δ_2 are symmetric systems, then the cascade connection $\Delta_1\Delta_2$ is a symmetric system and*

$$\Xi_{\Delta_1\Delta_2}(\lambda) = \Xi_{\Delta_1}(\lambda)\Xi_{\Delta_2}(\lambda), \qquad \lambda \in \mathbb{C}_-.$$

Let $\Delta = (T, Q, R; H, K)$ be a symmetric system, and let M be an invariant subspace of T. Consider the following partitionings:

$$T = \begin{bmatrix} T_{11} & T_{12} \\ 0 & T_{22} \end{bmatrix} : M \boxplus M^\perp \to M \boxplus M^\perp,$$

$$Q = \begin{bmatrix} Q_1 \\ Q_2 \end{bmatrix} : K \to M \boxplus M^\perp,$$

$$R = [R_1 \quad R_2] : M_1 \boxplus M_2 \to K.$$

The systems

$$\Delta_1 = (T_{11}, Q_1, R_1; M, K),$$
$$\Delta_2 = (T_{22}, Q_2, R_2; M^\perp, K),$$

are called, respectively, the *left projection* and the *right projection* of Δ associated with the invariant subspace M.

THEOREM 12.4. *Let Δ be a symmetric system, and let M be an invariant subspace for the state operator of Δ. Then the left projection Δ_1 and the right projection Δ_2 of Δ associated with M are symmetric systems, and $\Delta = \Delta_1\Delta_2$.*

Theorems 12.1–12.4 are proved in much the same way as the analogous results for unitary systems. They may also be derived from the latter results by applying an appropriate Möbius transformation. To do this, we need to make a few preparations.

Let T be the main operator of a symmetric system. From (4) we see that the imaginary part of T is a non-negative operator; in other words, in the terminology of Section XXI.2, the operator T is i-dissipative. But then we can apply Lemma X.3.2 to show that the spectrum of T lies in the closed upper half plane, and hence $\mathbb{C}_- \subset \rho(T)$, where $\rho(T)$ denotes the resolvent set of T. In particular, $iI + T$ is invertible. In the following lemma we relate T to the operator

$$(11) \qquad A = (iI - T)(iI + T)^{-1}.$$

LEMMA 12.5. *Let $T \in \mathcal{L}(H)$, and assume that $-i \in \rho(T)$. Define A by (11). Then $-1 \in \rho(A)$, and the operator A is a (completely non-unitary) contraction if and only if T is a (completely non-selfadjoint) i-dissipative operator.*

PROOF. Note that $A = M(T)$, where M is the Möbius transformation

$$(12) \qquad M(\lambda) = \frac{i-\lambda}{i+\lambda}, \qquad M^{-1}(z) = i\frac{1-z}{1+z},$$

which transforms the open upper half plane into $\mathbb{D}$, the real line into $\mathbb{T}$, and $\lambda = \infty$ into $z = -1$. Thus, by the spectral mapping theorem, $-1 \in \rho(A)$.

To prove the other statements, let us compute $I - A^*A$ and $I - AA^*$. We have

$$I - A^*A = I - (-iI - T^*)(-iI + T^*)^{-1}(iI + T)^{-1}(iI - T)$$
$$= (-iI + T^*)^{-1}\{(-iI + T^*)(iI + T) - (-iI - T^*)(iI - T)\}(iI + T)^{-1}$$
$$= (-iI + T^*)^{-1}\{-2i(T - T^*)\}(iI + T)^{-1}.$$

It follows that

$$(13a) \qquad\qquad I - A^*A = 4(iI + T)^{-*}T_{\Im}(iI + T)^{-1}.$$

Here S^{-*} denotes the adjoint of S^{-1}, and $T_{\Im}$ is the imaginary part of T. Since $M(-T^*) = M(T)^*$ and $(-T^*)_{\Im} = T_{\Im}$, we also have

$$(13b) \qquad\qquad I - AA^* = 4(iI + T)^{-1}T_{\Im}(iI + T)^{-*}.$$

From (13a) it is clear that A is a contraction if and only if T is i-dissipative. Now, assume that T is i-dissipative and not completely non-selfadjoint. Then, according to the definition in Section XXI.2, there exists a subspace $H_0 \neq \{0\}$, invariant under T and T^*, such that T and T^* coincide on H_0. In particular, $T_{\Im}$ acts as the zero operator on H_0. Since $-i$ is in the unbounded connected component of $\rho(T)$, the space H_0 is invariant under $(iI + T)^{-1}$. In a similar way, one shows that $(iI + T)^{-*}$ leaves H_0 invariant. It follows that H_0 is invariant under both A and A^*, and from the identities (13a), (13b) we see that A acts as a unitary operator on H_0. So A is not completely non-unitary. The reverse implication is proved analogously, using

$$(14) \qquad\qquad T = i(I - A)(I + A)^{-1}. \quad \square$$

LEMMA 12.6. *Let $\Delta = (T, Q, R; H, K)$ be a symmetric system. Put*

$$(15) \qquad A = (iI - T)(iI + T)^{-1}, \qquad C = R(iI + T)^{-1},$$

and let Σ be the standard unitary system with state operator A and output operator C. Then $-1 \in \rho(A)$ and

$$(16) \qquad\qquad \Xi_\Delta(\lambda) = \theta_\Sigma\left(\frac{i + \lambda}{i - \lambda}\right)\theta_\Sigma(-1)^*.$$

Conversely, let $\Sigma = (A, B, C, D; H, K, L)$ be a unitary system such that $-1 \in \rho(A)$, and put

$$(17) \qquad\qquad T = i(I - A)(I + A)^{-1}, \qquad R = 2iC(I + A)^{-1}.$$

Then $\Delta = (T, \frac{1}{2}iR^*, R; H, L)$ is a symmetric system and (16) holds.

PROOF. We start with the second part of the lemma. Let $\Sigma = (A, B, C, D; H, K, L)$ be a unitary system such that $-1 \in \rho(A)$. Then

$$\theta_\Sigma(-1) = \lim_{\lambda \to -1, \lambda \in \mathbb{D}} \theta_\Sigma(\lambda)$$

is well-defined. Let T and R be as in (17), and put $Q = \frac{1}{2}iR^*$. Then $R = 2iQ^*$, and by using the computations made in Part (a) of the proof of Theorem 10.2 one sees that

$$\theta_\Sigma\left(\frac{i+\lambda}{i-\lambda}\right)\theta_\Sigma(-1)^* = I + R(\lambda - T)Q.$$

Next, we show that $T - T^* = QR$. The latter identity follows from (13a). Indeed, since T is given by the first identity in (17), we have $M(T) = A$, where M is as in (12), and thus, by (13a),

$$\begin{aligned}
\frac{1}{2i}(T - T^*) &= \frac{1}{4}(iI + T)^*(I - A^*A)(iI + T) \\
&= (I + A)^{-*}C^*C(I + A)^{-1} \\
&= \frac{1}{2i}(I + A)^{-*}C^*R = \frac{1}{2i}QR.
\end{aligned}$$

Here we use that $iI + T = 2i(I + A)^{-1}$ and the second identity in formula (4b) of Section XXVIII.2. We have now proved that the system $(T, \frac{1}{2}iR^*, R; H, L)$ is symmetric and that its transfer function satisfies (16).

Next, let $\Sigma_0 = (A, B_0, C, D_0)$ be the standard unitary system associated with A and C, where A and C are defined by (15). From the first identity in (15) it follows that $-1 \in \rho(A)$ and T is given by the first identity in (17). Furthermore,

$$(18) \qquad\qquad I + A = 2i(iI + T)^{-1},$$

and hence R satisfies the second identity in (17). But then we can apply the first part of the proof to show that (16) holds for $\Sigma = \Sigma_0$. $\square$

Let $\Delta = (T, Q, R; H, K)$ be a unitary system, and let

$$\Sigma_0 = (A, B_0, C, D_0; H, K_0, K)$$

be the standard unitary system associated with the operators A and C defined by (15). Put $F = \theta_\Sigma(-1)^*$, and set

$$(19) \qquad\qquad \Sigma_\Delta = (A, B_0F^*, C, D_0F^*; H, K, K).$$

Since $\theta_\Sigma(-1)^*$ is unitary, the system Σ_Δ is unitary, and, by Lemma 10.6,

$$\Xi_\Delta(\lambda) = \theta_{\Sigma_\Delta}\left(\frac{i+\lambda}{i-\lambda}\right).$$

The map $\Delta \mapsto \Sigma_\Delta$ is one-one, and it is straightforward to check that this map preserves cascade connections, that is,

$$(20) \qquad\qquad \Sigma_{\Delta_1\Delta_2} = \Sigma_{\Delta_1}\Sigma_{\Delta_2}.$$

Let us return now to Theorems 12.1–12.4. Since the Möbius transformation

$$\lambda \mapsto \frac{i+\lambda}{i-\lambda}$$

maps the open lower half plane $\mathbb{C}_-$ onto $\mathbb{D}$ and the real line $\mathbf{R}$ onto $\mathbf{T}$, Theorem 12.1 is an immediate corollary of Lemma 12.6 and Theorem 2.1. Furthermore, by applying Lemma 12.6 and Theorem 5.2, we see that (for the case when $H = \mathbb{C}^k$) any operator function of the type appearing in Theorem 12.1 may be realized as the transfer function of a symmetric system. In a similar way, by applying Lemmas 12.5 and 12.6, one may derive Theorem 12.2 from its analogue for unitary systems. Theorems 12.3 and 12.4 are best proved by direct checking; we omit the details. Also the notion of regular factorization and Theorem 8.1 carry over to the class of operator-valued functions considered here. For the latter purpose formula (20) and Lemmas 12.5 and 12.6 are useful tools.

The main operator of a symmetric system is i-dissipative. Conversely, any i-dissipative operator appears in this way. Indeed, let $T \in \mathcal{L}(H)$ be i-dissipative. Then $T_\Im \geq 0$, and thus the square root of this operator is well-defined. Put $M = \overline{\operatorname{Im} T_\Im}$, and let $\tau_M \colon M \to H$ be the canonical embedding. Then

$$\Delta_T = (T, T_\Im^{1/2} \tau_M, \tau_M^* T_\Im^{1/2}; M, H)$$

is a symmetric system. The transfer function of the system Δ_T in (21) is called the *characteristic operator function* of the operator T and will be denoted by $V_T(\cdot)$. Thus

$$V_T(\lambda) = I + 2i T_\Im^{1/2}(\lambda - T)^{-1} T_\Im^{1/2}|_{\overline{\operatorname{Im} T_\Im}} : \overline{\operatorname{Im} T_\Im} \to \overline{\operatorname{Im} T_\Im}.$$

Now, consider $A = (iI - T)(iI + T)^{-1}$, where $T \in \mathcal{L}(H)$ is i-dissipative. Then the characteristic operator functions $V_T(\cdot)$ and $W_A(\cdot)$ of T and A, respectively, are well-defined. Put

$$C = \tau_M^* T_\Im^{1/2}(iI + T)^{-1} \colon H \to M,$$

where $M = \overline{\operatorname{Im} T_\Im}$. Then $D_A = UC$ for some unitary operator $U \colon M \to \overline{\operatorname{Im} D_A}$, and it follows from Lemma 12.6 that

$$(21) \qquad\qquad U V_T(\lambda) = W_A\left(\frac{i+\lambda}{i-\lambda}\right) W_A(-1)^*.$$

Finally, let us note that for the operator $\widetilde{A}$ defined by formula (1) of the previous section the transformation (14) yields the operator $T \colon L_2([0,1]) \to L_2([0,1])$ defined by

$$(22) \qquad\qquad (Tf)(t) = 2i \int\limits_t^1 f(s)\,ds, \qquad 0 \leq t \leq 1.$$

It follows that the characteristic operator function of T in (21) coincides with the scalar function $\exp(2i\lambda^{-1})$. Furthermore, the results of the previous section show that all regular factors of $\exp(2i\lambda^{-1})$ are of the form $\exp(2\tau i\lambda^{-1})$ for some $0 \leq \tau \leq 1$. We also see that, by applying the transformation (14), the results of the previous section could be derived directly from the last section of Chapter XXI.

COMMENTS ON PART VII

Chapter XXVI contains standard material on shift and block shift operators. The Von Neumann-Wold decomposition in Section XXVI.1 has its origin in the theory of stochastic processes (see Wold [1]). Theorem XXVI.2.1 is due to Rota [1] for strict contractions and the general version which appears here is from De Branges-Rovnyak [1]. The main theorem in Section XXVI.3 is P. Lax's generalization (see Lax [1], [2]) of Beurling's theorem which concerns the scalar case (Beurling [1]). For the more general operator version due to P. Halmos (Halmos [1]) we refer to Section IX.2 of the book Foias-Frazho [1]. References to other generalizations of Beurling's theorem may be found in the notes to Chapter 14 in Ball-Gohberg-Rodman [1]. The dilation theory in Chapter XXVII has its origin in Sz-Nagy [7]. The matrix representation of the unitary dilation in Section XXVII.4 is due to J.J. Schäffer (see Schäffer [1]). The connections between dilations and semi-invariant subspaces has been pointed out in Sarason [1]. Sources for the material in Section XXVII.5 are Davis-Kahan-Weinberger [1] and Parrott [1]; see also Chapter IV of Foias-Frazho [1] for other versions. We shall return to the topic of matrix completions in Part IX. The Douglas factorization lemma (Douglas [1]) is a standard tool in the theory of contractions. The commutant lifting theorem with its applications to interpolation was started by D. Sarason (Sarason [2]) and developed further to its present form by B. Sz-Nagy and C. Foias (see Sz-Nagy-Foias [2]). For a detailed and far reaching analysis of the commutant lifting theorem and its applications to various kinds of interpolation and extension problems the reader is referred to Foias-Frazho [1]. The characteristic operator functions for operators close to selfadjoint in the version as it appears in Section XXVII.12, but without the restriction to dissipative operators, was introduced by M.S. Livšic (see Livšic [1], [2]). For a comprehensive account of the theory of this type of characteristic operator functions and its applications, see M.S. Brodskii [1]. The characteristic operator function for arbitrary operators, using a signature matrix, originates in the papers Livšic-Potapov [1] and Smuljan [1]. For contractions, in the form as used in Chapter XXVIII, the characteristic operator function was introduced in Sz-Nagy-Foias [1], [3]. Another approach to characteristic operator functions and the associate model theorem was developed by L. de Branges and J. Rovnyak (see de Branges-Rovnyak [1], [2], de Branges [1] and the more recent de Branges [2]). Important contributions to the theory of characteristic operator functions also appear in Brodskii-Gohberg-Kreĭn [1], [2] and in Ball [1], [2]. In the sixties the state space theory was developed for input/output systems by Kalman [1], Youla [1], Weiss [1] and others, which allowed one to view the transfer function appearing in systems theory as a finite dimensional version of the theory of characteristic operator functions. Among the first to observe and to exploit the latter connection were P. Dewilde, P. Fuhrmann and J.W. Helton (see Dewilde [1], Fuhrmann [1], Helton [1]). See also Bart-Gohberg-Kaashoek [1], where the implications of this connection for the factorization theory are employed and developed further. For a detailed and up-to-date review of de Branges-Rovnyak operator models and mathematical systems theory we refer to Ball-Cohen [1]. The preliminaries about completely non-unitary operators in Section XXVIII.1 are taken from Gohberg-Kreĭn [3]. M.S. Brodskii [2] has been a source for the material in Sections XXVIII.2–XXVIII.8. A novelty in our presentation is the emphasis on and the terminology from mathematical systems theory. Sections XXVIII.9 and XXVIII.10 are based on Gohberg-Kreĭn [3] and

the papers Brodskii-Gohberg-Kreĭn [1], [2]. The multiplicative representation theorems
in Section XXVIII.10 appear for the first time in the work V.P. Potapov (see Potapov
[1]), these theorems also yield a new view on the multiplicative representations appearing in complex analysis, even for the scalar case. The factorization approach in Section
XXVIII.11 is basically as in M.S. Brodskii [1]; the latter book has also been the source for
Section XXVIII.12. The theorems about multiplicative representations of transfer functions of unitary systems are presented here in a restricted form. More general versions
may be found in Brodskii-Gohberg-Kreĭn [1], [2], Ginzburg [1] and Ginzburg-Zemskov
[1]. The latter two papers also contain results about the uniqueness of the multiplicative
representation, a topic which is not discussed in this part. A full treatment of the multiplicative representation theory, including the uniqueness, lies outside the scope of the
present book. For a recent review of the existing results we refer to Ginzburg-Shevchuk
[1].

EXERCISES TO PART VII

In the following exercises H and K are Hilbert spaces and all operators are assumed to be bounded linear operators acting between Hilbert spaces.

1. Show that a part of a block backward shift is unitarily equivalent to a block backward shift.

2. Show that $T \in \mathcal{L}(H, K)$ is an isometry if and only if T^* is a left inverse of T.

3. Determine the rotation operator associated with an isometry. Do the same for a co-isometry.

4. Show that $A \in \mathcal{L}(H)$ is an isometry if and only if A is a part of a unitary operator.

5. Show that a semi-invariant subspace of a selfadjoint operator A is invariant under A. Does this statement remain true if "selfadjoint" is replaced by "unitary"? What happens if A acts on a finite dimensional space?

6. Let $A \in \mathcal{L}(H)$ be unitary. Assume that $\sigma(A) \neq \mathbb{T}$. Show that any semi-invariant subspace of A is invariant under A. Conversely, if any semi-invariant subspace of the unitary operator A is invariant under A, does it follow that $\sigma(A) \neq \mathbb{T}$?

7. Let $A \in \mathcal{L}(H)$. Show that a dilation of a dilation of A is also a dilation of A.

8. Show that T is a dilation of A if and only if T^* is a dilation of A^*. What does this result say in terms of semi-invariant subspaces?

9. Let $A \in \mathcal{L}(H)$ be a contraction, and let T be its minimal isometric dilation. Show that

$$\operatorname{Ker} T^* = \overline{\{(I - TA^*)h \mid h \in H\}}.$$

10. Let A be the operator

$$A = \begin{bmatrix} 0 & 0 & \cdots & 0 & 0 \\ 1 & 0 & \cdots & 0 & 0 \\ 0 & 1 & \cdots & 0 & 0 \\ \vdots & \vdots & & \vdots & \vdots \\ 0 & 0 & \cdots & 1 & 0 \end{bmatrix} : \mathbb{C}^n \to \mathbb{C}^n.$$

Determine the minimal unitary dilation of A.

11. Let A be a contraction, and let $p(\lambda)$ be a polynomial in λ. Use dilation theory to prove Von Neumann's inequality:

$$\|p(A)\| \leq \sup\{|p(\lambda)| \mid \lambda \in \mathbb{D}\}.$$

12. Assume that

$$A = \begin{bmatrix} A_{11} & A_{12} \\ A_{21} & A_{22} \end{bmatrix} : H_1 \boxplus H_2 \to K_1 \boxplus K_2$$

is a strict contraction. Show that A_{11}, A_{12} and A_{22} are zero operators if and only if

$$A(I - A^*A)^{-1} = \begin{bmatrix} 0 & 0 \\ * & 0 \end{bmatrix}.$$

(Hint: use Corollaries XXVII.5.3 and XXVII.5.4).

The following three exercises lead to a proof of the so-called strong Parrott theorem (see Foias-Tannenbaum [1]). Given a subspace M of a Hilbert space H the operator $\tau_M: M \to H$ is the canonical embedding operator and $\pi_M = \tau_M^*: H \to M$ is the orthogonal projection of H onto M.

13. Let $H_0 \subset H$ and $K_0 \subset K$ be subspaces (i.e., closed linear manifolds) of H and K, respectively, and let

(1) $$T_0: H_0 \to K, \qquad T_0': H \to K_0$$

be contractions. Find a necessary and sufficient condition in order that there exists a contraction $T: H \to K$ such that

(a) $T_0 = T\tau_{H_0}$,

(b) $T_0' = \pi_{K_0} T$.

(Hint: see the problem as the geometric version of the problem discussed in Section XXVII.5.)

14. Let H_0 and H_1 be subspaces of H such that $H = \overline{H_0 + H_1}$. Let $T_0: H_0 \to K$ be a contraction, and let $V: H_1 \to K$ be an isometry. Find a necessary and sufficient condition in order that there exists a contraction S such that

$$T_0 = S\tau_{H_0}, \qquad V = S\tau_{H_1}.$$

(Hint: use the previous exercise with $K_0 = \operatorname{Im} V$ and $T_0' = \pi_{K_0} V \pi_{H_1}$.)

15. Let T_0 and T_0' be contractions as in (1). Let $H_1 \subset H$ and $K_1 \subset K$ be subspaces of H and K, respectively, and consider a unitary operator $V: H_1 \to K_1$. Show that there exists a contraction $T: H \to K$ such that

$$T_0 = T\tau_{H_0}, \qquad T_0' = \pi_{K_0} T, \qquad V = \pi_{K_1} T\tau_{H_1}$$

if and only if the following three conditions are satisfied:

(i) $\pi_{K_0} T_0 = T_0' \tau_{H_0}$,

(ii) $T_0' \tau_{H_1} = \pi_{K_0} \tau_{K_1} V$,

(iii) $\pi_{K_1} T_0 = V \pi_{H_1} \tau_{H_0}$.

(Hint: first consider the space $\overline{H_1 + H_0}$ and use the result of the previous exercise.)

16. Show that the conditions (4a), (4b), (4c) in Section XXVIII.2 are not independent. For example, prove that the second identity in (4b) may be derived from the first identity in (4b) and the four identities in (4a), (4c).

17. Let $\Sigma = (A, B, C, D; H, K, L)$ be a unitary system, and let

$$U = \begin{bmatrix} A & B \\ C & D \end{bmatrix} : H \boxplus K \to H \boxplus L.$$

Show that the excessive subspace of Σ is the largest subspace contained in H that is invariant under U and U^*.

18. Let $\theta: \mathbb{D} \to \mathcal{L}(\mathbb{C}^k, \mathbb{C}^\ell)$ be an analytic operator-valued function such that

$$\sup\{\|\theta(\lambda)\| \mid \lambda \in \mathbb{D}\} \leq 1.$$

Show that θ coincides with the characteristic operator function of some contraction if and only if $\theta(0)$ is a strict contraction.

19. Let $A \in \mathcal{L}(H)$ be compact and a contraction. Let P be the Riesz projection of A corresponding to the eigenvalues on the unit circle. Show that P is an orthogonal projection and that $A | \operatorname{Im} P$ is unitary.

20. Let $A \in \mathcal{L}(H)$ be the rank one operator given by $A = \langle \cdot, \varphi \rangle \psi$. Assume that $\|\varphi\| \|\psi\| \leq 1$. Compute the characteristic operator function $W_A(\cdot)$.

21. Let $A \in \mathcal{L}(H)$ be the finite rank operator given by

$$A = \sum_{j=1}^{n} \langle \cdot, \varphi_j \rangle \psi_j,$$

where $\sum_{j=1}^{n} \|\varphi_j\| \|\psi_j\| \leq 1$. Compute the characteristic operator function $W_A(\cdot)$.

22. Let $A: \ell_2 \to \ell_2$ be the rank one operator given by $A = c\langle \cdot, \varphi \rangle \psi$. Show that for c sufficiently small the lower triangular part of A (relative to the natural chain on ℓ_2) is a contraction and find its characteristic operator function.

23. Let $A: L_2([0, \tau]) \to L_2([0, \tau])$ be given by

$$(Af)(t) = 2i \sum_{j=1}^{n} \int_{t}^{\tau} \varphi_j(t) \overline{\varphi_j(s)} f(s)\, ds, \qquad 0 \leq t \leq \tau,$$

where $\varphi_1, \ldots, \varphi_n$ are given functions in $L_2([0, \tau])$. Let $\Phi(t)$ be the $1 \times n$ matrix $[\varphi_1(t) \cdots \varphi_n(t)]$, and for $0 \neq \lambda \in \mathbb{C}$ let $U(t; \lambda)$ be the fundamental matrix (normalized to $I_{\mathbb{C}^n}$ at $t = 0$) of the differential equation

$$x'(t) = \frac{2i}{\lambda} x(t) \Phi(t)^* \Phi(t), \qquad 0 \leq t \leq \tau.$$

Show that A is i-dissipative, and prove that its characteristic operator function $V_A(\cdot)$ coincides with $U(\tau; \cdot)$.

PART VIII

BANACH ALGEBRAS AND ALGEBRAS OF OPERATORS

This part presents an introduction to the theory of Banach algebras. During the past fifty years this field has grown considerably and is now an independent direction in mathematics with a wide range of applications. The choice of the material presented here is made with a view to applications in operator theory. The study of a single operator or a class of operators is sometimes easier and more transparent in the framework of a Banach algebra. Moreover, for some operator theory problems the only known solution is obtained by analyzing the single operator as a member of some Banach algebra.

This part consists of the four chapters XXIX–XXXII. Chapter XXIX has an introductory character and deals with invertibility and spectra. Chapter XXX contains the Gelfand theory of commutative Banach algebras. The last two chapters in this part are devoted mainly to applications, namely to the spectral theory of normal operators and the Fredholm theory of operators in a Banach algebra generated by Toeplitz operators defined by piecewise continuous functions. Elements of C^*-algebra theory are also included.

CHAPTER XXIX

GENERAL THEORY

This chapter contains the first basic properties of Banach algebras and deals with invertibility and spectra. Special attention is paid to the behaviour of the spectrum when the algebra is embedded isometrically into a larger Banach algebra. Also in this chapter we introduce a number of examples which will be used throughout this part to illustrate the general theory. Factorization of elements close to the identity is treated in the last section.

XXIX.1 DEFINITION AND EXAMPLES

A *Banach algebra* is a complex Banach space B on which a multiplication xy is defined with the following properties. For all x, y, z in B and $\alpha \in \mathbb{C}$,

(a) $(xy)z = x(yz)$,

(b) $x(y + z) = xy + xz$, $(y + z)x = yx + zx$,

(c) $\alpha(xy) = (\alpha x)y = x(\alpha y)$,

(d) $\|xy\| \leq \|x\|\|y\|$.

It is clear from the properties of the norm that the algebraic operations on B of addition and multiplication are continuous. If $xy = yx$ for all x, y, then B is called a *commutative* Banach algebra.

An element $e \in B$ is the *identity* or *unit* in B if $\|e\| = 1$ and $ex = xe = x$ for all $x \in B$. A Banach algebra with a unit is called a *unital* Banach algebra. A Banach algebra B without a unit can be embedded in a unital Banach algebra as follows. Put $B_1 = B \times \mathbb{C}$, and define

$$(x, \alpha) + (y, \beta) = (x + y, \alpha + \beta),$$
$$\gamma(x, \alpha) = (\gamma x, \gamma \alpha),$$
$$(x, \alpha)(y, \beta) = (xy + \beta x + \alpha y, \alpha\beta),$$
$$\|(x, \alpha)\| = \|x\| + |\alpha|,$$
$$e = (0, 1).$$

It is easy to check that B_1 is a Banach algebra with unit e and that B is isometrically isomorphic to $B \times \{0\}$ under the map $x \mapsto (x, 0)$.

A *subalgebra* A of a Banach algebra B is a linear manifold of B which is closed under multiplication, i.e., if x and y are in A, then xy is in A. A closed subalgebra of a Banach algebra B is a Banach algebra in its own right with the norm inherited from B.

We conclude this section with a first series of examples of Banach algebras.

(i) The space $C(S)$ of all bounded continuous complex valued functions on a topological space S is a commutative unital Banach algebra with respect to the usual product of functions and the norm

$$(1) \qquad \|f\| = \sup_{t \in S} |f(t)|.$$

The unit element is the constant function 1.

(ii) If S is a compact set in the complex plane, then the space $\mathcal{A}(S)$ of functions in $C(S)$ which are analytic on the interior of S is a closed subalgebra of $C(S)$.

(iii) The space $B(S)$ of all bounded complex valued functions on a set S is a commutative unital Banach algebra with respect to the usual product of functions and the norm (1). The unit element is again the constant function 1. If S is a topological space, then $C(S)$ is a closed subalgebra of $B(S)$.

(iv) Let $\mathcal{B} = PC([a, b])$ be the space of all complex valued functions f defined on $[a, b]$ such that for each $t \in [a, b]$ the right hand limit $f(t+)$ exists (and is finite) and the left hand limit $f(t-) = f(t)$. For the end point b we use the convention that $b+ = b$. The letters PC stand for piecewise continuous. Note that $PC([a, b])$ contains left continuous monotone functions and left continuous functions of bounded variation. It is easy to see that $PC([a, b])$ is a closed subalgebra of $B([a, b])$ with unit the constant function 1.

(v) The space $\mathcal{L}(X)$ of all bounded linear operators on a complex Banach space X is a unital non-commutative (if $\dim X > 1$) Banach algebra with respect to the usual sum and product of operators and the operator norm

$$\|T\| = \sup_{\|x\|=1} \|Tx\|.$$

The unit element is the identity operator I.

(vi) If $\mathcal{K}$ is the set of all compact operators on a Banach space X, then $\{\lambda I + K \mid \lambda \in \mathbb{C}, K \in \mathcal{K}\}$ is a closed subalgebra of $\mathcal{L}(X)$.

(vii) The Banach space S_1 of all trace class operators (endowed with the trace class norm) and the Banach space S_2 of all Hilbert-Schmidt operators on a Hilbert space are Banach algebras without unit (if H is infinite dimensional). In both cases the multiplication is the usual product of operators.

XXIX.2 WIENER ALGEBRAS

The four examples of Banach algebras presented in this section are called *Wiener algebras*.

(a) Let $\mathcal{W}$ be the space of all complex valued functions f which are defined on the unit circle $\mathbb{T} = \{\zeta \mid |\zeta| = 1\}$ and have the Fourier series expansion

$$f(\zeta) = \sum_{n=-\infty}^{\infty} a_n \zeta^n, \qquad \sum_{n=-\infty}^{\infty} |a_n| < \infty.$$

The space $\mathcal{W}$ is a normed linear space with respect to the usual definitions of addition and scalar multiplication and norm

$$\|f\| = \left\| \sum_{n=-\infty}^{\infty} a_n \zeta^n \right\| = \sum_{n=-\infty}^{\infty} |a_n|.$$

Moreover, $\mathcal{W}$ is a Banach space since it is linearly isometric to the Banach space

$$\ell_1(\mathbb{Z}) = \left\{ (\alpha_k)_{-\infty}^{\infty} \Big| \sum_{k=-\infty}^{\infty} |\alpha_k| < \infty \right\}.$$

Define multiplication on $\mathcal{W}$ to be the usual product of functions. If $f(\zeta) = \sum_{n=-\infty}^{\infty} a_n \zeta^n$ and $g(\zeta) = \sum_{n=-\infty}^{\infty} b_n \zeta^n$ are in $\mathcal{W}$, then

$$f(\zeta)g(\zeta) = \sum_{n=-\infty}^{\infty} c_n \zeta^n, \qquad \sum_{n=-\infty}^{\infty} |c_n| < \infty,$$

where

$$c_n = \sum_{k=-\infty}^{\infty} a_{n-k} b_k.$$

Indeed,

$$\|fg\| = \sum_{n=-\infty}^{\infty} |c_n| \leq \sum_{n=-\infty}^{\infty} \sum_{k=-\infty}^{\infty} |a_{n-k}||b_k|$$

$$= \sum_{k=-\infty}^{\infty} \left(\sum_{n=-\infty}^{\infty} |a_{n-k}| \right) |b_k|$$

$$= \sum_{j=-\infty}^{\infty} |a_j| \sum_{k=-\infty}^{\infty} |b_k| = \|f\|\|g\|.$$

We see that $\mathcal{W}$ is a commutative Banach algebra with unit the constant function 1.

(b) (Wiener algebra with weights). Let $\beta = (\beta_n)_{n=-\infty}^{\infty}$ be a sequence of positive numbers with the property that

$$(1) \qquad \beta_{j+k} \leq \beta_j \beta_k, \qquad -\infty < j, k < \infty.$$

Define $\mathcal{W}(\beta)$ to be the space of all formal power series $x = \sum_{n=-\infty}^{\infty} a_n z^n$ for which $\|x\| = \sum_{n=-\infty}^{\infty} |a_n|\beta_n < \infty$. The space $\mathcal{W}(\beta)$ may be identified with the weighted $\ell_1(\beta)$ space of all sequences $(a_n\beta_n)_{-\infty}^{\infty}$ which are in ℓ_1 and hence $\mathcal{W}(\beta)$ is a Banach space with respect to the norm defined above. Given $x = \sum_{n=-\infty}^{\infty} a_n z^n$ and $y = \sum_{n=-\infty}^{\infty} b_n z^n$ in $\mathcal{W}(\beta)$, define xy to be the usual product of power series, i.e.,

$$xy = \sum_{k=-\infty}^{\infty} \left(\sum_{j=-\infty}^{\infty} a_{k-j} b_j \right) z^k.$$

From (1) we get

$$\|xy\| = \sum_{k=-\infty}^{\infty} \left| \sum_{j=-\infty}^{\infty} a_{k-j} b_j \right| \beta_k \leq \sum_{k=-\infty}^{\infty} \sum_{j=-\infty}^{\infty} |a_{k-j}| \beta_{k-j} |b_j| \beta_j$$

$$= \sum_{j=-\infty}^{\infty} \left(\sum_{k=-\infty}^{\infty} |a_{k-j}| \beta_{k-j} \right) |b_j| \beta_j = \|x\| \|y\|.$$

Hence $\mathcal{W}(\beta)$ is a Banach algebra with unit the constant function 1.

(c) (Wiener convolution algebra). We shall now show that the Banach space $\mathcal{B} = L_1(\mathbf{R})$ is a commutative Banach algebra with convolution $*$ as multiplication, i.e.,

$$(f * g)(t) = \int_{-\infty}^{\infty} f(t - s)g(s)ds.$$

Here and in the sequel the equalities have to be understood as being valid almost everywhere on $\mathbf{R}$. First we need to prove that $h = f * g$ is in $\mathcal{B}$ whenever f and g are in $\mathcal{B}$. It follows from the measurability of f and g that $F(t, s) = f(t - s)g(s)$ is a measurable function on $\mathbf{R} \times \mathbf{R}$ (cf., [R], page 147). Since

$$(2) \qquad \int_{-\infty}^{\infty} \int_{-\infty}^{\infty} |F(t, s)| dt ds = \int_{-\infty}^{\infty} |g(s)| ds \int_{-\infty}^{\infty} |f(t - s)| dt = \|g\| \|f\|,$$

we can apply Fubini's theorem to show that $h = f * g$ is in $L_1(\mathbf{R})$ and by (2),

$$\|(f * g)\| = \int_{-\infty}^{\infty} |h(t)| dt \leq \int_{-\infty}^{\infty} \int_{-\infty}^{\infty} |F(t, s)| ds dt$$

$$= \int_{-\infty}^{\infty} \int_{-\infty}^{\infty} |F(t, s)| dt ds = \|f\| \|g\|.$$

It is clear that $f * (g + k) = f * g + f * k$. Also

$$[f * (g * k)](t) = \int_{-\infty}^{\infty} f(t - s) \left(\int_{-\infty}^{\infty} g(s - u)k(u)du \right) ds$$

$$= \int_{-\infty}^{\infty} \left(\int_{-\infty}^{\infty} f(t - s)g(s - u)ds \right) k(u)du$$

$$= \int_{-\infty}^{\infty} \left(\int_{-\infty}^{\infty} f(t - s - u)g(s)ds \right) k(u)du = [(f * g) * k](t),$$

and

$$(f * g)(t) = \int_{-\infty}^{\infty} f(t-s)g(s)ds = \int_{-\infty}^{\infty} f(\eta)g(t-\eta)d\eta = (g * f)(t).$$

Hence $\mathcal{B}$ is a commutative Banach algebra. It does not have a unit. To see this, we use the Fourier transform $\mathcal{F}$ on $L_1(\mathbb{R})$, which is defined by

$$(3) \qquad \widehat{f}(\lambda) = (\mathcal{F}f)(\lambda) = \int_{-\infty}^{\infty} e^{i\lambda t} f(t)dt, \qquad f \in L_1(\mathbb{R}).$$

(Note that here we depart from the notation in Section XII.1, where the Fourier transformation has an additional normalization term and λ is replaced by $-\lambda$. Throughout this Banach algebra part the *Fourier transform* $\widehat{f}$ and the *Fourier transformation* $\mathcal{F}$ are defined by (3).) Now, recall the fact that the Fourier transformation $\mathcal{F}$ is an algebra isomorphism from $L_1(\mathbb{R})$ into the space C_0 of all continuous complex valued functions on $\mathbb{R}$ which vanish at infinity (cf., [R], items 9.2 and 9.6). Therefore, if $\mathcal{B}$ has a unit e, then $\mathcal{F}(e) = 1$ which is impossible since $1 \notin C_0$.

(d) (Wiener convolution algebra with weights). Let $w(t)$ be a positive continuous function on $\mathbb{R}$ which satisfies

$$w(t+s) \le w(t)w(s).$$

Let $L_1(\mathbb{R}, w)$ be the space of all Lebesgue measurable complex valued functions f on $\mathbb{R}$ for which

$$(4) \qquad \|f\| = \int_{-\infty}^{\infty} |f(t)|w(t)dt < \infty.$$

An argument similar to the one given in example (b) shows that $L_1(\mathbb{R}, w)$ is a commutative Banach algebra with respect to the usual addition and scalar multiplication of functions and with multiplication $*$ given by

$$(f * g)(t) = \int_{-\infty}^{\infty} f(t-s)g(s)ds.$$

The norm $\|f\|$ is defined in (4).

XXIX.3 IDEALS AND QUOTIENT ALGEBRAS

The operations of addition and multiplication on a Banach algebra give rise, in a natural way, to special classes of linear functionals and subspaces, namely those linear functionals f which are multiplicative and those subspaces which are ideals.

A subset $\mathcal{I}$ of a Banach algebra $\mathcal{B}$ is called an *ideal* in $\mathcal{B}$ if $\mathcal{I}$ is a linear manifold in $\mathcal{B}$ and given any $x \in \mathcal{B}$ and $a \in \mathcal{I}$, both ax and xa are in $\mathcal{I}$. An ideal $\mathcal{I}$ in $\mathcal{B}$ is called a *proper* ideal if $\mathcal{I} \neq \{0\}$ and $\mathcal{I} \neq \mathcal{B}$. Let us consider a few examples.

(a) The set of all functions in $C(S)$ which vanish on a given set $K \subset S$ is a closed ideal in $C(S)$.

(b) The set of all trace class operators on an infinite dimensional Hilbert space H is a non-closed ideal in $\mathcal{L}(H)$. The same is true for the set of all Hilbert-Schmidt operators on H.

(c) The set of all compact operators on a complex Banach space X is a closed ideal in $\mathcal{L}(X)$.

(d) In the convolution algebra $\mathcal{B} = L_1(\mathbb{R})$, the set $\mathcal{I}$ of all functions $f \in \mathcal{B}$ with the property that its Fourier transform $\widehat{f}$ vanishes at a given fixed point $t_0 \in \mathbb{R}$ is a closed ideal in $\mathcal{B}$. Obviously $\mathcal{I}$ is a linear manifold of $\mathcal{B}$. Since $\widehat{f}$ is a continuous function, $\mathcal{I}$ is closed. Given $f \in \mathcal{I}$ and $g \in \mathcal{B}$, $\widehat{f * g}(t_0) = \widehat{f}(t_0)\widehat{g}(t_0) = 0$. Hence $f * g \in \mathcal{I}$ and $\mathcal{I}$ is an ideal in $\mathcal{B}$.

Let $\mathcal{I}$ be a closed ideal in a Banach algebra $\mathcal{B}$. Given $x \in \mathcal{B}$, let

$$[x] = \{x + m \mid m \in \mathcal{I}\}.$$

The set $[x]$ is called the *coset* containing x. The collection of all cosets is denoted by $\mathcal{B}/\mathcal{I}$. We define addition and multiplication on $\mathcal{B}/\mathcal{I}$ by

$$\alpha[x] = [\alpha x], \alpha \in \mathbb{C},$$
$$[x] + [y] = [x + y],$$
$$[x][y] = [xy].$$

The operations are unambiguously defined. Indeed, if $[x] = [x_1]$ and $[y] = [y_1]$, then $x_1 - x \in \mathcal{I}$ and $y_1 - y \in \mathcal{I}$. Hence it follows that

$$[x] + [y] = [x + y] = [x_1 + y_1] = [x_1] + [y_1].$$

Similarly,

$$\alpha[x] = [\alpha x] = [\alpha x_1] = \alpha[x_1].$$

Now $x_1 - x = u \in \mathcal{I}$ and $y_1 - y = v \in \mathcal{I}$. Since $\mathcal{I}$ is an ideal, $x_1 y_1 = xy + w$ for some $w \in \mathcal{I}$. Hence

$$[x][y] = [xy] = [x_1 y_1] = [x_1][y_1].$$

It is easy to verify that the algebraic operations in $\mathcal{B}/\mathcal{I}$ defined above satisfy conditions (a), (b) and (c) in Section XXIX.1. We call $\mathcal{B}/\mathcal{I}$ a *quotient algebra*.

The assumption that $\mathcal{I}$ is a closed linear manifold in $\mathcal{B}$ implies that

$$\|[x]\| = d(x, \mathcal{I}) = \inf\{\|x - z\| \mid z \in \mathcal{I}\}$$

defines a norm on the quotient algebra. This norm has the property that

$$\|[x][y]\| \leq \|[x]\|\|[y]\|.$$

Indeed for any $u, v \in \mathcal{I}$, we have $(x - u)(y - v) = xy + w$ for some $w \in \mathcal{I}$. Therefore

$$\|[x][y]\| = \|[xy]\| = d(xy, \mathcal{I}) \leq \|(x - u)(y - v)\| \leq \|x - u\|\|y - v\|.$$

Since u and v are arbitrary in $\mathcal{I}$, it follows that $\|[x][y]\| \leq \|[x]\|\|[y]\|$.

We shall prove that $\mathcal{B}/\mathcal{I}$ is a Banach algebra. In order to do this, it remains to show that $\mathcal{B}/\mathcal{I}$ is complete. So, suppose $([x_n])$ is a Cauchy sequence in $\mathcal{B}/\mathcal{I}$. There exists a subsequence $([x_{n_k}])$ such that $\sum_k \|[x_{n_{k+1}} - x_{n_k}]\| < \infty$. It follows from the definition of the norm of a coset that for any $v_1 \in [x_{n_1}]$ there exists $v_2 \in [x_{n_2}]$ such that

$$\|v_1 - v_2\| \leq 2\|[x_{n_1} - x_{n_2}]\|.$$

Similarly, there exists $v_3 \in [x_{n_3}]$ such that

$$\|v_2 - v_3\| \leq 2\|[x_{n_3} - x_{n_2}]\|.$$

Continuing in this manner, we obtain a Cauchy sequence (v_j) with $v_j \in [x_{n_j}]$. Since $\mathcal{B}$ is complete, (v_j) converges to some $v \in \mathcal{B}$. Hence

$$\|[x_{n_j}] - [v]\| = \|[v_j] - [v]\| \leq \|v_j - v\| \to 0,$$

and therefore the Cauchy sequence $([x_n])$ converges to $[v]$. We have established that $\mathcal{B}/\mathcal{I}$ is a Banach algebra.

If $\mathcal{B}$ is a unital Banach algebra with unit e, then $[e]$ is the unit in the quotient algebra $\mathcal{B}/\mathcal{I}$, provided $\mathcal{I}$ is a proper subset of $\mathcal{B}$. Note that $\|[e]\| = 1$, because

$$1 = \|e\| \geq \|[e]\|, \qquad \|[e]\|^2 \geq \|[e]\| \neq 0.$$

If $\mathcal{K}$ is the closed ideal of all compact operators on the complex Banach space X, then the quotient algebra $\mathcal{L}(X)/\mathcal{K}$ is the Calkin algebra which we used in Section XI.5. From what we proved above it follows that the Calkin algebra is a Banach algebra.

XXIX.4 INVERTIBILITY

An element x in an unital Banach algebra $\mathcal{B}$ with unit e is *invertible* in $\mathcal{B}$ if there exists $y \in \mathcal{B}$ such that $xy = yx = e$. Since there is only one y with this property, we call y the *inverse* of x and denote it by x^{-1}. Let us see what invertibility means for some concrete examples.

(a) Let $\mathcal{B}$ be any of the Banach algebras described in examples (i), (ii) and (iii) in Section XXIX.1. An element $f \in \mathcal{B}$ is invertible if and only if

$$\inf_{t \in S} |f(t)| > 0,$$

in which case $f^{-1}(t) = f(t)^{-1}$ for all $t \in S$.

(b) Let $\mathcal{B} = PC([a,b])$ as defined in Section XXIX.1, example (iv). A function $f \in \mathcal{B}$ is invertible if and only if $f(t) \neq 0$ and $f(t+) \neq 0$ for each $t \in [a,b]$ and in that case $f^{-1}(t) = f(t)^{-1}$.

(c) An operator $T \in \mathcal{L}(X)$ is invertible in the algebra $\mathcal{L}(X)$ if and only if T is bijective. The inverse of T in $\mathcal{L}(X)$ is the usual inverse operator T^{-1}.

(d) Let $\mathcal{B}$ be the closed subalgebra of $\mathcal{L}(X)$ consisting of all operators $\lambda I + K$, where X is an infinite dimensional Banach space, $\lambda \in \mathbb{C}$ and K is compact. An operator $\mu I + K \in \mathcal{B}$ is invertible in $\mathcal{B}$ if and only if $\mu \neq 0$ and $\mu I + K$ is injective or, equivalently, surjective (cf., Corollary 4.3 in Chapter XI). In this case the usual inverse operator $(\mu I + K)^{-1}$ is the inverse in $\mathcal{B}$ of $\mu I + K$, because

$$(\mu I + K)^{-1} = \frac{1}{\mu} I - \frac{1}{\mu}(\mu I + K)^{-1} K \in \mathcal{B}.$$

We shall show in Chapter XXX that a function f in the Wiener algebra $\mathcal{W}$ (example (a) in Section XXIX.2) is invertible if and only if $f(z) \neq 0$ for all z in the unit circle. In that case the inverse of f in $\mathcal{W}$ is the function $1/f$. For the weighted Wiener algebra $\mathcal{W}(\beta)$ (see Section XXIX.2, example (b)) an analogous result holds true. We shall prove (also in Chapter XXX) that $x = \sum_{n=-\infty}^{\infty} a_n \zeta^n \in \mathcal{W}(\beta)$ is invertible in $\mathcal{W}(\beta)$ if and only if $g(\zeta) = \sum_{n=-\infty}^{\infty} a_n \zeta^n \neq 0$ for all ζ in an annulus $0 < \rho_1 \leq |\zeta| \leq \rho_2 < \infty$ determined by the weights $\beta = (\beta_n)$. Furthermore, in this case, $x^{-1} = \sum_{n=-\infty}^{\infty} b_n \zeta^n$, where

$$1/g(\zeta) = \sum_{n=-\infty}^{\infty} b_n \zeta^n, \qquad \rho_1 \leq |\zeta| \leq \rho_2.$$

In Chapter XXX also other additional examples will be given.

THEOREM 4.1. *Suppose $x \in \mathcal{B}$ is invertible. If $\|x - y\| < \|x^{-1}\|^{-1}$, then y is invertible and*

$$(1) \qquad y^{-1} = \sum_{k=0}^{\infty} [x^{-1}(x - y)]^k x^{-1},$$

$$(2) \qquad \|x^{-1} - y^{-1}\| \leq \frac{\|x^{-1}\|^2 \|x - y\|}{1 - \|x^{-1}\|\|x - y\|}.$$

In particular, the set $\mathcal{G}$ of invertible elements in $\mathcal{B}$ is an open set in $\mathcal{B}$ and the map $x \mapsto x^{-1}$ is a homeomorphism from $\mathcal{G}$ onto $\mathcal{G}$.

PROOF. First we note that if $\|u\| < 1$, then $e - u$ is invertible and

$$(3) \qquad (e - u)^{-1} = \sum_{k=0}^{\infty} u^k.$$

This may be seen by setting $s_n = \sum_{k=0}^{n} u^k$ and noting that (s_n) converges and

$$(e - u)s_n = s_n(e - u) \to e.$$

Now $y = x[e - x^{-1}(x - y)]$ and $\|x^{-1}(x - y)\| < 1$. Hence it follows from (3) that y is invertible and y^{-1} is given by (1). But then

$$\|x^{-1} - y^{-1}\| = \left\| \sum_{k=1}^{\infty} [x^{-1}(x - y)]^k x^{-1} \right\|$$

$$\leq \|x^{-1}\| \sum_{k=1}^{\infty} \|x^{-1}\|^k \|x - y\|^k = \frac{\|x^{-1}\|^2 \|x - y\|}{1 - \|x^{-1}\|\|x - y\|}. \qquad \square$$

Recall that an element x in a ring $\mathcal{R}$ is a *divisor of zero* if $xy = yx = 0$ for some $y \neq 0$ in $\mathcal{R}$. We shall need the following extension of this notion. An element x in a Banach algebra $\mathcal{B}$ is called a *topological divisor of zero* if there exists a sequence (y_n) in $\mathcal{B}$, $\|y_n\| = 1$ for all n, such that

$$\lim_{n \to \infty} xy_n = \lim_{n \to \infty} y_n x = 0.$$

It is clear that a topological divisor of zero cannot be invertible. Let us consider a few examples.

(i) An example of a topological divisor of zero which is not a divisor of zero is the following. Let $f \in C([0, 1])$ vanish only at $t = 0$. Obviously, f is not a divisor of zero. However f is a topological divisor of zero. Indeed, let g_n be the function in $C([0, 1])$ whose graph is a line joining the points $(0, 1)$ to $(\frac{1}{n}, 0)$ and is zero on the interval $\frac{1}{n} \leq t \leq 1$. The continuity of f implies that $fg_n \to 0$ in $C([0, 1])$.

(ii) The forward shift operator $S \in \mathcal{L}(\ell_2)$ defined by $S(\alpha_1, \alpha_2, \ldots) = (0, \alpha_1, \alpha_2, \ldots)$ is not invertible, yet is not a topological divisor. To see this, recall that S has a left inverse L defined by $L(\alpha_1, \alpha_2, \ldots) = (\alpha_2, \alpha_3, \ldots)$. Thus $(T_n)_{n=1}^{\infty}$ in $\mathcal{L}(\ell_2)$ and $ST_n \to 0$ implies that $T_n = LST_n \to 0$.

(iii) Given a Banach space X, let $T \in \mathcal{L}(X)$ have the property that $\text{Im}\,T$ is neither dense nor closed in X. For example, if $X = \ell_2$ and

$$T(\alpha_1, \alpha_2, \ldots) = \left(0, \alpha_1, \frac{1}{2}\alpha_2, \frac{1}{3}\alpha_3, \ldots\right),$$

then T is compact and $\text{Im}\,T$ has the desired properties. We shall show that T is a topological divisor of zero in $\mathcal{L}(X)$. Since $\text{Im}\,T$ is not dense in X, there exists a continuous linear functional f on X such that $\|f\| = 1$ and $f(Tx) = 0$ for all $x \in X$. The fact that the range of T is not closed implies (cf., Theorem XI.2.1) that there exists a sequence (x_n) in X, $\|x_n\| = 1$ for all n, such that $Tx_n \to 0$. Define $A_n \in \mathcal{L}(X)$ by $A_n(x) = f(x)x_n$. Clearly, $\|A_n\| = 1$ for all n and

$$\|TA_n\| = \|Tx_n\| \to 0 \qquad (n \to \infty).$$

Furthermore, $A_n T = 0$ for $n = 1, 2, \ldots$.

THEOREM 4.2. *If x is in the boundary of the set of invertible elements in a unital Banach algebra $\mathcal{B}$, then x is a topological divisor of zero.*

PROOF. Recall that the boundary ∂A of a subset A of a topological space S consists of all points in the closure $\overline{A}$ that are not in the interior of A. Let $\mathcal{G}$ be the set of invertible elements in $\mathcal{B}$. Since $\mathcal{G}$ is an open set, x is not in $\mathcal{G}$ and there exists a sequence (x_n) in $\mathcal{G}$ such that $x_n \to x$. Now the sequence (x_n^{-1}) is unbounded, otherwise

$$x_n^{-1} x - e = x_n^{-1}(x - x_n) \to 0,$$

and thus by Theorem 4.1, the element $x_n^{-1} x$ is invertible for n sufficiently large, which is a contradiction. So, without loss of generality, we may assume $\|x_n^{-1}\| \to \infty$. Let $y_n = \|x_n^{-1}\|^{-1} x_n^{-1}$. Then $y_n \in \mathcal{G}$, $\|y_n\| = 1$ for all n, and

$$xy_n = (x - x_n)y_n + \frac{1}{\|x_n^{-1}\|} e \to 0,$$

$$y_n x = y_n(x - x_n) + \frac{1}{\|x_n^{-1}\|} e \to 0. \quad \square$$

COROLLARY 4.3. *Let $\mathcal{B}$ be a unital Banach algebra, and let $\mathcal{A}$ be a closed subalgebra of $\mathcal{B}$ with the same unit as $\mathcal{B}$. Then the boundary of the set of invertible elements in $\mathcal{A}$ is contained in the boundary of the set of invertible elements in $\mathcal{B}$.*

PROOF. We know that $\mathcal{A}$ is a Banach algebra in its own right. Let $\mathcal{G}_\mathcal{A}$ and $\mathcal{G}_\mathcal{B}$ denote the set of invertible elements in $\mathcal{A}$ and $\mathcal{B}$, respectively. Suppose x is in the boundary $\partial \mathcal{G}_\mathcal{A}$ of $\mathcal{G}_\mathcal{A}$. Then, by Theorem 4.2, the element x is a topological divisor of zero in $\mathcal{A}$ and therefore it is also a topological divisor of zero in $\mathcal{B}$. Hence x is not invertible in $\mathcal{B}$, that is, $x \notin \mathcal{G}_\mathcal{B}$. Since $x \in \partial \mathcal{G}_\mathcal{A} \subset \overline{\mathcal{G}}_\mathcal{A} \subset \overline{\mathcal{G}}_\mathcal{B}$ and $\mathcal{G}_\mathcal{B}$ is open, we have $x \in \overline{\mathcal{G}}_\mathcal{B} \backslash \mathcal{G}_\mathcal{B} = \partial \mathcal{G}_\mathcal{B}$. $\quad \square$

XXIX.5 SPECTRUM AND RESOLVENT

Given x in a unital Banach algebra $\mathcal{B}$ (with unit e) the *resolvent set* $\rho(x)$ is the set of those $\lambda \in \mathbb{C}$ for which $\lambda e - x$ is invertible. The complement in $\mathbb{C}$ of $\rho(x)$ is called the *spectrum* of x and is denoted by $\sigma(x)$.

For example, if f is in $C(S)$, where S is a compact Hausdorff space, then $\sigma(f) = \{f(t) \mid t \in S\}$. If S is an arbitrary set and $f \in B(S)$ (see example (iii) in Section XXIX.1), then

$$(1) \qquad \sigma(f) = \overline{\{f(t) \mid t \in S\}}.$$

With $S = [a, b]$ formula (1) also gives the spectrum of an element f in $PC([a, b])$. If T is a bounded linear operator on a complex Banach space, then the spectrum of T as defined in Section I.1 coincides with the spectrum of T as an element of the Banach algebra $\mathcal{L}(X)$. Other examples will appear in the next chapter.

THEOREM 5.1. *Given x in a unital Banach algebra $\mathcal{B}$, the resolvent set $\rho(x)$ is open and the map $\lambda \mapsto (\lambda e - x)^{-1}$ is analytic on $\rho(x)$.*

PROOF. Take $\mu \in \rho(x)$, and let $|\lambda - \mu| < \|(\mu e - x)^{-1}\|^{-1}$. Then

$$\|(\mu e - x) - (\lambda e - x)\| < \|(\mu e - x)^{-1}\|^{-1}.$$

So we can apply Theorem 4.1 to show that $\lambda e - x$ is invertible and

$$(2) \qquad (\lambda e - x)^{-1} = \sum_{k=0}^{\infty} (-1)^k (\lambda - \mu)^k (\mu e - x)^{-k-1}. \qquad \square$$

THEOREM 5.2. *The spectrum of an element in a unital Banach algebra B is non-empty and compact.*

PROOF. Given $x \in B$, Theorem 4.1 implies that $\lambda e - x = \lambda(e - \lambda^{-1}x)$ is invertible if $|\lambda| > \|x\|$. Hence $\sigma(x)$ is bounded and closed (because $\rho(x)$ is open). Thus $\sigma(x)$ is compact. Assume $\sigma(x) = \emptyset$. Then for each continuous linear functional f in the conjugate space B' we may conclude from Theorem 5.1 that $\varphi(\lambda) = f\big((\lambda e - x)^{-1}\big)$ is analytic on the entire complex plane. Moreover, the continuity of the inverse operation implies that

$$(3) \qquad \lim_{\lambda \to \infty} \varphi(\lambda) = \lim_{\lambda \to \infty} \lambda^{-1} f\big((e - \lambda^{-1}x)^{-1}\big) = 0.$$

Hence φ is a bounded entire function which must be constant by Liouville's theorem. But then we get from (3) that $\varphi(\lambda) = 0$ for all $\lambda \in \mathbb{C}$, that is, $f\big((\lambda e - x)^{-1}\big) = 0$ for all $\lambda \in \mathbb{C}$. Since f is arbitrary in B', we conclude that $(\lambda e - x)^{-1} = 0$, which is impossible. Hence $\sigma(x) \neq \emptyset$. $\square$

THEOREM 5.3. (Gelfand-Mazur). *If in a Banach algebra B with unit e every non-zero vector is invertible, then $B = \{\lambda e \mid \lambda \in \mathbb{C}\}$.*

PROOF. Given $x \in B$, there exists by Theorem 5.2 a $\lambda \in \mathbb{C}$ such that $\lambda e - x$ is not invertible. Hence $\lambda e - x = 0$ by hypothesis. $\square$

XXIX.6 SPECTRA RELATIVE TO SUBALGEBRAS

Let A be a closed subalgebra of the unital Banach algebra B, and assume that the unit e of B also belongs to A. Then each $x \in A$ has a spectrum $\sigma_A(x)$ with respect to the algebra A and a spectrum $\sigma_B(x)$ with respect to the algebra B. Obviously, if $\lambda e - x$ has no inverse in B, then it has no inverse in A. Therefore $\sigma_B(x) \subset \sigma_A(x)$. Equivalently, $\rho_A(x) \subset \rho_B(x)$, where $\rho_A(x)$ (resp., $\rho_B(x)$) is the resolvent set of x with respect to A (resp., B). The following example shows that the above inclusions may be strictly proper.

Let B be the Banach algebra $C(\mathbb{T})$, where $\mathbb{T}$ is the unit circle. Let A be the subalgebra of those $g \in B$ which can be extended continuously to a function which is analytic on the open unit disc $\mathbb{D}$. In other words,

$$A = \{g = h|\mathbb{T} \mid h \in A(\overline{\mathbb{D}})\},$$

where $A(\overline{\mathbb{D}})$ is the Banach algebra defined in Section XXIX.1, example (ii). By the maximum modulus principle every h in $A(\overline{\mathbb{D}})$ attains its maximum on $\mathbb{T}$. This fact implies that $\mathcal{A}(\overline{\mathbb{D}})$ is isometrically isomorphic to $\mathcal{A}$ via the restriction map $h \mapsto h|\mathbb{T}$. Hence $\mathcal{A}$ is a closed subalgebra of $\mathcal{B}$ with the same unit as $\mathcal{B}$, namely the function $e(\zeta) = 1$ for all $\zeta \in \mathbb{T}$. Let $f(\zeta) = \zeta$, $\zeta \in \mathbb{T}$. Then $\sigma_{\mathcal{B}}(f) = \mathbb{T}$ and $\sigma_{\mathcal{A}}(f) = \overline{\mathbb{D}}$, and hence $\sigma_{\mathcal{B}}(f)$ is a proper subset of $\sigma_{\mathcal{A}}(f)$. Note that $\sigma_{\mathcal{A}}(f)$ consists of $\sigma_{\mathcal{B}}(f)$ and a component (i.e., a maximal connected subset) of the resolvent set $\rho_{\mathcal{B}}(f)$. The next theorem shows that this is the general situation.

THEOREM 6.1. *Let $\mathcal{A}$ be a closed subalgebra of a Banach algebra $\mathcal{B}$ with the same unit as $\mathcal{B}$, and let x be in $\mathcal{A}$. Then each component in the resolvent set $\rho_{\mathcal{A}}(x)$ is a component in $\rho_{\mathcal{B}}(x)$. If U is a component of $\rho_{\mathcal{B}}(x)$ but U is not a component of $\rho_{\mathcal{A}}(x)$, then $U \subset \sigma_{\mathcal{A}}(x)$.*

PROOF. Clearly, $\rho_{\mathcal{A}}(x) \subset \rho_{\mathcal{B}}(x)$ and

$$(1) \qquad \rho_{\mathcal{B}}(x) = \rho_{\mathcal{A}}(x) \cup \big(\rho_{\mathcal{B}}(x) \cap \sigma_{\mathcal{A}}(x)\big).$$

It follows from Corollary 4.3 that the boundary $\partial \sigma_{\mathcal{A}}(x)$ of $\sigma_{\mathcal{A}}(x)$ is contained in the boundary $\partial \sigma_{\mathcal{B}}(x)$ of $\sigma_{\mathcal{B}}(x)$. It is now an easy matter to prove that $\mathcal{O} = \rho_{\mathcal{B}}(x) \cap \sigma_{\mathcal{A}}(x)$ is an open set in the complex plane. Suppose V is a component of $\rho_{\mathcal{A}}(x)$. Then V is contained in some component W of $\rho_{\mathcal{B}}(x)$. By (1)

$$(2) \qquad W = \big(W \cap \rho_{\mathcal{A}}(x)\big) \cup (W \cap \mathcal{O}),$$

which shows that W is the disjoint union of two open sets. Since W is connected and $V \subset W \cap \rho_{\mathcal{A}}(x)$, we may conclude that $W \cap \sigma_{\mathcal{A}}(x) = W \cap \mathcal{O} = \emptyset$. Hence $W \subset \rho_{\mathcal{A}}(x)$, and thus $V = W$. This proves the first assertion of the theorem.

Suppose U is a component of $\rho_{\mathcal{B}}(x)$ which is not a component of $\rho_{\mathcal{A}}(x)$. Then $U \cap \sigma_{\mathcal{A}}(x) \neq \emptyset$. If we replace W by U in (2) and use the fact that U is open and connected, then we may conclude that $U \cap \rho_{\mathcal{A}}(x) = \emptyset$ or $U \subset \sigma_{\mathcal{A}}(x)$. $\square$

Note that Theorem 6.1 implies that $\rho_{\mathcal{A}}(x)$ and $\rho_{\mathcal{B}}(x)$ have the same unbounded component.

COROLLARY 6.2. *Let $\mathcal{A}$ be a closed subalgebra of a Banach algebra $\mathcal{B}$ with the same unit as $\mathcal{B}$, and let x be in $\mathcal{A}$.*

(a) *If $\rho_{\mathcal{B}}(x)$ is connected, then $\sigma_{\mathcal{A}}(x) = \sigma_{\mathcal{B}}(x)$.*

(b) *If the interior of $\sigma_{\mathcal{A}}(x)$ is empty, then $\sigma_{\mathcal{A}}(x) = \sigma_{\mathcal{B}}(x)$.*

PROOF. (a) Since $\rho_{\mathcal{B}}(x)$ is unbounded, it is not contained in $\sigma_{\mathcal{A}}(x)$. Therefore Theorem 6.1 shows that $\rho_{\mathcal{B}}(x)$ is a component of $\rho_{\mathcal{A}}(x) \subset \rho_{\mathcal{B}}(x)$. Hence $\rho_{\mathcal{A}}(x) = \rho_{\mathcal{B}}(x)$.

(b) It was shown in the proof of Theorem 6.1 that $\rho_{\mathcal{B}}(x) \cap \sigma_{\mathcal{A}}(x)$ is open in $\mathbb{C}$. Since $\rho_{\mathcal{B}}(x) \cap \sigma_{\mathcal{A}}(x) \subset \sigma_{\mathcal{A}}(x)$ and the interior of $\sigma_{\mathcal{A}}(x)$ is empty, we have $\rho_{\mathcal{B}}(x) \cap \sigma_{\mathcal{A}}(x) = \emptyset$ or $\sigma_{\mathcal{A}}(x) \subset \sigma_{\mathcal{B}}(x) \subset \sigma_{\mathcal{A}}(x)$. $\square$

COROLLARY 6.3. *Let x be an element in a Banach algebra $\mathcal{B}$ with unit e such that the resolvent set $\rho(x)$ is connected. Then for each $\lambda \in \rho(x)$ the inverse $(\lambda e - x)^{-1}$ is the limit in $\mathcal{B}$ of a sequence of polynomials in x.*

PROOF. Let $\mathcal{A}$ be the closure in $\mathcal{B}$ of the set of all polynomials in x. Clearly, $\mathcal{A}$ is a closed subalgebra of $\mathcal{B}$ with the same unit as in $\mathcal{B}$. According to our hypothesis, $\rho_{\mathcal{B}}(x) = \rho(x)$ is connected, and hence $\rho_{\mathcal{B}}(x) = \rho_{\mathcal{A}}(x)$ by Corollary 6.2(a). It follows that for each $\lambda \in \rho(x)$ the inverse $(\lambda e - x)^{-1}$ is in $\mathcal{A}$, which proves the corollary. $\square$

Let K be a compact operator on the complex Banach space X. Then the resolvent set $\rho(K)$ of K is connected, and hence Corollary 6.3 implies that for each $\lambda \in \rho(K)$ the inverse $(\lambda I - K)^{-1}$ is the limit in the operator norm of a sequence of polynomials in K.

XXIX.7 SPECTRAL RADIUS

Given x in a unital Banach algebra $\mathcal{B}$, the *spectral radius* $r(x)$ is defined by

$$r(x) = \max\{|\lambda| \mid \lambda \in \sigma(x)\}.$$

In other words, the spectral radius of x is the radius of the smallest closed disc with centre at 0 containing $\sigma(x)$. If $|\lambda| > \|x\|$, then $\lambda \in \rho(x)$, and hence $r(x) \leq \|x\|$.

Let $\mathcal{B}$ be any of the Banach algebras appearing in the examples (i)–(iv) in Section XXIX.1. Then $r(f) = \|f\|$ for any $f \in \mathcal{B}$. For the Banach algebra $\mathcal{L}(H)$, where H is a Hilbert space, the situation is different. If $T \in \mathcal{L}(H)$ is selfadjoint, then $r(T) = \|T\|$, but for an arbitrary $T \in \mathcal{L}(H)$ the spectral radius may be strictly less than the norm. For example, if $T \in \mathcal{L}(\mathbb{C}^2)$ is given by

$$(1) \qquad\qquad T = \begin{bmatrix} 0 & 1 \\ 0 & 0 \end{bmatrix},$$

then $r(T) = 0$, but $\|T\| = 1$.

THEOREM 7.1. *The spectral radius of an element x in a unital Banach algebra $\mathcal{B}$ is given by*

$$(2) \qquad\qquad r(x) = \lim_{n \to \infty} \|x^n\|^{1/n}.$$

PROOF. Take $\mu \in \sigma(x)$. Then $\mu^n \in \sigma(x^n)$, because

$$\mu^n e - x^n = (\mu e - x)y = y(\mu e - x),$$

where $y = \mu^{n-1}e + \mu^{n-2}x + \cdots + x^{n-1}$. Hence $|\mu|^n \leq r(x^n) \leq \|x^n\|$. Since $\mu \in \sigma(x)$ was arbitrary, we get

$$(3) \qquad\qquad r(x) \leq \inf \|x^n\|^{1/n}.$$

To prove a reverse inequality, let f be a continuous linear functional on $\mathcal{B}$ and consider the function $\varphi(\lambda) = f\big((\lambda e - x)^{-1}\big)$. The function φ is analytic on the resolvent set $\rho(x)$ (cf., Theorem 5.1) and for $|\lambda| > \|x\|$,

$$(4) \qquad\qquad \varphi(\lambda) = \frac{1}{\lambda} f\left(\left(e - \frac{1}{\lambda}x\right)^{-1}\right) = \sum_{n=0}^{\infty} \frac{1}{\lambda^{n+1}} f(x^n).$$

Since φ is analytic for $|\lambda| > r(x)$, the series in (4) converges for $|\lambda| > r(x)$. Take $|\lambda| > r(x)$. Then $\left(\frac{1}{\lambda}\right)^{n+1} f(x^n) \to 0$ if $n \to \infty$, and hence

$$\sup_n \left| f\left(\frac{1}{\lambda^n} x^n\right) \right| < \infty.$$

But then the uniform boundedness principle implies that

$$m_\lambda := \sup_n \left\| \frac{1}{\lambda^n} x^n \right\| < \infty.$$

It follows that $\|x^n\|^{1/n} \le |\lambda| m_\lambda^{1/n}$, $n = 1, 2, \ldots$, which implies that $\limsup \|x^n\|^{1/n} \le r(x)$. This inequality combined with (3) proves the theorem. $\square$

Theorem 7.1 shows that the spectral radius of an element in a Banach algebra does not change if the algebra is enlarged. For example, the spectral radius of $f \in PC([a, b])$ is the same as the spectral radius of f in the algebra $B([a, b])$. Note that Corollary 4.3 also leads to the same remark.

COROLLARY 7.2. *Given x in a Banach algebra B with a unit e, the series expansion*

$$(5) \qquad (\lambda e - x)^{-1} = \sum_{n=0}^{\infty} \frac{1}{\lambda^{n+1}} x^n$$

holds for $|\lambda| > r(x)$.

PROOF. Formula (2) implies that the series in (5) converges for $|\lambda| > r(x)$. Let f be an arbitrary continuous linear functional on B. Then

$$(6) \qquad f\big((\lambda e - x)^{-1}\big) = f\left(\sum_{n=0}^{\infty} \frac{1}{\lambda^{n+1}} x^n \right)$$

for $|\lambda| > \|x\|$ (cf., formula (4)). Both sides of (6) are well-defined and analytic in λ for $|\lambda| > r(x)$. So (6) holds for $|\lambda| > r(x)$. But then the Hahn-Banach theorem implies (5). $\square$

COROLLARY 7.3. *If x and y are elements in a unital Banach algebra and $xy = yx$, then*

(a) $r(\alpha x) = |\alpha| r(x)$, $\alpha \in \mathbb{C}$,

(b) $r(x + y) \le r(x) + r(y)$,

(c) $r(xy) \le r(x) r(y)$.

PROOF. Equality (a) follows directly from formula (2). Since $xy = yx$, we have $(xy)^n = x^n y^n$. Thus

$$\|(xy)^n\|^{1/n} \le \|x^n\|^{1/n} \|y_n\|^{1/n} \to r(x) r(y),$$

which yields (c). It remains to prove (b). To do this, take $s > r(x)$ and $t > r(y)$, and put $u = \frac{1}{s}x$ and $v = \frac{1}{t}y$. From $xy = yx$ it follows that

$$(7) \qquad \|(x+y)^n\| = \left\|\sum_{k=0}^{n}\binom{n}{k}x^k y^{n-k}\right\| \le \sum_{k=0}^{n}\binom{n}{k}s^k t^{n-k}\|u^k\|\|v^{n-k}\|.$$

For each positive integer n choose a nonnegative integer $n' \le n$ so that

$$\|u^{n'}\|\|v^{n-n'}\| = \max_{0\le k\le n}\|u^k\|\|v^{n-k}\|.$$

It follows from (7) that for $n'' = n - n'$,

$$\|(x+y)^n\|^{1/n} \le (s+t)\|u^{n'}\|^{1/n}\|v^{n''}\|^{1/n}.$$

Hence by Theorem 7.1

$$(8) \qquad r(x+y) \le (s+t)\liminf_{n\to\infty}\|u^{n'}\|^{1/n}\|v^{n''}\|^{1/n}.$$

There exists a sequence $n_1 < n_2 < n_3 < \cdots$ such that $\ell := \lim_{\nu\to\infty} n'_\nu/n_\nu$ exists. If $\ell \ne 0$, then $n'_\nu \to \infty$ if $\nu \to \infty$ and in that case, because of (2),

$$\lim_{\nu\to\infty}\|u^{n'_\nu}\|^{1/n_\nu} = r(u)^\ell = \left(\frac{1}{s}r(x)\right)^\ell < 1.$$

If $\ell = 0$, then

$$\limsup_{\nu\to\infty}\|u^{n'_\nu}\|^{1/n_\nu} \le \lim_{\nu\to\infty}\|u\|^{n'_\nu/n_\nu} = 1.$$

In either case, $\limsup_{\nu\to\infty}\|u^{n'_\nu}\|^{1/n_\nu} \le 1$. Since $\lim_{\nu\to\infty} n''_\nu/n_\nu = 1-\ell$, a a similar argument shows that $\limsup_{\nu\to\infty}\|v^{n''_\nu}\|^{1/n_\nu} \le 1$. But then (8) implies that $r(x+y) \le s+t$. Since s and t were any numbers larger than $r(x)$ and $r(y)$, respectively, (b) follows. $\square$

Since $\sigma(x) \ne \emptyset$, the definition of the spectra radius implies that $r(x) = 0$ if and only if $\sigma(x) = \{0\}$. An element x with this property will be called quasi-nilpotent. More generally, an element x in a (not necessarily unital) Banach algebra is said to be *quasi-nilpotent* if $\lim_{n\to\infty}\|x^n\|^{1/n} = 0$. Obviously, if x is *nilpotent*, i.e., $x^n = 0$ for some positive integer n, then x is quasi-nilpotent. The operator $T \in \mathcal{L}(\mathbb{C}^2)$ defined by (1) is nilpotent, and the operator $T \in \mathcal{L}(\ell_2)$ defined by

$$T(\alpha_1, \alpha_2, \alpha_3, \ldots) = \left(0, \alpha_1, \frac{1}{2}\alpha_2, \ldots\right)$$

is quasi-nilpotent, but not nilpotent. We conclude this section with an example of a commutative Banach algebra in which each element is quasi-nilpotent.

Define on $L_1([0,1])$ the convolution

$$(f * g)(t) = \int_0^t f(t-s)g(s)ds, \qquad 0 \le t \le 1.$$

The equality has to be understood as being valid almost everywhere on $[0,1]$. With respect to this operation and the L_1-norm, $\mathcal{B} = L_1([0,1])$ is a commutative Banach algebra. This can be established in the same manner as the statement that the Wiener convolution algebra $L_1(\mathbb{R})$ (see Section XXIX.2) is a commutative Banach algebra. We shall now show that every $g \in \mathcal{B} = L_1([0,1])$ is quasi-nilpotent. Let $f_0 \in \mathcal{B}$ be the constant function 1. It is easy to verify that

$$f_0^n(t) = \frac{1}{(n-1)!}t^{n-1}.$$

Therefore every polynomial is in $\mathcal{B}$ and can be written in the form $\sum_{j=0}^p \alpha_j f_0^j$. Moreover $\|f_0^n\| = (n!)^{-1}$, which implies that $r(f_0) = 0$. It follows from Corollary 7.3 that every polynomial is quasi-nilpotent. Given $g \in \mathcal{B}$ and any $\varepsilon > 0$, there exists a polynomial $p \in \mathcal{B}$ such that $\|g - p\| < \varepsilon$. Hence (use Corollary 7.3)

$$r(g) \le r(g - p) + r(p) = r(g - p) \le \|g - p\| < \varepsilon,$$

which shows that $r(g) = 0$. Since every element in $\mathcal{B}$ is quasi-nilpotent, we have also shown that $\mathcal{B}$ does not have a unit.

Take $k \in L_1([0,1])$. The fact that each element in $\mathcal{B} = L_1([0,1])$ is quasi-nilpotent can be applied to show that for any $g \in L_1([0,1])$ the equation

$$(9) \qquad f(t) + \int_0^t k(t-s)f(s)ds = g(t), \qquad 0 \le t \le 1, \text{ a.e.,}$$

has a unique solution in $L_1([0,1])$. To see this, we adjoin a unit to $\mathcal{B} = L_1([0,1])$ as in Section XXIX.1 to obtain a unital Banach algebra $\widehat{\mathcal{B}}$, namely

$$\widehat{\mathcal{B}} = \{(f,\alpha) \mid \alpha \in \mathbb{C}, f \in \mathcal{B}\}.$$

Since $r(k) = 0$, the element $(k,1)$ is invertible in $\widehat{\mathcal{B}}$. But then there exists $\ell \in \mathcal{B}$ such that $k * \ell + \ell + k = 0$. Furthermore, equation (9) can be rewritten in the form $(k,1)(f,0) = (g,0)$. It follows that (9) has a unique solution f in $L_1([0,1])$, namely

$$f(t) = g(t) + \int_0^t \ell(t-s)g(s)ds, \qquad 0 \le t \le 1, \text{ a.e.} \,.$$

XXIX.8 MATRICES OVER BANACH ALGEBRAS

Let $\mathcal{B}$ be a Banach algebra. By $\mathcal{B}^{m \times m}$ we denote the set of all $m \times m$ matrices $[a_{ij}]_{i,j=1}^{m}$ with entries from $\mathcal{B}$. We endow $\mathcal{B}^{m \times m}$ with the usual matrix operations and the norm

$$(1) \qquad \|[a_{ij}]\| = \max_{1 \leq i \leq m} \sum_{j=1}^{m} \|a_{ij}\|.$$

Then $\mathcal{B}^{m \times m}$ is a Banach algebra. If $\mathcal{B}$ has unit e, say, then $\mathcal{B}^{m \times m}$ also has a unit, namely, the $m \times m$ matrix E with e on the main diagonal and zeros elsewhere.

The precise definition of the norm on $\mathcal{B}^{m \times m}$ is not so important, and for some cases there is a more natural norm than the one given in (1). For example, let $\mathcal{B} = C(S)$, where S is a compact Hausdorff space. For $\Phi = [\varphi_{ij}]_{i,j=1}^{m} \in C(S)^{m \times m}$ the natural norm is

$$(2) \qquad \|\Phi\| = \max_{t \in S} \left\| \begin{bmatrix} \varphi_{11}(t) & \cdots & \varphi_{1m}(t) \\ \vdots & & \vdots \\ \varphi_{m1}(t) & \cdots & \varphi_{mm}(t) \end{bmatrix} \right\|.$$

The norm in the right hand side of (2) is the usual operator norm on $\mathbb{C}^{m}$.

Given $b \in \mathcal{B}$, define $E_{ij}(b) \in \mathcal{B}^{m \times m}$ to be the $m \times m$ matrix with all entries zero except the (i,j)-th entry which is equal to b. With $\mathcal{B} = C(S)$ and the norm on $\mathcal{B}^{m \times m}$ as in (2),

$$(3) \qquad \|E_{ij}(b)\| = \|b\|, \qquad i,j = 1,\ldots,m.$$

In the sequal we shall require that the norm on $\mathcal{B}^{m \times m}$ is submultiplicative, satisfies condition (3) for each $b \in \mathcal{B}$, and if $\mathcal{B}$ has a unit e, then the unit E in $\mathcal{B}^{m \times m}$ should have norm one. One could work with weaker conditions on the norm on $\mathcal{B}^{m \times m}$, but for our purposes the three requirements mentioned in the previous sentence are good enough.

For $A = [a_{ij}]_{i,j=1}^{m} \in \mathcal{B}^{m \times m}$ the *determinant* of A is the element $\det A \in \mathcal{B}$ defined by

$$(4) \qquad \det A = \sum_{\sigma} (\operatorname{sign} \sigma) a_{1\sigma_1} a_{2\sigma_2} \cdots a_{m\sigma_m}.$$

In (4) the summation is over all permutations σ of the numbers $1, 2, \ldots, m$ and $\operatorname{sign} \sigma$ denotes the sign of the permutation σ. In the commutative case the determinant is a useful object.

THEOREM 8.1. *Let $\mathcal{B}$ be a unital Banach algebra, and let*

$$A = [a_{ij}]_{i,j=1}^{m} \in \mathcal{B}^{m \times m}.$$

Assume that the entries of A commute with one another. Then A is invertible in $\mathcal{B}^{m \times m}$ if and only if $\det A$ is invertible in $\mathcal{B}$, and in that case A^{-1} is given by Cramer's rule.

PROOF. The proof is the same as that of Proposition XI.7.2. $\square$

For later purposes (see Section XXXII.2) we mention the following lemma.

LEMMA 8.2. *Let $\mathcal{D}$ be a subset of $\mathcal{B}$ containing 0, and let $\mathcal{D}^{m \times m}$ be the set of all $m \times m$ matrices with entries from $\mathcal{D}$. Denote by $\mathcal{A}(\mathcal{D})$ (resp., $\mathcal{A}(\mathcal{D}^{m \times m})$) the smallest closed subalgebra of $\mathcal{B}$ (resp., $\mathcal{B}^{m \times m}$) containing all elements from $\mathcal{D}$ (resp., $\mathcal{D}^{m \times m}$). Then*

$$(5) \qquad \mathcal{A}(\mathcal{D}^{m \times m}) = \left\{ \begin{bmatrix} a_{11} & \cdots & a_{1m} \\ \vdots & & \vdots \\ a_{m1} & \cdots & a_{mm} \end{bmatrix} \mid a_{ij} \in \mathcal{A}(\mathcal{D}) \right\}.$$

PROOF. We begin with a few general observations. Let $A = [a_{ij}]_{i,j=1}^{m}$ be an arbitrary element in $\mathcal{B}^{m \times m}$. Since the norm on $\mathcal{B}^{m \times m}$ is submultiplicative, condition (3) implies that

$$(6) \qquad \begin{aligned} \|a_{ij}\| = \|E_{ij}(a_{ij})\| &= \|E_{ii}(e)AE_{jj}(e)\| \\ &\leq \|E_{ii}(e)\| \|A\| \|E_{jj}(e)\| \\ &= \|A\|, \qquad i,j = 1, \ldots, m. \end{aligned}$$

On the other hand,

$$(7) \qquad \|A\| = \left\| \sum_{i,j=1}^{m} E_{ij}(a_{ij}) \right\| \leq \sum_{i,j=1}^{m} \|E_{ij}(a_{ij})\| = \sum_{i,j=1}^{m} \|a_{ij}\|.$$

Write $\mathcal{A}(\mathcal{D})^{m \times m}$ for the right hand side of (5). Since $\mathcal{A}(\mathcal{D})$ is closed in $\mathcal{B}$, formula (6) implies that $\mathcal{A}(\mathcal{D})^{m \times m}$ is closed in $\mathcal{B}^{m \times m}$. Let $\mathcal{A}_0(\mathcal{D})$ be the smallest subalgebra of $\mathcal{B}$ containing all elements from $\mathcal{D}$, and define $\mathcal{A}_0(\mathcal{D}^{m \times m})$ analogously. Take $A = [a_{ij}]_{i,j=1}^{m}$ in $\mathcal{A}_0(\mathcal{D}^{m \times m})$. Then A is a sum of products of matrices from $\mathcal{D}^{m \times m}$. It follows that each entry a_{ij} is a sum of products of elements from $\mathcal{D}$. Hence $a_{ij} \in \mathcal{A}_0(\mathcal{D}) \subset \mathcal{A}(\mathcal{D})$ for all i and j. Thus $\mathcal{A}_0(\mathcal{D}^{m \times m}) \subset \mathcal{A}(\mathcal{D})^{m \times m}$. Since $\mathcal{A}(\mathcal{D}^{m \times m})$ is the closure of $\mathcal{A}_0(\mathcal{D}^{m \times m})$, we conclude that $\mathcal{A}(\mathcal{D}^{m \times m}) \subset \mathcal{A}(\mathcal{D})^{m \times m}$.

To prove the reverse inclusion, let $A = [a_{ij}]_{i,j=1}^{m}$ have all its entries in $\mathcal{A}_0(\mathcal{D})$. Since the set of all A's of this type is dense in $\mathcal{A}(\mathcal{D})^{m \times m}$ because of (7), it suffices to prove that $A \in \mathcal{A}_0(\mathcal{D}^{m \times m})$. Note that $A = \sum_{i,j=1}^{m} E_{ij}(a_{ij})$, where $E_{ij}(b)$ is the $m \times m$ matrix defined in the third paragraph of this section. We know that $\mathcal{A}_0(\mathcal{D}^{m \times m})$ is an algebra. So, without loss of generality, we may assume that $A = E_{ij}(b)$ with $b \in \mathcal{A}_0(\mathcal{D})$. In fact, we may take $b = b_1 \cdots b_r$ with $b_k \in \mathcal{D}$ for $k = 1, \ldots, r$. But then

$$A = E_{ij}(b) = E_{ij}(b_1)E_{jj}(b_2) \cdots E_{jj}(b_r).$$

Since $b_1, \ldots, b_r$ and 0 are in $\mathcal{D}$, the matrices $E_{ij}(b_1), E_{jj}(b_2), \ldots, E_{jj}(b_r)$ are in $\mathcal{D}^{m \times m}$. So $A \in \mathcal{A}_0(\mathcal{D}^{m \times m})$. $\square$

XXIX.9 FACTORIZATION IN BANACH ALGEBRAS

Throughout this section B is a Banach algebra with unit e. We call B a *decomposing Banach algebra* if B has closed subalgebras B_+ and B_-, both containing non-zero elements, such that

$$(1) \qquad B = B_- \oplus B_+.$$

For example, the algebra $\mathbb{C}^{m \times m}$ of the $m \times m$ matrices has a natural decomposition of this type, namely with

$$C_{-,0}^{m \times m} = \{[a_{ij}]_{i,j=1}^m \mid a_{ij} = 0 \text{ for } j - i \geq 0\},$$
$$\mathbb{C}_+^{m \times m} = \{[a_{ij}]_{i,j=1}^m \mid a_{ij} = 0 \text{ for } j - i < 0\}.$$

An important second example is the *matrix Wiener algebra* $\mathcal{W}^{m \times m}$. Indeed, put

$$(2a) \qquad \mathcal{W}_{-,0}^{m \times m} = \{\Phi \in \mathcal{W}^{m \times m} \mid \Phi_k = 0 \text{ for } k \geq 0\},$$

$$(2b) \qquad \mathcal{W}_+^{m \times m} = \{\Phi \in \mathcal{W}^{m \times m} \mid \Phi_k = 0 \text{ for } k < 0\},$$

where Φ_k denotes the k-th Fourier coefficient of Φ, i.e.,

$$\Phi_k = \frac{1}{2\pi} \int_{-\pi}^{\pi} e^{-ikt} \Phi(e^{it}) dt.$$

The algebra $\mathcal{W}^{m \times m}$ endowed with the norm

$$(3) \qquad \|\Phi\| = \sum_{k=-\infty}^{\infty} \|\Phi_k\|$$

is a unital Banach algebra. (The norm in the right hand side of (3) is the usual operator norm of a matrix, i.e., $\|\Phi_k\|$ is the square root of the largest eigenvalue of $\Phi_k^* \Phi_k$.) The spaces $\mathcal{W}_{-,0}^{m \times m}$ and $\mathcal{W}_+^{m \times m}$ defined in (2a), (2b) are subalgebras of $\mathcal{W}^{m \times m}$, and they are closed in the norm (3). Obviously,

$$(4) \qquad \mathcal{W}^{m \times m} = \mathcal{W}_{-,0}^{m \times m} \oplus \mathcal{W}_+^{m \times m},$$

and hence $\mathcal{W}^{m \times m}$ is a decomposing Banach algebra.

The following theorem is the main result of this section.

THEOREM 9.1. *Assume $B = B_- \oplus B_+$ is a decomposing Banach algebra. Let P be the projection of B onto B_+ along B_-, and put $Q = I - P$. If $a \in B$ and $\|a\| < \min\{\|P\|^{-1}, \|Q\|^{-1}\}$, then $e - a$ factorizes as*

$$(5) \qquad e - a = (e + b_-)(e + b_+)$$

with $e + b_\pm$ invertible in $\mathcal{B}$ and

$$(6) \qquad b_\pm \in \mathcal{B}_\pm, \qquad (e + b_\pm)^{-1} - e \in \mathcal{B}_\pm.$$

Furthermore, the elements b_+ and b_- are uniquely determined by a and are given by

$$(7) \qquad b_\pm = (e + x_\pm)^{-1} - e,$$

where $x_+ \in \mathcal{B}_+$ and $x_- \in \mathcal{B}_-$ are the unique solutions of the equations

$$(8) \qquad x_+ - P(ax_+) = Pa, \qquad x_- - Q(x_-a) = Qa.$$

The above theorem may be viewed as a generalization of Theorem XXII.8.2, which concerns factorization in a strongly decomposable algebra. Note that, in general, in a decomposing Banach algebra $y_\pm \in \mathcal{B}_\pm$ and $e + y_\pm$ invertible do not imply that

$$(e + y_\pm)^{-1} - e \in \mathcal{B}_\pm,$$

as is the case for strongly decomposable algebras. For that reason we have in Theorem 9.1 an additional condition on $\|a\|$. On the other hand, the equations in (8) are the analogues of the equations (4a) and (4b) in Theorem XXII.8.2. Furthermore, we shall see that, as in Theorem XXII.8.2,

$$(9) \qquad b_+ = x_- - x_-a - a, \qquad b_- = x_+ - x_+a - a,$$

where $x_+ \in \mathcal{B}_+$ is the solution of the first equation in (8) and $x_- \in \mathcal{B}_-$ is the solution of the second equation in (8).

For the proof of Theorem 9.1 we need the notion of left invertibility. An element x in $\mathcal{B}$ is said to be *left invertible* if there exists $y \in \mathcal{B}$ such that $yx = e$.

LEMMA 9.2. *Let $\alpha: [0, 1] \to \mathcal{B}$ be a continuous function. Assume that $\alpha(t)$ is left invertible for each $0 \leq t \leq 1$, and let $\alpha(0)$ be invertible. Then $\alpha(t)$ is invertible for each $0 \leq t \leq 1$.*

PROOF. First, let us show that the set of left invertible, non-invertible operators on a Banach space X is open. Let $A \in \mathcal{L}(X)$, and assume that $A^+ \in \mathcal{L}(X)$ is a left inverse of A. Take $B \in \mathcal{L}(X)$ such that $\|A - B\| < \|A^+\|^{-1}$. We claim that B is also left invertible. To see this note that

$$B = A - (A - B) = \left[I - (A - B)A^+ \right] A,$$

because A^+A is the identity operator on X. By our choice of B we have $\|(A-B)A^+\| < 1$. So $F = I - (A - B)A^+$ is invertible, and hence A^+F is a left inverse of B. Since F is invertible, $B = FA$ implies that B is non-invertible whenever A is non-invertible, which yields the desired result.

Next, we prove the lemma for the case when $\mathcal{B} = \mathcal{L}(X)$, where X is a Banach space. Consider the set

$$V = \left\{ t \in [0, 1] \mid \alpha(t) \text{ is invertible} \right\}.$$

By our hypotheses, $0 \in V$, and so V is not empty. Since the set of invertible operators is open, the continuity of $\alpha(\cdot)$ implies that V is open. By the result of the previous paragraph, we also have that $W = [0,1]\backslash V$ is open. So $[0,1]$ is the disjoint union of two open sets. But $[0,1]$ is connected. So one of the sets must be empty. We know that $V \neq \emptyset$. Hence $W = \emptyset$, and thus $V = [0,1]$.

To prove the lemma in general, let $A(t)\colon \mathcal{B} \to \mathcal{B}$ be the operator defined by

$$(10) \qquad A(t)x = \alpha(t)x, \qquad x \in \mathcal{B},\ 0 \leq t \leq 1.$$

Then $A\colon [0,1] \to \mathcal{L}(\mathcal{B})$ is continuous, $A(t)$ is a left invertible operator on $\mathcal{B}$ for each $0 \leq t \leq 1$ and $A(0)$ is an invertible operator on $\mathcal{B}$. So, by the result of the previous paragraph, $A(t)$ is invertible for $0 \leq t \leq 1$. Fix $0 \leq t \leq 1$. Since $A(t)$ is invertible as an operator on $\mathcal{B}$, we can find $x(t) \in \mathcal{B}$ so that $A(t)x(t) = e$. But then we see from (10) that $x(t)$ is a right inverse of $\alpha(t)$. So $\alpha(t)$ has a left and a right inverse, and hence $\alpha(t)$ is invertible.

PROOF OF THEOREM 9.1. We split the proof into four parts. The first part concerns the uniqueness statement.

Part (a). Assume $e-a$ admits the factorization (5) with $b_{\pm}$ as in (6). Consider a second factorization of this type, i.e.,

$$(11) \qquad e - a = (e + c_-)(e + c_+),$$

where $c_{\pm}$ have the same properties as $b_{\pm}$ in (6). We want to show that $c_{\pm} = b_{\pm}$. Put

$$y_- = (e + c_-)^{-1} - e \in \mathcal{B}_-, \qquad x_+ = (e + b_+)^{-1} - e \in \mathcal{B}_+.$$

From (5) and (11) we see that

$$(e + y_-)(e + b_-) = (e + c_+)(e + x_+).$$

Hence

$$y_- + b_- + y_- b_- = c_+ + x_+ + c_+ x_+ \in \mathcal{B}_- \cap \mathcal{B}_+,$$

and we can use (1) to show that $y_- + b_- + y_- b_- = 0$. Thus $(e + y_-)(e + b_-) = e$. It follows that

$$e + b_- = (e + y_-)^{-1} = e + c_-.$$

So $b_- = c_-$, which, by (5) and (11), also implies that $b_+ = c_+$.

Part (b). Let a satisfy the conditions mentioned in the theorem. In this part we show that the equations in (8) are uniquely solvable in $\mathcal{B}_+$ and $\mathcal{B}_-$, respectively. Put

$$(12) \qquad T_a x = x - P(ax), \quad S_a x = x - Q(xa) \qquad (x \in \mathcal{B}).$$

The fact that $\mathcal{B}_+$ and $\mathcal{B}_-$ are closed in $\mathcal{B}$ implies that P and Q are bounded linear operators on $\mathcal{B}$. Left multiplication and right multiplication by a also induce bounded linear operators on $\mathcal{B}$. It follows that T_a and S_a are well-defined bounded linear operators on $\mathcal{B}$. By our conditions on a, we have $\|I - T_a\| < 1$ and $\|I - S_a\| < 1$, and hence T_a and S_a are invertible. In fact,

$$(13) \qquad T_a^{-1} = I + \sum_{n=1}^{\infty}(I - T_a)^n, \qquad S_a^{-1} = I + \sum_{n=1}^{\infty}(I - S_a)^n.$$

Note that the equations in (8) may be rewritten as

$$(14) \qquad\qquad T_a x_+ = Pa, \qquad S_a x_- = Qa.$$

From (12) we see that $T_a \mathcal{B}_+ \subset \mathcal{B}_+$ and $S_a \mathcal{B}_- \subset \mathcal{B}_-$. These inclusions remain true if T_a is replaced by $I - T_a$ and S_a by $I - S_a$. Since $\mathcal{B}_+$ and $\mathcal{B}_-$ are closed, we see from (13) that T_a maps $\mathcal{B}_+$ in a one-one way onto $\mathcal{B}_+$ and S_a maps $\mathcal{B}_-$ in a one-one way onto $\mathcal{B}_-$. Thus the first equation in (14) is uniquely solvable in $\mathcal{B}_+$ and the second equation is uniquely solvable in $\mathcal{B}_-$. It follows that the same holds true for the equations in (8).

Part (c). Let $x_+ \in \mathcal{B}_+$ be the solution of the first equation in (8), and let $x_- \in \mathcal{B}_-$ be the solution of the second equation in (8). In this part we derive a number of useful formulas. Note that $P\big[a(e + x_+)\big] = x_+$. Thus

$$(15) \qquad \begin{aligned} (e - a)(e + x_+) &= e + x_+ - a(e + x_+) \\ &= e + x_+ - P\big[a(e + x_+)\big] - Q\big[a(e + x_+)\big] \\ &= e + b_-, \end{aligned}$$

with $b_- := -Q\big[a(e + x_+)\big] \in \mathcal{B}_-$. Similarly, since $Q\big[(e + x_-)a\big] = x_-$,

$$(16) \qquad \begin{aligned} (e + x_-)(e - a) &= e + x_- - P\big[(e + x_-)a\big] - Q\big[(e + x_-)a\big] \\ &= e + b_+, \end{aligned}$$

with $b_+ := -P\big[(e + x_-)a\big]$. The space $\mathcal{B}_+$ contains non-zero elements. Thus $\|P\| \geq 1$, and hence $\|P\|^{-1} \leq 1$, which implies that $\|a\| < 1$. So, by Theorem 4.1, the element $e - a$ is invertible, and the computations made above show that

$$\begin{aligned} (e + b_+)(e + x_+) &= (e + b_+)(e - a)^{-1}(e - a)(e + x_+) \\ &= (e + x_-)(e + b_-). \end{aligned}$$

Hence

$$b_+ + x_+ + b_+ x_+ = b_- + x_- + x_- b_- \in \mathcal{B}_- \cap \mathcal{B}_+.$$

But $\mathcal{B}_+ \cap \mathcal{B}_-$ consists of the zero element only. Therefore,

$$(17) \qquad\qquad (e + b_+)(e + x_+) = (e + x_-)(e + b_-) = e.$$

If $\mathcal{B}$ is commutative, then the identities in (16) and (17) provide the proof of the theorem.

Part (d). In this part we complete the proof for the non-commutative case. First, we repeat the above arguments with a replaced by ta, where $0 \le t \le 1$. This leads to the elements $x_\pm(t)$ and $b_\pm(t)$, which are related to ta in the same way as $x_\pm$ and $b_\pm$ are related to a. Note that

$$(T_{ta})^{-1} = I + \sum_{n=1}^{\infty} t^n (I - T_a)^n, \qquad (S_{ta})^{-1} = I + \sum_{n=1}^{\infty} t^n (I - S_a)^n$$

for $0 \le t \le 1$. It follows that the functions $x_\pm(\cdot)$ and $b_\pm(\cdot)$ are analytic on $[0,1]$. From (17) we know that

$$\big(e + b_+(t)\big)\big(e + x_+(t)\big) = \big(e + x_-(t)\big)\big(e + b_-(t)\big) = e, \qquad 0 \le t \le 1.$$

So $e + x_+(t)$ and $e + b_-(t)$ are left invertible for $0 \le t \le 1$. Obviously, $e + x_+(0)$ and $e + b_-(0)$ are equal to the unit e. So, we may apply Lemma 9.2 to conclude that $e + x_+(t)$ and $e + b_-(t)$ are invertible for $0 \le t \le 1$. In particular, $e + x_+$ and $e + b_-$ are invertible. By using (16) and (17) the theorem is now proved. $\square$

From the proof of Theorem 9.1 it is clear that the solutions of the equations in (8) are given by

$$x_+ = \lim_{n \to \infty} s_n, \qquad x_- = \lim_{n \to \infty} t_n,$$

where the sequences (s_n) and (t_n) are recursively defined by

$$s_0 = 0, \quad s_n = P(as_{n-1}) + Pa, \qquad n = 1, 2, \dots ,$$

$$t_0 = 0, \quad t_n = Q(t_{n-1}a) + Qa, \qquad n = 1, 2, \dots .$$

CHAPTER XXX
COMMUTATIVE BANACH ALGEBRAS

This chapter contains the Gelfand theory of commutative Banach algebras. The Gelfand spectrum and the Gelfand transform are introduced and analysed. The results are illustrated by various examples. In particular, it is explained in detail how under the Gelfand transformation piecewise continuous functions become continuous. Special attention is paid to finitely generated commutative Banach algebras and to the Banach algebra generated by a compact operator. The last two sections present applications to factorization of matrix functions and to Wiener-Hopf integral operators. The analysis starts with a study of multiplicative linear functionals.

XXX.1 MULTIPLICATIVE LINEAR FUNCTIONALS

A functional φ on a Banach algebra $\mathcal{B}$ is called *multiplicative* if φ is not the zero functional and $\varphi(xy) = \varphi(x)\varphi(y)$ for all x and y in $\mathcal{B}$. Clearly, if e is a unit in $\mathcal{B}$ and φ is multiplicative on $\mathcal{B}$, then $\varphi(e) = 1$.

THEOREM 1.1. *Every multiplicative linear functional on a unital Banach algebra is bounded and has norm 1.*

PROOF. Let φ be a multiplicative linear functional on a Banach algebra with unit e. Assume that there exists an $x \in \mathcal{B}$ such that $\|x\| = 1$ and $|\varphi(x)| > 1$. Then $\varphi(x)e - x$ is invertible and $\varphi(\varphi(x)e - x) = 0$. Since φ is multiplicative, this implies that

$$1 = \varphi(e) = \varphi(\varphi(x)e - x)\varphi((\varphi(x)e - x)^{-1}) = 0,$$

which is absurd. Hence $|\varphi(x)| \leq 1$ for all $x \in \mathcal{B}$ with $\|x\| = 1$. As $\varphi(e) = 1$, we get $\|\varphi\| = 1$. $\square$

In the remaining part of this section we describe the multiplicative linear functionals on the algebras which were discussed earlier in Sections XXIX.1 and XXIX.2.

(i) Let $\mathcal{B} = C(S)$, where S is a compact Hausdorff space. The unit of $\mathcal{B}$ is denoted by e. For each $t \in S$, the functional h_t defined on $\mathcal{B}$ by $h_t(f) = f(t)$ is linear and multiplicative. Note that h_t on f evaluates the function f at the point t. We shall prove that *every multiplicative linear functional φ on $\mathcal{B} = C(S)$ is an h_t for some t in S.*

In order to do this, it suffices to prove the existence of a $t \in S$ such that $\mathrm{Ker}\,\varphi \subset \mathrm{Ker}\,h_t$. For, if this is the case, then $\mathcal{B} = \mathrm{sp}\{e\} \oplus \mathrm{Ker}\,\varphi$, and therefore given $g \in \mathcal{B}$, $g = \alpha e + f$ for some $f \in \mathrm{Ker}\,\varphi$. Hence

$$\varphi(g) = \alpha = h_t(\alpha e) = h_t(g).$$

Assume that $\mathrm{Ker}\,\varphi \not\subset \mathrm{Ker}\,h_s$ for each $s \in S$. Thus, given an $s \in S$, there exists an $f_s \in \mathrm{Ker}\,\varphi$ such that $f_s(s) = h_s(f_s) \neq 0$. Since each f_s is continuous and S is compact, it follows that there exists $s_1, \ldots, s_n$ in S such that $f_{s_1}, \ldots, f_{s_n}$ are in $\operatorname{Ker} \varphi$ and

$$f(t) := \sum_{k=1}^{n} |f_{s_k}(t)|^2 > 0, \qquad t \in S.$$

Put $g(t) = 1/f(t)$. Then $g \in \mathcal{B}$, $g = f^{-1}$ and

$$1 = \varphi(fg) = \varphi\left(\sum_{k=1}^{n} f_{s_k} \overline{f}_{s_k} g\right) = \sum_{k=1}^{n} \varphi(f_{s_k}) \varphi(\overline{f}_{s_k}) \varphi(g) = 0.$$

Contradiction. Thus any multiplicative linear functional on $C(S)$ is an h_t for some $t \in S$.

(ii) Let $\mathcal{B} = \mathcal{A}(\overline{\mathbb{D}})$ denote the Banach algebra described in Section XXIX.1, example (ii), where $\overline{\mathbb{D}}$ is the closed unit disc in the complex plane. For each $\lambda \in \overline{\mathbb{D}}$ define h_λ on $\mathcal{B}$ by $h_\lambda(f) = f(\lambda)$. Clearly, h_λ is a multiplicative linear functional. We shall prove that *every multiplicative linear functional φ on $\mathcal{B} = \mathcal{A}(\overline{\mathbb{D}})$ is an h_λ for some $\lambda \in \overline{\mathbb{D}}$.*

In order to do this, let $g_0 \in \mathcal{B}$ be the function $g_0(\lambda) = \lambda$ and let $\lambda_0 = \varphi(g_0)$. Since $\|g_0\| = 1$ and $\|\varphi\| = 1$ (by Theorem 1.1), the point $\lambda_0 \in \overline{\mathbb{D}}$. We claim that $\varphi = h_{\lambda_0}$. From the assumption that φ is multiplicative and linear it follows that $\varphi(p) = p(\lambda_0)$ for every polynomial p. Thus φ and h_{λ_0} coincide on the polynomials. Since the polynomials are dense in $\mathcal{B}$ (use the second Weierstrass approximation theorem) and φ and h_{λ_0} are continuous on $\mathcal{B}$, we conclude that $\varphi = h_{\lambda_0}$.

(iii) Let $\mathcal{B} = PC([a, b])$ be the algebra of piecewise continuous functions defined in Section XXIX.1, example (iv). For each $t \in [a, b]$ define multiplicative linear functionals h_t and h_t^+ on $\mathcal{B}$ by

$$h_t(f) := f(t), \qquad h_t^+(f) := f(t+).$$

We shall show that *every multiplicative linear functional φ on $\mathcal{B} = PC([a, b])$ is either an h_t or an h_t^+ for some $t \in [a, b]$.*

As we noted in example (i) above, it suffices to prove that $\operatorname{Ker} \varphi \subset \operatorname{Ker} h_t$ or $\operatorname{Ker} \varphi \subset \operatorname{Ker} h_t^+$ for some t. Suppose this is not the case. Then for each $s \in [a, b]$ there exist f_s and g_s in $\operatorname{Ker} \varphi$ such that

$$f_s(s) = h_s(f_s) \neq 0, \qquad g_s(s+) = h_s^+(g_s) \neq 0.$$

Fix $s \in [a, b]$, and consider $d_s(t) = |f_s(t)|^2 + |g_s(t)|^2$. The function d_s is continuous from the left, $d_s(s) > 0$ and $d_s(s+) > 0$. Hence there exists $\delta > 0$ and an open neighbourhood U_s of s in $[a, b]$ such that $d_s(t) \geq \delta > 0$ for each $t \in U_s$. In particular, $d_s(t) > 0$ and $d_s(t+) > 0$ for $t \in U_s$. But then we can use the compactness of $[a, b]$ to show that there exist $s_1, \ldots, s_n$ in $[a, b]$ such that $f_{s_1}, \ldots, f_{s_n}, g_{s_1}, \ldots, g_{s_n}$ are in $\operatorname{Ker} \varphi$ and

$$f(t) := \sum_{k=1}^{n} \left(|f_{s_k}(t)|^2 + |g_{s_k}(t)|^2 \right) > 0, \qquad a \leq t \leq b,$$

$$f(t+) > 0, \qquad a \leq t \leq b.$$

It follows that f is invertible in $\mathcal{B}$ with inverse v, say. Hence

$$1 = \varphi(vf) = \varphi(v)\left\{\sum_{k=1}^{n}[\varphi(f_{s_k})\varphi(\overline{f}_{s_k}) + \varphi(g_{s_k})\varphi(\overline{g}_{s_k})]\right\} = 0,$$

which is absurd. Therefore $\varphi = h_t$ or $\varphi = h_t^+$ for some $a \leq t \leq b$.

(iv) Let $\mathcal{W}$ be the Wiener algebra described in Section XXIX.2, example (a). For each ζ in the unit circle $\mathbb{T}$, the functional h_ζ defined on $\mathcal{W}$ by $h_\zeta(f) = f(\zeta)$ is multiplicative and linear. We shall show that *every multiplicative linear functional φ on $\mathcal{W}$ is an h_ζ for some $\zeta \in \mathbb{T}$.*

Let $g \in \mathcal{W}$ be defined by $g(\zeta) = \zeta$. Note that g is invertible in $\mathcal{W}$ and $\|g\| = \|g^{-1}\| = 1$. From $1 = \varphi(gg^{-1}) = \varphi(g)\varphi(g^{-1})$ it follows that $\varphi(g^{-1}) = \varphi(g)^{-1}$. Hence $|\varphi(g)| = 1$, because

$$|\varphi(g)| \leq 1, \qquad |\varphi(g)|^{-1} = |\varphi(g^{-1})| \leq 1.$$

Put $\zeta_0 = \varphi(g)$. Then $\zeta_0 \in \mathbb{T}$. Take $f \in \mathcal{W}$. Recall that $f(\zeta) = \sum_{n=-\infty}^{\infty} a_n \zeta^n$, where $\sum_{n=-\infty}^{\infty} |a_n| < \infty$. So we may write $f = \sum_{n=-\infty}^{\infty} a_n g^n$ and the latter series converges in the norm of $\mathcal{W}$. The fact that φ is continuous, linear and multiplicative implies that

$$\varphi(f) = \sum_{n=-\infty}^{\infty} a_n \varphi(g)^n = f(\zeta_0) = h_{\zeta_0}(f).$$

(v) Let $\mathcal{B} = \mathcal{W}(\beta)$ be the Wiener algebra with weights $\beta = \{\beta_k\}_{-\infty}^{\infty}$ given in Section XXIX.2, example (b). In order to describe the multiplicative linear functionals on $\mathcal{B}$, we first show that

$$(1) \qquad \rho_1 := \sup_{n<0} \beta_n^{1/n} = \lim_{n \to -\infty} \beta_n^{1/n},$$

$$(2) \qquad \rho_2 := \inf_{n>0} \beta_n^{1/n} = \lim_{n \to \infty} \beta_n^{1/n},$$

$$(3) \qquad 0 < \rho_1 \leq \rho_2 < \infty.$$

We start with (2). Let k be any fixed positive integer. Given a positive integer n, there exist nonnegative integers m and r such that $n = mk + r$, $0 \leq r < k$. The assumption

$$(4) \qquad \beta_{j+k} \leq \beta_j \beta_k, \qquad -\infty < j, k < \infty,$$

implies

$$\beta_n^{1/n} \leq (\beta_k^m \beta_r)^{1/n} = \beta_k^{(1/k)-(r/kn)} \beta_r^{1/n} \to \beta_k^{1/k} \qquad (n \to \infty).$$

Hence $\limsup_{n\to\infty} \beta_n^{1/n} \leq \beta_k^{1/k}$. Since k was an arbitrary positive integer, it follows that $\lim_{n\to\infty} \beta_n^{1/n}$ exists and the second equality in (2) is proved. For $n \geq 0$ put $\gamma_n = \beta_{-n}$, and note that $\gamma_{j+k} \leq \gamma_j \gamma_k$ for $0 \leq j,\ k < \infty$. If the second equality in (2) is applied to $(\gamma_n)_{n=0}^{\infty}$, then one obtains the second equality in (1). Let ρ_1 and ρ_2 be defined by the first equalities in (1) and (2). If k is a positive integer, then $\beta_0 \leq \beta_k \beta_{-k}$, and therefore $\beta_0^{1/k} \beta_{-k}^{-1/k} \leq \beta_k^{1/k}$. Hence (1) and (2) imply $\rho_1 \leq \rho_2$. From (1) it also follows that $\rho_1 > 0$ and from (2) we see that $\rho_2 < \infty$. Thus (3) holds.

Let x be an arbitrary element in $\mathcal{W}(\beta)$. Recall that x is a formal power series $x = \sum_{n=-\infty}^{\infty} a_n \zeta^n$ for which $\|x\| = \sum_{n=-\infty}^{\infty} |a_n| \beta_n$ is finite. Take λ in the annulus $\rho_1 \leq |\lambda| \leq \rho_2$. Then (1) and (2) imply that

$$(5) \qquad h_\lambda(x) = \sum_{n=-\infty}^{\infty} a_n \lambda^n$$

is well-defined. In particular, the series in (5) is absolutely convergent. It is easy to see that h_λ is multiplicative and linear. We shall prove that *every multiplicative linear functional φ on $\mathcal{W}(\beta)$ is an h_λ for some λ in $\rho_1 \leq |\lambda| \leq \rho_2$.*

Let $y \in \mathcal{W}(\beta)$ be given by $y = \sum_{n=-\infty}^{\infty} \delta_{1,n} \zeta^n$, where $\delta_{1,n} = 0$ for $n \neq 1$ and $\delta_{1,1} = 1$. The element y is invertible in $\mathcal{W}(\beta)$ and $y^{-1} = \sum_{n=-\infty}^{\infty} \delta_{1,-n} \zeta^n$. For any positive integer k, we have $\|y^k\| = \beta_k$ and $\|y^{-k}\| = \beta_{-k}$. Since $\varphi(y^{-k}) = \varphi(y)^{-k}$, it follows that

$$\beta_{-k}^{-1/k} \leq |\varphi(y)| \leq \beta_k^{1/k}.$$

But then (1) and (2) imply that $\lambda_0 := \varphi(y)$ belongs to the annulus $\rho_1 \leq |\lambda| \leq \rho_2$. Finally, observe that $x = \sum_{n=-\infty}^{\infty} a_n \zeta^n \in \mathcal{W}(\beta)$ can be written as $x = \sum_{n=-\infty}^{\infty} a_n y^n$ and the latter series converges in the norm of $\mathcal{W}(\beta)$. Since φ is continuous, linear and multiplicative, we conclude that

$$\varphi(x) = \sum_{n=-\infty}^{\infty} a_n \varphi(y)^n = h_{\lambda_0}(x).$$

(vi) Let $\mathcal{B} = L_1(\mathbb{R})$ be the Wiener convolution algebra given in Section XXIX.2, example (c). For each $f \in \mathcal{B}$, let $\widehat{f}$ denote the Fourier transform of f, i.e.,

$$\widehat{f}(\lambda) = \int_{-\infty}^{\infty} f(t) e^{i\lambda t}\, dt.$$

It follows from the properties of the Fourier transform (see [R], Chapter 9) that for each $\lambda \in \mathbb{R}$, the functional h_λ defined on $\mathcal{B}$ by $h_\lambda(f) = \widehat{f}(\lambda)$ is multiplicative and linear. We shall prove that *every multiplicative linear functional on $\mathcal{B} = L_1(\mathbb{R})$ is an h_{λ_0} for some $\lambda_0 \in \mathbb{R}$*, i.e.,

$$(6) \qquad \varphi(f) = \widehat{f}(\lambda_0).$$

Since φ is bounded and linear, we know from integration theory that there exists $k \in L_\infty(\mathbb{R})$ such that $\|k\|_\infty = \|\varphi\| = 1$ and

$$(7) \qquad \varphi(f) = \int_{-\infty}^{\infty} f(s)k(s)ds, \qquad f \in L_1(\mathbb{R}).$$

Hence for f and g in $\mathcal{B} = L_1(\mathbb{R})$,

$$\varphi(f * g) = \int_{-\infty}^{\infty} \left(\int_{-\infty}^{\infty} f(t-s)g(s)ds \right) k(t)dt$$

$$= \int_{-\infty}^{\infty} \int_{-\infty}^{\infty} f(t-s)g(s)k(t)dsdt$$

$$= \int_{-\infty}^{\infty} \int_{-\infty}^{\infty} f(t)g(s)k(t+s)dsdt$$

$$= \int_{-\infty}^{\infty} f(t) \left(\int_{-\infty}^{\infty} k(t+s)g(s)ds \right) dt,$$

$$\varphi(f)\varphi(g) = \left(\int_{-\infty}^{\infty} f(t)k(t)dt \right) \left(\int_{-\infty}^{\infty} g(s)k(s)ds \right)$$

$$= \int_{-\infty}^{\infty} f(t) \left(\int_{-\infty}^{\infty} k(t)k(s)g(s)ds \right) dt.$$

Since $\varphi(f * g) = \varphi(f)\varphi(g)$ for all f, g in $\mathcal{B}$, there exists a set N_1 in $\mathbb{R}$ of measure zero such that for each $t \notin N_1$

$$(8) \qquad k(t+s) = k(t)k(s), \qquad s \in \mathbb{R}, \text{ a.e. }.$$

We shall show that there exists $\lambda_0 \in \mathbb{R}$ such that $k(s) = e^{i\lambda_0 s}$ a.e., and therefore by (7) formula (6) holds. The fact that $\|k\|_\infty = 1$ ensures the existence of a finite interval $[a, b]$ such that $\int_a^b k(s)ds = r \neq 0$. Moreover we can choose a and b such that

$$k(t+a) = k(t)k(a), \qquad k(t+b) = k(t)k(b)$$

for $t \notin N_1$. Formula (8) implies that

$$\int_{a+t}^{b+t} k(s)ds = \int_a^b k(t+s)ds = k(t) \int_a^b k(s)ds = rk(t)$$

for $t \notin N_1$. Define ℓ on $\mathbb{R}$ by

$$\ell(t) = \frac{1}{r} \int\limits_{a+t}^{b+t} k(s)\,ds, \qquad t \in \mathbb{R}.$$

Then $\ell = k$ a.e. . Furthermore, ℓ is absolutely continuous and there exists a set N_2 in $\mathbb{R}$ of measure zero such that $N_1 \subset N_2$ and for each $t \notin N_2$ we have

$$r\ell'(t) = k(b+t) - k(a+t)$$
$$= \big(k(b) - k(a)\big)k(t) = \big(k(b) - k(a)\big)\ell(t).$$

Since $\ell(0) = 1$, we conclude that $\ell(t) = e^{\alpha t}$ for some $\alpha \in \mathbb{C}$. Now use that

$$|e^{\alpha s}| \le \|\ell\|_\infty = \|k\|_\infty = 1, \qquad s \in \mathbb{R}.$$

It follows that $\Re\alpha = 0$, and thus $\ell(s) = e^{i\lambda_0 s}$ for some $\lambda_0 \in \mathbb{R}$.

(vii)   Let $\mathcal{B} = L_1(\mathbb{R}, w)$ be the weighted Wiener convolution algebra introduced in Section XXIX.2, example (d). By arguments analogous to those given in example (v), it can be shown that

$$(9) \qquad \tau_1 := \sup_{t>0}(-t)^{-1}\log w(t) = \lim_{t\to\infty}(-t)^{-1}\log w(t),$$

$$(10) \qquad \tau_2 := \inf_{t<0}(-t)^{-1}\log w(t) = \lim_{t\to-\infty}(-t)^{-1}\log w(t),$$

$$(11) \qquad -\infty < \tau_1 \le \tau_2 < \infty.$$

Furthermore, one can prove (see Loomis [1], page 74) that *every multiplicative linear functional φ on $\mathcal{B} = L_1(\mathbb{R}, w)$ is of the form*

$$\varphi(f) = \int\limits_{-\infty}^{\infty} f(t)e^{i\lambda t}\,dt,$$

where λ is a complex number such that $\tau_1 \le \Im\lambda \le \tau_2$.

(viii)   Let $\mathcal{B} = \mathbb{C}^{2\times 2}$ be the algebra of all 2×2 matrices with complex entries. We identify $\mathcal{B}$ in the usual way with the Banach algebra $\mathcal{L}(\mathbb{C}^2)$. Hence $\mathcal{B}$ is a non-commutative Banach algebra with a unit. We shall prove that *the set of multiplicative linear functionals on $\mathcal{B} = \mathbb{C}^{2\times 2}$ is empty.* To see this, assume φ is a multiplicative linear functional on $\mathcal{B}$. Then $M = \operatorname{Ker}\varphi$ is an ideal in $\mathcal{B}$. Since $\operatorname{codim}\operatorname{Ker}\varphi = 1$, the ideal $M \ne \{0\}$. Choose $A = \begin{bmatrix} a & b \\ c & d \end{bmatrix} \ne 0$ in M. Since the entries a, b, c and d are not all zero, we can choose 2×2 matrices B and C such that

$$\begin{bmatrix} 1 & 0 \\ 0 & 0 \end{bmatrix} = BAC \in M.$$

By multiplying $\left[\begin{smallmatrix} 1 & 0 \\ 0 & 0 \end{smallmatrix}\right]$ on the left and the right by a 2×2 permutation matrix we see that also the matrices $\left[\begin{smallmatrix} 0 & 1 \\ 0 & 0 \end{smallmatrix}\right]$, $\left[\begin{smallmatrix} 0 & 0 \\ 1 & 0 \end{smallmatrix}\right]$ and $\left[\begin{smallmatrix} 0 & 0 \\ 0 & 1 \end{smallmatrix}\right]$ are in M. But then $M = \mathcal{B}$, which contradicts the fact that $\varphi \neq 0$.

(ix) Let $\mathcal{B}$ be the commutative convolution algebra $L_1([0,1])$ introduced in Section XXIX.7. *The set of multiplicative linear functionals on $\mathcal{B} = L_1([0,1])$ is empty.* To see this, recall that any element f in $\mathcal{B}$ is quasi-nilpotent. So, if φ is a multiplicative linear functional on $\mathcal{B}$, then

$$|\varphi(f)| = |\varphi(f)^n|^{1/n} = |\varphi(f^n)|^{1/n} \le \|f^n\|^{1/n} \to 0$$

if $n \to \infty$ by the definition of quasi-nilpotency. Hence φ is the zero functional, which is impossible. The above argument also shows that the set of multiplicative linear functionals is empty for any Banach algebra $\mathcal{B}$ for which every element in $\mathcal{B}$ is quasi-nilpotent.

XXX.2 MAXIMAL IDEALS

Examples (viii) and (ix) in the preceding section show that if a Banach algebra $\mathcal{B}$ is not both commutative and unital, then $\mathcal{B}$ may fail to have any multiplicative linear functionals. In this section we prove by means of the existence of maximal ideals that the set of multiplicative linear functionals on a commutative unital Banach algebra is nonempty.

An ideal M in a Banach algebra $\mathcal{B}$ is called *maximal* if $M \neq \mathcal{B}$ and there does not exist an ideal $\mathcal{I}$ in $\mathcal{B}$ containing M such that $\mathcal{I} \neq M$ and $\mathcal{I} \neq \mathcal{B}$.

THEOREM 2.1. *Let $\mathcal{B}$ be a commutative unital Banach algebra. The kernel of a multiplicative linear functional on $\mathcal{B}$ is a maximal ideal in $\mathcal{B}$ and, conversely, every maximal ideal in $\mathcal{B}$ is the kernel of one and only one multiplicative linear functional on $\mathcal{B}$.*

PROOF. Let φ be a multiplicative linear functional on $\mathcal{B}$. Then $M = \operatorname{Ker}\varphi$ is a proper ideal in $\mathcal{B}$. Since $\operatorname{codim} M = 1$, the only strictly larger ideal which contains M is $\mathcal{B}$. Hence M is a maximal ideal in $\mathcal{B}$.

Conversely, let M be a maximal ideal in $\mathcal{B}$. We first prove that M is closed. Since $M \neq \mathcal{B}$, the unit e of $\mathcal{B}$ is not in M. More generally, $M \neq \mathcal{B}$ implies that M does not contain invertible elements. Hence the open ball $\|e - x\| < 1$ does not intersect with M. So e does not belong to the closure $\overline{M}$ of M. Obviously, $\overline{M}$ is an ideal and $M \subset \overline{M} \subset \mathcal{B}$. Since $e \notin \overline{M}$, we have $\overline{M} \neq \mathcal{B}$. Hence $M = \overline{M}$, because M is a maximal ideal.

So M is closed and $M \neq \mathcal{B}$. It follows that the quotient algebra $\mathcal{B}/M$ is a commutative unital Banach algebra with unit $[e]$. Let $[x]$ be a non-zero element in $\mathcal{B}/M$. We shall prove that $[x]$ is invertible in $\mathcal{B}/M$. Consider the set

$$J_x = \{xy + z \mid y \in \mathcal{B}, z \in M\}.$$

Obviously, J_x is an ideal, $M \subset J_x$ and $M \neq J_x$, because $x \notin M$. Since M is a maximal ideal, it follows that $J_x = \mathcal{B}$. Thus e can be represented in the form $e = xy_1 + z_1$ for some $y_1 \in \mathcal{B}$ and $z_1 \in M$. Hence in the quotient algebra $\mathcal{B}/M$

$$[x][y_1] = [xy_1] = [xy_1 + z_1] = [e].$$

Thus $[x]$ is invertible in $\mathcal{B}/M$. Since every non-zero element in $\mathcal{B}/M$ is invertible, we can apply Theorem XXIX.5.3 to show that

$$\mathcal{B}/M = \{\lambda[e] \mid \lambda \in \mathbb{C}\}.$$

Define $\varphi \colon \mathcal{B} \to \mathbb{C}$ as follows. Given $x \in \mathcal{B}$, let $\varphi(x) = \lambda$, where $[x] = \lambda[e]$. It is easy to see that φ is well-defined and φ is a multiplicative linear functional on $\mathcal{B}$ with kernel M. It was pointed out in Section XXX.1, example (i) that unequal linear functionals cannot have the same kernel. Thus φ is uniquely determined by M. $\quad\square$

The natural question now is whether multiplicative linear functionals exist or, equivalently, whether maximal ideals exist in every commutative unital Banach algebra. The following theorem answers this question in the affirmative.

THEOREM 2.2. *If $\mathcal{B}$ is a commutative unital Banach algebra, then the set of multiplicative linear functionals on $\mathcal{B}$ is nonempty.*

PROOF. It suffices, by Theorem 2.1, to show that there exists a maximal ideal in $\mathcal{B}$. Let y be any non-invertible vector in $\mathcal{B}$, e.g., $y = 0$. Let $\mathcal{P}$ be the set of all ideals in $\mathcal{B}$ which are not equal to $\mathcal{B}$ and contain y. The family $\mathcal{P}$ contains the ideal $\{xy \mid x \in \mathcal{B}\}$. Note that the latter ideal is different from $\mathcal{B}$ because y is not invertible. On $\mathcal{P}$ we define a partial ordering by set inclusion. Let $\mathcal{F}$ be a nonempty linearly ordered subset of $\mathcal{P}$. Then $J := \cup \mathcal{F}$ is an ideal which contains y. Since the unit e does not belong to any ideal in $\mathcal{F}$, also $e \notin J$ and hence J is different from $\mathcal{B}$. Thus $J \in \mathcal{P}$. A straightforward application of Zorn's lemma shows now that $\mathcal{P}$ has a maximal element M (with respect to inclusion). Obviously M is a maximal ideal in $\mathcal{B}$ which contains y. $\quad\square$

THEOREM 2.3. *An element y in a commutative unital Banach algebra $\mathcal{B}$ is invertible if and only if $\varphi(y) \neq 0$ for every multiplicative linear functional φ on $\mathcal{B}$.*

PROOF. If y is invertible, then for any multiplicative linear functional φ on $\mathcal{B}$, we have $1 = \varphi(y)\varphi(y^{-1})$ whence $\varphi(y) \neq 0$. On the other hand, if y is not invertible, then the last statement in the proof of Theorem 2.2 shows that there exists a maximal ideal M in $\mathcal{B}$ which contains y. But $M = \operatorname{Ker}\psi$ for some multiplicative linear functional ψ on $\mathcal{B}$ (by Theorem 2.1). In particular, $\psi(y) = 0$. $\quad\square$

COROLLARY 2.4. *Let $\mathcal{B}$ be a commutative unital Banach algebra, and let $\mathcal{M}$ denote the set of multiplicative linear functionals on $\mathcal{B}$. For every $x \in \mathcal{B}$,*

$$\sigma(x) = \{\varphi(x) \mid \varphi \in \mathcal{M}\}.$$

As a first application of the theory of commutative Banach algebras, we present Gelfand's proof of the Wiener theorem about inversion in the algebra $\mathcal{W}$ of functions with an absolutely convergent Fourier series.

THEOREM 2.5. *An element f in the Wiener algebra $\mathcal{W}$ is invertible in $\mathcal{W}$ if and only if $f(\zeta) \neq 0$ for each $\zeta \in \mathbb{T}$, and in that case the inverse element is the function $1/f(\zeta)$.*

PROOF. We apply Theorem 2.3. From example (iv) in Section XXX.1 we know that every multiplicative linear functional φ on $\mathcal{W}$ is of the form $\varphi = h_\zeta$ for some $\zeta \in \mathbb{T}$. Here $h_\zeta(f) = f(\zeta)$. So, by Theorem 2.3, the element $f \in \mathcal{W}$ is invertible in $\mathcal{W}$ if and only if $f(\zeta) = h_\zeta(f) \neq 0$ for each $\zeta \in \mathbb{T}$. This proves the if and only if part of the theorem.

Next, assume that $f(\zeta) \neq 0$ for all $\zeta \in \mathbb{T}$. Then f is invertible in $\mathcal{W}$ with inverse g, say. It follows that $g(\zeta)f(\zeta) = 1$ for all $\zeta \in \mathbb{T}$, and hence $g(\zeta) = \big(f(\zeta)\big)^{-1}$. $\quad\square$

Of course the "if part" in Theorem 2.5 is the surprising part. A continuous analogue is the following result.

THEOREM 2.6. *Suppose that k is in $L_1(\mathbb{R})$ and its Fourier transform $\widehat{k}(\lambda) \neq 1$ for all $\lambda \in \mathbb{R}$. Then there exists $\ell \in L_1(\mathbb{R})$ such that*

$$(1) \qquad \big(1 - \widehat{k}(\lambda)\big)^{-1} = 1 - \widehat{\ell}(\lambda), \qquad \lambda \in \mathbb{R}.$$

PROOF. Let $\mathcal{A}$ be the convolution algebra $L_1(\mathbb{R})$ with a unit e adjoined to it. We put $e = 1$, where 1 is the function identically equal to one on $\mathbb{R}$, and we identify $\mathcal{A}$ with the set of functions $\alpha + f$ with $f \in L_1(\mathbb{R})$ and $\alpha \in \mathbb{C}$. Let φ be a multiplicative linear functional on $\mathcal{A}$. Then its restriction to $L_1(\mathbb{R})$ is a multiplicative linear functional on $L_1(\mathbb{R})$, and hence we know from example (vi) in Section XXX.1 that there exists $\lambda_0 \in \mathbb{R}$ such that

$$\varphi(1 - k) = 1 - \widehat{k}(\lambda_0) \neq 0.$$

Hence $1 - k$ is invertible in $\mathcal{A}$ by Theorem 2.3, which means that there exists $\ell \in L_1(\mathbb{R})$ such that $1 - \ell = (1 - k)^{-1}$ or, equivalently, $\ell * k = \ell + k$. Taking Fourier transforms of both sides gives $\widetilde{\ell k} = \widehat{\ell} + \widehat{k}$, and hence (1) holds. $\quad\square$

COROLLARY 2.7. *Suppose that $k \in L_1(\mathbb{R})$. Then for each $g \in L_1(\mathbb{R})$ the equation*

$$(2) \qquad f(t) - \int_{-\infty}^{\infty} k(t - s)f(s)\,ds = g(t), \qquad t \in \mathbb{R}, \text{ a.e.,}$$

has a unique solution $f \in L_1(\mathbb{R})$ if and only if the Fourier transform $\widehat{k}(\lambda) \neq 1$ for all $\lambda \in \mathbb{R}$. In this case the unique solution of (2) is given by

$$(3) \qquad f(t) = g(t) - \int_{-\infty}^{\infty} \ell(t - s)g(s)\,ds, \qquad t \in \mathbb{R}, \text{ a.e.,}$$

where $\ell \in L_1(\mathbb{R})$ is given by (1).

PROOF. Let $\mathcal{A}$ be the Banach algebra introduced in the proof of Theorem 2.6. Equation (2) may be rewritten in the form $f - k * f = g$, and hence (2) is uniquely solvable in $L_1(\mathbb{R})$ for each right hand side g in $L_1(\mathbb{R})$ if and only if $1 - k$ is invertible in

$\mathcal{A}$. Assume $1 - k$ is invertible in $\mathcal{A}$ with $(1 - k)^{-1} = 1 - \ell \in \mathcal{A}$. Then $\ell * k = \ell + k$, and hence $\widehat{\ell k} = \widehat{\ell} + \widehat{k}$, which implies that $(1 - \widehat{\ell}(\lambda))(1 - \widehat{k}(\lambda)) = 1$. Thus $\widehat{k}(\lambda) \neq 1$ for each $\lambda \in \mathbb{R}$.

Conversely, assume that $\widehat{k}(\lambda) \neq 1$ for each $\lambda \in \mathbb{R}$. By Theorem 2.6 there exists $\ell \in L_1(\mathbb{R})$ such that (1) holds. It follows (cf., the proof of Theorem 2.6) that $1 - k$ is invertible in $\mathcal{A}$ with $(1 - k)^{-1} = 1 - \ell$. But then equation (2) has a unique solution $f \in L_1(\mathbb{R})$ for $g \in L_1(\mathbb{R})$, namely $f = g - \ell * g$. In other words, f is given by (3). $\square$

XXX.3 THE GELFAND TRANSFORM

Let $\mathcal{B}$ be a commutative unital Banach algebra. Let $\mathcal{M}$ denote the set of all multiplicative linear functionals on $\mathcal{B}$. We know from Theorem 2.2 that $\mathcal{M} \neq \emptyset$. Given an $x \in \mathcal{B}$ there is a natural way to associate with x a bounded complex valued function $\widehat{x}$ defined on $\mathcal{M}$, namely

$$\widehat{x}(\varphi) := \varphi(x), \qquad \varphi \in \mathcal{M}.$$

The function $\widehat{x}$ is called the *Gelfand transform* of x. From $|\widehat{x}(\varphi)| = |\varphi(x)| \leq \|x\|$ for $\varphi \in \mathcal{M}$, we see that $\widehat{x}$ is in the Banach algebra $B(\mathcal{M})$ (cf., example (iii) in Section XXIX.1) and

$$(1) \qquad \qquad \|\widehat{x}\| := \sup_{\varphi \in \mathcal{M}} |\widehat{x}(\varphi)| \leq \|x\|.$$

There exists a topology on $\mathcal{M}$, called the *Gelfand topology*, such that, with respect to this topology, $\mathcal{M}$ is a compact Hausdorff space and $\widehat{x}$ is in $C(\mathcal{M})$. The Gelfand topology is by definition the weakest topology on $\mathcal{M}$ such that every $\widehat{x}$ is continuous on $\mathcal{M}$. In other words, all sets of the form

$$\mathcal{N}(\varphi; x_1, \ldots, x_n, \varepsilon) = \bigcap_{i=1}^{n} \{\psi \in \mathcal{M} \mid |\widehat{x}_i(\varphi) - \widehat{x}_i(\psi)| < \varepsilon\},$$

$$= \bigcap_{i=1}^{n} \{\psi \in \mathcal{M} \mid |\varphi(x_i) - \psi(x_i)| < \varepsilon\},$$

where $\varphi \in \mathcal{M}$, $x_1, \ldots, x_n \in \mathcal{B}$ and $\varepsilon > 0$, constitute a basis for this topology. In particular, the Gelfand topology is a Hausdorff topology. The reader familiar with weak*-topologies will recognize the Gelfand topology as the topology on $\mathcal{M}$ inherited from the weak*-topology on the conjugate space $\mathcal{B}'$.

The set $\mathcal{M}$ of all multiplicative linear functionals on $\mathcal{B}$ endowed with the Gelfand topology is called the *Gelfand spectrum* of $\mathcal{B}$. Clearly, every $\widehat{x}$ is continuous on the Gelfand spectrum.

THEOREM 3.1. *The Gelfand spectrum of a commutative unital Banach algebra $\mathcal{B}$ is a compact Hausdorff space.*

PROOF. Let $\mathcal{M}$ denote the Gelfand spectrum of $\mathcal{B}$. Given $x \in \mathcal{B}$, define $C_x = \{\lambda \in \mathbb{C} \mid |\lambda| \leq \|x\|\}$, and let $S = \prod_{x \in \mathcal{B}} C_x$ with the product topology. Since each

C_x is compact, S is compact by the Tychonoff theorem (see [W], Theorem 17.8). Also, $\mathcal{M} \subset S$ since each $\varphi \in \mathcal{M}$ has norm 1. Recall that the sets

$$\mathcal{N}(h; x_1, \ldots, x_n, \varepsilon) = \bigcap_{i=1}^{n} \{g \in S \mid |h(x_i) - g(x_i)| < \varepsilon\},$$

where $h \in S$, $x_1, \ldots, x_n \in \mathcal{B}$ and $\varepsilon > 0$, constitute a basis for the product topology. It follows that the Gelfand topology on $\mathcal{M}$ is the topology on $\mathcal{M}$ inherited from S. Therefore to prove that $\mathcal{M}$ is compact, it suffices to show that $\mathcal{M}$ is closed in S.

Let h be in the closure of $\mathcal{M}$ in S. Given $\varepsilon > 0$ and x, y in $\mathcal{B}$, there exists

$$\varphi \in \mathcal{M} \cap \mathcal{N}(h; x, y, xy, e, \varepsilon),$$

where e is the unit of $\mathcal{B}$. Therefore,

$$\begin{aligned}
|h(xy) - h(x)h(y)| &\leq |h(xy) - \varphi(xy)| + |\varphi(x)\varphi(y) - h(x)h(y)| \\
&< \varepsilon + |(\varphi(x) - h(x))\varphi(y)| + |h(x)(\varphi(y) - h(y))| \\
&\leq \varepsilon + \varepsilon\|y\| + \varepsilon\|x\|, \\
|h(e) - 1| &= |h(e) - \varphi(e)| < \varepsilon.
\end{aligned}$$

Since $\varepsilon > 0$ is arbitrary, $h(xy) = h(x)h(y)$ and $h(e) = 1$. In a similar way, replacing xy by $\alpha x + \beta y$, one shows that h is linear. Hence $h \in \mathcal{M}$ which proves that $\mathcal{M}$ is closed in the compact Hausdorff space S. Thus $\mathcal{M}$ has the desired properties. $\square$

Let $\mathcal{M}$ be the Gelfand spectrum of the commutative unital Banach algebra $\mathcal{B}$. Now that we know that for each $x \in \mathcal{B}$ the Gelfand transform $\widehat{x}$ is in $C(\mathcal{M})$, we define

$$\Gamma \colon \mathcal{B} \to C(\mathcal{M}), \qquad \Gamma x = \widehat{x}.$$

The map Γ is called the *Gelfand representation* of $\mathcal{B}$. It is easy to see that Γ is linear and multiplicative, i.e., $\Gamma(xy) = \Gamma(x)\Gamma(y)$. In fact the latter equality follows from

$$\widehat{xy}(\varphi) = \varphi(xy) = \varphi(x)\varphi(y) = \widehat{x}(\varphi)\widehat{y}(\varphi) = (\widehat{x}\widehat{y})(\varphi).$$

A linear multiplicative map between algebras is called a *homomorphism*. An *isomorphism* is an injective homomorphism.

THEOREM 3.2. *The Gelfand representation Γ of a commutative unital Banach algebra $\mathcal{B}$ is a homomorphism and*

$$(2) \qquad \|\Gamma x\| = \lim_{n \to \infty} \|x^n\|^{1/n} \leq \|x\|.$$

PROOF. Let $\mathcal{M}$ denote the Gelfand spectrum of $\mathcal{B}$. The theorem follows from the equalities

$$\|\Gamma x\| = \|\widehat{x}\| = \max_{\varphi \in \mathcal{M}} |\widehat{x}(\varphi)| = \max_{\varphi \in \mathcal{M}} |\varphi(x)|,$$

Corollary 2.4 and Theorem XXIX.7.1. $\square$

XXX.4 EXAMPLES OF GELFAND SPECTRA AND TRANSFORMS

In this section the Gelfand spectrum and Gelfand transform are described explicitly for various concrete cases. All examples concern algebras of functions defined on some set S. It may happen that the functions are not continuous. In such a case we show how the Gelfand theory changes the underlying set S in order to make the functions continuous.

(i) Let $\mathcal{B} = C(S)$. In example (i) of Section XXX.1 it was shown that the Gelfand spectrum $\mathcal{M}$ of $\mathcal{B} = C(S)$ consists of all functionals of the form h_t, $t \in S$, where $h_t(f) = f(t)$. Therefore the map

$$\eta\colon S \to \mathcal{M}, \qquad \eta(t) = h_t$$

is surjective. It is also injective, for if $t_1 \neq t_2$, then by Urysohn's lemma (see [W], item 15.6) there exists $g \in C(S)$ such that $g(t_1) \neq g(t_2)$. Hence $\eta(t_1)(g) \neq \eta(t_2)(g)$. For $h_{t_0} \in \mathcal{M}$, $f_1, \ldots, f_n \in \mathcal{B}$ and $\varepsilon > 0$ we have

$$\eta^{-1}[\mathcal{N}(h_{t_0}; f_1, \ldots, f_n, \varepsilon)] = \bigcap_{j=1}^{n} \{ t \in S \mid |f_j(t_0) - f_j(t)| < \varepsilon \}.$$

It follows that η is continuous. Since S is compact and $\mathcal{M}$ is a Hausdorff space (see [W], Theorem 17.14), η is a homeomorphism. Given $f \in \mathcal{B}$,

$$\widehat{f}(h_t) = h_t(f) = f(t).$$

Therefore, if we identify S with $\mathcal{M}$ via the map η, then $\widehat{f}$ may be identified with f and the Gelfand representation $\Gamma f = \widehat{f}$ of $\mathcal{B} = C(S)$ may be considered as the identity map on $C(S)$.

(ii) Let $\mathcal{B} = \mathcal{A}(\overline{\mathbb{D}})$, where $\overline{\mathbb{D}}$ is the closed unit disc in the complex plane. Example (ii) in Section XXX.1 shows that the Gelfand spectrum $\mathcal{M}$ of $\mathcal{B} = \mathcal{A}(\overline{\mathbb{D}})$ consists of all linear functionals h_λ, $\lambda \in \overline{\mathbb{D}}$, where $h_\lambda(f) = f(\lambda)$, $f \in \mathcal{B}$. Arguments similar to the ones used in the above example show that

$$\eta\colon \overline{\mathbb{D}} \to \mathcal{M}, \qquad \eta(\lambda) = h_\lambda$$

is a homeomorphism and that the Gelfand representation of $\mathcal{B} = \mathcal{A}(\overline{\mathbb{D}})$ may be considered as the embedding map from $\mathcal{A}(\overline{\mathbb{D}})$ into $C(\overline{\mathbb{D}})$.

(iii) Let $\mathcal{B} = PC([a, b])$. This algebra contains discontinuous functions, which by the Gelfand representation are transformed into continuous functions. We shall show how the underlying interval $[a, b]$ and its topology have to be changed in order to deal with the piecewise continuous functions as functions that are continuous.

It was shown in example (iii) in Section XXX.1 that the Gelfand spectrum $\mathcal{M}$ of $\mathcal{B}$ consists of the functionals h_t and h_t^+ with $a \leq t \leq b$, where $h_t(f) = f(t)$ and $h_t^+(f) = f(t+)$. We shall identify $\mathcal{M}$ with the following topological space. Let

$$\Delta = [a, b] \times \{1\} \cup [a, b] \times \{2\},$$

and let a basis of open sets in Δ at the point $(t, 1)$ consist of all sets of the form

$$O^1_{t,\varepsilon} = \{(s,1) \mid t - \varepsilon < s \leq t\} \cup \{(s,2) \mid t - \varepsilon \leq s < t\},$$

and a basis of open sets for Δ at the point $(t, 2)$ consist of all sets of the form

$$O^2_{t,\varepsilon} = \{(s,1) \mid t < s \leq t + \varepsilon\} \cup \{(s,2) \mid t \leq s < t + \varepsilon\}.$$

Note that the topology on Δ is a Hausdorff topology. Next, define $\eta \colon \Delta \to \mathcal{M}$ by setting

$$\eta(t, 1) = h_t, \qquad a \leq t \leq b,$$
$$\eta(t, 2) = h_t^+, \qquad a \leq t < b.$$

Since $b+ = b$, and thus $h_b^+ = h_b$, the map η is surjective. As in example (i) above one proves that η is injective. Furthermore,

(1) $$\eta(O^1_{t,\varepsilon}) = \mathcal{N}(h_t; \chi_{(t-\varepsilon,t]}, 1),$$

(2) $$\eta(O^2_{t,\varepsilon}) = \mathcal{N}(h_t^+; \chi_{(t,t+\varepsilon]}, 1).$$

Here χ_E denotes the characteristic function of the set E (thus $\chi_E(t) = 1$ for $t \in E$ and $\chi_E(t) = 0$ otherwise). To prove (1) note that $h_s \in \mathcal{N}(h_t; \chi_{(t-\varepsilon,t]}, 1)$ if and only if

$$1 > |h_t(\chi_{(t-\varepsilon,t]}) - h_s(\chi_{(t-\varepsilon,t]})| = |1 - \chi_{(t-\varepsilon,t]}(s)|,$$

and hence

$$h_s \in \mathcal{N}(h_t; \chi_{(t-\varepsilon,t]}, 1) \Longleftrightarrow t - \varepsilon < s \leq t.$$

Similarly, $h_s^+ \in \mathcal{N}(h_t; \chi_{(t-\varepsilon,t]}, 1)$ if and only if $\chi_{(t-\varepsilon,t)}(s+) = 1$, and thus

$$h_s^+ \in \mathcal{N}(h_t; \chi_{(t-\varepsilon,t]}, 1) \Longleftrightarrow t - \varepsilon \leq s < t.$$

This proves the equality (1). The equality (2) is proved in an analogous way. Let $\tau \colon \mathcal{M} \to \Delta$ be the inverse map $\tau = \eta^{-1}$. Then (1) and (2) imply that τ is continuous. Since $\mathcal{M}$ is compact and Δ is a Hausdorff space, the map τ (and hence also η) is a homeomorphism (cf., [W], Theorem 17.14).

If we identify $\mathcal{M}$ with Δ via the map η, then the Gelfand representation $\Gamma \colon PC([a,b]) \to C(\Delta)$ is given by

$$(\Gamma f)(t, 1) = f(t), \quad a \leq t \leq b, \qquad (\Gamma f)(t, 2) = f(t+), \quad a \leq t < b.$$

Furthermore, $\|\Gamma f\| = \|f\|$. It is easily checked that Γ is surjective. In fact, if $g \in C(\Delta)$ and

$$f_1(t) = g(t, 1) \quad (a \leq t \leq b), \qquad f_2(t) = g(t, 2) \quad (a \leq t < b),$$

then $f_1(t) = f_1(t-) = f_2(t-)$ for $a < t \leq b$ and $f_2(t) = f_2(t+) = f_1(t+)$ for $a \leq t < b$. Thus $f_1 \in PC[a,b]$ and $\Gamma f_1 = g$. We conclude that $PC([a,b])$ and $C(\Delta)$ are isometrically isomorphic Banach algebras.

(iv) Let W be the Wiener algebra. We know from example (iv) in Section XXX.1 that every φ in the Gelfand spectrum $\mathcal{M}$ of W is of the form h_ζ for some ζ in the unit circle $\mathbf{T}$. Here $h_\zeta(f) = f(\zeta)$. The same arguments as in examples (i) and (ii) above show that $\mathbf{T}$ and $\mathcal{M}$ are homeomorphic under the map $\zeta \mapsto h_\zeta$. With this identification the Gelfand representation Γ may be considered as the identity map from W into $C(\mathbf{T})$. Note that $\operatorname{Im}\Gamma$ is dense in $C(\mathbf{T})$, but Γ is not surjective. In particular, $\operatorname{Im}\Gamma$ is not closed. This follows from the fact that the trigonometric polynomials are dense in $C(\mathbf{T})$ and the existence of functions in $C(\mathbf{T})$ which do not have absolutely convergent power series.

(v) Let $W(\beta)$ be the Wiener algebra with weights $\{\beta_k\}_{-\infty}^{\infty}$. We know from example (v) in Section XXX.1 that there exists an annulus $\Sigma = \{\lambda \mid \rho_1 \leq |\lambda| \leq \rho_2\}$, $0 < \rho_1 \leq \rho_2 < \infty$, determined by β, such that the Gelfand spectrum $\mathcal{M}$ of $W(\beta)$ consists of functionals φ_λ, $\lambda \in \Sigma$, defined by $\varphi_\lambda(x) = \sum_{n=-\infty}^{\infty} a_n \lambda^n$, where $x \in W(\beta)$ is the formal power series $x = \sum_{n=-\infty}^{\infty} a_n \zeta^n$. The map $\eta \colon \Sigma \to \mathcal{M}$ defined by $\eta(\lambda) = \varphi_\lambda$ is bijective and continuous. Since Σ is compact and $\mathcal{M}$ is a Hausdorff space (see [W], Theorem 17.14), the map η is a homeomorphism. Given $x = \sum_{n=-\infty}^{\infty} a_n \zeta^n \in W(\beta)$,

$$\widehat{x}(\varphi_\lambda) = \varphi_\lambda(x) = \sum_{n=-\infty}^{\infty} a_n \lambda^n.$$

Thus, if we identify $\mathcal{M}$ with Σ via the map η, then the Gelfand representation of $W(\beta)$ is the map $\Gamma \colon W(\beta) \to C(\Sigma)$ given by

$$(\Gamma x)(\lambda) = \sum_{n=-\infty}^{\infty} a_n \lambda^n, \qquad \lambda \in \Sigma.$$

Note that $\operatorname{Im}\Gamma \subset \mathcal{A}(\Sigma)$, the Banach algebra of continuous functions on Σ that are analytic in the interior of Σ.

(vi) Let $\mathcal{A}$ be the convolution algebra $L_1(\mathbb{R})$ with a unit adjoined to it. As in the proof of Theorem 2.6 we identify $\mathcal{A}$ with the set of functions $\alpha + f$ with $f \in L_1(\mathbb{R})$ and $\alpha \in \mathbb{C}$. It was shown in example (vi) of Section XXX.1 that the multiplicative linear functionals on $L_1(\mathbb{R})$ are of the form h_λ, $\lambda \in \mathbb{R}$, where $h_\lambda(f) = \widehat{f}(\lambda)$ and $\widehat{f}$ is the Fourier transform of f. Thus if $\mathcal{M}$ is the Gelfand spectrum of $\mathcal{A}$, then $\mathcal{M}$ consists of all functionals of the form φ_λ, $\lambda \in \mathbb{R}$, and φ_∞, where

$$\begin{aligned}
\varphi_\infty(\alpha + f) &= \alpha, & \alpha \in \mathbb{C}, f \in L_1(\mathbb{R}), \\
\varphi_\lambda(\alpha + f) &= \alpha + \widehat{f}(\lambda), & \alpha \in \mathbb{C}, f \in L_1(\mathbb{R}).
\end{aligned}$$

Let $\widetilde{\mathbb{R}} = \mathbb{R} \cup \{\infty\}$ be the usual compactification of $\mathbb{R}$; i.e., the open sets in $\widetilde{\mathbb{R}}$ are those open in $\mathbb{R}$ together with the neighbourhoods $\{t \mid |t| > c\}$ of ∞. For each $f \in L_1(\mathbb{R})$ the Fourier transform $\widehat{f}$ is a continuous function on $\mathbb{R}$ which by the continuous analogue of the Riemann-Lebesgue lemma (see [R], Theorem 9.6) vanishes at ∞ (i.e., $\widehat{f}(\lambda) \to 0$ if $|\lambda| \to \infty$). Using these properties it is not difficult to prove that $\eta \colon \widetilde{\mathbb{R}} \to \mathcal{M}$ defined by $\eta(\lambda) = \varphi_\lambda$ and $\eta(\infty) = \varphi_\infty$ is a homeomorphism. If $x = \alpha + f \in \mathcal{A}$, then

$$\widehat{x}(\varphi_\lambda) = \varphi_\lambda(x) = \alpha + \widehat{f}(\lambda), \quad \lambda \in \mathbb{R}, \qquad \widehat{x}(\varphi_\infty) = \varphi_\infty(x) = \alpha.$$

Therefore, if we identify $\mathcal{M}$ with $\widetilde{\mathbb{R}}$ via the map η, then given $g \in L_1(\mathbb{R})$, the Gelfand transform of g is just equal to the Fourier transform of g on $\mathbb{R}$ and equal to zero at ∞.

(vii) Let $\widetilde{\mathcal{B}}$ be the convolution algebra $L_1([0,1])$ with a unit adjoined to it. Thus

$$\widetilde{\mathcal{B}} = \{(f, \alpha) \mid \alpha \in \mathbb{C}, f \in L_1([0,1])\},$$

which is a commutative unital Banach algebra (see Section XXIX.7). From example (ix) in Section XXX.1 we know that there is no multiplicative linear functional on $\mathcal{B} = L_1([0,1])$. Hence the Gelfand spectrum $\mathcal{M}$ of $\widetilde{\mathcal{B}}$ is a singleton, namely $\mathcal{M} = \{\varphi\}$, where $\varphi(f, \alpha) = \alpha$. Thus, after a trivial identification, the Gelfand representation of $\widetilde{\mathcal{B}}$ is the map $\Gamma \colon \widetilde{\mathcal{B}} \to \mathbb{C}$ given by $\Gamma(f, \alpha) = \alpha$.

XXX.5 FINITELY GENERATED BANACH ALGEBRAS

Let $\mathcal{B}$ be a Banach algebra with unit e. Given $x_1, x_2, \ldots, x_n$ in $\mathcal{B}$, the smallest closed subalgebra $\mathcal{A}$ of $\mathcal{B}$ which contains these vectors as well as e is the algebra of those $x \in \mathcal{B}$ which are limits of polynomials in $x_1, x_2, \ldots, x_n$. If p is the constant polynomial a_0, then $p(y)$ is defined to be $a_0 e$ for all $y \in \mathcal{B}$. We say that $\mathcal{A}$ is the *unital closed subalgebra of $\mathcal{B}$ generated by* $x_1, \ldots, x_n$. If $\mathcal{A} = \mathcal{B}$, then $\mathcal{B}$ is called a *unital Banach algebra generated by* $x_1, \ldots, x_n$.

For example, the unital closed subalgebra of $B([a, b])$ generated by the function $f(t) = t$ is precisely $C([a, b])$. This is a consequence of the Weierstrass approximation theorem. The unital Banach algebra $\mathcal{A}(\overline{\mathbb{D}})$, where $\overline{\mathbb{D}}$ is the closed unit disc in the complex plane, is generated by the function $g_0(\lambda) = \lambda$. In example (iv) in Section XXX.1 we have shown that any f in the Wiener algebra $\mathcal{W}$ may be written in the form $f = \sum_{n=-\infty}^{\infty} a_n g^n$, where $g(\zeta) = \zeta$ and the series converges in the norm of $\mathcal{W}$. Thus $\mathcal{W}$ is generated by g and g^{-1}.

THEOREM 5.1. *Suppose $x_1, x_2, \ldots, x_n$ generate a commutative Banach algebra $\mathcal{B}$ with unit. Then the Gelfand spectrum $\mathcal{M}$ of $\mathcal{B}$ is homeomorphic to a compact subset of $\mathbb{C}^n$ under the map*

$$(1) \qquad \eta(\varphi) = (\varphi(x_1), \ldots, \varphi(x_n)), \qquad \varphi \in \mathcal{M}.$$

PROOF. It is clear from the topology on $\mathcal{M}$ that η is continuous. Moreover, η is injective. Indeed, suppose $\eta(\varphi) = \eta(\psi)$. Then $\varphi(x_i) = \psi(x_i)$ for $i = 1, \ldots, n$, and since φ and ψ are linear and multiplicative, φ and ψ agree on all polynomials in $x_1, \ldots, x_n$. The continuity of φ and ψ (Theorem 1.1) imply that $\varphi(y) = \psi(y)$ for all $y \in \mathcal{B}$. Hence η is injective. Put $V = \eta[\mathcal{M}]$. Then V is a subset of $\mathbb{C}^n$, and hence V is a Hausdorff space in its own right. It follows that $\eta \colon \mathcal{M} \to V$ is a homeomorphism (see [W], Theorem 17.14). $\square$

If the unital Banach algebra $\mathcal{B}$ is generated by only one element $x_1 = a$, say, then $\mathcal{B}$ is commutative and the map η in (1) is the Gelfand transform $\widehat{a}$ of a. Since the range of $\widehat{a}$ is the spectrum $\sigma(a)$ (Corollary 2.4) we have the following corollary to Theorem 5.1.

COROLLARY 5.2. *If the unital Banach algebra $\mathcal{B}$ is generated by an element a, then $\mathcal{B}$ is commutative and the Gelfand transform $\widehat{a}$ of a is a homeomorphism from the Gelfand spectrum of $\mathcal{B}$ onto $\sigma(a)$.*

The next theorem is a more general version of Corollary 5.2.

THEOREM 5.3. *Let $\mathcal{B}$ be a Banach algebra with unit e which is generated by $a, (z_1 e - a)^{-1}, \ldots, (z_n e - a)^{-1}$, where $z_1, \ldots, z_n$ are given points in the resolvent set of a. Then $\mathcal{B}$ is commutative, and the Gelfand transform $\widehat{a}$ is a homeomorphism from the Gelfand spectrum $\mathcal{M}$ of $\mathcal{B}$ onto $\sigma(a)$. Furthermore, if $\mathcal{M}$ and $\sigma(a)$ are identified via the map $\widehat{a}$, then for each $x \in \mathcal{B}$ the Gelfand transform $\widehat{x}$ is continuous on $\sigma(a)$ and analytic on the interior of $\sigma(a)$.*

PROOF. Since the elements $a, (z_1 e - a)^{-1}, \ldots, (z_n e - a)^{-1}$ are mutually commutative, $\mathcal{B}$ is a commutative algebra. The Gelfand transform $\widehat{a} \colon \mathcal{M} \to \sigma(a)$ is injective. Indeed, if $\widehat{a}(\varphi) = \widehat{a}(\psi)$, then $\varphi(a) = \psi(a)$, and therefore

$$\varphi[(z_j e - a)^{-1}] = \left(\varphi(z_j e - a)\right)^{-1} = \left(\psi(z_j e - a)\right)^{-1}$$
$$= \psi[(z_j e - a)^{-1}], \qquad j = 1, \ldots, n.$$

Next, use that φ and ψ are linear and multiplicative to show that φ and ψ coincide on all the polynomials in $a, (z_1 e - a)^{-1}, \ldots, (z_n e - a)^{-1}$. Thus by continuity of φ and ψ, the functionals φ and ψ coincide on $\mathcal{B}$. By Corollary 2.4, the range of $\widehat{a}$ is $\sigma(a)$. Since $\widehat{a} \in C(\mathcal{M})$, we conclude that $\widehat{a} \colon \mathcal{M} \to \sigma(a)$ is a continuous bijection, and thus $\widehat{a}$ is a homeomorphism.

Next, we identify $\mathcal{M}$ with $\sigma(a)$ via the map $\widehat{a}$. Thus for each element x of $\mathcal{B}$ the Gelfand transform $\widehat{x}$ is a continuous function on $\sigma(a)$. From $\widehat{e}(\lambda) = 1$ and $\widehat{a}(\lambda) = \lambda$ for each $\lambda \in \sigma(a)$, one sees that

$$(z_j e - a)^{-1}\widehat{}(\lambda) = \frac{1}{z_j - \lambda}, \qquad \lambda \in \sigma(a).$$

Now let $v = p(a, x_1, \ldots, x_n)$, where p is a polynomial in $n + 1$ variables and $x_j = (z_j e - a)^{-1}, j = 1, \ldots, n$. Then

$$\widehat{v}(\lambda) = p\left(\lambda, \frac{1}{z_1 - \lambda}, \ldots, \frac{1}{z_n - \lambda}\right), \qquad \lambda \in \sigma(a).$$

In particular, $\widehat{v}$ is continuous on $\sigma(a)$ and analytic in the interior of $\sigma(a)$. Given an $x \in \mathcal{B}$ and $\varepsilon > 0$, we may choose v so that $\|x - v\| < \varepsilon$. Therefore, $\|\widehat{x} - \widehat{v}\| \leq \|x - v\| < \varepsilon$ (cf., formula (1) in Section XXX.3). Hence $\widehat{x}$ is continuous on $\sigma(a)$ and analytic on the interior of $\sigma(a)$, because $\widehat{x}$ is the uniform limit of a sequence of functions of the form $\widehat{v}$. $\square$

Let us consider the following concrete example. Let $\mathbb{D}$ be the open unit disc in the complex plane, and let D_1 and D_2 be open discs such that $\overline{D}_1$ and $\overline{D}_2$ are disjoint and in $\mathbb{D}$. Take D_0 to be $\mathbb{D}$ with $\overline{D_1 \cup D_2}$ excluded. Let $\mathcal{A}(\overline{D}_0)$ be the Banach algebra (see Section XXIX, example (ii)) of all complex valued functions which are continuous on $\overline{D}_0$ and analytic on D_0 equipped with the usual supremum norm. Let $a \in \mathcal{A}(\overline{D}_0)$ be the function $a(\zeta) = \zeta$, and let $\mathcal{B}_0$ be the unital closed subalgebra of $\mathcal{A}(\overline{D}_0)$ generated by a. It is easy to see that $\mathcal{B}_0$ is the set of those functions in $\mathcal{A}(\overline{D}_0)$ that can be extended to be analytic in $\mathbb{D}$. In other words, $\mathcal{B}_0$ may be identified with $\mathcal{A}(\overline{\mathbb{D}})$. Next, choose any z_1 in D_1, and let $x_1 \in \mathcal{A}(\overline{D}_0)$ be the function $x_1(\zeta) = (z_1 - \zeta)^{-1}$. We denote by $\mathcal{B}_1$ the unital closed subalgebra of $\mathcal{A}(\overline{D}_0)$ generated by a and x_1. It follows from the theory of Laurent series expansions that $\mathcal{B}_1$ consists of those functions in $\mathcal{A}(\overline{D}_0)$ which can be extended to be analytic on $\mathbb{D}\backslash\overline{D}_1$, that is, $\mathcal{B}_1$ may be identified with $\mathcal{A}(\mathbb{D}\backslash\overline{D}_1)$. Similarly, let $x_2 \in \mathcal{A}(\overline{D}_0)$ be the function $x_2(\zeta) = (z_2 - \zeta)^{-1}$, where z_2 is some point in D_2. The unital closed subalgebra $\mathcal{B}_2$ of $\mathcal{A}(\overline{D}_0)$ generated by a, x_1 and x_2 is precisely equal to $\mathcal{A}(\overline{D}_0)$. Note that $x_1 = (z_1 e - a)^{-1}$ and $x_2 = (z_2 e - a)^{-1}$, where e is the unit in $\mathcal{B}_2$ and the inverses are taken in $\mathcal{B}_2$. Furthermore,

$$\sigma_{\mathcal{B}_0}(a) = \overline{\mathbb{D}}, \qquad \sigma_{\mathcal{B}_1}(a) = \overline{\mathbb{D}\backslash D_1}, \qquad \sigma_{\mathcal{B}_2}(a) = \overline{\mathbb{D}\backslash(D_1 \cup D_2)} = \overline{D}_0,$$

which corresponds to the statements in Theorem XXIX.6.1.

XXX.6 THE ALGEBRA GENERATED BY A COMPACT OPERATOR

Let K be a compact operator in $\mathcal{L}(X)$, where X is a complex Banach space, and let $\mathcal{A}$ be the unital closed subalgebra of $\mathcal{L}(X)$ generated by K. Since the resolvent set of K is connected, Corollary XXIX.6.2 shows that the spectrum $\sigma_{\mathcal{A}}(K)$ of K as an element of $\mathcal{A}$ coincides with the usual spectrum $\sigma(K)$ of K. Of course, $\mathcal{A}$ is a commutative Banach algebra, and we know from Corollary 5.2 that the Gelfand transform $\widehat{K}$ of K is a homeomorphism from the Gelfand spectrum of $\mathcal{A}$ onto $\sigma(K)$.

THEOREM 6.1. *Let K be a compact linear operator on a complex Banach space, and let $\mathcal{A}$ be the unital closed subalgebra of $\mathcal{L}(X)$ generated by K. Then the map*

$$(1) \qquad \Gamma: \mathcal{A} \to C(\sigma(K)), \qquad \Gamma(A) = \widehat{A} \circ \widehat{K}^{-1},$$

where $\widehat{T}$ denotes the Gelfand transform of $T \in \mathcal{A}$, is a homomorphism of norm 1,

$$(2) \qquad \Gamma(I)(\lambda) = 1, \qquad \Gamma(K)(\lambda) = \lambda, \quad (\lambda \in \sigma(K)),$$

and the image of Γ is dense in $C(\sigma(K))$.

PROOF. If we identify the Gelfand spectrum $\mathcal{M}$ of $\mathcal{A}$ with $\sigma(K)$ via the map $\widehat{K}$, then the map Γ defined by (1) is just the Gelfand representation of $\mathcal{A}$. Thus, except for the statement about $\operatorname{Im}\Gamma$, all other properties of Γ are clear.

Let $\mathcal{P}(\sigma(K))$ denote the set of polynomials restricted to $\sigma(K)$. From (2) it follows that $\mathcal{P}(\sigma(K)) \subset \operatorname{Im}\Gamma$. Thus to prove that $\operatorname{Im}\Gamma$ is dense in $C(\sigma(K))$, it suffices to prove that $\mathcal{P}(\sigma(K))$ is dense in $C(\sigma(K))$.

Take $f \in C(\sigma(K))$. If $\sigma(K)$ is finite, then there exists a polynomial which passes through the points of $f[\sigma(K)]$, and hence in that case $f \in \mathcal{P}(\sigma(K))$. Next, assume that $\sigma(K)$ is infinite. Take $\varepsilon > 0$. Since 0 is the only limit point of $\sigma(K)$ and f is continuous on $\sigma(K)$, there exists a closed disc S with centre at 0 such that

$$|f(\lambda) - f(0)| < \frac{1}{2}\varepsilon, \qquad \lambda \in S \cap \sigma(K).$$

There exists at most a finite number of points, say $\lambda_1, \dots, \lambda_N$ in $\sigma(K)$ which are not in S. Choose $q \in \mathcal{P}\big(\sigma(K)\big)$ such that $q(\lambda_i) = f(\lambda_i)$ for $i = 1, \dots, N$, and consider the function h on S defined by

$$h(\zeta) = (\zeta - \lambda_1)^{-1} \cdots (\zeta - \lambda_N)^{-1}\big(f(0) - q(\zeta)\big).$$

Obviously, h is continuous on S and analytic in the interior of S. Since S is a closed disc with centre at 0 there exists a sequence of polynomials $p_1, p_2, \dots$ such that $p_n(\zeta) \to h(\zeta)$ uniformly on S if $n \to \infty$. For $n = 1, 2, \dots$ put

$$q_n(\zeta) = (\zeta - \lambda_1) \cdots (\zeta - \lambda_N)p_n(\zeta) + q(\zeta).$$

Then $q_n(\lambda_i) = f(\lambda_i)$ for $i = 1, \dots, N$ and $q_n(\zeta) \to f(0)$ uniformly on S if $n \to \infty$. The latter property implies that there exists a positive integer k such that $|f(0) - q_k(\zeta)| < \frac{1}{2}\varepsilon$ for all $\zeta \in S$, and hence $|f(\lambda) - q_k(\lambda)| < \varepsilon$ for all $\lambda \in \sigma(K)$. Since $\varepsilon > 0$ was arbitrary, we conclude that $\mathcal{P}\big(\sigma(K)\big)$ is dense in $C\big(\sigma(K)\big)$. $\square$

XXX.7 THE RADICAL

We have seen that if $\mathcal{B}$ is a commutative unital Banach algebra, then its Gelfand representation Γ is a homomorphism from $\mathcal{B}$ into $C(\mathcal{M})$, where $\mathcal{M}$ is the Gelfand spectrum of $\mathcal{B}$. The map Γ need not be surjective or injective. For instance, it was shown in example (iv) of Section XXX.4 that if $\mathcal{B}$ is the Wiener algebra $\mathcal{W}$, then the image of Γ is a proper dense linear manifold of $C(\mathcal{M})$. In example (vii) of Section XXX.4 the kernel of Γ is the whole of $L_1([0,1])$.

By definition the *radical* of a commutative unital Banach algebra $\mathcal{B}$ is the kernel of its Gelfand representation Γ. An immediate consequence of Theorem 3.2 is the following description of the radical.

THEOREM 7.1. *Let $\mathcal{B}$ be a commutative unital Banach algebra. An element x is in the radical of $\mathcal{B}$ if and only if x is quasi-nilpotent.*

An element x is in the radical of $\mathcal{B}$ if and only if $\varphi(x) = (\Gamma x)(\varphi) = 0$ for all $\varphi \in \mathcal{M}$, where $\mathcal{M}$ is the Gelfand spectrum of $\mathcal{B}$. Hence the radical of $\mathcal{B}$ is equal to the set $\cap_{\varphi \in \mathcal{M}} \operatorname{Ker} \varphi$, which is precisely the intersection of the maximal ideals in $\mathcal{B}$.

It is clear from examples (i)–(vi) in Section XXX.4 that the radicals of the Banach algebras $C(S)$, $\mathcal{A}(\overline{\mathbb{D}})$, $PC([a,b])$, $\mathcal{W}$, $\mathcal{W}(\beta)$ and $L_1(\mathbb{R}) \oplus \operatorname{sp}\{1\}$ are all $\{0\}$. A commutative unital Banach algebra $\mathcal{B}$ whose radical is $\{0\}$ is called *semi-simple*. Thus $\mathcal{B}$ is semi-simple if and only if its Gelfand representation is injective.

The convolution algebra $L_1([0,1])$ with a unit e adjoined to it is not semi-simple (see example (vii) of Section XXX.4). Another example is the following. Let $\mathcal{T}_n$

consist of the $n \times n$ lower triangular Toeplitz matrices:

$$A = \begin{bmatrix} a_0 & & & \\ a_1 & a_0 & & \\ \vdots & \vdots & \ddots & \\ a_{n-1} & a_{n-2} & \cdots & a_0 \end{bmatrix}.$$

With respect to the usual matrix operations and norm $\|A\| = \sum_{j=0}^{n-1} |a_j|$, the set $\mathcal{T}_n$ is a commutative unital Banach algebra. The unit is equal to the $n \times n$ identity matrix I. Given A above, $A = a_0 I + N$, where N is nilpotent. Hence, by Theorem 7.1, the matrix N is in the radical of $\mathcal{T}_n$. Since the Gelfand transform of I is the constant function 1, the radical of $\mathcal{T}_n$ consists precisely of all lower triangular Toeplitz matrices with zeros on the main diagonal.

XXX.8 MATRICES OVER COMMUTATIVE BANACH ALGEBRAS

Let $\mathcal{B}$ be a commutative Banach algebra with unit e. As in Section XXIX.8 we denote by $\mathcal{B}^{m \times m}$ the Banach algebra of all $m \times m$ matrices $[a_{ij}]_{i,j=1}^{m}$ endowed with the usual matrix operations and the norm

$$\|[a_{ij}]\| = \max_{1 \le i \le m} \sum_{j=1}^{m} \|a_{ij}\|.$$

The unit E in $\mathcal{B}^{m \times m}$ is the $m \times m$ matrix with e on the main diagonal and zeros elsewhere. Note that $\mathcal{B}^{m \times m}$ is non-commutative for $m > 1$. We saw in example (viii) in Section XXX.1 that $\mathbb{C}^{2 \times 2}$ has no multiplicative linear functionals.

Let $\mathcal{M}$ denote the Gelfand spectrum of $\mathcal{B}$, and let $C(\mathcal{M})^{m \times m}$ denote the algebra all $m \times m$ matrices with entries from $C(\mathcal{M})$. It is easy to see that the map $\Gamma_m : \mathcal{B}^{m \times m} \to C(\mathcal{M})^{m \times m}$ defined by

$$(1) \qquad \Gamma_m\left([a_{ij}]_{i,j=1}^{m}\right) = [\widehat{a}_{ij}]_{i,j=1}^{m}$$

is a homomorphism. It follows from Theorem 7.1 that the kernel of Γ_m consists of those matrices in $\mathcal{B}^{m \times m}$ whose elements are quasi-nilpotent in $\mathcal{B}$. This observation together with Corollary XXIX.7.3 implies that every element A in the kernel of Γ_m is quasi-nilpotent in $\mathcal{B}^{m \times m}$. Indeed, let $\Gamma_m(A) = 0$ and $\lambda \ne 0$. Since $\mathcal{B}$ is commutative, $\det(\lambda E - A)$ is equal to $\lambda^m e + b$, where $b \in \mathcal{B}$ is quasi-nilpotent because of Corollary XXIX.7.3. It follows that $\det(\lambda E - A)$ is invertible in the commutative Banach algebra $\mathcal{B}$ and hence, by Theorem XXIX.8.1, the matrix $\lambda E - A$ is invertible in $\mathcal{B}^{m \times m}$. But $\lambda \ne 0$ is arbitrary. Thus A is quasi-nilpotent in $\mathcal{B}^{m \times m}$. The converse statement is false, i.e., if A is quasi-nilpotent in $\mathcal{B}^{m \times m}$, then it does not follow that $A \in \operatorname{Ker} \Gamma_m$. For example,

$$A = \begin{bmatrix} 0 & 0 \\ e & 0 \end{bmatrix} \in \mathcal{B}^{2 \times 2}$$

is not in the kernel of Γ_2, yet $A^2 = 0$.

THEOREM 8.1. *The matrix $A = [a_{ij}]_{i,j=1}^{m}$ in $\mathcal{B}^{m \times m}$ is invertible in $\mathcal{B}^{m \times m}$ if and only if*

$$(2) \qquad \det\left([\widehat{a}_{ij}(\varphi)]_{i,j=1}^{m}\right) \neq 0, \qquad \varphi \in \mathcal{M}.$$

Here $\mathcal{M}$ is the Gelfand spectrum of $\mathcal{B}$ and $\widehat{a}_{ij}$ is the Gelfand transform of a_{ij}.

PROOF. Suppose A is invertible in $\mathcal{B}^{m \times m}$ and $A^{-1} = [b_{ij}]_{i,j=1}^{m}$. Then for $\varphi \in \mathcal{M}$

$$\varphi\left(\sum_{k=1}^{m} a_{ik}b_{kj}\right) = \delta_{ij}, \qquad 1 \leq i,j \leq m,$$

where δ_{ij} is the Kronecker delta. It follows that

$$[\widehat{a}_{ij}(\varphi)]_{i,j=1}^{m} \cdot [\widehat{b}_{ij}(\varphi)]_{i,j=1}^{m} = I,$$

the $m \times m$ identity matrix in $\mathbb{C}^{m \times m}$. Hence (2) holds.

Conversely, assume (2) holds. Since $\mathcal{B}$ is commutative, $\det A$ is a well-defined element in $\mathcal{B}$. According to (2)

$$\varphi(\det A) = \det\left([\varphi(a_{ij})]_{i,j=1}^{m}\right) \neq 0$$

for each $\varphi \in \mathcal{M}$. But then we can apply Theorem 2.3 to obtain that $\det A$ is invertible in the commutative algebra $\mathcal{B}$. It follows from Theorem XXIX.8.1 that A is invertible in $\mathcal{B}^{m \times m}$. $\square$

If $\mathcal{B} = \mathcal{W}$ is the Wiener algebra, then it follows from (2) and example (iv) in Section XXX.1 that

$$A = [f_{ij}]_{i,j=1}^{m} \in \mathcal{W}^{m \times m}$$

is invertible in $\mathcal{W}^{m \times m}$ if and only if

$$\det\left([f_{ij}(\lambda)]_{i,j=1}^{m}\right) \neq 0, \qquad |\lambda| = 1,$$

and in this case $A^{-1} = [g_{ij}]_{i,j=1}^{m}$ is given by

$$[g_{ij}(\lambda)]_{i,j=1}^{m} = \left([f_{ij}(\lambda)]_{i,j=1}^{m}\right)^{-1}, \qquad |\lambda| = 1.$$

This is the matrix version of (Wiener's) Theorem 2.5. The continuous analogue of the above remark yields the following system version of Corollary 2.7.

THEOREM 8.2. *Suppose that $k = [k_{ij}]_{i,j=1}^{m}$ is an $m \times m$ matrix with entries in $L_1(\mathbb{R})$, and let $\widehat{h}$ denote the Fourier transform of an element $h \in L_1(\mathbb{R})$. Then for $g_1, \ldots, g_m$ in $L_1(\mathbb{R})$ the system of equations*

$$(3) \qquad f_i(t) - \sum_{j=1}^{m} \int_{-\infty}^{\infty} k_{ij}(t-s)f_j(s)ds = g_i(t), \qquad t \in \mathbb{R}, \text{ a.e.} \quad (i = 1, \ldots, m),$$

has a unique solution $f_1, \ldots, f_m$ in $L_1(\mathbb{R})$ if and only if $\det\big(I - [\widehat{k}_{ij}(\lambda)]_{i,j=1}^m\big) \neq 0$ for each $\lambda \in \mathbb{R}$, where I is the $m \times m$ identity matrix. In this case the unique solution $f_1, \ldots, f_m$ of (4) is given by

$$
(4) \qquad f_i(t) = g_i(t) - \sum_{j=1}^m \int_{-\infty}^{\infty} \ell_{ij}(t-s)g_j(s)ds, \quad t \in \mathbb{R}, \text{ a.e.} \qquad (i = 1, \ldots, m),
$$

where $\ell_{ij} \in L_1(\mathbb{R})$, $i, j = 1, \ldots, m$, is determined by

$$
[\widehat{\ell}_{ij}(\lambda)]_{i,j=1}^m = I - \big(I - [\widehat{k}_{ij}(\lambda)]_{i,j=1}^m\big)^{-1}, \qquad \lambda \in \mathbb{R}.
$$

XXX.9 FACTORIZATION OF MATRIX FUNCTIONS

In this section our aim is to prove two factorization theorems. The first concerns functions Φ in the $m \times m$ matrix Wiener algebra $\mathcal{W}^{m \times m}$ (see Section XXIX.9), and may be viewed as a generalization of Theorem XXIV.3.1. The second, which generalizes Theorem XIII.2.1, deals with matrix functions on the real line of the form

$$
(1) \qquad W(\lambda) = I_m - \int_{-\infty}^{\infty} e^{i\lambda t} k(t)dt, \qquad \lambda \in \mathbb{R},
$$

where I_m is the $m \times m$ identity matrix and k is an $m \times m$ matrix function with entries in $L_1(\mathbb{R})$.

THEOREM 9.1. *Let Φ be in the matrix Wiener algebra $\mathcal{W}^{m \times m}$, and assume that $\det \Phi(\zeta) \neq 0$ for each $\zeta \in \mathbb{T}$. Then there exist Φ_+ and Φ_- in $\mathcal{W}^{m \times m}$ and integers $\kappa_1 \leq \kappa_2 \leq \cdots \leq \kappa_m$ such that*

$$
(2) \qquad \Phi(\zeta) = \Phi_-(\zeta) \begin{bmatrix} \zeta^{\kappa_1} & & & \\ & \zeta^{\kappa_2} & & \\ & & \ddots & \\ & & & \zeta^{\kappa_m} \end{bmatrix} \Phi_+(\zeta), \qquad \zeta \in \mathbb{T},
$$

and

    (i)   *Φ_+ is continuous on $\overline{\mathbb{D}}$, analytic on $\mathbb{D}$ and $\det \Phi_+(\zeta) \neq 0$ for $\zeta \in \overline{\mathbb{D}}$,*

    (ii) *Φ_- is continuous on $\mathbb{C}_\infty \backslash \mathbb{D}$, analytic on $\mathbb{C}_\infty \backslash \overline{\mathbb{D}}$ and $\det \Phi_-(\zeta) \neq 0$ for $\zeta \in \mathbb{C}_\infty \backslash \mathbb{D}$.*

Furthermore, the indices $\kappa_1, \ldots, \kappa_m$ are uniquely determined by Φ.

THEOREM 9.2. *Let W be as in (1), and assume $\det W(\lambda) \neq 0$ for each $\lambda \in \mathbb{R}$. Then there exist matrix functions W_+ and W_-,*

$$
(3a) \qquad W_+(\lambda) = I_m - \int_0^{\infty} e^{i\lambda t} k_+(t)dt, \qquad \Im \lambda \geq 0,
$$

$$(3b) \qquad W_-(\lambda) = I_m - \int_{-\infty}^{0} e^{i\lambda t} k_-(t)dt, \qquad \Im\lambda \le 0,$$

where k_+ and k_- are $m \times m$ matrix functions with entries in $L_1([0,\infty])$ and $L_1((-\infty,0])$, respectively, and there exist an invertible $m \times m$ matrix D and integers $\kappa_1 \ge \kappa_2 \ge \cdots \ge \kappa_m$ such that

$$(4) \quad W(\lambda) = W_-(\lambda)D \begin{bmatrix} \left(\frac{\lambda-i}{\lambda+i}\right)^{\kappa_1} & & & \\ & \left(\frac{\lambda-i}{\lambda+i}\right)^{\kappa_2} & & \\ & & \ddots & \\ & & & \left(\frac{\lambda-i}{\lambda+i}\right)^{\kappa_m} \end{bmatrix} D^{-1}W_+(\lambda), \qquad \lambda \in \mathbb{R},$$

and

(j) $\det W_+(\lambda) \ne 0$ *for* $\Im\lambda \ge 0,$

(jj) $\det W_-(\lambda) \ne 0$ *for* $\Im\lambda \le 0.$

Furthermore, the indices $\kappa_1,\ldots,\kappa_m$ are uniquely determined by W.

The factorizations (2) and (4) are called (*right*) *Wiener-Hopf factorizations*. In (2) the factorization is relative to the circle $\mathbb{T}$ and in (4) relative to $\mathbb{R}$. One refers to the indices $\kappa_1,\ldots,\kappa_m$ as the (*right*) *factorization indices*. If these indices are all equal to zero, then the corresponding factorization is called a (*right*) *canonical factorization*.

PROOF OF THEOREM 9.1.   Let Φ_ν be the ν-th Fourier coefficient of Φ, and consider

$$R_N(\zeta) = \sum_{\nu=-N}^{N} \zeta^\nu \Phi_\nu.$$

Since $\Phi \in \mathcal{W}^{m \times m}$, we have

$$(5) \qquad \|\!|\Phi|\!\| := \sum_{\nu=-\infty}^{\infty} \|\Phi_\nu\| < \infty,$$

and hence

$$\|\!|R_N - \Phi|\!\| = \sum_{|\nu|>N} \|\Phi_\nu\| \to 0 \qquad (N \to \infty).$$

By our hypotheses on Φ the function Φ is an invertible element of $\mathcal{W}^{m \times m}$ (see the previous section). Thus for N sufficiently large R_N^{-1} exists and belongs to $\mathcal{W}^{m \times m}$. Furthermore,

$$\|\!|(R_N - \Phi)R_N^{-1}|\!\| \to 0 \qquad (N \to \infty).$$

Next recall (see Section XXIX.9) that $\mathcal{W}^{m \times m}$ is a decomposing Banach algebra with respect to the decomposition

$$\mathcal{W}^{m \times m} = \mathcal{W}_{-,0}^{m \times m} \oplus \mathcal{W}_+^{m \times m}.$$

Here

$$\mathcal{W}^{m\times m}_{-,0} = \{\Phi \in \mathcal{W}^{m\times m} \mid \Phi_k = 0 \text{ for } k \geq 0\},$$

$$\mathcal{W}^{m\times m}_{+} = \{\Phi \in \mathcal{W}^{m\times m} \mid \Phi_k = 0 \text{ for } k < 0\}.$$

Let P be the projection of $\mathcal{W}^{m\times m}$ onto $\mathcal{W}^{m\times m}_{+}$ along $\mathcal{W}^{m\times m}_{-,0}$, and put $Q = I - P$. Since the norm $\|\| \cdot \|\|$ on $\mathcal{W}^{m\times m}$ is given by (5), we have $\|P\| = \|Q\| = 1$. Now, choose N so that

$$\|\|(R_N - \Phi)R_N^{-1}\|\| < 1,$$

and put $H = E - (R_N - \Phi)R_N^{-1}$. Then $\Phi = HR_N$. Furthermore, by Theorem XXIX.9.1, H factorizes as $H = H_-H_+$, where H_- and H_+ are invertible elements of $\mathcal{W}^{m\times m}$ such that $H_{\pm}^{\pm 1} \in \mathcal{W}^{m\times m}_{+}$ and

$$H_{-}^{\pm 1} \in \mathcal{W}^{m\times m}_{-} := \{\Phi \in \mathcal{W}^{m\times m} \mid \Phi_k = 0 \text{ for } k > 0\}.$$

In particular, $\Phi = H_-H_+R_N$.

Note that statement (ii) of the theorem holds for H_- in place of Φ_-. So in order to establish the desired Wiener-Hopf factorization for Φ it suffices to show that H_+R_N admits such a factorization. To do this choose a scalar polynomial q such that $q(\zeta) \neq 0$ for $\zeta \in \mathbb{T}$ and $q(\zeta)R_N(\zeta)$ is a matrix polynomial. Put

$$\Psi(\zeta) = H_+(\zeta)\big(q(\zeta)R_N(\zeta)\big).$$

Then Ψ is analytic on the open unit disc $\mathbb{D}$, continuous on $\overline{\mathbb{D}}$ and det $\Psi(\zeta) \neq 0$ for $\zeta \in \mathbb{T}$. In particular, det $\Psi(\zeta)$ has a finite number of zeros in $\mathbb{D}$. But then we can repeat the arguments of the last two paragraphs of the proof of Theorem XIII.2.1 (with Ψ in place of Φ and polynomials replaced by analytic functions on $\mathbb{D}$) to show that Ψ admits a factorization,

$$\Psi(\zeta) = \Psi_-(\zeta)\begin{bmatrix} \zeta^{\kappa_1} & & & \\ & \zeta^{\kappa_2} & & \\ & & \ddots & \\ & & & \zeta^{\kappa_m} \end{bmatrix}\Psi_+(\zeta),$$

where $\kappa_1, \ldots, \kappa_m$ are integers, Ψ_- is a rational $m \times m$ matrix function such that

(α) Ψ_- has no poles on $|\zeta| \geq 1$ and det $\Psi_-(\zeta) \neq 0$ for $|\zeta| \geq 1$,

(β) Ψ_- has no pole at infinity and det $\Psi_-(\infty) \neq 0$,

and Ψ_+ is analytic on $\mathbb{D}$, continuous on $\overline{\mathbb{D}}$ with det $\Psi_+(\zeta) \neq 0$ for $\zeta \in \overline{\mathbb{D}}$. We also know (see the third paragraph of the proof of Theorem XIII.2.1) that $q(\cdot)^{-1}$ has a Wiener-Hopf factorization,

$$q(\zeta)^{-1} = \varphi_-(\zeta)\zeta^{\kappa}\varphi_+(\zeta),$$

with scalar rational factors φ_- and φ_+. It follows that (2) holds with

$$\Phi_- := \varphi_-H_-\Psi_-, \qquad \Phi_+ := \Psi_+.$$

Since φ_- and Ψ_- are rational and $H_- \in \mathcal{W}^{m \times m}$, we also hae $\Phi_- \in \mathcal{W}^{m \times m}$. But then we can use (2) to conclude that $\Phi_+ \in \mathcal{W}^{m \times m}$ too. So Φ has the required factorization.

To prove the uniqueness of the indices one first proves (cf., the next section) the analogue of Theorem XXIV.4.2 for the function Φ considered here, and next one uses the arguments appearing in the last paragraph of Section XXIV.4. $\square$

Since Φ_+ and Φ_- in (2) are in $\mathcal{W}^{m \times m}$, the analyticity conditions in (i) and (ii) imply that $\Phi_+ \in \mathcal{W}_+^{m \times m}$ and $\Phi_- \in \mathcal{W}_-^{m \times m}$. Furthermore, because of the conditions on the determinants in (i) and (ii), we can use the matrix version of Wiener's theorem (see the previous section) to show that $\Phi_+^{-1} \in \mathcal{W}_+^{m \times m}$ and $\Phi_-^{-1} \in \mathcal{W}_-^{m \times m}$.

PROOF OF THEOREM 9.2. Let $\mathcal{B}$ be the algebra of all $m \times m$ matrix functions F on $\mathbb{R}$ of the form

$$(6) \qquad F(\lambda) = D + \int_{-\infty}^{\infty} e^{i\lambda t} f(t)\,dt, \qquad \lambda \in \mathbb{R},$$

where D is a constant $m \times m$ matrix and f is an $m \times m$ matrix function with entries in $L_1(\mathbb{R})$. The algebra $\mathcal{B}$ is a unital Banach algebra. The norm $\|\|\cdot\|\|$ on $\mathcal{B}$ is given by

$$\|\|F\|\| = \|D\| + \int_{-\infty}^{\infty} \|k(t)\|\,dt.$$

Here $\|A\|$ is the usual operator norm of the $m \times m$ matrix considered as an operator on $\mathbb{C}^m$. The unit E of $\mathcal{B}$ is the function which is identically equal to the $m \times m$ identity matrix. The algebra $\mathcal{B}$, which is called the $m \times m$ *matrix Wiener algebra on the line*, is also a decomposing algebra. Indeed, let $\mathcal{B}_+$ be the set of all F of the form (6) with the additional property that $f(t) = 0$ a.e. on $-\infty < t < 0$, and let $\mathcal{B}_{-,0}$ be the set of all F such that

$$(7) \qquad F(\lambda) = \int_{-\infty}^{0} e^{i\lambda t} f(t)\,dt,$$

with f an $m \times m$ matrix function with entries in $L_1((-\infty, 0])$. Then $\mathcal{B}_+$ and $\mathcal{B}_{-,0}$ are closed subalgebras of $\mathcal{B}$ and

$$(8) \qquad \mathcal{B} = \mathcal{B}_{-,0} \oplus \mathcal{B}_+.$$

By $\mathcal{R}$ we denote the set of all F of the form (6) such that each entry of f is a finite linear combination of functions of the form

$$h(t) = \begin{cases} a_1 t^{n_1} e^{-i\alpha_1 t}, & t > 0, \\ a_2 t^{n_2} e^{-i\alpha_2 t}, & t < 0, \end{cases}$$

where a_1, a_2 are arbitrary complex numbers, $\Im \alpha_1 < 0$ and $\Im \alpha_2 > 0$. The arguments used in Part (v) of the proof of Theorem XII.3.1 show that $\mathcal{R}$ is dense in $\mathcal{B}$. Furthermore, the elements of $\mathcal{R}$ are rational matrix functions.

Now, let W be as in the theorem. Since $\mathcal{R}$ is dense in $\mathcal{B}$, we can find $R \in \mathcal{R}$ such that $\|\|R - W\|\|$ is as small as desired. Note that

$$\|R(\lambda) - W(\lambda)\| \leq \|\|R - W\|\|, \qquad \lambda \in \mathbb{R} \cup \{\infty\}.$$

Thus for $\|\|R - W\|\|$ sufficiently small, we have $\det R(\lambda) \neq 0$ for each $\lambda \in \mathbb{R} \cup \{\infty\}$, and in that case $R^{-1} \in \mathcal{B}$. Hence we may choose $R \in \mathcal{R}$ such that

$$(9) \qquad \|\|(R - W)R^{-1}\|\| < 1.$$

Let $R \in \mathcal{R}$ be as in (9). Note that the natural projections associated with the decomposition (8) have norm equal to one. So we may apply Theorem XXIX.9.1 to show that $V := E - (R - W)R^{-1}$ factorizes as $V = V_- V_+$, where V_- and V_+ are invertible elements of $\mathcal{B}$ such that $V_+^{\pm 1} \in \mathcal{B}_+$ and $V_-^{\pm 1} \in \mathcal{B}_-$. Here $\mathcal{B}_-$ is the set of all F of the form (6) such that $f(t) = 0$ a.e. on $0 < t < \infty$. Note that $W = V_-(V_+ R)$.

Next, consider $V_+ R$. To factorize $V_+ R$ we apply the Möbius transform

$$(10) \qquad z \mapsto \lambda(z) = i\frac{1 + z}{1 - z}.$$

Put

$$\widetilde{V}_+(z) = V_+\left(i\frac{1 + z}{1 - z}\right), \qquad \widetilde{R}(z) = R\left(i\frac{1 + z}{1 - z}\right).$$

Then $\widetilde{V}_+$ is analytic on the open unit disc $\mathbb{D}$, continuous on $\overline{\mathbb{D}}$ and $\det \widetilde{V}_+(z) \neq 0$ for $z \in \overline{\mathbb{D}}$. Since R is rational, the same holds for $\widetilde{R}$. Furthermore, $\det \widetilde{R}(z) \neq 0$ for $z \in \mathbb{T}$, because $\det R(\lambda) \neq 0$ for $\lambda \in \mathbb{R} \cup \{\infty\}$. The arguments used in the one but last paragraph of the proof of Theorem 9.1 show that we have a factorization

$$\widetilde{V}_+(z)\widetilde{R}_+(z) = \widetilde{G}_-(z) \begin{bmatrix} z^{\kappa_1} & & & \\ & z^{\kappa_2} & & \\ & & \ddots & \\ & & & z^{\kappa_m} \end{bmatrix} \widetilde{G}_+(z), \qquad z \in \mathbb{T},$$

where $\kappa_1 \geq \kappa_2 \geq \cdots \geq \kappa_m$ are integers, $\widetilde{G}_-$ is a rational matrix function such that

(α) $\widetilde{G}_-$ has no poles on $|z| \geq 1$ and $\det \widetilde{G}_-(z) \neq 0$ for $|z| \geq 1$,

(β) $\widetilde{G}_-$ has no pole at infinity and $\det \widetilde{G}_-(\infty) \neq 0$,

and $\widetilde{G}_+$ is analytic on $\mathbb{D}$, continuous on $\overline{\mathbb{D}}$ with $\det \widetilde{G}_+(z) \neq 0$ for $z \in \overline{\mathbb{D}}$. Now, use the inverse of the transformation (10), and put

$$G_-(\lambda) = \widetilde{G}_-\left(\frac{\lambda - i}{\lambda + i}\right), \qquad G_+(\lambda) = \widetilde{G}_+\left(\frac{\lambda - i}{\lambda + i}\right).$$

Then W admits the following factorization:

$$W(\lambda) = V_-(\lambda)G_-(\lambda) \begin{bmatrix} \left(\frac{\lambda - i}{\lambda + i}\right)^{\kappa_1} & & \\ & \ddots & \\ & & \left(\frac{\lambda - i}{\lambda + i}\right)^{\kappa_m} \end{bmatrix} G_+(\lambda).$$

Here G_- is a rational matrix function which is analytic on the closed lower half plane (including the point infinity) and $\det G_-(\lambda) \neq 0$ for $\Im\lambda \leq 0$ (including $\lambda = \infty$). It follows that $G_- \in \mathcal{B}_-$, and thus $V_-G_- \in \mathcal{B}_-$. Hence we may choose $W_-(\lambda)$ as in (3b) and an invertible $m \times m$ matrix D such that

$$V_-(\lambda)G_-(\lambda) = W_-(\lambda)D$$

and such that condition (jj) is fulfilled. Since

$$G_+(\lambda) = \begin{bmatrix} \left(\frac{\lambda-i}{\lambda+i}\right)^{-\kappa_1} & & \\ & \ddots & \\ & & \left(\frac{\lambda-i}{\lambda+i}\right)^{-\kappa_m} \end{bmatrix} D^{-1}W_-(\lambda)^{-1}W(\lambda),$$

we see that $G_+ \in \mathcal{B}$ and

$$\lim_{\lambda\in\mathbb{R},\ |\lambda|\to\infty} G_+(\lambda) = D^{-1}.$$

But then we may write $G_+(\lambda) = D^{-1}W_+(\lambda)$, where W_+ is as in (3a) and satisfies condition (j).

To prove the uniqueness of the indices one first proves the analogue of Theorem XIII.3.2 for the case considered here (see the next section), and next one uses the same arguments as in the one but last paragraph of Section XIII.3. $\square$

Let W_+ and W_- be as in Theorem 9.2. The inverses $W_+(\lambda)^{-1}$ and $W_-(\lambda)^{-1}$, which exist for each $\lambda \in \mathbb{R}$, admit the following representations:

$$(11) \qquad W_+(\lambda)^{-1} = I_m + \int_0^\infty e^{i\lambda t}\gamma_+(t)dt, \qquad \lambda \in \mathbb{R},$$

$$(12) \qquad W_-(\lambda)^{-1} = I_m + \int_{-\infty}^0 e^{i\lambda t}\gamma_-(t)dt, \qquad \lambda \in \mathbb{R},$$

where γ_+ and γ_- are $m \times m$ matrix functions with entries in $L_1(\mathbb{R})$. Let us prove this for W_+. Since $\det W_+(\lambda) \neq 0$ for each $\lambda \in \mathbb{R}$, Theorem 8.1 implies that

$$W_+(\lambda)^{-1} = I_m + \int_{-\infty}^\infty e^{i\lambda t}\omega(t)dt, \qquad \lambda \in \mathbb{R},$$

for some $m \times m$ matrix function ω with entries in $L_1(\mathbb{R})$. Now, put

$$V_+(\lambda) = \int_0^\infty e^{i\lambda t}\omega(t)dt, \qquad V_-(\lambda) = \int_{-\infty}^0 e^{i\lambda t}\omega(t)dt.$$

Then

$$V_-(\lambda) = W_+(\lambda)^{-1} - I_m - V_+(\lambda), \qquad \lambda \in \mathbb{R}.$$

Condition (j) in Theorem 9.2 implies that $W_+(\cdot)^{-1}$ is analytic on the open upper half plane $\mathbb{C}_+$ and continuous on its closure $\overline{\mathbb{C}}_+$. The same holds true for $V_+(\cdot)$. On the other hand, $V_-(\cdot)$ is analytic on the open lower half plane $\mathbb{C}_-$ and continuous on its closure $\overline{\mathbb{C}}_-$. It follows that V_- extends to an analytic function on the entire complex plane. Furthermore, by the continuous analogue of the Riemann-Lebesgue lemma (cf., [R], item 9.6), we have

$$\lim_{\lambda \in \overline{\mathbb{C}}_+,\ \lambda \to \infty} \left(W_+(\lambda)^{-1} - I_m - V_+(\lambda) \right) = 0,$$

$$\lim_{\lambda \in \overline{\mathbb{C}}_-,\ \lambda \to \infty} V_-(\lambda) = 0.$$

Thus Liouville's theorem implies that $V_- = 0$. Therefore, $W_+(\cdot)^{-1} = I + V_+(\cdot)$, and hence $W_+(\cdot)^{-1}$ has the desired representation (11). In an analogous way one proves (12).

Theorems 9.1 and 9.2 can be put into a more general context. To get a factorization of the type (2) one needs a Banach algebra $\mathcal{A}$ with norm $\| \cdot \|_{\mathcal{A}}$ consisting of continuous functions on $\mathbb{T}$ such that the identity $e(\lambda) \equiv 1$ belongs to $\mathcal{A}$ and $\mathcal{A}$ is continuously embedded in $C(\mathbb{T})$. Let $\mathcal{A}$ be such an algebra. Define $\mathcal{A}_+$ to be the subalgebra consisting of those $f \in \mathcal{A}$ that have an extension which is continuous on $\overline{\mathbb{D}}$ and analytic on $\mathbb{D}$, and let $\mathcal{A}_-^0$ consist of those $f \in \mathcal{A}$ that have an extension which is continuous on $\mathbb{C}_\infty \backslash \mathbb{D}$, analytic on $\mathbb{C}_\infty \backslash \overline{\mathbb{D}}$ and have the value 0 at infinity. One calls (see Clancey-Gohberg [1], Section II.2) the algebra $\mathcal{A}$ a *decomposing Banach algebra of continuous functions* on $\mathbb{T}$ if, in addition,

$$(13) \qquad\qquad \mathcal{A} = \mathcal{A}_+ \oplus \mathcal{A}_-^0.$$

Since $\mathcal{A}$ is continuously embedded in $C(\mathbb{T})$, the subalgebras $\mathcal{A}_+$ and $\mathcal{A}_-^0$ are always closed. Obviously, $\mathcal{W}$ is a decomposing Banach algebra of continuous functions on $\mathbb{T}$; one can show that $C(\mathbb{T})$ is not decomposing.

Now, let $\mathcal{A}$ be a decomposing Banach algebra of continuous functions with the additional property that the rational functions with no poles on $\mathbb{T}$ are dense in $\mathcal{A}$. Then arguments similar to those used in the proof of Theorem 9.1 show that any $\Phi \in \mathcal{A}^{m \times m}$ satisfying $\det \Phi(\lambda) \neq 0$ for each $\lambda \in \mathbb{T}$ admits a factorization of the type (2) such that the factors $\Phi_+(\cdot)^{\pm 1}$ are in $\mathcal{A}_+$ and $\Phi_-(\cdot)^{\pm 1}$ in $\mathcal{A}_-$, where

$$\mathcal{A}_- = \{ \alpha e + f \mid f \in \mathcal{A}_-^0 \}.$$

Furthermore, one can show that to obtain such a factorization result the decomposition (13) is necessary provided $\mathcal{A}$ is inverse closed. The latter means that any $f \in \mathcal{A}$ satisfying $f(\lambda) \neq 0$ for each $\lambda \in \mathbb{T}$ is invertible in $\mathcal{A}$. It follows that, in general, for $\Phi \in C(\mathbb{T})^{m \times m}$ one cannot obtain a factorization as in Theorem 9.1. For these and related results we refer to Chapter II in Clancey-Gohberg [1].

One cannot derive Theorem 9.2 from Theorem 9.1 by applying directly the

Möbius transformation (10). Indeed, consider

$$(14) \qquad \widetilde{\mathcal{B}} = \{ \widetilde{F} \mid \widetilde{F}(z) = F\left(i\frac{1+z}{1-z}\right), \ z \in \mathbb{T}, \ F \in \mathcal{B} \},$$

where $\mathcal{B}$ is the $m \times m$ matrix Wiener algebra on the line. Then $\widetilde{\mathcal{B}} \neq \mathcal{W}^{m \times m}$. On the other hand, $\widetilde{\mathcal{B}}$ is an algebra of continuous functions on $\mathbb{T}$, and with $\|\|\widetilde{F}\|\| = \|\|F\|\|$, where $\widetilde{F}$ and F are related as in (14), we have a Banach algebra norm on $\widetilde{\mathcal{B}}$ which is stronger than the norm on $C(\mathbb{T})^{m \times m}$. In fact, $\widetilde{\mathcal{B}} = \mathcal{A}^{m \times m}$, where $\mathcal{A}$ is some decomposing Banach algebra of continuous functions on $\mathbb{T}$, and hence one may use the Möbius transformation (10) and the factorization result mentioned in the previous paragraph to obtain Theorem 9.2.

XXX.10 WIENER-HOPF INTEGRAL OPERATORS REVISITED

In this section we extend the inversion and Fredholm theorems of Section XIII.3, which concern Wiener-Hopf operators with a rational matrix symbol, to a larger class of Wiener-Hopf operators, namely to Wiener-Hopf operators T whose kernel functions k belong to $L_1(\mathbb{R})^{m \times m}$. Thus in this section T will be an operator on $L_2^m([0, \infty))$ of the form

$$(1) \qquad (T\varphi)(t) = \varphi(t) - \int_0^\infty k(t-s)\varphi(s)ds, \qquad 0 \le t < \infty,$$

where k is an $m \times m$ matrix function on the real line whose entries are in $L_1(\mathbb{R})$. In this case we shall say that the symbol W of T,

$$(2) \qquad W(\lambda) = I_m - \int_{-\infty}^\infty e^{i\lambda t} k(t)dt, \qquad \lambda \in \mathbb{R},$$

belongs to the $m \times m$ matrix Wiener algebra on the line. (The latter algebra is the Banach algebra $\mathcal{B}$ introduced in the first paragraph of the proof of Theorem 9.2.)

Using the Wiener-Hopf factorization result stated in Theorem 9.2 the following generalizations of Theorems 3.1 and 3.2 in Section XIII.3 may be derived.

THEOREM 10.1. *Let T be a Wiener-Hopf operator on $L_2^m([0, \infty))$ with symbol W from the $m \times m$ matrix Wiener algebra on the line. Then T is invertible if and only if*

 (i) $\det W(\lambda) \neq 0$ *for all $\lambda \in \mathbb{R}$,*

 (ii) *W admits a (right) canonical factorization relative to the real line.*

In this case the inverse of T is obtained in the following way. Construct a right canonical factorization $W(\lambda) = W_-(\lambda)W_+(\lambda)$, $\lambda \in \mathbb{R}$, and choose $m \times m$ matrix functions γ_- and

γ_+ *with entries in* $L_1(\mathbb{R})$ *such that*

$$(3a) \qquad W_-(\lambda)^{-1} = I_m + \int_{-\infty}^{0} e^{i\lambda t}\gamma_-(t)dt, \qquad \lambda \in \mathbb{R},$$

$$(3b) \qquad W_+(\lambda)^{-1} = I_m + \int_{0}^{\infty} e^{i\lambda t}\gamma_+(t)dt, \qquad \lambda \in \mathbb{R}.$$

Then

$$(4) \qquad (T^{-1}\psi)(t) = \psi(t) + \int_{0}^{\infty} \gamma(t,s)\psi(s)ds, \qquad t \geq 0,$$

where

$$(5) \qquad \gamma(t,s) = \begin{cases} \gamma_+(t-s) + \int_{0}^{s} \gamma_+(t-\alpha)\gamma_-(\alpha-s)d\alpha, & 0 \leq s < t < \infty, \\[3mm] \gamma_-(t-s) + \int_{0}^{t} \gamma_+(t-\alpha)\gamma_-(\alpha-s)d\alpha, & 0 \leq t < s < \infty. \end{cases}$$

THEOREM 10.2. *Let* T *be a Wiener-Hopf operator on* $L_2^m([0,\infty))$ *with symbol* W *from the* $m \times m$ *matrix Wiener algebra on the line. Then* T *is Fredholm if and only if* $\det W(\lambda) \neq 0$ *for all* $\lambda \in \mathbb{R}$. *Assume the latter condition to be fulfilled, and let*

$$(6) \qquad W(\lambda) = W_-(\lambda)D\left(\left[\delta_{kj}\left(\frac{\lambda-i}{\lambda+i}\right)^{\kappa_j}\right]_{k,j=1}^{m}\right)D^{-1}W_+(\lambda), \qquad \lambda \in \mathbb{R},$$

be a right Wiener-Hopf factorization of W *relative to the real line. Then*

$$(7) \qquad n(T) = \sum_{\kappa_j \leq 0} -\kappa_j, \qquad d(T) = \sum_{\kappa_j \geq 0} \kappa_j,$$

and a generalized inverse of T *is given by*

$$(8) \qquad T^+ = V_1 M \begin{bmatrix} S_2^{-\kappa_1} & & & & & & & \\ & \ddots & & & & & & \\ & & S_2^{-\kappa_r} & & & & & \\ & & & I & & & & \\ & & & & \ddots & & & \\ & & & & & I & & \\ & & & & & & S_1^{\kappa_{s+1}} & \\ & & & & & & & \ddots & \\ & & & & & & & & S_1^{\kappa_m} \end{bmatrix} M^{-1} V_2,$$

where V_1 and V_2 are the Wiener-Hopf operators on $L_2^m([0,\infty))$ with symbols $W_+(\cdot)^{-1}$ and $W_-(\cdot)^{-1}$, respectively, S_1 and S_2 are the Wiener-Hopf operators on $L_2([0,\infty))$ defined by

$$(S_1 f)(t) = f(t) - 2 \int_t^\infty e^{t-s} f(s)ds, \qquad t \geq 0,$$

$$(S_2 f)(t) = f(t) - 2 \int_0^t e^{s-t} f(s)ds, \qquad t \geq 0,$$

M is the operator on $L_2^m([0,\infty))$ defined by $(M\varphi)(t) = D\varphi(t)$, the numbers $\kappa_1, \ldots, \kappa_r$ in (8) are the negative factorization indices, $\kappa_{s+1}, \ldots, \kappa_m$ in (8) are the positive factorization indices, and the unspecified entries of the $m \times m$ operator matrix in (8) are zero.

Theorems 10.1 and 10.2 are proved in much the same way as Theorems 3.1 and 3.2 in Section XIII.3. Note that the first part of Theorem 10.2 (i.e., the if and only if statement) is covered by Theorem XII.3.1. Now, assume that we have the factorization (6). Let T_1 and T_2 be the Wiener-Hopf operators with symbols W_+ and W_-, respectively. We have to show that T_1 and T_2 are invertible.

From the remark made in the first paragraph after the proof of Theorem 9.2 we know that

$$W_+(\lambda)^{-1} = I_m + \int_0^\infty e^{i\lambda t} \gamma_+(t)dt, \qquad \lambda \in \mathbb{R},$$

$$W_-(\lambda)^{-1} = I_m + \int_{-\infty}^0 e^{i\lambda t} \gamma_-(t)dt, \qquad \lambda \in \mathbb{R},$$

where γ_+ and γ_- are $m \times m$ matrix functions with entries in $L_1(\mathbb{R})$. Consider the Wiener-Hopf operators V_1 and V_2,

$$(V_1\varphi)(t) = \varphi(t) + \int_0^t \gamma_+(t-s)\varphi(s)ds, \qquad t \geq 0,$$

$$(V_2\varphi)(t) = \varphi(t) + \int_t^\infty \gamma_-(t-s)\varphi(s)ds, \qquad t \geq 0.$$

Note that T_ν has a similar representation as V_ν ($\nu = 1, 2$). Thus we can apply Theorem 8.2 and the arguments used in the proof of Theorem XII.2.3 (cf., Theorem XIII.1.1) to show that T_1 and T_2 are invertible, $T_1^{-1} = V_1$ and $T_2^{-1} = V_2$.

All other arguments necessary to prove Theorems 10.1 and 10.2 are now precisely the same as those used in Section XIII.3. We omit further details.

COROLLARY 10.3. *Let W be the element in the $m \times m$ matrix Wiener algebra on the line given by (2), and assume that $W(\lambda)$ is positive definite for each $\lambda \in \mathbb{R}$. Then the factorization indices of W relative to the real line are all equal to zero and W factorizes as*

$$(9) \qquad W(\lambda) = W_+(\lambda)^* W_+(\lambda), \qquad \lambda \in \mathbb{R},$$

where $\det W_+(\lambda) \neq 0$ *for each* $\Im \lambda \geq 0$,

$$(10) \qquad W_+(\lambda) = I_m - \int_0^\infty e^{i\lambda t} k_+(t) dt, \qquad \Im \lambda \geq 0,$$

$$(11) \qquad W_+(\lambda)^{-1} = I_m + \int_0^\infty e^{i\lambda t} \gamma_+(dt, \qquad \Im \lambda \geq 0,$$

and k_+ and γ_+ are $m \times m$ matrix functions with entries in $L_1([0,\infty))$.

PROOF. Since $W(\cdot)$ is continuous on $\mathbb{R}$ and

$$\lim_{\lambda \in \mathbb{R}, \ |\lambda| \to \infty} W(\lambda) = I_m,$$

our hypotheses on $W(\cdot)$ imply that there exists $\varepsilon > 0$ such that $\langle W(\lambda)x, x \rangle \geq \varepsilon \|x\|^2$ for each $x \in \mathbb{C}^m$ and $\lambda \in \mathbb{R}$. Thus, for $\varphi \in L_2^m(\mathbb{R})$,

$$\int_0^\infty \langle W(\lambda)\varphi(\lambda), \varphi(\lambda) \rangle d\lambda \geq \varepsilon \|\varphi\|^2,$$

which shows that the operator M_W on $L_2^m(\mathbb{R})$ of multiplication by $W(\cdot)$ is strictly positive. Now, let T be the Wiener-Hopf operator on $L_2^m([0,\infty))$ with symbol W. According to Theorem XII.2.1 the operator T is unitarily equivalent to the compression of M_W to $L_2^m([0,\infty))$, and hence T is also strictly positive. In particular, T is invertible, and so we can apply Theorem 10.1 to show that all factorization indices are equal to zero.

Theorem 10.1 also implies that W admits a right canonical factorization relative to the real line,

$$(12) \qquad W(\lambda) = W_-(\lambda) W_+(\lambda), \qquad \lambda \in \mathbb{R}.$$

Here $\det W_+(\lambda) \neq 0$ for $\Im \lambda \geq 0$, and for $W_+(\cdot)$ and $W_+(\cdot)^{-1}$ we have the representations (10) and (11). To complete the proof it remains to show that in (12) we may take $W_-(\lambda) = W_+(\lambda)^*$.

For $\lambda \in \mathbb{R}$, put $V_+(\lambda) = W_-(\lambda)^*$ and $V_-(\lambda) = W_+(\lambda)^*$. Thus, by (10) and (11),

$$V_-(\lambda) = I_m - \int_{-\infty}^0 e^{i\lambda t} k_+(-t)^* dt, \qquad \lambda \in \mathbb{R}$$

$$V_-(\lambda)^{-1} = I_m + \int_{-\infty}^{0} e^{i\lambda t}\gamma_+(-t)^* dt, \qquad \lambda \in \mathbb{R}.$$

Since $W(\lambda) = W(\lambda)^*$ for each $\lambda \in \mathbb{R}$, we have $W(\lambda) = V_-(\lambda)V_+(\lambda)$ for $\lambda \in \mathbb{R}$, and hence

$$(13) \qquad V_-(\lambda)^{-1}W_-(\lambda) = V_+(\lambda)W_+(\lambda)^{-1}.$$

The right hand side of (13) is analytic on the open upper half plane $\mathbb{C}_+$ and continuous on its closure $\overline{\mathbb{C}}_+$. The left hand side of (13) has the same properties with the open lower half plane $\mathbb{C}_-$ in place of $\mathbb{C}_+$ and its closure $\overline{\mathbb{C}}_-$ in place of $\overline{\mathbb{C}}_+$. It follows that $V_-(\cdot)W_-(\cdot)^{-1}$ extends to an analytic function on the entire complex plane. Furthermore, by the continuous analogue of the Riemann-Lebesgue lemma (cf., [R], item 9.6),

$$\lim_{\lambda\in\overline{\mathbb{C}}_-,\ \lambda\to\infty} V_-(\lambda)^{-1}W_-(\lambda) = I_m, \qquad \lim_{\lambda\in\overline{\mathbb{C}}_+,\ \lambda\to\infty} V_+(\lambda)W_+(\lambda)^{-1} = I_m,$$

and therefore $V_-(\lambda)^{-1}W_-(\lambda) \equiv I_m$. It follows that

$$W_-(\lambda) = V_-(\lambda) = W_+(\lambda)^*, \qquad \lambda \in \mathbb{R}. \quad \square$$

The arguments given in the last paragraph of the proof of Corollary 10.3 show that the factors W_+ and W_- in a canonical factorization are essentially unique. One may multiply $W_+(\cdot)$ on the left by a constant invertible matrix D in which case $W_-(\cdot)$ has to be replaced by $W_-(\cdot)D^{-1}$, and this is all the freedom one has.

By employing Theorem 9.1 one can show that Theorems 4.1 and 4.2 in Section XXIV remain valid for a block Toeplitz operator on ℓ_2^m defined by an $m \times m$ matrix function from the matrix Wiener algebra $\mathcal{W}^{m\times m}$. The line of arguments is the same as in Section XXIV; the changes are similar to those for the Wiener-Hopf case indicated in the two paragraphs after Theorem 10.2. We omit the details. Let us mention the following corollary, which is the discrete analogue of Corollary 10.3.

COROLLARY 10.4. *Let* $\Phi \in \mathcal{W}^{m\times m}$, *and assume that* $\Phi(\lambda)$ *is positive definite for each* $\lambda \in \mathbb{T}$. *Then the factorization indices of* Φ *relative to* $\mathbb{T}$ *are all equal to zero and* Φ *factorizes as*

$$(14) \qquad \Phi(\lambda) = \Phi_+(\lambda)^*\Phi_+(\lambda), \qquad \lambda \in \mathbb{T},$$

where $\det \Phi_+(\lambda) \neq 0$ *for* $|\lambda| \leq 1$,

$$(15) \qquad \Phi_+(\lambda) = \sum_{n=0}^{\infty} \lambda^n\Phi_n, \qquad |\lambda| \leq 1,$$

$$(16) \qquad \Phi_+(\lambda)^{-1} = \sum_{n=0}^{\infty} \lambda^n\Phi_n^{\times}, \qquad |\lambda| \leq 1,$$

and the series in (15) *and* (16) *are absolutely convergent.*

One proves Corollary 10.4 in a similar way as Corollary 10.3. We omit the details. The factorizations in (9) and (14) are called (right) *spectral factorizations*; (9) is a spectral factorization relative to the line, and (14) is one relative to the circle. Spectral factorizations will play an important role in Part IX.

CHAPTER XXXI
ELEMENTS OF C^*-ALGEBRA THEORY

This chapter contains the Stone-Weierstrass theorem and the Gelfand-Naimark representation theorem for commutative Banach algebras with an involution. The latter theorem is used to derive the spectral theorem for a normal operator. The Gelfand-Naimark characterization of a C^*-algebra as a closed $*$-subalgebra of operators is also presented.

XXXI.1 PRELIMINARIES ABOUT C^*-ALGEBRAS

In Section 7 of the preceding chapter we have seen that the Gelfand representation Γ of a commutative unital Banach algebra need not be injective or surjective. In this chapter we meet an important class of algebras whose representation Γ is an isometric isomorphism. The result we present is the Gelfand-Naimark representation theorem which requires that the algebra have an involution defined on it.

An *involution* on a Banach algebra $\mathcal{B}$ is a mapping which takes each $x \in \mathcal{B}$ to an element $x^* \in \mathcal{B}$ and has the following properties:

(a) $(x + y)^* = x^* + y^*$,

(b) $(\alpha x)^* = \overline{\alpha} x^*$,

(c) $(xy)^* = y^* x^*$,

(d) $x^{**} = x$.

Here x and y are in $\mathcal{B}$ and $\alpha \in \mathbb{C}$. If e is the unit in $\mathcal{B}$, then $e^* = e$. This follows from

$$e^* = e^* e = (e^* e)^* = e^{**} = e.$$

A Banach algebra $\mathcal{B}$ with an involution is called a C^*-*algebra* if, in addition, $\|x^* x\| = \|x\|^2$ for each $x \in \mathcal{B}$.

The most important examples of C^*-algebras are $C(S)$ with $f^* = \overline{f}$, the complex conjugate of f, and $\mathcal{L}(H)$, with T^* the usual adjoint of T. Clearly, the adjoint operation satisfies (a)–(d). Moreover, since $\|T^*\| = \|T\|$,

$$\|Tx\|^2 = \langle T^* T x, x \rangle \leq \|T^* T\| \leq \|T\|^2$$

for $\|x\| = 1$, and hence $\|T\|^2 = \|T^* T\|$. The Calkin algebra $\mathcal{L}(H)/\mathcal{K}(H)$, where H is a Hilbert space, provides another example of a C^*-algebra. More generally, if $\mathcal{J}$ is a closed ideal in a C^*-algebra $\mathcal{B}$ with the property that $x \in \mathcal{J}$ implies $x^* \in \mathcal{J}$, then the quotient algebra $\mathcal{B}/\mathcal{J}$ is a C^*-algebra with involution defined by

$$(x + \mathcal{J})^* = x^* + \mathcal{J}.$$

The Banach algebras $B(S)$ and $PC([a,b])$ (see Section XXIX.1 for the definitions) are also C^*-algebras with complex conjugation as the involution. Complex conjugation defines also an involution on the Wiener algebra $\mathcal{W}$ but under this operation $\mathcal{W}$ is not a C^*-algebra. For example, take $f(\zeta) = \zeta + 1 - \zeta^{-1}$. Then $\|f\| = \|\overline{f}\| = 3$, but $\|\overline{f}f\| = 5$.

THEOREM 1.1. *Let $\mathcal{B}$ be a unital C^*-algebra. Then*

(i) $\|x\| = \|x^*\|$ *for all* $x \in \mathcal{B}$,

(ii) $x = x^*$ *implies that* $\sigma(x) \subset \mathbb{R}$.

PROOF. (i) Take $x \in \mathcal{B}$. Then

$$\tag{1} \|x\|^2 = \|x^*x\| \le \|x^*\|\|x\|.$$

Since $x^{**} = x$, formula (1) with x replaced by x^*, yields $\|x^*\|^2 \le \|x^*\|\|x\|$. Together the inequalities show that $\|x\| = \|x^*\|$.

(ii) Suppose $x = x^*$. Let $\mathcal{B}_0$ be the unital closed subalgebra of $\mathcal{B}$ generated by x, i.e., $\mathcal{B}_0$ is the closure in $\mathcal{B}$ of all polynomials in x. Since $x = x^*$, the algebra $\mathcal{B}_0$ is also closed under the involution operation, and hence $\mathcal{B}_0$ is a commutative C^*-algebra in its own right. Given $\lambda = a + ib \in \sigma(x)$, with a and b real, it is obvious that λ is in $\sigma_0(x)$, where $\sigma_0(x)$ is the spectrum of x with respect to $\mathcal{B}_0$. Hence (see Corollary XXX.5.2) there exists a functional φ in the Gelfand spectrum of $\mathcal{B}_0$ such that $\varphi(x) = \lambda$. Take $y = x + ire$ with r real. Now $y \in \mathcal{B}_0$, $y^*y = x^2 + r^2e \in \mathcal{B}_0$ and

$$a^2 + (b+r)^2 = |\lambda + ir|^2 = |\varphi(y)|^2 \le \|y\|^2$$
$$= \|y^*y\| \le \|x\|^2 + r^2.$$

Hence $2br \le \|x\|^2 - |\lambda|^2$ for all $r \in \mathbb{R}$, which implies that $b = 0$.

If $x^* = x$, then the element x will be called *selfadjoint*. For the C^*-algebra $\mathcal{L}(H)$ this corresponds with the usual terminology.

THEOREM 1.2. *If $\mathcal{B}$ is a commutative unital C^*-algebra and $\widehat{x}$ is the Gelfand transform of $x \in \mathcal{B}$, then*

$$\|\widehat{x}\| = \lim_{n\to\infty} \|x^n\|^{1/n} = \|x\|.$$

Hence the Gelfand representation of $\mathcal{B}$ is an isometry.

PROOF. From the equalities

$$\|x^2\|^2 = \|(x^2)^*x^2\| = \|(x^*x)(x^*x)\| = \|x^*x\|^2 = \|x\|^4,$$

we get $\|x^2\| = \|x\|^2$. By induction, if $n = 2^k$, where k is a positive integer, then $\|x^n\| = \|x\|^n$. The theorem now follows immediately from Theorem XXX.3.2. $\square$

COROLLARY 1.3. *Let x be an element in a unital C^*-algebra $\mathcal{B}$ such that $xx^* = x^*x$. Then the spectral radius of x is equal to $\|x\|$.*

PROOF. Let $\mathcal{A}$ be the unital closed subalgebra of $\mathcal{B}$ generated by x and x^*. From $xx^* = x^*x$ it follows that $\mathcal{A}$ is a commutative C^*-algebra in its own right. But then we can apply Theorem 1.2 to get the desired result. $\square$

XXXI.2 THE STONE-WEIERSTRASS THEOREM

In this book we have already used a few times a classical theorem due to Weierstrass which states that every continuous function defined on a compact interval $[a, b]$ is the uniform limit of a sequence of polynomials. In terms of algebras the Weierstrass theorem tells us that the subalgebra of polynomials is dense in the Banach algebra $C([a, b])$. In 1937 M.H. Stone produced a wonderful generalization of this result, which is known as the Stone-Weierstrass theorem. We start with the version which we shall need in this chapter. First a few preliminaries.

In what follows, S is a compact Hausdorff space. As before $C(S)$ denotes the Banach algebra of all complex-valued continuous functions on S. A subset $\mathcal{F}$ of $C(S)$ is said to *separate the points* of S if for every pair of points s, t in S, $s \neq t$, there exists $f \in \mathcal{F}$ such that $f(s) \neq f(t)$. The set $\mathcal{F}$ is *closed under complex conjugation* if $f \in \mathcal{F}$ implies that its complex conjugate $\overline{f}$ is also in $\mathcal{F}$.

THEOREM 2.1 (Stone-Weierstrass). *Let $\mathcal{A}$ be a subalgebra of $C(S)$, and assume that $\mathcal{A}$ contains the constant function 1, separates the points of S and is closed under complex conjugation. Then $\mathcal{A}$ is dense in $C(S)$.*

The proof of Theorem 2.1 is based on the "real" version of the theorem. We denote by $C_R(S)$ the real Banach space of continuous real-valued functions on S endowed with the supremum norm. A linear manifold $\mathcal{A}$ in $C_R(S)$ is called a *subalgebra* of $C_R(S)$ if the product of two functions in $\mathcal{A}$ is again in $\mathcal{A}$.

THEOREM 2.2. *Let $\mathcal{A}$ be a subalgebra of $C_R(S)$ which separates the points of S and contains the constant function 1. Then $\mathcal{A}$ is dense in $C_R(S)$.*

PROOF. First we show that if f and g are in the closure $\overline{\mathcal{A}}$ of $\mathcal{A}$, then $f \vee g$ and $f \wedge g$ are in $\overline{\mathcal{A}}$, where

$$(f \vee g)(s) = \max\big(f(s), g(s)\big), \qquad (f \wedge g)(s) = \min\big(f(s), g(s)\big).$$

Since

$$f \vee g = \frac{1}{2}(f + g + |f - g|), \qquad f \wedge g = \frac{1}{2}(f + g - |f - g|),$$

it suffices to prove that $|h| \in \overline{\mathcal{A}}$ whenever h is in $\overline{\mathcal{A}}$ and $\|h\| \leq 1$. From calculus we know that the binomial series expansion

$$(1 - t)^{1/2} = 1 - \frac{1}{2}t - \sum_{n=1}^{\infty} a_n t^{n+1}, \qquad a_n = \frac{1 \cdot 3 \cdots (2n - 1)}{2 \cdot 4 \cdots 2(n + 1)},$$

is valid on $[0, 1]$ and that the series converges uniformly on this interval. Hence, if $y = 1 - h^2$, then $0 \leq y(s) \leq 1$ for each $s \in S$, and

$$|h| = (1 - y)^{1/2} = 1 - \frac{1}{2}y - \sum_{n=1}^{\infty} a_n y^{n+1} \in \overline{\mathcal{A}}.$$

The next step in the proof is to show that given $F \in C_R(S)$ and s, t in S, there is $f_{st} \in \mathcal{A}$ with $f_{st}(s) = F(s)$ and $f_{st}(t) = F(t)$. Indeed, if $s = t$, take f_{st} to be the function identically equal to $F(s)$. If $s \neq t$, there exists $g \in \mathcal{A}$ with $g(s) \neq g(t)$. Take $f_{st} = ag + b$, where a and b satisfy

$$ag(s) + b = F(s), \qquad ag(t) + b = F(t).$$

Then f_{st} has the required properties. Now for the final step of the proof of the theorem, let $F \in C_R(S)$ and $\varepsilon > 0$ be given. For each pair s, t in S let f_{st} be as above. Define

$$U_{st} = \{x \in S \mid f_{st}(x) < F(x) + \varepsilon\}.$$

For each t, we have $S \subset \cup_{s \in S} U_{st}$ and U_{st} is an open set. Since S is compact, we may find $s_1, s_2, \ldots, s_N$ so that $S \subset \cup_{j=1}^{N} U_{s_j t}$. Define $f_t = f_{s_1 t} \wedge f_{s_2 t} \wedge \cdots \wedge f_{s_N t}$. Then $f_t \in \overline{\mathcal{A}}$ and

$$(1) \qquad\qquad f_t(x) < F(x) + \varepsilon, \qquad t, x \in S.$$

Define

$$V_t = \{x \in S \mid f_t(x) > F(x) - \varepsilon\}.$$

Since $f_{st}(t) = F(t)$ for all $s \in S$, it follows that $f_t(t) = F(t)$. Hence t is in the open set V_t and $S \subset \cup_{t \in S} V_t$. Therefore $S \subset \cup_{i=1}^{M} V_{t_i}$ for some $t_1, \ldots, t_M$ in S. Let $f = f_{t_1} \vee \cdots \vee f_{t_M}$. Then $f \in \overline{\mathcal{A}}$ and $f(x) > F(x) - \varepsilon$ for all $x \in S$. Also, from (1) we have that $f(x) < F(x) + \varepsilon$ for each $x \in S$. Hence $\|f - F\| < \varepsilon$. $\square$

    COROLLARY 2.3. *Every real-valued function which is continuous on a compact set in $\mathbb{R}^n$ is the uniform limit of a sequence of polynomials in n variables.*

    PROOF OF THEOREM 2.1. Let $\mathcal{A}_R$ consist of the real-valued functions in $\mathcal{A}$. Clearly, $\mathcal{A}_R$ is a subalgebra of $C_R(S)$ which contains the constant function 1. Moreover, $\mathcal{A}_R$ separates points of S. Indeed, if s, t are in S, $s \neq t$, then there exists $f \in \mathcal{A}$ such that $f(s) \neq f(t)$. Therefore $\Re f(s) \neq \Re f(t)$ or $\Im f(s) \neq \Im f(t)$. Also, $\Re f = \frac{1}{2}(f + \overline{f}) \in \mathcal{A}_R$ and $\Im f = \frac{1}{2}(f - \overline{f}) \in \mathcal{A}_R$. Hence $\mathcal{A}_R$ is dense in $C_R(S)$ by Theorem 2.2. Since each $g \in C(S)$ is a linear combination of $\Re g$ and $\Im g$, the theorem follows. $\square$

XXXI.3 THE GELFAND-NAIMARK REPRESENTATION THEOREM

    THEOREM 3.1 (Gelfand-Naimark). *Let $\mathcal{B}$ be a commutative unital C^*-algebra with Gelfand spectrum $\mathcal{M}$. The Gelfand representation Γ of $\mathcal{B}$ is an isometric isomorphism from $\mathcal{B}$ onto $C(\mathcal{M})$ and $\Gamma(x^*) = \overline{\Gamma(x)}$.*

    PROOF. We know from Theorem 1.2 that the homomorphism Γ is an isometry. Therefore $\operatorname{Im}\Gamma$ is a closed subalgebra of $C(\mathcal{M})$. In order to prove that $\operatorname{Im}\Gamma = C(\mathcal{M})$ we apply the Stone-Weierstrass theorem (see the preceding section). Let e be the unit of $\mathcal{B}$. Then $(\Gamma e)(\varphi) = \varphi(e) = 1$, and hence $\operatorname{Im}\Gamma$ contains the constant function 1. Next, note that $\operatorname{Im}\Gamma$ separates the points of $\mathcal{M}$. Indeed, if φ_1 and φ_2 are in $\mathcal{M}$ and $\varphi_1 \neq \varphi_2$, then there exists an $x \in \mathcal{B}$ such that

$$(\Gamma x)(\varphi_1) = \varphi_1(x) \neq \varphi_2(x) = (\Gamma x)(\varphi_2).$$

Since $\operatorname{Im}\Gamma$ is a closed subalgebra, it remains to show that $\operatorname{Im}\Gamma$ is closed under complex conjugation. To do this we prove that $\overline{\Gamma x} = \Gamma x^*$. Given $x \in \mathcal{B}$, the vectors $u = \frac{1}{2}(x + x^*)$ and $v = \frac{1}{2i}(x - x^*)$ are selfadjoint and $x = u + iv$. By Corollary XXX.5.2, the Gelfand transforms Γu and Γv have ranges $\sigma(u)$ and $\sigma(v)$, respectively. Hence Γu and Γv are real-valued by Theorem 1.1(ii). Therefore,

$$\Gamma x^* = \Gamma u - i\Gamma v = \overline{\Gamma(u + iv)} = \overline{\Gamma x},$$

which completes the proof of the theorem. $\square$

A homomorphism between Banach algebras with involutions is called a $*$-*homomorphism* if it preserves the involutions, i.e., if h is a homomorphism from a Banach algebra $\mathcal{B}_1$ with involution $*$ into a Banach algebra $\mathcal{B}_2$ with involution $\#$, then

$$(1) \qquad h(x^*) = h(x)^{\#}, \qquad x \in \mathcal{B}_1.$$

If h is an isomorphism and (1) holds, then h is called a $*$-*isomorphism*. The Gelfand-Naimark theorem shows that the Gelfand representation of a commutative unital C^*-algebra is a $*$-isomorphism.

THEOREM 3.2. *Let $\mathcal{B}$ be a commutative unital C^*-algebra, and assume $\mathcal{B}$ is generated by a and a^*. Then the Gelfand transform $\widehat{a}$ of a is a homeomorphism from the Gelfand spectrum $\mathcal{M}$ of $\mathcal{B}$ onto $\sigma(a)$. Moreover, $\mathcal{B}$ is isometrically $*$-isomorphic to $C(\sigma(a))$ under the map*

$$(2) \qquad x \mapsto \eta(x) = (\Gamma x) \circ \widehat{a}^{-1},$$

where Γ is the Gelfand representation of $\mathcal{B}$. In particular, if e is the unit in $\mathcal{B}$, then $\eta(e)$ is the constant function 1, and

$$(3) \qquad \eta(a)(\lambda) = \lambda, \qquad \eta(a^*)(\lambda) = \overline{\lambda} \qquad (\lambda \in \sigma(a)).$$

PROOF. Let us start with the proof that $\widehat{a}$ is injective. Suppose $\widehat{a}(\varphi) = \widehat{a}(\psi)$ for some φ and ψ in $\mathcal{M}$. Then $\varphi(a) = \psi(a)$ and by Theorem 3.1

$$\varphi(a^*) = (\Gamma a^*)(\varphi) = \overline{(\Gamma a)}(\varphi) = \overline{\varphi(a)} = \overline{\psi(a)} = \psi(a^*).$$

We see that φ and ψ coincide on all polynomials in a and a^*. By continuity it follows that $\varphi = \psi$. Thus $\widehat{a}$ is an injective continuous map from the compact space $\mathcal{M}$ onto the Hausdorff space $\sigma(a)$. So (see [W], Theorem 17.14) the Gelfand transform $\widehat{a}$ is a homeomorphism. The other statements of the theorem follow readily from Theorem 3.1. $\square$

For later purposes we note that the inverse of the map η in (2) is given by

$$(4) \qquad \eta^{-1}(f) = \Gamma^{-1}(f \circ \widehat{a}), \qquad f \in C(\sigma(a)).$$

A subalgebra $\mathcal{A}$ of a Banach algebra with involution is called a $*$-*subalgebra* if $\mathcal{A}$ is closed under the operation of involution, i.e., $x^* \in \mathcal{A}$ whenever $x \in \mathcal{A}$.

THEOREM 3.3. *Let $\mathcal{A}$ be a closed $*$-subalgebra of a C^*-algebra $\mathcal{B}$ with the same unit e as $\mathcal{B}$. Then $\sigma_{\mathcal{A}}(x) = \sigma_{\mathcal{B}}(x)$ for all $x \in \mathcal{A}$.*

PROOF. It suffices to prove that if $y \in \mathcal{A}$ has an inverse $z \in \mathcal{B}$, then $z \in \mathcal{A}$. From $zy = yz = e$ and $e^* = e$ it follows that z^* is the inverse of y^* in $\mathcal{B}$. Hence y^*y is invertible in $\mathcal{B}$ with inverse zz^*. Since y^*y is selfadjoint, Theorem 1.1(ii) implies that $\sigma_{\mathcal{B}}(y^*y)$ is on the real axis. In particular, $\rho_{\mathcal{B}}(y^*y)$ is connected. Note that y^*y also belongs to $\mathcal{A}$, because $\mathcal{A}$ is a $*$-subalgebra. Thus $0 \in \rho_{\mathcal{B}}(y^*y) = \rho_{\mathcal{A}}(y^*y)$ by Corollary XXIX.6.2. We conclude that the inverse zz^* of y^*y belongs to $\mathcal{A}$, and hence $z = (zz^*)y^* \in \mathcal{A}$. $\square$

XXXI.4 FUNCTIONAL CALCULUS FOR NORMAL ELEMENTS

An element x in a Banach algebra with involution $*$ is called *normal* if $xx^* = x^*x$. In this section a functional calculus is developed for normal elements in a unital C^*-algebra.

Let x be a normal element in a unital C^*-algebra $\mathcal{B}$ with unit e. Given $f \in C(\sigma(x))$, we define $f(x)$ by means of the Gelfand-Naimark representation theorem as follows. Let $\mathcal{A}_x$ be the unital closed subalgebra of $\mathcal{B}$ generated by x and x^*. Since x is normal, $\mathcal{A}_x$ is a commutative unital C^*-algebra in its own right. By Theorem 3.3, the spectrum of x as an element of $\mathcal{A}_x$ is equal to $\sigma(x)$ (the spectrum of x as an element of $\mathcal{B}$), and therefore, by Theorem 3.2, the algebra $C(\sigma(x))$ is isometrically $*$-isomorphic to $\mathcal{A}_x$ under the map $f \mapsto \Gamma^{-1}(f \circ \widehat{x})$. Here Γ is the Gelfand representation of $\mathcal{A}_x$ and $\widehat{x} = \Gamma(x)$. We define (cf., formula (4) in the preceding section)

$$(1) \qquad f(x) = \Gamma^{-1}(f \circ \widehat{x}), \qquad f \in C(\sigma(x)).$$

It is clear that $f(x) = x$ if $f(\lambda) = \lambda$, $\lambda \in \sigma(x)$, and $g(x) = e$ if $g(\lambda) = 1$, $\lambda \in \sigma(x)$. To summarize, we have the following result.

THEOREM 4.1. *Let x be a normal element in a unital C^*-algebra $\mathcal{B}$ with unit e. Then the map $f \mapsto f(x)$ defined by (1) is an isomorphic $*$-isomorphism from $C(\sigma(x))$ onto the unital closed subalgebra of $\mathcal{B}$ generated by x and x^*. Moreover, $f(x) = x$ if $f(\lambda) = \lambda$ for $\lambda \in \sigma(x)$, and $g(x) = e$ if $g(\lambda) = 1$ for $\lambda \in \sigma(x)$.*

In the event that f is analytic on an open neighbourhood of $\sigma(x)$, then the definition (1) of $f(x)$ yields an integral representation of the type appearing in the functional calculus of Section I.3. This is the content of the next theorem.

THEOREM 4.2. *Let x be a normal element in a unital C^*-algebra $\mathcal{B}$ with unit e, and let f be analytic on an open neighbourhood of $\sigma(x)$. If $f(x)$ is defined by (1), then*

$$(2) \qquad f(x) = \frac{1}{2\pi i} \int_{\gamma} f(\lambda)(\lambda e - x)^{-1} d\lambda,$$

where γ is a Cauchy contour around $\sigma(x)$ in the domain of f.

PROOF. Let y be the element defined by the integral in (2). As before let $\mathcal{A}_x$ be the unital closed subalgebra of $\mathcal{B}$ generated by x and x^*. Since $\rho(x)$ is equal to the resolvent set of x as an element of $\mathcal{A}_x$ (by Theorem 3.3), the element $(\lambda e - x)^{-1} \in \mathcal{A}_x$ for each $\lambda \in \rho(x)$. It therefore follows from the Riemann-Stieltjes sums used to define the integral in (2) that y is in $\mathcal{A}_x$.

We shall now find a formula for the Gelfand transform $\widehat{y}$ of y with respect to the algebra $\mathcal{A}_x$. Take φ in the Gelfand spectrum $\mathcal{M}$ of $\mathcal{A}_x$. Then

$$\widehat{y}(\varphi) = \varphi(y) = \frac{1}{2\pi i} \int_\gamma f(\lambda)\varphi\big[(\lambda e - x)^{-1}\big]d\lambda$$

$$= \frac{1}{2\pi i} \int_\gamma \frac{f(\lambda)}{\lambda - \varphi(x)}d\lambda.$$

Since $\varphi(x) \in \sigma(x)$, it is obvious that $\varphi(x)$ is inside the curve γ, whence the Cauchy integral formula implies that

$$\widehat{y}(\varphi) = f\big(\varphi(x)\big) = f\big(\widehat{x}(\varphi)\big) = (f \circ \widehat{x})(\varphi),$$

where $\widehat{x}$ is the Gelfand transform of x relative to $\mathcal{A}_x$. Thus $\Gamma(y) = \widehat{y} = f \circ \widehat{x}$ or, equivalently, $y = \Gamma^{-1}(f \circ \widehat{x})$. But then $y = f(x)$, by (1). $\square$

For the functional calculus defined by (1) we have the following spectral mapping theorem.

THEOREM 4.3 (Spectral mapping theorem). *Let x be a normal element in a unital C^*-algebra $\mathcal{B}$. If $f \in C\big(\sigma(x)\big)$, then*

$$f\big(\sigma(x)\big) = \sigma\big(f(x)\big).$$

PROOF. By Theorem 3.3, $\sigma\big(f(x)\big)$ is also the spectrum of $f(x)$ with respect to the C^*-algebra generated by x and x^*. It now follows from Theorem 3.2 that the spectrum of f as an element of $C\big(\sigma(x)\big)$ is the same as $\sigma\big(f(x)\big)$. Since $\sigma(f) = \{f(t) \mid t \in \sigma(x)\}$, the theorem is proved. $\square$

XXXI.5 NON-NEGATIVE ELEMENTS AND POSITIVE LINEAR FUNCTIONALS

This section has an auxiliary character. The results that are presented will be used in the next section to prove the Gelfand-Naimark characterization of C^*-algebras.

Let $\mathcal{B}$ be a unital C^*-algebra. An element $x \in \mathcal{B}$ is called *non-negative* if x is selfadjoint and $\sigma(x) \subset \mathbb{R}^+ = [0, \infty)$. The set of non-negative elements of $\mathcal{B}$ is denoted by $\mathcal{B}^+$.

Assume $\mathcal{B}$ is a closed $*$-subalgebra of the C^*-algebra of all bounded linear operators on a Hilbert space H such that the identity operator is in $\mathcal{B}$. Then $T \in \mathcal{B}^+$ if and only if T is a non-negative operator, i.e.,

$$(1) \qquad\qquad \langle Tu, u \rangle \geq 0 \qquad (u \in H).$$

To see this, note that, by Theorem 3.3, the spectrum of T as an element of $\mathcal{B}$ is equal to the spectrum of T as an operator on H. Thus, if (1) holds, then $T = T^*$ and

$$\sigma_{\mathcal{B}}(T) = \sigma(T) \subset [0, \infty),$$

by Theorem V.2.1, and hence T is a non-negative element of $\mathcal{B}$. Conversely, if $T = T^*$ and $\sigma_{\mathcal{B}}(T) \subset [0, \infty)$, then $\sigma(T) \subset [0, \infty)$, and (1) follows by using Theorem V.2.1 again.

Let $\mathcal{B}$ be a unital C^*-algebra, and take $x \in \mathcal{B}^+$. Let $f(t) = \sqrt{t}$ for $t \geq 0$. Then $f \in C(\sigma(x))$, and hence $y = f(x)$ is a well-defined element of $\mathcal{B}$. According to the functional calculus of Theorem 4.1, we have $y^* = y$ and $x = y^2$. Thus a non-negative element x in $\mathcal{B}$ factorizes as $x = y^*y$ for some $y \in \mathcal{B}$. We shall see later (Theorem 5.4 below) that, conversely, $y^*y \in \mathcal{B}^+$ for any $y \in \mathcal{B}$.

In the following lemma we collect together a number of properties of non-negative elements that will be useful later.

LEMMA 5.1. *For a selfadjoint element x in a unital C^*-algebra $\mathcal{B}$ (with unit e) the following holds.*

(a) *If $f \in C(\sigma(x))$, then $f(x) \in \mathcal{B}^+$ if and only if $f(t) \geq 0$ for each $t \in \sigma(x)$.*

(b) *There exist x_1 and x_2 in $\mathcal{B}^+$ such that*

$$(2) \qquad x = x_1 - x_2, \qquad x_1 x_2 = x_2 x_1 = 0.$$

(c) *$x \in \mathcal{B}^+$ if and only if for some $\mu \geq 0$ one has $\|\mu e - x\| \leq \mu$, in which case $\|\lambda e - x\| \leq \lambda$ for all $\lambda \geq \|x\|$.*

PROOF. (a) If $f(t) \geq 0$ for each $t \in \sigma(x)$, then, by Theorems 4.1 and 4.3, we have $f(x)^* = \overline{f}(x) = f(x)$ and $\sigma\big(f(x)\big) = f(\sigma(x)) \subset \mathbb{R}^+$. Thus $f(x) \in \mathcal{B}^+$. Conversely, if $f(x) \in \mathcal{B}^+$, then

$$f\big(\sigma(x)\big) = \sigma\big(f(x)\big) \subset \mathbb{R}^+,$$

and thus $f(t) \geq 0$ for each $t \in \sigma(x)$.

(b) Let $h(t) = t$, and define

$$h_1(t) = \max\{t, 0\}, \qquad h_2 = \max\{-t, 0\}$$

for $t \in \mathbb{R}$. Now $h = h_1 - h_2$ and $h_1 h_2 = 0$. Thus

$$x = h(x) = h_1(x) - h_2(x), \qquad h_1(x)h_2(x) = h_2(x)h_1(x) = 0,$$

by Theorem 4.1. Also, $x_1 := h_1(x)$ and $x_2 := h_2(x)$ are in $\mathcal{B}^+$ by (a).

(c) Since $x = x^*$, we know from Theorem 1.1(ii) and Corollary 1.3 that $\sigma(x) \subset [-\|x\|, \|x\|]$. Hence, by Theorem 4.1, for $\lambda \geq \|x\|$ we have

$$(3) \qquad \|\lambda e - x\| = \sup_{t \in \sigma(x)} (\lambda - t).$$

Thus, if $x \in \mathcal{B}^+$, then $\|\lambda e - x\| \leq \lambda$. Conversely, assume $\|\mu e - x\| \leq \mu$ for some $\mu \geq 0$. Note that

$$\sigma(\mu e - x) = \{\mu - t \mid t \in \sigma(x)\}.$$

It follows that $\mu - t \leq \|\mu e - x\| \leq \mu$ for each $t \in \sigma(x)$, which implies that $\mu \geq \mu - t$ for each $t \in \sigma(x)$. Hence $\sigma(x) \subset \mathbb{R}^+$, and therefore x is non-negative. $\sqsubset$

COROLLARY 5.2. *The set of non-negative elements in a unital C^*-algebra is closed under addition and multiplication by non-negative real numbers.*

PROOF. Let x and y be non-negative elements in the unital C^*-algebra $\mathcal{B}$. Note that $x + y$ is selfadjoint. Put $\lambda = \|x\|$ and $\mu = \|y\|$. Then $\lambda + \mu \geq \|x + y\|$. By Lemma 5.1(c) we have

$$\lambda \geq \|\lambda e - x\|, \qquad \mu \geq \|\mu e - x\|,$$

and thus

$$\|(\lambda + \mu)e - (x + y)\| \leq \|\lambda e - x\| + \|\mu e - y\| \leq \lambda + \mu.$$

It follows (again apply Lemma 5.1(c)) that $x + y \in \mathcal{B}^+$.

Let $0 \leq \alpha \leq \mathbb{R}$. Then αx is selfadjoint, and from $\sigma(\alpha x) = \alpha \sigma(x) \subset \mathbb{R}^+$ we see that αx is non-negative. $\square$

LEMMA 5.3. *Let $\mathcal{B}$ be a unital C^*-algebra. If $x \in \mathcal{B}$ and $-x^*x \in \mathcal{B}^+$, then $x = 0$.*

PROOF. First note (cf., the proof of Corollary VII.6.2) that

$$\sigma(-xx^*) \subset \sigma(-x^*x) \cup \{0\}. \tag{4}$$

Indeed, if $\lambda \neq 0$ and $\lambda \notin \sigma(-x^*x)$, then a direct computation shows that

$$(\lambda e + xx^*)\{\lambda^{-1}e - \lambda^{-2}x(e + \lambda^{-1}x^*x)^{-1}x^*\}$$
$$= \{\lambda^{-1}e - \lambda^{-2}x(e + \lambda^{-1}x^*x)^{-1}x^*\}(\lambda e + xx^*) = e,$$

and hence $\lambda \notin \sigma(-xx^*)$. Since $-x^*x \in \mathcal{B}^+$, the right hand side of (4) is a subset of $\mathbb{R}^+$, and therefore $\sigma(-xx^*) \subset \mathbb{R}^+$. Moreover, $-xx^*$ is selfadjoint. So $-xx^*$ is a non-negative too.

Write $x = a_1 + ia_2$, where a_1 and a_2 are selfadjoint. The elements a_1^2, a_2^2 are in $\mathcal{B}^+$ because they are selfadjoint and (by Theorem 4.3)

$$\sigma(a_i^2) = \{\lambda^2 \mid \lambda \in \sigma(a_i)\} \subset \mathbb{R}^+, \qquad i = 1, 2.$$

Here we used that the spectrum of a selfadjoint element is real (Theorem 1.1(ii)). Now

$$x^*x + xx^* = (a_1 - ia_2)(a_1 + ia_2) + (a_1 + ia_2)(a_1 - ia_2)$$
$$= 2a_1^2 + 2a_2^2,$$

whence $x^*x = 2a_1^2 + 2a_2^2 + (-xx^*)$. Thus $x^*x \in \mathcal{B}^+$ by Corollary 5.2. Therefore, x^*x and $-x^*x$ are non-negative, which implies that

$$\sigma(x^*x) \subset [0, \infty) \cap (-\infty, 0] = \{0\}.$$

So $\|x\|^2 = \|x^*x\| = 0$, by Corollary 1.3. $\square$

THEOREM 5.4. *An element x is a unital C^*-algebra $\mathcal{B}$ is non-negative if and only if $x = y^*y$ for some $y \in \mathcal{B}$.*

PROOF. We have already seen (in the fourth paragraph of this section) that each $x \in \mathcal{B}^+$ admits a factorization $x = y^*y$ for some $y \in \mathcal{B}$. So we have to prove the converse.

Let $x = y^*y$ for some $y \in \mathcal{B}$. Since y^*y is selfadjoint, there exist, by Lemma 5.1(b), elements a_1, a_2 in $\mathcal{B}^+$ such that $x = a_1 - a_2$ and $a_1 a_2 = a_2 a_1 = 0$. Put $h = ya_2$. Then

$$h^*h = a_2 y^* y a_2 = a_2(a_1 - a_2)a_2 = -(a_2)^3.$$

Since $\sigma(a_2) \subset \mathbb{R}^+$, the spectral mapping theorem (Theorem 4.3) implies that

$$\sigma(-h^*h) = \sigma(a_2^3) = \{\lambda^3 \mid \lambda \in \sigma(a_2)\} \subset \mathbb{R}^+.$$

Therefore, $-h^*h \in \mathcal{B}^+$, and from Lemma 5.3 it follows that $h = 0$. So $a_2^3 = 0$. But then $a_2 = 0$ because, by Corollary 1.3,

$$\|a_2\| = \lim_{n \to \infty} \|a_2^n\|^{1/n} = \lim_{k \to \infty} \|a_2^{3k}\|^{1/3k} = 0.$$

Thus $x = a_1 \in \mathcal{B}^+$. $\square$

Let $\mathcal{B}$ be a unital C^*-algebra. A linear functional f on $\mathcal{B}$ is said to be *positive* if f maps $\mathcal{B}^+$ into $\mathbb{R}^+ = [0, \infty)$.

PROPOSITION 5.5. *Let f be a linear functional on a unital C^*-algebra $\mathcal{B}$ (with unit e). Then*

(i) $f(x^*) = \overline{f(x)}$ *for each $x \in \mathcal{B}$ whenever f is positive;*

(ii) f *is positive if and only if f is bounded and $\|f\| = f(e)$.*

PROOF. (i) Let f be positive. If x is selfadjoint, then $x = x_1 - x_2$ for some x_1, x_2 in $\mathcal{B}^+$ by Lemma 5.1(b). Therefore, $f(x) = f(x_1) - f(x_2)$ is real. In the general case, write $x = a + ib$ where a and b are selfadjoint. It follows that

$$f(x^*) = f(a) - if(b) = \overline{f(a) + if(b)} = \overline{f(x)}.$$

(ii) Suppose f is positive. Given $a \in \mathcal{B}$, set $|f(a)| = e^{i\theta} f(a)$ for some $\theta \in \mathbb{R}$. By (i),

$$|f(a)| = f(e^{i\theta} a) = f(e^{-i\theta} a^*) = f\left[\frac{1}{2}(e^{-i\theta} a^* + e^{-i\theta} a^*)\right]$$

$$= f\left[\frac{1}{2}(e^{i\theta} a + e^{-i\theta} a^*)\right].$$

Let $u = \frac{1}{2}(e^{i\theta}a + e^{-i\theta}a^*)$. Then $u = u^*$ and $\|u\| \leq \|a\|$. Therefore, $\|a\|e - u \in \mathcal{B}^+$ by Lemma 5.1(c). Hence

$$0 \leq \|a\|f(e) - f(u) = \|a\|f(e) - |f(a)|.$$

Thus $|f(a)| \leq \|a\|f(e)$. This holds for each $a \in \mathcal{B}$. Hence $\|f\| \leq f(e) \leq \|f\|$.

Conversely, suppose f is a bounded linear functional on $\mathcal{B}$ with $\|f\| = f(e) \neq 0$. Let $g = \{f(e)\}^{-1}f$. Then $g(e) = \|g\| = 1$. It suffices to prove that g is positive on $\mathcal{B}$. Given $a \in \mathcal{B}^+$, $g(a) = \alpha + i\beta$ for some α, β in $\mathbb{C}$. We need to prove that $\alpha \geq 0$ and $\beta = 0$. Since $\sigma(a) \in \mathbb{R}^+$, we have that for all sufficiently small positive numbers s,

$$\sigma(e - sa) = \{1 - s\lambda \mid \lambda \in \sigma(a)\} \subset [0,1].$$

Therefore $\|e - sa\| = r(e - sa) \leq 1$, by Corollary 1.3. Hence

$$1 - s\alpha \leq |1 - s(\alpha + i\beta)| = |g(e - sa)| \leq \|e - sa\| \leq 1.$$

Thus $\alpha \geq 0$. Let $x_n = a - \alpha e + in\beta e$, $n = 1, 2, \ldots$. Then

$$\beta^2(1+n)^2 = |g(x_n)|^2 \leq \|x_n\|^2 = \|x_n^* x_n\| = \|(a - \alpha e)^2 + n^2\beta^2 e\| \leq \|a - \alpha e\|^2 + n^2\beta^2$$

which implies that $\beta = 0$.

The next theorem shows that a unital C^*-algebra has many positive linear functionals.

THEOREM 5.6. *Let $\mathcal{B}$ be a unital C^*-algebra (with unit e). Given $a \in \mathcal{B}$ and $\lambda \in \sigma(a)$, there exists a positive linear functional f on $\mathcal{B}$ such that $f(a) = \lambda$ and $f(e) = 1$.*

PROOF. Suppose $a \notin \mathrm{span}\{e\}$. Define f on $\mathrm{span}\{a, e\}$ by $f(\alpha a + \beta e) = \alpha\lambda + \beta$. Clearly, $\alpha\lambda + \beta \in \sigma(\alpha a + \beta e)$. Hence

$$|f(\alpha a + \beta e)| = |\alpha\lambda + \beta| \leq \|\alpha a + \beta e\|.$$

Therefore, $\|f\| \leq 1$. Since $f(e) = 1$, we have $\|f\| = 1$.

If $a \in \mathrm{span}\{e\}$, define f on $\mathrm{span}\{e\}$ by $f(\gamma e) = \gamma$. Then $\|f\| = 1 = f(e)$. In either case, f may be extended to a bounded linear functional F on $\mathcal{B}$ with $\|F\| = 1 = F(e)$. Thus F is positive, by Proposition 5.5(ii), and $F(a) = f(a) = \lambda$. $\square$

XXXI.6 CHARACTERIZATION OF C^*-ALGEBRAS

Throughout this section $\mathcal{B}$ is a unital C^*-algebra with unit e. Our aim is to show that $\mathcal{B}$ is isometrically $*$-isomorphic to a subalgebra of $\mathcal{L}(H)$ for some Hilbert space H (depending on $\mathcal{B}$). To state the precise result we use the following definition. A $*$-homomorphism φ mapping $\mathcal{B}$ into $\mathcal{L}(H)$ with $\varphi(e) = I$ is called a *representation* of $\mathcal{B}$ on H.

THEOREM 6.1 (Gelfand-Naimark). *Given a unital C^*-algebra $\mathcal{B}$, there exists a Hilbert space H and an isometric representation of $\mathcal{B}$ on H.*

Let $\varphi: \mathcal{B} \to \mathcal{L}(H)$ be a representation of $\mathcal{B}$ on H. Then for each $u \in H$,

$$f_u(x) = \langle \varphi(x)u, u \rangle, \qquad x \in \mathcal{B},$$

defines a positive linear functional on $\mathcal{B}$. Indeed, f_u is linear, and if $x \in \mathcal{B}^+$, then $x = y^*y$ for some $y \in \mathcal{B}$, by Theorem 5.4, and hence

$$f_u(x) = \langle \varphi(y^*y)u, u \rangle = \langle \varphi(y)^*\varphi(y)u, u \rangle = \|\varphi(y)u\|^2 \geq 0.$$

It turns out that, conversely, positive linear functionals may be used to construct representations. This fact will be one of our main tools in the proof of Theorem 6.1.

LEMMA 6.2. *Let f be a positive linear functional on the C^*-algebra $\mathcal{B}$ with $f(e) = 1$. Then there exist an element $u \in \mathcal{B}$, $\|u\| = 1$, a Hilbert space H and a representation φ of $\mathcal{B}$ on H (all three depending on f) such that*

$$f(a) = \langle \varphi(a)u, u \rangle, \quad \|\varphi(a)\| \leq \|a\| \qquad (a \in \mathcal{B}).$$

PROOF. Define $\langle a, b \rangle_1 = f(b^*a)$ for $a, b \in \mathcal{B}$. Since f is a positive linear functional and $a^*a \in \mathcal{B}^+$ for each $a \in \mathcal{B}$ (by Theorem 5.4), it follows that $\langle \cdot, \cdot \rangle_1$ is a sesquilinear form (i.e., linear in the first variable and conjugate linear in the second variable) on $\mathcal{B}$ and $\langle a, a \rangle_1 \geq 0$ for $a \in \mathcal{B}$. Thus the Cauchy-Schwartz inequality holds (same proof as for an inner product), i.e.,

$$|\langle a, b \rangle_1| \leq \langle a, a \rangle_1^{1/2} \langle b, b \rangle_1^{1/2} \qquad (a, b \in \mathcal{B}).$$

To obtain the corresponding inner product space, let $M = \{a \in \mathcal{B} \mid \langle a, a \rangle_1 = 0\}$. Now M is a linear manifold in $\mathcal{B}$. Indeed, if a, b are in M and $\alpha \in \mathbb{C}$, then

$$\begin{aligned}
0 \leq \langle \alpha a + b, \alpha a + b \rangle_1 &= |\alpha|^2 \langle a, a \rangle_1 + 2\Re\alpha \langle a, b \rangle_1 + \langle b, b \rangle_1 \\
&= 2\Re\alpha \langle a, b \rangle_1 \\
&\leq 2|\alpha|\langle a, a \rangle_1^{1/2} \langle b, b \rangle_1^{1/2} = 0.
\end{aligned}$$

Thus $\alpha a + b \in M$. Since a positive linear functional is bounded (Proposition 5.5(ii)), the space M is closed. Indeed, if $u_n \to u$, $u_n \in M$, then

$$0 = \langle u_n, u_n \rangle_1 = f(u_n^* u_n) \to f(u^* u) = \langle u, u \rangle_1.$$

Hence $\langle u, u \rangle_1 = 0$, and thus $u \in M$. The space M has the additional property that if $a \in \mathcal{B}$ and $u \in M$, then $au \in M$, i.e., M is a left ideal, since

$$\begin{aligned}
0 \leq \langle au, au \rangle_1 &= f((u^*a^*a)u) = f((a^*au)^*u) = \langle u, a^*au \rangle_1 \\
&\leq \langle u, u \rangle_1^{1/2} \langle a^*au, a^*au \rangle_1 = 0.
\end{aligned}$$

Let us now consider the quotient space $\mathcal{B}/M$. Define $\langle [a], [b] \rangle = \langle a, b \rangle_1$ for $[a]$, $[b]$ in $\mathcal{B}/M$. It is easy to see that the above defines an inner product on $\mathcal{B}/M$. Let H be a Hilbert space completion of this inner product space.

For each $a \in \mathcal{B}$ define $L_a : \mathcal{B}/M \to \mathcal{B}/M$ by $L_a([v]) = [av]$. This map is unambiguously defined for if $[v] = [v_1]$, then $a(v - v_1) \in M$, since M is a left ideal. Thus $[av] = [av_1]$. Clearly, L_a is linear. To see that L_a is bounded, we note that

$$(1) \qquad \|L_a[v]\|^2 = \langle [av], [av] \rangle = \langle av, av \rangle_1 = f(v^* a^* av)$$

and

$$(2) \qquad v^* a^* av = \|a\|^2 v^* v - v^*(\|a\|^2 e - a^* a)v.$$

Now $v^*(\|a\|^2 e - a^* a)v \in \mathcal{B}^+$. To see this, note that $\|a\|^2 e - a^* a$ is selfadjoint. Furthermore,

$$\sigma(\|a\|^2 e - a^* a) = \{ \|a\|^2 - \lambda \mid \lambda \in \sigma(a^* a) \} \subset \mathbb{R}^+,$$

because $\sigma(a^* a) \subset \mathbb{R}^+$ and the spectral radius of $a^* a$ is equal to $\|a^* a\| = \|a\|^2$ (by Corollary 1.3). Thus $\|a\|^2 e - a^* a \in \mathcal{B}^+$. But then Theorem 5.4 ensures the existence of a $w \in \mathcal{B}$ such that $\|a\|^2 e - a^* a = w^* w$. Hence

$$(3) \qquad v^*(\|a\|^2 e - a^* a)w = v^* w^* wv = (wv)^*(wv) \in \mathcal{B}^+,$$

by Theorem 5.4. From (1), (2), (3) and the assumption that f is positive, we get

$$\begin{aligned}
\big\|L_a([v])\big\|^2 &= \|a\|^2 f(v^* v) - f\big(v^*(\|a\|^2 e - a^* a)v\big) \\
&\leq \|a\|^2 f(v^* v) = \|a\|^2 \langle v, v \rangle_1 = \|a\|^2 \|[v]\|^2.
\end{aligned}$$

Thus $\|L_a\| \leq \|a\|$. We may therefore extend L_a (uniquely) to H without increasing its norm. For each $a \in \mathcal{B}$, define $\varphi(a)$ to be this extension of L_a. We shall denote this extension again by L_a. So,

$$(4) \qquad \|\varphi(a)\| = \|L_a\| \leq \|a\|.$$

We shall show that φ is a representation of $\mathcal{B}$ on H. Clearly, $\varphi(e) = I$. It is easy to check that φ is a homomorphism on $\mathcal{B}$. Also, $\varphi(a^*) = \varphi(a)^*$, $a \in \mathcal{B}$, since

$$\begin{aligned}
\langle \varphi(a^*)[u], [v] \rangle &= \langle a^* u, v \rangle_1 = f(v^* a^* u) = f\big((av)^* u\big) \\
&= \langle u, av \rangle_1 = \langle [u], [av] \rangle = \langle [u], \varphi(a)[v] \rangle \\
&= \big\langle (\varphi(a))^* [u], [v] \big\rangle, \qquad u, v \in \mathcal{B}.
\end{aligned}$$

It now follows from the boundedness of φ and the denseness of $\mathcal{B}/M$ in H, that φ is a representation of $\mathcal{B}$ on H.

Finally, Let $u = [e]$. Then

$$\langle \varphi(a)u, u \rangle = \langle [a], [e] \rangle = \langle a, e \rangle_1 = f(a), \qquad a \in \mathcal{B},$$

and $\|u\|^2 = \|[e]\|^2 = f(e) = 1$. $\square$

PROOF OF THEOREM 6.1. Let $\mathcal{P}$ be the set of all positive linear functionals f on $\mathcal{B}$ such that $f(e) = 1$. According to Lemma 6.2, for each $f \in \mathcal{P}$, we may choose an element $u_f \in \mathcal{B}$, $\|u_f\| = 1$, a Hilbert space H_f and a representation φ_f of $\mathcal{B}$ on H_f such that

$$(5) \qquad f(a) = \langle \varphi_f(a)u_f, u_f \rangle, \quad \|\varphi_f(a)\| \le \|a\| \qquad (a \in \mathcal{B}).$$

Let H be the Hilbert space direct sum $\bigoplus_{f \in \mathcal{P}} H_f$. Thus H consists of all $F \colon \mathcal{P} \to \cup_{f \in \mathcal{P}} H_f$ with

$$(6) \qquad \|\!|F|\!\| = \left(\sum_{f \in \mathcal{P}} \|F(f)\|^2 \right)^{1/2} < \infty.$$

With $\|\!|\cdot|\!\|$ defined by (6) the space H is a Hilbert space. Define $\varphi \colon \mathcal{B} \to \mathcal{L}(H)$ by setting

$$(\varphi(a)F)(f) = \varphi_f(a)F(f), \qquad f \in \mathcal{P},$$

where F is an arbitrary element of H. It is readily checked that φ is a representation of $\mathcal{B}$ on H. From the second inequality it follows that

$$\|\!|\varphi(a)F|\!\|^2 = \sum_{f \in \mathcal{P}} \|\varphi_f(a)F(f)\|^2 \le \|a\|^2 \sum_{f \in \mathcal{P}} \|F(f)\|^2 = \|a\|^2 \|\!|F|\!\|^2,$$

and hence $\|\varphi(a)\| \le \|a\|$. On the other hand,

$$\|\varphi_f(a)\|^2 \ge \|\varphi_f(a)u_f\|^2 = \langle \varphi_f(a)u_f, \varphi_f(a)u_f \rangle = \langle \varphi_f(a^*a)u_f, u_f \rangle = f(a^*a).$$

Given $\lambda \in \sigma(a)$ there exists, by Theorem 5.6, a functional $f_0 \in \mathcal{P}$ such that $f_0(a) = \lambda$, and hence $f_0(a^*a) = |\lambda|^2$. It follows that

$$\begin{aligned}
\|\varphi(a)\|^2 &= \sup\{ \|\varphi_f(a)\|^2 \mid f \in \mathcal{P} \} \\
&\ge \sup\{ f(a^*a) \mid f \in \mathcal{P} \} \ge r(a^*a),
\end{aligned}$$

where $r(a^*a)$ is the spectral radius of a^*a. Now, $r(a^*a) = \|a^*a\| = \|a\|^2$, by Corollary 1.3. Thus $\|\varphi(a)\| = \|a\|$ for each $a \in \mathcal{B}$. $\square$

We conclude this section with a corollary about strictly positive elements. Let $\mathcal{A}$ be a unital C^*-algebra with unit e. An element $a \in \mathcal{A}$ is called *strictly positive* if $a - re$ is in $\mathcal{A}^+$ for some $r > 0$.

COROLLARY 6.3. *For an element a in a unital C^*-algebra $\mathcal{A}$ the following statements are equivalent:*

(i) *a is strictly positive,*

(ii) *$a \in \mathcal{A}^+$ and a is invertible,*

(iii) *$a = v^*v$ for some invertible v in $\mathcal{A}$.*

PROOF. First we note that (i) is equivalent to (ii). Indeed, if $a - re$ is in $\mathcal{A}^+$ for some $r > 0$, then a is selfadjoint and

$$\{\lambda - r \mid \lambda \in \sigma(a)\} = \sigma(a - re) \subset [0, \infty).$$

Thus $\sigma(a) \subset [r, \infty)$ and a is invertible. Conversely, if $a \in \mathcal{A}^+$ and a is invertible, then $0 \notin \sigma(a)$ and since $\sigma(a)$ is compact, we have $\sigma(a) \subset [r, \infty)$ for some $r > 0$. Thus $a - re \in \mathcal{A}^+$.

By Theorem 5.4, statement (iii) implies (ii). Assume (ii) holds, and let us prove (iii). According to Theorem 6.1, we may assume without loss of generality that $\mathcal{A}$ is a closed *-subalgebra of $\mathcal{L}(H)$ for some Hilbert space H such that the identity operator on H is the unit of $\mathcal{A}$. Then (ii) implies that a is a non-negative operator on H (cf., the third paragraph in the previous section). Hence there exists a (unique) $b \in \mathcal{L}(H)$ such that $b^2 = a$. From the proofs of Theorem V.6.1 and Lemma V.6.2 it follows that b is the limit in the operator norm of polynomials in a. Thus $b \in \mathcal{A}$. Since a is invertible, $a = b^2$ implies that b is invertible. Thus (iii) holds. $\square$

Strictly positive elements will also be called *positive definite*. Later (in Section XXXIV.1) we shall use item (iii) in Corollary 6.3 to define positive definite elements in an arbitrary algebra with an involution.

Let $\mathcal{A}$ be a unital C^*-algebra with unit e. Since $\mathcal{A}^+$ is closed under addition (Corollary 5.2), it is clear from the definition that the sum of two strictly positive elements in $\mathcal{A}$ is again strictly positive. For later purposes (see Chapter XXXIV) we also mention that $g \in \mathcal{A}$ has $\|g\| < 1$ if and only if $e - g^*g$ is strictly positive. To prove the latter result one applies Theorem 6.1 and uses the analogous result for operators on a Hilbert space.

XXXI.7 THE SPECTRAL THEORY FOR NORMAL OPERATORS

We know from linear algebra that if T is a normal operator defined on a finite dimensional inner product space H, then there exist an orthonormal basis $\varphi_1, \ldots, \varphi_n$ of H and complex numbers $\lambda_1, \ldots, \lambda_n$ such that

$$Tx = \sum_{i=1}^{n} \lambda_i \langle x, \varphi_i \rangle \varphi_i, \qquad x \in H.$$

Let P_k be the orthogonal projection from H onto $\mathrm{span}\{\varphi_1, \ldots, \varphi_k\}$, i.e., $P_k x = \sum_{i=1}^{k} \langle x, \varphi_i \rangle \varphi_i$. Take $P_0 = 0$. It is clear that $\Delta P_j = P_j - P_{j-1}$ ($j = 1, \ldots, n$) is an orthogonal projection, $\Delta P_j \Delta P_k = 0$ for $j \neq k$, the sum $\sum_{j=1}^{n} \Delta P_j = I$ and

$$(1) \qquad\qquad T = \sum_{j=1}^{n} \lambda_j \Delta P_j.$$

In this section we shall extend this result to normal operators on infinite dimensional Hilbert space. The above formula suggests a Riemann-Stieltjes integral

$$T = \int\limits_{\sigma(T)} \lambda dE,$$

which will be a limit in norm of linear combinations of orthogonal projections. As a clue to how this representation may be found, let us return to the case where H is finite dimensional. Suppose $\sigma(T) = \{\lambda_1, \ldots, \lambda_n\}$. For each j, let χ_j be the characteristic function of the set $\{\lambda_j\}$. Since $\sigma(T)$ is a discrete set, every complex valued function defined on $\sigma(T)$ is in $C(\sigma(T))$. In particular, $\chi_1, \ldots, \chi_n$ are in $C(\sigma(T))$. Now we use the functional calculus introduced in Section XXXI.4 (with $\mathcal{B} = \mathcal{L}(H)$ and $x = T$), and define $E_j = \chi_j(T)$ for $j = 1, \ldots, n$. It follows from Theorem 4.1 that each E_j is an orthogonal projection, $E_j E_k = 0$ for $j \neq k$ and $\sum_{j=1}^{n} E_j = I$. Also, since $\lambda = \sum_{j=1}^{n} \lambda_j \chi_j(\lambda)$ for each $\lambda \in \sigma(T)$, Theorem 4.1 implies that

$$T = \left(\sum_{j=1}^{n} \lambda_j \chi_j \right)(T) = \sum_{j=1}^{n} \lambda_j E_j.$$

We have therefore shown that the spectral theorem for normal operators on a finite dimensional Hilbert space is a corollary of Theorem 4.1. But what is more important, we see that the desired projections E_j are characteristic functions of λ_i evaluated at T. If we now wish to extend the result to infinite dimensional H, one might try to obtain the desired projections by taking characteristic functions of certain subsets of $\sigma(T)$ and evaluate them at T. Immediately we reach an obstacle, namely, a characteristic function χ need not be in $C(\sigma(T))$ and therefore $\chi(T)$ is not defined. So, the first major step in the proof of the spectral theorem is to extend the functional calculus for T normal from $C(\sigma(T))$ to the space of bounded Borel measurable functions on $\sigma(T)$. Our projections E will then be defined as the characteristic functions on these Borel sets evaluated at T. The next major step is to define a Riemann-Stieltjes integral with respect to these projections and to show that $T = \int_{\sigma(T)} \lambda dE$. For the definitions of Borel sets, Borel measurable functions and complex Borel measures we refer to [R].

We start with a lemma. Recall that a functional $\psi(x, y)$ on $H \times H$ is *sesquilinear* if ψ is linear in x for each y and $\overline{\psi(x, y)}$ is linear in y for each x. The sesquilinear functional ψ is *bounded* if

$$\sup_{\|x\| = \|y\| = 1} |\psi(x, y)| = \|\psi\| < \infty.$$

LEMMA 7.1. *Let H be a Hilbert space. Every bounded sesquilinear functional Ψ on $H \times H$ is of the form $\Psi(x, y) = \langle Ax, y \rangle$, where A is a bounded linear operator on H uniquely determined by Ψ.*

PROOF. Since for each $x \in H$ the functional $\overline{\Psi(x, \cdot)}$ is a bounded linear functional on H, there exists a unique $v_x \in H$ such that $\overline{\psi(x, y)} = \langle y, v_x \rangle$ by the Riesz representation theorem (see [GG], Theorem II.5.2). Define $A\colon H \to H$ by $Ax = v_x$. It is easy to verify that A is linear and

$$\|Ax\|^2 = \langle Ax, Ax \rangle = \langle Ax, v_x \rangle = \overline{\Psi(x, Ax)} \le \|\psi\| \|Ax\| \|x\|.$$

Hence $\|Ax\| \le \|\psi\| \cdot \|x\|$, and thus A is bounded. Clearly, A is uniquely determined by Ψ. $\square$

THEOREM 7.2. *Let T be a normal operator in $\mathcal{L}(H)$. There exists a mapping E from the σ-algebra of all Borel subsets of $\sigma(T)$ into $\mathcal{L}(H)$ with the following properties:*

(a) *$E(\Delta)$ is an orthogonal projection for every Borel set $\Delta \subset S = \sigma(T)$;*

(b) *$E(\emptyset) = 0$, $E(S) = I$;*

(c) *$E(\Delta_1 \cap \Delta_2) = E(\Delta_1)E(\Delta_2)$;*

(d) *$E(\Delta_1 \cup \Delta_2) = E(\Delta_1) + E(\Delta_2)$ if $\Delta_1 \cap \Delta_2 = \emptyset$;*

(e) *for each x, y in H, the function $E_{x,y}$ defined by $E_{x,y}(\Delta) = \langle E(\Delta)x, y \rangle$ is a regular complex Borel measure on $\sigma(T)$.*

The operator valued function E is called a *spectral measure* on $\sigma(T)$.

PROOF. By $\mathrm{Bor}(S)$ we denote the Banach algebra of all bounded complex-valued Borel measurable functions on $S = \sigma(T)$ endowed with the supremum norm. Note that $\mathrm{Bor}(S)$ is a C^*-algebra with complex conjugation as the involution. Recall (see Theorem 4.1) that the functional calculus

$$(2) \qquad\qquad f \mapsto f(T), \qquad f \in C(S),$$

defines a $*$-isomorphism which maps $C(S)$ isometrically into $\mathcal{L}(H)$. The existence of a spectral measure E on S is shown by first proving that the map (2) can be extended to a $*$-homomorphism Φ from $\mathrm{Bor}(S)$ into $\mathcal{L}(H)$. The spectral measure E is then defined by $E(\Delta) = \Phi(\chi_\Delta)$, where χ_Δ is the characteristic function of Δ. Let us start with the construction of Φ.

For x, y fixed in H, the map $f \mapsto \langle f(T)x, y \rangle$ is a bounded linear functional with norm at most $\|x\| \|y\|$. Hence by the Riesz representation theorem for continuous linear functionals on $C(S)$ (see [R], Theorem 6.19) there exists a unique regular complex Borel measure $\mu_{x,y}$ such that the total variation $|\mu_{x,y}| \le \|x\| \|y\|$ and

$$(3) \qquad\qquad \langle f(T)x, y \rangle = \int_S f\, d\mu_{x,y}, \qquad f \in C(S).$$

It follows from (3) that $\mu_{\alpha x, y} = \alpha \mu_{x,y}$, $\mu_{x+y,z} = \mu_{x,z} + \mu_{y,z}$ and $\mu_{x,\alpha y} = \overline{\alpha}\mu_{x,y}$. Hence for $g \in \mathrm{Bor}(S)$, the functional

$$\Psi_g(x,y) = \int_S g\, d\mu_{x,y}$$

is bounded and sesquilinear. Therefore Lemma 7.1 ensures the existence of a unique $\Phi(g) \in \mathcal{L}(H)$ such that for all x, y in H

$$(4) \qquad \langle \Phi(g)x, y \rangle = \int_S g \, d\mu_{x,y}.$$

It is easy to see that Φ is a linear map from $\mathrm{Bor}(S)$ into $\mathcal{L}(H)$ with $\|\Phi\| \leq 1$. Moreover (3) and (4) imply that $\Phi(f) = f(T)$ for $f \in C(S)$. We shall now prove that Φ is multiplicative and $*$-preserving.

Take $g \in C(S)$, and put $v = g(T)x$, where x is some vector in H. Let f be an arbitrary element in $C(S)$. Since $(fg)(T) = f(T)g(T)$, formula (3) implies that

$$(5) \qquad \int_S fg \, d\mu_{x,y} = \langle f(T)g(T)x, y \rangle = \int_S f \, d\mu_{v,y},$$

where y is an arbitrary vector in H. Now use that $C(S)$ is dense in $L_1(S, \mu)$, where μ is the positive measure $\mu = |\mu_{x,y}| + |\mu_{v,y}|$. Also, $\mathrm{Bor}(S) \subset L_1(S, \mu)$. By continuity it follows from (5) that

$$(6) \qquad \int_S fg \, d\mu_{x,y} = \int_S f \, d_{v,y}, \qquad f \in \mathrm{Bor}(S).$$

Next, take a fixed $f \in \mathrm{Bor}(S)$, and put $z = \Phi(f)^*y$. Then, because of (6),

$$\int_S fg \, d\mu_{x,y} = \langle \Phi(f)v, y \rangle = \langle g(T)x, z \rangle = \int_S g \, d\mu_{x,z}.$$

Since g is an arbitrary element of $C(S)$, a continuity argument similar to the one used to prove (6) yields

$$\int_S fg \, d\mu_{x,y} = \int_S g \, d\mu_{x,z}$$

for each g in $\mathrm{Bor}(S)$. Now fix also g in $\mathrm{Bor}(S)$. Then

$$\langle \Phi(fg)x, y \rangle = \langle \Phi(g)x, z \rangle = \langle \Phi(f)\Phi(g)x, y \rangle$$

for all x and y in H. Hence $\Phi(fg) = \Phi(f)\Phi(g)$.

To show that Φ is $*$-preserving, we first note that $\mu_{x,y} = \overline{\mu}_{y,x}$ for each x and y in H. Indeed, if $f \in C(S)$ is real-valued, then $f(T)$ is selfadjoint (because $f \mapsto f(T)$ is $*$-preserving), and thus

$$(7) \qquad \int_S f \, d\mu_{x,y} = \langle f(T)x, y \rangle = \overline{\langle f(T)y, x \rangle} = \int_S \overline{f} \, d\overline{\mu}_{y,x}.$$

If we apply (7) to the real and imaginary parts of f in $C(S)$, then we may conclude that the integrals in (7) are equal for each $f \in C(S)$. Hence $\mu_{x,y} = \overline{\mu}_{y,x}$. Given $f \in \mathrm{Bor}(S)$

$$\langle \Phi(f)x, y \rangle = \int_S f d\mu_{x,y} = \overline{\int_S \overline{f} d\mu_{y,x}} = \overline{\langle \Phi(\overline{f})y, x \rangle} = \langle x, \Phi(\overline{f})y \rangle.$$

Thus $\Phi(f)^* = \Phi(\overline{f})$. We have shown that Φ is a $*$-homomorphism of $\mathrm{Bor}(S)$ into $\mathcal{L}(H)$.

Given a Borel set $\Delta \subset S$, let χ_Δ denote the characteristic function of Δ and define $E(\Delta) = \Phi(\chi_\Delta)$. Since Φ is a $*$-homomorphism, $E(\Delta)$ is an orthogonal projection. Also, $E(\emptyset) = \Phi(0) = 0$ and $E(S) = \Phi(\chi_S) = \chi_S(T) = I$. From (4) we get

$$(8) \qquad E_{x,y}(\Delta) = \langle E(\Delta)x, y \rangle = \int_S \chi_\Delta d\mu_{x,y} = \mu_{x,y}(\Delta),$$

and hence $E_{x,y}$ is a regular complex Borel measure. For any two Borel sets Δ_1 and Δ_2, we have $\chi_{\Delta_1} \chi_{\Delta_2} = \chi_{\Delta_1 \cap \Delta_2}$. Furthermore, $\chi_{\Delta_1 \cup \Delta_2} = \chi_{\Delta_1} + \chi_{\Delta_2}$ provided $\Delta_1 \cap \Delta_2 = \emptyset$. Since Φ is a homomorphism, statements (c) and (d) hold true. The proof of the theorem is complete. $\square$

We shall now define the integral of a function in $\mathrm{Bor}(S)$ with respect to a spectral measure E on $S = \sigma(T)$. First let us consider a simple function f of the form $f = \sum_{i=1}^{n} \alpha_i \chi_{\Delta_i}$, where the Δ_i's are mutually disjoint Borel sets of S and the α_i's are complex numbers. Define the integral $\mathcal{J}(f)$ by

$$\mathcal{J}(f) = \int_S f dE = \sum_{i=1}^{n} \alpha_i E(\Delta_i) \in \mathcal{L}(H).$$

It follows from the properties of E that the integral $\mathcal{J}(f)$ is independent of the representation of f. Now

$$(9) \qquad \langle \mathcal{J}(f)x, y \rangle = \sum_{i=1}^{n} \alpha_i \langle E(\Delta_i)x, y \rangle = \sum_{i=1}^{n} \alpha_i E_{x,y}(\Delta_i) = \int_S f dE_{x,y},$$

$$(10) \qquad \sum_{i=1}^{n} E_{x,x}(\Delta_i) = \left\langle E\left(\bigcup_{i=1}^{n} \Delta_i\right)x, x \right\rangle \leq \|x\|^2,$$

where the Δ_i's are mutually disjoint Borel sets. From (9) and (10) we obtain

$$(11) \qquad \langle \mathcal{J}(f)x, x \rangle \leq \|f\| \|x\|^2,$$

where $\|f\|$ is the norm of f as an element of $\mathrm{Bor}(S)$. The so-called polarization identity for inner products (which expresses the inner product in terms of the corresponding quadratic form) and the parallellogram law applied to (11) yield

$$(12) \qquad \left\| \int_S f dE \right\| \leq 2\|f\|.$$

Next we shall use that any function f in $\mathrm{Bor}(S)$ is the limit in the supremum norm of a sequence of simple functions. To prove the latter fact, we may assume without loss of generality that f is real. Since f is bounded, there exist real numbers a and b such that $f[S] \subset [a, b]$. Let $\varepsilon > 0$ be given, and choose a partition $a = a_0 < a_1 < \cdots < a_n = b$ such that $|a_j - a_{j-1}| < \varepsilon$ for $j = 1, \ldots, n$. Put $\Delta_0 = \{t \in S \mid f(t) = a\}$, and for $j = 1, \ldots, n$ let $\Delta_j = \{t \in S \mid a_{j-1} < f(t) \leq a_j\}$. The fact that f is Borel measurable implies that $\Delta_0, \ldots, \Delta_n$ are disjoint Borel subsets of S. From our construction it follows that

$$\left\| f - \sum_{j=0}^{n} a_j \chi_{\Delta_j} \right\| = \sup_{t \in S} \left| f(t) - \sum_{j=0}^{n} a_j \chi_{\Delta}(t) \right| \leq \varepsilon$$

for each $t \in S$, which yields the desired result.

Since the simple functions are dense in $\mathrm{Bor}(S)$, it follows from (12) and the completeness of $\mathcal{L}(H)$ that for any $f \in \mathrm{Bor}(S)$ there exists a sequence (f_n) of simple functions such that $\|f - f_n\| \to 0$ and $\lim_{n \to \infty} \int_S f_n dE$ exists in $\mathcal{L}(H)$ independent of the choice of (f_n). We define $\int_S f dE$ to be this limit. Note that in the definition of the integral we did not use that $S = \sigma(T)$; only the properties (a)–(e) of a spectral measure and the compactness of S were employed.

LEMMA 7.3. *Let E be a spectral measure defined on a compact set $S \subset \mathbb{C}$, and let $\int_S f dE$ be defined as above for each $f \in \mathrm{Bor}(S)$. Then the map $f \mapsto \int_S f dE$ is a $*$-homomorphism from $\mathrm{Bor}(S)$ into $\mathcal{L}(H)$.*

PROOF. It suffices to check that the map $\mathcal{J}$, defined by $\mathcal{J}(f) = \int_S f dE$, is a $*$-homomorphism on the $*$-subalgebra of all simple functions. Let $f = \sum_{j=1}^{n} \alpha_j \chi_{\Delta_j}$, where the Δ_j's are mutually disjoint Borel subsets of S and the α_j's are complex numbers. Then $\mathcal{J}(f) = \sum_{j=1}^{n} \alpha_j E(\Delta_j)$. Since each $E(\Delta_j)$ is selfadjoint,

$$\mathcal{J}(f)^* = \sum_{j=1}^{n} \overline{\alpha}_j E(\Delta_j) = \mathcal{J}(\overline{f}).$$

Thus $\mathcal{J}$ is $*$-preserving. Clearly, $\mathcal{J}$ is linear. To prove that $\mathcal{J}$ is multiplicative, let $g = \sum_{i=1}^{k} \beta_i \chi_{\Delta_i'}$ be another simple function in $\mathrm{Bor}(S)$. Then

$$\mathcal{J}(f)\mathcal{J}(g) = \sum_{j=1}^{n} \sum_{i=1}^{k} \alpha_j \beta_i E(\Delta_j) E(\Delta_i')$$

$$= \sum_{j=1}^{n} \sum_{i=1}^{k} \alpha_j \beta_i E(\Delta_j \cap \Delta_i')$$

$$= \mathcal{J}\left(\sum_{j=1}^{n} \sum_{i=1}^{k} \alpha_j \beta_i \chi_{\Delta_j \cap \Delta_i'} \right) = \mathcal{J}(fg),$$

which proves the lemma. $\square$

THEOREM 7.4 (Spectral Theorem). *Let T be a normal operator in $\mathcal{L}(H)$. Then there exists a unique spectral measure E on $\sigma(T)$ such that*

$$(13) \qquad T = \int_{\sigma(T)} \lambda dE.$$

Furthermore, the map $f \mapsto \int_{\sigma(T)} fdE$ is a continuous $*$-homomorphism from $\mathrm{Bor}(\sigma(T))$ into $\mathcal{L}(H)$,

$$(14) \qquad f(T) = \int_{\sigma(T)} fdE, \qquad f \in C(\sigma(T)),$$

and an operator $A \in \mathcal{L}(H)$ commutes with T and T^* if and only if A commutes with every $E(\Delta)$.

PROOF. Let E be the spectral measure which was defined in the proof of Theorem 7.2. Put $S = \sigma(T)$, and define $\mathcal{J}(f) = \int_S fdE$ for f in $\mathrm{Bor}(S)$. From (4), (8) and (9) it follows that

$$(15) \qquad \langle \mathcal{J}(f)x, y \rangle = \int_S fdE_{x,y} = \int_S fd\mu_{x,y} = \langle \Phi(f)x, y \rangle$$

for each simple function f in $\mathrm{Bor}(S)$, where Φ is the continuous $*$-homomorphism introduced in the proof of Theorem 5.2. By continuity, it follows that (15) holds for each f in $\mathrm{Bor}(S)$, and hence

$$(16) \qquad \Phi(f) = \int_S fdE, \qquad f \in \mathrm{Bor}(S).$$

Recall that $\Phi(f) = f(T)$ for each $f \in C(\sigma(T))$. Thus (14) holds true. If we take $f(\lambda) = \lambda$, then $f(T) = T$ and hence (14) implies (13).

Suppose the E' is another spectral measure on S such that $T = \int_S \lambda dE'$. It follows from Lemma 7.3 that for any polynomial P in two variables,

$$(17) \qquad P(T, T^*) = \int_S P(\lambda, \overline{\lambda})dE = \int_S P(\lambda, \overline{\lambda})dE'.$$

By the Stone-Weierstrass theorem, the functions $\lambda \mapsto P(\lambda, \overline{\lambda})$ are dense in $C(S)$. Hence

$$\int_S fdE = \int_S fdE'$$

for every $f \in C(S)$. Since

$$\left\langle \left(\int_S fdE \right)x, y \right\rangle = \int_S fdE_{x,y},$$

$$\left\langle \left(\int_S fdE' \right)x, y \right\rangle = \int_S fdE'_{x,y}, \quad f \in \mathrm{Bor}(S),$$

we conclude that

$$\int_S f\, dE_{x,y} = \int_S f\, dE'_{x,y}$$

for every $f \in C(S)$. Since $E_{x,y}$ and $E'_{x,y}$ are complex regular Borel measures, the uniqueness assertion in the Riesz representation theorem shows that $E_{x,y} = E'_{x,y}$ for all x, y, i.e., $\langle E(\Delta)x, y \rangle = \langle E'(\Delta)x, y \rangle$ for all Borel subsets Δ of S. Hence $E = E'$.

Since Φ is a continuous $*$-homomorphism, (16) implies that the map

$$f \mapsto \int_S f\, dE$$

is a continuous $*$-homomorphism.

It remains to prove the last statement of the theorem. Suppose that $A \in \mathcal{L}(H)$ commutes with T and T^*. Then for any polynomial P in two variables, formulas (17) and (15) imply

$$(18) \qquad \langle AP(T, T^*)x, y \rangle = \langle P(T, T^*)x, A^*y \rangle = \int_S P(\lambda, \overline{\lambda})\, dE_{x, A^*y},$$

$$(19) \qquad \langle P(T, T^*)Ax, y \rangle = \int_S P(\lambda, \overline{\lambda})\, dE_{Ax, y}.$$

Since $AP(T, T^*) = P(T, T^*)A$ and the set of functions $\lambda \mapsto P(\lambda, \overline{\lambda})$ is dense in $C(S)$, it follows from (18) and (19) that

$$\int_S f\, dE_{x, A^*y} = \int_S f\, dE_{Ax, y}$$

for all $f \in C(S)$. Hence $E_{x, A^*y} = E_{Ax, y}$, i.e., $\langle E(\Delta)x, A^*y \rangle = \langle E(\Delta)Ax, y \rangle$ for all Borel subsets $\Delta \subset S$. Therefore, $AE(\Delta) = E(\Delta)A$. The converse is obtained by reversing the steps of the argument. $\square$

We shall refer to the spectral measure E in Theorem 7.4 as the *spectral measure associated with* T. In view of formula (14) in Theorem 7.4, we define

$$(20) \qquad f(T) = \int_{\sigma(T)} f\, dE, \qquad f \in \mathrm{Bor}(\sigma(T)).$$

The next theorem contains further information about the spectral measure associated with a normal operator T and gives an analysis of the eigenvalues of T in terms of this measure.

THEOREM 7.5. *Let T be a normal operator in $\mathcal{L}(H)$, and let E be a spectral measure associated with T. Then*

(a) *$E(\cdot)x$ is countably additive for each $x \in H$, i.e., if Δ is the union of mutually disjoint Borel subsets $\Delta_1, \Delta_2, \ldots$ in $\sigma(T)$, then*

$$E(\Delta)x = \sum_{j=1}^{\infty} E(\Delta_j)x;$$

(b) *$E(\Delta) \neq 0$ for any nonempty open subset Δ of $\sigma(T)$;*

(c) *$\mathrm{Ker}(\lambda I - T) = \mathrm{Im}\, E(\{\lambda\})$ for every $\lambda \in \sigma(T)$;*

(d) *every isolated point λ of $\sigma(T)$ is an eigenvalue of T with eigenspace $\mathrm{Im}\, E(\{\lambda\})$;*

(e) *for T compact*

$$Tx = \sum_{j} \lambda_j E(\{\lambda_j\})x, \qquad x \in H,$$

where $\lambda_1, \lambda_2, \ldots$ are the non-zero eigenvalues of T.

PROOF. (a) Let $\Delta = \cup_{j=1}^{\infty}\Delta_j$ be a disjoint union of Borel sets in $\sigma(T)$. Put $\Delta'_n = \Delta \setminus \cup_{j=1}^{n} \Delta_n$. It suffices to show that $E(\Delta'_n) \to 0$ for $n \to \infty$. Note that $\Delta'_1 \supset \Delta'_2 \supset \cdots$ and $\cap_{n=1}^{\infty}\Delta'_n = \emptyset$. Since $E_{x,x}$ is countably additive, this implies that

$$\|E(\Delta'_n)x\|^2 = \langle E(\Delta'_n)x, E(\Delta'_n)x \rangle$$
$$= \langle E(\Delta'_n)x, x \rangle = E_{x,x}(\Delta'_n) \to 0 \qquad (n \to \infty).$$

(b) Suppose Δ is a nonempty open subset of $\sigma(T)$ and $E(\Delta) = 0$. By Urysohn's lemma (see [W], item 5.6) there exists $f \neq 0$ in $C(\sigma(T))$ such that $f(t) = 0$ for $t \notin \Delta$. For every Borel set $\Delta' \subset \Delta$

$$E(\Delta') = E(\Delta' \cap \Delta) = E(\Delta')E(\Delta) = 0.$$

Hence it follows from (14) that $f(T) = 0$. But this is impossible, because the map $g \mapsto g(T)$ is injective on $C(\sigma(T))$ by Theorem 4.1.

(c) Note that $E(\{\lambda_0\}) = \int_{\sigma(T)} \chi_{\lambda_0}\, dE$, where χ_{λ_0} is the characteristic function of the set $\{\lambda_0\}$. Since $\lambda\chi_{\lambda_0}(\lambda) = \lambda_0\chi_{\lambda_0}(\lambda)$, we can use Theorem 7.4 to show that

$$TE(\{\lambda_0\}) = \int_{\sigma(T)} \lambda\chi_{\lambda_0}\, dE = \lambda_0 E(\{\lambda_0\}).$$

Hence $\mathrm{Im}\, E(\{\lambda_0\}) \subset \mathrm{Ker}(\lambda_0 I - T)$. Suppose that $x \in \mathrm{Ker}(\lambda_0 I - T)$. Let

$$S_n := \left\{ \lambda \in \sigma(T) \mid\, |\lambda - \lambda_0| > \frac{1}{n} \right\},$$

and define

$$f_n(\lambda) = \begin{cases} (\lambda_0 - \lambda)^{-1}, & \lambda \in S_n, \\ 0, & \lambda \in \sigma(T)\backslash S_n. \end{cases}$$

Then by (20) and Theorem 7.4

$$(21) \qquad 0 = f_n(T)(\lambda_0 I - T)x = \chi_{S_n}(T)(x) = E(S_n)x,$$

where χ_{S_n} is the characteristic function of the set S_n. Clearly

$$(22) \qquad \sigma(T)\backslash\{\lambda_0\} = \bigcup_{n=1}^{\infty} S_n, \qquad S_1 \subset S_2 \subset S_3 \subset \cdots .$$

Now $E(\cdot)x$ is countably additive on $\sigma(T)$ by (a). So equations (21) and (22) imply that

$$x - E(\{\lambda_0\})x = E(\sigma(T))x - E(\{\lambda_0\})x = E\left(\bigcup_{n=1}^{\infty} S_n\right)x = \lim_{n\to\infty} E(S_n)x = 0.$$

Hence $\operatorname{Ker}(\lambda_0 I - T) \subset \operatorname{Im} E(\{\lambda_0\})$.

(d) If λ_0 is an isolated point in $\sigma(T)$, then $\{\lambda_0\}$ is an open set in $\sigma(T)$. Thus (d) is an immediate consequence of (b) and (c).

(e) Since $E(\cdot)x$ is countably additive,

$$x = E(\{0\})x + \sum_{j} E(\{\lambda_i\})x,$$

and thus by (c)

$$Tx = \sum_{j} TE(\{\lambda_j\})x = \sum_{j} \lambda_j E(\{\lambda_j\})x. \quad \square$$

The map $f \mapsto f(T)$ defined by (20) is an isometry from $C(\sigma(T))$ into $\mathcal{L}(H)$, but in general not from $\operatorname{Bor}(\sigma(T))$ into $\mathcal{L}(H)$. The first statement follows from Theorem 4.1. To prove the second statement, take $\lambda_0 \in \sigma(T)$ such that λ_0 is not an eigenvalue of T. Let f be the characteristic function of the set $\{\lambda_0\}$. Then $f \neq 0$, but $f(T) = E(\{\lambda_0\}) = 0$ by Theorem 7.5(c).

Recall that a subspace M of H is said to *reduce* an operator $T \in \mathcal{L}(H)$ if $TM \subset M$ and $TM^{\perp} \subset M^{\perp}$. If T is normal, then we have the following result.

THEOREM 7.6. *Let $T \in \mathcal{L}(H)$ be normal, and let E be the spectral measure associated with T. Then for every Borel subset Δ of $\sigma(T)$, the space $\operatorname{Im} E(\Delta)$ reduces T.*

PROOF. The theorem follows readily from the equalities

$$TE(\Delta) = E(\Delta)T, \qquad (\operatorname{Im} E(\Delta))^{\perp} = \operatorname{Im}(I - E(\Delta)). \quad \square$$

We note that the normal operator T always has a non-trivial reducing subspace if $\dim H > 1$. For if $E(\Delta) = 0$ or I for every Borel subset Δ of $\sigma(T)$, then

$$T = \int_{\sigma(T)} \lambda \, dE = \mu I$$

for some $\mu \in \mathbb{C}$ and therefore our assertion is obvious.

THEOREM 7.7. *Let T be a normal operator in $\mathcal{L}(H)$ with corresponding spectral measure $E(\cdot)$. Then for each closed set Δ in the complex plane,*

$$\text{(23)} \qquad \operatorname{Im} E(\Delta) = \bigcap_{\lambda \notin \Delta} \operatorname{Im}(\lambda - T).$$

In particular,

$$\bigcap_{\lambda \in \mathbb{C}} \operatorname{Im}(\lambda - T) = \{0\}.$$

PROOF. Suppose $x \in \operatorname{Im} E(\Delta)$ and $\lambda_0 \notin \Delta$. Let

$$f(\lambda) = \begin{cases} (\lambda - \lambda_0)^{-1} , & \lambda \notin \Delta, \\ 0 & , \quad \lambda \in \Delta. \end{cases}$$

It follows from Theorem 7.4 (the spectral theorem) that $x = (\lambda_0 - T)f(T)x$. Hence $\operatorname{Im} E(\Delta) \subset \operatorname{Im}(\lambda_0 - T)$ for each $\lambda_0 \notin \Delta$, and thus $\operatorname{Im} E(\Delta)$ is contained in the right hand side of (23). It remains to prove the reverse inclusion:

$$\text{(24)} \qquad \bigcap_{\lambda \notin \Delta} \operatorname{Im}(\lambda - T) \subset \operatorname{Im} E(\Delta).$$

Set $\mathcal{O} = \mathbb{C} \backslash \Delta$, and identify $\mathbb{C}$ in the usual way with $\mathbb{R}^2$. Since $\mathcal{O}$ is open in the plane, $\mathcal{O}$ is the union of countably many disjoint bounded squares S_j whose sides are parallel to the coordinate axis and whose closures are contained in $\mathcal{O}$. To prove (24) it suffices to show that

$$\text{(25)} \qquad \bigcap_{\lambda \in \mathcal{O}} \operatorname{Im}(\lambda - T) \subset \operatorname{Ker} E(S_j), \qquad j = 1, 2, \dots .$$

Indeed, if (25) holds and x belongs to the left hand side of (24), then $E(S_j)x = 0$ for $j \geq 1$, and hence

$$E(\mathcal{O})x = \sum_{j=1}^{\infty} E(S_j)x = 0.$$

Thus $x \in \operatorname{Ker} E(\mathcal{O}) = \operatorname{Im} E(\Delta)$, which proves (24).

To show that (25) holds, fix $S = S_j$, and take x in the left hand side of (25). By induction, we construct a sequence of squares as follows. Take $R_0 = S$. Let

$\Delta_j = 1, 2, 3, 4$ be the mutually disjoint squares contained in R_0 which one obtaines by halving the sides of the closure $\overline{R}_0$ of R_0. Since $\|E(R_0)x\|^2 = \sum_{j=1}^{4} \|E(\Delta_j)x\|^2$, it follows that for some i, $\|E(\Delta_i)x\|^2 \geq \frac{1}{4}\|E(R_0)x\|^2$. Let R_1 be this Δ_i. Thus $\|E(R_0)x\|^2 \leq 4\|E(R_1)x\|^2$. Now let R_2 be obtained from R_1 in the same way that R_1 was obtained from R_0. Continuing in this manner we obtain a sequence $R_0 \supset R_1 \supset \cdots$ of squares such that

$$\|E(S)x\|^2 \leq 4^n\|E(R_n)x\|^2$$

and $d_n = 2^{-n}d_0$, where d_n is the diameter of R_n, $n = 1, 2, \dots$. Since $\overline{R}_0 \supset \overline{R}_1 \supset \cdots$ and $d_n \to 0$, we have $\bigcap_n \overline{R}_n = \{\lambda_0\}$ for some $\lambda_0 \in \mathbb{R}^2$. In particular, $\lambda_0 \in \overline{S} \subset \mathcal{O}$ and therefore there exists $y \in H$ such that $(\lambda_0 - T)y = x$. By Theorem 7.4, the operator T^* commutes with $E(\cdot)$. Hence,

$$\|E(S)x\|^2 \leq 4^n\|E(R_n)x\|^2 = 4^n\langle E(R_n)(\lambda_0 - T)y, (\lambda_0 - T)y\rangle$$

$$= 4^n\langle E(R_n)(\lambda_0 - T)^*(\lambda_0 - T)y, y\rangle = 4^n\int_{R_n} |\lambda_0 - \lambda|^2 d\langle E(\cdot)y, y\rangle$$

$$= 4^n\int_{R_n\setminus\{\lambda_0\}} |\lambda_0 - \lambda|^2 d\langle E(\cdot)y, y\rangle \leq 4^n d_n^2 v(R_n\setminus\{\lambda_0\}) \leq d_0^2 v(R_n\setminus\{\lambda_0\}) \to 0,$$

where $v(K)$ is the variation of the measure $\langle E(\cdot)y, y\rangle$ on the set K. Hence $E(S)x = 0$ for $x \in \bigcap_{\lambda \in S} \mathrm{Im}(\lambda - T)$. Thus (25) holds. $\square$

As a corollary of the above result we obtain Fuglede's theorem (Theorem 7.9 below). First we prove the following lemma.

LEMMA 7.8. *Let T be a normal operator in $\mathcal{L}(H)$. For each $\lambda \in \mathbb{C}$ there exists a unitary operator $U_\lambda \in \mathcal{L}(H)$ such that $\lambda - T = U_\lambda(\lambda - T)^*$ and U_λ commutes with T and T^*.*

PROOF. We have $H = \mathrm{Im}(\lambda - T)^* \oplus \mathrm{Ker}(\lambda - T)$. Since T is normal, $\|(\lambda - T)^*x\| = \|(\lambda - T)x\|$ for each $x \in H$. Thus the map

$$\varphi_\lambda\big((\lambda - T)^*x\big) = (\lambda - T)x$$

is a linear isometry from $\overline{\mathrm{Im}(\lambda - T)^*}$ onto $\mathrm{Im}(\lambda - T)$. Let $\widetilde{\varphi}_\lambda$ be the unique continuous extension of φ_λ to all of $\overline{\mathrm{Im}(\lambda - T)^*}$. Define U_λ on H by

$$U_\lambda(x + y) = \widetilde{\varphi}_\lambda(x) + y, \qquad x \in \overline{\mathrm{Im}(\lambda - T)^*}, \quad y \in \mathrm{Ker}(\lambda - T).$$

Clearly, U_λ is unitary and

(26) $$U_\lambda(\lambda - T)^* = \lambda - T.$$

It remains to show that U_λ commutes with T and T^*. Since $U_\lambda^{-1} = U_\lambda^*$, formula (26) yields $(\lambda - T)^* = U_\lambda^*(\lambda - T)$. Taking adjoints of both sides gives $\lambda - T = (\lambda - T)^*U_\lambda$. Thus (see (26))

(27) $$U_\lambda(\lambda - T)^* = (\lambda - T)^*U_\lambda,$$

and we see that U_λ commutes with T^*. Furthermore,

$$U_\lambda(\lambda - T) = U_\lambda^2(\lambda - T)^* = U_\lambda(\lambda - T)^*U_\lambda = (\lambda - T)U_\lambda,$$

by (26) and (27). It follows that U_λ commutes with T. $\square$

THEOREM 7.9. *Let $T \in \mathcal{L}(H)$ be normal. If T commutes with $B \in \mathcal{L}(H)$, then T^* commutes with B.*

PROOF. For each $\lambda \in \mathbb{C}$, let U_λ be as in the preceding lemma. Then

$$(\lambda - T)^* = (\lambda - T)U_\lambda^* = U_\lambda^*(\lambda - T).$$

Hence

$$T^*B - BT^* = B(\lambda - T)^* - (\lambda - T)^*B = B(\lambda - T)U_\lambda^* - (\lambda - T)U_\lambda^*B$$
$$= (\lambda - T)\{BU_\lambda^* - U_\lambda^*B\},$$

because $TB = BT$. It follows that

$$\mathrm{Im}(T^*B - BT^*) \subset \bigcap_{\lambda \in \mathbb{C}} \mathrm{Im}(\lambda - T) = \{0\},$$

by Theorem 7.7. $\square$

CHAPTER XXXII
BANACH ALGEBRAS GENERATED
BY TOEPLITZ OPERATORS

In this chapter the Banach algebras generated by Toeplitz operators defined by continuous and, more generally, by piecewise continuous functions are studied. These algebras are not commutative, but they contain the compact operators as a proper closed ideal and in the scalar case the corresponding quotient algebras turn out to be commutative. The Gelfand spectra and transforms of these quotient algebras are described and analyzed.

XXXII.1 ALGEBRAS OF TOEPLITZ OPERATORS DEFINED BY CONTINUOUS FUNCTIONS (SCALAR CASE)

By T_φ we denote the Toeplitz operator defined by the (scalar) function φ. Thus (see Chapter XXIII)

$$(1) \qquad T_\varphi = \begin{bmatrix} a_0 & a_{-1} & a_{-2} & \cdots \\ a_1 & a_0 & a_{-1} & \cdots \\ a_2 & a_1 & a_0 & \\ \vdots & \vdots & & \ddots \end{bmatrix}$$

is a bounded linear operator on ℓ_2 and φ is a measurable essentially bounded function on $\mathbb{T}$ of which the n-th Fourier coefficient is equal to a_n. The representation (1) means that for a sequence $y = (\eta_0, \eta_1, \ldots)$ in ℓ_2 the i-th entry of $T_\varphi y$ is given by

$$(2) \qquad (T_\varphi y)_i = \sum_{j=0}^{\infty} a_{i-j}\eta_j, \qquad i = 0, 1, 2, \ldots \ .$$

In this section we treat the case when the defining function φ is continuous on $\mathbb{T}$.

In the sequel $\mathcal{T}(C)$ denotes the smallest closed subalgebra of $\mathcal{L}(\ell_2)$ containing all operators T_φ with φ from $C(\mathbb{T})$. Obviously, the identity operator on ℓ_2 is in $\mathcal{T}(C)$ (take $\varphi(\zeta) \equiv 1$ on $\mathbb{T}$). From (1) it is easy to see that

$$(3) \qquad (T_\varphi)^* = T_{\overline{\varphi}},$$

and hence $\mathcal{T}(C)$ is closed under the adjoint operation. It follows that $\mathcal{T}(C)$ is a unital closed $*$-subalgebra of $\mathcal{L}(\ell_2)$, and thus $\mathcal{T}(C)$ is a unital C^*-algebra in its own right.

THEOREM 1.1. *The C^*-algebra $\mathcal{T}(C)$ contains the set $\mathcal{K}$ of all compact operators on ℓ_2 as a proper closed ideal and the quotient algebra $\mathcal{T}(C)/\mathcal{K}$ is isometrically $*$-isomorphic to $C(\mathbb{T})$.*

PROOF. First we prove that $\mathcal{K} \subset \mathcal{T}(C)$. Let K be an arbitrary compact operator on ℓ_2. Let P_ν be the orthogonal projection on ℓ_2 defined by

$$(4) \qquad P_\nu(\eta_0, \eta_1, \eta_2, \ldots) = (\eta_0, \ldots, \eta_\nu, 0, 0, \ldots).$$

Since K is compact, $P_\nu K \to K$ $(\nu \to \infty)$ in the operator norm (see [GG], Section XI.3). Also $P_\nu K P_\mu \to P_\nu K$ in the operator norm if $\mu \to \infty$. So it suffices to show that $P_\nu K P_\mu \in \mathcal{T}(C)$ for each $\mu, \nu \geq 0$. Since $\mathcal{T}(C)$ is a linear space, it follows that we may assume that in the matrix $[k_{ij}]_{i,j=0}^\infty$ of K (relative to the standard orthogonal basis in ℓ_2) all entries are zero except one. So, let a be an arbitrary complex number, fix two nonnegative integers p and q, and assume that $K = [k_{ij}]_{i,j=0}^\infty$, where $k_{pq} = a$ and all other entries are zero. Let

$$(5) \qquad S_\ell = \begin{bmatrix} 0 & 1 & & & \\ & 0 & 1 & & \\ & & 0 & 1 & \\ & & & \ddots & \ddots \\ & & & & \ddots \end{bmatrix}, \qquad S_r = \begin{bmatrix} 0 & & & & \\ 1 & 0 & & & \\ & 1 & 0 & & \\ & & 1 & \ddots & \\ & & & \ddots & \ddots \end{bmatrix}$$

be the backward and forward shifts on ℓ_2, respectively. Obviously, S_ℓ and S_r are in $\mathcal{T}(C)$. So $P_0 = S_\ell S_r - S_r S_\ell \in \mathcal{T}(C)$, and hence

$$K = a S_r^p (S_\ell S_r - S_r S_\ell) S_\ell^q \in \mathcal{T}(C).$$

Thus $\mathcal{K} \subset \mathcal{T}(C)$. Note that $\mathcal{K}$ is closed under the operation of taking adjoints. Thus $\mathcal{T}(C)/\mathcal{K}$ is a well-defined unital C^*-algebra.

Next, we prove that for $\varphi \in C(\mathbb{T})$,

$$(6) \qquad \inf_{K \in \mathcal{K}} \|T_\varphi + K\| = \|T_\varphi\| = \max_{|\zeta|=1} |\varphi(\zeta)|.$$

The second identity follows from Corollary XXIII.3.2 (and the continuity of φ). Let $\mathcal{B}$ be the Calkin algebra $\mathcal{B} = \mathcal{L}(\ell_2)/\mathcal{K}$, and denote by $[T]$ the coset $T + \mathcal{K}$. From Theorem XI.5.2 we know that $\lambda[I] - [T_\varphi]$ is invertible in $\mathcal{B}$ if and only if $\lambda I - T_\varphi$ is Fredholm. On the other hand, since φ is continuous, the operator $\lambda I - T_\varphi$ is Fredholm if and only if $\lambda \neq \varphi(\zeta)$ for all $\zeta \in \mathbb{T}$ (see Theorem XXIII.4.3). It follows that $\sigma([T_\varphi])$ coincides with the range of φ. In particular, the spectral radius $r([T_\varphi])$ is equal to the third term in (6). But then

$$\|T_\varphi\| \geq \inf_{K \in \mathcal{K}} \|T_\varphi + K\| = \|[T_\varphi]\| \geq r([T_\varphi]) = \max_{|\zeta|=1} |\varphi(\zeta)| = \|T_\varphi\|,$$

and (6) is proved.

The third step of the proof consists of showing that

$$(7) \qquad \mathcal{T}(C) = \{T_\varphi + K \mid \varphi \in C(\mathbb{T}), K \in \mathcal{K}\}.$$

Let $\mathcal{A}$ denote the right hand side of (7). According to the first part of the proof, $\mathcal{K} \subset \mathcal{T}(C)$ and hence $\mathcal{A} \subset \mathcal{T}(C)$. By Corollary XXIII.4.2,

$$(8) \qquad T_{\varphi_1} T_{\varphi_2} - T_{\varphi_1 \varphi_2} \in \mathcal{K},$$

whenever $\varphi_1, \varphi_2 \in C(\mathbb{T})$. Since $\mathcal{K}$ is an ideal in $\mathcal{L}(\ell_2)$, this implies that $\mathcal{A}$ is an algebra. Hence to prove (7), it remains to show that $\mathcal{A}$ is closed. Assume that $T_{\varphi_n} + K_n \to T$ in the operator norm. Then (6) implies that the sequence (φ_n) converges in $C(\mathbb{T})$ to a function φ, say. It follows that $T_{\varphi_n} \to T_\varphi$ in the operator norm (by Corollary XXIII.3.2), and therefore $K_n \to T - T_\varphi$ if $n \to \infty$. Since the K_n's are assumed to be compact, the operator $K = T - T_\varphi$ is also compact. Hence $T = T_\varphi + K \in \mathcal{A}$. Thus $\mathcal{A}$ is closed and (7) is proved.

Finally, define $\mathcal{J} \colon C(\mathbb{T}) \to \mathcal{T}(C)/\mathcal{K}$ by setting $\mathcal{J}(\varphi) = T_\varphi + \mathcal{K}$. Obviously, $\mathcal{J}$ is a linear map. From (3) and (8) it follows that $\mathcal{J}$ is a $*$-homomorphism. Formulas (6) and (7) show that $\mathcal{J}$ is an isometry from $C(\mathbb{T})$ onto $\mathcal{T}(C)/\mathcal{K}$. $\quad\square$

COROLLARY 1.2. *The Gelfand spectrum of the commutative unital C^*-algebra $\mathcal{T}(C)/\mathcal{K}$ is homeomorphic to the unit circle $\mathbb{T}$. Here $\mathcal{K}$ is the set of all compact operators on ℓ_2.*

PROOF. Theorem 1.1 implies that the Gelfand spectra of $\mathcal{T}(C)/\mathcal{K}$ and $C(\mathbb{T})$ are homeomorphic. We know (see example (i) in Section XXX.4) that the Gelfand spectrum of $C(\mathbb{T})$ is homeomorphic to $\mathbb{T}$. These two remarks prove the corollary. $\quad\square$

If we identify the Gelfand spectrum of $\mathcal{T}(C)/\mathcal{K}$ with $\mathbb{T}$, then the Gelfand representation of $\mathcal{T}(C)/\mathcal{K}$ may be identified with the map

$$\Gamma \colon \mathcal{T}(C)/\mathcal{K} \to C(\mathbb{T}), \qquad \Gamma([T_\varphi]) = \varphi.$$

Note that Γ is an isometric $*$-isomorphism.

XXXII.2 ALGEBRAS OF TOEPLITZ OPERATORS DEFINED BY CONTINUOUS FUNCTIONS (MATRIX CASE)

Let Φ be a measurable essentially bounded $m \times m$ matrix function on the unit circle $\mathbb{T}$. Thus

$$(1) \qquad \Phi(\zeta) = \begin{bmatrix} \varphi_{11}(\zeta) & \cdots & \varphi_{1m}(\zeta) \\ \vdots & & \vdots \\ \varphi_{m1}(\zeta) & \cdots & \varphi_{mm}(\zeta) \end{bmatrix}, \qquad \zeta \in \mathbb{T},$$

where for each i and j the scalar function φ_{ij} is measurable and essentially bounded on $\mathbb{T}$. The block Toeplitz operator defined by Φ is the bounded linear operator on ℓ_2^m given by

$$(2) \qquad T_\Phi = \begin{bmatrix} A_0 & A_{-1} & A_{-2} & \cdots \\ A_1 & A_0 & A_{-1} & \cdots \\ A_2 & A_1 & A_0 & \\ \vdots & \vdots & & \ddots \end{bmatrix}.$$

Here A_n is the n-th Fourier coefficient of Φ. Recall (see Section XXIII.1) that ℓ_2^m is the Hilbert space consisting of all square summable sequences $y = (\eta_0, \eta_1, \ldots)$ with entries η_j $(j \geq 0)$ in $\mathbb{C}^m$. The representation (2) means that the i-th vector in the sequence $T_\Phi y$ is given by

$$(T_\Phi y)_i = \sum_{j=0}^\infty A_{i-j} \eta_j, \qquad i = 0, 1, 2, \ldots .$$

The coefficient A_n is an $m \times m$ matrix, and as usual we identify an $m \times m$ matrix with the operator on $\mathbb{C}^m$ defined by the canonical action of the matrix on the vectors of $\mathbb{C}^m$. The space ℓ_2^m is also equal to an orthogonal direct sum of m copies of ℓ_2, and hence T_Φ may be written as an $m \times m$ operator matrix. In fact

$$(3) \qquad T_\Phi = \begin{bmatrix} T_{\varphi_{11}} & \cdots & T_{\varphi_{1m}} \\ \vdots & & \vdots \\ T_{\varphi_{m1}} & \cdots & T_{\varphi_{mm}} \end{bmatrix},$$

where $T_{\varphi_{ij}}$ is the Toeplitz operator on ℓ_2 defined by the (i,j)-th entry of the right hand side of (1).

In this section we deal with block Toeplitz operators defined by Φ from $C^{m \times m}(\mathbb{T})$, the set of all $m \times m$ matrix functions which are continuous on $\mathbb{T}$. Note that $C^{m \times m}(\mathbb{T})$ is a unital C^*-algebra. The algebraic operations are defined in the usual way, the involution is given by $\Phi^*(\zeta) = \Phi(\zeta)^*$ and

$$(4) \qquad \|\Phi\| = \max_{|\zeta|=1} \|\Phi(\zeta)\|.$$

The norm in the right hand side of (4) is the operator on $\mathcal{L}(\mathbb{C}^m)$ (in other words, $\|\Phi(\zeta)\|$ is the largest singular value of the matrix $\Phi(\zeta)$). The unit in $C^{m \times m}(\mathbb{T})$ is the $m \times m$ matrix function E on $\mathbb{T}$ which is identically equal to the $m \times m$ identity matrix.

We denote by $\mathcal{T}_m(C)$ the smallest closed subalgebra of $\mathcal{L}(\ell_2^m)$ containing all operators T_Φ with Φ from $C^{m \times m}(\mathbb{T})$. Since T_E is the identity operator on ℓ_2^m and

$$(5) \qquad (T_\Phi)^* = T_{\Phi^*},$$

the algebra $\mathcal{T}_m(C)$ is a unital closed $*$-subalgebra of $\mathcal{L}(\ell_2^m)$, and hence $\mathcal{T}_m(C)$ is a unital C^*-algebra in its own right.

THEOREM 2.1. *The C^*-algebra $\mathcal{T}_m(C)$ contains the set $\mathcal{K}$ of all compact operators on ℓ_2^m as a proper closed ideal and the quotient algebra $\mathcal{T}_m(C)/\mathcal{K}$ is isometrically $*$-isomorphic to $C^{m \times m}(\mathbb{T})$.*

PROOF. Let $\mathcal{D}$ be the set of all Toeplitz operators T_φ on ℓ_2 defined by a continuous function φ on $\mathbb{T}$, and let $\mathcal{D}^{m \times m}$ be the set of all $m \times m$ matrices with entries from $\mathcal{D}$. According to (3) the set $\mathcal{D}^{m \times m}$ consists of all block Toeplitz operators T_Φ with Φ from $C^{m \times m}(\mathbb{T})$. An application of Lemma XXIX.8.2 yields that

$$(6) \qquad \mathcal{T}_m(C) = \left\{ \begin{bmatrix} T_{11} & \cdots & T_{1m} \\ \vdots & & \vdots \\ T_{m1} & \cdots & T_{mm} \end{bmatrix} \,\middle|\, T_{ij} \in \mathcal{T}(C) \right\}.$$

Next observe that

$$T = \begin{bmatrix} T_{11} & \cdots & T_{1m} \\ \vdots & & \vdots \\ T_{m1} & \cdots & T_{mm} \end{bmatrix}$$

is a compact linear operator on ℓ_2^m if and only if each entry T_{ij} is compact on ℓ_2. But then we can use (6) and formula (7) in the previous section to show that

$$(7) \qquad \mathcal{T}_m(C) = \{T_\Phi + K \mid \Phi \in C^{m \times m}(\mathbb{T}), K \text{ compact}\}.$$

Next, we show that

$$(8) \qquad \inf_{K \in \mathcal{K}} \|T_\Phi + K\| = \|T_\Phi\| = \max_{|\zeta|=1} \|\Phi(\zeta)\|$$

for each $\Phi \in C^{m \times m}(\mathbb{T})$. Note that (8) is the matrix analogue of formula (6) in the previous section. To obtain (8) we have to modify the proof of the scalar case. Let $[T]$ denote the coset $T + \mathcal{K}$. From Theorems XI.5.2 and XXIII.4.3 it follows that $\lambda[I] - [T_\Phi]$ is invertible in the Calkin algebra $\mathcal{L}(\ell_2^m)/\mathcal{K}$ if and only if $\det(\lambda I - \Phi(\zeta)) \neq 0$ for all $\zeta \in \mathbb{T}$. According to Theorem XXIII.2.4 this implies that the spectrum $\sigma([T_\Phi])$ of $[T_\Phi]$ (relative to the Calkin algebra) is equal to the spectrum of L_Φ, where L_Φ is the block Laurent operator on $\ell_2^m(\mathbb{Z})$ defined by Φ. In particular, the corresponding spectral radii are equal, i.e., $r([T_\Phi]) = r(L_\Phi)$. Since $T_\Phi T_{\Phi^*} - T_{\Phi\Phi^*}$ is compact (Corollary XXIII.4.2), it follows that

$$\begin{aligned} \|T_\Phi\|^2 = \|T_\Phi T_\Phi^*\| &= \|T_\Phi T_{\Phi^*}\| \\ &\geq \inf_{K \in \mathcal{K}} \|T_{\Phi\Phi^*} + K\| = \|[T_{\Phi\Phi^*}]\| \\ &\geq r([T_{\Phi\Phi^*}]) = r(L_{\Phi\Phi^*}). \end{aligned}$$

But $L_{\Phi\Phi^*}$ is selfadjoint. Hence $r(L_{\Phi\Phi^*}) = \|L_{\Phi\Phi^*}\|$ by Corollary XXXI.1.3. Thus

$$\|T_\Phi\|^2 \geq \|L_{\Phi\Phi^*}\| = \|L_\Phi L_{\Phi^*}\| = \|L_\Phi\|^2 = \left(\max_{|\zeta|=1} \|\Phi(\zeta)\|\right)^2 = \|\Phi\|^2.$$

We already know (see Corollary XXIII.3.2) that $\|T_\Phi\| = \|\Phi\|$. So (8) is proved.

For the remaining part of the proof one can use the same arguments as in the proof of Theorem 1.1. Since $T_{\Phi_1} T_{\Phi_2} - T_{\Phi_1\Phi_2}$ is compact whenever Φ_1 and Φ_2 are continuous on $\mathbb{T}$ (Corollary XXIII.4.2), the map $\mathcal{J}: C^{m \times m}(\mathbb{T}) \to \mathcal{T}_m(C)/\mathcal{K}$, defined by $\mathcal{J}(\Phi) = T_\Phi + \mathcal{K}$, is a homomorphism. From (5) it follows that $\mathcal{J}$ is a $*$-homomorphism, (7) implies that $\mathcal{J}$ is surjective, and from (8) we conclude that $\mathcal{J}$ is an isometry. $\quad\square$

XXXII.3 ALGEBRAS OF TOEPLITZ OPERATORS DEFINED BY PIECEWISE CONTINUOUS FUNCTIONS (FINITELY MANY DISCONTINUITIES)

Our aim is to extend the theorem of the previous section to block Toeplitz operators defined by piecewise continuous matrix functions. First we consider the case when the defining functions have their discontinuities in a fixed finite set. In what follows we assume the reader to be familiar with the contents of Chapter XXV.

Let $\zeta_1, \ldots, \zeta_k$ be k different points on $\mathbb{T}$. By $\mathcal{T}_m(PC; \zeta_1, \ldots, \zeta_k)$ we denote the smallest closed subalgebra of $\mathcal{L}(\ell_2^m)$ containing all block Toeplitz operators T_Φ defined by $m \times m$ matrix functions from $PC^{m \times m}(\mathbb{T}; \zeta_1, \ldots, \zeta_k)$. If $m = 1$, we drop the index m and just write $\mathcal{T}(PC; \zeta_1, \ldots, \zeta_k)$. A sum of products of block Toeplitz operators defined by functions from $PC^{m \times m}(\mathbb{T}; \zeta_1, \ldots, \zeta_k)$ is an operator in $\mathcal{T}_m(PC; \zeta_1, \ldots, \zeta_k)$. In fact, the set of all operators of the latter type is a subalgebra which is dense in $\mathcal{T}_m(PC; \zeta_1, \ldots, \zeta_k)$. This remark allows us to use the results of Section XXV.5 to study the algebra $\mathcal{T}_m(PC; \zeta_1, \ldots, \zeta_k)$.

Let $\Phi \in PC^{m \times m}(\mathbb{T}; \zeta_1, \ldots, \zeta_k)$. We denote by Φ_* the unique function in $PC^{m \times m}(\mathbb{T}; \zeta_1, \ldots, \zeta_k)$ such that

$$(1) \qquad \Phi_*(\zeta) = \Phi(\zeta)^*, \qquad \zeta \in \mathbb{T}, \zeta \neq \zeta_j \quad (j = 1, \ldots, k).$$

Since $\Phi_*(\zeta)$ and $\Phi(\zeta)^*$ differ only in a finite number of points, these two functions define the same block Toeplitz operator, and hence

$$(2) \qquad (T_\Phi)^* = T_{\Phi_*}.$$

It follows that $\mathcal{T}_m(PC; \zeta_1, \ldots, \zeta_k)$ is a closed $*$-subalgebra of $\mathcal{L}(\ell_2^m)$. Of course the identity operator on ℓ_2^m is in $\mathcal{T}_m(PC; \zeta_1, \ldots, \zeta_k)$, and thus $\mathcal{T}_m(PC; \zeta_1, \ldots, \zeta_k)$ is a unital C^*-algebra in its own right.

We write $C^{m \times m}[\mathbb{T}[\zeta_1, \ldots, \zeta_k]]$ for the set of all $m \times m$ matrix functions F on $\mathbb{T} \times [0, 1]$ of the form

$$(3) \qquad F(\zeta, \mu) = \begin{cases} \Phi(\zeta) & \text{if } \zeta \neq \zeta_j \quad (j = 1, \ldots, k), \\ G_j(\mu) & \text{if } \zeta = \zeta_j. \end{cases}$$

Here Φ is an arbitrary element of $PC^{m \times m}(\mathbb{T}; \zeta_1, \ldots, \zeta_k)$ and $G_1, \ldots, G_k$ are continuous $m \times m$ matrix functions on $[0, 1]$ such that

$$(4) \qquad G_j(0) = \Phi(\zeta_j), \qquad G_j(1) = \Phi(\zeta_j+), \qquad j = 1, \ldots, k.$$

In (3) the variable $\zeta \in \mathbb{T}$ and $\mu \in [0, 1]$. We define on $C^{m \times m}[\mathbb{T}[\zeta_1, \ldots, \zeta_k]]$ an involution by setting

$$(5) \qquad F^*(\zeta, \mu) = F(\zeta, \mu)^*, \qquad (\zeta, \mu) \in \mathbb{T} \times [0, 1].$$

Note that for F given by (3)

$$(6) \qquad F^*(\zeta, \mu) = \begin{cases} \Phi_*(\zeta) & \text{if } \zeta \neq \zeta_j \quad (j = 1, \ldots, k), \\ G_j(\mu)^* & \text{if } \zeta = \zeta_j. \end{cases}$$

It follows that F^* belongs to $C^{m \times m}[\mathbb{T}[\zeta_1, \ldots, \zeta_k]]$ indeed. With the usual algebraic operations, the involution defined by (5) and with the norm

$$(7) \qquad \|F\| = \max_{\zeta \in \mathbb{T}, 0 \leq \mu \leq 1} \|F(\zeta, \mu)\|$$

the set $C^{m\times m}\big[\mathbb{T}[\zeta_1,\ldots,\zeta_k]\big]$ is a unital C^*-algebra. The norm in the right hand side of (7) is the usual operator norm on $\mathcal{L}(\mathbb{C}^m)$. The unit in $C^{m\times m}\big[\mathbb{T}[\zeta_1,\ldots,\zeta_k]\big]$ is the $m\times m$ matrix function E which is equal to the $m\times m$ identity matrix at each point of $\mathbb{T}\times[0,1]$. We may also identify $C^{m\times m}\big[\mathbb{T}[\zeta_1,\ldots,\zeta_k]\big]$ with the algebra of all continuous $m\times m$ matrix functions on the deformed circle $\widetilde{\mathbb{T}}(\zeta_1,\ldots,\zeta_k)$ defined in Section XXV.1. We shall prove the following theorem.

THEOREM 3.1. *The C^*-algebra $\mathcal{T}_m(PC;\zeta_1,\ldots,\zeta_k)$ contains the set $\mathcal{K}$ of all compact operators on ℓ_2^m as a proper closed ideal and the quotient algebra*

$$\mathcal{T}_m(PC;\zeta_1,\ldots,\zeta_k)/\mathcal{K}$$

is isometrically $*$-isomorphic to the C^*-algebra $C^{m\times m}\big[\mathbb{T}[\zeta_1,\ldots,\zeta_k]\big]$.

To prove Theorem 3.1 we need the following lemma, which may be viewed as a corollary of the Fredholm theory developed in Section XXV.5. First a few preparations. Let S be a sum of products of block Toeplitz operators,

$$(8) \qquad\qquad S = \sum_{i=1}^{p} T_{\Phi_{i1}} T_{\Phi_{i2}} \cdots T_{\Phi_{iq}}$$

with $\Phi_{ij}\in PC^{m\times m}(\mathbb{T};\zeta_1,\ldots,\zeta_k)$ for each i and j. Recall that the symbol of S is the function

$$(9) \qquad\qquad \Omega(\zeta,\mu) = \sum_{i=1}^{p} \widehat{\Phi}_{i1}(\zeta,\mu)\widehat{\Phi}_{i2}(\zeta,\mu)\cdots\widehat{\Phi}_{iq}(\zeta,\mu)$$

where $\widehat{\Phi}_{ij}$ is the symbol associated to Φ_{ij} (see Sections XXV.3 and XXV.5). Obviously, Ω belongs to $C^{m\times m}\big[\mathbb{T}[\zeta_1,\ldots,\zeta_k]\big]$.

LEMMA 3.2. *Let S be a sum of products of block Toeplitz operators defined by functions from $PC^{m\times m}(\mathbb{T};\zeta_1,\ldots,\zeta_k)$, and let Ω be its symbol. Denote by $\mathcal{K}$ the set of all compact operators on ℓ_2^m. Then*

$$(10) \qquad\qquad \inf_{K\in\mathcal{K}}\|S+K\| = \max_{\zeta\in\mathbb{T},0\le\mu\le 1}\|\Omega(\zeta,\mu)\|.$$

PROOF. Let S be as in (8). Then

$$(11) \qquad\qquad S^* = \sum_{i=1}^{p} T_{\Psi_{iq}} T_{\Psi_{i,q-1}} \cdots T_{\Psi_{i1}},$$

with $\Psi_{ij} = (\Phi_{ij})_*$. It follows that SS^* is also a sum of products of block Toeplitz operators defined by functions from $PC^{m\times m}(\mathbb{T};\zeta_1,\ldots,\zeta_k)$ and its symbol is given by $\Omega\Omega^*$, where $*$ is the involution defined by (5). Given $T\in\mathcal{L}(\ell_2^m)$, let $[T]$ denote the coset $T+\mathcal{K}$. Consider the spectrum of $[SS^*]$ in the Calkin algebra $\mathcal{L}(\ell_2^m)/\mathcal{K}$. We write $r([SS^*])$ for the corresponding spectral radius. From Theorems XI.5.2 and XXV.5.1 it follows that $\lambda \in \sigma([SS^*])$ if and only if λ is an eigenvalue of $\Omega(\zeta_0, \mu_0)\Omega(\zeta_0, \mu_0)^*$ for some $(\zeta_0, \mu_0) \in \mathbb{T} \times [0, 1]$. Since Ω is an element of $C^{m \times m}[\mathbb{T}[\zeta_1, \ldots, \zeta_k]]$, the maximum in the right hand side of (10) is well-defined, and thus

$$(12) \qquad r([SS^*]) = \max_{\zeta \in \mathbb{T}, 0 \leq \mu \leq 1} \|\Omega(\zeta, \mu)\|^2.$$

Next we use the fact that the Calkin algebra $\mathcal{L}(\ell_2^m)/\mathcal{K}$ is a unital C^*-algebra. So Corollary XXXI.1.3 yields

$$(13) \qquad \|[S]\|^2 = \|[S][S]^*\| = \|[SS^*]\| = r([SS^*]).$$

According to the definition of the norm in the Calkin algebra,

$$(14) \qquad \inf_{K \in \mathcal{K}} \|S + K\| = \|[S]\|.$$

Together, formulas (12), (13) and (14) prove the lemma. $\square$

Lemma 3.2 allows us to define the symbol of an operator A in

$$\mathcal{T}_m(PC; \zeta_1, \ldots, \zeta_k)$$

as follows. Given A, we may choose a sequence $S_1, S_2, \ldots$ such that each S_j, $j = 1, 2, \ldots$, is a sum of products of block Toeplitz operators defined by functions from $PC^{m \times m}(\mathbb{T}; \zeta_1, \ldots, \zeta_k)$ and $S_n \to A$ $(n \to \infty)$ in the operator norm. Let Ω_j be the symbol of S_j for $1, 2, \ldots$ (see Section XXV.5). Formula (10) implies that the sequence $\Omega_1, \Omega_2, \ldots$ converges uniformly on $\mathbb{T} \times [0, 1]$ to a function Ω, say. We call Ω the *symbol* of A. Since each Ω_j, $j = 1, 2, \ldots$, is in $C^{m \times m}[\mathbb{T}[\zeta_1, \ldots, \zeta_k]]$, the same is true for Ω. From formula (10) we may also conclude that the definition of the symbol of A does not depend on the particular choice of the sequence $S_1, S_2, \ldots$. In particular, if A itself is a sum of products of block Toeplitz operators defined by functions from $PC^{m \times m}(\mathbb{T}; \zeta_1, \ldots, \zeta_k)$, then the symbol of A as defined above coincides with the symbol defined in Section XXV.5.

PROOF OF THEOREM 3.1. Since $\mathcal{T}_m(C) \subset \mathcal{T}_m(PC; \zeta_1, \ldots, \zeta_k)$, the set $\mathcal{K}$ is a proper closed ideal in $\mathcal{T}_m(PC; \zeta_1, \ldots, \zeta_k)$. Let $\mathcal{B}$ denote the corresponding quotient algebra. Since $K \in \mathcal{K}$ implies $K^* \in \mathcal{K}$, the algebra $\mathcal{B}$ is a unital C^*-algebra. We denote the coset $A + \mathcal{K}$ by $[A]$. Define

$$(15) \qquad \mathcal{J}: \mathcal{B} \to C^{m \times m}[\mathbb{T}[\zeta_1, \ldots, \zeta_k]], \qquad \mathcal{J}[A] = \Omega,$$

where Ω is the symbol of A. First let us show that $\mathcal{J}$ is well-defined. Take $\widetilde{A} \in [A]$, and let $\widetilde{\Omega}$ be the symbol of $\widetilde{A}$. We have to prove that $\Omega = \widetilde{\Omega}$. Choose a sequence $S_1, S_2, \ldots$ such that each S_j $(j = 1, 2, \ldots)$ is a sum of products of block Toeplitz operators defined by functions from $PC^{m \times m}(\mathbb{T}; \zeta_1, \ldots, \zeta_k)$ and $S_n \to A$ $(n \to \infty)$ in the operator norm. Let $\widetilde{S}_1, \widetilde{S}_2, \ldots$ be a corresponding sequence for $\widetilde{A}$. For each j let Ω_j be the symbol of S_j and $\widetilde{\Omega}_j$ for $\widetilde{S}_j$ $(j = 1, 2, \ldots)$. Then $\Omega_j - \widetilde{\Omega}_j$ is the symbol of $S_j - \widetilde{S}_j$. Note that $K := A - \widetilde{A} \in \mathcal{K}$ and $S_j - \widetilde{S}_j - K \to 0$ if $j \to \infty$. Formula (10) implies that

$$\max_{\zeta \in \mathbb{T}, 0 \leq \mu \leq 1} \|\Omega_j(\zeta, \mu) - \widetilde{\Omega}_j(\zeta, \mu)\| \leq \|S_j - \widetilde{S}_j - K\|.$$

Thus $\Omega_j - \widetilde{\Omega}_j \to 0$ uniformly on $\mathbf{T} \times [0,1]$, which implies that $\Omega = \widetilde{\Omega}$. Hence the map $\mathcal{J}$ in (15) is well-defined.

Let $\mathcal{B}_0$ be the subalgebra of $\mathcal{B}$ consisting of cosets that contain an operator which can be written as a sum of products of block Toeplitz operators defined by functions from $PC^{m \times m}(\mathbf{T}; \zeta_1, \ldots, \zeta_k)$. Obviously, $\mathcal{B}_0$ is dense in $\mathcal{B}$. From (11) it is clear that $\mathcal{B}_0$ is a $*$-subalgebra of $\mathcal{B}_0$. Lemma 3.2 tells us that

$$(16) \qquad \|\mathcal{J}[S]\| = \|[S]\|, \qquad [S] \in \mathcal{B}_0.$$

The first norm in (16) is the norm on $C^{m \times m}\big[\mathbf{T}[\zeta_1, \ldots, \zeta_k]\big]$, which is given by (7). From the definition of the symbol we know that for each $[A] \in \mathcal{B}$ there exists a sequence $[S_1], [S_2], \ldots$ in $\mathcal{B}_0$ such that $[S_n] \to [A]$ $(n \to \infty)$ in the norm of $\mathcal{B}$ and $\mathcal{J}[S_n] \to \mathcal{J}[A]$ $(n \to \infty)$ in the norm of $C^{m \times m}\big[\mathbf{T}[\zeta_1, \ldots, \zeta_k]\big]$. Thus (16) implies that

$$(17) \qquad \|\mathcal{J}[A]\| = \|[A]\|, \qquad [A] \in \mathcal{B},$$

and hence $\mathcal{J}$ is an isometry.

It is easily checked that $\mathcal{J}$ is a $*$-preserving homomorphism from $\mathcal{B}_0$ into $C^{m \times m}\big[\mathbf{T}[\zeta_1, \ldots, \zeta_k]\big]$. Using (17) we conclude that $\mathcal{J}$ is a $*$-isomorphism from $\mathcal{B}$ into $C^{m \times m}\big[\mathbf{T}[\zeta_1, \ldots, \zeta_k]\big]$. It remains to prove that $\mathcal{J}$ is onto. Note that $\operatorname{Im}\mathcal{J}$ is closed and contains all sum of products of function $\widehat{\Phi}$ with Φ from $PC^{m \times m}(\mathbf{T}; \zeta_1, \ldots, \zeta_k)$. So to prove that $\mathcal{J}$ is onto it suffices to prove the next lemma. $\square$

LEMMA 3.3.   *The set of all sums of products of symbols $\widehat{\Phi}$ with Φ from $PC^{m \times m}(\mathbf{T}; \zeta_1, \ldots, \zeta_k)$ is dense in $C^{m \times m}\big[\mathbf{T}[\zeta_1, \ldots, \zeta_k]\big]$.*

PROOF.   Let $\mathcal{D}$ be the set of all (scalar) functions $\widehat{\varphi}$ with φ from $PC(\mathbf{T}; \zeta_1, \ldots, \zeta_k)$, and define $\mathcal{A}(\mathcal{D})$ to be the smallest closed subalgebra of $C\big[\mathbf{T}[\zeta_1, \ldots, \zeta_k]\big]$ containing all functions from $\mathcal{D}$. Lemma XXV.1.3 tells us that $\mathcal{A}(\mathcal{D})$ is equal to $C\big[\mathbf{T}[\zeta_1, \ldots, \zeta_k]\big]$. Let $\mathcal{E}$ be the set of all functions $\widehat{\Phi}$ with Φ from $PC^{m \times m}(\mathbf{T}; \zeta_1, \ldots, \zeta_k)$, and define $\mathcal{A}(\mathcal{E})$ be the smallest closed subalgebra in $C^{m \times m}\big[\mathbf{T}[\zeta_1, \ldots, \zeta_k]\big]$ containing the functions from $\mathcal{E}$. Of course, $\mathcal{E}$ is equal to the set $\mathcal{D}^{m \times m}$ of all $m \times m$ matrix functions with entries from $\mathcal{D}$. So we can apply Lemma XXIX.8.2 to show that

$$\mathcal{A}(\mathcal{E}) = \left\{ \begin{bmatrix} \omega_{11} & \cdots & \omega_{1m} \\ \vdots & & \vdots \\ \omega_{m1} & \cdots & \omega_{mm} \end{bmatrix} \;\Big|\; \omega_{ij} \in \mathcal{A}(\mathcal{D}) \right\}.$$

Since $\mathcal{A}(\mathcal{D}) = C\big[\mathbf{T}[\zeta_1, \ldots, \zeta_k]\big]$, this proves the lemma. $\square$

COROLLARY 3.4.   *The Gelfand spectrum of the commutative unital C^*-algebra $\mathcal{T}(PC; \zeta_1, \ldots, \zeta_k)/\mathcal{K}$ is homeomorphic to the deformed circle $\widetilde{\mathbf{T}}(\zeta_1, \ldots, \zeta_k)$. Here $\mathcal{K}$ is the set of all compact operators on ℓ_2.*

PROOF.   Apply Theorem 3.1 with $m = 1$, and recall (see Section XXV.1) that $C\big[\mathbf{T}[\zeta_1, \ldots, \zeta_k]\big]$ may be identified with the Banach algebra of all continuous functions on $\widetilde{\mathbf{T}}(\zeta_1, \ldots, \zeta_k)$. $\square$

XXXII.4 ALGEBRAS OF TOEPLITZ OPERATORS DEFINED BY PIECEWISE CONTINUOUS MATRIX FUNCTIONS (GENERAL CASE)

Recall (see Section XXV.3) that $PC^{m \times m}(\mathbb{T})$ stands for the set of all $m \times m$ matrix functions Φ that are piecewise continuous on $\mathbb{T}$ and continuous from the left. In this section we do not require the functions to have a finite number of discontinuities. By $\mathcal{T}_m(PC)$ we denote the smallest closed subalgebra of $\mathcal{L}(\ell_2^m)$ containing all block Toeplitz operators T_Φ with Φ from $PC^{m \times m}(\mathbb{T})$. If $m = 1$, we shall drop the index m and just write $\mathcal{T}(PC)$. Endowed with the operator norm, $\mathcal{T}_m(PC)$ is a unital Banach algebra in its own right. We shall see later (see the proof of Theorem 4.2 below) that $\mathcal{T}_m(PC)$ is closed under taking adjoints, and hence $\mathcal{T}_m(PC)$ is a C^*-algebra. According to Theorem 2.1 the algebra $\mathcal{T}_m(PC)$ contains the set $\mathcal{K}$ of all compact operators on ℓ_2^m as a proper closed ideal. We shall study the quotient algebra $\mathcal{T}_m(PC)/\mathcal{K}$.

First a few preparations. With $\Phi \in PC^{m \times m}(\mathbb{T})$ we associate the function

$$(1) \qquad \widehat{\Phi}(\zeta, \mu) = \mu \Phi(\zeta+) + (1 - \mu)\Phi(\zeta), \qquad \zeta \in \mathbb{T}, \ 0 \leq \mu \leq 1.$$

We call $\widehat{\Phi}$ the *symbol* of Φ (or of the block Toeplitz operator defined by Φ). We would like to think about $\widehat{\Phi}$ as a continuous function on the cylinder $\mathbb{T} \times [0,1]$. But this is not true if $\mathbb{T} \times [0,1]$ is endowed with the usual (product) topology. Therefore we shall define a new topology on $\mathbb{T} \times [0,1]$.

Take $\zeta = e^{it} \in \mathbb{T}$ and $0 < \mu_0 < 1$. Furthermore, let $0 < \varepsilon < 1$. Consider the following sets:

$$O_\varepsilon(\zeta, 0) = \{(e^{is}, \lambda) \mid t - \varepsilon < s < t, 0 \leq \lambda \leq 1\} \cup \{(e^{it}, \lambda) \mid 0 \leq \lambda < \varepsilon\},$$
$$O_\varepsilon(\zeta, 1) = \{(e^{is}, \lambda) \mid t < s < t + \varepsilon, 0 \leq \lambda \leq 1\} \cup \{(e^{it}, \lambda) \mid 1 - \varepsilon < \lambda \leq 1\},$$
$$O_\varepsilon(\zeta, \mu_0) = \{(\zeta, \lambda) \mid 0 < \lambda < 1, |\lambda - \mu_0| < \varepsilon\}.$$

We shall refer to these sets as the ε-neighborhoods of $(\zeta, 0)$, $(\zeta, 1)$ and (ζ, μ_0), respectively. They are sketched in the pictures on the next page. The ε-neighborhoods form a basis for a topology on $\mathbb{T} \times [0,1]$ which we shall call the *PC-topology* on $\mathbb{T} \times [0,1]$. Thus a non-empty subset U of $\mathbb{T} \times [0,1]$ is open in the PC-topology if and only if for each (ζ, μ) in U there exists $\varepsilon > 0$ such that the ε-neighborhood $O_\varepsilon(\zeta, \mu)$ lies entirely in U. Obviously, the PC-topology is a Hausdorff topology.

Take $\Phi \in PC^{m \times m}(\mathbb{T})$, and let us check that $\widehat{\Phi}$ is continuous at $(\zeta_0, 0)$. Note that

$$(2) \qquad \widehat{\Phi}(\zeta, \mu) - \widehat{\Phi}(\zeta_0, 0) = \mu\big(\Phi(\zeta+) - \Phi(\zeta_0)\big) + (1 - \mu)\big(\Phi(\zeta) - \Phi(\zeta_0)\big).$$

If $(\zeta, \mu) \to (\zeta_0, 0)$ in the PC-topology, then $\zeta \to \zeta_0$ from the left and $\mu \to 0$. Hence the right hand side of (2) goes to zero, which proves that $\widehat{\Phi}$ is continuous at $(\zeta_0, 0)$. In a similar way one proves the continuity of $\widehat{\Phi}$ at other points of $\mathbb{T} \times [0,1]$ endowed with the PC-topology.

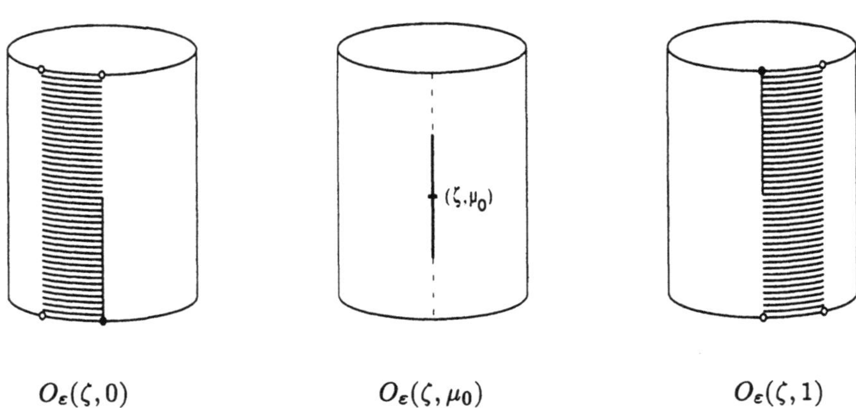

$$O_\varepsilon(\zeta, 0) \qquad\qquad O_\varepsilon(\zeta, \mu_0) \qquad\qquad O_\varepsilon(\zeta, 1)$$

LEMMA 4.1. *The set* $\mathbb{T} \times [0, 1]$ *endowed with the PC-topology is a compact Hausdorff space.*

PROOF. We already know that the PC-topology is a Hausdorff topology. Let us prove that $\mathbb{T} \times [0, 1]$ is compact in the PC-topology. Consider a family $\mathcal{P}$ of sets which are open in the PC-topology and cover $\mathbb{T} \times [0, 1]$. We have to show that $\mathcal{P}$ has a finite subset which also covers $\mathbb{T} \times [0, 1]$. By contradiction, suppose that such a finite subset does not exist. Consider the following closed arcs:

$$\Delta = \{e^{it} \mid -\pi \leq t \leq 0\}, \qquad \Delta' = \{e^{it} \mid 0 \leq t \leq \pi\}.$$

Since $\mathbb{T} \times [0, 1]$ is not covered by a finite subset of $\mathcal{P}$, the same is true for one (or perhaps for both) of the sets $\Delta \times [0, 1]$ and $\Delta' \times [0, 1]$. Put $\Delta_1 = \Delta$ if $\Delta \times [0, 1]$ is not covered by a finite subset of $\mathcal{P}$ and take $\Delta_1 = \Delta'$ otherwise. Now, repeat the argument for Δ_1. This yields an arc $\Delta_2 \subset \Delta_1$ of length $\frac{1}{2}\pi$ which is a closed subset of $\mathbb{T}$ (in the usual topology) such that $\Delta_2 \times [0, 1]$ is not covered by a finite subset of $\mathcal{P}$. Proceeding in this way we obtain a sequence $\Delta_1 \supset \Delta_2 \supset \Delta_3 \supset \cdots$ of arcs of $\mathbb{T}$, which are closed in the usual topology on $\mathbb{T}$, such that Δ_k has length $2^{-k+1}\pi$ and $\Delta_k \times [0, 1]$ is not covered by a finite subset of $\mathcal{P}$ $(k = 1, 2, \ldots)$. Let ζ_0 be the point on $\mathbb{T}$ which belongs to all arcs $\Delta_1, \Delta_2, \ldots$, and consider the set $\ell = \{\zeta_0\} \times [0, 1]$. Obviously, ℓ is covered by the family $\mathcal{P}$. Now use that the relative PC-topology on ℓ coincides with the usual product topology on ℓ. In the latter topology ℓ is compact. So there exist $U_1, \ldots, U_r$ in $\mathcal{P}$ such that ℓ is covered by the sets $U_1, \ldots, U_r$. In particular, there exist p and q in $\{1, 2, \ldots, r\}$ such that $(\zeta_0, 0) \in U_p$ and $(\zeta_0, 1) \in U_q$. Since U_p and U_q are open in the PC-topology, we can find $\varepsilon > 0$ such that $O_\varepsilon(\zeta_0, 0) \subset U_p$ and $O_\varepsilon(\zeta_0, 1) \subset U_q$. But this implies that for k sufficiently large the set $\Delta_k \times [0, 1]$ is covered by $U_1, \ldots, U_r$, which is impossible. So $\mathbb{T} \times [0, 1]$ is compact in the PC-topology. $\square$

Let us denote by $C^{m \times m}(\mathbb{T} \times [0, 1]; PC)$ the space of all $m \times m$ matrix functions that are continuous on $\mathbb{T} \times [0, 1]$ endowed with the PC-topology. With the usual algebraic operations and the norm defined by

$$(3) \qquad \|F\| = \max_{\zeta \in \mathbf{T}, 0 \le \mu \le 1} \|F(\zeta, \mu)\|$$

the space $C^{m \times m}(\mathbf{T} \times [0,1]; PC)$ is a unital C^*-algebra. The unit is the function on $\mathbf{T} \times [0,1]$ which is identically equal to the $m \times m$ identity matrix and the involution $*$ is defined by

$$(4) \qquad F^*(\zeta, \mu) = F(\zeta, \mu)^*, \qquad (\zeta, \mu) \in \mathbf{T} \times [0,1].$$

The norm in the right hand side of (3) is the usual operator norm on $\mathbb{C}^{m \times m}$. For $m = 1$ we drop the index m and write $C(\mathbf{T} \times [0,1]; PC)$ instead of $C^{1 \times 1}(\mathbf{T} \times [0,1]; PC)$.

THEOREM 4.2. *The C^*-algebra $\mathcal{T}_m(PC)$ contains the set $\mathcal{K}$ of all compact operators on ℓ_2^m as a proper closed ideal and the quotient algebra $\mathcal{T}_m(PC)/\mathcal{K}$ is isometrically $*$-isomorphic to the C^*-algebra $C^{m \times m}(\mathbf{T} \times [0,1]; PC)$.*

To prove Theorem 4.2 we use the following lemma which allows us to reduce the proof to the case considered in the preceding section.

LEMMA 4.3. *A continuous $m \times m$ matrix function on $\mathbf{T} \times [0,1]$ endowed with the PC-topology is the limit in the supremum norm of a sequence of sums of products of symbols $\widehat{\Phi}$ with Φ from $PC^{m \times m}(\mathbf{T})$ and Φ has a finite number of discontinuities.*

PROOF. First we consider the case $m = 1$. Let $\mathcal{D}$ be the family of all symbols $\widehat{\varphi}$ with φ from $PC(\mathbf{T})$ and φ has a finite number of discontinuities. From the remark made in the paragraph preceding Lemma 4.1 we know that $\mathcal{D}$ is contained in $C(\mathbf{T} \times [0,1]; PC)$. Let $\mathcal{A}_0(\mathcal{D})$ be the set of all sums of products of functions from $\mathcal{D}$, and let $\mathcal{A}(\mathcal{D})$ be the closure of $\mathcal{A}_0(\mathcal{D})$ in $C(\mathbf{T} \times [0,1]; PC)$. Of course, $\mathcal{A}(\mathcal{D})$ is the smallest closed subalgebra of $C(\mathbf{T} \times [0,1]; PC)$ containing the functions from $\mathcal{D}$. We have to show that $\mathcal{A}(\mathcal{D})$ is equal to $C(\mathbf{T} \times [0,1]; PC)$. To do this we apply the Stone-Weierstrass theorem (Theorem XXXI.2.1). It is easily checked that $\mathcal{A}(\mathcal{D})$ contains the constant function 1 and separates the points of $\mathbf{T} \times [0,1]$. In fact, $\mathcal{D}$ has already these two properties. Take $\omega \in \mathcal{A}_0(\mathcal{D})$. Then we can find $\zeta_1, \ldots, \zeta_k$ in $\mathbf{T}$, depending on ω, such that $\omega \in C\big[\mathbf{T}[\zeta_1, \ldots, \zeta_k]\big]$. It follows (cf., formulas (5) and (6) in the previous section) that

$$\overline{\omega} \in C\big[\mathbf{T}[\zeta_1, \ldots, \zeta_k]\big] \subset \mathcal{A}(\mathcal{D}),$$

where $\overline{\omega}$ is the complex conjugate of ω. Since $\overline{\mathcal{A}_0(\mathcal{D})} = \mathcal{A}(\mathcal{D})$, we conclude that $\mathcal{A}(\mathcal{D})$ is closed under complex conjugation. The Stone-Weierstrass Theorem implies that $\mathcal{A}(\mathcal{D}) = C(\mathbf{T} \times [0,1]; PC)$, and the lemma is proved for $m = 1$.

Take $m > 1$. Define $\mathcal{E}$ to be the family of all symbols $\widehat{\Phi}$ with Φ from $PC^{m \times m}(\mathbf{T})$ and Φ has a finite number of discontinuities. We know that $\mathcal{E}$ is contained in $C^{m \times m}(\mathbf{T} \times [0,1]; PC)$. Let $\mathcal{A}(\mathcal{E})$ be the smallest closed subalgebra of

$$C^{m \times m}(\mathbf{T} \times [0,1]; PC)$$

containing $\mathcal{E}$. We have to prove that $\mathcal{A}(\mathcal{E})$ is equal to $C^{m \times m}(\mathbf{T} \times [0,1]; PC)$. To do this, note that $\mathcal{E}$ is precisely the set $\mathcal{D}^{m \times m}$ of all $m \times m$ matrix functions with entries from

$\mathcal{D}$, where $\mathcal{D}$ is as in the first part of the proof. But then we can apply Lemma XXIX.8.2 to show that

$$\mathcal{A}(\mathcal{E}) = \left\{ \begin{bmatrix} \omega_{11} & \cdots & \omega_{1m} \\ \vdots & & \vdots \\ \omega_{m1} & \cdots & \omega_{mm} \end{bmatrix} \mid \omega_{ij} \in \mathcal{A}(\mathcal{D}) \right\}.$$

Since $\mathcal{A}(\mathcal{D}) = C(\mathbb{T} \times [0,1]; PC)$, this implies that $\mathcal{A}(\mathcal{E}) = C^{m \times m}(\mathbb{T} \times [0,1]; PC)$, and the lemma is proved. $\square$

PROOF OF THEOREM 4.2. Let T be a block Toeplitz operator defined by a function Φ from $PC^{m \times m}(\mathbb{T})$. Then $\widehat{\Phi}$ is continuous on $\mathbb{T} \times [0,1]$ endowed with the PC-topology. So there exist a sequence $\Omega_1, \Omega_2, \ldots$ of $m \times m$ matrix functions on $\mathbb{T} \times [0,1]$ such that $\Omega_\nu \to \widehat{\Phi}$ ($\nu \to \infty$) uniformly on $\mathbb{T} \times [0,1]$ and for each ν the function Ω_ν is a sum of products of functions $\widehat{\Psi}$ with Ψ from $PC^{m \times m}(\mathbb{T})$ and Ψ has a finite number of discontinuities. For $\nu = 1, 2, \ldots$ put $\Phi_\nu(\zeta) = \Omega_\nu(\zeta, 0)$. Then $\Phi_\nu \in PC^{m \times m}(\mathbb{T})$ and Φ_ν has a finite number of discontinuities for each ν. Let T_ν be the block Toeplitz operator on ℓ_2^m defined by Φ_ν. From

$$\begin{aligned} \|T - T_\nu\| &= \operatorname*{ess\,sup}_{-\pi \leq t \leq \pi} \|\Phi(e^{it}) - \Phi_\nu(e^{it})\| \\ &\leq \sup_{\zeta \in \mathbb{T}} \|\Phi(\zeta) - \Phi_\nu(\zeta)\| \\ &= \sup_{\zeta \in \mathbb{T}} \|\widehat{\Phi}(\zeta, 0) - \Omega_\nu(\zeta, 0)\| \\ &\leq \max_{\zeta \in \mathbb{T}, 0 \leq \mu \leq 1} \|\widehat{\Phi}(\zeta, \mu) - \Omega_\nu(\zeta, \mu)\|, \end{aligned}$$

it follows that $T_\nu \to T$ if $\nu \to \infty$.

Let $\mathcal{C}_0$ be the subset of $\mathcal{T}_m(PC)$ consisting of all sums of products of block Toeplitz operators defined by functions from $PC^{m \times m}(\mathbb{T})$ with a finite number of discontinuities. Of course, $\mathcal{C}_0$ is a subalgebra of $\mathcal{T}_m(PC)$, and the same is true for its closure $\overline{\mathcal{C}}_0$. The remark made in the preceding paragraph implies that all block Toeplitz operators defined by functions from $PC^{m \times m}(\mathbb{T})$ are in $\overline{\mathcal{C}}_0$. We conclude that $\mathcal{C}_0$ is dense in $\mathcal{T}_m(PC)$.

Take $S \in \mathcal{C}_0$. Thus

$$(5) \qquad\qquad S = \sum_{i=1}^{p} T_{\Phi_{i1}} T_{\Phi_{i2}} \cdots T_{\Phi_{iq}}$$

with $\Phi_{ij} \in PC^{m \times m}(\mathbb{T})$ and Φ_{ij} has a finite number of discontinuities for each i and j. It follows that we can find $\zeta_1, \ldots, \zeta_k$ in $\mathbb{T}$ such that all Φ_{ij} are in $PC^{m \times m}(\mathbb{T}; \zeta_1, \ldots, \zeta_k)$. This allows us to apply the result of the previous section to S. Note that formula (11) in the previous section implies that S^* is also in $\mathcal{C}_0$. Thus $\mathcal{C}_0$ is closed under the operation of taking adjoints. Since $\mathcal{C}_0$ is dense in $\mathcal{T}_m(PC)$, we see that $\mathcal{T}_m(PC)$ is a C^*-algebra.

Let $\mathcal{B}$ be the quotient algebra $\mathcal{T}_m(PC)/\mathcal{K}$, and given $T \in \mathcal{T}_m(PC)$ let $[T]$

denote the coset $T + \mathcal{K}$. Put

$$\mathcal{B}_0 = \{[S] \mid S \in \mathcal{C}_0\}.$$

Then $\mathcal{B}_0$ is a $*$-subalgebra of $\mathcal{B}$ which is dense in $\mathcal{B}$. Define

$$\mathcal{J} : \mathcal{B}_0 \to C^{m \times m}(\mathbb{T} \times [0,1]; PC), \qquad \mathcal{J}[S] = \Omega,$$

where Ω is the symbol of S. From the results of the previous section we know that $\mathcal{J}$ is a well-defined isometric $*$-isomorphism from $\mathcal{B}_0$ onto a $*$-subalgebra of

$$C^{m \times m}(\mathbb{T} \times [0,1]; PC).$$

By continuity $\mathcal{J}$ extends to an isometric $*$-isomorphism, also denoted by $\mathcal{J}$, from $\mathcal{B}$ onto a closed $*$-subalgebra of $C^{m \times m}(\mathbb{T} \times [0,1]; PC)$. Lemma 4.3 implies that $\mathcal{J}[\mathcal{B}_0]$ is dense in $C^{m \times m}(\mathbb{T} \times [0,1]; PC)$. Hence $\mathcal{J}$ is surjective and the theorem is proved. $\square$

For $m = 1$ Theorem 4.2 implies that $\mathcal{T}(PC)/\mathcal{K}$ is commutative and the Gelfand spectra of $\mathcal{T}(PC)/\mathcal{K}$ and $C(\mathbb{T} \times [0,1]; PC)$ are homeomorphic. This yields the following corollary.

COROLLARY 4.3.   *The Gelfand spectrum of the commutative unital C^*-algebra $\mathcal{T}(PC)/\mathcal{K}$ is homeomorphic to the space $\mathbb{T} \times [0,1]$ endowed with the PC-topology. Here $\mathcal{K}$ is the set of all compact operators on ℓ_2.*

If we identify the Gelfand spectrum of $\mathcal{T}(PC)/\mathcal{K}$ with $\mathbb{T} \times [0,1]$ endowed with the PC-topology, then the map $\mathcal{J}$ defined in the proof of Theorem 4.2 may be identified with the Gelfand representation of $\mathcal{T}(PC)/\mathcal{K}$.

XXXII.5   THE FREDHOLM INDEX

Let A be an operator in the C^*-algebra $\mathcal{T}_m(PC)$. By definition the *symbol* of A is the $m \times m$ matrix function F on $\mathbb{T} \times [0,1]$ given by $F = \mathcal{J}[A]$ where $\mathcal{J}$ is the isometric $*$-isomorphism,

$$(1) \qquad \qquad \mathcal{J} : \mathcal{T}_m(PC)/\mathcal{K} \to C^{m \times m}(\mathbb{T} \times [0,1]; PC)$$

defined in the proof of Theorem 4.2. If A is an operator in $\mathcal{T}_m(PC; \zeta_1, \ldots, \zeta_k)$, then the symbol of A (as defined above) coincides with the symbol of A introduced in Section XXXII.3. In this section we shall prove that A is a Fredholm operator if and only if the determinant of its symbol F does not vanish on $\mathbb{T} \times [0,1]$, and we shall identify the Fredholm index of A in terms of a winding number associated with $\det F$. Our first concern will be to define this winding number.

Let $\mathcal{A}_0$ be the family of scalar functions that are equal to a sum of products of functions $\hat{\varphi}$ with φ from $PC(\mathbb{T})$ and φ has a finite number of discontinuities. Of course, $\mathcal{A}_0$ is an algebra which, by Lemma 4.3, is dense in $C(\mathbb{T} \times [0,1]; PC)$. Take $\omega \in \mathcal{A}_0$, and assume $\omega(\zeta, \mu) \neq 0$ for $(\zeta, \mu) \in \mathbb{T} \times [0,1]$. From the definition of $\mathcal{A}_0$ it follows that we can find $\zeta_1, \ldots, \zeta_k$ in $\mathbb{T}$ (depending on ω) such that ω is a sum of products of functions $\hat{\varphi}$ with φ from $PC(\mathbb{T}; \zeta_1, \ldots, \zeta_k)$. So we know from Section XXV.1 that the winding number

$n(\omega; 0)$ of ω relative to 0 is well-defined. Moreover, the definition does not depend on the choice of $\zeta_1, \ldots, \zeta_k$. If ω_1, ω_2 are in $\mathcal{A}_0$, then

$$(2) \quad |\omega_2(\zeta, \mu) - \omega_1(\zeta, \mu)| < |\omega_1(\zeta, \mu)| \qquad (\zeta \in \mathbb{T}, 0 \le \mu \le 1) \Longrightarrow n(\omega_1; 0) = n(\omega_2; 0).$$

To see this, choose $\zeta_1, \ldots, \zeta_k$ in $\mathbb{T}$ such that both ω_1 and ω_2 are equal to a sum of products of functions $\widehat{\varphi}$ with φ from $PC(\mathbb{T}; \zeta_1, \ldots, \zeta_k)$ and apply the results of Section XXV.1. The property described in (2) allows us to extend the definition of the winding number of functions that are continuous on $\mathbb{T} \times [0, 1]$ endowed with the PC-topology.

Take $f \in C(\mathbb{T} \times [0, 1]; PC)$, and assume that $f(\zeta, \mu) \ne 0$ for (ζ, μ) in $\mathbb{T} \times [0, 1]$. Since $\mathcal{A}_0$ is dense in $C(\mathbb{T} \times [0, 1]; PC)$, there exists a sequence $\omega_1, \omega_2, \ldots$ in $\mathcal{A}_0$ such that $\omega_j \to f$ $(j \to \infty)$ uniformly on $\mathbb{T} \times [0, 1]$. Because of our hypothesis on f we may assume that $\omega_j(\zeta, \mu)$ does not vanish on $\mathbb{T} \times [0, 1]$ for $j = 1, 2, \ldots$. Hence $n(\omega_j; 0)$ is well-defined for each j. We set

$$(3) \qquad\qquad n(f; 0) = \lim_{j \to \infty} n(\omega_j; 0),$$

and we call $n(f; 0)$ the *winding number of f relative to* 0. Formula (2) implies that the limit in the right hand side of (3) exists and does not depend on the special choice of the sequence $\omega_1, \omega_2, \ldots$.

THEOREM 5.1. *Let A be an operator in the C^*-algebra $\mathcal{T}_m(PC)$, and let F be its symbol. Then A is a Fredholm operator if and only if $\det F(\zeta, \mu) \ne 0$ for all $\zeta \in \mathbb{T}$ and $0 \le \mu \le 1$. In that case*

$$(4) \qquad\qquad \operatorname{ind} A = -n(\det F; 0).$$

PROOF. Let $\mathcal{K}$ denote the set of all compact operators on ℓ_2^m. Given T in $\mathcal{L}(\ell_2^m)$ we write $[T]$ for the coset $T + \mathcal{K}$. Let $A \in \mathcal{T}_m(PC)$. First, note that $\mathcal{T}_m(PC)/\mathcal{K}$ is a closed $*$-subalgebra of the Calkin algebra $\mathcal{L}(\ell_2^m)/\mathcal{K}$ and both algebras have the same unit. So Theorems XXXI.3.3 and XI.5.2 imply that A is a Fredholm operator if and only if $[A]$ is invertible in the quotient algebra $\mathcal{T}_m(PC)/\mathcal{K}$. But then we can use the isometric $*$-isomorphism $\mathcal{J}$ introduced in the proof of Theorem 4.2 (see also formula (1)) to show that A is Fredholm if and only if $F = \mathcal{J}[A]$ is invertible in $C^{m \times m}(\mathbb{T} \times [0, 1]; PC)$. Since the latter property is equivalent to $\det F(\zeta, \mu) \ne 0$ for each $(\zeta, \mu) \in \mathbb{T} \times [0, 1]$, the first part of the theorem is proved.

Next, assume that A is Fredholm. We have to prove the index formula (4). Let $\mathcal{C}_0$ be the subset of $\mathcal{T}_m(PC)$ consisting of all sums of products of block Toeplitz operators defined by functions from $PC^{m \times m}(\mathbb{T})$ with a finite number of discontinuities. We know that $\mathcal{C}_0$ is dense in $\mathcal{T}_m(PC)$, and hence there exists a sequence $S_1, S_2, \ldots$ in $\mathcal{C}_0$ such that $S_\nu \to A$ $(\nu \to \infty)$ in the operator norm. Without loss of generality we may assume (see Theorem XI.4.1) that S_ν is a Fredholm operator and $\operatorname{ind} S_\nu = \operatorname{ind} A$ for each ν. Let Ω_ν be the symbol of S_ν $(\nu = 1, 2, \ldots)$. Thus $\Omega_\nu = \mathcal{J}[S_\nu]$, where $\mathcal{J}$ is the isometric $*$-isomorphism defined in the proof of Theorem 4.2. It follows that

$$F = \mathcal{J}(A) = \lim_{\nu \to \infty} \mathcal{J}(S_\nu) = \lim_{\nu \to \infty} \Omega_\nu.$$

In other words, the sequence $\Omega_1, \Omega_2, \ldots$ converges to F uniformly on $\mathbb{T} \times [0, 1]$. But then $\det F$ is the limit in the supremum norm of the sequence $\det \Omega_1, \det \Omega_2, \ldots$. Thus $n(\det F; 0) = n(\det \Omega_\nu; 0)$ for ν sufficiently large (see formula (2)). From Theorem XXV.5.1 we know that $\operatorname{ind} S_\nu = -n(\det \Omega_\nu; 0)$. Thus (4) also holds true. $\square$

COMMENTS ON PART VIII

In this part the main emphasis is on the theory of commutative Banach algebras and commutative C^*-algebras. This theory, which has its origin in papers of I.M. Gelfand, M.A. Naimark, G.E. Shilov and others produced in the forties, is presented here as an important tool in operator theory. Also some results of the non-commutative theory are included; the latter concern mainly algebras of matrices over commutative Banach algebras. With the exception of the last two sections, Chapter XXIX contains standard elements of the general theory. The factorization theorem in the last section is a special case of a more general result in Gohberg-Kreĭn [4] (see also Section II.1 in Clancey-Gohberg [1]), which has many predecessors with less sharp evaluations of the size of the perturbation (see Section IV.4 in Gohberg-Kreĭn [4] for some historical remarks). Theorem XXIX.8.1 is due to Krupnik [1]. Section XXIX.6 follows the corresponding material in Gohberg-Krupnik [2]. Sections 1–8 in Chapter XXX contain the basic theory of commutative Banach algebras. Some examples, for instance those about piecewise continuous functions and about the algebra generated by a compact operator, may be difficult to find in other texts. The last two sections of Chapter XXX, which concern applications to systems of Wiener-Hopf integral equations, consist of basic material in the latter area and have their roots in Gohberg-Kreĭn [2] (see also Gohberg-Fel'dman [1]). Chapter XXXI contains the standard theory of commutative C^*-algebras; the general Gelfand-Naimark theorem about representations of C^*-algebras as algebras of operators is also included. As an application of the general theory the spectral theorem of normal operators is derived and the Pták-Vroba description (see Pták-Vroba [1]) of the range of the spectral projection is given (see also Putnam [1] for a more general version of the latter result). The proofs of Lemma XXXI.7.8 and Theorem XXXI.7.9 are due to R.W. Whitley [1]. Chapter XXXII contains applications of Banach algebra theory to algebras generated by Toeplitz and block Toeplitz operators. The first two sections have their origin in Gohberg [1], which deals with the singular integral version; see also Coburn [1], [2] and Gohberg [2]. The results in Sections XXXII.3–XXXII.5 are due to Gohberg-Krupnik [2]. For further developments in this area we refer the reader to Böttcher-Silbermann [1] and Krupnik [1].

EXERCISES TO PART VIII

1. Let $\mathcal{A}$ be an algebra satisfying all the axioms of a Banach algebra except the requirement that $\mathcal{A}$ has a unit. Let $\mathcal{B} = \mathbb{C} \times \mathcal{A}$. Define the operations on $\mathcal{B}$ by

$$(\alpha, x) + (\beta, y) = (\alpha + \beta, x + y)$$

$$\lambda(\alpha, x) = (\lambda\alpha, \lambda x)$$

$$(\alpha, x)(\beta, y) = (\alpha\beta, \alpha y + \beta x + xy)$$

$$\|(\alpha, x)\| = |\alpha| + \|x\|.$$

Show that $\mathcal{B}$ is a Banach algebra with unit and that $\mathcal{A}$ is isometrically isomorphic to a subalgebra of codimension 1 in $\mathcal{B}$.

2. Let $\mathcal{M}$ be the set of doubly infinite matrices $A = (a_{ij})_{i,j=-\infty}^{\infty}$ with the property that

$$(1) \qquad \|A\| = \sum_{r=-\infty}^{\infty} \sup_j |a_{j,j+r}| < \infty.$$

Show that $\mathcal{M}$, with the usual definitions of matrix multiplication and addition, together with the norm defined by (1) is a Banach algebra.

3. Let $\{\mathcal{A}_\alpha \mid \alpha \in S\}$ be a family of Banach algebras. Define $\mathcal{B}$ to be the set of functions f on S such that $F(\alpha) \in \mathcal{A}_\alpha$ and

$$\|f\| = \sup_{\alpha \in S} \|f(\alpha)\| < \infty.$$

(a) Show that $\mathcal{B}$ with the natural operations of addition, multiplication and scalar multiplication for functions is a Banach algebra with the above norm.

(b) Take S to be the set of positive integers and define

$$M = \big\{ f \in \mathcal{B} \mid \lim_{n \to \infty} \|f(n)\| = 0 \big\}.$$

Determine whether or not M is a maximal ideal in $\mathcal{B}$.

In the next seven exercises $\mathcal{A}$ is a unital Banach algebra with unit e.

4. Assume $x \in \mathcal{A}$ is left invertible, but not right invertible. Show that the left inverse of $\mathcal{A}$ is not unique.

5. Let x be a left invertible element in $\mathcal{A}$, and let y be an arbitrary element in $\mathcal{A}$. Determine the largest radius $\rho > 0$ such that $x - \lambda y$ is left invertible for all $|\lambda| < \rho$.

6. Let

(2)
$$X = \begin{bmatrix} a & 0 \\ c & d \end{bmatrix} \in \mathcal{A}^{2\times 2},$$

and assume X is invertible in $\mathcal{A}^{2\times 2}$. Show that a is right invertible and d is left invertible in $\mathcal{A}$.

7. Give an example of a unital Banach algebra $\mathcal{A}$ and an invertible element X as in (2) so that a is not left invertible and d is not right invertible in $\mathcal{A}$.

8. Derive a formula for the radius of convergence of a power series with coefficients in $\mathcal{A}$.

9. Let x and y be arbitrary elements of $\mathcal{A}$. How are the spectra of xy and yx related? What do you know about the spectral radii of xy and yx?

10. Let x be an element of $\mathcal{A}$, and let $p(\lambda)$ be a polynomial with complex coefficients. Show that

$$\sigma\big(p(x)\big) = \{p(z) \mid z \in \sigma(z)\}.$$

What do you know about $\sigma(x)$ when $p(x) = 0$?

11. Let $S = \{\lambda \in \mathbb{C} \mid 1 \leq |\lambda| \leq 2\}$. Let $\mathcal{A}$ be the smallest closed subalgebra of $C(S)$ which contains 1 and the function $f(\lambda) = \lambda$. Describe the spectrum $\sigma_{\mathcal{A}}(f)$. Next, describe the spectrum $\sigma_{\mathcal{B}}(1/f)$, where $\mathcal{B}$ is the smallest closed subalgebra of $C(S)$ generated by 1 and $1/f$.

12. Let $\mathcal{A} = C'([0,1])$ be the subalgebra of $C([0,1])$ consisting of the continuously differentiable complex-valued functions on $[0,1]$. Consider $\mathcal{A}$ with the norm

$$\|f\| = \max_{0\leq t\leq 1} |f(t)| + \max_{0\leq t\leq 1} |f'(t)|.$$

Show that $\mathcal{A}$ is a semi-simple commutative Banach algebra and find its maximal ideal space.

13. Let $\mathcal{A} = C^{(n)}([0,1])$ be the subalgebra of $C([0,1])$ consisting of the n-times continuously differentiable complex-valued functions. Consider $\mathcal{A}$ with the norm

$$\|f\| = \max_{0\leq t\leq 1} \sum_{k=0}^{n} \frac{|f^{(k)}(t)|}{k!}.$$

Show that $\mathcal{A}$ is a commutative Banach algebra and find its maximal ideal space.

14. Let $\mathcal{A}$ be the algebra of all functions on $\mathbb{R}^2$ of the form

$$f(t,s) = \sum_{m,n=-\infty}^{\infty} a_{mn} e^{i(mt+ns)},$$

endowed with the norm

$$\|f\| = \sum_{m,n=-\infty}^{\infty} |a_{mn}|.$$

(a) Show that $\mathcal{A}$ is a commutative Banach algebra.

(b) Show that the Gelfand spectrum of $\mathcal{A}$ can be identified with the two-dimensional torus

$$\mathbb{T}^2 = \{(e^{it}, e^{is}) \mid t, s \in \mathbb{R}\}.$$

15. Let $\mathcal{A} = C(S)$ with S given by

$$S = \{\lambda \in \mathbb{C} \mid |\lambda| = 1\} \cup \{\lambda \in \mathbb{C} \mid |\lambda| = 2\}.$$

Describe the Gelfand spectrum and the Gelfand representation of the closed subalgebra of $\mathcal{A}$ generated by the functions f, g (and h) for each of the following cases:

(a) $f(z) = 1$, $g(z) = z$;

(b) $f(z) = 1$, $g(z) = z^{-1}$;

(c) $f(z) = z$, $g(z) = z^{-1}$;

(d) $f(z) = z$, $g(z) = (z - \alpha)^{-1}$, where $1 < |\alpha| < 2$;

(e) $f(z) = z$, $g(z) = (z - \alpha)^{-1}$, $h(z) = (z - \beta)^{-1}$, where $\alpha, \beta \in \mathbb{R}$;

(f) $f(z) = 1$, $g(z) = (z - \alpha)^{-1}$, $h(z) = (z - \beta)^{-1}$, where $\alpha, \beta \in \mathbb{R}$.

Note that in problems (e) and (f) the answer will depend on the location of α and β on the real line. One is asked to consider all possible locations.

16. Let $\mathcal{A}$ be a commutative unital Banach algebra with unit e, and suppose that $A = \mathcal{A}_1 + \mathcal{A}_2$, where $\mathcal{A}_1$, $\mathcal{A}_2$ are closed subalgebras of $\mathcal{A}$ with $\mathcal{A}_1 \cap \mathcal{A}_2 = \mathrm{span}\{e\}$. Let $\mathcal{M}_i$ be the set of maximal ideals of $\mathcal{A}_i$, $i = 1, 2$. Show that the set of maximal ideals of $\mathcal{A}$ can be identified with $\mathcal{M}_1 \times \mathcal{M}_2$.

17. Let $\mathcal{A} = C(S)$, where S is a square in $\mathbb{R}^2$. Let $\mathcal{A}_1$ be the subalgebra of $\mathcal{A}$ of functions $f = f(x, y)$ which depend on x only, and let $\mathcal{A}_2$ be the subalgebra of $\mathcal{A}$ of functions $f = f(x, y)$, which depend on y only.

(a) Show that the set of maximal ideals of $\mathcal{A}$ can be identified with $\mathcal{M}_1 \times \mathcal{M}_2$, where $\mathcal{M}_i$ is the set of maximal ideals of $\mathcal{A}_i$, $i = 1, 2$.

(b) If the square is replaced by a circle, is the above conclusion still true?

18. Let $\mathcal{A}$ be the algebra of all functions f on $\mathbb{R}$ of the form

$$f(\lambda) = d + \int_0^\infty e^{i\lambda t} k(t)\, dt, \qquad \lambda \in \mathbb{R},$$

where d is an arbitrary complex number and k is an arbitrary element of $L_1([0, \infty))$. The norm on $\mathcal{A}$ is defined by

$$\|f\| = |d| + \int_0^\infty |k(t)|\, dt.$$

Show that $\mathcal{A}$ is a commutative Banach algebra and find its Gelfand spectrum.

19. Let $\mathcal{B}$ be the algebra of all $m \times m$ matrix functions F on $\mathbb{R}$ of the form

$$F(\lambda) = D + \int_0^\infty e^{i\lambda t} K(t)dt, \qquad \lambda \in \mathbb{R},$$

where D is an arbitrary $m \times m$ complex matrix and K is an arbitrary $m \times m$ matrix function with entries in $L_1([0,\infty))$. Given $F \in \mathcal{B}$, find necessary and sufficient conditions for F to be invertible in $\mathcal{B}$, and derive a formula for F^{-1}.

20. Let $\mathcal{A}$ be a commutative unital Banach algebra, and let Γ be its Gelfand representation. Assume $\mathcal{A}$ has an involution $*$. Show

(a) $\|\Gamma x\| = \|\Gamma x^*\|$;

(b) if $x^* = x$, then $\|x\| = \|\Gamma x\|$;

(c) if $\|x^* x\| = \|x^*\|\|x\|$ for each $x \in \mathcal{A}$, then $\mathcal{A}$ is a C^*-algebra (hint: consider $x = u^* u$).

21. Show that any non-empty compact subset of $\mathbb{C}$ can appear as the spectrum of a normal operator.

22. Fix a positive real number ε, and let $\ell_2(\varepsilon)$ be the Hilbert space of all sequences $x = (x_0, x_1, x_2, \ldots)$ with complex entries such that

$$\|x\| = \left(\sum_{j=0}^\infty |\varepsilon^j x_j|^2\right)^{1/2} < \infty.$$

Let $\mathcal{A}$ be the smallest closed subalgebra of $\mathcal{L}(\ell_2(\varepsilon))$ containing all operators T on $\ell_2(\varepsilon)$ of the form

$$(Tx)_i = \sum_{\substack{j=-N+i \\ j \geq 0}}^{N+i} \tau_{i-j} x_j, \qquad i = 0, 1, 2, \ldots ,$$

where N is a positive integer depending on T. Show that $\mathcal{A}$ is a C^*-algebra with respect to the operator norm, and develop for $\mathcal{A}$ the analogue of the theory in Section XXXII.1.

23. Let P be the orthogonal projection of $\ell_2(\mathbb{Z})$ defined by

$$P(\ldots, x_{-1}, x_0, x_1, \ldots) = (\ldots, 0, 0, x_0, x_1, \ldots).$$

Let $\mathcal{A}$ be the smallest closed subalgebra of $\mathcal{L}(\ell_2(\mathbb{Z}))$ containing P and all Laurent operators on $\ell_2(\mathbb{Z})$ defined by continuous functions on $\mathbb{T}$.

(a) Show that $\mathcal{A}$ is a C^*-algebra with respect to the operator norm.

(b) Prove that $\mathcal{A}$ contains the set $\mathcal{K}$ of all compact operators on $\ell_2(\mathbb{Z})$.

(c) Prove that $\mathcal{A}/\mathcal{K}$ is a commutative unital C^*-algebra.

(d) Find the Gelfand spectrum and Gelfand representation of $\mathcal{A}/\mathcal{K}$.

(e) Let L_φ and L_ψ be the Laurent operators defined by the continuous functions φ and ψ. Determine the Gelfand transform of $L_\varphi P + L_\psi Q$, where $Q = I - P$.

24. Let $\mathcal{A}$ be the set of all 2×2 matrix functions M on $S = \{(t, \mu) \mid a \leq t \leq b,\ 0 \leq \mu \leq 1\}$ defined by

$$M(t, \mu) = \begin{bmatrix} a(t+)\mu + a(t)(1 - \mu) & \sqrt{\mu(1 - \mu)}\,[a(t+) - a(t)] \\ \sqrt{\mu(1 - \mu)}\,[a(t+) - a(t)] & a(t+)(1 - \mu) + a(t)\mu \end{bmatrix},$$

where a is an arbitrary function in $PC([a, b])$. The norm on $\mathcal{A}$ is given by

$$(3) \qquad \|\|M\|\| = \sup_{(t,\mu) \in S} \|M(t, \mu)\|,$$

where the norm in the right hand side of (3) is the norm of a 2×2 matrix as an operator on $\mathbb{C}^2$.

(a) Prove that $\mathcal{A}$ is a commutative unital C^*-algebra.

(b) Prove that $\mathcal{A}$ is isometrically $*$-isomorphic to $PC([a, b])$.

(c) Generalize the above for the case when a is an $m \times m$ matrix function with entries in $PC([a, b])$.

PART IX

EXTENSION AND COMPLETION PROBLEMS

This part presents an operator theoretical method for solving a general class of extension and completion problems. It contains the abstract theory, which is based on ideas and results of various previous parts in the present volume, and applications of the abstract theory to a number of matrix-valued versions of classical interpolation problems, such as those of Carathéodory-Toeplitz and Nehari. There are three chapters. The first treats extension problems for finite matrices by a one-step method. This chapter serves as an introduction, and it provides the motivation for the more advanced topics in the other two chapters. The second chapter presents the band method, a general scheme, which allows one to treat problems of extension and completion with constraints from one point of view. The last chapter specifies the results for finite operator matrices and contains the applications referred to above. Here also a tangential matrix-version of the Nevanlinna-Pick interpolation problem is treated.

CHAPTER XXXIII

COMPLETIONS OF MATRICES

This chapter, which has an introductory character, deals with completion problems for finite matrices with scalar entries. Here we introduce the simplest versions of the problems that will be treated in the next chapters with more powerful tools. This first chapter presents in a nutshell some of the main features of completion problems, and the results obtained here serve as a motivation for the more sophisticated theorems that will follow later. This chapter consists of three sections. In the first section we analyze the possibilities of one step extensions for invertible selfadjoint matrices. In the second section the results are developed further for the positive definite case. Here notions such as band extension and maximum entropy solution appear. In the last section analogous results are derived for contractive completion problems by reduction to the positive definite case.

XXXIII.1 ONE STEP EXTENSIONS

In this section we deal with the following completion problem. Consider the matrix

$$
(1) \qquad A(z) = \begin{bmatrix}
a_{00} & a_{01} & \cdots & a_{0,n-1} & \overline{z} \\
a_{10} & a_{11} & \cdots & a_{1,n-1} & a_{1n} \\
\vdots & \vdots & & \vdots & \vdots \\
a_{n-1,0} & a_{n-1,1} & \cdots & a_{n-1,n-1} & a_{n-1,n} \\
z & a_{n1} & \cdots & a_{n,n-1} & a_{nn}
\end{bmatrix}.
$$

Here the entries a_{ij} are given complex numbers such that $a_{ji} = \overline{a_{ij}}$ for $|i - j| \leq n - 1$, and z is an unknown. So we may think of $A(z)$ as a partially given selfadjoint matrix which we want to complete. In fact, the problem treated in the present section asks to determine $z \in \mathbb{C}$ in such a way that the resulting selfadjoint completion $A(z)$ is invertible. In this case we shall refer to $A(z)$ as a *one-step extension* of $A := A(0)$. We shall give the solution of the above problem assuming additionally that the following three submatrices are non-singular:

$$
(2) \qquad L = [a_{ij}]_{i,j=0}^{n-1}, \qquad M = [a_{ij}]_{i,j=1}^{n-1}, \qquad R = [a_{ij}]_{i,j=1}^{n}.
$$

In the next section we shall be interested in one-step extensions that are positive definite. If such an extension exists, then the three matrices L, M and R in (2) are positive definite, and hence in that case the invertibility condition on the matrices in (2) is natural.

THEOREM 1.1. *Let $A = A(0) = [a_{ij}]_{i,j=0}^{n}$ be a selfadjoint matrix with $a_{n0} = \overline{a_{0n}} = 0$, and assume that the submatrices L, M and R defined in (2) are non-singular. Let $A(z)$ be as in (1).*

(α) *If* $\det L$ *and* $\det R$ *have opposite signs, then* $A(z)$ *is invertible for each* $z \in \mathbb{C}$. *Furthermore, in this case,* $\det A(z)$ *and* $\det M$ *have opposite signs too.*

(β) *If* $\det L$ *and* $\det R$ *have the same sign, then* $A(z)$ *is invertible for all* $z \in \mathbb{C}$ *except those on the circle with radius* ρ *and center* z_0, *where*

$$(3a) \qquad \rho = \frac{\sqrt{(\det L)(\det R)}}{|\det M|},$$

$$(3b) \qquad z_0 = -\frac{1}{y_0} \sum_{j=1}^{n-1} a_{nj} y_j,$$

with

$$(4) \qquad \begin{bmatrix} y_0 \\ y_1 \\ \vdots \\ y_{n-1} \end{bmatrix} = L^{-1} \begin{bmatrix} 1 \\ 0 \\ \vdots \\ 0 \end{bmatrix}.$$

PROOF. We split the proof into four parts. In the following, z is an arbitrary complex number.

Part (a). We may partition $A(z)$ as follows:

$$(5) \qquad A(z) = \begin{bmatrix} L & c(z)^* \\ c(z) & a_{nn} \end{bmatrix},$$

where $c(z) = [z \quad a_{n1} \cdots a_{n,n-1}]$. Since $\det L \neq 0$, the partitioning in (5) allows us to produce an LU-factorization of $A(z)$, namely

$$(6) \qquad A(z) = \begin{bmatrix} I & 0 \\ c(z)L^{-1} & 1 \end{bmatrix} \begin{bmatrix} L & 0 \\ 0 & a_{nn} - c(z)L^{-1}c(z)^* \end{bmatrix} \begin{bmatrix} I & L^{-1}c(z)^* \\ 0 & 1 \end{bmatrix}.$$

Here I is the $n \times n$ identity matrix. Take the determinant of the left and of the right hand side of $A(z)$. This yields

$$\det A(z) = \left(a_{nn} - c(z)L^{-1}c(z)^*\right) \det L.$$

Rewrite $c(z)^*$ as follows:

$$c(z)^* = \bar{z}e_0 + c^*, \qquad e_0 = \begin{bmatrix} 1 \\ 0 \\ \vdots \\ 0 \end{bmatrix}, \qquad c^* = \begin{bmatrix} 0 \\ a_{1n} \\ \vdots \\ a_{n-1,n} \end{bmatrix},$$

and put

$$\alpha = \langle L^{-1}e_0, e_0 \rangle, \qquad \beta = \langle L^{-1}e_0, c^* \rangle, \qquad \gamma = \langle L^{-1}c^*, c^* \rangle,$$

$$\delta = \frac{a_{nn} - \gamma}{\alpha} + \frac{|\beta|^2}{\alpha^2}.$$

By Cramer's rule, $\alpha = \det M / \det L \neq 0$. Also, α is real, because M and L are selfadjoint. Now,

$$\begin{aligned}
c(z)L^{-1}c(z)^* &= \langle L^{-1}c(z)^*, c(z)^* \rangle \\
&= \langle L^{-1}(\bar{z}e_0 + c^*), (\bar{z}e_0 + c^*) \rangle \\
&= \alpha z\bar{z} + \bar{z}\beta + z\bar{\beta} + \gamma \\
&= \alpha|z + \frac{\beta}{\alpha}|^2 + \gamma - \frac{|\beta|^2}{\alpha}.
\end{aligned}$$

So we have proved that

$$(7) \qquad \frac{\det A(z)}{\det M} = \left(q_{nn} - c(z)L^{-1}c(z)^*\right)\frac{\det L}{\det M} = \delta - |z + \frac{\beta}{\alpha}|^2.$$

This holds for each $z \in \mathbb{C}$. If the number $\delta < 0$, then $A(z)$ is invertible for each $z \in \mathbb{C}$; if $\delta \geq 0$, then $A(z)$ is invertible for all $z \in \mathbb{C}$ except those on the circle with center $z_0 = -\alpha^{-1}\beta$ and radius $\rho = \sqrt{\delta}$.

      Part (b). In this part we prove (3b) and derive an auxiliary formula. Let $y_0, y_1, \ldots, y_{n-1}$ be defined by (4). Note that $y_0 = \langle L^{-1}e_0, e_0 \rangle = \alpha$, and so $y_0 \neq 0$. Furthermore,

$$\begin{aligned}
z_0 = -\alpha^{-1}\beta &= -\frac{1}{y_0}\langle L^{-1}e_0, c^* \rangle \\
&= -\frac{1}{y_0}\langle \begin{bmatrix} y_0 \\ y_1 \\ \vdots \\ y_{n-1} \end{bmatrix}, \begin{bmatrix} 0 \\ a_{1n} \\ \vdots \\ a_{n-1,n} \end{bmatrix} \rangle \\
&= -\frac{1}{y_0}\sum_{j=1}^{n-1} y_j\overline{a_{jn}} = -\frac{1}{y_0}\sum_{j=1}^{n-1} a_{nj}y_j,
\end{aligned}$$

which proves (3b). From (3b) it follows that

$$c(z)\begin{bmatrix} y_0 \\ \vdots \\ y_{n-1} \end{bmatrix} = zy_0 + \sum_{j=1}^{n-1} a_{nj}y_j = (z - z_0)y_0,$$

and thus

$$(8) \qquad A(z)\begin{bmatrix} y_0 \\ \vdots \\ y_{n-1} \\ 0 \end{bmatrix} = \begin{bmatrix} 1 \\ 0 \\ \vdots \\ 0 \\ (z - z_0)y_0 \end{bmatrix}.$$

Part (c). In this part we replace the partitioning (5) by

$$(9) \qquad A(z) = \begin{bmatrix} a_{00} & b(z) \\ b(z)^* & R \end{bmatrix},$$

where $b(z) = [a_{01} \cdots a_{0,n-1} \ \bar{z}]$. Since R is invertible, the matrix $A(z)$ has a UL-factorization, and we can repeat the arguments of Part (a) to show that

$$(10) \qquad \frac{\det A(z)}{\det M} = \widetilde{\delta} - |z + \widetilde{\alpha}^{-1}\widetilde{\beta}|^2,$$

where

$$\widetilde{\alpha} = \left\langle R^{-1} \begin{bmatrix} 0 \\ \vdots \\ 0 \\ 1 \end{bmatrix}, \begin{bmatrix} 0 \\ \vdots \\ 0 \\ 1 \end{bmatrix} \right\rangle, \qquad \widetilde{\beta} = \left\langle \begin{bmatrix} a_{10} \\ \vdots \\ a_{n-1,0} \\ 0 \end{bmatrix}, R^{-1} \begin{bmatrix} 0 \\ \vdots \\ 0 \\ 1 \end{bmatrix} \right\rangle,$$

and $\widetilde{\delta}$ is a complex number which we shall not specify further. By comparing the right hand sides of (10) and (7), we see that z_0 in (3b) is also given by $-\widetilde{\alpha}^{-1}\widetilde{\beta}$. Let $x_1, \ldots, x_n$ be defined by

$$(11) \qquad \begin{bmatrix} x_1 \\ \vdots \\ x_n \end{bmatrix} = R^{-1} \begin{bmatrix} 0 \\ \vdots \\ 0 \\ 1 \end{bmatrix}.$$

Note that $x_n = \widetilde{\alpha} \neq 0$. By using the same arguments as in Part (b) one now shows that

$$(12) \qquad z_0 = -\widetilde{\alpha}^{-1}\widetilde{\beta} = -\frac{1}{x_n} \overline{\sum_{j=1}^{n-1} a_{0j} x_j}.$$

It follows that

$$b(z) \begin{bmatrix} x_1 \\ \vdots \\ x_n \end{bmatrix} = \sum_{j=1}^{n-1} a_{0j} x_j + \bar{z} x_n = (\bar{z} - \bar{z}_0) x_n,$$

and so we derive the following analogue of (8):

$$(13) \qquad A(z) \begin{bmatrix} 0 \\ x_1 \\ \vdots \\ x_n \end{bmatrix} = \begin{bmatrix} (\bar{z} - \bar{z}_0)x_n \\ 0 \\ \vdots \\ 0 \\ 1 \end{bmatrix}.$$

Part (d). In this part we derive an alternative expression for the right hand side of (7). From (8) and (13) we see that the following identities hold:

$$(14a) \qquad A(z)\left\{ \begin{bmatrix} y_0 \\ \vdots \\ y_{n-1} \\ 0 \end{bmatrix} - (z - z_0)y_0 \begin{bmatrix} 0 \\ x_1 \\ \vdots \\ x_n \end{bmatrix} \right\} = \begin{bmatrix} r(z) \\ 0 \\ \vdots \\ 0 \end{bmatrix},$$

$$
(14b) \qquad A(z)\left\{ -(\bar{z} - \bar{z}_0)x_n \begin{bmatrix} y_0 \\ \vdots \\ y_{n-1} \\ 0 \end{bmatrix} + \begin{bmatrix} 0 \\ x_1 \\ \vdots \\ x_n \end{bmatrix} \right\} = \begin{bmatrix} 0 \\ \vdots \\ 0 \\ r(z) \end{bmatrix},
$$

where

$$
(15) \qquad r(z) = 1 - |z - z_0|^2 y_0 x_n.
$$

If $A(z)$ is invertible, then (14a) and (14b) yield

$$
(16) \qquad \frac{\det A(z)}{\det R} = \frac{1}{y_0} r(z), \qquad \frac{\det A(z)}{\det L} = \frac{1}{x_n} r(z).
$$

By continuity, it follows that the identities in (16) hold for each $z \in \mathbb{C}$. Recall that

$$
(17) \qquad y_0 = \frac{\det M}{\det L}, \qquad x_n = \frac{\det M}{\det R}.
$$

By using the latter three identities, we obtain that

$$
(18) \qquad \frac{\det A(z)}{\det M} = \frac{1}{y_0 x_n} - |z - z_0|^2.
$$

> Part (e). In this part we prove the theorem. From (17) we see that

$$
(19) \qquad \frac{1}{y_0 x_n} = \frac{(\det L)(\det R)}{|\det M|^2}.
$$

If $\det L$ and $\det R$ have opposite signs, then the left hand side of (19) is strictly negative, and hence the same is true for the right hand side of (18). It follows that in this case $\det A(z) \neq 0$ for each $z \in \mathbb{C}$. Next, assume that $\det L$ and $\det R$ have the same sign. Then we see from (19) that $(y_0 x_n)^{-1} > 0$, and hence $\det A(z) \neq 0$ for all z except those on the circle with center z_0 and radius

$$
\rho = \sqrt{\frac{1}{y_0 x_n}} = \frac{\sqrt{(\det L)(\det R)}}{|\det M|},
$$

which completes the proof. $\quad \Box$

> For later purposes we note that if $\det L$ and $\det R$ in Theorem 1.1 have the same sign, then $\det A(z)$ is given by the following expression:

$$
(20) \qquad \det A(z) = \frac{(\det L)(\det R)}{\det M}\left(1 - \frac{|z - z_0|^2}{\rho^2}\right),
$$

where ρ and z_0 are given by (3a) and (3b), respectively. Formula (20) is an immediate corollary of (17), (18), and the formula for ρ.

> Let $A = A(0)$ be as in Theorem 1.1. The number z_0 given by (3b) is called the *center of extension* of A. The *radius of extension* ρ of A is defined by (3a) if $\det L$

and $\det R$ have the same sign and otherwise is defined to be ∞. The extension $A(z_0)$ is called the *central (one step) extension* of A.

COROLLARY 1.2. *Let $A = A(0) = [a_{ij}]_{i,j=0}^n$ be a selfadjoint matrix with $a_{n0} = \overline{a}_{0n} = 0$, and such that the submatrices L, M and R in (2) are invertible. Then the central extension $A(z_0)$ is the unique one-step extension of A with the property that the $(n,0)$-th entry of $A(z_0)^{-1}$ is equal to zero.*

PROOF. Formula (8) shows that the $(n,0)$-th entry of $A(z_0)^{-1}$ is equal to zero. Conversely, assume that the $(n,0)$-th entry of $A(z)^{-1}$ is zero. We have to show $z = z_0$. Let $r(z)$ be as in (15). Since $A(z)$ is invertible, we have $r(z) \neq 0$, and hence (14a) shows that the $(n,0)$-th entry of $A(z)^{-1}$ is given by $-(z - z_0)y_0 x_n r(z)^{-1}$. According to our assumption, the latter number is zero. This can only happen when $z = z_0$. $\square$

XXXIII.2 POSITIVE COMPLETIONS

Let $0 \le m < n$ be integers. An $(n+1) \times (n+1)$ matrix $A = [a_{ij}]_{i,j=0}^n$ is called an *m-band matrix* if $a_{ij} = 0$ for $|i - j| > m$. Let A be such a matrix. We call an $(n+1) \times (n+1)$ matrix $F = [f_{ij}]_{i,j=0}^n$ an *extension* (or *completion*) of A if

$$(1) \qquad f_{ij} = a_{ij}, \qquad |i - j| \le m.$$

The one-step extensions considered in the previous section are extensions of $(n-1)$-band matrices. In this section we are interested in *positive extensions* of A, i.e., in extensions of A that are positive definite.

Recall that a $k \times k$ matrix F is said to be *positive definite* (notation: $F > 0$) if F is selfadjoint and all the eigenvalues of F are strictly positive. In other words, F is positive definite if (and only if) the operator on $\mathbb{C}^k$ induced by the canonical action of F on the standard basis of $\mathbb{C}^k$ is a strictly positive (or positive definite) operator. Here $\mathbb{C}^k$ is assumed to be endowed with its standard Hilbert space structure. From the operator description of positive definiteness it follows that the *central submatrices* (i.e., the square submatrices that are located symmetrically around the main diagonal) of a positive definite matrix are again positive definite.

THEOREM 2.1. *The m-band matrix $A = [a_{ij}]_{i,j=0}^n$ has a positive extension if and only if*

$$(2) \qquad \begin{bmatrix} a_{ii} & \cdots & a_{i,m+i} \\ \vdots & & \vdots \\ a_{m+i,i} & \cdots & a_{m+i,m+i} \end{bmatrix} > 0, \qquad i = 0, \ldots, n - m.$$

In this case there exists a unique positive extension $C = [c_{ij}]_{i,j=0}^n$ of A with the additional property that for all $m + 1 \le t \le n$ and $0 \le s \le t - m - 1$ the submatrix

$$(3) \qquad C([s,t]) = \begin{bmatrix} c_{ss} & \cdots & c_{st} \\ \vdots & & \vdots \\ c_{ts} & \cdots & c_{tt} \end{bmatrix}$$

is the central one-step extension of the corresponding $(t - s - 1)$-band matrix.

PROOF. Assume that A has a positive extension, F say. From the remarks made in the paragraph preceding the theorem we know that the central submatrices of F are positive definite. Since F is an extension of A, the matrices in (2) are central submatrices of F, and hence they have to be positive definite.

Next, consider the case when (2) is fulfilled, and let us construct a positive extension C with the stated properties. First we consider the case $m = n - 1$. In this case we take $C = A(z_0)$, where $A(z_0)$ is the central one-step extension of the matrix $A = A(0)$. We have only to show that $A(z_0)$ is well-defined and positive definite. To do this put

$$L = [a_{ij}]_{i,j=0}^{n-1}, \qquad M = [a_{ij}]_{i,j=1}^{n-1}, \qquad R = [a_{ij}]_{i,j=1}^{n}.$$

Condition (2) implies that the matrices L, M and R are positive definite. In particular, L, M and R are invertible, and, by Theorem 1.1, the central one-step extension $A(z_0)$ of $A = A(0)$ exists. From formulas (17) and (18) in the previous section we see that

$$(4) \qquad \det A(z_0) = \frac{(\det L)(\det R)}{\det M} > 0,$$

because L, M and R are positive definite. Now, recall (see (6) in the previous section) that $A(z_0)$ factors as

$$(5) \qquad A(z_0) = \begin{bmatrix} I & 0 \\ e^* & 1 \end{bmatrix} \begin{bmatrix} L & 0 \\ 0 & d \end{bmatrix} \begin{bmatrix} I & e \\ 0 & 1 \end{bmatrix}$$

for some $e \in \mathbb{C}^n$ and some $d \in \mathbb{C}$. Since L is positive definite, (4) implies that $d > 0$, and hence the second factor in the right hand side of (5) is a positive definite matrix. But then (5) shows that $A(z_0) > 0$. We have now proved the theorem for $n - m = 1$.

We treat the general case by induction on $n - m$. Fix $0 \leq m < m + 1 < n$, and assume that the theorem is proved for $n - m - 1$. Consider the matrices

$$A\big([i, m + i + 1]\big) = [a_{jk}]_{j,k=i}^{m+i+1}, \qquad i = 0, \ldots, m - n - 1.$$

Note that $A\big([i, m+i+1]\big)$ is an m-band matrix of size $m + 1$. Furthermore, condition (2) holds for $A\big([i, m + i + 1]\big)$ in place of A. So, by the result of the previous paragraph, the central one-step extension of $A\big([i, m + i + 1]\big)$ exists and is positive definite. Let z_i be the corresponding center of extension. Introduce a new selfadjoint matrix $\widetilde{A} = [\alpha_{jk}]_{j,k=0}^{n}$ by setting

$$\alpha_{jk} = a_{jk}, \qquad |j - k| \leq m,$$
$$\alpha_{jk} = \overline{\alpha}_{kj} = \overline{z}_j, \qquad k - j = m + 1,$$
$$\alpha_{jk} = 0, \qquad |j - k| > m + 1.$$

Note that $\widetilde{A}$ is an $(m+1)$-band matrix. Furthermore, condition (2) holds for $\widetilde{A}$ in place of A. In fact the matrix $[\alpha_{jk}]_{j,k=i}^{m+i+1}$ is positive definite, because this matrix is precisely the central one-step extension of $A([i, m+i+1])$. Here $i = 0, \ldots, n-m-1$. It follows that our inductive hypothesis applies to $\widetilde{A}$. Let C be the corresponding positive extension, with the additional properties as stated in the theorem (with $\widetilde{A}$ in place of A). Since $\widetilde{A}$ is an extension of A, the same holds true for C. From the construction of $\widetilde{A}$ and the properties of C derived so far it is clear that C is the desired positive extension. The uniqueness of C is an immediate consequence of the uniqueness of the central one-step extension. $\square$

The positive extension C with the additional properties stated in Theorem 2.1 is called the *central extension* of A. From its properties (see also the last paragraph of the proof of Theorem 2.1) it is clear that the central extension is obtained by repeatedly applying central one-step extensions.

A positive extension B of the m-band matrix A is called a *band extension* if B^{-1} is an m-band matrix, i.e., the (i,j)-th entry of B^{-1} is zero if $|i-j| \geq m+1$. Note that, by definition, a band extension is positive definite.

THEOREM 2.2. *Let $A = [a_{ij}]_{i,j=0}^{n}$ be an m-band matrix such that (2) holds. Then the central extension of A is a band extension, and A has no other band extensions.*

PROOF. We shall first prove, by induction on n, that the central extension C of A is a band extension. For $n = m+1$ the result follows from Corollary 1.2. Next, take $n > m+1$, and let us assume that the result holds for an m-band matrix of size $n \times n$. Let Γ be the $n \times n$ matrix in the left upper corner of C. Note that Γ is positive definite and an extension of the m-band matrix $A([0, n-1]) = [a_{ij}]_{i,j=0}^{n-1}$. From the properties of the central extension (mentioned in Theorem 2.1) it follows that Γ is the central extension of $A([0, n-1])$. The latter matrix has size $n \times n$, an so, by our induction hypothesis, Γ^{-1} is an m-band matrix. A similar reasoning may be applied to the $n \times n$ matrix Δ in the right lower corner of C. This matrix Δ is the central extension of the m-band matrix $A([1, n]) = [a_{ij}]_{i,j=1}^{n}$, and hence, by our induction hypotheses, Δ^{-1} is an m-band matrix. Now, $C = [c_{ij}]_{i,j=0}^{n}$ is the central one-step extension of the matrix $C(0)$, which is obtained from C by replacing the entries c_{0n} and c_{n0} by zero. It follows that we can use formula (8) in the previous section to show that

$$\begin{bmatrix} y_0 \\ \vdots \\ y_{n-1} \end{bmatrix} = \Gamma^{-1} \begin{bmatrix} 1 \\ 0 \\ \vdots \\ 0 \end{bmatrix} \Rightarrow C \begin{bmatrix} y_0 \\ \vdots \\ y_{n-1} \\ 0 \end{bmatrix} = \begin{bmatrix} 1 \\ 0 \\ \vdots \\ 0 \end{bmatrix}.$$

Since Γ^{-1} is an m-band matrix, we have $y_j = 0$ for $j = m+1, \ldots, n-1$. Next, note that

$$C \begin{bmatrix} 0 & 0 \\ 0 & \Delta^{-1} \end{bmatrix} = \begin{bmatrix} 0 & a \\ 0 & I \end{bmatrix},$$

where a is some $1 \times n$ matrix and I is the $n \times n$ identity matrix. It follows that

$$C \begin{bmatrix} y_0 \\ \vdots \\ y_m \\ 0 \\ \vdots \\ 0 \end{bmatrix} \begin{bmatrix} 1 & -a \end{bmatrix} = \begin{bmatrix} 1 & -a \\ 0 & 0 \\ \vdots & \vdots \\ 0 & 0 \end{bmatrix},$$

and hence

$$(6) \qquad C\left\{ \begin{bmatrix} 0 & 0 \\ 0 & \Delta^{-1} \end{bmatrix} + \begin{bmatrix} y_0 \\ \vdots \\ y_m \\ 0 \\ \vdots \\ 0 \end{bmatrix} \begin{bmatrix} 1 & -a \end{bmatrix} \right\} = \begin{bmatrix} 1 & 0 \\ 0 & I \end{bmatrix}.$$

Thus the factor between brackets in the left hand side of (6) is equal to C^{-1}. Since Δ^{-1} is an m-band matrix and C is selfadjoint, we see that C^{-1} is an m-band matrix. Thus C is a band extension of A.

Next, we prove the uniqueness of the band extension. So, let B be a second band extension of A. Since both B and C are positive definite, we may construct factorizations

$$B = U^* D U, \qquad C = L^* E L,$$

where U is an $(n+1) \times (n+1)$ upper triangular matrix with diagonal entries equal to one, L is an $(n+1) \times (n+1)$ lower triangular matrix with diagonal entries equal to one, and D and E are diagonal matrices with positive diagonal entries. Note that

$$(7) \qquad U^{-1} = B^{-1} U^* D.$$

Since $U^* D$ is lower triangular and B^{-1} is an m-band matrix, it follows that U^{-1} is an upper triangular m-band matrix. In a similar way one shows that L^{-1} is a lower triangular m-band matrix. Consider the index set

$$\mathcal{E} = \left\{ (i,j) \mid i - j \geq m, \; 0 \leq i \leq n, \; 0 \leq j \leq n \right\}.$$

Since B and C are positive extensions of the same m-band matrix, we may write $B - C = R + R^*$, where R is an $(n+1) \times (n+1)$ matrix such that the (i,j)-th entry of R is zero whenever $(i,j) \in \mathcal{E}$. Now $C^{-1} - B^{-1} = B^{-1}(R + R^*)C^{-1}$, and hence

$$(8) \qquad U(C^{-1} - B^{-1})L^* = R_1 + S_1,$$

where $R_1 = D^{-1}(U^*)^{-1} R L^{-1} E^{-1}$ and $S_1 = D^{-1}(U^*)^{-1} R^* L^{-1} E^{-1}$. Since L^{-1} and $(U^*)^{-1}$ are lower triangular m-band matrices and R^* is upper triangular, the (i,j)-th entry of S_1 is zero if $(i,j) \notin \mathcal{E}$. The left hand side of (8) has the same property. Indeed, $C^{-1} - B^{-1}$ is an m-band matrix, and the matrices U and L^* are upper triangular. So the (i,j)-th entry of $U(C^{-1} - B^{-1})L^*$ is zero if $(i,j) \notin \mathcal{E}$. But then we see from (8) that the (i,j)-th entry of R_1 is zero if $(i,j) \notin \mathcal{E}$. On the other hand, since L^{-1} and $(U^*)^{-1}$ are lower triangular, R_1 is a matrix of the same type as R, i.e., the (i,j)-th entry of R_1 is zero if $(i,j) \in \mathcal{E}$. So we have proved that $R_1 = 0$. It follows that $R = 0$, and hence $B = C$. $\square$

Theorem 2.2 yields the following *permanence principle* for band extensions.

COROLLARY 2.3. *Let* $A = [a_{ij}]_{i,j=0}^{n}$ *be an* m-*band matrix satisfying condition* (2), *and let* $B = [b_{ij}]_{i,j=0}^{n}$ *be its band extension. Then for all* $m + 1 \leq t \leq n$ *and* $0 \leq s \leq t - m - 1$ *the matrix* $B([s,t]) = [b_{ij}]_{i,j=s}^{t}$ *is the band extension of the* m-*band matrix* $A([s,t]) = [a_{ij}]_{i,j=s}^{t}$.

PROOF. By Theorem 2.2 it suffices to prove the corollary with "central extension" in place of "band extension". But for central extensions the result is true by definition. $\square$

Let $A = [a_{ij}]_{i,j=0}^{n}$ be an m-band matrix satisfying condition (2). The central extension C has another interesting property, namely among all positive extensions of A the central extension is the one with maximal determinant. In other words, if F is a positive extension of A, then

$$(9) \qquad\qquad \det F \leq \det C$$

and equality holds in (9) if and only if $C = F$. For the case when $m = n - 1$ this so-called *maximum entropy principle* is an immediate consequence of formula (20) in the previous section. (Recall that under condition (2) the numbers $\det L$, $\det R$ and $\det M$ appearing in formula (20) of the previous section are all strictly positive.) The proof for the general case will be given in Section XXXV.1 using the abstract maximum entropy principle derived in Section XXXIV.4.

As with the central extension, arbitrary positive extensions of A may also be obtained by repeatedly applying one-step extensions. In fact, in this way a full description of all positive extensions may be obtained (see Woerdeman [1]). In the next chapter we shall develop another, more transparent and powerful method, which yields a linear fractional description of all positive extensions in terms of UL- and LU-factorizations of the central extension.

XXXIII.3 STRICTLY CONTRACTIVE COMPLETIONS

Consider the matrix

$$(1) \qquad\qquad \Phi = \begin{bmatrix} \varphi_{11} & \cdots & \varphi_{1r} \\ \vdots & & \vdots \\ \varphi_{n1} & \cdots & \varphi_{nr} \end{bmatrix}.$$

Fix an integer p such that $-n < p < r$. In this section we assume that the entries φ_{ij} with $1 \leq i \leq n$, $1 \leq j \leq r$ and $j - i \leq p$ are given, and the problem is to determine the remaining entries in such a way that the norm of the complete matrix is strictly less than one. Recall that the norm of an $n \times r$ matrix Φ is equal to the norm of the operator from $\mathbb{C}^r$ into $\mathbb{C}^n$ induced by the canonical action of Φ on the standard bases of $\mathbb{C}^r$. Here $\mathbb{C}^r$ and $\mathbb{C}^n$ are endowed with their standard Hilbert space structure.

We say that Φ in (1) is *p-lower triangular* if for $1 \leq i \leq n$, $1 \leq j \leq r$ and $j - i > p$ the entries φ_{ij} are zero. An $n \times r$ matrix $\Psi = [\psi_{ij}]_{i=1,j=1}^{n \ \ r}$ is called an *extension* (or *completion*) of the p-lower triangular matrix Φ if

$$(2) \qquad \psi_{ij} = \varphi_{ij}, \qquad 1 \leq i \leq n, \ 1 \leq j \leq r, \ j - i \leq p.$$

The *strictly contractive* extensions are, by definition, the extensions with norm strictly less than one.

The strictly contractive extension problem may be reduced to a positive extension problem. For this purpose we need the following notation. Given an $n \times r$ matrix Ψ we write A_Ψ for the $(n+r) \times (n+r)$ matrix defined by

$$(3) \qquad A_\Psi = \begin{bmatrix} I_n & \Psi \\ \Psi^* & I_r \end{bmatrix}.$$

Here I_n and I_r are identity matrices of sizes $n \times n$ and $r \times r$, respectively.

LEMMA 3.1. *Let Φ be a p-lower triangular $n \times r$ matrix. Then A_Φ is an $(n+p)$-band matrix, and Ψ is a strictly contractive extension of Φ if and only if A_Ψ is a positive extension of A_Φ. Furthermore, for $\max\{0, r-n\} \leq p < r$ any positive extension F of A_Φ is of the form A_Ψ for some strictly contractive extension Ψ of Φ.*

PROOF. Obviously, A_Φ is an $(n+p)$-band matrix. Note that A_Ψ factorizes as

$$(4) \qquad A_\Psi = \begin{bmatrix} I_n & 0 \\ \Psi^* & I_r \end{bmatrix} \begin{bmatrix} I_n & 0 \\ 0 & I_r - \Psi^*\Psi \end{bmatrix} \begin{bmatrix} I_n & \Psi \\ 0 & I_r \end{bmatrix}.$$

It follows that A_Ψ is positive definite if and only if Ψ is a strict contraction. Here we used the fact that $\|\Psi\| < 1$ is equivalent to the requirement that $I_r - \Psi^*\Psi$ is positive definite. From (3) we see that A_Ψ is an extension of A_Φ if and only if Ψ is an extension of Φ. It remains to prove the last part of the lemma. The additional condition $p \geq \max\{0, r-n\}$ guarantees that any extension of the $(n+p)$-band matrix A_Φ is of the form A_Ψ for some extension Ψ of Φ, and hence we can use the first part of the lemma to complete the proof. $\square$

THEOREM 3.2. *Let $\Phi = [\varphi_{ij}]_{i=1,j=1}^{n \ \ r}$ be a p-lower triangular matrix. Then Φ has a strictly contractive extension if and only if the matrices*

$$(5) \qquad \Gamma_k = \begin{bmatrix} \varphi_{k,1} & \cdots & \varphi_{k,k+p} \\ \vdots & & \vdots \\ \varphi_{n,1} & \cdots & \varphi_{n,k+p} \end{bmatrix}, \qquad \max\{1, 1-p\} \leq k \leq \min\{n, r-p\}$$

have norm strictly less than one. In this case there exists a unique strictly contractive extension Δ of Φ such that the (i,j)-th entry of $\Delta(I - \Delta^*\Delta)^{-1}$ is zero for $j - i > p$.

PROOF. If Φ has a strictly contractive extension, then, obviously, the matrices Γ_k in (5) have norm strictly less than one (because they are compressions of strict contractions). So in the following we assume that $\|\Gamma_k\| < 1$ for the given k, and we prove that Φ has a unique strictly contractive extension Δ with the additional properties stated in the theorem.

First, we assume that $\max\{0, r - n\} \le p \le r$. Then k in (5) runs from $1, \ldots, r - p$. Put

$$(6) \qquad A_\Phi = \begin{bmatrix} I_n & \Phi \\ \Phi^* & I_r \end{bmatrix},$$

$$(7) \qquad A_{\Gamma_k} = \begin{bmatrix} I_{n-k-1} & \Gamma_k \\ \Gamma_k^* & I_{k+p} \end{bmatrix}, \qquad k = 1, \ldots, r - p.$$

The matrix A_Φ is an $(n + p)$-band matrix. According to our hypotheses, A_{Γ_k} is positive definite for $k = 1, \ldots, r - p$, and it follows that condition (2) in Theorem 2.1 is fulfilled for the case considered here. So, by Theorem 2.2, the matrix A_Φ has a band extension, B say. But then we can apply Lemma 3.1 to show that $B = A_\Delta$ for some strictly contractive extension Δ of Φ. We claim that Δ has the desired additional properties. To see this, we use (4) to compute A_Δ^{-1}. It follows that

$$(8) \qquad A_\Delta^{-1} = \begin{bmatrix} I + \Delta(I - \Delta^*\Delta)^{-1}\Delta^* & -\Delta(I - \Delta^*\Delta)^{-1} \\ -(I - \Delta^*\Delta)^{-1}\Delta^* & (I - \Delta^*\Delta)^{-1} \end{bmatrix}.$$

Now $A_\Delta^{-1} = B^{-1}$ is an $(n+p)$-band matrix. It follows that for $j - i > p$ the (i, j)-th entry of $\Delta(I - \Delta^*\Delta)^{-1}$ is zero. Conversely, if the latter holds, then A_Δ^{-1} is an $(n + p)$-band matrix, and hence A_Δ is a band extension of A_Φ. So the uniqueness statement about Δ follows from the uniqueness of the band extension (Theorem 2.2).

The general case, without restrictions on p, is reduced to the case when $p \ge \max\{0, r - n\}$. To see this, assume, for example, that $-n < p < 0$ and $p < r - n$. Put $q = -p$, and consider the square matrix

$$\Phi_0 = [\varphi_{ij}]_{i=q+1, j=1}^{n \quad n-q},$$

where φ_{ij} is the (i, j)-th entry of Φ. Since $\varphi_{ij} = 0$ for $j - i > p$, the matrix Φ_0 is a 0-lower triangular matrix, and we can apply the results proved so far to Φ_0. Next, one uses the following fact. If

$$(9) \qquad H = \begin{bmatrix} H_{11} & H_{12} \\ H_{21} & H_{22} \end{bmatrix} : \mathbb{C}^k \oplus \mathbb{C}^\ell \to \mathbb{C}^{k'} \oplus \mathbb{C}^{\ell'}$$

is strictly contractive and

$$H(I - H^*H)^{-1} = \begin{bmatrix} 0 & 0 \\ * & 0 \end{bmatrix},$$

then the block entries H_{11}, H_{12} and H_{22} in (9) are identically equal to zero (see Exercise 12 to Part VII). With these remarks it is now straightforward to complete the proof. $\square$

The special extension Δ in Theorem 3.2 is called the *triangular extension* of Φ. As we shall see later the role of triangular extensions in theory of contractive extensions is analogous to the one of band extensions in the theory of positive extensions. The triangular extension is also characterized by a *maximum entropy principle*, as follows. If Δ is the triangular extension of Φ and Ψ is an arbitrary contractive extension, then

$$\det(I - \Psi^*\Psi) \leq \det(I - \Delta^*\Delta),$$

and equality holds if and only if $\Psi = \Delta$. This principle may be derived directly from the analogous principle for positive extensions. To see this one applies Lemma 3.1, and uses the equality

$$\det A_\Psi = \det(I - \Psi^*\Psi),$$

which follows from (4).

The results of this section are related to those of Section XXVII.5; we shall return to this connection in Section XXXV.2.

CHAPTER XXXIV
A GENERAL SCHEME FOR COMPLETION AND EXTENSION PROBLEMS

This chapter presents an abstract scheme, called the *band method*, which allows one to deal with various positive and strictly contractive extension problems from one point of view. The theory developed here consists of three main elements. The first reduces the problem of finding a band extension to one of solving linear equations. The second is that all solutions of a positive extension problem are obtained via a linear fractional transformation of which the coefficients can be read off from a left and a right spectral factorization of the band extension. The third identifies the band extension in terms of an abstract maximum entropy principle. For strictly contractive extension problems this abstract approach has the same features, with triangular extensions in place of band extensions. This chapter consists of four sections. The first two develop the band method for positive extension problems, the third for strictly contractive extension problems, and the fourth presents the abstract maximum entropy principle. Applications appear in the next chapter.

XXXIV.1 BAND EXTENSIONS

Throughout this section $\mathcal{M}$ is an algebra with a unit e and an involution $*$. The latter is a mapping which takes each $a \in \mathcal{M}$ to an element $a^* \in \mathcal{M}$ and has the properties (a)–(d) mentioned in the second paragraph of Section XXXI.1. In particular, $e^* = e$. We shall assume that $\mathcal{M}$ admits a direct sum decomposition

$$(1) \qquad \mathcal{M} = \mathcal{M}_1 \oplus \mathcal{M}_2^0 \oplus \mathcal{M}_d \oplus \mathcal{M}_3^0 \oplus \mathcal{M}_4,$$

where $\mathcal{M}_1$, $\mathcal{M}_2^0$, $\mathcal{M}_d$, $\mathcal{M}_3^0$ and $\mathcal{M}_4$ are linear manifolds of $\mathcal{M}$. We say that $\mathcal{M}$ is an *algebra with band structure* (1) if, in addition, the following three conditions are satisfied

(E1) $e \in \mathcal{M}_d$,

(E2) $\mathcal{M}_1^* = \mathcal{M}_4$, $(\mathcal{M}_2^0)^* = \mathcal{M}_3^0$, $\mathcal{M}_d^* = \mathcal{M}_d$,

(E3) the following multiplication table describes some additional restrictions on the multiplication in $\mathcal{M}$:

$$(2) \quad
\begin{array}{c|ccccc}
 & \mathcal{M}_1 & \mathcal{M}_2^0 & \mathcal{M}_d & \mathcal{M}_3^0 & \mathcal{M}_4 \\
\hline
\mathcal{M}_1 & \mathcal{M}_1 & \mathcal{M}_1 & \mathcal{M}_1 & \mathcal{M}_+^0 & \mathcal{M} \\
\mathcal{M}_2^0 & \mathcal{M}_1 & \mathcal{M}_+^0 & \mathcal{M}_2^0 & \mathcal{M}_c & \mathcal{M}_-^0 \\
\mathcal{M}_d & \mathcal{M}_1 & \mathcal{M}_2^0 & \mathcal{M}_d & \mathcal{M}_3^0 & \mathcal{M}_4 \\
\mathcal{M}_3^0 & \mathcal{M}_+^0 & \mathcal{M}_c & \mathcal{M}_3^0 & \mathcal{M}_-^0 & \mathcal{M}_4 \\
\mathcal{M}_4 & \mathcal{M} & \mathcal{M}_-^0 & \mathcal{M}_4 & \mathcal{M}_4 & \mathcal{M}_4
\end{array}
$$

where

$$(3) \quad \mathcal{M}_+^0 := \mathcal{M}_1 \oplus \mathcal{M}_2^0, \qquad \mathcal{M}_-^0 := \mathcal{M}_3^0 \oplus \mathcal{M}_4,$$

$$(4) \quad \mathcal{M}_c := \mathcal{M}_2^0 \oplus \mathcal{M}_d \oplus \mathcal{M}_3^0.$$

Given $a \in \mathcal{M}$, we call a^* the *adjoint* of a. The first identity in (E2) requires the adjoint of an element in $\mathcal{M}_1$ to be in $\mathcal{M}_4$ and, conversely, the adjoint of an element in $\mathcal{M}_4$ belongs to $\mathcal{M}_1$. The other identities in (E2) have an analogous meaning. The multiplication table (2) tells us to which space a product ab belongs if a is in one of the spaces appearing in the left column of (2) and b belongs to one of the spaces in the top row of (2). For example, $ab \in \mathcal{M}_-^0$ if $a \in \mathcal{M}_2^0$ and $b \in \mathcal{M}_4$. Note that the table (2) is symmetric with respect to its main diagonal; thus $a \in \mathcal{M}_4$ and $b \in \mathcal{M}_2^0$ also imply that $ab \in \mathcal{M}_-^0$. The space $\mathcal{M}_c$ defined by (4) is called the *band* of $\mathcal{M}$, and an element in $\mathcal{M}_d$ will be called a *diagonal*.

In the sequel we put

$$(5) \quad
\begin{aligned}
\mathcal{M}_+ &:= \mathcal{M}_+^0 \oplus \mathcal{M}_d, & \mathcal{M}_- &:= \mathcal{M}_-^0 \oplus \mathcal{M}_d, \\
\mathcal{M}_2 &:= \mathcal{M}_2^0 \oplus \mathcal{M}_d, & \mathcal{M}_3 &:= \mathcal{M}_3^- \oplus \mathcal{M}_d.
\end{aligned}
$$

The multiplication table (2) implies that $\mathcal{M}_\pm^0$ as well as $\mathcal{M}_\pm$ are subalgebras of $\mathcal{M}$. Furthermore, $\mathcal{M}_1$ is a two-sided ideal of $\mathcal{M}_+^0$ and also of $\mathcal{M}_+$. A similar statement holds for $\mathcal{M}_4$ with respect to $\mathcal{M}_-^0$ and $\mathcal{M}_-$.

We shall use the symbols P_1, P_2^0, P_d, P_3^0 and P_4 to denote the natural projections associated with the decomposition (1). Thus, for example, P_2^0 is the projection of $\mathcal{M}$ onto $\mathcal{M}_2^0$ along the sum of the other spaces in the direct sum decomposition (1). We put

$$(6) \quad P_c = P_2^0 + P_d + P_3^0,$$

and hence P_c is the projection of $\mathcal{M}$ onto the band $\mathcal{M}_c$ along the sum of $\mathcal{M}_1$ and $\mathcal{M}_4$.

In the next chapter we shall meet several examples of algebras with a band structure. For the sake of illustration we mention here the following. Let $\mathcal{L}$ be the algebra of all $(n+1) \times (n+1)$ matrices $A = [a_{ij}]_{i,j=0}^n$ with entries a_{ij} in $\mathbb{C}$. Fix an integer m such that $0 \leq m < n$, and consider the following subsets of $\mathcal{L}$:

$$(7a) \qquad \mathcal{L}_1 = \left\{ A = [a_{ij}]_{i,j=0}^n \mid a_{ij} = 0 \text{ for } j - i \leq m \right\},$$

$$(7b) \qquad \mathcal{L}_2^0 = \left\{ A = [a_{ij}]_{i,j=0}^n \mid a_{ij} = 0 \text{ for } j - i > m \text{ or } j - i \leq 0 \right\},$$

$$(7c) \qquad \mathcal{L}_d = \left\{ A = [a_{ij}]_{i,j=0}^n \mid a_{ij} = 0 \text{ for } i \neq j \right\},$$

$$(7d) \qquad \mathcal{L}_3^0 = \left\{ A = [a_{ij}]_{i,j=0}^n \mid a_{ij} = 0 \text{ for } j - i \geq 0 \text{ or } j - i < -m \right\},$$

$$(7e) \qquad \mathcal{L}_4 = \left\{ A = [a_{ij}]_{i,j=0}^n \mid a_{ij} = 0 \text{ for } j - i \geq -m \right\}.$$

Then

$$(8) \qquad \mathcal{L} = \mathcal{L}_1 \oplus \mathcal{L}_2^0 \oplus \mathcal{L}_d \oplus \mathcal{L}_3^0 \oplus \mathcal{L}_4,$$

and it is straightforward to check that (8) defines a band structure in $\mathcal{L}$. Note that in this case the band $\mathcal{L}_c$ is precisely the set of all m-band matrices appearing in Section XXXIII.2, and the projection P_c is the map which assigns to $A = [a_{ij}]_{i,j=0}^n$ the matrix which one obtains from A by replacing the entries a_{ij} with $|i - j| > m$ by zero. Furthermore, $\mathcal{L}_d$ is the set of all $(n+1) \times (n+1)$ diagonal matrices and $\mathcal{L}_+^0 = \mathcal{L}_1 \oplus \mathcal{L}_2^0$ consists of all strictly upper triangular $(n+1) \times (n+1)$ matrices.

Let $\mathcal{M}$ be an algebra with band structure (1), and let $b \in \mathcal{M}$. We say that b admits a *right spectral factorization* (relative to the decomposition (1)) if b factorizes as $b = b_+^* b_+$, where b_+ is an invertible element of $\mathcal{M}$ such that b_+ and its inverse b_+^{-1} are both in $\mathcal{M}_+$. Analogously, b is said to have a *left spectral factorization* (relative to (1)) if $b = b_-^* b_-$ with b_- an invertible element of $\mathcal{M}$ and $b_-^{\pm 1}$ in $\mathcal{M}_-$. From the symmetry conditions on $\mathcal{M}$ it follows that b admits a right spectral factorization if b^{-1} admits a left spectral factorization, and conversely. Note that $B \in \mathcal{L}$ admits a right spectral factorization relative to (8) if and only if B factorizes as $B = C^*C$, where C is an invertible upper triangular $(n+1) \times (n+1)$ matrix. By replacing "upper triangular" by "lower triangular" one obtains a left spectral factorization of B.

Let $\mathcal{A}$ be an algebra with a unit and an involution $*$. An element $a \in \mathcal{A}$ is called *selfadjoint* if $a^* = a$, and a is said to be *positive definite* in $\mathcal{A}$ if there exists an invertible element c in $\mathcal{A}$ such that $a = c^*c$. If $b \in \mathcal{M}$ admits a right (or left) spectral factorization relative to (1), then b is automatically positive definite in $\mathcal{M}$. If $\mathcal{R}$ is a unital C^*-algebra with unit element e, then $g \in \mathcal{R}$ has $\|g\| < 1$ if and only if $e - g^*g$ is positive definite in $\mathcal{R}$. Furthermore, the sum of two positive definite elements in $\mathcal{R}$ is again positive definite. The latter two facts, which are proved in Section XXXI.6, will turn out to be very useful.

We are now ready to state the extension problem studied in this section. In the sequel $\mathcal{M}$ will be an *algebra with band structure* (1) *in a unital C^*-algebra* $\mathcal{R}$. The latter means that $\mathcal{M}$ is a $*$-subalgebra of a unital C^*-algebra $\mathcal{R}$ and the unit e of $\mathcal{M}$ is also the unit of $\mathcal{R}$. Let k be an element in the band $\mathcal{M}_c$. An element $b \in \mathcal{M}$ is called an *$\mathcal{R}$-positive extension* of k if b is positive definite in $\mathcal{R}$ and $P_c b = k$. The latter identity means that

$$(9) \qquad\qquad b = m_1 + k + m_4$$

for some elements m_1 in $\mathcal{M}_1$ and m_4 in $\mathcal{M}_4$. A *band extension* of k is an $\mathcal{R}$-positive extension b of k with the additional property that $b^{-1} \in \mathcal{M}_c$. Since an $\mathcal{R}$-positive extension b is selfadjoint, property (E2) of an algebra with band structure implies that we must have $m_1^* = m_4$ and $k = k^*$ in (9). So it is natural to assume that k is a selfadjoint element in $\mathcal{M}_c$. We are interested in finding all $\mathcal{R}$-positive extensions of k.

If b satisfies (9) and b is positive definite in $\mathcal{M}$, then we drop the prefix $\mathcal{R}$ and call b a *positive extension* of k. In this case b is an $\mathcal{R}$-positive extension whatever the choice of $\mathcal{R}$ is. Positive extensions that have a right or left spectral factorization relative to (1) are of special interest.

The next two theorems reduce the problem of finding band extensions which have spectral factorizations to solving certain linear equations. In the sequel we write c^{-*} for $(c^{-1})^*$ or $(c^*)^{-1}$.

THEOREM 1.1. *Let $\mathcal{M}$ be an algebra with band structure (1), and let $k = k^* \in \mathcal{M}_c$. Then k has a band extension b with a right spectral factorization relative to (1) if and only if the equation*

$$(10) \qquad\qquad P_2(kx) = e$$

has a solution x with the following properties:

    (i) $x \in \mathcal{M}_2$,

    (ii) x *is invertible and* $x^{-1} \in \mathcal{M}_+$,

    (iii) $P_d x = d^* d$ *for some* $d \in \mathcal{M}_d$ *which is invertible in* $\mathcal{M}_d$.

Furthermore, in this case such an element b is obtained by taking

$$(11) \qquad\qquad b = u^{-*} u^{-1}, \qquad u := x d^{-1},$$

where x is any solution of (10) satisfying (i)–(iii).

THEOREM 1.2. *Let $\mathcal{M}$ be an algebra with band structure (1), and let $k = k^* \in \mathcal{M}_c$. Then k has a band extension b with a left spectral factorization relative to (1) if and only if the equation*

$$(12) \qquad\qquad P_3(ky) = e$$

has a solution y with the following properties:

    (i) $y \in \mathcal{M}_3$,

    (ii) y *is invertible and* $y^{-1} \in \mathcal{M}_-$,

(iii) $P_d y = g^* g$ *for some* $g \in \mathcal{M}_d$ *which is invertible in* $\mathcal{M}_d$.

Furthermore, in this case such an element b is obtained by taking

$$(13) \qquad b = v^{-*}v^{-1}, \qquad v := yg^{-1},$$

where y is any solution of (12) *satisfying* (i)–(iii).

The procedures described in Theorems 1.1 and 1.2 lead to the same band extension. This is a corollary of the next theorem.

THEOREM 1.3. *Let $\mathcal{M}$ be an algebra with band structure* (1), *and let $k = k^* \in \mathcal{M}_c$. If k has a band extension b with a right spectral factorization relative to* (1) *and a band extension $\widetilde{b}$ with a left spectral factorization relative to* (1), *then $b = \widetilde{b}$.*

For the proof of Theorems 1.1–1.3 we need the following lemma.

LEMMA 1.4. *Let $\mathcal{M}$ be an algebra with band structure* (1). *Let $a \in \mathcal{M}_\pm$ be invertible, and assume $a^{-1} \in \mathcal{M}_\pm$. Then $P_d a$ is invertible and $(P_d a)^{-1} = P_d a^{-1}$. Furthermore, if $a^* a \in \mathcal{M}_c$, then $a \in \mathcal{M}_c \cap \mathcal{M}_\pm$.*

PROOF. Let us take $a^{\pm 1} \in \mathcal{M}_+$. Write $a = d + m$ and $a^{-1} = d^\times + m^\times$, with d and $d^\times$ in $\mathcal{M}_d$ and m and $m^\times$ in $\mathcal{M}_+^0$. In particular, $d = P_d a$ and $d^\times = P_d a^{-1}$. Note that

$$(14) \qquad e = aa^{-1} = dd^\times + (dm^\times + md^\times + mm^\times).$$

From the multiplication table (2) we see that $dm^\times + md^\times + mm^\times \in \mathcal{M}_+^0$ and $dd^\times \in \mathcal{M}_d$. Now let us apply the projection P_d to the right hand side of (14). We get $e = P_d e = dd^\times$. In a similar way, using $a^{-1} a = e$, we obtain $d^\times d = e$. Hence d is invertible, and $d^{-1} = d^\times$.

Next, assume additionally that $c := a^* a \in \mathcal{M}_c$. We have to show that $a \in \mathcal{M}_2$. But

$$(15) \qquad a = a^{-*}c \in (\mathcal{M}_+)^* \mathcal{M}_c = \mathcal{M}_- \mathcal{M}_c \subset \mathcal{M}_c + \mathcal{M}_4,$$

by virtue of the multiplication table (2). On the other hand, by our hypotheses, $a \in \mathcal{M}_+$. So

$$a = P_+ a \in P_+(\mathcal{M}_c + \mathcal{M}_4) = \mathcal{M}_2.$$

The proofs for $a^{\pm 1} \in \mathcal{M}_-$ are analogous. $\square$

PROOF OF THEOREM 1.1. We split the proof into two parts. In the first part we prove the necessity of the conditions (i)–(iii). The sufficiency and formula (11) are proved in the second part.

Part (a). Let b be a band extension of k with right spectral factorization $b = c^* c$. Then $b^{-1} = uu^*$ with

$$(u^*)^{\pm 1} = (u^{\pm 1})^* = ((c^{-1})^{\pm 1})^* \in \mathcal{M}_-,$$

and we can apply the second part of Lemma 1.4 to show that $u^* \in \mathcal{M}_3$. Put $x = u(P_d u^*)$. Then $x \in \mathcal{M}_2$, by condition (E2) and the multiplication table (2). We shall prove that x has the desired properties.

Since $b \in \mathcal{M}$ is an extension of k, we have $b = m_1 + k + m_4$ with $m_1 \in \mathcal{M}_1$ and $m_4 \in \mathcal{M}_4$. The multiplication table (2) shows that

$$(16) \qquad m_1 x \in \mathcal{M}_1 \mathcal{M}_2 \subset \mathcal{M}_1, \qquad m_4 x \in \mathcal{M}_4 \mathcal{M}_2 \subset \mathcal{M}_-^0.$$

So $P_2(kx) = P_2(bx)$. Now,

$$(17) \qquad bx = u^{-*} u^{-1}(u P_d u^*) = (u^*)^{-1}(P_d u^*).$$

The first part of Lemma 1.4 shows that $(u^*)^{-1} = (P_d u^*)^{-1} + m_-^0$, with m_-^0 in $\mathcal{M}_-^0$, and it follows from the multiplication table (2) that $bx = e + \widetilde{m}_-^0$ with $\widetilde{m}_-^0 \in \mathcal{M}_-^0$. So $P_2(bx) = e$, and we have proved that x satisfies the equation (10). Note that the preceding arguments also show that $P_1(bx) = 0$, and hence in the decomposition $b = m_1 + k + m_4$ the element m_1 satisfies the identity $m_1 x = -P_1(kx)$.

Let us check the properties (ii) and (iii). We already noted that $(P_d u^*)^{-1} = P_d u^{-*}$. Hence x is invertible and, by virtue of the multiplication table (2),

$$x^{-1} = (P_d u^{-*}) u^{-1} \in \mathcal{M}_d \mathcal{M}_+ \subset \mathcal{M}_+.$$

To get (iii), we use $\mathcal{M}_+^0 \mathcal{M}_d \subset \mathcal{M}_+^0$ to show that $P_d x = (P_d u)(P_d u^*)$. Condition (E2) implies that $(P_d u^*)^* = P_d u$, and so (iii) holds with $d = P_d u^*$. For the sake of completeness, we note that $u = x d^{-1}$, and hence b is given by (11).

Part (b). Assume equation (10) has a solution x with the properties (i)–(iii). We want to show that k has a band extension b with a right spectral factorization. Of course, b must be of the form $b = m_1 + k + m_4$ with $m_1 \in \mathcal{M}_1$ and $m_4 \in \mathcal{M}_4$. The remark made at the end of the second paragraph of Part (a) of the proof suggests taking

$$(18) \qquad m_1 = -P_1(kx) x^{-1}.$$

Note that the right hand side of (18) is in $\mathcal{M}_1$ because $\mathcal{M}_1 \mathcal{M}_+ \subset \mathcal{M}_1$. So let us define m_1 by (18), and set $b = m_1 + k + m_4$ with $m_4 = m_1^*$. Then b is an extension of k. Since $x \in \mathcal{M}_2$, we can use (16) and (18) to show that $P_1(bx) = 0$ and $P_2(bx) = e$. It follows that $bx = e + m_-^0$ for some $m_-^0 \in \mathcal{M}_-^0$. Now $x^* \in \mathcal{M}_-$. So, by $\mathcal{M}_- \mathcal{M}_-^0 \subset \mathcal{M}_-^0$,

$$(19) \qquad x^* bx = x^* + x^* m_-^0 = P_d(x^*) + \widetilde{m}_-^0,$$

for some $\widetilde{m}_-^0 \in \mathcal{M}_-^0$. By (iii) the element $P_d x$ is selfadjoint, and hence $P_d(x^*) = P_d(x)^* = P_d(x)$. Also, $x^* bx$ is selfadjoint. So we see from (19) that $\widetilde{m}_-^0$ is selfadjoint, and hence $\widetilde{m}_-^0 \in \mathcal{M}_-^0 \cap \mathcal{M}_+^0 = \{0\}$. We conclude that $\widetilde{m}_-^0 = 0$, and formula (19) yields the identity (11). Put $c = dx^{-1}$, where d is as in (iii). Then $b = c^* c$ and $c \in \mathcal{M}_+$. Also $c^{-1} = x d^{-1} \in \mathcal{M}_+$, because the inverse of d is again in $\mathcal{M}_d$, and so $c^{-1} \in \mathcal{M}_2 \mathcal{M}_d \subset \mathcal{M}_+$. Thus b admits a right spectral factorization relative to (1). In particular, b is an $\mathcal{R}$-positive extension of k.

The proof of Theorem 1.2 is analogous to that of Theorem 1.1. One has only to interchange subscripts at appropriate places. We omit the details.

PROOF OF THEOREM 1.3. The proof will follow in an abstract setting the same line of reasoning as the one used in the uniqueness part of the proof of Theorem XXXIII.2.2. Let b and $\widetilde{b}$ be band extensions of k, let $b = u^{-*}u^{-1}$ be a right spectral factorization of b relative to (1), and let $\widetilde{b} = v^{-*}v^{-1}$ be a left spectral factorization of $\widetilde{b}$ relative to (1). Here $u^{\pm 1} \in \mathcal{M}_+$ and $v^{\pm 1} \in \mathcal{M}_-$. Since b and $\widetilde{b}$ are both band extensions, we know from the second part of Lemma 1.4 that $u \in \mathcal{M}_2$ and $v \in \mathcal{M}_3$. Both $P_c b$ and $P_c \widetilde{b}$ are equal to k. So we may write $b - \widetilde{b} = m + m^*$ with $m \in \mathcal{M}_1$. Now $\widetilde{b}^{-1} - b^{-1} = b^{-1}(m + m^*)\widetilde{b}^{-1}$, and hence

$$(20) \qquad u^{-1}(\widetilde{b}^{-1} - b^{-1})v^{-*} = u^* m v + u^* m^* v.$$

By the multiplication table (2) we have

$$u^{-1}(\widetilde{b}^{-1} - b^{-1})v^{-*} \in \mathcal{M}_+ \mathcal{M}_c \mathcal{M}_+ \subset \mathcal{M}_1 + \mathcal{M}_c,$$

$$u^* m v \in \mathcal{M}_3 \mathcal{M}_1 \mathcal{M}_3 \subset \mathcal{M}_+ \mathcal{M}_3 \subset \mathcal{M}_1 + \mathcal{M}_c,$$

$$u^* m^* v \in \mathcal{M}_3 \mathcal{M}_4 \mathcal{M}_3 \subset \mathcal{M}_4 \mathcal{M}_3 \subset \mathcal{M}_4.$$

So, by applying the projection P_4 to both sides of (20), we obtain that $u^* m v = 0$. Recall that u and v are invertible. Thus $m = 0$, and we have proved that $b = \widetilde{b}$. $\square$

Let us illustrate Theorems 1.1–1.3 on the algebra $\mathcal{L}$, which has the band structure (8). Note that $\mathcal{L}$ is a unital C^*-algebra in its own right. Let A be a selfadjoint element in the band $\mathcal{L}_c$. This means that $A = [a_{ij}]_{i,j=0}^n$ is a selfadjoint m-band matrix. Since a positive definite element in $\mathcal{L}$ admits a left and right spectral factorization, we know from Theorem 1.3 that the band extension of A (assuming it exists) is unique. By Theorem 1.1, to find a band extension of A we have to solve the equation

$$(21) \qquad P_2(AX) = I,$$

where the unknown X has to satisfy the properties (i)–(iii) in Theorem 1.1. In this case this means that the solution X has to be an upper triangular m-band matrix with strictly positive diagonal entries. The right hand side of (21) is the identity matrix of order $n+1$, and P_2 is the mapping which assigns to a matrix $T = [t_{ij}]_{i,j=0}^n$ the matrix which one obtains by replacing the entry t_{ij} by zero whenever $j - i > m$ or $j - i < 0$. It follows that (21) is a short hand notation for the following set of equations

$$(22) \qquad \begin{bmatrix} a_{\alpha(j),\alpha(j)} & \cdots & a_{\alpha(j),j} \\ \vdots & & \vdots \\ a_{j,\alpha(j)} & \cdots & a_{jj} \end{bmatrix} \begin{bmatrix} x_{\alpha(j),j} \\ \vdots \\ x_{jj} \end{bmatrix} = \begin{bmatrix} 0 \\ \vdots \\ 0 \\ 1 \end{bmatrix}, \qquad j = 0, \ldots, n.$$

Here $\alpha(j) = \max\{0, j - m\}$ for $j = 0, \ldots, n$, and x_{ij} is the (i,j)-th entry of X for $0 \le j - i \le m$. In (22) we look for solutions with $x_{jj} > 0$ for $j = 0, \ldots, n$.

We already know (see Theorem XXXIII.2.1) that for A to have a positive extension it is necessary that

$$(23) \qquad \begin{bmatrix} a_{ii} & \cdots & a_{i,m+i} \\ \vdots & & \vdots \\ a_{m+i,i} & \cdots & a_{m+i,m+i} \end{bmatrix} > 0, \qquad i = 0, \ldots, n - m.$$

Assume (23) holds. Since a central submatrix of a positive definite matrix is again positive definite, we see that for each j the square matrix appearing in the left hand side of (22) is positive definite. It follows that (22) is uniquely solvable. Moreover, x_{jj} is a diagonal entry of a positive definite matrix (because the inverse of a positive definite matrix is again positive definite), and so $x_{jj} > 0$. So (23) is the necessary and sufficient condition for A to have a band extension, and if (23) holds, then A has a unique band extension B, namely

$$(24a) \qquad\qquad\qquad B = U^{-*}U^{-1},$$

where U is the upper triangular m-band matrix whose (i,j)-th entry u_{ij} is given by

$$(24b) \qquad\qquad u_{ij} = \begin{cases} x_{ij}x_{jj}^{-1/2}, & i = \max\{0, j-m\}, \ldots, j, \\ 0 & , \quad \text{otherwise}, \end{cases}$$

where the x_{ij} are determined by (22).

Applying Theorem 1.2 leads to an alternative set of equations, namely,

$$(25) \qquad \begin{bmatrix} a_{ii} & \cdots & a_{i,\beta(i)} \\ \vdots & & \vdots \\ a_{\beta(i),i} & \cdots & a_{\beta(i),\beta(i)} \end{bmatrix} \begin{bmatrix} y_{ii} \\ \vdots \\ y_{\beta(i),i} \end{bmatrix} = \begin{bmatrix} 1 \\ 0 \\ \vdots \\ 0 \end{bmatrix}, \qquad i = 0, \ldots, n.$$

Here $\beta(i) = \min\{n, i+m\}$ for $i = 0, \ldots, n$. It follows that the unique band extension of A (if it exists) is also given by

$$(26a) \qquad\qquad\qquad B = V^{-*}V^{-1},$$

where V is the lower triangular m-band matrix whose (i,j)-th entry v_{ij} is given by

$$(26b) \qquad\qquad v_{ij} = \begin{cases} y_{ij}y_{jj}^{-1/2}, & i = j, \ldots, \min\{n, j+m\}, \\ 0 & , \quad \text{otherwise}. \end{cases}$$

Note that (24a) and (26a) provide explicit formulas for the central extension in Theorem XXXIII.2.2, which up to now we could only find by repeatedly applying one step extensions.

XXXIV.2 POSITIVE EXTENSIONS

Let $\mathcal{M}$ be an algebra with band structure

$$(1) \qquad\qquad \mathcal{M} = \mathcal{M}_1 \oplus \mathcal{M}_2^0 \oplus \mathcal{M}_d \oplus \mathcal{M}_3^0 \oplus \mathcal{M}_4$$

in a unital C^*-algebra $\mathcal{R}$, and let k be a selfadjoint element in the band $\mathcal{M}_c$. In this section the element k is assumed to have a band extension with a left and right spectral factorization relative to (1). From the results in the previous section we know how to construct such an extension b. Our aim here is to describe all $\mathcal{R}$-positive extensions of k in $\mathcal{M}$. This requires extra structure. In the sequel $\|\cdot\|_{\mathcal{R}}$ denotes the norm of the C^*-algebra $\mathcal{R}$. We shall assume that the following axiom holds:

AXIOM (A). *If $g \in \mathcal{M}_+$ and $\|g\|_{\mathcal{R}} < 1$, then $(e - g)^{-1} \in \mathcal{M}_+$.*

For g as in axiom (A) we have $(e-g)^{-1} = \sum_{n=0}^{\infty} g^n$, where the series converges in $\mathcal{R}$. Since the partial sums are in $\mathcal{M}_+$, it follows that $(e - g)^{-1}$ belongs to the closure of $\mathcal{M}_+$. Thus axiom (A) is automatically fulfilled if $\mathcal{M}_+$ is closed in $\mathcal{R}$.

Let $\mathcal{L}$ be the algebra of all $(n + 1) \times (n + 1)$ matrices (with entries in $\mathbb{C}$) with band structure

$$(2) \qquad \mathcal{L} = \mathcal{L}_1 \oplus \mathcal{L}_2^0 \oplus \mathcal{L}_d \oplus \mathcal{L}_3^0 \oplus \mathcal{L}_4,$$

the various terms in this decomposition being given by (7a)–(7e) in the previous section. For $\mathcal{L}$ the axiom above is fulfilled. Indeed, in this case $\mathcal{R}$ is just $\mathcal{L}$ endowed with the operator norm, i.e., for $T \in \mathcal{L}$ we define $\|T\|$ to be the norm of the operator on $\mathbb{C}^{n+1}$ induced by the canonical action of T on the standard basis of $\mathbb{C}^{n+1}$. Here $\mathbb{C}^{n+1}$ is equipped with its standard Hilbert space structure, and thus $\mathcal{L}$ with the operator norm is a unital C^*-algebra. Since $\mathcal{L}_+$ consists of all upper triangular $(n+1) \times (n+1)$ matrices, $\mathcal{L}_+$ is closed, and hence axiom (A) is fulfilled.

THEOREM 2.1. *Let $\mathcal{M}$ be an algebra with band structure (1) in a unital C^*-algebra $\mathcal{R}$, and assume that axiom (A) holds. Let $k = k^* \in \mathcal{M}_c$, and suppose that k has a band extension b which admits a right and left spectral factorization relative to (1):*

$$(3) \qquad b = u^{-*}u^{-1} = v^{-*}v^{-1}, \qquad u^{\pm 1} \in \mathcal{M}_+, \ v^{\pm 1} \in \mathcal{M}_-.$$

Then each $\mathcal{R}$-positive extension of k is of the form

$$(4) \qquad \mathcal{F}(g) = (vg + u)^{-*}(e - g^*g)(vg + u)^{-1},$$

where the free parameter g is an arbitrary element in $\mathcal{M}_1$ such that $\|g\|_{\mathcal{R}} < 1$. Moreover, the map $\mathcal{F}$ provides a one-one correspondence between all such g and all $\mathcal{R}$-positive extensions of k.

In the above theorem the map $\mathcal{F}$ may be replaced by

$$\mathcal{F}'(h) = (uh + v)^{-*}(e - h^*h)(uh + v)^{-1},$$

where now the free parameter h is an arbitrary element of $\mathcal{M}_4$ such that $\|h\|_{\mathcal{R}} < 1$.

From the proof of Theorem 2.1 it will follow that (4) yields a positive extension (i.e., an extension which is positive definite in $\mathcal{M}$) if and only if the free parameter $g \in \mathcal{M}_1$ is such that $e - g^*g$ is positive definite in $\mathcal{M}$.

In the proof of Theorem 2.1 we shall use the following lemma.

LEMMA 2.2. *Let $\mathcal{M}$ be an algebra with band structure (1) in a unital C^*-algebra $\mathcal{R}$, and assume axiom (A) holds. Let $a \in \mathcal{M}_+$ be such that $a + a^*$ is positive definite in $\mathcal{R}$. Then a is invertible and $a^{-1} \in \mathcal{M}_+$.*

PROOF. Write $a + a^* = c^* c$, with $c \in \mathcal{R}$ and c invertible in $\mathcal{R}$. For $0 < t \in \mathbb{R}$ consider $g(t) = e - ta$. We have $g(t) \in \mathcal{M}_+$ and

$$
\begin{aligned}
e - g(t)^* g(t) &= e - \{e - t(a^* + a) + t^2 a^* a\} \\
&= t(a^* + a) - t^2 a^* a \\
&= tc^* \{e - tc^{-*} a^* a c^{-1}\} c.
\end{aligned}
$$

Now fix $t_0 > 0$ so that $\sqrt{t_0}\|ac^{-1}\| < 1$, and recall that in the C^*-algebra $\mathcal{R}$ an element x has $\|x\|_{\mathcal{R}} < 1$ if and only if $e - x^* x$ is positive definite in $\mathcal{R}$. It follows that $e - t_0 c^{-*} a^* a c^{-1}$ is positive definite in $\mathcal{R}$. But then, since c is invertible, $e - g(t_0)^* g(t_0)$ is positive definite in $\mathcal{R}$. So $\|g(t_0)\|_{\mathcal{R}} < 1$, and we may apply axiom (A) to conclude that $\left(e - g(t_0)\right)^{-1} \in \mathcal{M}_+$. Finally, note that $a = t_0^{-1}\left(e - g(t_0)\right)$. So a has the desired properties. $\square$

From the symmetry in the band structure (see property (E2) in the previous section) it is clear that Lemma 2.2 remains true if $\mathcal{M}_+$ in Lemma 2.2 is replaced by $\mathcal{M}_-$.

PROOF OF THEOREM 2.1. We split the proof into four parts. In the first part we rewrite the map $\mathcal{F}$.

Part (a). Since $b = b^*$, we may choose $c \in \mathcal{M}_+$ such that $b = c^* + c$. Fix such an element c, and consider

$$
(5) \qquad \widetilde{\mathcal{F}}(g) = (-c^* vg + cu)(vg + u)^{-1}
$$

for all $g \in \mathcal{M}$ for which $vg + u$ is invertible in $\mathcal{M}$. Here u and v are as in (3). It follows that

$$
\begin{aligned}
\widetilde{\mathcal{F}}(g) + \widetilde{\mathcal{F}}(g)^* &= (vg + u)^{-*}\{(g^* v^* + u^*)(-c^* vg + cu) \\
&\qquad + (-g^* v^* c + u^* c^*)(vg + u)\}(vg + u)^{-1} \\
&= (vg + u)^{-*}\{u^*(c^* + c)u - g^* v^*(c^* + c)vg\}(vg + u)^{-1} \\
&= (vg + u)^{-*}(e - g^* g)(vg + u)^{-1} = \mathcal{F}(g),
\end{aligned}
$$

whenever $u + vg$ is invertible. If $u + vg$ is invertible, then the same is true for $e + gu^{-1}v$, and in this case

$$
\begin{aligned}
(6) \qquad \widetilde{\mathcal{F}}(g) &= \{-c^* vg + c(vg + u - vg)\}(vg + u)^{-1} \\
&= c - (c + c^*)vg(vg + u)^{-1} \\
&= c - v^{-*}g(e + u^{-1}vg)^{-1}u^{-1} \\
&= c - v^{-*}(e + gu^{-1}v)^{-1}gu^{-1}.
\end{aligned}
$$

Part (b). Assume $g \in \mathcal{M}_1$ and $\|g\|_{\mathcal{R}} < 1$. Then $e - g^* g$ is positive definite in $\mathcal{R}$. In this part we show that $\mathcal{F}(g)$ is well-defined and an $\mathcal{R}$-positive extension of k. By the second part of Lemma 1.4, the element v is in $\mathcal{M}_3$. So (use the multiplication table (2) in the previous section) we have $vg \in \mathcal{M}_+$, and hence $u^{-1}vg \in \mathcal{M}_+$. Furthermore, since

$$
e - (u^{-1}vg)^*(u^{-1}vg) = e - g^* v^* bvg = e - g^* g
$$

is positive definite in $\mathcal{R}$, we have $\|u^{-1}vg\| < 1$, and hence we can use axiom (A) to show that $e + u^{-1}vg$ is invertible and $(e + u^{-1}vg)^{-1} \in \mathcal{M}_+$. In particular, $\mathcal{F}(g)$ and $\widetilde{\mathcal{F}}(g)$ are well-defined elements of $\mathcal{M}$ and $\mathcal{F}(g)$ is positive definite in $\mathcal{R}$. It remains to show that $\mathcal{F}(g)$ is an extension of k. For this purpose, note that

$$(e + gu^{-1}v)^{-1} = e - g(e + u^{-1}vg)^{-1}u^{-1}v$$
$$\in e + \mathcal{M}_1\mathcal{M}_+\mathcal{M}_+\mathcal{M}_3$$
$$\subset e + \mathcal{M}_1\mathcal{M}_3 \subset \mathcal{M}_+,$$

by the multiplication table for $\mathcal{M}$. So, by virtue of (6),

$$\widetilde{\mathcal{F}}(g) \in c + \mathcal{M}_+\mathcal{M}_+\mathcal{M}_1\mathcal{M}_+ \subset c + \mathcal{M}_1,$$

and thus

$$\mathcal{F}(g) = \widetilde{\mathcal{F}}(g) + \widetilde{\mathcal{F}}(g)^* \in b + \mathcal{M}_1 + \mathcal{M}_4,$$

which shows that $P_c\big(\mathcal{F}(g)\big) = k$.

Part (c). This part concerns the uniqueness of the representation $f = \mathcal{F}(g)$. Assume $g \in \mathcal{M}_1$ and $\|g\|_\mathcal{R} < 1$. Let us show that in this case g is uniquely determined by $\mathcal{F}(g)$. Put $m = \widetilde{\mathcal{F}}(g)$. Then, by (6),

$$v^*(m - c)v = -(e + gu^{-1}v)^{-1}(gu^{-1}v + e - e)$$
$$= (e + gu^{-1}v)^{-1} - e.$$

It follows that

(7)
$$g = \big[e + v^*(m - c)v\big]^{-1}v^{-1}u - v^{-1}u,$$

and thus g is uniquely determined by $\mathcal{F}(g)$. Note that (7) also gives a hint of how to find an element g such that $f = \mathcal{F}(g)$ for a given element f.

Part (d). Suppose $f \in \mathcal{M}$ is an arbitrary $\mathcal{R}$-positive extension of k. In this part we show that f has the desired representation. Because of selfadjointness, we may write $f = m + m^*$ with $m \in \mathcal{M}_+$. Since both f and b are extensions of k, we have $P_d(f) = P_d(b)$, and thus

$$d := P_d m - P_d c = -P_d m^* + P_d c^* = -d^*.$$

So without loss of generality we may assume that $P_d m = P_d c$ (replace m by $m - d$ if necessary). Put $w = m - c$. Then $w \in \mathcal{M}_+^0$, and $w + w^* = f - b$ implies that $w \in \mathcal{M}_1$. Since the sum of two positive definite elements in $\mathcal{R}$ is again a positive definite element in $\mathcal{R}$, the element $b + f$ is positive definite in $\mathcal{R}$. Now

$$v^*(b + f)v = v^*(b + b + w^* + w)v = 2e + v^*w^*v + v^*wv,$$

and so $(e + v^*wv) + (e + v^*wv)^*$ is positive definite in $\mathcal{R}$. Note

$$v^*wv \in \mathcal{M}_2\mathcal{M}_1\mathcal{M}_3 \subset \mathcal{M}_1\mathcal{M}_3 \subset \mathcal{M}_+.$$

But then we can apply Lemma 2.2 to show that $e + v^* w v$ is invertible and $(e + v^* w v)^{-1} \in \mathcal{M}_+$. Put $g := -(e + v^* w v)^{-1} v^* w u$. We know that v^* and u are in $\mathcal{M}_2$, by virtue of the second part of Lemma 1.4. So $g \in \mathcal{M}_+ \mathcal{M}_2 \mathcal{M}_1 \mathcal{M}_2 \subset \mathcal{M}_1$. Furthermore,

$$
\begin{aligned}
vg + u &= -v(e + v^* w v)^{-1} v^* w u + u \\
&= -v(e + v^* w v)^{-1} \{v^* w v + e - e\} v^{-1} u + u \\
&= v(e + v^* w v)^{-1} v^{-1} u,
\end{aligned}
$$

and so $vg + u$ is invertible in $\mathcal{M}$. Also, note

$$
v(e + v^* w v)^{-1} = (e + v v^* w)^{-1} v.
$$

Hence $vg + u = (e + v v^* w)^{-1} u$, and we see that

$$
\begin{aligned}
\widetilde{\mathcal{F}}(g) &= \{c^* v(e + v^* w v)^{-1} v^* w u + c u\} u^{-1}(e + v v^* w) \\
&= \{c^*(e + v v^* w)^{-1}(v v^* w + e - e) + c\}(e + v v^* w) \\
&= \{c^* + c - c^*(e + v v^* w)^{-1}\}(e + v v^* w) \\
&= b(e + v v^* w) - c^* \\
&= b + w - c^* = c + w = m.
\end{aligned}
$$

Thus $f = m + m^* = \widetilde{\mathcal{F}}(g) + \widetilde{\mathcal{F}}(g)^* = \mathcal{F}(g)$. Since f is positive definite in $\mathcal{R}$, it follows that $e - g^* g$ is positive definite, and hence $\|g\|_{\mathcal{R}} < 1$. Thus f has the desired representation. $\square$

Let $A = [a_{ij}]_{i,j=0}^m$ be an m-band matrix, and assume that the necessary and sufficient condition for A to have a positive extension is fulfilled, i.e., assume that

$$
(8) \qquad \begin{bmatrix} a_{ii} & \cdots & a_{i,m+i} \\ \vdots & & \vdots \\ a_{m+i,i} & \cdots & a_{m+i,m+i} \end{bmatrix} > 0, \qquad i = 0, \ldots, n - m.
$$

We can now use Theorem 2.1 to given a full parametrization of all positive extensions of A. To see this, let $\mathcal{L}$ be the algebra of all $(n+1) \times (n+1)$ matrices endowed with the band structure (2). We know that $\mathcal{L}$ equipped with the operator norm is a unital C^*-algebra for which axiom (A) holds. Note that A is a selfadjoint element in the band $\mathcal{L}_c$. By the results mentioned at the end of the previous paragraph, A has a unique band extension B, and the formulas (24a,b), (26a,b) in the previous section present in an explicit form a right and a left spectral factorization of B. So all the conditions of Theorem 2.1 are fulfilled for this case. Hence all positive extensions F of A are given by

$$
(9) \qquad F = (VG + U)^{-*}(I - G^* G)(VG + U)^{-1},
$$

where the free parameter G runs over all matrices $G = [g_{ij}]_{i,j=0}^n$ with $g_{ij} = 0$ for $j - i \leq m$ and $\|G\| < 1$. The coefficients U and V are the matrices $U = [u_{ij}]_{i,j=0}^n$ and $V = [v_{ij}]_{i,j=0}^n$ whose entries appear in formulas (24b) and (26b) of the previous section.

XXXIV.3 STRICTLY CONTRACTIVE EXTENSIONS

In this section we develop a general scheme to treat strictly contractive extension problems. As in the previous sections $\mathcal{M}$ is an algebra with a unit e and an involution $*$, and $\mathcal{M}$ is equipped with the band structure

$$(1) \qquad \mathcal{M} = \mathcal{M}_1 \oplus \mathcal{M}_2^0 \oplus \mathcal{M}_d \oplus \mathcal{M}_3^0 \oplus \mathcal{M}_4.$$

Throughout this section we assume that $\mathcal{M}$ is a $*$-subalgebra of a unital C^*-algebra $\mathcal{R}$ and the unit e of $\mathcal{M}$ is also the unit of $\mathcal{R}$. Put

$$(2) \qquad \mathcal{M}_\ell = \mathcal{M}_4 \oplus \mathcal{M}_3^0 \oplus \mathcal{M}_d \oplus \mathcal{M}_2^0,$$

and let $\varphi \in \mathcal{M}_\ell$ be given. We say that $\psi \in \mathcal{M}$ is a *strictly contractive extension* (or *completion*) of φ if $\varphi - \psi \in \mathcal{M}_1$ and $\|\psi\|_\mathcal{R} < 1$. In this section our aim is to find all solutions of this extension problem.

To give a simple example (more involved ones will appear in the next chapter), let us again consider the algebra $\mathcal{L}$ of all $(n+1) \times (n+1)$ matrices with complex entries. For the band structure

$$(3) \qquad \mathcal{L} = \mathcal{L}_1 \oplus \mathcal{L}_2^0 \oplus \mathcal{L}_d \oplus \mathcal{L}_3^0 \oplus \mathcal{L}_4$$

defined by formulas (7a)–(7e) in the first section of this chapter, the space $\mathcal{L}_\ell$ consists of all m-lower triangular matrices. It follows that for this case the strictly contractive extension problem is a problem of the type considered in Section XXXIII.3.

Let $\varphi \in \mathcal{M}_\ell$ be given. A *triangular extension* of φ is a strictly contractive extension g of φ with the additional property that $g(e - g^*g)^{-1}$ belongs to $\mathcal{M}_\ell$. We shall see that the triangular extensions play a special role, similar to the one of band extensions in the positive extension problem. The first theorem of this section reduces the problem of finding a triangular extension to solving certain linear equations.

To state the first theorem, we need certain projections associated with decompositions of $\mathcal{M}$. Put

$$\mathcal{M}_u := \mathcal{M}_\ell^* = \mathcal{M}_1 \oplus \mathcal{M}_2^0 \oplus \mathcal{M}_d \oplus \mathcal{M}_3^0.$$

Then

$$(4) \qquad \mathcal{M} = \mathcal{M}_\ell \oplus \mathcal{M}_1, \qquad \mathcal{M} = \mathcal{M}_u \oplus \mathcal{M}_4.$$

The projection of $\mathcal{M}$ onto $\mathcal{M}_\ell$ along $\mathcal{M}_1$ is denoted by P_ℓ. We write P_u for the projection of $\mathcal{M}$ onto $\mathcal{M}_u$ along $\mathcal{M}_4$. The natural projections of $\mathcal{M}$ onto the spaces $\mathcal{M}_-$, $\mathcal{M}_+$ and $\mathcal{M}_d$ are denoted by P_-, P_+ and P_d, respectively. Recall (see the first paragraph of Section XXXIV.1) that

$$\mathcal{M}_- = \mathcal{M}_d \oplus \mathcal{M}_3^0 \oplus \mathcal{M}_4, \qquad \mathcal{M}_+ = \mathcal{M}_1 \oplus \mathcal{M}_2^0 \oplus \mathcal{M}_d.$$

So, for example, P_- is equal to $P_d + P_3^0 + P_4$, where P_d, P_3^0 and P_4 are defined in Section XXXIV.1.

THEOREM 3.1. *Let $\mathcal{M}$ be an algebra with band structure (1) in a unital C^*-algebra $\mathcal{R}$, and let $\varphi \in \mathcal{M}_\ell$. The element φ has a triangular extension g in $\mathcal{M}$ such that $e - g^*g$ admits a right and $e - gg^*$ admits a left spectral factorization relative to (1) if and only if the equations*

$$(5) \qquad a - P_-\big(\varphi(P_u(\varphi^*a))\big) = e, \qquad d - P_+\big(\varphi^*(P_\ell(\varphi d))\big) = e$$

have solutions a and d with the properties:

(i) $a \in \mathcal{M}_-,\ d \in \mathcal{M}_+$,

(ii) *a and d are invertible,* $a^{-1} \in \mathcal{M}_-,\ d^{-1} \in \mathcal{M}_+$,

(iii) $P_d a$ *and* $P_d d$ *are positive definite in* $\mathcal{M}_d$.

In this case φ has a unique triangular extension g in $\mathcal{M}$ such that $e - g^*g$ admits a right and $e - gg^*$ admits a left spectral factorization relative to (1) , and this g is given by

$$(6) \qquad g = bd^{-1} = a^{-*}c^*,$$

where

$$(7) \qquad b = P_\ell(\varphi d), \qquad c = P_u(\varphi^*a).$$

We shall prove Theorem 3.1 by transforming the corresponding extension problem into a positive extension problem. For this purpose we need the algebra $\mathcal{A} = \mathcal{M}^{2\times 2}$ of 2×2 block matrices with entries in $\mathcal{M}$, i.e.,

$$\mathcal{A} = \Big\{ A = \begin{bmatrix} a & b \\ c & d \end{bmatrix} \ \big|\ a, b, c, d \in \mathcal{M} \Big\}.$$

The algebraic operations in $\mathcal{A}$ are given by the usual rules for addition and multiplication of matrices. Clearly, the matrix

$$(8) \qquad E = \begin{bmatrix} e & 0 \\ 0 & e \end{bmatrix}$$

is the unit in $\mathcal{A}$. In $\mathcal{A}$ we define an involution by setting

$$(9) \qquad \begin{bmatrix} a & b \\ c & d \end{bmatrix}^* = \begin{bmatrix} a^* & c^* \\ b^* & d^* \end{bmatrix}.$$

The next step is to turn $\mathcal{A}$ into an algebra with band structure. To do this, consider the following subsets of $\mathcal{A}$:

$$\mathcal{A}_1 = \begin{bmatrix} 0 & \mathcal{M}_1 \\ 0 & 0 \end{bmatrix} = \Big\{ \begin{bmatrix} 0 & b \\ 0 & 0 \end{bmatrix} \ \big|\ b \in \mathcal{M}_1 \Big\},$$

$$\mathcal{A}_2^0 = \begin{bmatrix} \mathcal{M}_+^0 & \mathcal{M}_\ell \\ 0 & \mathcal{M}_+^0 \end{bmatrix} = \Big\{ \begin{bmatrix} a & b \\ 0 & d \end{bmatrix} \ \big|\ a, d \in \mathcal{M}_+^0,\ b \in \mathcal{M}_\ell \Big\},$$

$$\mathcal{A}_d = \begin{bmatrix} \mathcal{M}_d & 0 \\ 0 & \mathcal{M}_d \end{bmatrix} = \Big\{ \begin{bmatrix} a & 0 \\ 0 & d \end{bmatrix} \ \big|\ a, d \in \mathcal{M}_d \Big\},$$

$$\mathcal{A}_3^0 = \begin{bmatrix} \mathcal{M}_-^0 & 0 \\ \mathcal{M}_u & \mathcal{M}_-^0 \end{bmatrix} = \Big\{ \begin{bmatrix} a & 0 \\ c & d \end{bmatrix} \ \big|\ a, d \in \mathcal{M}_-^0,\ c \in \mathcal{M}_u \Big\},$$

$$\mathcal{A}_4 = \begin{bmatrix} 0 & 0 \\ \mathcal{M}_4 & 0 \end{bmatrix} = \Big\{ \begin{bmatrix} 0 & 0 \\ c & 0 \end{bmatrix} \ \big|\ c \in \mathcal{M}_4 \Big\}.$$

From (1) and the decompositions in (4) it follows that

$$(10) \qquad \mathcal{A} = \mathcal{A}_1 \oplus \mathcal{A}_2^0 \oplus \mathcal{A}_d \oplus \mathcal{A}_3^0 \oplus \mathcal{A}_4.$$

We already know that $\mathcal{A}$ is an algebra with a unit and an involution (see (8) and (9)). So to prove that (10) defines a band structure on $\mathcal{A}$ we have to show that the conditions (E1)–(E3) in Section XXXIV.1 hold for $\mathcal{A}$ in place of $\mathcal{M}$. In this case

$$(11) \qquad \mathcal{A}_+^0 = \begin{bmatrix} \mathcal{M}_+^0 & \mathcal{M} \\ 0 & \mathcal{M}_+^0 \end{bmatrix}, \qquad \mathcal{A}_-^0 = \begin{bmatrix} \mathcal{M}_-^0 & 0 \\ \mathcal{M} & \mathcal{M}_-^0 \end{bmatrix},$$

$$(12) \qquad \mathcal{A}_c = \begin{bmatrix} \mathcal{M} & \mathcal{M}_\ell \\ \mathcal{M}_u & \mathcal{M} \end{bmatrix}.$$

Note that the unit E of $\mathcal{A}$ is in $\mathcal{A}_d$. Furthermore, since $\mathcal{M}_u = \mathcal{M}_\ell^*$, the symmetry conditions on (1) imply that (E2) holds for $\mathcal{A}$. By using the multiplication table for (1) it is straightforward to check that this table also holds for $\mathcal{A}$ in place of $\mathcal{M}$. Thus with (10) the algebra $\mathcal{A}$ is an algebra with band structure.

Let $\mathcal{R}^{2\times 2}$ be the algebra of 2×2 block matrices with entries in $\mathcal{R}$. We may view $\mathcal{R}$ as a norm closed $*$-subalgebra of $\mathcal{L}(H)$ for some Hilbert space H such that $\|\cdot\|_\mathcal{R}$ is equal to the operator norm (see Section XXXI.6). In this case $\mathcal{R}^{2\times 2}$ consists of operators on the Hilbert space direct sum $H \oplus H$, and hence $\mathcal{R}^{2\times 2}$ has a natural C^*-algebra structure which is inherited from $\mathcal{L}(H \oplus H)$. In the sequel we assume that $\mathcal{R}^{2\times 2}$ is endowed with this C^*-algebra structure. From our hypotheses on $\mathcal{M}$ and $\mathcal{R}$ it is clear that $\mathcal{A} = \mathcal{M}^{2\times 2}$ is a $*$-subalgebra of $\mathcal{R}^{2\times 2}$ and the unit E of $\mathcal{A}$ is the unit of $\mathcal{R}^{2\times 2}$. We are now ready to prove Theorem 3.1.

PROOF OF THEOREM 3.1. We split the proof into four parts. The first part contains a general remark related to Lemma XXXIII.3.1.

Part (a). For each $\psi \in \mathcal{M}$ set

$$(13) \qquad A_\psi = \begin{bmatrix} e & \psi \\ \psi^* & e \end{bmatrix} \in \mathcal{A} = \mathcal{M}^{2\times 2}.$$

Here e is the unit of $\mathcal{M}$. Note that A_ψ admits the following factorizations:

$$(14a) \qquad A_\psi = \begin{bmatrix} e & 0 \\ \psi^* & e \end{bmatrix} \begin{bmatrix} e & 0 \\ 0 & e - \psi^*\psi \end{bmatrix} \begin{bmatrix} e & \psi \\ 0 & e \end{bmatrix},$$

$$(14b) \qquad A_\psi = \begin{bmatrix} e & \psi \\ 0 & e \end{bmatrix} \begin{bmatrix} e - \psi\psi^* & 0 \\ 0 & e \end{bmatrix} \begin{bmatrix} e & 0 \\ \psi^* & e \end{bmatrix}.$$

We set $K = A_\varphi$, where φ is the given element in $\mathcal{M}_\ell$. From (12), (13) and $\mathcal{M}_u = \mathcal{M}_\ell^*$ it is clear that K is a selfadjoint element in $\mathcal{A}_c$.

Part (b). Suppose that g is a triangular extension of φ such that

$$(15) \qquad e - g^*g = r^*r, \qquad e - gg^* = s^*s,$$

where $r^{\pm 1} \in \mathcal{M}_+$ and $s^{\pm 1} \in \mathcal{M}_-$. In this part we show that A_g is a band extension of K and that A_g admits a right and a left spectral factorization relative to the decomposition (10). Note that the first equality in (15) is a right spectral factorization of $e - g^*g$ and the second gives a left spectral factorization for $e - gg^*$ (in both cases relative to (1)). By using the factorizations (15) in (14a) and (14b), we see that

$$(16a) \qquad A_g = \begin{bmatrix} e & g \\ 0 & r \end{bmatrix}^* \begin{bmatrix} e & g \\ 0 & r \end{bmatrix},$$

$$(16b) \qquad A_g = \begin{bmatrix} s & 0 \\ g^* & e \end{bmatrix}^* \begin{bmatrix} s & 0 \\ g^* & e \end{bmatrix}.$$

Note that

$$(17) \qquad \begin{bmatrix} e & g \\ 0 & r \end{bmatrix}^{-1} = \begin{bmatrix} e & -gr^{-1} \\ 0 & r^{-1} \end{bmatrix}, \qquad \begin{bmatrix} s & 0 \\ g^* & e \end{bmatrix}^{-1} = \begin{bmatrix} s^{-1} & 0 \\ -g^*s^{-1} & e \end{bmatrix}.$$

Now use the fact that $r^{\pm 1} \in \mathcal{M}_+$ and $s^{\pm 1} \in \mathcal{M}_-$. Since

$$(18) \qquad \mathcal{A}_+ = \begin{bmatrix} \mathcal{M}_+ & \mathcal{M} \\ 0 & \mathcal{M}_+ \end{bmatrix}, \qquad \mathcal{A}_- = \begin{bmatrix} \mathcal{M}_- & 0 \\ \mathcal{M} & \mathcal{M}_- \end{bmatrix},$$

equality (16a) (resp., (16b)) is a right (resp., left) spectral factorization of A_g relative to the decomposition (10). In particular, A_g is positive definite in $\mathcal{A}$, and hence in $\mathcal{R}^{2\times 2}$. Since g is an extension of φ, we have $g - \varphi \in \mathcal{M}_1$, and hence

$$A_g - K = \begin{bmatrix} 0 & g - \varphi \\ (g - \varphi)^* & 0 \end{bmatrix} \in \mathcal{A}_1 + \mathcal{A}_4.$$

Thus A_g is a positive extension of K. Furthermore, from (14a) and (12) we see that

$$(19) \qquad A_g^{-1} = \begin{bmatrix} e + g(e - g^*g)^{-1}g^* & -g(e - g^*g)^{-1} \\ -(e - g^*g)^{-1}g^* & (e - g^*g)^{-1} \end{bmatrix} \in \mathcal{A}_c,$$

because g is a triangular extension of φ and $\mathcal{M}_u = \mathcal{M}_\ell^*$. So A_g has the desired properties.

      Part (c). In this part we prove the necessity of the conditions in Theorem 3.1. Let g be as in Part (b). The result proved in Part (b) allows us to apply Theorems 1.1 and 1.2 to A_g. Let Q_2, Q_3 and Q_d be the natural projections of $\mathcal{A}$ onto the spaces $\mathcal{A}_2$, $\mathcal{A}_3$ and $\mathcal{A}_d$, respectively. By Theorems 1.1 and 1.2 there exist

$$X = \begin{bmatrix} \delta & -b \\ 0 & d \end{bmatrix} \in \mathcal{A}_2, \qquad Y = \begin{bmatrix} a & 0 \\ -c & \alpha \end{bmatrix} \in \mathcal{A}_3$$

such that X is invertible and $X^{-1} \in \mathcal{A}_+$, the element Y is invertible and $Y^{-1} \in \mathcal{A}_-$, the diagonals Q_dX and Q_dY are positive definite in $\mathcal{A}_d$ and

$$(20) \qquad Q_2(KX) = E, \qquad Q_3(KY) = E.$$

Now

$$(21a) \qquad Q_2(KX) = \begin{bmatrix} \delta & -b + P_\ell(\varphi d) \\ 0 & d - P_+(\varphi^* b) \end{bmatrix},$$

$$(21b) \qquad Q_3(KY) = \begin{bmatrix} a - P_-(\varphi c) & 0 \\ P_u(\varphi^* a) - c & \alpha \end{bmatrix}.$$

Here we used that δ and d are in $\mathcal{M}_+$, $b \in \mathcal{M}_\ell$, α and a in $\mathcal{M}_-$ and $c \in \mathcal{M}_\ell^* = \mathcal{M}_u$. Thus the two equalities in (20) yield:

$$(22a) \qquad \delta = e, \qquad b = P_\ell(\varphi d), \qquad d - P_+(\varphi^* b) = e,$$

$$(22b) \qquad \alpha = e, \qquad c = P_u(\varphi^* a), \qquad a - P_-(\varphi c) = e.$$

Since

$$X^{-1} = \begin{bmatrix} e & bd^{-1} \\ 0 & d^{-1} \end{bmatrix} \in \mathcal{A}_+, \qquad Y^{-1} = \begin{bmatrix} a^{-1} & 0 \\ ca^{-1} & e \end{bmatrix} \in \mathcal{A}_-,$$

we see that $d^{-1} \in \mathcal{M}_+$ and $a^{-1} \in \mathcal{M}_-$. Furthermore,

$$Q_d X = \begin{bmatrix} e & 0 \\ 0 & P_d d \end{bmatrix}, \qquad Q_d Y = \begin{bmatrix} P_d a & 0 \\ 0 & e \end{bmatrix},$$

and these elements are positive definite in $\mathcal{A}_d$. So both $Q_d X$ and $Q_d Y$ are of the form

$$\begin{bmatrix} p & 0 \\ 0 & q \end{bmatrix}^* \begin{bmatrix} p & 0 \\ 0 & q \end{bmatrix} \left(= \begin{bmatrix} p^* p & 0 \\ 0 & q^* q \end{bmatrix} \right)$$

for some p and q in $\mathcal{M}_d$ with p and q invertible in $\mathcal{M}_d$. It follows that $P_d a$ and $P_d d$ are positive definite in $\mathcal{M}_d$. We have now proved that a and d are solutions of the equations in (5) and have the properties (i)–(iii) stated in the theorem.

Part (d). In this part we prove the sufficiency of the conditions and the uniqueness. Suppose that a and d exist as in the theorem. Define b and c by (7), and put

$$X = \begin{bmatrix} e & -b \\ 0 & d \end{bmatrix}, \qquad Y = \begin{bmatrix} a & 0 \\ -c & e \end{bmatrix}.$$

Then, by the properties (i)–(iii), we have $X \in \mathcal{A}_2$ and $Y \in \mathcal{A}_3$,

$$(23) \qquad X^{-1} = \begin{bmatrix} e & bd^{-1} \\ 0 & d^{-1} \end{bmatrix} \in \mathcal{A}_+, \qquad Y^{-1} = \begin{bmatrix} a^{-1} & 0 \\ ca^{-1} & e \end{bmatrix} \in \mathcal{A}_-,$$

and $P_d X$ and $P_d Y$ are positive definite in $\mathcal{A}_d$. Furthermore, the two equalities in (5) imply that

$$Q_2(KX) = E, \qquad Q_3(KY) = E.$$

So we can apply Theorems 1.1–1.3 to show that K has a unique band extension B, namely

$$(24) \qquad B = X^{-*}(Q_d X)X^{-1} = Y^{-*}(Q_d Y)Y^{-1}.$$

Here

$$(25) \qquad Q_d X = \begin{bmatrix} e & 0 \\ 0 & P_d d \end{bmatrix}, \qquad Q_d Y = \begin{bmatrix} P_d a & 0 \\ 0 & e \end{bmatrix}.$$

Since $B - K \in \mathcal{A}_1 + \mathcal{A}_4$, we have

$$(26) \qquad B = A_g = \begin{bmatrix} e & g \\ g^* & e \end{bmatrix},$$

where $g \in \mathcal{M}$ and $g - \varphi \in \mathcal{M}_\ell$. From $B^{-1} \in \mathcal{A}_c$ and (19) we see that $g(e - g^* g)^{-1}$ is in $\mathcal{M}_\ell$. The identities in (23), (24) and (25) imply that (6) holds. From (14a) applied to $\psi = g$, the first identity in (6) and the first identity in (24), we get

$$
\begin{aligned}
(27) \qquad e - g^* g &= [-g^* \quad e]B \begin{bmatrix} -g \\ e \end{bmatrix} \\
&= [-g^* \quad e] \begin{bmatrix} e & 0 \\ d^{-*}b^* & d^{-*} \end{bmatrix}(Q_d X)\begin{bmatrix} e & bd^{-1} \\ 0 & d^{-1} \end{bmatrix}\begin{bmatrix} -g \\ e \end{bmatrix} \\
&= [0 \quad d^{-*}](Q_d X)\begin{bmatrix} 0 \\ d^{-1} \end{bmatrix} \\
&= d^{-*}(P_d d)d^{-1}.
\end{aligned}
$$

In a similar way, using (14b) with g in place of ψ, the second identity in (6) and the second identity in (24), one shows that

$$(28) \qquad e - gg^* = a^{-*}(P_d a)a^{-1}.$$

According to our hypotheses, $P_d a = p^* p$ and $P_d d = q^* q$ for some invertible p and q in $\mathcal{M}_d$ with inverses also in $\mathcal{M}_d$. Since $d^{\pm 1} \in \mathcal{M}_+$ and $a^{\pm 1} \in \mathcal{M}_-$, it follows that $e - g^* g$ and $e - gg^*$ have the desired factorizations. In particular, $e - g^* g$ is positive definite in $\mathcal{R}$, and hence $\|g\|_{\mathcal{R}} < 1$. So g is a triangular extension of φ with the desired properties. From the result of Part (b) and the uniqueness of the band extension of K, we obtain the uniqueness of g. $\square$

      To derive the description of all strictly contractive extensions we need axiom (A) introduced in the previous section.

      THEOREM 3.2. *Let $\mathcal{M}$ be an algebra with band structure (1) in a unital C^*-algebra $\mathcal{R}$, and assume that axiom (A) holds. Let $\varphi \in \mathcal{M}_\ell$, and suppose that φ has a triangular extension g such that $e - gg^*$ admits a left and $e - g^* g$ admits a right spectral factorization relative to (1). More precisely, let*

$$(29) \qquad e - gg^* = \alpha^{-*}\alpha^{-1} \ (\alpha^{\pm 1} \in \mathcal{M}_-), \qquad e - g^* g = \delta^{-*}\delta^{-1} \ (\delta^{\pm 1} \in \mathcal{M}_+).$$

Put

$$(30) \qquad \beta = P_\ell(\varphi\delta), \qquad \gamma = P_u(\varphi^*\alpha).$$

Then each strictly contractive extension of φ in $\mathcal{M}$ is given by

$$(31) \qquad \mathcal{F}(h) = (\alpha h + \beta)(\gamma h + \delta)^{-1},$$

where the free parameter h is an arbitrary element in $\mathcal{M}_1$ such that $\|h\|_\mathcal{R} < 1$. Moreover, the map $\mathcal{F}$ provides a one-one correspondence between all such h and all strictly contractive extensions of φ.

The coefficients α, β, γ and δ of the linear fractional representation (31) may be constructed by applying the method of Theorem 3.1. Indeed, let a, b, c and d be as in Theorem 3.1. By property (iii) in Theorem 3.1 we may choose invertible elements p and q in $\mathcal{M}_d$ (with inverses in $\mathcal{M}_d$) such that $P_d a = p^* p$ and $P_d(d) = q^* q$. Then we know from Part (d) of the proof of Theorem 3.1 (see formulas (27) and (28)) that

$$(32) \qquad e - g^* g = (qd^{-1})^*(qd^{-1}), \qquad e - gg^* = (pa^{-1})^*(pa^{-1}).$$

Now put

$$(33) \qquad \alpha = ap^{-1}, \qquad \beta = bq^{-1}, \qquad \gamma = cp^{-1}, \qquad \delta = dq^{-1}.$$

Then (29) holds by virtue of (32), and (30) holds, because of (7) and the fact that $\mathcal{M}_d \mathcal{M}_\ell \subset \mathcal{M}_\ell$. So the elements α, β, γ and δ in (33) have the desired properties.

The linear fractional map $\mathcal{F}$ in Theorem 3.1 may be replaced by

$$\mathcal{F}'(f) = (\alpha^* + f\beta^*)^{-1}(\gamma^* + f\delta^*),$$

where the free parameter f now runs over all f in $\mathcal{M}_1$ such that $\|f\|_\mathcal{R} < 1$.

For the proof of Theorem 3.2 we need the following lemma, which may be viewed as an abstract version of Lemma XXXIII.3.1.

LEMMA 3.3. *Let $\mathcal{M}$ be an algebra with band structure (1) in a unital C^*-algebra $\mathcal{R}$, and let $\varphi \in \mathcal{M}_\ell$. For each $\psi \in \mathcal{M}$ let A_ψ be defined by (13). Then $\psi \in \mathcal{M}$ is a strictly contractive extension of φ if and only if A_ψ is a $\mathcal{R}^{2\times2}$-positive extension of $K := A_\varphi$. Furthermore, if $F \in \mathcal{M}^{2\times2}$ is an $\mathcal{R}^{2\times2}$-positive extension of K, then $F = A_\psi$ for some strictly contractive extension ψ of φ in $\mathcal{M}$.*

PROOF. Assume ψ in $\mathcal{M}$ is a strictly contractive extension of φ. So $\psi - \varphi \in \mathcal{M}_1$ and $e - \psi^*\psi = r^*r$ for some invertible r in $\mathcal{R}$. From (13) we see that $A_\psi - K \in \mathcal{A}_1 \oplus \mathcal{A}_4$, and (14a) implies that

$$(34) \qquad A_\psi = \begin{bmatrix} e & \psi \\ 0 & r \end{bmatrix}^* \begin{bmatrix} e & \psi \\ 0 & r \end{bmatrix}.$$

Since r is invertible in $\mathcal{R}$, the factors in the right hand side of (34) are invertible in $\mathcal{R}^{2\times2}$, and A_ψ is an $\mathcal{R}^{2\times2}$-positive extension of K.

Conversely, let F be an $\mathcal{R}^{2\times 2}$-positive extension of K. Then $F-K \in \mathcal{A}_1 \oplus \mathcal{A}_4$. It follows that $F = A_\psi$ for some $\psi \in \mathcal{M}$ such that $\psi - \varphi \in \mathcal{M}_1$. It remains to prove that $e - \psi^*\psi$ is positive definite in $\mathcal{R}$. Since A_ψ is positive definite in $\mathcal{R}^{2\times 2}$, we have $A_\psi = C^*C$ for some invertible C in $\mathcal{R}^{2\times 2}$, and hence, by (14a),

$$(35) \qquad \begin{bmatrix} e & 0 \\ 0 & e - \psi^*\psi \end{bmatrix} = (C\begin{bmatrix} e & -\psi \\ 0 & e \end{bmatrix})^*(C\begin{bmatrix} e & -\psi \\ 0 & e \end{bmatrix}).$$

It follows that the left hand side of (35) is positive definite in $\mathcal{R}^{2\times 2}$. From the definition of the C^*-algebra structure on $\mathcal{R}^{2\times 2}$ it follows that the diagonal entries of a positive definite element in $\mathcal{R}^{2\times 2}$ are positive definite in $\mathcal{R}$. So we may conclude that $e - \psi^*\psi$ is positive definite in $\mathcal{R}$, and hence $\|\psi\|_{\mathcal{R}} < 1$. $\square$

PROOF OF THEOREM 3.2. We split the proof into two parts. In the first part we show that axiom (A) holds for $\mathcal{A} = \mathcal{M}^{2\times 2}$ relative to $\mathcal{R}^{2\times 2}$.

Part (a). Put $\mathcal{A} = \mathcal{M}^{2\times 2}$, and decompose $\mathcal{A}$ as in (10). Then (see the paragraph preceding the proof of Theorem 3.1) we may view $\mathcal{A}$ as an algebra with band structure (10) in the unital C^*-algebra $\mathcal{R}^{2\times 2}$. Now take $G \in \mathcal{A}_+$ with $\|G\|_{\mathcal{R}^{2\times 2}} < 1$. Thus

$$G = \begin{bmatrix} g_{11} & g_{12} \\ 0 & g_{22} \end{bmatrix}, \qquad g_{\nu\nu} \in \mathcal{M}_+ \quad (\nu = 1,2).$$

The norm condition on G implies that $\|g_{\nu\nu}\|_{\mathcal{R}} < 1$ for $\nu = 1,2$. Since axiom (A) holds for $\mathcal{M}$ relative to $\mathcal{R}$, we see that $(e - g_{\nu\nu})^{-1} \in \mathcal{M}_+$ for $\nu = 1,2$. It follows that

$$(E - G)^{-1} = \begin{bmatrix} (e - g_{11})^{-1} & -(e - g_{11})^{-1}g_{12}(e - g_{22})^{-1} \\ 0 & (e - g_{22})^{-1} \end{bmatrix} \in \mathcal{A}_+,$$

and hence axiom (A) holds for $\mathcal{A}$ relative to $\mathcal{R}^{2\times 2}$.

Part (b). Let $K = A_\varphi$, where φ is as in the theorem, and put $r = \delta^{-1}$, $s = \alpha^{-1}$. From Part (b) of the proof of Theorem 3.1 we know that A_g is a band extension of K and

$$(36a) \qquad A_g = U^{-*}U^{-1}, \qquad U = \begin{bmatrix} e & -gr^{-1} \\ 0 & r^{-1} \end{bmatrix},$$

$$(36b) \qquad A_g = V^{-*}V^{-1}, \qquad V = \begin{bmatrix} s^{-1} & 0 \\ -g^*s^{-1} & e \end{bmatrix},$$

where (36a) is a right and (36b) is a left spectral factorization relative to (10). So we may apply Theorem 2.1 to $\mathcal{A}$ and to $K = K^* \in \mathcal{A}_c$. So each $\mathcal{R}^{2\times 2}$-positive extension F of K is of the form

$$(37) \qquad F = (V\Delta + U)^{-*}(E - \Delta^*\Delta)(V\Delta + U)^{-1},$$

where Δ is an arbitrary element in $\mathcal{A}_1$ such that $\|\Delta\|_{\mathcal{R}^{2\times 2}} < 1$. From the definition of $\mathcal{A}_1$ we see that

$$(38) \qquad \Delta = \begin{bmatrix} 0 & -h \\ 0 & 0 \end{bmatrix},$$

where $h \in \mathcal{M}_1$. It follows that h in (38) is an arbitrary element of $\mathcal{M}_1$ such that $\|h\|_{\mathcal{R}} < 1$.

Since A_g is a band extension, $U \in \mathcal{A}_2$ and hence $gr^{-1} \in \mathcal{M}_\ell$. Furthermore, $g = \varphi + m$ with $m \in \mathcal{M}_1$. So $gr^{-1} = g\delta = \varphi\delta + m\delta$ with $m\delta \in \mathcal{M}_1\mathcal{M}_+ \subset \mathcal{M}_1$. Hence $g\delta = P_\ell(\varphi\delta)$, and so $g = \beta\delta^{-1}$, by the first identity in (30). In a similar way one shows that $g^* = \gamma\alpha^{-1}$, where γ is given by the second identity in (30).

By inserting these expressions for g and g^* into U and V, respectively, it is straightforward to check that

$$V\Delta + U = \begin{bmatrix} e & -(\beta + \alpha h) \\ 0 & \delta + \gamma h \end{bmatrix},$$

and hence F in (37) is equal to

$$F = \begin{bmatrix} e & \mathcal{F}(h) \\ \mathcal{F}(h)^* & e \end{bmatrix},$$

where $\mathcal{F}(h)$ is given by (31). But then, by Lemma 3.3, we may conclude that the set of strictly contractive extensions of φ consists of all $\mathcal{F}(h)$ with $h \in \mathcal{M}_1$ such that $\|h\|_{\mathcal{R}} < 1$. Since Δ in (37) is uniquely determined by F (by virtue of Theorem 2.1), the element $\mathcal{F}(h)$ determines h uniquely. Thus the map $\mathcal{F}$ gives the desired one-one correspondence. $\square$

Let us apply Theorems 3.1 and 3.2 to the algebra $\mathcal{L}$. Fix $0 \leq m < n$, and let $\Phi = [\varphi_{ij}]_{i,j=0}^n$ be an m-lower triangular matrix. Thus the entries φ_{ij} are zero for $j - i > m$. From Theorem XXXIII.3.2 we know that Φ has a strictly contractive extension if and only if the matrices

$$(39) \qquad \Gamma_k = \begin{bmatrix} \varphi_{k0} & \cdots & \varphi_{k,k+m} \\ \vdots & & \vdots \\ \varphi_{n0} & \cdots & \varphi_{n,k+m} \end{bmatrix}, \qquad k = 0, \ldots, n-m,$$

have norm strictly less than one. Moreover, if these norm conditions are fulfilled, then Φ has a unique triangular extension T. Theorem 3.1 allows us to determine T explicitly, and Theorem 3.2 provides a method to find all strictly contractive extensions.

The first step is to solve the equations in (5) for the case considered here. This means that we have to find a lower triangular matrix $A = [a_{ij}]_{i,j=0}^n$ and an upper triangular matrix $D = [d_{ij}]_{i,j=0}^n$ such that the diagonal entries a_{ii} and d_{ii} are strictly positive for $i = 0, 1, \ldots, n$ and

$$(40a) \qquad A - P_-\big(\Phi P_u(\Phi^* A)\big) = I,$$

$$(40b) \qquad D - P_+\big(\Phi^* P_\ell(\Phi D)\big) = I.$$

Here I is the $(n+1) \times (n+1)$ identity matrix, and P_- is the map which assigns to a matrix $\Psi = [\psi_{ij}]_{i,j=0}^n$ the matrix which one obtains from Ψ by replacing the entries ψ_{ij}

with $j - i > 0$ by zero. The maps P_u, P_+, P_ℓ in (40a) and (40b) have a natural analogous interpretation. Fix $0 \leq k \leq n - m$, and let us compute the k-th column of A by using (40a). Recall that A is lower triangular, and so we are interested in a_{ik} for $k \leq i \leq n$ only. Now, by (40a),

$$(41) \qquad a_{ik} - \sum_{\nu=k}^{n} \left(\sum_{j=0}^{k+m} \varphi_{ij}\overline{\varphi_{\nu j}} \right) a_{\nu k} = \delta_{ik}, \qquad i = k, \ldots, n,$$

where δ_{ik} is the Kronecker delta. For $n - m < k \leq n$ equation (40a) yields

$$(42) \qquad a_{ik} - \sum_{\nu=k}^{n} \left(\sum_{j=0}^{n} \varphi_{ij}\overline{\varphi_{\nu j}} \right) a_{\nu k} = \delta_{ik}, \qquad i = k, \ldots, n.$$

The equations (41) and (42) can be rewritten in a more concise form as follows

$$(43) \qquad (I_{n-k+1} - \Gamma_k \Gamma_k^*) \begin{bmatrix} a_{kk} \\ \vdots \\ a_{nk} \end{bmatrix} = \begin{bmatrix} 1 \\ 0 \\ \vdots \\ 0 \end{bmatrix}, \qquad k = 0, \ldots, n.$$

Here I_{n-k+1} is the identity matrix of order $n - k + 1$, the matrix Γ_k is given by (39) for $k = 0, \ldots, n - m$ and

$$(44) \qquad \Gamma_k = \begin{bmatrix} \varphi_{k0} & \cdots & \varphi_{kn} \\ \vdots & & \vdots \\ \varphi_{n0} & \cdots & \varphi_{nn} \end{bmatrix}, \qquad k = n - m + 1, \ldots, n.$$

Since the matrices in (39) have norm strictly less than one, the same is true for the matrices in (44), and hence $I_{n-k+1} - \Gamma_k \Gamma_k^*$ is positive definite for each k. So

$$\begin{bmatrix} a_{kk} \\ \vdots \\ a_{nk} \end{bmatrix}, \qquad k = 0, \ldots, n,$$

is precisely the first column of $(I_{n-k+1} - \Gamma_k \Gamma_k^*)^{-1}$. Since the latter matrix is positive definite, its entry in the left upper corner is strictly positive, and thus $a_{kk} > 0$. To solve (40b) we need the matrices

$$\Lambda_k = \begin{bmatrix} \varphi_{00} & \cdots & \varphi_{0k} \\ \vdots & & \vdots \\ \varphi_{n0} & \cdots & \varphi_{nk} \end{bmatrix}, \qquad k = 0, \ldots, m,$$

$$\Lambda_k = \begin{bmatrix} \varphi_{k-m,0} & \cdots & \varphi_{k-m,k} \\ \vdots & & \vdots \\ \varphi_{n0} & \cdots & \varphi_{nk} \end{bmatrix}, \qquad k = m + 1, \ldots, n.$$

The fact that the matrices in (39) have norm strictly less than one implies that the same is true for the matrices $\Lambda_0, \ldots, \Lambda_n$. One computes that (40b) is equivalent to

$$(45) \qquad (I_k - \Lambda_k^* \Lambda_k) \begin{bmatrix} d_{0k} \\ \vdots \\ d_{kk} \end{bmatrix} = \begin{bmatrix} 0 \\ \vdots \\ 0 \\ 1 \end{bmatrix}, \qquad k = 0, \ldots, n.$$

It follows that

$$\begin{bmatrix} d_{0k} \\ \vdots \\ d_{kk} \end{bmatrix}, \qquad k = 0, \ldots, n,$$

is the last column of $(I_k - \Lambda_k^* \Lambda_k)^{-1}$. The latter matrix is positive definite. So we may conclude that $d_{kk} > 0$.

The next step is to determine the entries of the matrices

$$(46) \qquad B = P_\ell(\Phi D), \qquad C = P_u(\Phi^* A).$$

Put $B = [b_{ij}]_{i,j=0}^n$ and $C = [c_{ij}]_{i,j=0}^n$. Fix $0 \le k \le n$, and let us compute c_{ik}. By definition, $c_{ik} = 0$ for $k - i < -m$, and thus we are only interested in c_{ik} for $i = 0, \ldots, \nu(k)$, where $\nu(k) = \min\{n, m + k\}$. Now, by the second equation in (46),

$$c_{ik} = \sum_{j=k}^n \overline{\varphi_{ji}} a_{jk}, \qquad i = 0, \ldots, \nu(k),$$

which we can rewrite as

$$(47) \qquad \begin{bmatrix} c_{0k} \\ \vdots \\ c_{\nu(k),k} \end{bmatrix} = \Gamma_k^* \begin{bmatrix} a_{kk} \\ \vdots \\ a_{nk} \end{bmatrix}, \qquad k = 0, \ldots, n.$$

In a similar way, using the first identity in (46), we have $b_{ik} = 0$ for $k - i > m$ and

$$(48) \qquad \begin{bmatrix} b_{\mu(k),k} \\ \vdots \\ b_{nk} \end{bmatrix} = \Lambda_k \begin{bmatrix} d_{0k} \\ \vdots \\ d_{kk} \end{bmatrix}, \qquad k = 0, \ldots, n.$$

Here $\mu(k) = \max\{0, k - m\}$ for $k = 0, \ldots, n$. We can now apply Theorem 3.1 to show that the unique triangular extension T of Φ is given by

$$T = BD^{-1} = A^{-*}C^*.$$

To get the description of all positive extensions it remains to apply Theorem 3.2. We already know that our algebra $\mathcal{L}$ satisfies axiom (A) relative to $\mathcal{R} = \mathcal{L}$. So, by Theorem 3.2, all positive extensions F of Φ are given by a linear fractional representation,

$$(49) \qquad F = (\widetilde{A}H + \widetilde{B})(\widetilde{C}H + \widetilde{D})^{-1},$$

where $H = [h_{ij}]_{i,j=0}^{n}$ is an arbitrary matrix of norm strictly less than one such that $h_{ij} = 0$ for $j - i \leq m$. Furthermore, by the remark made in the first paragraph after Theorem 3.2, the coefficients in (49) are given by

$$\widetilde{A} = AP^{-1}, \qquad \widetilde{B} = BQ^{-1}, \qquad \widetilde{C} = CP^{-1}, \qquad \widetilde{D} = DQ^{-1},$$

where A, B, C and D are the matrices constructed in the previous two paragraphs and the matrices P and Q are $(n+1) \times (n+1)$ diagonal matrices with i-th diagonal entry equal to $\sqrt{a_{ii}}$ and $\sqrt{d_{ii}}$, respectively.

For the algebra $\mathcal{L}$ with band structure (3), properties (i) and (iii) in Theorem 3.1 imply property (ii). In general, this is not true, and in concrete applications often the main difficulty is to check property (ii). For this purpose the following proposition is useful. In this proposition we assume that $\mathcal{M}$ is a Banach algebra and we employ a new axiom.

AXIOM (B). *If A is a positive definite element in $\mathcal{A} = \mathcal{M}^{2\times 2}$, then its diagonal entries are positive definite in $\mathcal{M}$.*

PROPOSITION 3.4. *Let $\mathcal{M}$ be an algebra with band structure (1) satisfying axiom (B). Assume that $\mathcal{M}$ is a Banach algebra and that the linear spaces in (1) are closed in $\mathcal{M}$. Let $\varphi \in \mathcal{M}_\ell$, and suppose that there exist analytic functions*

$$\alpha\colon [0,1] \to \mathcal{M}_-, \qquad \delta\colon [0,1] \to \mathcal{M}_+$$

such that

$$(50a) \qquad \alpha(t) - t^2 P_-\big(\varphi(P_u(\varphi^* \alpha(t)))\big) = e, \qquad 0 \leq t \leq 1,$$

$$(50b) \qquad \delta(t) - t^2 P_+\big(\varphi^*(P_\ell(\varphi \delta(t)))\big) = e, \qquad 0 \leq t \leq 1,$$

and $P_d\big(\alpha(t)\big)$ and $P_d\big(\delta(t)\big)$ are positive definite in $\mathcal{M}_d$. Then $a = \alpha(1)$ and $d = \delta(1)$ are invertible, $a^{-1} \in \mathcal{M}_-$ and $d^{-1} \in \mathcal{M}_+$.

PROOF. Let $\mathcal{A} = \mathcal{M}^{2\times 2}$, and endow $\mathcal{A}$ with the band structure defined by (10). By Q_d we denote the natural projection of $\mathcal{A}$ onto $\mathcal{A}_d$. Consider

$$(51) \qquad X(t) = \begin{bmatrix} e & -\beta(t) \\ 0 & \delta(t) \end{bmatrix}, \qquad Y(t) = \begin{bmatrix} \alpha(t) & 0 \\ -\gamma(t) & e \end{bmatrix}, \qquad 0 \leq t \leq 1,$$

where

$$(52) \qquad \beta(t) = t P_\ell\big(\varphi \delta(t)\big), \quad \gamma(t) = t P_u\big(\varphi^* \alpha(t)\big), \qquad 0 \leq t \leq 1.$$

Note that $\mathcal{M}_-$ and $\mathcal{M}_+$ are Banach algebras in their own right, and recall that the set of invertible elements in a Banach algebra is open. Since $\alpha(0) = e$ and $\delta(0) = e$, the continuity of the functions α and δ implies that there exists $0 < \varepsilon \leq 1$ such that for $0 \leq t < \varepsilon$ the elements $\alpha(t)$ and $\delta(t)$ are invertible in $\mathcal{M}_-$ and $\mathcal{M}_+$, respectively. Thus for $0 \leq t < \varepsilon$ the elements $\alpha(t)$ and $\delta(t)$ satisfy equation (5) with $t\varphi$ in place of φ and have the properties (i)–(iii) stated in Theorem 3.1. So we know from Part (d) of the proof of Theorem 3.1 that

$$K(t) = \begin{bmatrix} e & t\varphi \\ t\varphi^* & e \end{bmatrix} \in \mathcal{A}_c$$

has (relative to the band structure (10)) a unique band extension $B(t)$ for $0 \leq t < \varepsilon$. In fact, according to formula (24) we have

$$(53) \qquad B(t) = X(t)^{-*}(Q_d X(t)) X(t)^{-1} = Y(t)^{-*}(Q_d Y(t)) Y(t)^{-1}, \qquad 0 \leq t < \varepsilon.$$

Here $T^{-*} = (T^{-1})^*$. From (51) we see that

$$Q_d X(t) = \begin{bmatrix} e & 0 \\ 0 & P_d \delta(t) \end{bmatrix}, \qquad Q_d Y(t) = \begin{bmatrix} P_d \alpha(t) & 0 \\ 0 & e \end{bmatrix}$$

for each $0 \leq t \leq 1$. Furthermore, by our hypotheses on $P_d \delta(t)$ and $P_d \alpha(t)$, the block matrices $Q_d X(t)$ and $Q_d Y(t)$ are invertible in $\mathcal{A}$ for each $0 \leq t \leq 1$. But then we see from (53) that

$$(54) \qquad X(t)(Q_d X(t))^{-1} X(t)^* = Y(t)(Q_d Y(t))^{-1} Y(t)^*, \qquad 0 \leq t < \varepsilon.$$

By computing the $(1,1)$ entry in the left hand side and in the right hand side of (54), we find that

$$(55) \qquad e + \beta(t)(P_d \delta(t))^{-1} \beta(t)^* = \alpha(t)(P_d \alpha(t))^{-1} \alpha(t)^*$$

for $0 \leq t < \varepsilon$. Now, use the fact that the projections P_ℓ and P_d are continuous linear operators. It follows that the functions

$$t \mapsto e + \beta(t)(P_d \delta(t))^{-1} \beta(t)^*, \qquad t \mapsto \alpha(t)(P_d \alpha(t))^{-1} \alpha(t)^*$$

are analytic in the real variable $t \in [0,1]$. Thus, by analytic continuation, (55) holds true for each $0 \leq t \leq 1$. Write $P_d \delta(t) = q(t)^* q(t)$, $q(t)^{\pm 1} \in \mathcal{M}_d$. Then the left hand side of (55) is the $(1,1)$ entry of

$$(56) \qquad \begin{bmatrix} e & \beta(t)q(t)^{-1} \\ 0 & e \end{bmatrix} \begin{bmatrix} e & \beta(t)q(t)^{-1} \\ 0 & e \end{bmatrix}^* .$$

The product in (56) is positive definite in $\mathcal{A}$. So we can apply axiom (B) to show that the left hand side of (55) is positive definite in $\mathcal{M}$. In particular, the left hand side of (55) is invertible in $\mathcal{M}$. The latter holds for each $0 \leq t \leq 1$.

Put

$$\alpha^\times(t) = (P_d \alpha(t))^{-1} \alpha(t)^* \{ e + \beta(t)(P_d \delta(t))^{-1} \beta(t)^* \}^{-1}, \qquad 0 \leq t \leq 1.$$

From what has been proved so far we know that $\alpha(t)\alpha^\times(t) = e$ for $0 \leq t \leq 1$. Furthermore, since $\alpha(t)$ is invertible for $0 \leq t < \varepsilon$, we also have

$$(57) \qquad \alpha^\times(t)\alpha(t) = e, \qquad 0 \leq t < \varepsilon.$$

Obviously, $t \mapsto \alpha^{\times}(t)$ is analytic in the real variable $t \in [0,1]$. So, by analytic continuation, we see that the equality in (57) holds for each $0 \le t \le 1$. We have now proved that $\alpha(t)$ is invertible for $0 \le t \le 1$. Since $\alpha(t)^{-1} \in \mathcal{M}_-$ for $0 \le t < \varepsilon$, we have $P_+^0 \alpha(t)^{-1} = 0$ for $0 \le t < \varepsilon$. The projection P_+^0 is continuous, and so $t \mapsto P_+^0 \alpha(t)^{-1}$ is analytic on $0 \le t \le 1$. Therefore, by analytic continuation, $P_+^0 \alpha(t)^{-1} = 0$ for each $t \in [0,1]$, and thus $\alpha(t)^{-1} \in \mathcal{M}_-$ for $0 \le t \le 1$. So $a = \alpha(1)$ has the desired properties.

The proof for $d = \delta(1)$ follows in an analogous way by computing the $(2,2)$ entries of the block matrices in the left and right hand side of (54). $\square$

We conclude this section with two remarks. It may happen that in (1) the space $\mathcal{M}_2^0$ (and hence $\mathcal{M}_3^0$ too) consists of the zero element only. In fact, this case is of special interest. For example, for our algebra $\mathcal{L}$ it means that we seek strictly contractive extensions of a lower triangular matrix. In general, if $\mathcal{M}_2^0 = \{0\}$, then $\mathcal{M}_\ell = \mathcal{M}_-$ and $\mathcal{M}_u = \mathcal{M}_+$, and hence the projections P_ℓ and P_u appearing in (7) are equal to P_- and P_+, respectively.

The second remark concerns contractive extension problems for strictly lower triangular patterns or for rectangular non-square matrices (for instance, of the type considered in Section XXXIII.3 with $p < 0$ and/or $n \ne r$). To treat such problems in the general setting of the band method requires a modification of the set up of the present section. Consider a direct sum of linear spaces, $\mathcal{M} = \mathcal{M}_- \oplus \mathcal{M}_+$, and let $\varphi \in \mathcal{M}_-$ be given. We assume that $\mathcal{M}$ appears as the set of the $(1,2)$-elements in an algebra of 2×2 block matrices:

$$(58) \qquad \mathcal{A} = \left\{ \begin{bmatrix} a & b \\ c & d \end{bmatrix} \mid a \in \mathcal{H},\ b \in \mathcal{M},\ c \in \mathcal{N},\ d \in \mathcal{K} \right\},$$

where $\mathcal{H}$ and $\mathcal{K}$ are $*$-algebras with units $e_{\mathcal{H}}$ and $e_{\mathcal{K}}$, respectively, and $\mathcal{N}$ is a linear space which is isomorphic to $\mathcal{M}$ via an operator $*$ whose inverse is also denoted by $*$. For example, if $\mathcal{M}$ is the linear space of all $n \times r$ matrices, then we take $\mathcal{H}$ to be the algebra of all $n \times n$ matrices, $\mathcal{K}$ the algebra of all $r \times r$ matrices, and $\mathcal{N}$ the linear space of all $r \times n$ matrices. Note that in this case the operation of taking adjoints provides the desired isomorphism between $\mathcal{M}$ and $\mathcal{N}$. Since $\mathcal{A}$ is assumed to be an algebra with respect to the usual rules of addition and multiplication of matrices, we have $\mathcal{N}\mathcal{M} \subset \mathcal{K}$ and $\mathcal{M}\mathcal{N} \subset \mathcal{H}$. We shall also assume that $\mathcal{A}$ is a $*$-subalgebra of a unital C^*-algebra $\mathcal{B}$. To state the extension problem for φ, let us say that $\psi \in \mathcal{M}$ is a *strictly contractive extension* of φ if $\psi - \varphi \in \mathcal{M}_+$ and $\|\psi\| < 1$. Here $\|\cdot\|$ denotes the norm on $\mathcal{M}$ which $\mathcal{M}$ inherits from the norm on $\mathcal{B}$.

Given direct sum decompositions

$$\mathcal{H} = \mathcal{H}_-^0 \oplus \mathcal{H}_d \oplus \mathcal{H}_+^0, \qquad \mathcal{K} = \mathcal{K}_-^0 \oplus \mathcal{K}_d \oplus \mathcal{K}_+^0,$$

and

$$\mathcal{N}_- = \mathcal{M}_+^*, \qquad \mathcal{N}_+ = \mathcal{M}_-^*,$$

it is not difficult to give a set of natural conditions which guarantee that $\mathcal{A}$ in (58) becomes an algebra with band structure

$$(59) \qquad \mathcal{A} = \mathcal{A}_1 \oplus \mathcal{A}_2^0 \oplus \mathcal{A}_d \oplus \mathcal{A}_3^0 \oplus \mathcal{A}_4,$$

if we take

$$\mathcal{A}_1 = \{ \begin{bmatrix} 0 & b \\ 0 & 0 \end{bmatrix} \mid b \in \mathcal{M}_+ \},$$

$$\mathcal{A}_2^0 = \{ \begin{bmatrix} a & b \\ 0 & d \end{bmatrix} \mid a \in \mathcal{H}_+^0,\ d \in \mathcal{K}_+^0,\ b \in \mathcal{M}_- \},$$

$$\mathcal{A}_d = \{ \begin{bmatrix} a & 0 \\ 0 & d \end{bmatrix} \mid a \in \mathcal{H}_d,\ d \in \mathcal{K}_d \},$$

$$\mathcal{A}_3^0 = \{ \begin{bmatrix} a & 0 \\ c & d \end{bmatrix} \mid a \in \mathcal{H}_-^0,\ d \in \mathcal{K}_-^0,\ c \in \mathcal{N}_+ \},$$

$$\mathcal{A}_4 = \{ \begin{bmatrix} 0 & 0 \\ c & 0 \end{bmatrix} \mid c \in \mathcal{N}_- \}.$$

As soon as one has the band structure (59) one can solve the strictly contractive extension problem for $\varphi \in \mathcal{M}_-$ by using the same type of arguments as employed in the proofs of Theorems 3.1 and 3.2. For further details we refer to Section II.1 of Gohberg-Kaashoek-Woerdeman [1], where such problems are considered in a slightly different setting. See also Section II.1 in Gohberg-Kaashoek-Woerdeman [3] and the exercises.

XXXIV.4 MAXIMUM ENTROPY PRINCIPLES

In this section an abstract maximum entropy principle is derived in the general setting of the band method. This principle provides an alternative way to identify the band extension (for positive extension problems) or the triangular extension (for contractive extension problems). Throughout this section, $\mathcal{M}$ is an algebra with involution * and unit e. Furthermore, $\mathcal{M}$ is endowed with the band structure

$$(1) \qquad \mathcal{M} = \mathcal{M}_1 \oplus \mathcal{M}_2^0 \oplus \mathcal{M}_d \oplus \mathcal{M}_3^0 \oplus \mathcal{M}_4.$$

We begin with a lemma.

LEMMA 4.1. *Let b be an element in $\mathcal{M}$ with a right spectral factorization relative to* (1). *Then b factorizes as*

$$(2) \qquad b = (e + m_+^0)^* \Delta (e + m_+^0),$$

where $\Delta \in \mathcal{M}_d$ and the element $m_+^0 + e$ is invertible with

$$(3) \qquad (e + m_+^0)^{\pm 1} \in e + \mathcal{M}_+^0.$$

Moreover, the factors in (2) *are uniquely determined by b.*

PROOF. Let $b = b_+^* b_+$ be a right spectral factorization of b relative to (1). Thus $b_+^{\pm 1} \in \mathcal{M}_+$. Put $d = P_d(b_+)$. By the first part of Lemma 1.4, the element d is invertible and $d^{-1} = P_d(b_+^{-1})$. Note that $b_+ = d + b_+^0$, with $b_+^0 \in \mathcal{M}_+^0$. It follows that (2) holds with $m_+^0 = d^{-1} b_+ - e$ and $\Delta = d^* d$. Furthermore, by the multiplication table for (1),

$$m_+^0 = d^{-1}(d + b_+^0) - e = d^{-1} b_+^0 \in \mathcal{M}_+^0.$$

Since $(e + m_+^0)^{-1} = b_+^{-1} d$, we have

$$(e + m_+^0)^{-1} - e = b_+^{-1}(d - b_+) = -b_+^{-1} b_+^0 \in \mathcal{M}_+^0,$$

because $\mathcal{M}_+ \mathcal{M}_+^0 \subset \mathcal{M}_+^0$. Also, $\Delta \in \mathcal{M}_d$, because $d^* \in \mathcal{M}_d$ and $\mathcal{M}_d \mathcal{M}_d \subset \mathcal{M}_d$.

To prove the uniqueness of the factorization, consider a second factorization,

$$(4) \qquad\qquad b = (e + n_+^0)^* \Lambda (e + n_+^0),$$

where $\Lambda \in \mathcal{M}_d$ and (3) holds with n_+^0 in place of m_+^0. From (2) and (4) we see that

$$(5) \qquad\qquad (e + n_+^0)^{-*}(e + m_+^0)^* \Delta = \Lambda(e + n_+^0)(e + m_+^0)^{-1}.$$

Here c^{-*} stands for $(c^*)^{-1}$ for any invertible c. Now use the multiplication table for an algebra with band structure to show that the set $e + \mathcal{M}_+^0$ is closed under multiplication. Since $\mathcal{M}_d \mathcal{M}_+^0 \subset \mathcal{M}_+^0$ and $\mathcal{M}_+^0 \mathcal{M}_d \subset \mathcal{M}_+^0$, we conclude that the right hand side of (5) belongs to $\Lambda + \mathcal{M}_+^0$. In a similar way, using $\mathcal{M}_-^0 = (\mathcal{M}_+^0)^*$, one sees that the left hand side of (5) is in $\Delta + \mathcal{M}_-^0$. It follows that $\Delta = \Lambda$ and

$$(6) \qquad\qquad \Lambda = \Lambda(e + n_+^0)(e + m_+^0)^{-1}.$$

All the terms in (6) are invertible, and hence $n_+^0 = m_+^0$. $\quad\square$

The element Δ in (2) is called the *right multiplicative diagonal* of b (relative to (1)) and is denoted by $\Delta_r(b)$. From its construction in the first paragraph of the proof of Lemma 4.1 it is clear that $\Delta_r(b)$ is a positive definite element in $\mathcal{M}_d$. In fact, $\Delta_r(b) = P_d(b_+)^* P_d(b_+)$, where $b = b_+^* b_+$ is a right spectral factorization of b.

Let $\mathcal{L}$ be the algebra of all $(n + 1) \times (n + 1)$ matrices with complex entries and with band structure

$$(7) \qquad\qquad \mathcal{L} = \mathcal{L}_1 \oplus \mathcal{L}_2^0 \oplus \mathcal{L}_d \oplus \mathcal{L}_3^0 \oplus \mathcal{L}_4$$

defined by formulas (7a)–(7e) in the first section of this chapter. For a positive definite element B in $\mathcal{L}$ the factorization (2) is just an LU-factorization (see Chapter XXII) with respect to the finite chain $\mathbb{P} = \{P_0, P_1, \ldots, P_n\}$, where P_j is the projection of $\mathbb{C}^{n+1}$ given by

$$P_j(x_0, \ldots, x_n) = (x_0, \ldots, x_{j-1}, 0, \ldots, 0), \qquad j = 0, \ldots, n.$$

It follows that we can apply the LU-factorization theorem in Section XXII.1 to get explicit formulas for the factors in (2) for the case considered here. We shall return to this connection (with more details) in Section XXXV.1. Here we only note that for the algebra $\mathcal{L}$ the first (resp., third) factor in the right hand side of (2) is a lower (resp., upper) triangular matrix with diagonal entries equal to one, and hence the positive definite matrix B and its right multiplicative diagonal $\Delta_r(B)$ have the same determinant, i.e.,

$$(8) \qquad\qquad \det B = \det \Delta_r(B).$$

Assume that b admits a left spectral factorization relative to (1), $b = b_-^* b_-$ say. Then Lemma 4.1 applies to b^{-1}, and it follows from the symmetry conditions on the band structure (1) that b has a unique factorization

$$(9) \qquad b = (e + m_-^0)^* \widetilde{\Delta}(e + m_-^0),$$

such that $\widetilde{\Delta} \in \mathcal{M}_d$ and $m_-^0 + e$ is invertible with

$$(10) \qquad (e + m_-^0)^{\pm 1} \in e + \mathcal{M}_-^0.$$

The element $\widetilde{\Delta}$ in (9) is called the *left multiplicative diagonal* of b (relative to (1)) and is denoted by $\Delta_\ell(b)$. We have $\Delta_\ell(b) = P_d(b_-)^* P_d(b_-)$, and thus $\Delta_\ell(b)$ is also positive definite in $\mathcal{M}_d$ (by virtue of the first part of Lemma 1.4).

In what follows we assume that $\mathcal{M}$ is an algebra with band structure (1) in a unital C^*-algebra $\mathcal{R}$. An element $h \in \mathcal{R}$ is said to be *positive semidefinite* or *non-negative* (cf., Theorem XXXI.5.4) in $\mathcal{R}$ if $h = g^* g$ for some $g \in \mathcal{R}$. For a and b in $\mathcal{M}$ we write $a \geq_\mathcal{R} b$ whenever $a - b$ is positive semidefinite in $\mathcal{R}$.

To obtain a maximum entropy principle in the present abstract setting, $\mathcal{M}$ will be required to satisfy the following two axioms relative to $\mathcal{R}$.

AXIOM (C1). *If $a \in \mathcal{M}$ is positive semidefinite in $\mathcal{R}$, then $P_d(a)$ is positive semidefinite in $\mathcal{R}$.*

AXIOM (C2). *If $a \in \mathcal{M}$ is positive semidefinite in $\mathcal{R}$ and $P_d(a) = 0$, then $a = 0$.*

Axioms (C1) and (C2) are fulfilled for the algebra $\mathcal{L}$ of all $(n+1) \times (n+1)$ matrices with band structure (7). In this case $\mathcal{R}$ is just $\mathcal{L}$ endowed with the operator norm. Furthermore, $A \in \mathcal{L}$ is positive semidefinite in $\mathcal{L}$ if and only if

$$(11) \qquad \langle Ax, x \rangle \geq 0, \qquad x \in \mathbb{C}^{n+1},$$

where $\langle \cdot, \cdot \rangle$ is the standard inner product on $\mathbb{C}^{n+1}$. Thus, if $A = C^* C$ for some $C \in \mathcal{L}$, then the diagonal entries of A are nonnegative, and hence $P_d(A)$ is positive semidefinite in $\mathcal{L}$. So axiom (C1) is fulfilled. Next, assume that the diagonal entries of $A = C^* C$ are zero. Since the central submatrices of a positive semidefinite matrix in $\mathcal{L}$ are again positive semidefinite, it follows that for each $0 \leq i < j \leq n$ the matrix

$$\begin{bmatrix} 0 & a_{ij} \\ \overline{a}_{ij} & 0 \end{bmatrix}$$

is positive semidefinite, which can only happen when $a_{ij} = 0$. Thus axiom (C2) also holds for $\mathcal{L}$.

The next theorem provides the maximum entropy principle in the general setting of the band method.

THEOREM 4.2. *Let $\mathcal{M}$ be an algebra with band structure (1) in a unital C^*-algebra $\mathcal{R}$, and assume that $\mathcal{M}$ satisfies axioms (C1) and (C2). Let $k = k^* \in \mathcal{M}_c$*

have a band extension b with a right spectral factorization relative to (1). Then for any positive extension a of k such that a admits a right spectral factorization relative to (1), we have

$$(12) \qquad \Delta_r(b) \geq_{\mathcal{R}} \Delta_r(a),$$

and in (12) equality holds if and only if $a = b$.

PROOF. Since a and b both admit a right spectral factorization, the right multiplicative diagonals of a and b are well-defined, and we may write

$$(13) \qquad a = (e + a_+)^* \Delta_r(a)(e + a_+), \qquad b = (e + b_+)^* \Delta_r(b)(e + b_+),$$

with $(e + a_+)^{\pm 1}$ and $(e + b_+)^{\pm 1}$ in $e + \mathcal{M}_+^0$. By our hypotheses, b is a band extension. Thus $b^{-1} \in \mathcal{M}_c$. Furthermore, we have $\Delta_r(b) = d^* d$ for some $d^{\pm 1} \in \mathcal{M}_d$. Hence $b^{-1} = hh^*$ with $h = (e + b_+)^{-1} d^{-1}$, and we can use the second part of Lemma 1.4 to conclude that $h \in \mathcal{M}_2$. It follows that

$$(14) \qquad c_+ := (e + b_+)^{-1} - e \in \mathcal{M}_2^0.$$

By using the identities in (13) we see that

$$(15) \qquad \Delta_r(b) = (e + c_+)^*(b - a)(e + c_+) + (e + h_+)^* \Delta_r(a)(e + h_+),$$

where

$$(16) \qquad e + h_+ = (e + a_+)(e + c_+) \in e + \mathcal{M}_+^0,$$

because $e + \mathcal{M}_+^0$ is closed under multiplication. Both b and a are positive extensions of k. It follows that $b - a \in \mathcal{M}_1 + \mathcal{M}_4$, and hence formula (14), the multiplication table for (1) and the symmetry properties of (1) imply that

$$(b - a)c_+ \in \mathcal{M}_1 + \mathcal{M}_-^0, \qquad c_+^*(b - a) \in \mathcal{M}_+^0 + \mathcal{M}_4,$$

$$c_+^*(b - a)c_+ \in \mathcal{M}_-^0 + \mathcal{M}_+^0.$$

In particular,

$$P_d\big((e + c_+)^*(b - a)(e + c_+)\big) = 0.$$

Next, note that $\Delta_r(a)h_+ \in \mathcal{M}_+^0$ and $h_+^* \Delta_r(a) \in \mathcal{M}_-^0$. So

$$(17) \qquad \Delta_r(b) = P_d\big((e + h_+)^* \Delta_r(a)(e + h_+)\big) = \Delta_r(a) + P_d\big(h_+^* \Delta_r(a)h_+\big).$$

We know that $\Delta_r(a)$ is positive definite in $\mathcal{M}_d$. In other words, $\Delta_r(a) = d^* d$ for some invertible d in $\mathcal{M}_d$. It follows that $h_+^* \Delta_r(a)h_+ = c^* c$ with $c = dh_+$. But then, by axiom (C1), the element $P_d\big(h_+^* \Delta_r(a)h_+\big)$ is positive semidefinite in $\mathcal{R}$, and we have proved (12).

Now, assume that we have equality in (12). Then we see from (17) that $P_d\big(h_+^* \Delta_r(a)h_+\big) = 0$. So, by axiom (C2), we have $h_+^* d^* dh_+ = 0$. It follows that

$$\|dh_+\|_{\mathcal{R}}^2 = \|h_+^* d^* dh_+\|_{\mathcal{R}} = 0,$$

and hence $dh_+ = 0$. The element d is invertible. Therefore $h_+ = 0$, which implies that $a_+ = b_+$ (see (16)), and thus $a = b$. $\square$

In concrete cases one can often find a natural strictly monotone function on the positive definite elements. This allows one to replace (12) by a numerical condition. To make this statement more precise, consider the set $M_d^>$ consisting of all elements in $\mathcal{M}_d$ that are positive definite in $\mathcal{M}$. A function $F\colon M_d^> \to \mathbb{R}$ is said to be *strictly $\mathcal{R}$-monotone* if

$$d_1, d_2 \in \mathcal{M}_d^>, \qquad d_1 \geq_{\mathcal{R}} d_2, d_1 \neq d_2 \Rightarrow F(d_1) > F(d_2).$$

With such a function Theorem 4.2 can be rephrased as follows.

THEOREM 4.3. *Let $\mathcal{M}$ be an algebra with band structure (1) in a unital C^*-algebra $\mathcal{R}$, and assume that $\mathcal{M}$ satisfies axioms (C1) and (C2). Let $F\colon \mathcal{M}_d^> \to \mathbb{R}$ be strictly $\mathcal{R}$-monotone. Suppose that $k = k^* \in \mathcal{M}_c$ has a band extension with a right spectral factorization relative to (1). Then for any positive extension a of k such that a admits a right spectral factorization relative to (1), we have*

(18) $$F\big(\Delta_r(b)\big) \geq F\big(\Delta_r(a)\big),$$

and in (18) equality holds if and only if $a = b$.

The analogues of Theorems 4.2 and 4.3 with left spectral factorization in place of right spectral factorization and with left multiplicative diagonals in place of right multiplicative diagonals also hold true.

Let us specify Theorems 4.2 and 4.3 for the algebra $\mathcal{L}$ with band structure (7). We already know that $\mathcal{L}$ satisfies the axioms (C1) and (C2). Let K be a (selfadjoint) m-band matrix with a positive extension A, and let B be the (unique) band extension of K. Since positive definite matrices have spectral factorizations (left and right), the multiplicative diagonals $\Delta_r(B)$ and $\Delta_r(A)$ are well-defined. Now use (8) for both B and A, and note that the determinant is a strictly monotone function on the positive definite diagonal matrices. We conclude that $\det B \geq \det A$, and equality holds if and only if $B = A$. Thus the band extension B is the unique positive extension that maximizes the determinant, which is precisely the maximum entropy principle appearing in Section XXXIII.2.

We turn now to the contractive case and derive a maximum entropy principle for strictly contractive extensions in the general setting of the previous section.

THEOREM 4.4. *Let $\mathcal{M}$ be an algebra with band structure (1) in a unital C^*-algebra $\mathcal{R}$, and assume that $\mathcal{M}$ satisfies the axioms (C1) and (C2). Let*

$$\varphi \in \mathcal{M}_\ell = \mathcal{M}_4 \oplus \mathcal{M}_3^0 \oplus \mathcal{M}_d \oplus \mathcal{M}_2^0$$

have a triangular extension g such that $e - g^*g$ has a right spectral factorization relative to (1). Then for any strictly contractive extension $\psi \in \mathcal{M}$ of φ such that $e - \psi^*\psi$ admits a right spectral factorization relative to (1) we have

$$(19) \qquad \Delta_r(e - g^*g) \geq_{\mathcal{R}} \Delta_r(e - \psi^*\psi),$$

and in (19) equality holds if and only if $g = \psi$.

PROOF. We split the proof into three parts. As in the previous section, $\mathcal{A} = \mathcal{M}^{2\times 2}$ is the $*$-algebra of 2×2 block matrices with entries in $\mathcal{M}$. Recall that the unit E in $\mathcal{A}$ and the involution $*$ are given by formulas (8) and (9) in the previous section, respectively. We equip $\mathcal{A}$ with the band structure

$$(20) \qquad \mathcal{A} = \mathcal{A}_1 \oplus \mathcal{A}_2^0 \oplus \mathcal{A}_d \oplus \mathcal{A}_3^0 \oplus \mathcal{A}_4,$$

which is defined in the paragraph after the statement of Theorem 3.1. By Q_d we denote the projection of $\mathcal{A}$ onto $\mathcal{A}_d$ along the other spaces in the decomposition (20).

Part (a). In this part we show that $\mathcal{A} = \mathcal{M}^{2\times 2}$ satisfies axioms (C1) and (C2) relative to $\mathcal{R}^{2\times 2}$, where $\mathcal{R}^{2\times 2}$ is endowed with the C^*-algebra structure considered in the paragraph preceding the proof of Theorem 3.1. Assume $A \in \mathcal{M}^{2\times 2}$ is positive semidefinite in $\mathcal{R}^{2\times 2}$. Thus $A = C^*C$ for some $C \in \mathcal{R}^{2\times 2}$.

First we show $Q_d(A)$ is positive semidefinite in $\mathcal{R}^{2\times 2}$. Let

$$A = \begin{bmatrix} a & b \\ c & d \end{bmatrix}, \qquad C = \begin{bmatrix} \alpha & \beta \\ \gamma & \delta \end{bmatrix}.$$

Then $a = \alpha^*\alpha + \gamma^*\gamma$ and $d = \beta^*\beta + \delta^*\delta$. Since the sum of two positive semidefinite elements in a C^*-algebra is again positive semidefinite (see Corollary XXXI.5.2), we see that a and d are positive semidefinite in $\mathcal{R}$. The elements a and d are in $\mathcal{M}$. So we can apply axiom (C1) to show that

$$P_d(a) = p^*p, \qquad P_d(d) = q^*q$$

for some p and q in $\mathcal{R}$. It follows that

$$Q_d(A) = \begin{bmatrix} P_d(a) & 0 \\ 0 & P_d(d) \end{bmatrix} = \begin{bmatrix} p^* & 0 \\ 0 & q^* \end{bmatrix} \begin{bmatrix} p & 0 \\ 0 & q \end{bmatrix},$$

and hence $Q_d(A)$ is positive semidefinite in $\mathcal{R}^{2\times 2}$.

Next, assume that $Q_d(A) = 0$. We have to show that $A = 0$. Since axiom (C2) holds for $\mathcal{M}$ relative to $\mathcal{R}$, we conclude that a and d are zero. Indeed, from the previous paragraph we know that a and d belong to $\mathcal{M}$ and are positive semidefinite in $\mathcal{R}$. Furthermore, $P_d(a)$ and $P_d(d)$ are both zero, because $Q_d(A) = 0$. So, by axiom (C2), we have $a = d = 0$. Since $A^* = A$, we also have $c = b^*$. We may view $\mathcal{R}^{2\times 2}$ as a C^*-algebra of operators on the Hilbert space direct sum $H \oplus H$. Thus

$$A = \begin{bmatrix} 0 & b \\ b^* & 0 \end{bmatrix} : H \oplus H \to H \oplus H$$

is a non-negative operator. It follows that for each $x \in H$,

$$0 \leq \left\langle A \begin{bmatrix} -bx \\ x \end{bmatrix}, \begin{bmatrix} -bx \\ x \end{bmatrix} \right\rangle = -2\|bx\|^2.$$

This can only happen when $b = 0$. So we have shown that $A = 0$.

Part (b). Let ψ be as in the theorem, and set

$$A_\psi = \begin{bmatrix} e & \psi \\ \psi^* & e \end{bmatrix}.$$

In this part we show that A_ψ has a right spectral factorization relative to (20). Recall (see formula (14a) in the previous section) that A_ψ factorizes as

$$(21) \qquad A_\psi = \begin{bmatrix} e & 0 \\ \psi^* & e \end{bmatrix} \begin{bmatrix} e & 0 \\ 0 & e - \psi^*\psi \end{bmatrix} \begin{bmatrix} e & \psi \\ 0 & e \end{bmatrix}.$$

Since $e - \psi^*\psi$ admits a right spectral factorization (relative to (1)), the right multiplicative diagonal of $e - \psi^*\psi$ is well-defined, and we may write

$$(22) \qquad e - \psi^*\psi = (e + m_+)^* \Delta_r(e - \psi^*\psi)(e + m_+),$$

where $(e + m_+)^{\pm 1} \in e + \mathcal{M}_+^0$. Inserting (22) into (21) yields

$$(23) \qquad A_\psi = (E + M_+)^* \begin{bmatrix} e & 0 \\ 0 & \Delta_r(e - \psi^*\psi) \end{bmatrix} (E + M_+),$$

where

$$E + M_+ = \begin{bmatrix} e & \psi \\ 0 & e + m_+ \end{bmatrix}, \qquad (E + M_+)^{-1} = \begin{bmatrix} e & -\psi(e + m_+)^{-1} \\ 0 & (e + m_+)^{-1} \end{bmatrix}.$$

From the definitions of the various spaces in the band structure (20) we conclude that $(E + M_+)^{\pm 1} \in E + \mathcal{A}_+^0$. The middle factor in the right hand side of (23) may be written as D^*D, where D is an invertible element in $\mathcal{A}_d$. In fact,

$$D = \begin{bmatrix} e & 0 \\ 0 & d \end{bmatrix}.$$

Here d is an invertible element in $\mathcal{M}_d$ such that $\Delta_r(e - \psi^*\psi) = d^*d$. It follows that A_ψ has a right spectral factorization relative to (20) and

$$(24) \qquad \Delta_r(A_\psi) = \begin{bmatrix} e & 0 \\ 0 & \Delta_r(e - \psi^*\psi) \end{bmatrix}.$$

By repeating the above reasoning with g in place of ψ we see that A_g has a right spectral factorization relative to (20) and

$$(25) \qquad \Delta_r(A_g) = \begin{bmatrix} e & 0 \\ 0 & \Delta_r(e - g^*g) \end{bmatrix}.$$

In particular, both A_ψ and A_g are positive definite in $\mathcal{A}$.

Part (c). In this part we derive the maximum entropy principle. Put $K = A_\varphi$, where φ is as in the theorem. According to our hypotheses, ψ is a strictly contractive extension of φ. This implies that $\psi - \varphi \in \mathcal{M}_1$, and hence

$$A_\psi - K = \begin{bmatrix} 0 & \psi - \varphi \\ (\psi - \varphi)^* & 0 \end{bmatrix} \in \mathcal{A}_1 + \mathcal{A}_4.$$

So A_ψ is a positive extension of K. Similarly, A_g is a positive extension of K. Moreover, by formula (19) in the previous section, A_g is a band extension of K. We now apply Theorem 4.2 to $\mathcal{A}$ and K which is allowed because the algebra $\mathcal{A}$ satisfies axioms (C1) and (C2) relative to $\mathcal{R}^{2\times 2}$. We conclude that $\Delta_r(A_g) \geq_{\mathcal{R}^{2\times 2}} \Delta_r(A_\psi)$, and equality holds if and only if $A_g = A_\psi$. By (24) and (25), we have

$$(26) \qquad \Delta_r(A_g) - \Delta_r(A_\psi) = \begin{bmatrix} 0 & 0 \\ 0 & \Delta_r(e - g^*g) - \Delta_r(e - \psi^*\psi) \end{bmatrix}.$$

Since $\mathcal{R}^{2\times 2}$ is a C^*-algebra of operators on a Hilbert space direct sum $H \oplus H$, the left hand side of (26) is a non-negative operator on $H \oplus H$. But then the $(2,2)$-entry in the right hand side of (26) is a non-negative operator on H, and hence (19) holds. Finally, if we have equality in (19), then (26) yields $\Delta_r(A_g) = \Delta_r(A_\psi)$, and thus $A_g = A_\psi$. But the latter is equivalent to $g = \psi$. $\square$

As in Theorem 4.3, any strictly $\mathcal{R}$-monotone function on $\mathcal{M}_d^>$ can be used to pick out the triangular extension.

THEOREM 4.5.   *Let $\mathcal{M}$ be an algebra with band structure (1) in a unital C^*-algebra $\mathcal{R}$, and assume that $\mathcal{M}$ satisfies the axioms (C1) and (C2). Let $F: \mathcal{M}_d^> \to \mathbb{R}$ be strictly $\mathcal{R}$-monotone. Suppose that $\varphi \in \mathcal{M}_\ell$ has a triangular extension g such that $e - g^*g$ has a right spectral factorization relative to (1). Then for any strictly contractive extension ψ of φ such that $e - \psi^*\psi$ admits a right spectral factorization relative to (1), we have*

$$(27) \qquad F\big(\Delta_r(g)\big) \geq F\big(\Delta_r(\psi)\big),$$

and equality holds in (27) if and only if $\psi = g$.

Theorems 4.4 and 4.5 have analogues with left spectral factorizations for $e - gg^*$ and $e - \psi\psi^*$ instead of right spectral factorizations of $e - g^*g$ and $e - \psi^*\psi$, and with the left multiplicative diagonals of $e - gg^*$ and $e - \psi\psi^*$ in place of the right multiplicative diagonals of $e - g^*g$ and $e - \psi^*\psi$.

Since $\mathcal{L}$ in (7) satisfies the axioms (C1) and (C2), the strict monotonicity of the determinant on diagonal matrices with positive diagonal entries allows us to apply Theorem 4.5 to $\mathcal{L}$ with $F(A) = \det A$. Doing so and using (8), we conclude that for an m-band $(n+1) \times (n+1)$ matrix Φ the triangular extension T is the unique strictly contractive extension that maximizes the expression $\det(I - T^*T)$. This criterion, which singles out the triangular extension, is the maximum entropy principle referred to in Section XXXIII.3.

CHAPTER XXXV

APPLICATIONS OF THE BAND METHOD

In this chapter the general scheme for dealing with positive and strictly contractive extension problems developed in the previous chapter is used to solve a number of concrete problems. The chapter consists of seven sections. The first two treat, respectively, the positive and the strictly contractive completion problem for operator matrices. Sections 3 and 4 concern matrix-valued functions on the unit circle, and solve the Carathéodory-Toeplitz problem (Section 3) and the Nehari problem (Section 4). Sections 5 and 6 concern two different versions for matrix-valued functions of the Nevanlinna-Pick interpolation problem considered in Section XXVII.7. We deal with these interpolation problems as extension problems. The last section treats the Nehari problem for rational matrix functions in terms of realization.

XXXV.1 POSITIVE COMPLETIONS OF OPERATOR MATRICES

Let n be a non-negative integer, and let $H_0, \ldots, H_n$ be Hilbert spaces. Consider an $(n+1) \times (n+1)$ operator matrix

$$
(1) \qquad F = \begin{bmatrix} F_{00} & \cdots & F_{0n} \\ \vdots & & \vdots \\ F_{n0} & \cdots & F_{nn} \end{bmatrix} : H_0 \oplus \cdots \oplus H_n \to H_0 \oplus \cdots \oplus H_n.
$$

For each i and j the (i,j)-th entry F_{ij} is a bounded linear operator from H_j into H_i, and hence F acts as a bounded linear operator on the Hilbert space direct sum $H_0 \oplus \cdots \oplus H_n$. The set of all operator matrices (1) will be denoted by $\mathcal{L}(H_0, \ldots, H_n)$.

Fix $0 \le m < n$, and let

$$
(2) \qquad A_{ij} \in \mathcal{L}(H_j, H_i), \qquad |j - i| \le m
$$

be a given indexed set of operators. We say that F in (1) is a *positive completion* of the *operator band* (2) if

(a) $F_{ij} = A_{ij}$ for $|j - i| \le m$,

(b) $F > 0$.

Condition (b) means that F is strictly positive as an operator on $H_0 \oplus \cdots \oplus H_n$. The latter implies that the (central) submatrix $[F_{ij}]_{i,j=k}^{\ell}$ of F is a strictly positive operator on $H_k \oplus \cdots \oplus H_\ell$ for each $0 \le k \le \ell \le n$. Hence, because of (a), in order that the operator band (2) have a positive completion it is necessary that

$$
(3) \qquad \begin{bmatrix} A_{kk} & \cdots & A_{k,m+k} \\ \vdots & & \vdots \\ A_{m+k,k} & \cdots & A_{m+k,m+k} \end{bmatrix} > 0, \qquad k = 0, \ldots, n - m.
$$

These conditions are also sufficient.

THEOREM 1.1. *The operator band* (2) *has a positive completion if and only if* (3) *holds. In this case the band* (2) *has a unique positive completion B with the additional property that the (i,j)-th entry of B^{-1} is zero for $|j - i| > m$. Furthermore, B is constructed in the following way. For $j = 0, \ldots, n$ let*

$$(4) \qquad \begin{bmatrix} X_{\alpha(j),j} \\ \vdots \\ X_{jj} \end{bmatrix} = \begin{bmatrix} A_{\alpha(j),\alpha(j)} & \cdots & A_{\alpha(j),j} \\ \vdots & & \vdots \\ A_{j,\alpha(j)} & \cdots & A_{jj} \end{bmatrix}^{-1} \begin{bmatrix} 0 \\ \vdots \\ 0 \\ I_j \end{bmatrix},$$

$$(5) \qquad \begin{bmatrix} Y_{jj} \\ \vdots \\ Y_{\beta(j),j} \end{bmatrix} = \begin{bmatrix} A_{jj} & \cdots & A_{j,\beta(j)} \\ \vdots & & \vdots \\ A_{\beta(j),j} & \cdots & A_{\beta(j),\beta(j)} \end{bmatrix}^{-1} \begin{bmatrix} I_j \\ 0 \\ \vdots \\ 0 \end{bmatrix},$$

where I_j is the identity operator on H_j and

$$(6) \qquad \alpha(j) = \max\{0, j - m\}, \qquad \beta(j) = \min\{n, j + m\}.$$

Define $U = [U_{ij}]_{i,j=0}^n$ and $V = [V_{ij}]_{i,j=0}^n$ in $\mathcal{L}(H_0, \ldots, H_n)$ by setting

$$(7) \qquad U_{ij} = \begin{cases} X_{ij} X_{jj}^{-1/2}, & i = \alpha(j), \cdots, j, \\ 0, & \text{otherwise}; \end{cases}$$

$$(8) \qquad V_{ij} = \begin{cases} Y_{ij} Y_{jj}^{-1/2}, & i = j, \ldots, \beta(j), \\ 0, & \text{otherwise}. \end{cases}$$

Then B is given by the following factorizations

$$(9) \qquad B = (U^{-1})^* U^{-1} = (V^{-1})^* V^{-1}.$$

For $H_0 = H_1 = \cdots = H_n = \mathbb{C}$ the above theorem is proved in the last three paragraphs of Section XXXIV.1. As we shall see the more general case considered here is proved in an analogous way.

In (9) the star denotes the usual adjoint for Hilbert space operators. Thus for F in (1) we have

$$(10) \qquad F^* = \begin{bmatrix} F_{00}^* & \cdots & F_{n0}^* \\ \vdots & & \vdots \\ F_{0n}^* & \cdots & F_{nn}^* \end{bmatrix}.$$

Since $\mathcal{L}(H_0, \ldots, H_n)$ coincides with the algebra of all bounded linear operators on $H_0 \oplus \cdots \oplus H_n$, the set $\mathcal{L}(H_0, \ldots, H_n)$ is a unital C^*-algebra with the involution $*$ as in (10) and the unit equal to the identity operator I on $H_0 \oplus \cdots \oplus H_n$. It follows that $F \in \mathcal{L}(H_0, \ldots, H_n)$ is positive definite in $\mathcal{L}(H_0, \ldots, H_n)$ if and only if $F > 0$. To prove Theorem 1.1 we equip $\mathcal{L}(H_0, \ldots, H_n)$ with a band structure by introducing the following subsets:

$$
\begin{aligned}
\mathcal{M}_1 &= \{F \in \mathcal{L}(H_0, \ldots, H_n) \mid F_{ij} = 0 \text{ for } j - i \le m\}, \\
\mathcal{M}_2^0 &= \{F \in \mathcal{L}(H_0, \ldots, H_n) \mid F_{ij} = 0 \text{ for } j - i > m \text{ or } j - i \le 0\}, \\
\mathcal{M}_d &= \{F \in \mathcal{L}(H_0, \ldots, H_n) \mid F_{ij} = 0 \text{ for } i \ne j\}, \\
\mathcal{M}_3^0 &= \{F \in \mathcal{L}(H_0, \ldots, H_n) \mid F_{ij} = 0 \text{ for } j - i \ge 0 \text{ or } j - i < -m\}, \\
\mathcal{M}_4 &= \{F \in \mathcal{L}(H_0, \ldots, H_n) \mid F_{ij} = 0 \text{ for } j - i \ge -m\}.
\end{aligned}
$$

Then

$$
(11) \qquad \mathcal{L}(H_0, \ldots, H_n) = \mathcal{M}_1 \oplus \mathcal{M}_2^0 \oplus \mathcal{M}_d \oplus \mathcal{M}_3^0 \oplus \mathcal{M}_4,
$$

and it is straightforward to check that $\mathcal{L}(H_0, \ldots, H_n)$ with the decomposition (11) is an algebra with band structure. Note that $F = C^*C$ is a right (left) spectral factorization relative to (11) if and only if C is an upper (lower) triangular $(n+1) \times (n+1)$ operator matrix with invertible diagonal entries. We are now ready to prove the theorem.

PROOF OF THEOREM 1.1. The necessity has already been established. So in the following we assume that (3) holds. Let $K = [K_{ij}]_{i,j=0}^n$ be defined by

$$
(12) \qquad K_{ij} = \begin{cases} A_{ij}, & |j - i| \le m, \\ 0, & \text{otherwise.} \end{cases}
$$

From (3) it follows that $A_{ij} = A_{ji}^*$ for $|j - i| \le m$. Hence

$$
K = K^* \in \mathcal{M}_c = \mathcal{M}_2^0 \oplus \mathcal{M}_d \oplus \mathcal{M}_3^0.
$$

So K is a selfadjoint element in the band $\mathcal{M}_c$ of (11). Note that $B \in \mathcal{L}(H_0, \ldots, H_n)$ is a positive completion of the band (2) if and only if B is a positive extension of K (relative to (11)). Moreover, B is a positive completion of (2) with the additional property that $(B^{-1})_{ij} = 0$ for $|j - i| > m$ if and only if B is a band extension of K. So we have to prove that K has a unique band extension. For this purpose we apply Theorems 1.1–1.3 in Section XXXIV.1 with $\mathcal{M} = \mathcal{L}(H_0, \ldots, H_n)$, with band structure (11) and with $\mathcal{R} = \mathcal{M}$.

From condition (3) it follows that the inverse matrices appearing in (4) and (5) exist and are strictly positive operators on the corresponding Hilbert spaces. Thus the left hand sides of (4) and (5) are well-defined. Note that X_{jj} is a diagonal entry of a strictly positive operator matrix. So X_{jj} is strictly positive as an operator on H_j. Similarly, Y_{jj} is strictly positive as an operator on H_j. Introduce operator matrices $X = [X_{ij}']_{i,j=0}^n$ and $Y = [Y_{ij}']_{i,j=0}^n$ by setting

$$
X_{ij}' = \begin{cases} X_{ij}, & i = \alpha(j), \ldots, j, \\ 0, & \text{otherwise;} \end{cases}
$$

$$Y'_{ij} = \begin{cases} Y_{ij}, & i = j, \ldots, \beta(j), \\ 0, & \text{otherwise.} \end{cases}$$

From the definitions of $\alpha(j)$ and $\beta(j)$ in (6) we see that $X \in \mathcal{M}_2$ and $Y \in \mathcal{M}_3$. (Here and in remaining part of the proof we use the notations introduced in Section XXXIV.1.) A straightforward computation shows that

$$P_2(KX) = I, \qquad P_3(KY) = I.$$

Since X_{jj} and Y_{jj} are strictly positive operators for each j, the operators X and Y are invertible, X^{-1} is upper triangular and Y^{-1} is lower triangular. In other words, $X^{-1} \in \mathcal{M}_+$ and $Y^{-1} \in \mathcal{M}_-$. Also, $P_d(X)$ and $P_d(Y)$ are positive definite in $\mathcal{M}_d$. So we can apply Theorems 1.1–1.3 in Section XXXIV.1 to show that K has a unique band extension B which is given by

$$(13) \qquad B = X^{-*}P_d(X)X^{-1} = Y^{-*}P_d(Y)Y^{-1},$$

where T^{-*} stands for $(T^{-1})^*$. Now, put

$$U = X\big(P_d(X)\big)^{-1/2}, \qquad V = Y\big(P_d(Y)\big)^{-1/2}.$$

Then the entries of U and V are given by (7) and (8), respectively, and (9) holds. $\square$

The next theorem gives a full description of all positive completions of (2).

THEOREM 1.2. *In order that the operator band* (2) *have a positive completion it is necessary and sufficient that* (3) *holds. In this case each positive completion F of* (2) *is of the form*

$$(14) \qquad F = (G^*V^* + U^*)^{-1}(I - G^*G)(VG + U)^{-1},$$

where $U = [U_{ij}]_{i,j=0}^n$ *and* $V = [V_{ij}]_{i,j=0}^n$ *are the operator matrices defined by* (7) *and* (8), *respectively, and the free parameter* $G = [G_{ij}]_{i,j=0}^n$ *is a strict contraction on*

$$H_0 \oplus \cdots \oplus H_n$$

such that $G_{ij} = 0$ *for* $j - i \leq m$. *Furthermore, formula* (14) *gives a one-one correspondence between all such G and all positive completions F of* (2).

PROOF. The necessity and sufficiency of (3) has already been proved (Theorem 1.1). It remains to derive the parametrization (14). So, in the following we assume that (3) holds. To obtain (14) we apply Theorem XXXIV.2.1 to $\mathcal{M} = \mathcal{L}(H_0, \ldots, H_n)$ with band structure (11). If $F^{(k)} = [F_{ij}^{(k)}]_{i,j=0}^n$ and $F = [F_{ij}]_{i,j=0}^n$, then $F^{(k)} \to F$ $(k \to \infty)$ in the operator norm implies

$$F_{ij}^{(k)} \to F_{ij} \qquad (k \to \infty)$$

in the operator norm for each i and j. It follows that set $\mathcal{M}_+$, which consists of all upper triangular $(n+1) \times (n+1)$ operator matrices, is closed in the operator norm, and hence axiom (A) is fulfilled. Let K and B be as in the proof of Theorem 1.1, and let U and V be as in (9). The first identity in (9) gives a right spectral factorization of B and the second a left. So, by Theorem XXXIV.2.1, each positive extension of K is given by (14), where G is an arbitrary element in $\mathcal{M}_1$ (with $\mathcal{M}_1$ as in (11)) such that $\|G\| < 1$. So, Theorem XXXIV.2.1 yields the desired parametrization of the positive completions F of (2). $\square$

The special completion B constructed in Theorem 1.1 is called the *band completion* (or *band extension*) of the operator band (2). The band completion may be identified by a maximum entropy principle. For this purpose we need the notion of a multiplication diagonal for operator matrices.

Let $F = [F_{ij}]_{i,j=0}^n \in \mathcal{L}(H_0,\ldots,H_n)$ be strictly positive as an operator on $H_0 \oplus \cdots \oplus H_n$. Then F factorizes as $F = U^* D U$, where $U \in \mathcal{L}(H_0,\ldots,H_n)$ is an upper triangular operator matrix with j-th diagonal entry equal to the identity operator on H_j and $D \in \mathcal{L}(H_0,\ldots,H_n)$ is a diagonal operator matrix with j-th diagonal entry $\Delta_j^{(r)}(F)$ being given by

$$(15a) \qquad \Delta_0^{(r)}(F) = F_{00},$$

$$(15b) \qquad \Delta_j^{(r)}(F) = F_{jj} - [F_{j0} \quad \cdots \quad F_{j,j-1}] \begin{bmatrix} F_{00} & \cdots & F_{0,j-1} \\ \vdots & & \vdots \\ F_{j-1,1} & \cdots & F_{j-1,j-1} \end{bmatrix}^{-1} \begin{bmatrix} F_{0j} \\ \vdots \\ F_{j-1,j} \end{bmatrix},$$
$$j = 1,\ldots,n.$$

To obtain such a factorization one applies the LU-factorization theorem for finite chains in Section XXII.1. Indeed, let Q_j be the orthogonal projection of $H = H_0 \oplus \cdots \oplus H_n$ onto the subspace

$$H_0 \oplus \cdots \oplus H_{j-1} \oplus \{0\} \oplus \cdots \oplus \{0\}.$$

Then $\mathbb{Q} = \{Q_0,\ldots,Q_{n+1}\}$ is a finite chain on H. Since F is strictly positive, the operators $I - Q_j + Q_j F Q_j$ are invertible for $j = 0,\ldots,n+1$. So we may apply Theorem XXII.1.1 with $\mathbb{P} = \mathbb{Q}$ and $K = I - F$ to obtain the factorization $F = U^* D U$ mentioned above. The operator matrix $D = \operatorname{diag}(\Delta_j^{(r)}(F))_{j=0}^n$ is called the *right multiplicative diagonal* of F.

LEMMA 1.3. *Assume that the operator band (2) satisfies condition (3), and let B be the band completion of (2). For $j = 0,\ldots,n$ put*

$$(16) \qquad M_j = [0 \quad \cdots \quad 0 \quad I_j] \begin{bmatrix} A_{\alpha(j),\alpha(j)} & \cdots & A_{\alpha(j),j} \\ \vdots & & \vdots \\ A_{j,\alpha(j)} & \cdots & A_{jj} \end{bmatrix}^{-1} \begin{bmatrix} 0 \\ \vdots \\ 0 \\ I_j \end{bmatrix},$$

where $\alpha(j) = \max\{0, j - m\}$ and I_j is the identity operator on H_j. Then the diagonal entries of the right multiplicative diagonal of B are given by

$$(17) \qquad \Delta_j^{(r)}(B) = M_j^{-1}, \qquad j = 0,\ldots,n.$$

PROOF. To prove (17) we use the first identity in (13) and the uniqueness of the LU-factorization. It follows that

$$\mathrm{diag}(\Delta_j^{(r)}(B))_{j=0}^n = (P_d(X))^{-1}.$$

Now, $P_d(X) = \mathrm{diag}(X_{jj})_{j=0}^n$, where X_{jj} is given by (4). Thus X_{jj} is precisely the operator M_j defined by (16), and thus (17) holds. $\square$

In the next theorem $A \geq B$ means that $A - B$ is a non-negative operator (see Section V.6).

THEOREM 1.4. *Assume that the operator band (2) satisfies condition (3). Then for the right multiplicative diagonal* $\mathrm{diag}(\Delta_j^{(r)}(F))_{j=0}^n$ *of any positive completion F of (2) the following inequalities hold:*

$$(18) \qquad \Delta_j^{(r)}(F) \leq M_j^{-1}, \qquad j = 0, \ldots, n,$$

where $M_0, \ldots, M_n$ are as in (16). Furthermore, equality holds in (18) for all j if and only if F is the band completion B of (2).

PROOF. We use the band structure (11), and apply Theorem XXXIV.4.2 to $\mathcal{M} = \mathcal{L}(H_0, \ldots, H_n)$ with band structure (11) and with $\mathcal{R} = \mathcal{M}$. The first step is to check that $\mathcal{L}(H_0, \ldots, H_n)$ satisfies the axioms (C1) and (C2). Note that $F \in \mathcal{M} = \mathcal{L}(H_0, \ldots, H_n)$ is positive semidefinite in $\mathcal{R} = \mathcal{M}$ if and only if F is a non-negative operator on $H = H_0 \oplus \cdots \oplus H_n$ (cf., Theorem XXXI.5.4). Furthermore, in this setting $P_d(F)$ is the diagonal matrix which one obtains from F by replacing the off diagonal entries of F by a zero operator. Now, take $A \in \mathcal{L}(H_0, \ldots, H_n)$, and let $A \geq 0$. Since the diagonal entries of a non-negative operator matrix are also non-negative, $P_d(A) = D^*D$, where D is the diagonal matrix whose j-th diagonal entry is equal to the square root of the j-th diagonal entry of A. Thus axiom (C1) is fulfilled. Next, assume additionally that $P_d(A) = 0$, i.e., the diagonal entries of A are zero. Fix $0 \leq i < j \leq n$. Since A is a non-negative operator, it follows that $A_{ji} = A_{ij}^*$ and

$$\Phi = \begin{bmatrix} 0 & A_{ij} \\ A_{ji} & 0 \end{bmatrix} = \begin{bmatrix} A_{ii} & A_{ij} \\ A_{ji} & A_{jj} \end{bmatrix} \geq 0.$$

So for each $x \in H_j$ we have

$$0 \leq \left\langle \Phi \begin{bmatrix} -A_{ij}x \\ x \end{bmatrix}, \begin{bmatrix} -A_{ij}x \\ x \end{bmatrix} \right\rangle = -2\|A_{ij}x\|^2,$$

which implies that $A_{ij}x = 0$. Therefore we may conclude that $A_{ij} = 0$ for $i \neq j$, and hence axiom (C2) is fulfilled.

Next, let B be the band completion of (2), and let F be an arbitrary positive completion of (2). Furthermore, let K be as in the proof of Theorem 1.1. Then B is the band extension of K relative to the band structure (11) and the right multiplicative diagonal of B relative to (11) is precisely $\mathrm{diag}(\Delta_j^{(r)}(B))_{j=0}^n$. Similarly, F is a positive extension of K relative to (11), and $\mathrm{diag}(\Delta_j^{(r)}(F))_{j=0}^n$ is the right multiplicative diagonal of F relative to (11). So we may apply Theorem XXXIV.4.2 and Lemma 1.3 to complete the proof. $\square$

If the spaces $H_0, H_1, \ldots, H_n$ are finite dimensional, then the determinant provides a strictly monotone function on the set of strictly positive diagonal operator matrices. This observation yields the following corollary.

COROLLARY 1.5. *Assume that the operator band (2) satisfies condition (3), and let the spaces $H_0, \ldots, H_n$ in (2) be finite dimensional. Then for any positive completion F of (2)*

$$(19) \qquad \det F \le \left(\prod_{j=0}^n \det M_j \right)^{-1},$$

where $M_0, \ldots, M_n$ are as in (16). Furthermore, equality holds in (19) if and only if F is equal to the band completion B of (2).

PROOF. We apply Theorem XXXIV.4.3 to $\mathcal{M} = \mathcal{L}(H_0, \ldots, H_n)$ with band structure (11) and with $\mathcal{R} = \mathcal{M}$. Let $\mathcal{M}_d$ be as in (11), and consider

$$\mathcal{M}_d^{>} = \{ F \in \mathcal{M}_d \mid F_{jj} > 0 \ (j = 0, \ldots, n) \}.$$

Then $\mathcal{F}(F) = \det F$ defines a strictly monotone function $\mathcal{F}$ on $\mathcal{M}_d^{>}$. Since the determinant of a product is the product of the determinants, the determinant of a strictly positive operator matrix in $\mathcal{L}(H_0, \ldots, H_n)$ is equal to the determinant of its right multiplicative diagonal, and hence the corollary follows by applying Theorem XXXIV.4.3 and Lemma 1.3. $\square$

Formula (14) in Theorem 1.2 may be replaced by

$$F = (G^*U^* + V^*)^{-1}(I - G^*G)(UG + V),$$

where now the free parameter G is a strictly contractive operator matrix, $G = [G_{ij}]_{i,j=0}^n$, such that $G_{ij} = 0$ for $j - i \ge -m$.

Lemma 1.3, Theorem 1.4 and Corollary 1.5 have "left" analogues in which the role of the right multiplicative diagonal is taken over by the left multiplicative diagonal.

XXXV.2 STRICTLY CONTRACTIVE COMPLETIONS OF OPERATOR MATRICES

Let $0 \le m \le n$ be integers, and let $H_0, \ldots, H_n$ be Hilbert spaces. In this section we treat the following problem. Given

$$(1) \qquad \Phi_{ij} \in \mathcal{L}(H_j, H_i), \qquad i \ge 0, \ j \ge 0, \ j - i \le m,$$

find all operator matrices

$$(2) \qquad F = \begin{bmatrix} F_{00} & \cdots & F_{0n} \\ \vdots & & \vdots \\ F_{n0} & \cdots & F_{nn} \end{bmatrix} : H_0 \oplus \cdots \oplus H_n \to H_0 \oplus \cdots \oplus H_n$$

such that

(a) $F_{ij} = \Phi_{ij}$ for $j - i \leq m$,

(b) $\|F\| < 1$.

We shall refer to such an operator F as a *strictly contractive completion* of the *operator lower triangular part* (1). The norm in (b) is the norm of F as an operator on the Hilbert space direct sum $H_0 \oplus \cdots \oplus H_n$. Since compressions of a strict contraction are also strict contractions, it follows that if (1) has a strictly contractive completion, then necessarily

$$(3) \qquad \left\| \begin{bmatrix} \Phi_{k0} & \cdots & \Phi_{k,k+m} \\ \vdots & & \vdots \\ \Phi_{n0} & \cdots & \Phi_{n,k+m} \end{bmatrix} \right\| < 1, \qquad k = 0, \ldots, n - m.$$

The norm in (3) is the operator norm for operators from $H_0 \oplus \cdots \oplus H_{k+m}$ into $H_k \oplus \cdots \oplus H_n$. We shall see that (3) is not only necessary but also sufficient.

To give the solution of the strictly contractive completion problem we associate with (1) the following operator matrices:

$$(4) \qquad \Gamma_k = \begin{bmatrix} \Phi_{k0} & \cdots & \Phi_{k,\nu(k)} \\ \vdots & & \vdots \\ \Phi_{n0} & \cdots & \Phi_{n,\nu(k)} \end{bmatrix} : H_0 \oplus \cdots \oplus H_{\nu(k)} \to H_k \oplus \cdots \oplus H_n,$$

$$(5) \qquad \Lambda_k = \begin{bmatrix} \Phi_{\mu(k),0} & \cdots & \Phi_{\mu(k),k} \\ \vdots & & \vdots \\ \Phi_{n,0} & \cdots & \Phi_{n,k} \end{bmatrix} : H_0 \oplus \cdots \oplus H_k \to H_{\mu(k)} \oplus \cdots \oplus H_n.$$

Here $\nu(k) = \min\{k + m, n\}$ and $\mu(k) = \max\{0, k - m\}$. Note that condition (3) implies that the operators in (4) and (5) are strict contractions.

THEOREM 2.1. *The operator lower triangular part (1) has a strictly contractive completion if and only if (3) holds. In this case there exists a unique strictly contractive completion T of (1) with the additional property that the (i,j)-th operator entry of $T(I - T^*T)^{-1}$ is a zero operator for $j - i > m$. This completion T is constructed in the following way. Introduce operator matrices*

$$A = [A_{ij}]_{i,j=0}^n, \qquad B = [B_{ij}]_{i,j=0}^n, \qquad C = [C_{ij}]_{i,j=0}^n, \qquad D = [D_{ij}]_{i,j=0}^n,$$

of which the entries are given by the following formulas:

$$(6a) \qquad \begin{bmatrix} A_{kk} \\ \vdots \\ A_{nk} \end{bmatrix} = (I - \Gamma_k \Gamma_k^*)^{-1} \begin{bmatrix} I_k \\ 0 \\ \vdots \\ 0 \end{bmatrix}, \qquad k = 0, \ldots, n,$$

$$(6b) \qquad A_{ik} = 0, \qquad 0 \le i < k \le n;$$

$$(7a) \qquad \begin{bmatrix} D_{0k} \\ \vdots \\ D_{kk} \end{bmatrix} = (I - \Lambda_k^* \Lambda_k)^{-1} \begin{bmatrix} 0 \\ \vdots \\ 0 \\ I_k \end{bmatrix}, \qquad k = 0, \ldots, n,$$

$$(7b) \qquad D_{ik} = 0, \quad 0 \le k < i < n;$$

$$(8a) \qquad \begin{bmatrix} B_{\mu(k),k} \\ \vdots \\ B_{nk} \end{bmatrix} = \Lambda_k \begin{bmatrix} D_{0k} \\ \vdots \\ D_{kk} \end{bmatrix}, \qquad k = 0, \ldots, n,$$

$$(8b) \qquad B_{ik} = 0, \qquad 0 \le i < \mu(k);$$

$$(9a) \qquad \begin{bmatrix} C_{0k} \\ \vdots \\ C_{\nu(k),k} \end{bmatrix} = \Gamma_k^* \begin{bmatrix} A_{kk} \\ \vdots \\ A_{nk} \end{bmatrix}, \qquad k = 0, \ldots, n,$$

$$(9b) \qquad C_{ik} = 0, \qquad \nu(k) < i \le n.$$

Here $\nu(k)$ and $\mu(k)$ are as in (4) and (5), respectively, and I_k denotes the identity operator on H_k. Then

$$(10) \qquad T = BD^{-1} = (A^{-1})^* C^*.$$

Furthermore, all strictly contractive completions of (1) are of the form

$$(11) \qquad F = (\widetilde{A}G + \widetilde{B})(\widetilde{C}G + \widetilde{D})^{-1}.$$

Here the free parameter $G = [G_{ij}]_{i,j=0}^n$ is a strict contraction on $H_0 \oplus \cdots \oplus H_n$ such that $G_{ij} = 0$ for $j - i \le m$, and

$$(12) \qquad \widetilde{A} = AP^{-1}, \qquad \widetilde{B} = BQ^{-1}, \qquad \widetilde{C} = CP^{-1}, \qquad \widetilde{D} = DQ^{-1},$$

where P and Q are the $(n{+}1){\times}(n{+}1)$ diagonal operator matrices with k-th diagonal entry equal to $A_{kk}^{1/2}$ and $D_{kk}^{1/2}$, respectively. Moreover, (11) defines a one-one correspondence between the set of all free parameters G and all strictly contractive extensions F of (1).

Assume condition (3) is fulfilled. Then the operators Γ_k and Λ_k in (4) and (5) are strict contractions. Hence $I - \Gamma_k \Gamma_k^*$ and $I - \Lambda_k^* \Lambda_k$ are invertible for $k = 0, \ldots, n,$

and thus the left hand sides of (6a), (7a), (8a) and (9a) are well-defined. Note that $I - \Gamma_k \Gamma_k^*$ and $I - \Lambda_k^* \Lambda_k$ are strictly positive operators. It follows that $(I - \Gamma_k \Gamma_k^*)^{-1}$ and $(I - \Lambda_k^* \Lambda_k)^{-1}$ are strictly positive, and hence the same is true for A_{kk} and D_{kk}. In particular, A_{kk} and D_{kk} are invertible. Since A and D are lower and upper triangular, respectively, we see that A and D are invertible, and thus the operators BD^{-1} and $(A^{-1})^* C^*$ in (10) are well-defined. From $A_{kk} > 0$ and $D_{kk} > 0$ it also follows that the square roots $A_{kk}^{1/2}$ and $D_{kk}^{1/2}$ exist and are invertible. So the operators P^{-1} and Q^{-1} in (12) are well-defined.

For the case when $H_0 = H_1 = \cdots = H_n = \mathbb{C}$ the proof of Theorem 2.1 appears in Section XXXIV.3 (in the first four paragraphs after the proof of Theorem XXXIV.3.2). To prove the general case considered here, one applies Theorems XXXIV.3.1 and XXXIV.3.2 to $\mathcal{M} = \mathcal{L}(H_0, \ldots, H_n)$ endowed with the band structure

$$(13) \qquad \mathcal{L}(H_0, \ldots, H_n) = \mathcal{M}_1 \oplus \mathcal{M}_2^0 \oplus \mathcal{M}_d \oplus \mathcal{M}_3^0 \oplus \mathcal{M}_4$$

introduced in the previous section (see (11) in the previous section), and with $\mathcal{R} = \mathcal{M}$. From the previous section we know that axiom (A) (see Section XXXIV.2) is fulfilled for $\mathcal{M} = \mathcal{L}(H_0, \ldots, H_n)$ relative to $\mathcal{R} = \mathcal{M}$. Let $\Phi = [\Phi_{ij}]_{i,j=0}^n$, where Φ_{ij} is as in (1) for $j - i \leq m$ and $\Phi_{ij} = 0$ otherwise. Then

$$(14) \qquad \Phi \overset{\cdot}{\in} \mathcal{M}_\ell := \mathcal{M}_4 \oplus \mathcal{M}_3^0 \oplus \mathcal{M}_d \oplus \mathcal{M}_2^0,$$

and Φ plays the role of φ in Section XXXIV.3. The further reasoning is the same as for the scalar case in Section XXIV.3, and with the remarks made in the previous paragraph it is straightforward to provide all the details.

In Theorem 2.1 formula (11) may be replaced by

$$F = (\widetilde{A}^* + H\widetilde{B}^*)^{-1}(C^* + HD^*)^{-1},$$

where now the free parameter $H = [H_{ij}]_{i,j=0}^n$ is a strict contraction on $H_0 \oplus \cdots \oplus H_n$ such that $H_{ij} = 0$ for $i - j \leq m$.

The special completion T given by (10) is called the *triangular completion* (or *triangular extension*) of (1). The triangular completion is characterized by the following maximum entropy principle.

THEOREM 2.2. *Assume that the operator lower triangular part* (1) *satisfies condition* (3), *and put*

$$(15) \qquad L_k = [0 \quad \cdots \quad 0 \quad I_k](I - \Lambda_k^* \Lambda_k)^{-1} \begin{bmatrix} 0 \\ \vdots \\ 0 \\ I_k \end{bmatrix}, \qquad k = 0, \ldots, n,$$

where Λ_k is as in (5) *and I_k is the identity operator on H_k. Then for any strictly contractive completion F of* (1) *the right multiplicative diagonal* $\mathrm{diag}\big(\Delta_j^{(r)}(I - F^* F)\big)_{j=0}^n$ *of $I - F^* F$ satisfies the following inequalities:*

$$(16) \qquad \Delta_j^{(r)}(I - F^* F) \leq L_j^{-1}, \qquad j = 0, \ldots, n.$$

Furthermore, equality holds in (16) *for all j if and only if F is the unique triangular completion of* (1).

PROOF. We apply Theorem XXXIV.4.4 to the algebra $\mathcal{M} = \mathcal{L}(H_0, \ldots, H_n)$ endowed with the band structure (13) and with $\mathcal{R} = \mathcal{M}$. From the previous section we know that $\mathcal{M} = \mathcal{L}(H_0, \ldots, H_n)$ satisfies axioms (C1) and (C2) relative to $\mathcal{R} = \mathcal{M}$. Let Φ be as in (14). Note that F is a strictly contractive completion of (1) if and only if F is a strictly contractive extension of Φ relative to the band structure (13). Since any positive definite element in $\mathcal{M} = \mathcal{L}(H_0, \ldots, H_n)$ has a right (and a left) spectral factorization relative to (13), we may apply Theorem XXXIV.4.4 to any strictly contractive extension of (1). Next, observe that the triangular completion T of (1) is precisely the triangular extension of Φ relative to (13). We have already seen (in the previous section) that the right multiplicative diagonal of a strictly positive operator matrix in $\mathcal{L}(H_0, \ldots, H_n)$ coincides with its right multiplicative diagonal relative to the band structure (13). So to complete the proof it remains to show that for the triangular extension T

$$\Delta_j^{(r)}(I - T^*T) = L_j^{-1}, \qquad j = 0, \ldots, n,$$

where $L_0, \ldots, L_n$ are as in (15). But these equalities are a direct consequence of the identity

$$\begin{bmatrix} I & T \\ T^* & I \end{bmatrix} = \begin{bmatrix} I & 0 \\ T^* & I \end{bmatrix} \begin{bmatrix} I & 0 \\ 0 & I - T^*T \end{bmatrix} \begin{bmatrix} I & T \\ 0 & I \end{bmatrix}$$

and Lemma 1.3. $\square$

If the spaces $H_0, H_1, \ldots, H_n$ are finite dimensional, then the determinant can be used to identify the triangular completion of a given lower triangular part. We have the following corollary (the proof is analogous to that of Corollary 1.5).

COROLLARY 2.3. *Assume that the operator lower triangular part* (1) *satisfies condition* (3), *and let the spaces $H_0, \ldots, H_n$ be finite dimensional. If F is a strictly contractive completion of* (1), *then*

$$(17) \qquad \det(I - F^*F) \le \left(\prod_{j=0}^{n} L_j \right)^{-1}.$$

where $L_0, \ldots, L_n$ are as in (15). *Furthermore, equality holds in* (17) *if and only if F is the unique triangular completion of* (1).

Note that the results of this section are related to those in Section XXVII.5.

XXXV.3 THE CARATHÉODORY-TOEPLITZ EXTENSION PROBLEM

In this section we treat the matrix-valued version of the classical Carathéodory-Toeplitz extension problem appearing in complex function theory. The problem is as follows. One has given a trigonometric polynomial

$$(1) \qquad \sum_{k=-p}^{p} e^{ikt} A_k,$$

of which the coefficients $A_{-p}, \ldots, A_p$ are $m \times m$ matrices, and one seeks $m \times m$ matrix-valued functions Φ on the unit circle $\mathbb{T}$ such that

(CT1) $\Phi(\zeta)$ is a positive definite matrix for each $\zeta \in \mathbb{T}$,

(CT2) $\Phi_k = A_k$ for $|k| \leq p$,

(CT3) $\sum_{k=-\infty}^{\infty} \|\Phi_k\| < \infty$.

Here Φ_k is k-th Fourier coefficient of Φ, i.e.,

$$\Phi_k = \frac{1}{2\pi} \int_{-\pi}^{\pi} e^{-ikt} \Phi(e^{it}) dt.$$

The third condition means that Φ is required to be an element of the matrix Wiener algebra $\mathcal{W}^{m \times m}$.

Assume we have a solution Φ of the above problem. Property (CT3) implies that the entries of $\Phi(\cdot)$ are continuous functions on $\mathbb{T}$. Let L be the block Laurent operator on $\ell_2^m(\mathbb{Z})$ defined by Φ, and let M be the operator of multiplication by Φ on $L_2^m(\mathbb{T})$. Property (CT1) implies that $\det \Phi(\zeta) \neq 0$ for each $\zeta \in \mathbb{T}$. Since Φ is continuous, it follows that L is invertible, by virtue of Theorem XXIII.2.4. Next, observe that

$$\langle Mf, f \rangle = \frac{1}{2\pi} \int_{-\pi}^{\pi} \langle \Phi(e^{it}) f(e^{it}), f(e^{it}) \rangle dt,$$

and hence we can use (CT1) to show that M is non-negative. From Theorem XXIII.2.1 we know that L and M are unitarily equivalent. We conclude that L is an invertible non-negative operator, and thus L is strictly positive (by Theorem V.2.1). The latter implies that the compression of L to a subspace of $\ell_2^m(\mathbb{Z})$ is again strictly positive. It follows that for each n the block matrix

$$(2) \qquad \begin{bmatrix} \Phi_0 & \Phi_{-1} & \cdots & \Phi_{-n} \\ \Phi_1 & \Phi_0 & \cdots & \Phi_{-n+1} \\ \vdots & \vdots & & \vdots \\ \Phi_n & \Phi_{n-1} & \cdots & \Phi_0 \end{bmatrix}$$

is positive definite. In particular, this holds true for $n = p$. But in the latter case, because of (CT2), the block matrix (2) is just equal to

$$(3) \qquad \Gamma = \begin{bmatrix} A_0 & A_{-1} & \cdots & A_{-p} \\ A_1 & A_0 & \cdots & A_{-p+1} \\ \vdots & \vdots & & \vdots \\ A_p & A_{p-1} & \cdots & A_0 \end{bmatrix}.$$

So, for the *Carathéodory-Toeplitz extension* problem (CT1)–(CT3) to be solvable, the block matrix (3) must be positive definite. This condition turns out to be sufficient too.

THEOREM 3.1. *The Carathéodory-Toeplitz extension problem* (CT1)–(CT3) *for the trigonometric $m \times m$ matrix polynomial* (1) *is solvable if and only if the block matrix Γ in* (3) *is positive definite. In this case there exists a unique solution $\Phi_b(\cdot)$ of* (CT1)–(CT3) *with the additional property that the j-th Fourier coefficient of $\Phi_b(\cdot)^{-1}$ is equal to the zero $m \times m$ matrix for $|j| > p$, and this solution is obtained in the following way. Introduce*

$$
(4) \qquad
\begin{bmatrix} X_0 \\ X_1 \\ \vdots \\ X_p \end{bmatrix}
= \Gamma^{-1}
\begin{bmatrix} I \\ 0 \\ \vdots \\ 0 \end{bmatrix}, \qquad
\begin{bmatrix} Y_{-p} \\ \vdots \\ Y_{-1} \\ Y_0 \end{bmatrix}
= \Gamma^{-1}
\begin{bmatrix} 0 \\ \vdots \\ 0 \\ I \end{bmatrix},
$$

and put

$$
(5a) \qquad U(\lambda) = (X_0 + \lambda X_1 + \cdots + \lambda^p X_p) X_0^{-1/2},
$$

$$
(5b) \qquad V(\lambda) = (Y_0 + \lambda^{-1} Y_{-1} + \cdots + \lambda^{-p} Y_{-p}) Y_0^{-1/2}.
$$

Then $\det U(\lambda) \neq 0$ for $|\lambda| \leq 1$, $\det V(\lambda) \neq 0$ for $|\lambda| \geq 1$ including $\lambda = \infty$, and

$$
(6) \qquad \Phi_b(\lambda) = (U(\lambda)^*)^{-1} U(\lambda)^{-1} = (V(\lambda)^*)^{-1} V(\lambda)^{-1}, \qquad \lambda \in \mathbf{T}.
$$

Furthermore, each solution Φ of the problem (CT1)–(CT3) *is of the form*

$$
(7) \quad \Phi(\lambda) = (G(\lambda)^* V(\lambda)^* + U(\lambda)^*)^{-1} (I - G(\lambda)^* G(\lambda)) (V(\lambda) G(\lambda) + U(\lambda))^{-1}, \qquad \lambda \in \mathbf{T},
$$

where G is an arbitrary element of $\mathcal{W}^{m \times m}$ such that $\|G(\lambda)\| < 1$ for $\lambda \in \mathbf{T}$ and the j-th Fourier coefficient of G is the zero $m \times m$ matrix for $j \leq p$. Moreover, (7) *gives a one-one correspondence between all such G and all solutions Φ.*

PROOF. The necessity of the condition $\Gamma > 0$ has already been established. Here we prove the sufficiency and construct the solutions by applying the band method from Sections XXXIV.1 and XXXIV.2. We split the proof into four parts.

Part (a). We begin with some preparations. Note that the algebra $\mathcal{W}^{m \times m}$ has a natural involution, namely

$$
(8) \qquad \Phi^*(\zeta) = \Phi(\zeta)^*, \qquad \zeta \in \mathbf{T}.
$$

The * in the right hand side of (8) is the usual adjoint of a matrix. The function $E(\zeta) = I$, where I is the $m \times m$ identity matrix, is the unit of $\mathcal{W}^{m \times m}$. Consider the following subspaces:

$$
(9a) \qquad \mathcal{M}_1 = \{\Phi \in \mathcal{W}^{m \times m} \mid \Phi_j = 0 \text{ for } j \leq p\},
$$

$$
(9b) \qquad \mathcal{M}_2^0 = \{\Phi \in \mathcal{W}^{m \times m} \mid \Phi_j = 0 \text{ for } j \leq 0 \text{ or } j > p\},
$$

$$(9c) \qquad \mathcal{M}_d = \{\Phi \in \mathcal{W}^{m \times m} \mid \Phi_j = 0 \text{ for } j \neq 0\},$$

$$(9d) \qquad \mathcal{M}_3^0 = \{\Phi \in \mathcal{W}^{m \times m} \mid \Phi_j = 0 \text{ for } j \geq 0 \text{ or } j < -p\},$$

$$(9e) \qquad \mathcal{M}_4 = \{\Phi \in \mathcal{W}^{m \times m} \mid \Phi_j = 0 \text{ for } j \geq -p\}.$$

Then

$$(10) \qquad \mathcal{W}^{m \times m} = \mathcal{M}_1 \oplus \mathcal{M}_2^0 \oplus \mathcal{M}_d \oplus \mathcal{M}_3^0 \oplus \mathcal{M}_4,$$

and it is straightforward to check that $\mathcal{W}^{m \times m}$ with the direct sum decomposition (10) is an algebra with band structure. The corresponding band $\mathcal{M}_c$ is precisely the set of all trigonometric polynomials (1). Furthermore,

$$\mathcal{M}_+ = \{\Phi \in \mathcal{W}^{m \times m} \mid \Phi_j = 0 \text{ for } j < 0\},$$

$$\mathcal{M}_- = \{\Phi \in \mathcal{W}^{m \times m} \mid \Phi_j = 0 \text{ for } j > 0\}.$$

Thus, if $\Phi \in \mathcal{W}^{m \times m}$ satisfies condition (CT1), then Φ admits a left and right spectral factorization relative to (10), by virtue of the factorization theorems in Section XXX.10 (see Corollary XXX.10.4). In particular, $\Phi(\zeta) > 0$, $\zeta \in \mathbb{T}$, implies that Φ is positive definite in $\mathcal{W}^{m \times m}$. The converse of the latter statement is trivially true.

Let $C(\mathbb{T})^{m \times m}$ be the algebra of all $m \times m$ matrix functions that are continuous on $\mathbb{T}$. The algebra $C(\mathbb{T})^{m \times m}$ is a unital C^*-algebra with the involution $*$ as in (8) and with the norm $\|\| \cdot \|\|$ given by

$$\|\|\Phi\|\| = \max_{\zeta \in \mathbb{T}} \|\Phi(\zeta)\|.$$

Here $\|\cdot\|$ is the usual operator norm of a matrix. Obviously, $\mathcal{W}^{m \times m}$ is a $*$-subalgebra of $C^{m \times m}(\mathbb{T})$ and the unit E of $\mathcal{W}^{m \times m}$ is the unit of $C^{m \times m}(\mathbb{T})$.

If $\Phi \in \mathcal{W}^{m \times m}$ is positive definite in the larger C^*-algebra $C^{m \times m}(\mathbb{T})$, then $\Phi(\zeta) > 0$ for each $\zeta \in \mathbb{T}$, and hence, by the remarks made above, Φ is positive definite in $\mathcal{W}^{m \times m}$. It follows that for $\Phi \in \mathcal{W}^{m \times m}$ condition (CT1) is also equivalent to the requirement that Φ is positive definite in $C^{m \times m}(\mathbb{T})$.

Part (b). This part concerns the uniqueness statement. Let K be the trigonometric matrix polynomial defined by (1). Then $K \in \mathcal{M}_c$, and the fact that Γ in (3) is positive definite implies that $K = K^*$. Assume Φ is a solution of the problem (CT1)–(CT3) which has the additional property that for $|j| > p$ the j-th Fourier coefficient of $\Phi(\cdot)^{-1}$ is equal to zero. Then Φ is a band extension of K relative to the band structure (10). We also know that Φ has a left and right spectral factorization relative to (10). But then we can apply Theorem XXXIV.1.3 to show that there exists atmost one such Φ.

Part (c). In this part we apply Theorems 1.1 and 1.2 in Section XXXIV.1 to construct the desired solution Φ_b. Consider the equations

$$(11) \qquad P_2(KX) = E, \qquad P_3(KY) = E,$$

where E is the unit of $\mathcal{W}^{m \times m}$. We seek solutions

$$(12) \qquad \begin{aligned} X(\zeta) &= X_0 + \zeta X_1 + \cdots + \zeta^p X_p \in \mathcal{M}_2, \\ Y(\zeta) &= Y_0 + \zeta^{-1} Y_{-1} + \cdots + \zeta^{-p} Y_{-p} \in \mathcal{M}_3. \end{aligned}$$

Note that in this case P_2 (resp., P_3) is the map which assigns to Φ, $\Phi(\zeta) = \sum_{\nu=-\infty}^{\infty} \zeta^\nu \Phi_\nu$, the function $\sum_{\nu=0}^{p} \zeta^\nu \Phi_\nu$ (resp., $\sum_{\nu=0}^{p} \zeta^{-\nu} \Phi_{-\nu}$). It follows that in terms of Fourier coefficients the equations (11) can be rewritten in the following equivalent form:

$$(13) \qquad \Gamma \begin{bmatrix} X_0 \\ X_1 \\ \vdots \\ X_p \end{bmatrix} = \begin{bmatrix} I \\ 0 \\ \vdots \\ 0 \end{bmatrix}, \qquad \Gamma \begin{bmatrix} Y_{-p} \\ \vdots \\ Y_{-1} \\ Y_0 \end{bmatrix} = \begin{bmatrix} 0 \\ \vdots \\ 0 \\ I \end{bmatrix},$$

where Γ is given by (3). Since Γ is positive definite, Γ^{-1} exists and its block diagonal entries are positive definite. We conclude that there exist $X(\cdot)$ and $Y(\cdot)$ as in (12) such that (11) holds. Moreover, X_0 and Y_0 are positive definite.

Next, let us show that $\det X(\lambda) \neq 0$ for all $|\lambda| \leq 1$. Put $\Gamma_{p-1} = [A_{i-j}]_{i,j=0}^{p-1}$. The Toeplitz structure of Γ implies that Γ partitions as

$$(14) \qquad \Gamma = \begin{bmatrix} A_0 & B \\ B^* & \Gamma_{p-1} \end{bmatrix},$$

where $B = [A_{-1} \ \cdots \ A_{-p}]$. For $j = 1, \ldots, p$ put $S_j = X_j X_0^{-1}$, and set

$$\Sigma = \begin{bmatrix} S_1 \\ S_2 \\ \vdots \\ S_p \end{bmatrix}.$$

From the partitioning in (14) and the first identity in (13) it follows that

$$A_0 + B\Sigma = X_0^{-1}, \qquad B^* + \Gamma_{p-1}\Sigma = 0,$$

and therefore

$$(15) \qquad A_0 - \Sigma^* \Gamma_{p-1} \Sigma = X_0^{-1}.$$

Introduce the following companion-type block matrices:

$$T = \begin{bmatrix} -S_1 & I & 0 & \cdots & 0 \\ -S_2 & 0 & I & & 0 \\ \vdots & \vdots & & \ddots & \\ -S_{p-1} & 0 & 0 & & I \\ -S_p & 0 & 0 & \cdots & 0 \end{bmatrix}, \qquad S = \begin{bmatrix} 0 & I & 0 & \cdots & 0 \\ 0 & 0 & I & & 0 \\ \vdots & \vdots & & \ddots & \\ 0 & 0 & 0 & & I \\ -S_p & -S_{p-1} & -S_{p-2} & \cdots & -S_1 \end{bmatrix}.$$

Here the blank spots denote zero entries. We have $TG = GS$, where G is the lower triangular block Toeplitz matrix given by

$$G = \begin{bmatrix} I & 0 & 0 & \cdots & 0 & 0 \\ S_1 & I & 0 & \cdots & 0 & 0 \\ S_2 & S_1 & I & & 0 & 0 \\ \vdots & \vdots & & \ddots & & \vdots \\ S_{p-2} & S_{p-3} & S_{p-4} & & I & 0 \\ S_{p-1} & S_{p-2} & S_{p-3} & \cdots & S_1 & I \end{bmatrix}.$$

The identity (15) and the block Toeplitz structure of Γ imply that

$$(16) \qquad \Gamma_{p-1} - T^*\Gamma_{p-1}T = \begin{bmatrix} X_0^{-1} & & & \\ & 0 & & \\ & & \ddots & \\ & & & 0 \end{bmatrix}.$$

The off diagonal elements in the block matrix in the right hand side of (16) are all zero. It follows that

$$(17) \qquad G^*\Gamma_{p-1}G - S^*(G^*\Gamma_{p-1}G)S = \begin{bmatrix} X_0^{-1} & & & \\ & 0 & & \\ & & \ddots & \\ & & & 0 \end{bmatrix}.$$

Note that $\det X(0) \neq 0$. Assume $0 < |\lambda_0| \leq 1$ and $\det X(\lambda_0) = 0$. Then there exists $x_0 \neq 0$ such that $X(\lambda_0)x_0 = 0$. Put $z_0 = \lambda_0^{-1}$ and $y_0 = X_0^{1/2}x_0$. Then

$$(S_p + z_0 S_{p-1} + \cdots + z_0^{p-1}S_1 + z_0^p I)y_0 = 0,$$

and hence (cf., formula (4) in Section III.1) we have $Sv = z_0 v$, where v is the column whose i-th entry $(i = 1, \ldots, p)$ is equal to $z_0^{i-1}y_0$. Now use (17) to conclude that

$$(18) \qquad (1 - |z_0|^2)v^*(G^*\Gamma_{p-1}G)v = y_0^* X_0^{-1} y_0.$$

The left hand side of (18) is not positive, and the right hand side is not negative. It follows that both sides of (18) must be zero. Since X_0^{-1} is positive definite, we see that $y_0 = 0$, which is a contradiction. Thus $\det X(\lambda) \neq 0$ for all $|\lambda| \leq 1$. In a similar way one shows that $\det Y(\lambda) \neq 0$ for $|\lambda| \geq 1$ (including $\lambda = \infty$).

The results of the previous two paragraphs, together with the matrix version of Wiener's theorem (see Section XXX.8) show that the equations in (11) have solutions $X \in \mathcal{M}_2$ and $Y \in \mathcal{M}_3$, respectively, such that

$$X^{-1} \in \mathcal{M}_+, \qquad Y^{-1} \in \mathcal{M}_-,$$

and the elements $P_d X$ and $P_d Y$ are positive definite in $\mathcal{M}_d$. Put

$$U(\lambda) = X(\lambda)X_0^{-1/2}, \qquad V(\lambda) = Y(\lambda)Y_0^{-1/2}.$$

Then $U(\cdot)$ and $V(\cdot)$ have the desired properties, and we can apply Theorems 1.1 and 1.2 in Section XXXIV.1 to show that Φ_b and $\widetilde{\Phi}_b$,

$$\Phi_b(\lambda) = \left(U(\lambda)^*\right)^{-1}U(\lambda)^{-1}, \qquad \widetilde{\Phi}_b(\lambda) = \left(V(\lambda)^*\right)^{-1}V(\lambda)^{-1},$$

are band extensions of K relative to (10). By Theorem XXXIV.1.3 we have $\Phi_b = \widetilde{\Phi}_b$, and we see that Φ_b has all the properties mentioned in the theorem.

Part (d). In this part we apply Theorem XXXIV.2.1 to get all solutions of the problem (CT1)–(CT3). For $\mathcal{M}$ we take the algebra $\mathcal{W}^{m\times m}$ with band structure (10), and the role of the C^*-algebra $\mathcal{R}$ is played by $C(\mathbb{T})^{m\times m}$. First, we show that axiom (A) is fulfilled in the present context. Take $G \in \mathcal{M}_+$, and assume that $\|\!|G|\!\| < 1$. Note that G is analytic on $|\lambda| < 1$ and continuous on $|\lambda| \leq 1$. So, by the maximum modulus principle,

$$\|G(\lambda)\| \leq \max_{\zeta \in \mathbb{T}} \|G(\zeta)\| = \|\!|G|\!\| < 1, \qquad |\lambda| \leq 1.$$

It follows that $\det\!\left(I - G(\lambda)\right) \neq 0$ for each $|\lambda| \leq 1$, and thus, by the matrix version of Wiener's theorem (see Section XXX.8), we have $(E - G)^{-1} \in \mathcal{W}^{m\times m}$. Since $\left(I - G(\cdot)\right)^{-1}$ extends to an analytic function on $|\lambda| < 1$, we obtain $(E - G)^{-1} \in \mathcal{M}_+$, and hence axiom (A) is fulfilled. According to what has been proved so far, the band extension Φ_b of K admits a right and a left spectral factorization relative to (10). So all the conditions of Theorem XXXIV.2.1 are satisfied for the case considered here. Note that in this case the $\mathcal{R}$-positive extensions coincide with the positive extensions. Thus we can apply Theorem XXXIV.2.1 to get the desired description of all solutions. $\square$

The solution Φ_b constructed in Theorem 3.1 will be called the *band extension* of (1). The next theorem presents the so-called maximum entropy characterization of the band extension Φ_b.

THEOREM 3.2. *Assume that the block matrix Γ in (3) is positive definite. Then the band extension of (1) is the unique solution Φ of the Carathéodory-Toeplitz problem* (CT1)–(CT3) *for (1) that maximizes the integral*

$$\frac{1}{2\pi} \int_{-\pi}^{\pi} \log \det \Phi(e^{it})\,dt.$$

PROOF. We apply Theorem XXXIV.4.2 to the algebra $\mathcal{W}^{m\times m}$ endowed with the band structure (10) and with $\mathcal{R} = C^{m\times m}(\mathbb{T})$. First, let us show that axioms (C1) and (C2) are fulfilled for the case considered here.

Take $\Phi \in \mathcal{W}^{m\times m}$, and assume Φ is positive semidefinite in $\mathcal{R} = C^{m\times m}(\mathbb{T})$. Then $\Phi(\zeta) \geq 0$ for each $\zeta \in \mathbb{T}$. For each $x \in \mathbb{C}^m$ we have

$$(19) \qquad \langle \Phi_0 x, x \rangle = \frac{1}{2\pi} \int_{-\pi}^{\pi} \langle \Phi(e^{it})x, x \rangle\,dt.$$

The integrand is a continuous function of t whose values are non-negative real numbers. It follows that $\langle \Phi_0 x, x \rangle \geq 0$ for each x. So the matrix Φ_0 is positive semidefinite. Hence $\Phi_0 = F_0^* F_0$ with $F_0 = \Phi_0^{1/2}$. Since $P_d(\Phi) = \Phi_0$, we see that axiom (C1) is fulfilled.

Next, assume additionally that $P_d(\Phi) = 0$. So $\Phi_0 = 0$, and hence for each $x \in \mathbb{C}^m$ the integral in (19) is zero. Since the integrand is continuous, we conclude that $\langle \Phi(\zeta)x, x \rangle = 0$ for each $x \in \mathbb{C}^m$ and each $\zeta \in \mathbb{T}$. This implies that $\Phi = 0$, and hence axiom (C2) is fulfilled.

Let $\Phi \in \mathcal{W}^{m \times m}$ be a solution of the Carathéodory-Toeplitz problem (CT1)–(CT3) for (1). So Φ is a positive extension relative to (10) of K, where K is the trigonometric polynomial (1). We know that Φ admits a right spectral factorization relative to (10), and hence (see Lemma XXXIV.4.1) the function Φ factorizes as

$$\Phi = (E + F_+)^* \Delta(\Phi)(E + F_+),$$

where $\Delta(\Phi)$ is the right multiplicative diagonal of Φ relative to (10) and $F_+ \in \mathcal{W}^{m \times m}$ is of the form

$$F_+(\zeta) = \sum_{n=1}^{\infty} \zeta^n F_n, \qquad \zeta \in \mathbb{T}.$$

It follows that $\det(I + F_+(\cdot))$ is analytic on $|\lambda| < 1$, and $\det(I + F_+(\cdot))$ is continuous and non-zero on $|\lambda| \leq 1$. From these properties we conclude that $\log \det(I + F_+(\cdot))$ is analytic on $|\lambda| < 1$ and continuous on $|\lambda| \leq 1$. Therefore, by Cauchy's theorem,

$$\frac{1}{2\pi} \int_{-\pi}^{\pi} \log \det\left(I + F_+(e^{it})\right) dt = 0.$$

But then we see that

$$\frac{1}{2\pi} \int_{-\pi}^{\pi} \log \det \Phi(e^{it}) dt = \log \det \Delta(\Phi).$$

Next use the fact that the log function is a monotonically strictly increasing function on $(0, \infty)$ and the determinant is a strictly monotone function on the positive definite matrices. Furthermore, note that for the case considered here $\mathcal{R} = C^{m \times m}(\mathbb{T})$, and hence

$$\Delta(\Phi_b) \geq_{\mathcal{R}} \Delta(\Phi)$$

is equivalent to the requirement that $\Delta(\Phi_b) - \Delta(\Phi)$ is a positive semidefinite matrix. A straightforward application of Theorem XXXIV.4.3 gives now the desired result. $\square$

XXXV.4 THE NEHARI EXTENSION PROBLEM

In this section we treat a matrix-valued version of an extension problem which has its origin in a classical paper of Z. Nehari (see Nehari [1]). Let $\Psi_0, \Psi_{-1}, \Psi_{-2}, \ldots$ be a given sequence of $m \times m$ matrices such that

$$(1) \qquad \sum_{j=-\infty}^{0} \|\Psi_j\| < \infty.$$

Here, as well as in the sequel, the norm of an $m \times m$ matrix T is the norm of T viewed as an operator on $\mathbb{C}^m$. The problem is to find $m \times m$ matrix-valued functions Φ on the unit circle $\mathbb{T}$ with the following properties:

(N1) $\|\Phi(\zeta)\| < 1$ for each $\zeta \in \mathbb{T}$,

(N2) $\Phi_k = \Psi_k$ for $k = 0, -1, -2, \ldots$,

(N3) $\sum_{k=-\infty}^{\infty} \|\Phi_k\| < \infty$.

Here Φ_k is the k-th Fourier coefficient of Φ. Thus condition (N3) requires that $\Phi \in \mathcal{W}^{m \times m}$, and hence Φ is continuous on $\mathbb{T}$.

Assume the *Nehari extension problem* (N1)–(N3) has a solution Φ, and let L be the Laurent operator on $\ell_2^m(\mathbb{Z})$ defined by Φ. Since Φ is continuous on $\mathbb{T}$, condition (N1) implies (cf., Corollary XXIII.2.2) that

$$\|L\| = \max_t \|\Phi(e^{it})\| < 1.$$

It follows that any compression of L is a strict contraction. In particular (cf., formula (1) in Section XXIII.4), the Hankel operator

$$(2) \qquad \Gamma = \begin{bmatrix} \Psi_0 & \Psi_{-1} & \Psi_{-2} & \cdots \\ \Psi_{-1} & \Psi_{-2} & \Psi_{-3} & \cdots \\ \Psi_{-2} & \Psi_{-3} & \Psi_{-4} & \cdots \\ \vdots & \vdots & \vdots & \end{bmatrix} : \ell_2^m \to \ell_2^m$$

must have norm strictly less than one. This condition is not only necessary but also sufficient.

THEOREM 4.1. *In order that the Nehari extension problem (N1)–(N3) be solvable it is necessary and sufficient that the Hankel operator Γ in (2) be a strict contraction. In this case there exists a unique solution $\Phi_\tau(\cdot)$ of (N1)–(N3) with the additional property that the j-th Fourier coefficient of $\left(I - \Phi_\tau(\cdot)^* \Phi_\tau(\cdot)\right)^{-1}$ is equal to the zero $m \times m$ matrix for $j = 1, 2, \ldots$, and this solution is obtained in the following way. Put*

$$(3a) \qquad \begin{bmatrix} a_0 \\ a_{-1} \\ a_{-2} \\ \vdots \end{bmatrix} := (I - \Gamma\Gamma^*)^{-1} \begin{bmatrix} I_m \\ 0 \\ 0 \\ \vdots \end{bmatrix}, \qquad \begin{bmatrix} c_0 \\ c_1 \\ c_2 \\ \vdots \end{bmatrix} := \Gamma^* \begin{bmatrix} a_0 \\ a_1 \\ a_2 \\ \vdots \end{bmatrix},$$

$$(3b) \qquad \begin{bmatrix} d_0 \\ d_1 \\ d_2 \\ \vdots \end{bmatrix} := (I - \Gamma^*\Gamma)^{-1} \begin{bmatrix} I_m \\ 0 \\ 0 \\ \vdots \end{bmatrix}, \qquad \begin{bmatrix} b_0 \\ b_{-1} \\ b_{-2} \\ \vdots \end{bmatrix} := \Gamma \begin{bmatrix} d_0 \\ d_1 \\ d_2 \\ \vdots \end{bmatrix},$$

and let

$$(4a) \qquad \alpha(\zeta) = \sum_{j=-\infty}^{0} a_j a_0^{-1/2} \zeta^j, \qquad \gamma(\zeta) = \sum_{j=0}^{\infty} c_j a_0^{-1/2} \zeta^j,$$

$$(4b) \qquad \delta(\zeta) = \sum_{j=0}^{\infty} d_j d_0^{-1/2} \zeta^j, \qquad \beta(\zeta) = \sum_{j=-\infty}^{0} b_j d_0^{-1/2} \zeta^j.$$

Then

$$(5) \qquad \Phi_\tau(\zeta) = \beta(\zeta)\delta(\zeta)^{-1} = \left(\alpha(\zeta)^{-1}\right)^* \gamma(\zeta)^*, \qquad \zeta \in \mathbf{T}.$$

Furthermore, each solution Φ of the problem (N1)–(N3) is of the form

$$(6) \qquad \Phi(\zeta) = \left(\alpha(\zeta)H(z) + \beta(\zeta)\right)\left(\gamma(\zeta)H(\zeta) + \delta(\zeta)\right)^{-1}, \qquad \zeta \in \mathbf{T},$$

where H is an arbitrary element of $\mathcal{W}^{m \times m}$ such that $\|H(\zeta)\| < 1$ for $\zeta \in \mathbf{T}$ and the j-th Fourier coefficient of H is the zero $m \times m$ matrix for $j = 0, -1, -2, \ldots$. Moreover, (6) gives a one-one correspondence between all such H and all solutions Φ.

PROOF. The necessity of the condition $\|\Gamma\| < 1$ has already been established. To prove the sufficiency and to construct all solutions we apply the main result of Section XXXIV.3. We divide the proof into four parts.

Part (a). In the sequel we use the notations introduced in the proof of Theorem XXXV.3.1. Thus $\mathcal{W}^{m \times m}$ is the $m \times m$ matrix Wiener algebra, the involution $*$ on $\mathcal{W}^{m \times m}$ is defined by formula (8) in the previous section, and E is the unit of $\mathcal{W}^{m \times m}$. Put

$$\mathcal{M}_-^0 = \{\Phi \in \mathcal{W}^{m \times m} \mid \Phi_j = 0, \ j \geq 0\},$$

$$\mathcal{M}_d = \{\Phi \in \mathcal{W}^{m \times m} \mid \Phi_j = 0, \ j \neq 0\},$$

$$\mathcal{M}_+^0 = \{\Phi \in \mathcal{W}^{m \times m} \mid \Phi_j = 0, \ j \leq 0\}.$$

Then

$$(7) \qquad \mathcal{W}^{m \times m} = \mathcal{M}_+^0 \oplus \{0\} \oplus \mathcal{M}_d \oplus \{0\} \oplus \mathcal{M}_-^0,$$

and it is straightforward to check that $\mathcal{W}^{m \times m}$ with the direct sum decomposition (7) is an algebra with band structure. Note that

$$\mathcal{M}_\ell := \mathcal{M}_- = \mathcal{M}_-^0 \oplus \mathcal{M}_d, \qquad \mathcal{M}_u := \mathcal{M}_+ = \mathcal{M}_+^0 \oplus \mathcal{M}_d.$$

Since (1) holds,

$$(8) \qquad \Psi(\zeta) = \sum_{j=-\infty}^{0} \zeta^j \Psi_j \in \mathcal{M}_\ell.$$

The algebra $\mathcal{W}^{m \times m}$ is a $*$-subalgebra of the unital C^*-algebra $C^{m \times m}(\mathbb{T})$ and the unit E of $\mathcal{W}^{m \times m}$ is also the unit of $C^{m \times m}(\mathbb{T})$. In what follows the role of $\mathcal{R}$ in Section XXXIV.3 is played by $C^{m \times m}(\mathbb{T})$. Thus

$$\|\Phi\|_{\mathcal{R}} = \|\!|\Phi|\!\| = \max_{\zeta \in \mathbb{T}} \|\Phi(\zeta)\|.$$

It follows that Φ is a solution of the Nehari problem (N1)–(N3) if and only if $\Phi \in \mathcal{W}^{m \times m}$ is a strictly contractive extension of Ψ relative to the band structure (7). Here Ψ is defined by (8).

Part (b). In what follows we assume $\|\Gamma\| < 1$. Condition (1) on the data implies that Γ and Γ^* both act as compact operators on ℓ_1^m. Thus $I - \Gamma\Gamma^*$ is a Fredholm operator on ℓ_1^m of index equal to zero. Now, $\ell_1^m \subset \ell_2^m$ and on ℓ_2^m the operator $I - \Gamma\Gamma^*$ is invertible. In particular, $I - \Gamma\Gamma^*$ is an injective operator on ℓ_1^m. Since its index is zero, we conclude that $I - \Gamma\Gamma^*$ acts as an invertible operator on ℓ_1^m. By interchanging the roles of Γ and Γ^* we see that the same holds true for $I - \Gamma^*\Gamma$. It follows that the sequences of $m \times m$ matrices defined in (3a) and (3b) are all summable in norm. Hence we have

$$(9a) \qquad a(\zeta) := \sum_{j=-\infty}^{0} \zeta^j a_j \in \mathcal{M}_-, \qquad b(\zeta) := \sum_{j=-\infty}^{0} \zeta^j b_j \in \mathcal{M}_-,$$

$$(9b) \qquad c(\zeta) := \sum_{j=0}^{\infty} \zeta^j c_j \in \mathcal{M}_+, \qquad d(\zeta) := \sum_{j=0}^{\infty} \zeta^j d_j \in \mathcal{M}_+.$$

Part (c). In this part we apply Theorem XXXIV.3.1 to $\mathcal{M} = \mathcal{W}^{m \times m}$ endowed with the band structure (7) and with $\mathcal{R} = C^{m \times m}(\mathbb{T})$. Let P_+, P_- and P_d be the natural projections of $\mathcal{W}^{m \times m}$ associated with the spaces $\mathcal{M}_+$, $\mathcal{M}_-$ and $\mathcal{M}_d$, respectively. Since Γ is the Hankel operator corresponding to the function Ψ in (8), the identities in the left hand side of (3a) and (3b) show that the matrix functions $a(\cdot)$ and $d(\cdot)$ in (9a) and (9b), respectively, satisfy the following equations

$$a - P_-\big(\Psi(P_+(\Psi^* a))\big) = E, \qquad d - P_+\big(\Psi^*(P_-(\Psi d))\big) = E.$$

Note that $P_d a = a_0$ and $P_d d = d_0$. Thus we see from (3a) and (3b) that $P_d a$ and $P_d d$ are positive definite in $\mathcal{M}_d$.

Since the algebra of 2×2 matrices with entries in $\mathcal{W}^{m \times m}$ may be identified with $\mathcal{W}^{2m \times 2m}$, it is straightforward to check that axiom (B) (see Section XXXIV.3) holds for $\mathcal{W}^{m \times m}$. Furthermore, $\mathcal{W}^{m \times m}$ endowed with the norm

$$(10) \qquad \|\Phi(\cdot)\| = \sum_{j=-\infty}^{\infty} \|\Phi_j\|$$

is a Banach algebra, and the spaces appearing in the right hand side of (7) are closed in the topology defined by the norm (10). So we may apply Proposition XXXIV.3.4 to show that $a(\cdot)$ and $d(\cdot)$ are invertible in $\mathcal{W}^{m \times m}$, and $a(\cdot)^{-1} \in \mathcal{M}_-$, $d(\cdot)^{-1} \in \mathcal{M}_+$.

Indeed, if one replaces Ψ by $t\Psi$, then the corresponding Hankel operator Γ is replaced by $t\Gamma$. Let $0 \leq t \leq 1$. Then $I - t^2\Gamma\Gamma^*$ and $I - t^2\Gamma^*\Gamma$ are invertible operators on ℓ_1^m. Hence $(I - t^2\Gamma\Gamma^*)^{-1}$ and $(I - t^2\Gamma^*\Gamma)^{-1}$ are bounded linear operators on ℓ_1^m which depend analytically on t. Put

$$
\begin{bmatrix} a_0(t) \\ a_{-1}(t) \\ a_{-2}(t) \\ \vdots \end{bmatrix} = (I - t^2\Gamma\Gamma^*)^{-1} \begin{bmatrix} I_m \\ 0 \\ 0 \\ \vdots \end{bmatrix},
$$

$$
\begin{bmatrix} d_0(t) \\ d_1(t) \\ d_2(t) \\ \vdots \end{bmatrix} = (I - t^2\Gamma^*\Gamma)^{-1} \begin{bmatrix} I_m \\ 0 \\ 0 \\ \vdots \end{bmatrix},
$$

and set

$$
a_t(\zeta) = \sum_{j=-\infty}^{0} a_j(t)\zeta^j, \qquad d_t(\zeta) = \sum_{j=0}^{\infty} d_j(t)\zeta^j.
$$

Then $a_t(\cdot) \in \mathcal{M}_-$ and $d_t(\cdot) \in \mathcal{M}_+$ and these functions depend analytically on t with respect to the norm (10). From the result in the first paragraph of the present part of the proof we see that the functions $t \mapsto a_t$ and $t \mapsto d_t$ have the properties described in Proposition XXXIV.3.4 for the algebra $\mathcal{W}^{m \times m}$ with band structure (7). So we may apply Proposition XXXIV.3.4. It follows that $a(\cdot)$ and $d(\cdot)$ are invertible in $\mathcal{W}^{m \times m}$, and $a(\cdot)^{-1} \in \mathcal{M}_-$ and $d(\cdot)^{-1} \in \mathcal{M}_+$.

Thus $a = a(\cdot)$ and $d = d(\cdot)$ satisfy conditions (i), (ii) and (iii) in Theorem XXXIV.3.1 for the case considered here. Hence, by Theorem XXXIV.3.1, the given Ψ has a unique strictly contractive extension Φ_τ with the additional property that

$$
\Phi_\tau(E - \Phi_\tau^*\Phi_\tau)^{-1} \in \mathcal{M}_-.
$$

Furthermore, Φ_τ is given by

$$
\Phi_\tau = b(\cdot)d(\cdot)^{-1} = \big(a(\cdot)^{-1}\big)^* c(\cdot)^*,
$$

where $a(\cdot)$, $b(\cdot)$, $c(\cdot)$ and $d(\cdot)$ are defined by (9a) and (9b). Here we used that $\mathcal{M}_\ell = \mathcal{M}_-$, $\mathcal{M}_u = \mathcal{M}_+$, and the identities in the right hand side of (3a) and (3b) which imply that

$$
b(\cdot) = P_-\big(\Psi(\cdot)d(\cdot)\big), \qquad c(\cdot) = P_+\big(\Psi(\cdot)^* a(\cdot)\big).
$$

From (4a) and (4b) we see that

$$
(11a) \qquad\qquad \alpha(\cdot) = a(\cdot)a_0^{-1/2}, \qquad \gamma(\cdot) = c(\cdot)a_0^{-1/2},
$$

$$
(11b) \qquad\qquad \delta(\cdot) = d(\cdot)d_0^{-1/2}, \qquad \beta(\cdot) = b(\cdot)d_0^{-1/2}.
$$

So we have proved the first part of the theorem up to and including formula (5).

Part (d). It remains to derive the description of all solutions. To this end we apply Theorem XXXIV.3.2 with $\mathcal{M} = \mathcal{W}^{m \times m}$ and $\mathcal{R} = C^{m \times m}(\mathbb{T})$. Note that $\mathcal{M}_+$ is the same $\mathcal{M}_+$ as considered in the previous section, and hence we know from Part (d) of the proof of Theorem XXXV.3.1 that axiom (A) holds in the present context. Also we know that Ψ has a triangular extension with the desired factorization properties. So all conditions of Theorem XXXIV.3.2 are fulfilled for the case considered here. Together with the remarks made in the paragraph directly after Theorem XXXIV.3.2 this yields the desired description of all solutions. $\square$

The solution Φ_τ constructed in Theorem 4.1 will be called the *triangular extension* of (8). The next theorem presents the so-called maximum entropy characterization of the triangular extension Φ_τ.

THEOREM 4.2. *Assume that the Hankel operator (2) is a strict contraction. Then the triangular extension of (8) is the unique solution Φ of the Nehari extension problem (N1)–(N3) that maximizes the integral*

$$\frac{1}{2\pi} \int_{-\pi}^{\pi} \log \det\big(I - \Phi(e^{it})^* \Phi(e^{it})\big) dt.$$

PROOF. We apply Theorem XXXIV.4.4 with $\mathcal{M} = \mathcal{W}^{m \times m}$ and $\mathcal{R} = C^{m \times m}(\mathbb{T})$. The band structure in $\mathcal{M} = \mathcal{W}^{m \times m}$ is given by (7). We already know (see the proof of Theorem XXXV.3.2) that axioms (C1) and (C2) are fulfilled in the present context.

Let $\Phi \in \mathcal{W}^{m \times m}$ be a solution of the Nehari problem (N1)–(N3). So Φ is a strictly contractive extension relative to (7) of Ψ, where Ψ is given by (8). Since $E - \Phi^* \Phi$ is positive definite in $\mathcal{R} = C^{m \times m}(\mathbb{T})$, we know (see the last paragraph of Part (a) of the proof of Theorem XXXV.3.1) that $E - \Phi^* \Phi$ admits a right spectral factorization relative to (7). Hence $E - \Phi^* \Phi$ has a well-defined right multiplicative diagonal $\Delta(E - \Phi^* \Phi)$. As in the proof of Theorem XXXV.3.2 one shows that

$$\frac{1}{2\pi} \int_{-\pi}^{\pi} \log \det\big(I - \Phi(e^{it})^* \Phi(e^{it})\big) dt = \log \det \Delta(E - \Phi^* \Phi).$$

A straightforward application of Theorem XXXIV.4.5, using the monotonicity of $\log \det$ on the positive definite matrices, yields now the desired result (cf., the last paragraph of the proof of Theorem XXXV.3.2). $\square$

XXXV.5 THE NEVANLINNA-PICK INTERPOLATION PROBLEM REVISITED

Let $z_1, \ldots, z_N$ be points in the open unit disc $\mathbb{D}$, $z_i \neq z_j$ for $i \neq j$, and let $y_1, \ldots, y_N$ be arbitrary complex numbers. Consider the following version of the

Nevanlinna-Pick interpolation problem (cf., Section XXVII.7). Find

$$\varphi \in \mathcal{W}_+ := \left\{ f \in \mathcal{W} \mid \frac{1}{2\pi} \int_{-\pi}^{\pi} e^{-ikt} f(e^{it})dt = 0, \ k < 0 \right\}$$

such that

(NP1) $\varphi(z_j) = y_j$, $j = 1, \ldots, N$,

(NP2) $\sup_{|z|<1} |\varphi(z)| < 1$.

Here $\mathcal{W}$ is the Wiener algebra on $\mathbb{T}$. The above problem may be restated as a Nehari extension problem. To see this, let $\widetilde{\varphi}$ be an arbitrary function in $\mathcal{W}_+$ satisfying (NP1). For example, we may use the Lagrange interpolation formula, and assume that

$$(1) \qquad \widetilde{\varphi}(\lambda) = \sum_{j=1}^{N} y_j \left(\prod_{\substack{i=1 \\ i \neq j}}^{N} \frac{\lambda - z_i}{z_j - z_i} \right).$$

Next, consider the following Blaschke type product

$$(2) \qquad u(\lambda) = \prod_{j=1}^{N} \frac{\lambda - z_j}{1 - \lambda \overline{z}_j}.$$

Note that $u \in \mathcal{W}_+$. Furthermore, $u(z_j) = 0$ for $j = 1, \ldots, N$ and $|u(e^{it})| = 1$ for $-\pi \leq t \leq \pi$.

LEMMA 5.1. *The function $\varphi \in \mathcal{W}_+$ is a solution of the Nevanlinna-Pick interpolation problem (NP1), (NP2) if and only if $\varphi = \psi u$, where $\psi \in \mathcal{W}$ is a solution of a Nehari extension problem, namely*

(Ne1) $|\psi(\zeta)| < 1$ *for each $\zeta \in \mathbb{T}$,*

(Ne2)   *for $j = -1, -2, \ldots$ the functions ψ and $\widetilde{\varphi}u^{-1}$ have the same j-th Fourier coefficient.*

PROOF. Assume $\varphi \in \mathcal{W}_+$ satisfies (NP1), (NP2). Put $\psi = \varphi u^{-1}$. Then $\psi \in \mathcal{W}$ and

$$(3) \qquad \psi - \widetilde{\varphi}u^{-1} = (\varphi - \widetilde{\varphi})u^{-1} \in \mathcal{W}_+.$$

In particular, (Ne2) holds. Furthermore,

$$(4) \qquad \sup_{\zeta \in \mathbb{T}} |\psi(\zeta)| = \sup_{\zeta \in \mathbb{T}} |\varphi(\zeta)u(\zeta)^{-1}| = \sup_{\zeta \in \mathbb{T}} |\varphi(\zeta)| = \sup_{z \in \mathbb{D}} |\varphi(z)| < 1,$$

because φ is analytic on $\mathbb{D}$ and continuous on $\overline{\mathbb{D}}$. It follows that (Ne1) holds.

Conversely, assume $\psi \in \mathcal{W}$ and ψ satisfies (Ne1), (Ne2). Put $\varphi = \psi u$. Then $\varphi \in \mathcal{W}$, and according to (Ne2) we have $\psi - \widetilde{\varphi}u^{-1} \in \mathcal{W}_+$. It follows that $\varphi - \widetilde{\varphi} = hu$ for some $h \in \mathcal{W}_+$. In particular, $\varphi \in \mathcal{W}_+$. Furthermore,

$$\varphi(z_j) = \widetilde{\varphi}(z_j) + h(z_j)u(z_j) = \widetilde{\varphi}(z_j) = y_j, \qquad j = 1, \ldots, N,$$

and thus φ satisfies (NP1). Since $\varphi \in \mathcal{W}_+$, the identities in (4) show that (NP2) is also fulfilled. $\square$

The connection between Nevanlinna-Pick interpolation and Nehari extension allows us to obtain a linear fractional representation of all solutions of the (strict) Nevanlinna-Pick interpolation problem considered here. In fact, as we shall see below, this connection also holds for matrix-valued functions. Indeed, let $z_1, \ldots, z_N$ be N different points in $\mathbb{D}$, and let $Y_1, \ldots, Y_N$ be $m \times m$ matrices. The matrix version of the Nevanlinna-Pick interpolation problem is to determine all

$$\Phi \in \mathcal{W}_+^{m \times m} := \left\{ F \in \mathcal{W}^{m \times m} \mid \frac{1}{2\pi} \int_{-\pi}^{\pi} e^{-ikt} F(e^{it}) dt = 0, \ k < 0 \right\}$$

such that

(MNP1) $\Phi(z_j) = Y_j, j = 1, \ldots, N,$

(MNP2) $\sup_{|z| < 1} \|\Phi(z)\| < 1.$

The norm in (MNP2) is the usual norm of an $m \times m$ matrix. Now consider

$$(5) \qquad \widetilde{\Phi}(\lambda) = \sum_{j=1}^{N} Y_j \left(\prod_{\substack{i=1 \\ i \neq j}}^{N} \frac{\lambda - z_i}{z_j - z_i} I \right),$$

$$(6) \qquad U(\lambda) = \prod_{j=1}^{N} \frac{\lambda - z_j}{1 - \lambda \overline{z}_j} I,$$

where I is the $m \times m$ identity matrix.

THEOREM 5.2. *Let $\widetilde{\Phi}$ and U be as in (5) and (6), respectively, and let Γ be the block Hankel operator*

$$(7) \qquad \Gamma = \begin{bmatrix} \Lambda_{-1} & \Lambda_{-2} & \Lambda_{-3} & \cdots \\ \Lambda_{-2} & \Lambda_{-3} & \Lambda_{-4} & \cdots \\ \Lambda_{-3} & \Lambda_{-4} & \Lambda_{-5} & \cdots \\ \vdots & \vdots & \vdots & \end{bmatrix} : \ell_2^m \to \ell_2^m$$

where Λ_k is the k-th Fourier coefficient of $\widetilde{\Phi}U^{-1}$ ($k < 0$). Then the matrix Nevanlinna-Pick interpolation problem (MNP1), (MNP2) has a solution Φ if and only if $\|\Gamma\| < 1$, and in this case all solutions are obtained as follows. Put

$$(8a) \qquad \begin{bmatrix} a_{-1} \\ a_{-2} \\ a_{-3} \\ \vdots \end{bmatrix} := (I - \Gamma\Gamma^*)^{-1} \begin{bmatrix} I_m \\ 0 \\ 0 \\ \vdots \end{bmatrix}, \qquad \begin{bmatrix} c_0 \\ c_1 \\ c_2 \\ \vdots \end{bmatrix} := \Gamma^* \begin{bmatrix} a_{-1} \\ a_{-2} \\ a_{-3} \\ \vdots \end{bmatrix},$$

$$(8b) \qquad \begin{bmatrix} d_0 \\ d_1 \\ d_2 \\ \vdots \end{bmatrix} := (I - \Gamma^*\Gamma)^{-1} \begin{bmatrix} I_m \\ 0 \\ 0 \\ \vdots \end{bmatrix}, \qquad \begin{bmatrix} b_{-1} \\ b_{-2} \\ b_{-3} \\ \vdots \end{bmatrix} := \Gamma \begin{bmatrix} d_0 \\ d_1 \\ d_2 \\ \vdots \end{bmatrix},$$

and let

$$(9a) \qquad \alpha(\zeta) = \sum_{j=-\infty}^{-1} a_j a_{-1}^{-1/2} \zeta^j, \qquad \gamma(\zeta) = \sum_{j=0}^{\infty} c_j a_{-1}^{-1/2} \zeta^j,$$

$$(9b) \qquad \delta(\zeta) = \sum_{j=0}^{\infty} d_j d_0^{-1/2} \zeta^j, \qquad \beta(\zeta) = \sum_{j=-\infty}^{-1} b_j d_0^{-1/2} \zeta^j.$$

Then all solutions Φ are given by

$$(10) \qquad \Phi = (\alpha H + \beta)(\gamma H + \delta)^{-1} U,$$

where H is an arbitrary element of $\mathcal{W}^{m \times m}$ such that $\|H(\zeta)\| < 1$ for $\zeta \in \mathbb{T}$ and the j-th Fourier coefficient of H is zero for $j = 0, -1, -2, \ldots$.

PROOF. By using the same arguments as in the proof of Lemma 5.1 one shows that Φ is a solution of the matrix Nevanlinna-Pick interpolation problem (MN1), (MN2) if and only if $\Phi = \Psi U$, where $\Psi \in \mathcal{W}^{m \times m}$ and Ψ has the following properties:

    (MNe1) $\|\Psi(\zeta)\| < 1$ for $\zeta \in \mathbb{T}$,

    (MNe2)   for $j = -1, -2, \ldots$ the j-th Fourier coefficient of Ψ is equal to Λ_j.

Note that there exists a solution $\Psi \in \mathcal{W}^{m \times m}$ to the problem (MNe1), (MNe2) if and only if for $\Psi_k = \Lambda_{k-1}$, $k = 0, -1, -2, \ldots$, the Nehari extension problem considered in the previous section is solvable, and in this case all $\Psi \in \mathcal{W}^{m \times m}$ satisfying (MNe1), (MNe2) are of the form

$$\Psi(\zeta) = \zeta^{-1} \widetilde{\Psi}(\zeta), \qquad \zeta \in \mathbb{T},$$

where $\widetilde{\Psi}$ is an arbitrary solution of the Nehari extension problem referred to above. A straightforward application of Theorem XXXV.4.1 now gives the desired result. $\square$

The Hankel operator Γ in (7) does not depend on the particular choice of the interpolant $\widetilde{\Phi}$. Indeed, let $\widetilde{\Phi}'$ be another matrix function in $\mathcal{W}_+^{m \times m}$ such that

$$\widetilde{\Phi}'(z_j) = Y_j, \qquad j = 1, \ldots, N.$$

Then $(\widetilde{\Phi} - \widetilde{\Phi}')U^{-1}$ is analytic on $\mathbb{D}$, and hence for $j = -1, -2, -3, \ldots$ the j-th Fourier coefficients of $\widetilde{\Phi}U^{-1}$ and $\widetilde{\Phi}'U^{-1}$ coincide.

From Theorem 5.2 and the results in Section XXVII.7 it follows that for the scalar case ($m = 1$) the condition $\|\Gamma\| < 1$ in Theorem 5.2 is equivalent to the requirement that the classical Pick matrix

$$\left[\frac{1 - \overline{y}_i y_j}{1 - \overline{z}_i z_j} \right]_{i,j=1}^{N}$$

is positive definite. For the matrix case this equivalence remains true provided one replaces $\overline{y}_i y_j$ by $Y_i^* Y_j$.

From the connection between the Nevanlinna-Pick interpolation problem and the Nehari extension problem it follows that also in the Nevanlinna-Pick problem there is a special solution from which all other solutions may be read off, namely the solution corresponding to the triangular extension in the Nehari problem (MNe1), (MNe2).

XXXV.6 TANGENTIAL NEVANLINNA-PICK INTERPOLATION

In this section we treat the tangential version of the (strict) Nevanlinna-Pick interpolation problem considered in the previous section. Let $z_1, \ldots, z_N$ be points in the open unit disc $\mathbb{D}$, and let $X_1, \ldots, X_N$ and $Y_1, \ldots, Y_N$ be vectors in $\mathbb{C}^m$. We assume that the points $z_1, \ldots, z_N$ are different and that the vectors $X_1, \ldots, X_N$ are non-zero. We seek $\Phi \in \mathcal{W}_+^{m \times m}$ such that

(TNP1) $\Phi(z_j) X_j = Y_j, \ j = 1, \ldots, N,$

(TNP2) $\sup_{|z|<1} \|\Phi(z)\| < 1.$

Here $\mathcal{W}_+^{m \times m}$ is as in the previous paragraph.

Our aim is to solve this tangential interpolation problem by using an appropriate modification of the method employed in the previous section.

The first step is to construct a $\widetilde{\Phi} \in \mathcal{W}_+^{m \times m}$ satisfying (TNP1). For this purpose we need the following auxiliary matrices:

$$(1) \qquad C_- = [X_1 \quad \cdots \quad X_N], \qquad C_+ = [Y_1 \quad \cdots \quad Y_N]$$

$$(2) \qquad A = \begin{bmatrix} z_1 & & \\ & \ddots & \\ & & z_N \end{bmatrix}.$$

In (2) the off diagonal blank spots denote zero entries. Thus A is a diagonal $N \times N$ matrix. We also need the matrix

$$(3) \qquad \Omega = \begin{bmatrix} C_- \\ C_- A \\ \vdots \\ C_- A^{N-1} \end{bmatrix} = \begin{bmatrix} X_1 & \cdots & X_N \\ z_1 X_1 & \cdots & z_N X_N \\ \vdots & & \vdots \\ z_1^{N-1} X_1 & \cdots & z_N^{N-1} X_N \end{bmatrix}.$$

Since the points $z_1, \ldots, z_N$ are different, the $N \times N$ Vandermonde matrix

$$\begin{bmatrix} 1 & \cdots & 1 \\ z_1 & \cdots & z_N \\ \vdots & & \vdots \\ z_1^{N-1} & \cdots & z_N^{N-1} \end{bmatrix}$$

is non-singular. But then we may use the fact that the vectors $X_1, \ldots, X_N$ are non-zero to show that $\operatorname{Ker} \Omega$ consists of the zero vector only. It follows that Ω has a left inverse Ω^+ which we partition as

$$(4) \qquad \Omega^+ = [H_0 \quad H_1 \quad \cdots \quad H_{N-1}],$$

where $H_0, \ldots, H_{N-1}$ are $N \times m$ matrices.

LEMMA 6.1. *Let Ω^+ be a left inverse of the matrix Ω defined by (3), and let Ω^+ be partitioned as in (4). Let C_+ be as in (1). Then the matrix polynomial*

$$(5) \qquad \widetilde{\Phi}(\lambda) = \sum_{k=0}^{N-1} \lambda^k C_+ H_k$$

satisfies the interpolation condition (TNP1).

PROOF. Fix $1 \le j \le N$. Then

$$\widetilde{\Phi}(z_j)X_j = \sum_{k=0}^{N-1} z_j^k C_+ H_k X_j = C_+\left(\sum_{k=0}^{N-1} H_k z_j^k X_j \right) = C_+ \Omega^+ \begin{bmatrix} X_j \\ z_j X_j \\ \vdots \\ z_j^{N-1} X_j \end{bmatrix} = C_+ \Omega^+ \Omega e_j,$$

where e_j is the j-th vector in the standard basis of $\mathbb{C}^N$. Since $\Omega^+ \Omega = I_N$, we see that

$$C_+ \Omega^+ \Omega e_j = C_+ e_j = Y_j,$$

which completes the proof. $\square$

The second step in our solution of the problem (TNP1), (TNP2) involves a homogeneous interpolation problem, namely we seek all $F \in \mathcal{W}_+^{m \times m}$ such that

$$(6) \qquad F(z_j)X_j = 0, \qquad j = 1, \ldots, N.$$

To solve the latter problem we shall employ some elements of the theory of unitary systems developed in Chapter XXXVIII.

Put $S = \sum_{\nu=0}^{\infty} (A^*)^\nu C_-^* C_- A^\nu$, where C_- and A are as in (1) and (2). Note that

$$(7) \qquad A^* S A + C_-^* C_- = S.$$

Let Ω be as in (3). Since $S \ge \Omega^* \Omega$ and Ω is left invertible, the $N \times N$ matrix S is positive definite. Define

$$(8) \qquad \alpha := S^{-1/2} A^* S^{1/2}, \qquad \beta = S^{-1/2} C_-^*.$$

Then $\alpha\alpha^* + \beta\beta^* = I_N$, because of (7), and we may choose (cf., Lemma XXVIII.7.1) matrices γ and δ, of sizes $m \times N$ and $m \times m$, respectively, such that

$$(9) \qquad \begin{bmatrix} \alpha & \beta \\ \gamma & \delta \end{bmatrix}$$

is unitary. Let $U(\cdot)$ be the transfer function of the corresponding unitary system $\Sigma = (\alpha, \beta, \gamma, \delta)$, i.e.,

$$(10) \qquad U(\lambda) = \delta + \lambda\gamma(I_N - \lambda\alpha)^{-1}\beta.$$

Since Σ is a finite dimensional system, we know that $U(\cdot)$ is a rational matrix function, which does not have poles on $\overline{\mathbb{D}}$, and $U(\lambda)$ is unitary for each $\lambda \in \mathbb{T}$ (see Corollary XXVIII.2.2). It follows that

$$(11) \qquad U(\lambda)^{-1} = \delta^* + \beta^*(\lambda - \alpha^*)^{-1}\gamma^*.$$

LEMMA 6.2. *The $m \times m$ matrix $F \in \mathcal{W}_+^{m \times m}$ is a solution of the homogeneous interpolation problem* (6) *if and only if* $F(\lambda) = Q(\lambda)U(\lambda)$ *for some* $Q \in \mathcal{W}_+^{m \times m}$. *Here $U(\cdot)$ is as in* (10).

PROOF. Take $F \in \mathcal{W}_+^{m \times m}$, and let e_j be the j-th vector in the standard basis of $\mathbb{C}^N$. Then

$$(12) \qquad F(\lambda)C_-(\lambda - A)^{-1}e_j = \frac{1}{\lambda - z_j}F(\lambda)X_j.$$

Thus (6) holds if and only if the left hand side of (12) is analytic on $\mathbb{D}$ for $j = 1, \ldots, N$. From the definition of α and β in (8) it follows that

$$C_-(\lambda - A)^{-1} = \beta^*(\lambda - \alpha^*)^{-1}S^{-1/2}.$$

So $F \in \mathcal{W}_+^{m \times m}$ satisfies (6) if and only if the function $F(\lambda)\beta^*(\lambda - \alpha^*)^{-1}$ is analytic on $\mathbb{D}$.

Since $\beta\beta^* = I_N - \alpha\alpha^*$ and

$$\lambda(I_N - \alpha\alpha^*) = \lambda(I_N - \lambda\alpha) + (\lambda - \alpha^*) - (I_N - \lambda\alpha)(\lambda - \alpha^*),$$

we have

$$U(\lambda)\beta^*(\lambda - \alpha^*)^{-1} = \delta\beta^*(\lambda - \alpha^*)^{-1} + \gamma(I_N - \lambda\alpha)^{-1}(\lambda - \lambda\alpha\alpha^*)(\lambda - \alpha^*)^{-1}$$
$$= \delta\beta^*(\lambda - \alpha^*)^{-1} + \lambda\gamma(\lambda - \alpha^*)^{-1} + \gamma(I_N - \lambda\alpha)^{-1} - \gamma.$$

The matrix in (9) is unitary. So $\delta\beta^* = -\gamma\alpha^*$, and hence

$$\delta\beta^*(\lambda - \alpha^*)^{-1} = \gamma(\lambda - \alpha^* - \lambda)(\lambda - \alpha^*)^{-1} = \gamma - \lambda\gamma(\lambda - \alpha^*)^{-1}.$$

We conclude that

$$U(\lambda)\beta^*(\lambda - \alpha^*)^{-1} = \gamma(I_N - \lambda\alpha)^{-1}.$$

The right hand side of the last formula is analytic on $\mathbb{D}$, and so, by the result of the previous paragraph, $U(z_j)X_j = 0$ for $j = 1, \ldots, N$. In particular, if $F = QU$ for some $Q \in \mathcal{W}_+^{m \times m}$, then $F \in \mathcal{W}_+^{m \times m}$, because $U \in \mathcal{W}_+^{m \times m}$, and

$$F(z_j)X_j = Q(z_j)U(z_j)X_j = 0, \qquad j = 1, \ldots, m.$$

To prove the converse implication, take $F \in \mathcal{W}_+^{m \times m}$, and assume (6) holds. Put $Q = FU^{-1}$. Then $Q \in \mathcal{W}^{m \times m}$. From (11) it follows that

$$Q(z) = F(z)U(z)^{-1} = F(z)\delta^* + \left\{ F(z)\beta^*(z - \alpha^*)^{-1} \right\}\gamma^*$$

for each $z \in \mathbb{T}$. Now, use again the result of the first paragraph of the proof. We see that Q extends to a continuous function on $\overline{\mathbb{D}}$ which is analytic on $\mathbb{D}$. So $Q \in \mathcal{W}_+^{m \times m}$. $\square$

LEMMA 6.3. *Let $\widetilde{\Phi}$ and U be given by (5) and (10), respectively. Then $\Phi \in \mathcal{W}_+^{m \times m}$ is a solution of the tangential Nevanlinna-Pick interpolation problem (TNP1), (TNP2) if and only if $\Phi = \Psi U$, where $\Psi \in \mathcal{W}^{m \times m}$ is a solution of the following Nehari extension problem:*

(MNe1) $\|\Psi(\zeta)\| < 1$ *for each $\zeta \in \mathbb{T}$,*

(MNe2) *for $j = -1, -2, \ldots$ the functions Ψ and $\widetilde{\Phi}U^{-1}$ have the same j-th Fourier coefficient.*

PROOF. Assume $\Phi \in \mathcal{W}_+^{m \times m}$ satisfies (TNP1), (TNP2). Put $\Psi = \Phi U^{-1}$. Then $\Psi \in \mathcal{W}^{m \times m}$. Since $F = \Phi - \widetilde{\Phi}$ satisfies (6), we can apply Lemma 6.2 to show that

$$\Psi - \widetilde{\Phi}U^{-1} = (\Phi - \widetilde{\Phi})U^{-1} = FU^{-1} \in \mathcal{W}_+^{m \times m}.$$

In particular, (MNe2) holds. Next, we use the fact that $U(z)$ is unitary for each $z \in \mathbb{T}$. This implies that (MNe1) is fulfilled, because

$$(13) \qquad \sup_{z \in \mathbb{T}} \|\Psi(z)\| = \sup_{z \in \mathbb{T}} \|\Phi(z)U(z)^{-1}\| = \sup_{z \in \mathbb{T}} \|\Phi(z)\| = \sup_{\lambda \in \mathbb{D}} \|\Phi(\lambda)\| < 1.$$

The last equality follows from the fact that Φ is analytic on $\mathbb{D}$ and continuous on $\overline{\mathbb{D}}$.

To prove the converse, assume that $\Psi \in \mathcal{W}^{m \times m}$ and Ψ satisfies (MNe1), (MNe2). Put $\Phi = \Psi U^{-1}$. Then $\Phi \in \mathcal{W}^{m \times m}$, and by (MNe2) we have $\Psi - \Phi U^{-1} \in \mathcal{W}_+^{m \times m}$. It follows that $\Phi - \widetilde{\Phi} = QU$ for some $Q \in \mathcal{W}_+^{m \times m}$, and we apply Lemma 6.2 to show that $\Phi - \widetilde{\Phi} \in \mathcal{W}_+^{m \times m}$ and $\Phi(z_j)X_j = \widetilde{\Phi}(z_j)X_j$ for $j = 1, \ldots, N$. Thus $\Phi \in \mathcal{W}_+^{m \times m}$ and Φ satisfies (TNP1). From the equalities in (13) we may conclude that Φ also satisfies (TNP2). $\square$

THEOREM 6.4. *Let $\widetilde{\Phi}$ and U be given by (5) and (10), respectively, and let Γ be the block Hankel operator*

$$(14) \qquad \Gamma = \begin{bmatrix} \Lambda_{-1} & \Lambda_{-2} & \Lambda_{-3} & \cdots \\ \Lambda_{-2} & \Lambda_{-3} & \Lambda_{-4} & \cdots \\ \Lambda_{-3} & \Lambda_{-4} & \Lambda_{-5} & \cdots \\ \vdots & \vdots & \vdots & \end{bmatrix} : \ell_2^m \to \ell_2^m,$$

where Λ_k is the k-th Fourier coefficient of $\widetilde{\Phi}U^{-1}$ $(k < 0)$. Then the tangential Nevanlinna-Pick interpolation problem (TNP1), (TNP2) has a solution if and only if $\|\Gamma\| < 1$, and in this case all solutions Φ are given by

$$(15) \qquad \Phi = (\widetilde{\alpha}H + \widetilde{\beta})(\widetilde{\gamma}H + \widetilde{\delta})^{-1}U,$$

where H is an arbitrary element of $\mathcal{W}^{m\times m}$ such that $\|H(\zeta)\| < 1$ for $\zeta \in \mathbb{T}$ and the j-th Fourier coefficient of H is zero for $j = 1, 2, \ldots$. The coefficients $\widetilde{\alpha}$, $\widetilde{\beta}$, $\widetilde{\gamma}$ and $\widetilde{\delta}$ in the linear fractional representation (15) are computed by using the formulas (9a), (9b) of the previous section with Γ as in (14).

The proof of Theorem 6.4 proceeds in the same way as that of Theorem 5.2. We omit the details. It can be shown (see, e.g., Theorem 18.2.2 in Ball-Gohberg-Rodman [1]) that the condition $\|\Gamma\| < 1$ in Theorem 6.4 is equivalent to the requirement that the matrix

$$\left[\frac{X_i^* X_j - Y_i^* Y_j}{1 - \overline{z}_i z_j}\right]_{i,j=1}^N$$

is positive definite.

XXXV.7 RATIONAL CONTRACTIVE INTERPOLANTS

In applications, for example to problems in system theory and control, the Nehari extension problem is of special interest for the case when the given data come from a rational matrix function. In this case one seeks solutions that are also rational matrix functions. Furthermore, in this context rational matrix functions are often transfer functions of systems. In other words, the data are given in realized form (cf., Section XXIV.5), and the solutions should appear in this form too. In this section we treat this rational matrix version of the Nehari extension problem.

Throughout this section, M is an $m \times m$ rational matrix function which has its poles in the open unit disc $\mathbb{D}$. We also assume that M is *strictly proper*, i.e., the function M is analytic at infinity and its value at infinity is equal to the zero matrix. We call $F \in \mathcal{W}^{m\times m}$ a *strictly contractive interpolant* of M if

(CI1) $\|F(\zeta)\| < 1$ for each $\zeta \in \mathbb{T}$,

(CI2) F and M have the same k-th Fourier coefficient for $k = -1, -2, \ldots$.

If F is a rational matrix function, then $F \in \mathcal{W}^{m\times m}$ is equivalent to the requirement that F has no poles on $\mathbb{T}$, and in this case condition (CI2) requires $F - M$ to be analytic on $\mathbb{D}$.

Since M is a strictly proper rational matrix function we know from the first paragraph of the proof of Theorem XXIV.5.1 that M may be represented in the form

$$(1) \qquad\qquad M(\lambda) = C(\lambda - A)^{-1}B,$$

where A is a square matrix of order n say, B and C are matrices of sizes $n \times m$ and $m \times n$, respectively, and A has all its eigenvalues in $\mathbb{D}$. In this case we refer to (1) as a *stable realization* of M. Given a stable realization of M we are interested in finding all rational strictly contractive interpolants of M in terms of the matrices A, B and C appearing in the realization (1).

With a stable realization (1) we associate the following two matrices:

$$(2) \qquad G_c = \sum_{j=0}^{\infty} A^j BB^*(A^*)^j, \qquad G_o = \sum_{j=0}^{\infty} (A^*)^j C^* C A^j.$$

The fact that the eigenvalues of A are in $\mathbb{D}$ implies that the series in (2) converge in any matrix norm. The matrices G_c and G_o are called, respectively, the *controllability* and *observability Gramians* of (1). We are now ready to state the first main result of this section.

THEOREM 7.1. *Let $M(\lambda) = C(\lambda - A)^{-1}B$ be a stable realization, and let G_c and G_o be the corresponding controllability and observability Gramians. Then M has a strictly contractive interpolant if and only if all the eigenvalues of $G_c G_o$ are in $\mathbb{D}$, and in this case all strictly contractive interpolants of M are given by*

$$(3) \qquad F_E(\lambda) = C(\lambda - A)^{-1}B + \Omega_{21}(\lambda) + \Omega_{22}(\lambda)E(\lambda)\big\{ I_m - \Omega_{12}(\lambda)E(\lambda) \big\}^{-1}\Omega_{11}(\lambda),$$

where the free parameter E is any $m \times m$ matrix function in $\mathcal{W}_+^{m \times m}$ such that $\|E(\zeta)\| < 1$ for each $\zeta \in \mathbb{T}$. Furthermore, the coefficients Ω_{ij} are rational matrix functions which have no poles on $\overline{\mathbb{D}}$ and are given by the following realizations:

$$(4) \qquad \Omega_{11}(\lambda) = Y - \lambda Y B^* A^*(I - \lambda R_1^{-1} R_0 A^*)^{-1} R_1^{-1} G_o B,$$

$$(5) \qquad \Omega_{12}(\lambda) = -\lambda Y B^* R_0^{-1} C^* X - \lambda^2 Y B^* A^*(I - \lambda R_1^{-1} R_0 A^*) R_1^{-1} C^* X,$$

$$(6) \qquad \Omega_{21}(\lambda) = -C G_c A^*(I - \lambda R_1^{-1} R_0 A^*)^{-1} R_1^{-1} G_o B,$$

$$(7) \qquad \Omega_{22}(\lambda) = X - \lambda C G_c A^*(I - \lambda R_1^{-1} R_0 A^*)^{-1} R_1^{-1} C^* X.$$

Here

$$(8) \qquad R_0 = I - G_o G_c, \qquad R_1 = I - G_o A G_c A^*,$$

$$(9) \qquad \begin{aligned} X &= \big\{ I - C(I - G_c A^* G_o A)^{-1} G_c C^* \big\}^{1/2}, \\ Y &= \big\{ I - B^*(I - G_o A G_c A^*)^{-1} G_o B \big\}^{1/2}. \end{aligned}$$

Moreover, (3) defines a 1-1 correspondence between all strictly contractive interpolants of M and the $m \times m$ matrix functions E in $\mathcal{W}_+^{m \times m}$ with the property that $\|E(\zeta)\| < 1$ for each $\zeta \in \mathbb{T}$. In particular, all rational strictly contractive interpolants of M are obtained by choosing the free parameter E to be any rational $m \times m$ matrix function without poles on $\overline{\mathbb{D}}$ and such that $\|E(\zeta)\| < 1$ for each $\zeta \in \mathbb{T}$.

The second main result of this section concerns the maximum entropy principle. Assume M has a strictly contractive interpolant. From the results in Section XXXV.4 it follows that among all strictly contractive interpolants F of M there is precisely one that maximizes the integral

$$(10) \qquad \frac{1}{2\pi} \int_{-\pi}^{\pi} \log \deg \left(I - F(e^{it})^* F(e^{it})\right) dt,$$

which we shall call the *maximum entropy interpolant* of M.

THEOREM 7.2. *Let $M(\lambda) = C(\lambda - A)^{-1}B$ be a stable realization, and let G_c and G_o be the corresponding controllability and observability Gramians. Assume that the eigenvalues of $G_c G_o$ are in $\mathbb{D}$. Then the maximum entropy interpolant of M is the rational matrix function*

$$(11) \qquad H(\lambda) = C(\lambda - A)^{-1}B - C_-(I - \lambda A_-)^{-1}B_-,$$

where

$$A_- = (I - G_o A G_c A^*)^{-1}(I - G_o G_c)A^*,$$
$$B_- = (I - G_o A G_c A^*)^{-1}G_o B,$$
$$C_- = C G_c A^*.$$

In addition, for each strictly contractive interpolant F of M we have

$$(12) \quad \frac{1}{2\pi} \int_{-\pi}^{\pi} \log \det\left(I - F(e^{it})^* F(e^{it})\right) dt \leq \log \det \left\{ I - B^*(I - G_o A G_c A^*)^{-1} G_o B \right\},$$

and equality holds in (12) *if and only if $F = H$.*

PROOF OF THEOREM 7.1. We split the proof into six parts. In the first two parts we apply Theorem 4.1 with

$$(13) \qquad \Psi_{-k} = C A^k B, \qquad k = 0, 1, 2, \dots .$$

Part (a). In this part we derive the necessary and sufficient condition for the existence of a strictly contractive interpolant. Let Ψ_{-k} be as in (13). Since A has its eigenvalues in $\mathbb{D}$, formula (1) in Section XXXV.4 holds. Note that $\Phi \in \mathcal{W}^{m \times m}$ is a solution of the corresponding Nehari extension problem (N1), (N2), (N3) if and only if $F(\lambda) = \lambda^{-1}\Phi(\lambda)$ is a strictly contractive interpolant of M. It follows that such an interpolant exists if and only if the operator

$$(14) \qquad \Gamma = \begin{bmatrix} CB & CAB & CA^2B & \cdots \\ CAB & CA^2B & CA^3B & \cdots \\ CA^2B & CA^3B & CA^4B & \cdots \\ \vdots & \vdots & \vdots & \end{bmatrix} : \ell_2^m \to \ell_2^m$$

has norm strictly less than one. Now, consider the following auxiliary operators

$$\Sigma : \ell_2^m \to \mathbb{C}^n, \qquad \Sigma\big((u_j)_{j=0}^\infty\big) = \sum_{j=0}^\infty A^j B u_j,$$

$$\Lambda : \mathbb{C}^n \to \ell_2^m, \qquad \Lambda x = (C A^j x)_{j=0}^\infty.$$

The stability of the realization implies that Σ and Λ, are well-defined bounded linear operators. In terms of Σ and Λ the controllability and observability Gramians in (2) have the following representation:

$$G_c = \Sigma\Sigma^*, \qquad G_o = \Lambda^*\Lambda.$$

Furthermore, the operator Γ in (14) is given by $\Gamma = \Lambda\Sigma$. Thus $\|\Gamma\| < 1$ is equivalent to the requirement that the Hermitian matrix $\Sigma^*\Lambda^*\Lambda\Sigma$ have all its eigenvalues in the open unit disc $\mathbb{D}$. Since $\Sigma^*\Lambda^*\Lambda\Sigma$ and $G_c G_o = \Sigma\Sigma^*\Lambda^*\Lambda$ have the same non-zero eigenvalues (cf., the proof of Corollary VII.6.2), we conclude that M has a strictly contractive interpolant if and only if the eigenvalues of $G_c G_o$ are in $\mathbb{D}$.

        Part (b). In this and the remaining parts of the proof we assume that $G_c G_o$ has all its eigenvalues in $\mathbb{D}$. Thus $\|\Gamma\| < 1$, and we can apply Theorem 4.1 to show that all strictly contractive interpolants F of M are given by $F(\lambda) = \lambda^{-1}\Phi(\lambda)$, where

$$(15) \qquad \Phi(\lambda) = \big(\alpha(\lambda)H(\lambda) + \beta(\lambda)\big)\big(\gamma(\lambda)H(\lambda) + \delta(\lambda)\big)^{-1}, \qquad \lambda \in \mathbf{T}.$$

Here H is an arbitrary element of $\mathcal{W}^{m\times m}$ such that $\|H(\zeta)\| < 1$ for $\zeta \in \mathbf{T}$ and the j-th Fourier coefficient of H is zero for $j = 0, -1, -2, \ldots$. Moreover, the correspondence between H and F is one-one. The coefficients α, β, γ and δ in the linear fractional representation (15) are given by (4a) and (4b) in Section XXXV.4. To find these coefficients we have to analyse the operators $(I - \Gamma\Gamma^*)^{-1}$ and $(I - \Gamma^*\Gamma)^{-1}$ for the case considered here.

        If T, S and D are operators such that D and $I - TD^{-1}S$ are invertible, then (cf., the proof of Lemma XXIV.6.2) the operator $D - ST$ is invertible and

$$(16) \qquad (I - TD^{-1}S)^{-1} = I + T(D - ST)^{-1}S.$$

By applying this identity twice, one obtains:

$$(17a) \qquad (I - \Gamma\Gamma^*)^{-1} = I + \Lambda(I - G_c G_o)^{-1} G_c \Lambda^*,$$

$$(17b) \qquad (I - \Gamma^*\Gamma)^{-1} = I + \Sigma^*(I - G_o G_c)^{-1} G_o \Sigma.$$

For example, to get (17a) one applies (16) with $T = \Lambda$, $S = G_c\Lambda^*$ and $D = I$, and one uses $\Gamma\Gamma^* = \Lambda\Sigma\Sigma^*\Lambda^* = \Lambda G_c\Lambda^*$. Formula (17b) follows in an analogous way from $\Gamma^*\Gamma = \Sigma^* G_o\Sigma$.

Part (c). In this part we compute realizations for the functions α and δ in (15). Put

$$(18) \qquad \begin{bmatrix} a_0 \\ a_{-1} \\ a_{-2} \\ \vdots \end{bmatrix} = (I - \Gamma\Gamma^*)^{-1} \begin{bmatrix} I_m \\ 0 \\ 0 \\ \vdots \end{bmatrix}, \qquad \begin{bmatrix} d_0 \\ d_1 \\ d_2 \\ \vdots \end{bmatrix} = (I - \Gamma^*\Gamma)^{-1} \begin{bmatrix} I_m \\ 0 \\ 0 \\ \vdots \end{bmatrix}.$$

Since $I - \Gamma\Gamma^*$ is a strictly positive operator, the same holds true for $(I - \Gamma\Gamma^*)^{-1}$, and hence the matrix a_0 is positive definite. In particular, a_0 is invertible. From (17a) and the definition of Λ we see that

$$a_0 = I_m + C(I - G_c G_o)^{-1} G_c C^*.$$

Note that $G_c G_o - G_c C^* C = G_c A^* G_o A$. So we can apply (16) to show that the matrix $I - G_c A^* G_o A$ is invertible and

$$(19) \qquad a_0^{-1} = I_m - C(I - G_c A^* G_o A)^{-1} G_c C^*.$$

The right hand side of (19) is positive definite, because a_0 is positive definite. It follows that the matrix X in (9) is well-defined and $a_0^{-1/2} = X$. From (17a) and the definition of Λ we also see that $a_{-j} = C A^j (I - G_c G_o)^{-1} G_c C^*$ for $j \geq 1$, and hence, according to the first identity in formula (4a) of Section XXXV.4, we have

$$(20) \qquad \alpha(\lambda) = \left\{ I_m + \lambda C(\lambda - A)^{-1}(I - G_c G_o)^{-1} G_c A^* \right\} X.$$

Similarly, using (17b) in place of (17a) and Σ in place of Λ, one may show that the matrix Y in (9) is well-defined, that $d_0^{-1/2} = Y$ and that

$$(21) \qquad \delta(\lambda) = \left\{ I_m + B^*(I - \lambda A^*)^{-1}(I - G_o G_c)^{-1} G_o B \right\} Y.$$

Part (d). In this part we compute the functions β and γ. Put

$$(22) \qquad \begin{bmatrix} b_0 \\ b_{-1} \\ b_{-2} \\ \vdots \end{bmatrix} = \Gamma \begin{bmatrix} d_0 \\ d_1 \\ d_2 \\ \vdots \end{bmatrix}, \qquad \begin{bmatrix} c_0 \\ c_1 \\ c_2 \\ \vdots \end{bmatrix} = \Gamma^* \begin{bmatrix} a_0 \\ a_1 \\ a_2 \\ \vdots \end{bmatrix}.$$

From (17a) and $\Gamma = \Lambda\Sigma$ we see that

$$\Gamma(I - \Gamma^*\Gamma)^{-1} = (I - \Gamma\Gamma^*)^{-1}\Gamma = \left\{ \Lambda + \Lambda(I - G_c G_o)^{-1} G_c \Lambda^* \Lambda \right\} \Sigma$$
$$= \Lambda \left\{ I + (I - G_c G_o)^{-1} G_c G_o \right\} \Sigma = \Lambda(I - G_c G_o)^{-1} \Sigma.$$

From this calculation and the definition of the column $(d_j)_{j=0}^{\infty}$ in (18) we see that $b_{-j} = C A^j (I - G_c G_o)^{-1} B$ for $j \geq 0$, and thus, by the second identity in formula (4b) of Section XXXV.4, we have

$$(23) \qquad \beta(\lambda) = \lambda C(\lambda - A)^{-1}(I - G_c G_o)^{-1} B Y.$$

In a similar way, replacing (17a) by (17b) and Λ by Σ, one may show that

$$\Gamma^*(I - \Gamma\Gamma^*)^{-1} = \Sigma^*(I - G_oG_c)^{-1}\Lambda^*,$$

and it follows that

$$(24) \qquad \gamma(\lambda) = B^*(I - \lambda A^*)^{-1}(I - G_oG_c)^{-1}C^*X.$$

Part (e). From Part (b) we know that all strictly contractive interpolants of M are given by

$$(25) \qquad F(\lambda) = \left(\alpha(\lambda)E(\lambda) + \widetilde{\beta}(\lambda)\right)\left(\widetilde{\gamma}(\lambda)E(\lambda) + \delta(\lambda)\right)^{-1}, \qquad \lambda \in \mathbb{T},$$

where $\widetilde{\beta}(\lambda) = \lambda^{-1}\beta(\lambda)$ and $\widetilde{\gamma}(\lambda) = \lambda\gamma(\lambda)$. Moreover, the free parameter E in (25) is an arbitrary element of $\mathcal{W}_+^{m\times m}$ satisfying $\|E(\zeta)\| < 1$ for each $\zeta \in \mathbb{T}$. Since $\delta(\lambda)$ is invertible for each $\lambda \in \mathbb{T}$, formula (25) may be rewritten as

$$(26) \qquad F(\lambda) = \theta_{21}(\lambda) + \theta_{22}(\lambda)E(\lambda)\{I - \theta_{12}(\lambda)E(\lambda)\}^{-1}\theta_{11}(\lambda), \qquad \lambda \in \mathbb{T},$$

with

$$(27) \qquad \begin{aligned} \theta(\lambda) &= \begin{bmatrix} \theta_{11}(\lambda) & \theta_{12}(\lambda) \\ \theta_{21}(\lambda) & \theta_{22}(\lambda) \end{bmatrix} \\ &= \begin{bmatrix} \delta(\lambda)^{-1} & -\delta(\lambda)^{-1}\widetilde{\gamma}(\lambda) \\ \widetilde{\beta}(\lambda)\delta(\lambda)^{-1} & \alpha(\lambda) - \widetilde{\beta}(\lambda)\delta(\lambda)^{-1}\widetilde{\gamma}(\lambda) \end{bmatrix}, \qquad \lambda \in \mathbb{T}. \end{aligned}$$

In this part we compute realizations for the functions θ_{ij}.

To compute a realization for $\theta_{11}(\lambda) = \delta(\lambda)^{-1}$ we apply (16) to (21). This yields

$$(28) \qquad \theta_{11}(\lambda) = Y^{-1} - Y^{-1}B^*[R_0^{-1}R_1 - \lambda A^*]^{-1}R_0^{-1}G_oB, \qquad \lambda \in \mathbb{T},$$

where R_0 and R_1 are the matrices defined by (8). To see this, we first note that

$$(29) \qquad I + R_0^{-1}G_oBB^* = R_0^{-1}\{I - G_oG_c + G_oBB^*\} = R_0^{-1}R_1.$$

It follows that

$$I - \lambda A^* + (I - G_oG_c)^{-1}G_oBB^* = R_0^{-1}R_1 - \lambda A^*,$$

and hence (28) follows by applying (16) to (21). Recall that $Y^{-2} = d_0$, and thus we see from (17b) that $Y^{-2} = I_m + B^*R_0^{-1}G_oB$, and hence

$$Y^{-2}B^* = B^* + B^*R_0^{-1}G_oBB^* = B^*\{I + R_0^{-1}G_0BB^*\} = B^*R_0^{-1}R_1,$$

because of (29). By using these identities we see that

$$(30) \qquad \theta_{11}(\lambda) = Y - \lambda Y B^*A^*(R_0^{-1}R_1 - \lambda A^*)^{-1}R_0^{-1}G_oB, \qquad \lambda \in \mathbb{T}.$$

Next, we compute

$$\theta_{12}(\lambda) = -\delta(\lambda)^{-1}\widetilde{\gamma}(\lambda) = -\lambda\theta_{11}(\lambda)\gamma(\lambda), \qquad \lambda \in \mathbb{T}.$$

To do this we use the realizations (24) and (30). It follows that

$$\theta_{12}(\lambda) = -\lambda Y\gamma(\lambda) + \lambda^2 Y B^* A^*(R_0 R_1^{-1} - \lambda A)^* R_0^{-1} G_o BB^*(I - \lambda A^*)^{-1} R_0^{-1} C^* X.$$

According to (29) we may write

$$(31) \qquad R_0^{-1} G_o BB^* = (R_0^{-1} R_1 - \lambda A^*) - (I - \lambda A^*),$$

and thus we obtain that

$$(32) \quad \theta_{12}(\lambda) = -\lambda Y B^* R_0^{-1} C^* X - \lambda^2 Y B^* A^*(R_0^{-1} R_1 - \lambda A^*)^{-1} R_0^{-1} C^* X, \qquad \lambda \in \mathbb{T}.$$

To find a realization for $\theta_{21}(\lambda) = \lambda^{-1}\beta(\lambda)\theta_{11}(\lambda)$, $\lambda \in \mathbf{T}$, we use (23) and (28). Thus

$$\theta_{21}(\lambda) = C(\lambda - A)^{-1}(I - G_c G_o)^{-1} B - C(\lambda - A)^{-1}(I - G_c G_o)^{-1}$$
$$\cdot BB^*(R_0^{-1} R_1 - \lambda A^*)^{-1} R_0^{-1} G_o B.$$

Formula (16) implies that $(I - G_c G_o)^{-1} = I + G_c R_0^{-1} G_o$. But then we can use (29) to show that

$$(I - G_c G_o)^{-1} BB^* = BB^* + G_c R_0^{-1} G_o BB^* = BB^* + G_c R_0^{-1} R_1 - G_c$$
$$= G_c R_0^{-1} R_1 - AG_c A^*,$$

and hence

$$(I - G_c G_o)^{-1} BB^* = (\lambda - A)G_c A^* + G_c(R_0^{-1} R_1 - \lambda A^*).$$

It follows that

$$(33) \qquad \theta_{21}(\lambda) = C(\lambda - A)^{-1} B - CG_c A^*(R_0^{-1} R_1 - \lambda A^*)^{-1} R_0^{-1} G_o B, \qquad \lambda \in \mathbb{T}.$$

Finally, we compute a realization for $\theta_{22}(\lambda) = \alpha(\lambda) - \theta_{21}(\lambda)\widetilde{\gamma}(\lambda)$, $\lambda \in \mathbf{T}$, by using (20), (24) and (33). Since $\widetilde{\gamma}(\lambda) = \lambda\gamma(\lambda)$ and

$$\lambda B^* B = (\lambda - A)G_c + AG_c(I - \lambda A^*),$$

we have

$$C(\lambda - A)^{-1} B\widetilde{\gamma}(\lambda) = C(\lambda - A)^{-1}(\lambda BB^*)(I - \lambda A^*)^{-1} R_0^{-1} C^* X$$
$$= CG_c(I - \lambda A^*)^{-1} R_0^{-1} C^* X - CG_c R_0^{-1} C^* X + \lambda C(\lambda - A)^{-1} G_c R_0^{-1} C^* X.$$

Furthermore, by applying (31) we see that

$$CG_c A^*(R_0^{-1} R_1 - \lambda A^*)^{-1} R_0^{-1} G_o B\widetilde{\gamma}(\lambda)$$
$$= -CG_c R_0^{-1} C^* X + CG_c(I - \lambda A^*)^{-1} R_0^{-1} C^* X -$$
$$- \lambda CG_c A^*(R_0^{-1} R_1 - \lambda A^*)^{-1} R_0^{-1} C^* X.$$

It follows that

$$\theta_{21}(\lambda)\widetilde{\gamma}(\lambda) = \lambda C(\lambda - A)^{-1}G_c R_0^{-1}C^*X + \lambda CG_c A^*(R_0^{-1}R_1 - \lambda A^*)^{-1}R_0^{-1}C^*X.$$

Since $(I - G_c G_o)^{-1}G_c = G_c R_0^{-1}$, we conclude that

$$(34) \qquad \theta_{22}(\lambda) = X - \lambda CG_c A^*(R_0^{-1}R_1 - \lambda A^*)^{-1}R_0^{-1}C^*X, \qquad \lambda \in \mathbb{T}.$$

Part (f). In this part we finish the proof. Introduce the following matrix functions:

$$\Omega_{11}(\lambda) = \theta_{11}(\lambda), \qquad \Omega_{12}(\lambda) = \theta_{12}(\lambda),$$
$$\Omega_{21}(\lambda) = \theta_{21}(\lambda) - C(\lambda - A)^{-1}B,$$
$$\Omega_{22}(\lambda) = \theta_{22}(\lambda).$$

Then (according to (26)) all strictly contractive interpolants of M are given by (3) with a free parameter E of the desired type. From the general theory of strictly contractive extensions (see Theorem XXXIV.3.1) we know that $\delta^{\pm 1} \in \mathcal{W}_+^{m \times m}$. Thus $\delta(\lambda)$ is invertible for each $\lambda \in \overline{\mathbb{D}}$. The same holds true for $I - \lambda A^*$, because all the eigenvalues of A are in $\mathbb{D}$. But then we can apply (16) and the realization (21) to show that $R_0^{-1}R_1 - \lambda A^*$ is invertible for each $\lambda \in \overline{\mathbb{D}}$. In particular, $R_0^{-1}R_1$ is invertible. It follows that (4)–(7) hold true. Since $I - \lambda R_1^{-1}R_0 A^*$ is invertible for each $\lambda \in \overline{\mathbb{D}}$, we conclude that the functions Ω_{ij} have no poles on $\overline{\mathbb{D}}$.

If E in (3) is rational, then the resulting interpolant F_E is also rational. Conversely, if $F = F_E$ is rational, then by solving E from (25) it follows that E is rational. Indeed,

$$E(\lambda) = \alpha(\lambda)^{-1}\big(I_m - F(\lambda)\widetilde{\gamma}(\lambda)\alpha(\lambda)^{-1}\big)^{-1}\big(F(\lambda)\delta(\lambda) - \widetilde{\beta}(\lambda)\big), \qquad \lambda \in \mathbb{T},$$

and the inverse exists because for each $\lambda \in \mathbb{T}$ we have $\|F(\lambda)\| < 1$ (by assumption) and $\|\widetilde{\gamma}(\lambda)\alpha(\lambda)^{-1}\| < 1$ (by Theorem XXXIV.3.1). $\square$

PROOF OF THEOREM 7.2. Let H be the maximum entropy interpolant of M. Put $\widetilde{H}(\lambda) = \lambda H(\lambda)$, and let $\Psi_0, \Psi_{-1}, \Psi_{-2}, \ldots$ be as in (3). Note that

$$F(e^{it})^* F(e^{it}) = \Phi(e^{it})^* \Phi(e^{it}),$$

whenever $\Phi(\lambda) = \lambda F(\lambda)$ for $\lambda \in \mathbb{T}$. Hence among all solutions Φ of the Nehari extension problem (N1), (N2), (N3) for $\Psi_0, \Psi_{-1}, \Psi_{-2}, \ldots$, the function $\widetilde{H}$ is the one that maximizes the integral

$$\frac{1}{2\pi} \int_{-\pi}^{\pi} \log \det\big(I - \Phi(e^{it})^* \Phi(e^{it})\big) dt.$$

It follows that $\widetilde{H}$ is the triangular extension of $\lambda M(\lambda)$, by Theorem 4.2, and thus $\widetilde{H}(\lambda) = \beta(\lambda)\delta(\lambda)^{-1}$, where β and δ are as in (15). But then we may conclude (see the proof of Theorem 7.1) that

$$H(\lambda) = \lambda^{-1}\widetilde{H}(\lambda) = \lambda\beta(\lambda)\delta(\lambda)^{-1} = \theta_{21}(\lambda) = C(\lambda - A)^{-1}B + \Omega_{21}(\lambda),$$

where $\Omega_{21}(\lambda)$ is given by (6).

To complete the proof it remains to show that

$$(35) \quad \frac{1}{2\pi} \int_{-\pi}^{\pi} \log \det\left(I - H(e^{it})^* H(e^{it})\right) dt = \log \det\{I - B^*(I - G_o A G_c A^*)^{-1} G_o B\}.$$

In the left hand side of (35) we may replace H by $\widetilde{H}$. In this case we know from the proof of Theorem 4.2 that the left hand side of (35) is equal to $\log \det \Delta$, where Δ is the right multiplicative diagonal of $E - \widetilde{H}^* \widetilde{H}$, relative to the decomposition (7) in Section XXXV.4. Since $\widetilde{H}$ is the triangular extension, we can use the general theory developed in Section XXXIV.3 to show that $E - \widetilde{H}^* \widetilde{H}$ factorizes as

$$E - \widetilde{H}^* \widetilde{H} = \delta^{-*}(d_0)\delta^{-1},$$

where $\delta(\lambda) = \sum_{j=0}^{\infty} \lambda^j d_j$ with $d_0, d_1, d_2, \ldots$ given by the second identity in (18). It follows that $\Delta = d_0^{-1}$. From the proof of Theorem 7.1 we know that $d_0^{-1} = Y^2$, where Y is given by (9). Thus $\log \det \Delta$ is equal to the right hand side of (35). $\quad\square$

Let $M(\lambda) = C(\lambda - A)^{-1} B$ be a stable realization, and consider the description of all strictly contractive interpolants of M in (3). To get rational interpolants the free parameter E has to be an $m \times m$ rational matrix function without poles on $\overline{\mathbb{D}}$ and with the property that $\|E(\zeta)\| < 1$ for $\zeta \in \mathbb{T}$. The fact that E has no poles on $\mathbb{D}$ implies that E may be represented in the form

$$(36) \qquad E(\lambda) = C_E(I - \lambda A_E)^{-1} B_E$$

where A_E is a square matrix of order ℓ say, which has all its eigenvalues in $\mathbb{D}$, and B_E and C_E are matrices of sizes $\ell \times m$ and $m \times \ell$, respectively. Inserting (36) into (3) and using the realizations (4)–(7) for the coefficients Ω_{ij} yields a formula for the corresponding interpolant F_E directly in terms of the data in the realizations of M and E. In fact, we have

$$(37) \qquad F_E(\lambda) = C(\lambda - A)^{-1} B + \widetilde{C}_-(I - \lambda \widetilde{A}_-)^{-1} \widetilde{B}_-$$

with

$$\widetilde{A}_- = \begin{bmatrix} 0 & 0 \\ 0 & A_E \end{bmatrix} + \begin{bmatrix} I & 0 \\ 0 & B_E \end{bmatrix} \begin{bmatrix} R_1^{-1} \\ -YB^* R_0^{-1} \end{bmatrix} [R_0 A^* \quad C^* X] \begin{bmatrix} I & 0 \\ 0 & C_E \end{bmatrix},$$

$$\widetilde{B}_- = \begin{bmatrix} I & 0 \\ 0 & B_E \end{bmatrix} \begin{bmatrix} R_1^{-1} G_o B \\ Y \end{bmatrix},$$

$$\widetilde{C}_- = [-CG_c A^* \quad X] \begin{bmatrix} I & 0 \\ 0 & C_E \end{bmatrix}.$$

Here G_o and G_c are the controllability and observability Gramians of $C(\lambda - A)^{-1} B$ and the matrices R_0, R_1, X and Y are as in (8) and (9). To prove (37) one may use

Theorem 7.6.1 in Gohberg-Lancaster-Rodman [2], which describes a realization of a linear fractional transformation given realizations of the coefficients W_{ij} in the transformation and given a realization of the matrix function V that is being transformed. Actually, one applies this theorem with

$$W_{11}(\lambda) = \Omega_{11}(\lambda^{-1}), \qquad W_{12}(\lambda) = \lambda\Omega_{12}(\lambda^{-1}),$$

$$W_{21}(\lambda) = \lambda^{-1}\Omega_{21}(\lambda^{-1}), \qquad W_{22}(\lambda) = \Omega_{22}(\lambda^{-1}),$$

and $V(\lambda) = C_E(\lambda - A_E)^{-1}B_E$.

One may use Theorem 7.1 to derive explicit realization formulas for the rational matrix solutions of the tangential Nevanlinna-Pick problem (TNP1), (TNP2) considered in the previous section. To see this note that the function $\widetilde{\Phi}U^{-1}$ (see formulas (5) and (10) in the previous section) is a rational matrix function which does not have poles on $\mathbb{T}$. So we may choose matrices A, B and C, where A is a square matrix whose eigenvalues are in $\mathbb{D}$, such that

$$\widetilde{\Phi}(\lambda)U(\lambda)^{-1} - C(\lambda - A)^{-1}B$$

has no poles on $\mathbb{D}$. In particular, $M(\lambda) = C(\lambda - A)^{-1}B$ is a stable realization. Since U is a regular matrix function, Lemma 6.3 tells us that all rational matrix functions Φ satisfying (TNP1), (TNP2) are of the form $\Phi = FU$, where F is a rational strictly contractive interpolant of $M(\lambda) = C(\lambda - A)^{-1}B$. In particular, F may be written in the form (37). Finally, since U is also given in a realized form (see formula (10) in the previous section), it remains to compute a product realization to obtain a realization for the solution Φ of the tangential Nevanlinna-Pick interpolation problem.

COMMENTS ON PART IX

For the positive definite case the material in Section XXXIII.1 is standard; the more general version considered here is taken from Ellis-Gohberg-Lay [1]. Section XXXIII.2 also follows from the corresponding material in Ellis-Gohberg-Lay [1]. The reduction of a strictly contractive extension problem to a positive definite one, described in Section XXXIII.3, is well-known and appears, for example, in Dym-Gohberg [3]. The band method, described in Chapter XXXIV, has its origin in the papers Dym-Gohberg [1], [2], and has been developed to a more final form in the papers Gohberg-Kaashoek-Woerdeman [1], [2], [3]. Further results in this direction may be found in Gohberg-Kaashoek-Woerdeman [4], [5] and [6]. The presentation in Chapter XXXIV follows mainly Gohberg-Kaashoek-Woerdeman [2], [3]; the modifications are inspired by a different version of the abstract band method presented in Ellis-Gohberg-Lay [2]. For block matrices Theorem XXXV.1.1, the construction of the triangular strictly contractive extension in Theorem XXXV.2.1, and the maximum entropy results in Sections XXXV.1 and XXXV.2 are due to Dym-Gohberg [4]. Linear fractional representations of all positive and all strictly contractive extensions were obtained for the block matrix case in

Ball-Gohberg [1] by applying a finite dimensional version of the method of Ball-Helton [1]. For the (block) matrix case linear fractional description of all positive extensions appear in a somewhat less explicit form also in Dewilde-Deprettere [1], Dym [1] and Woerdeman [1]; see also Section I.6 in Woerdeman [2]. The latter two references also treat the strictly contractive case. The construction of the band extension in Theorem XXXV.3.1 is due to Dym-Gohberg [5]; see also the papers Morf-Vieira-Kailath [1] and Morf-Vieira-Lee-Kailath [1], which also contain applications to estimation theory. The maximum entropy results in Section XXXV.3 have their origin in Burg [1] and Dym-Gohberg [5]. The linear fractional description of all solutions is mentioned (without proof) in Dym [1] and is derived in Gohberg-Kaashoek-Woerdeman [2] for the more general case of functions in the operator Wiener algebra; the analogous result for the matrix Wiener algebra is established in Gohberg-Kaashoek-Woerdeman [1]. The linear fractional description of all solutions of the Nehari problem in Section XXXV.4 is due to Adamjan-Arov-Kreĭn [1] (see also Dym-Gohberg [6], [7]). The maximum entropy result in Section XXXV.4 is due to Dym-Gohberg [7]; the analogous result for functions in the operator Wiener algebra may be found in Gohberg-Kaashoek-Woerdeman [3]. Other applications of the band method appear in Bakonyi-Woerdeman [1] and Spitkovsky-Woerdeman [1]. For the scalar case the idea on which Sections XXXV.5 and XXXV.6 are based goes back to Sarason [2] (see also Section 8 in Dym-Gohberg [7]). The material in Section XXXV.7 is taken from Gohberg-Kaashoek-Van Schagen [2]. Besides the band method there are several other methods for dealing with extension and interpolation problems of the type considered here. For an extended review of the various approaches we refer to "Notes for Part V" in Ball-Gohberg-Rodman [1] (see also Foias-Frazho [1]). Connections with and motivation from engineering problems are discussed in Helton [1].

EXERCISES TO PART IX

1. For each of the following three matrices A, find complex numbers x and y such that A^{-1} exists and is a 2-band matrix. Discuss the uniqueness of the solution.

(a)
$$A = \begin{bmatrix} 0 & 1 & x \\ 2 & 3 & 1 \\ y & 4 & 1 \end{bmatrix},$$

(b)
$$A = \begin{bmatrix} 3 & 1 & x \\ 2 & 0 & 1 \\ y & 4 & 1 \end{bmatrix},$$

(c)
$$A = \begin{bmatrix} 1 & 0 & x \\ 0 & 0 & 4 \\ y & 4 & 1 \end{bmatrix}.$$

2. Consider the 3×3 matrix

$$A(x,y) = \begin{bmatrix} a_{11} & a_{12} & x \\ a_{21} & a_{22} & a_{23} \\ y & a_{32} & a_{33} \end{bmatrix}.$$

Find necessary and sufficient conditions on the entries a_{ij}, $|i-j| \leq 1$, such that $A(x,y)$ admits LU- and UL-factorizations and $A(x,y)^{-1}$ is a 1-band matrix for some complex numbers x and y. In this case find such x and y, and discuss uniqueness.

3. Generalize the previous problem for $n \times n$ matrices.

4. Consider the block matrix

$$A(X) = \begin{bmatrix} A_{11} & A_{12} \\ A_{12}^* & X \end{bmatrix},$$

where A_{11} and X are square matrices. Find necessary and sufficient conditions on the given entries A_{11} and A_{12} such that $A(X)$ is positive definite for some X, and in this case find all such X.

5. Consider the 3×3 block matrix:

$$A(X,Y) = \begin{bmatrix} A_{11} & A_{12} & Y^* \\ A_{12}^* & X & A_{23} \\ Y & A_{23}^* & A_{33} \end{bmatrix},$$

where A_{11}, X and A_{33} are square matrices. Find necessary and sufficient conditions on the given entries A_{ij} such that $A(X, Y)$ is positive definite for some X and Y, and in this case describe all such X and Y.

6. Generalize the previous problem to 3×3 operator matrices with entries acting on Hilbert spaces.

7. Consider an $n \times n$ matrix of which the first column, the first row and the main diagonal are given. All other entries are unspecified. Find necessary and sufficient conditions in order that the matrix can be completed to a matrix A which is positive definite. Does there exist a completion A such that the entries of A^{-1} outside the first column, first row and main diagonal are zero?

8. Can the result of the previous exercise be obtained by applying the band method?

9. Given complex numbers $a_0, \ldots, a_n$, consider the problem to find all complex functions φ,

$$\varphi(\lambda) = \sum_{\nu=0}^{\infty} \lambda^{\nu} \varphi_{\nu},$$

satisfying the following three conditions:

   (i) $\varphi_j = a_j$ for $j = 0, \ldots, n$,

   (ii) $\sum_{\nu=0}^{\infty} |\varphi_{\nu}| < \infty$,

   (iii) $\sup_{\lambda \in \mathbb{D}} |\varphi(\lambda)| < 1$.

(a) Find necessary and sufficient conditions in order that the problem has a solution.

(b) Show that all solutions (when they exist) can be described by a linear fractional transformation, and find formulas for the coefficients in this transformation.

(c) What is in this case the analogue of the maximum entropy principle?

Hint: reduce the problem to a Nehari extension problem.

10. Do the previous exercise for matrix-valued functions.

11. Solve the following generalization of the Carathéodory-Toeplitz extension problem. Let $A_{-p}, \ldots, A_p$ be given $m \times m$ matrices. Find (if possible) $m \times m$ matrix-valued functions Φ on the unit circle $\mathbb{T}$,

$$\Phi(\zeta) = \sum_{\nu=-\infty}^{\infty} \zeta^{\nu} \Phi_{\nu}, \qquad \zeta \in \mathbb{T},$$

such that

   (i) $\sum_{\nu=-\infty}^{\infty} \|\Phi_{\nu}\| < \infty$,

   (ii) $\Phi_k = A_k$ for $|k| \leq p$,

   (iii) Φ admits a left and right canonical factorization relative to $\mathbb{T}$.

12. Show that the solution of the positive completion problem for finite band matrices can be derived from the solution of the Carathéodory-Toeplitz extension problem. Hint:

replace an $n \times n$ matrix $[a_{ij}]_{i,j=0}^{n}$ by the following matrix-valued function on $\mathbb{T}$:

$$\Phi(\zeta) = \begin{bmatrix} 1 & & & \\ & \zeta & & \\ & & \ddots & \\ & & & \zeta^n \end{bmatrix} \begin{bmatrix} a_{00} & a_{01} & \cdots & a_{0n} \\ a_{10} & a_{11} & \cdots & a_{1n} \\ \vdots & \vdots & & \vdots \\ a_{n0} & a_{n1} & \cdots & a_{nn} \end{bmatrix} \begin{bmatrix} 1 & & & \\ & \zeta^{-1} & & \\ & & \ddots & \\ & & & \zeta^{-n} \end{bmatrix}, \qquad \zeta \in \mathbb{T}.$$

In the next two exercises $\mathcal{M}$ is an algebra with a band structure defined by

(1)
$$\mathcal{M} = \mathcal{M}_1 \oplus \mathcal{M}_2^0 \oplus \mathcal{M}_d \oplus \mathcal{M}_3^0 \oplus \mathcal{M}_4.$$

13. Put $\mathcal{B}_- = \mathcal{M}_4$ and $\mathcal{B}_+ = \mathcal{M}_1 \oplus \mathcal{M}_2^0 \oplus \mathcal{M}_d \oplus \mathcal{M}_3^0$, and let $\varphi \in \mathcal{B}_-$ be given. Formulate the strictly contractive extension problem for φ relative to the decomposition

$$\mathcal{M} = \mathcal{B}_- \oplus \mathcal{B}_+.$$

Reduce the problem to a positive extension problem in $\mathcal{M}^{2 \times 2}$. Derive the analogues of Theorems XXXIV.3.1 and XXXIV.3.2 for the present setting.

14. Assume $\mathcal{M}$ admits the direct sum decomposition

(2)
$$\mathcal{M} = \mathcal{B}_- \oplus \mathcal{B}_+,$$

where $\mathcal{B}_-$ and $\mathcal{B}_+$ are linear spaces which are related to the band structure (1) in the following way:

$$\mathcal{B}_\pm \mathcal{M}_\pm^0 \subset \mathcal{B}_\pm, \qquad \mathcal{M}_\pm^0 \mathcal{B}_\pm \subset \mathcal{B}_\pm,$$
$$\mathcal{B}_\pm \mathcal{M}_d \subset \mathcal{B}_\pm, \qquad \mathcal{M}_d \mathcal{B}_\pm \subset \mathcal{B}_\pm,$$
$$\mathcal{B}_+ \mathcal{B}_-^* \subset \mathcal{M}_+^0, \qquad \mathcal{B}_-^* \mathcal{B}_+ \subset \mathcal{M}_+^0.$$

Let $\varphi \in \mathcal{B}_-$ be given. Formulate the strictly contractive extension problem for φ relative to the decomposition (2), and derive the analogues of Theorems XXXIV.3.1 and XXXIV.3.2 for the present setting.

15. Let $A = [a_{ij}]_{i,j=0}^{n}$ be a p-lower triangular matrix with $p < 0$. Use the results of Exercise 13 to find the triangular extension of A. Also, describe all strictly contractive extensions of A.

16. Do the previous exercise for the case when a_{ij} is a bounded linear operator from a Hilbert space H_j into a Hilbert space H_i for each i and j.

17. Prove that the coefficient matrix of the linear fractional map $\mathcal{F}$ in Theorem XXXIV.3.2 is J-unitary with $J = \begin{bmatrix} e & 0 \\ 0 & -e \end{bmatrix}$, i.e., show that

$$\begin{bmatrix} \alpha & \beta \\ \gamma & \delta \end{bmatrix} \begin{bmatrix} e & 0 \\ 0 & -e \end{bmatrix} \begin{bmatrix} \alpha^* & \gamma^* \\ \beta^* & \delta^* \end{bmatrix} = \begin{bmatrix} e & 0 \\ 0 & -e \end{bmatrix},$$

$$\begin{bmatrix} \alpha^* & \gamma^* \\ \beta^* & \delta^* \end{bmatrix} \begin{bmatrix} e & 0 \\ 0 & -e \end{bmatrix} \begin{bmatrix} \alpha & \beta \\ \gamma & \delta \end{bmatrix} = \begin{bmatrix} e & 0 \\ 0 & -e \end{bmatrix}.$$

STANDARD REFERENCES TEXTS

[C]   Conway, J.B., *Functions of One Complex Variable*, Graduate Texts in Mathematics, Springer-Verlag, New York, 1973.

[GG] Gohberg, I., Goldberg, S., *Basic Operator Theory*, Birkhäuser Verlag, Boston, 1981.

[R]   Rudin, W., *Real and Complex Analysis*, McGraw-Hill, New York, 1966.

[W]   Willard, S., *General Topology*, Addison Wesley, Reading, 1970.

BIBLIOGRAPHY

Adamjan, V.M., Arov, D.Z., Kreĭn, M.G.

[1] Infinite Hankel block matrices and related extension problems, *Izv. Akad. Nauk SSSR. Ser. Math.* **6** (1971), 87–112 (Russian); English Transl., *Amer. Math. Soc. Transl.* **111** (1978), 133–156.

Apostol, C.

[1] On a spectral equivalence of operators, in: *Topics in Operator Theory (Constantin Apostol Memorial Issue)*, Operator Theory: Advances and Applications, Vol. 32, Birkhäuser Verlag, Basel, 1988, pp. 15–35.

Bakonyi, M., Woerdeman, H.J.

[1] The central method for positive semi-definite, contractive and strong Parrott type completion problems, in *Operator Theory and Complex Analysis* (Ando, T. and Gohberg, I., eds.), Operator Theory: Advances and Applications, Vol. 59, Birkhäuser Verlag, Basel, 1992, pp. 78–95.

Ball, J.A.

[1] *Unitary Perturbations of Contractions*, Ph.D. Thesis, University of Virginia at Charlottesville, 1973.

[2] *Factorization and model theory for contraction operators with unitary part*, Memoirs Amer. Math. Soc., Vol. 198, Amer. Math. Soc., Providence, R.I., 1978.

Ball, J.A., Cohen, N.

[1] De Branges-Rovnyak operator models and systems theory: a survey, in: *Topics in Matrix and Operator Theory* (Bart, H., Gohberg, I. and Kaashoek, M.A., eds.), Operator Theory: Advances and Applications, Vol. 50, Birkhäuser Verlag, Basel, 1991, pp. 93–136.

Ball, J.A., Gohberg, I.

[1] Shift invariant subspaces, factorization, and interpolation for matrices, I: The canonical case, *Linear Alg. Appl.* **74** (1986), 87–150.

Ball, J.A., Gohberg, I., Rodman, L.

[1] *Interpolation of rational matrix functions*, Operator Theory: Advances and Applications, Vol. 45, Birkhäuser Verlag, Basel, 1990.

Ball, J.A., Helton, J.W.

[1] Factorization results related to shifts in an indefinite metric, *Integral Equations and Operator Theory* **5** (1982), 632–658.

Barkar, M.A., Gohberg, I.C.

[1] On factorizations of operators relative to a discrete chain of projectors in a Banach space, *Mat. Issled.* **1** (1966), 32–54 (Russian); English Transl., *Amer. Math. Soc. Transl.* (Series 2) **90** (1970), 81–103.

Bart, H., Gohberg, I., Kaashoek, M.A.

[1] *Minimal factorization of matrix and operator functions*, Operator Theory: Advances and Applications, Vol. 1, Birkhäuser Verlag, Basel, 1979.

[2] Wiener-Hopf integral equations, Toeplitz matrices and linear systems, in: *Toeplitz Centennial*, Operator Theory: Advances and Applications, Vol. 4, Birkhäuser Verlag, Basel, 1982, pp. 85–135.

[3] Convolution equations and linear systems, *Integral Equations and Operator Theory* **5** (1982), 283–340.

[4] The coupling method for solving integral equations, in: *Topics in Operator Theory and Networks, the Rehovot Workshop*, Operator Theory: Advances and Applications, Vol. 12, Birkhäuser Verlag, Basel, 1984, pp. 39–73.

[5] Explicit Wiener-Hopf factorization and realization, in: *Constructive Methods for Wiener-Hopf Factorization*, Operator Theory: Advances and Applications, Vol. 21, Birkhäuser Verlag, Basel, 1986, pp. 235–316.

[6] Fredholm theory of Wiener-Hopf equations in terms of realization of their symbols, *Integral Equations and Operator Theory* **8** (1985), 590–613.

[7] Wiener-Hopf factorization, inverse Fourier transforms and exponentially dichotomous operators, *J. Funct. Analysis* **68** (1986), 1–42.

[8] Wiener-Hopf equations with symbols analytic in a strip, in: *Constructive Methods of Wiener-Hopf Factorization* (Gohberg, I. and Kaashoek, M.A., eds.), Operator Theory: Advances and Applications, Vol. 21, Birkhäuser Verlag, Basel, 1986, pp. 39–74.

Bart, H., Kaashoek, M.A., Lay, D.C.

[1] The integral formula for the reduced algebraic multiplicity of meromorphic operator functions, *Proc. Edinburgh Math. Soc.* **21** (1978), 65–72.

Beauzamy, B.

[1] Un operateur sans sous-espace invariant: simplification de l'exemple de P. Enflo, *Integral Equations and Operator Theory* **8** (1985), 314–384.

[2] *Introduction to Operator Theory and Invariant Subspaces*, North-Holland Math. Library 42, Elsevier Science Publ., Amsterdam, 1988.

Bellman, R.

[1] Functional equations in the theory of dynamic programming – VII, A partial differential equation for the Fredholm resolvent, *Proc. Amer. Math.* **8** (1957), 435–440.

Beurling, A.

[1] On two problems concerning linear transformations in Hilbert space, *Acta Math.* **81** (1949), 239–255.

Böttcher, A., Silbermann, B.

[1] *Analysis of Toeplitz Operators*, Springer-Verlag, Berlin, 1990.

Boutet de Monvel, L., Guillemin, V.

[1] *The spectral theory of Toeplitz operators*, Annals of Mathematics Studies, Princeton, N.J., 1981.

Branges de, L.

[1] Factorization and invariant subspaces, *J. Math. Anal. Appl.* **29** (1970), 163–200.

[2] Unitary linear systems whose transfer functions are Riemann mapping functions, in: *Operator Theory and Systems* (Bart, H., Gohberg, I. and Kaashoek, M.A., eds.), Operator Theory: Advances and Applications, Vol. 19, Birkhäuser Verlag, Basel, 1986, pp. 105–124.

Branges de, L., Rovnyak, J.

[1] *Square Summable Power Series*, Holt, Rinehart and Winston, New York, 1966.

[2] Appendix on square summable power series, canonical models in quantum scattering theory, in: *Perturbation Theory and Its Applications in Quantum Mechanics* (Wilcox, C.H., ed.), Wiley, New York, 1966.

Brodskii, M.S.

[1] *Triangular and Jordan Representations of Linear Operators*, Transl. Math. Monographs, Vol. 32, Amer. Math. Soc., Providence, R.I., 1970.

[2] Unitary operator colligations and their characteristic functions, *Uspekhi Math. Nauk* **33**(4) (1978), 141–168 (Russian); English Transl., *Russian Math. Surveys* **33**(4) (1987), 159–191.

Brodskii, V.M., Gohberg, I.C., Kreĭn, M.G.

[1] General theorems on triangular representations of linear operators and multiplications of their characteristic functions, *Funk. Anal. i Priloz* **3**(4) (1969), 1–27 (Russian); English Transl., *Funct. Anal. Appl.* **3** (1969), 255–276.

[2] On characteristic functions of an invertible operator, *Acta Sci. Math. (Szeged)* **32** (1971), 141–164 (Russian).

Burg, J.P.

[1] *Maximum Entropy Spectral Analysis*, Doctoral Dissertation, Department of Geophysics, Stanford University, 1975.

Calderon, A.P.

[1] The analytic calculation of the index of elliptic equations, *Proc. Nat. Acad. Sci. USA* **57** (1967), 1193–1194.

Carleman, T.

[1] Zur Theorie der linearen Integralgleichunger, *Math. Zeilsch.* **9** (1921), 196–217.

Clancey, K, Gohberg, I.

[1] *Factorization of matrix functions and singular integral operators.* Operator Theory: Advances and Applications, Vol. 3, Birkhäuser Verlag, Basel, 1981.

Coburn, L.A.

[1] The C*-algebra generated by an isometry, I, *Bull. Amer. Math. Soc.* **73** (1967), 722–726.

[2] The C*-algebra generated by an isometry, II, *Trans. Amer. Math. Soc.* **137** (1969), 211–217.

Colojoara, L., Foiaş, C.

[1] *The Theory of Generalized Spectral Operators*, Gordon and Breach, New York, 1968.

Conway, J.B.

[1] *A Course in Functional Analysis*, Springer-Verlag, Berlin, 1985.

Daleckii, Ju.L., Kreĭn, M.G.

[1] *Stability of Solutions of Differential Equations in Banach Space*, Transl. Math. Monographs, Vol. 43, Amer. Math. Soc., Providence, R.I., 1974.

Davidson, K.R.

[1] *Nest algebras*, Pitman Research Notes in Mathematics Series 191, Pitman, London, 1988.

Davies, E.B.

[1] *One-Parameter Semigroups*, London Math. Soc. Monographs, No. 15, Academic Press, London, 1980.

Davis, C., Kahan, W.M., Weinberger, H.F.

[1] Norm-preserving dilations and their applications to optimal error bounds, *SIAM J. Numer. Anal.* **19** (1982), 445–469.

Dewilde, P.

[1]   Input-output description of roomy systems, *SIAM J. Control* **14** (1976), 712–736.

Dewilde, P., Deprettere, E.F.A.

[1]   The generalized Schur algorithm: approximation and hierarchy, in: *Topics in Operator Theory and Interpolation* (Gohberg, I., ed.), Operator Theory: Advances Applications, Vol. 28, Birkhäuser Verlag, Basel, 1988, pp. 97–116.

Douglas, R.G.

[1]   On majorization, factorization, and range inclusion of operators in Hilbert space, *Proc. Amer. Math. Soc.* **17** (1966), 413–415.

[2]   *Banach Algebra Techniques in Operator Theory*, Academic Press, New York, 1972.

Dunford, N., Schwartz, J.T.

[1]   *Linear Operators*, Part II: *Spectral Theory (Self Adjoint Operators in Hilbert Space)*, Wiley-Interscience, New York, 1963.

[2]   *Linear Operators*, Part III: *Spectral Operators*, Wiley-Interscience, New York, 1971.

Dym, H.

[1]   *J Contractive Matrix Functions, Reproducing Kernel Hilbert Spaces and Interpolation*, CBMS Regional Conference Series in Mathematics 71, Amer. Math. Soc., Providence, R.I., 1989.

Dym, H , Gohberg, I.

[1]   Extensions of kernels of Fredholm operators, *J. Anal. Math.* **42** (1982/83), 51–97.

[2]   A new class of contractive interpolants and maximum entropy principles, in: *Topics in Operator Theory and Interpolation* (Gohberg, I., ed.), Operator Theory: Advances Applications, Vol. 29, Birkhäuser Verlag, 1988, pp. 117–150.

[3]   Extensions of matrix valued functions and block matrices, *Indiana Univ. Math. J.* **31**(5) (1982), 733–765.

[4]   Extensions of band matrices with band inverses, *Linear Alg. Appl.* **36** (1981), 1–24.

[5]   Extensions of matrix valued functions with rational polynomial inverse, *Integral Equations and Operator Theory* **2** (1979), 503–528.

[6]   Unitary interpolants, factorization indices and infinite Hankel block matrices, *J. Funct. Anal.* **54** (1983), 229–289.

[7]   A maximum entropy principle for contractive interpolants, *J. Funct. Anal.* **65** (1986), 83–125.

Ellis, R.L., Gohberg, I., Lay, D.C

[1] Invertible selfadjoint extensions of band matrices and their entropy, *SIAM J. Alg. Disc. Meth.* **8**(3) (1987), 483–500.

[2] Extensions with positive real part, A new version of the abstract band method with applications, *Integral Equations and Operator Theory* **16** (1993), 360–384.

Enflo, P.H.

[1] On the invariant subspace problem in Banach spaces, *Acta Mathematica* **158** (1987), 213–313.

Feintuch, A., Saeks, R.

[1] *System Theory: A Hilbert Space Approach*, Academic Press, New York, 1982.

Foias, C., Frazho, A.E.

[1] *The commutant lifting approach to interpolation problems*, Operator Theory: Advances and Applications, Vol. 44, Birkhäuser Verlag, Basel, 1990.

Foias, C., Tannenbaum, A.

[1] The strong Parrott theorem, *Proc. Amer. Math. Soc.* **106** (1989), 777–784.

Fredholm, I.

[1] Sur une classe d'équations fonctionnelles, *Acta Math.* **27** (1903), 365–390.

Friedman, A.

[1] *Partial Differential Equations*, Krieger Publ. Co., New York, 1976.

Fuhrmann, P.

[1] *Linear Systems and Operators in Hilbert Space*, McGraw Hill, New York, 1981.

Gel'fand, I.M.

[1] *Lectures on Linear Algebra*, Interscience Publ., New York, 1961.

Gilburg, D., Trudinger, N.S.

[1] *Elliptic Partial Differential Equations of the Second Order* (2nd edition), Springer-Verlag, New York, 1984.

Ginzburg, Yu.P.

[1] Multiplicative representations and minorants of bounded analytic operator-functions, *Funk. Anal. i Priloz* **1**(3) (1967), 9–23 (Russian); English Transl., *Funct. Anal. Appl.* **1** (1976), 180–192.

Ginzburg, Yu.P., Shevchuk, L.V.

[1] On the Potapov theory of multiplicative representations, Operator Theory: Advances and Applications, to appear.

Ginzburg, Yu.P., Zemskov, L.M.

[1] On the multiplicative representations of operator functions of bounded form, *Teor. Funkcii, Funk. Anal. i Priloz* **53** (1990) 108–118.

Gohberg, I.C.

[1] On an application of the theory of normed rings to singular integral equations, *Uspekhi Math. Nauk* **7** (1952), 149–156 (Russian).

[2] On a factorization problem in normed rings, functions of isometric and symmetric operators and singular integral equations, *Uspekhi Math. Nauk* **19**(1) (1964) 71–124 (Russian); English Transl., *Russian Math. Surveys* **19**(1) (1964), 63–144.

Gohberg, I.C., Fel'dman, I.A.

[1] *Convolution Equations and Projection Methods for Their Solution*, Transl. Math. Monographs, Vol. 41, Amer. Math. Soc., Providence, R.I., 1974.

Gohberg, I., Goldberg, S.

[1] Semi-separable operators along chains of projections and systems, *J. Math. Anal.* **125**(1) (1987), 124–140.

[2] Finite dimensional Wiener-Hopf equations and factorization of matrices, *Linear Alg. Appl.* **48** (1982), 219–236.

[3] Factorization of semi-separable operators along continuous chains of projections, *J. Math. Anal.* **133**(1) (1988), 27–43.

Gohberg, I., Kaashoek, M.A.

[1] Time varying linear systems with boundary conditions and integral operators, I. The transfer operator and its properties, *Integral Equations and Operator Theory* **7** (1984), 325–391.

[2] Minimal factorizations of integral operators and cascade decompositions of systems, in: *Constructive Methods of Wiener-Hopf Factorization* (Gohberg, I. and Kaashoek, M.A., eds.), Operator Theory: Advances and Applications, Vol. 21, Birkhäuser Verlag, Basel, 1986, pp. 157–230.

[3] Block Toeplitz operators with rational symbols, in: *Contributions to Operator Theory and Its Applications* (Gohberg, I., Helton, J.W. and Rodman, L., eds.), Operator Theory: Advances and Applications, Vol. 35, 1988, pp. 385–440.

[4] Asymptotic formulas of Szegö-Kac-Achiezer type, *Asymptotic Analysis* **5** (1992), 187–220.

Gohberg, I., Kaashoek, M.A., Lay, D.C.

[1] Spectral classification of operators and operator functions, *Bull. Amer. Math. Soc.* **82** (1976), 587–589.

[2] Equivalence, linearization and decompositions of holomorphic operator functions, *J. Funct. Anal.* **28** (1978), 102–144.

Gohberg, I., Kaashoek, M.A., Schagen, F. van

[1] Szegö-Kac-Achiezer formulas in terms of realization of the symbol, *J. Funct. Anal.* **74** (1987), 24–51.

[2] Rational contractive and unitary interpolants in realized form, *Integral Equations and Operator Theory* **11** (1988), 105–127.

Gohberg, I., Kaashoek, M.A., Woerdeman, H.J.

[1] The band method for positive and contractive extension problems, *J. Operator Theory* **22** (1989), 109–155.

[2] The band method for positive and contractive extension problems: An alternative version and new applications, *Integral Equations and Operator Theory* **12** (1989), 343–382.

[3] A maximum entropy principle in the general framework of the band method, *J. Funct. Anal.* **95** (1991), 231–254.

[4] A note on extensions of band matrices with maximal and submaximal invertible blocks, *Linear Alg. Appl.* **150** (1991), 157–166.

[5] The time variant versions of the Nehari and four block problems, in: H_∞-*control theory* (Mosca, E., Pandolfi, L., eds.), Springer Lecture Notes in Math 1496, Springer-Verlag, Berlin, 1991.

[6] The band method for several positive extension problems of non-band type, Operator Theory: Advances and Applications, to appear.

Gohberg, I., Koltracht, I.

[1] Numerical solutions of integral equations, fast algorithms and Kreĭn-Sobolev equation, *Numer. Math.* **47** (1985), 237–288.

Gohberg, I.C., Kreĭn, M.G.

[1] The basic propositions on defect numbers, root numbers and indices of linear operators, *Uspekhi Math. Nauk* **12**, 2(74) (1957), 43–118 (Russian); English Transl., *Amer. Math. Soc. Transl.* (Series 2) **13** (1960), 185–265.

[2] Systems of integral equations on a half line with kernels depending on the difference of arguments, *Uspekhi Math. Nauk* **13**, 2(80) (1958), 3–72 (Russian); English Transl., *Amer. Math. Soc. Transl.* (Series 2) **14** (1960), 217–287.

[3] *Introduction to the Theory of Linear Nonselfadjoint Operators*, Transl. Math. Monographs, Vol. 18, Amer. Math. Soc., Providence, R.I., 1969.

[4]   *Theory and Applications of Volterra Operators in Hilbert Space*, Transl. Math. Monographs, Vol. 24, Amer. Math. Soc., Providence, R.I., 1970.

[5]   On the problem of factoring operators in a Hilbert space, *Dokl. Akad. Nauk SSSR* **147** (1962), 279–282 (Russian); English Transl.,*Amer. Math. Soc. Transl.* (Series 2) **51** (1966), 155–188.

[6]   Description of contraction operators which are similar to unitary operators, *Funct. Anal. Appl.* **1**(1) (1967), 33–52.

Gohberg, I., Krupnik, N.Ya.

[1]   *Einführung in die Theorie der Eindimensionalen Singulären Integraloperatoren*, Mathematische Reihe, Band 63, Birkhäuser Verlag, Basel, 1979.

[2]   On an algebra generated by Toeplitz matrices, *Funk. Anal. i Priloz* **3** (1969), 46–56 (Russian); English Transl., *Funct. Anal. Appl.* **3** (1969), 119–127.

[3]   On the algebra generated by one-dimensional singular integral operators with piecewise continuous coefficients, *Funk. Anal. i Priloz* **4** (1970), 26–36 (Russian); English Transl.,*Funct. Anal. Appl.* **4** (1970), 193–201.

[4]   *One-dimensional linear singular integral equations, I. Introduction*, Operator Theory: Advances and Applications, Vol. 53, Birkhäuser Verlag, Basel, 1992.

[5]   *One-dimensional linear singular integral equations, II. General theory and applications*, Operator Theory: Advances and Applications, Vol. 54, Birkhäuser Verlag, Basel, 1992.

Gohberg, I., Lancaster, P., Rodman, L.

[1]   *Matrix Polynomials*, Academic Press, New York, 1982.

[2]   *Invariant Subspaces of Matrices with Applications*, Wiley, New York, 1986.

Gohberg, I.C., Sigal, E.I.

[1]   An operator generalization of the logarithmic residue theorem and the theorem of Rouché, *Mat. Sbornik* **84**(1260) (1971), 607–629 (Russian); English Transl., *Math. USSR Sbornik* **13** (1971), 603–625.

Goldberg, S.

[1]   *Unbounded Linear Operators, Theory and Applications*, McGraw-Hill, New York, 1966.

Greenberg, W., Mee, C.V.M. van der, Protopopescu, V.

[1]   *Boundary Value Problems in Abstract Kinetic Theory*, Operator Theory: Advances and Applications, Vol. 23, Birkhäuser Verlag, Basel, 1987.

Guillemin, V.

[1] Toeplitz operators in n dimensions, *Integral Equations and Operator Theory* **7** (1984), 145–205.

Halmos, P.R.

[1] Shifts on Hilbert spaces, *J. reine angew. Math.* **208** (1961), 102–112.

Heinig, G., Rost, K.

[1] *Algebraic methods for Toeplitz-like matrices and operators*, Operator Theory: Advances and Applications, Vol. 13, Birkhäuser Verlag, Basel, 1984.

Helton, J.W.

[1] Systems with infinite-dimensional state space: the Hilbert space approach, *Proc. IEEE* **64**(1) (1976), 145–160.

[2] *Operator Theory, Analytic Functions, Matrices, and Electrical Engineering*, CBMS Regional Conference Series in Mathematics 68, Amer. Math. Soc., Providence, R.I., 1987.

Hilbert, D.

[1] Grundzüge einer allgemeinen Theorie der Linearen Integralgleichungen (Erste Mitteilung), *Nachr. Wiss. Gesell. Göttingen, Math.-Phys. Kl.* (1904), 49–91.

Hille, E., Phillips, R.S.

[1] *Functional Analysis and Semigroups*, Amer. Math. Soc. Colloquim Publ., Vol. XXXI, Amer. Math. Soc., Providence, R.I., 1957.

Horn, A.

[1] On the eigenvalues of a matrix with prescribed singular values, *Proc. Math. Soc.* **5** (1954), 4–7.

Kaashoek, M.A.

[1] Closed linear operators on Banach spaces, *Proc. Acad. Sci. Amsterdam* **A68** (1965), 405–414.

Kaashoek, M.A., Mee, C.V.M. van der, Rodman, L.

[1] Analytic operator functions with compact spectrum, I. Spectral nodes, linearization and equivalence, *Integral Equations and Operator Theory* **4** (1981), 504–547.

Kailath, T.

[1] *Linear Systems*, Prentice-Hall, Inc., Englewood Cliffs, New Jersey, 1980.

[2] Fredholm resolvents, Wiener-Hopf equations, and Ricatti differential equations, *IEEE Trans. Inform. Theory* **I.T.-15** (1969), 665–672.

[3]   A RKHS approach to detection and estimation problems, Part I, *IEEE Trans. Inform. Theory* **I.T.-17** (1971), 530–549.

Kailath, T, Duttweiler, D.

[1]   A RKHS approach to detection and estimation problems, Part III, *IEEE Trans. Inform. Theory* **I.T.-18** (1972), 730–745.

Kalman, R.E.

[1]   Mathematical description of linear dynamical systems, *SIAM J. Control* **1**(2) (1963), 152–192.

Kalman, R.E., Falb, P.L., Arbib, M.A.

[1]   *Topics in Mathematical System Theory*, McGraw-Hill, New York, 1969.

Kaper, H.G., Lekkerkerker, C.G., Hejtmanek, J.

[1]   *Spectral Methods in Linear Transport Theory*, Operator Theory: Advances and Applications, Vol. 5, Birkhäuser Verlag, Basel, 1982.

Kato, T.

[1]   *Perturbation Theory for Linear Operators*, Grundlehren, Band 132, Springer-Verlag, Berlin, 1966.

Keldysh, V.M.

[1]   On the eigenvalues and eigenfunctions of certain classes of nonselfadjoint equations, *Dokl. Akad. Nauk SSSR* **77** (1951), 11–14 (Russian); English Transl. in Markus [1].

Knuth, D.E.

[1]   *The Art of Computer Programming*, Addison Wesley, Reading, M.A., 1980.

König, H.

[1]   *Eigenvalue Distribution of Compact Operators*, Operator Theory: Advances and Applications, Vol. 16, Birkhäuser Verlag, Basel, 1986.

Kreĭn, M.G.

[1]   On inverse problems for nonhomogeneous cord, *Dokl. Akad. Nauk SSSR* **82** (1952), 669–672 (Russian).

[2]   On integral equations generating differential equations of second-order, *Dokl. Akad. Nauk SSSR* **97** (1954), 21–24 (Russian).

[3]   Integral equations on a half-line with kernel depending upon the difference of the arguments, *Uspekhi Math. Nauk* **13**(5) (1958), 3–120 (Russian); English Transl., *Amer. Math. Soc. Transl.* (Series 2) **22** (1962), 163–288.

Krupnik, N.Ya.

[1] *Banach Algebras with Symbol and Singular Integral Operators*, Operator Theory:
     Advances and Applications, Vol. 26, Birkhäuser Verlag, Basel, 1987.

Larson, D.R.

[1] Nest algebras and similarity transformations, *Ann. Math.* **121** (1985), 409–427.

Lax, P.

[1] Translation invariant spaces, *Acta Math.* **101** (1959), 163–178.

[2] Translation invariant spaces, *Proc. Int. Symp. Linear Spaces, Jerusalem (1960)*,
     New York, 1961, pp. 299–306.

Leiterer, J.

[1] Local and global equivalence of meromorphic operator functions, I, *Math. Nachr.*
     83 (1978), 7–29.

[2] Local and global equivalence of meromorphic operator functions, II, *Math. Nachr.*
     84 (1978), 145–170.

Lidskii, V.B.

[1] Theorems on the completeness of the system of eigenelements and adjoined elements
     of operators having a discrete spectrum, *Dokl. Akad. Nauk SSSR* **119** (1958), 1088–
     1091 (Russian); English Transl., *Amer. Math. Soc. Transl.* (Series 2) **47** (1965),
     37–41.

[2] Non-selfadjoint operators with a trace, *Dokl. Akad. Nauk SSSR* **125** (1959), 485–587
     (Russian); English Transl., *Amer. Math. Soc. Transl.* (Series 2) **47** (1965), 43–46.

Livšic, M.S.

[1] On a class of linear operators in a Hilbert space, *Mat. Sbornik* **19**(61) (1946), 239–
     262 (Russian); English Transl., *Amer. Math. Soc. Transl.* (Series 2) **13** (1960),
     61–83.

[2] On the spectral resolution of linear non-selfadjoint operators, *Mat. Sbornik* **34**(76)
     (1954), 145–199 (Russian); English Transl., *Amer. Math. Soc. Transl.* (Series 2) **5**
     (1957), 67–114.

Livšic, M.S., Potapov, V.P.

[1] Theorem of multiplication of characteristic matrix functions, *Dokl. Akad. Nauk* **72**
     (1950), 625–628.

Loomis, L.H.

[1] *Abstract Harmonic Analysis*, Van Nostrand, New York, 1953.

Lorch, E.R.

[1] Return to the self-adjoint transformation, *Acta Sci. Math. (Szeged)* **12**(B) (1950), 137–144.

[2] *Spectral Theory*, University Texts in the Mathematical Sciences, Oxford University Press, New York, 1962.

Lumer, G., Rosenblum, M.

[1] Linear operator equations, *Proc. Amer. Math. Soc.* **10** (1959), 32–41.

Lyubich, Ju., Macaev, V.I.

[1] On operators with a separable spectrum, *Mat. Sb.* **56** (1962), 433–468 (Russian); English Transl., *Amer. Math. Soc. Transl.* (Series 2) **47** (1965), 89–129.

Macaev, V.I.

[1] A class of completely continuous operators, *Dokl. Akad. Nauk SSSR* **139** (1961), 548–551 (Russian); English Transl., *Soviet Math. Dokl.* **1** (1961), 972–975.

[2] Several theorems on completeness of root subspaces of completely continuous operators, *Dokl. Akad. Nauk SSSR* **155** (1964), 273–276 (Russian), English Transl., *Soviet Math. Dokl.* **5** (1964), 396–399.

Markus, A.S.

[1] *Introduction to Spectral Theory of Polynomial Operator Pencils*, Transl. Math. Monographs, Vol. 71, Amer. Math. Soc., Providence, R.I., 1988.

Markus, A.S., Fel'dman, I.A.

[1] Index of an operator matrix, *Funct. Anal. Appl.* **11** (1977), 149–151.

Markus, A.S., Sigal, E.I.

[1] On the multiplicity of a characteristic value of an analytic operator-function, *Mat. Issled* **5**, No. 3(17) (1970), 129–147 (Russian).

Mee, C.V.M. van der

[1] *Semigroup and Factorization Methods in Transport Theory*, Thesis Vrije Universiteit, Amsterdam, 1981 = Mathematical Centre Tracts **146**, Mathematisch Centrum, Amsterdam, 1981.

Mikhlin, S.G.

[1] *Integral Equations* (2nd Revised Edition), The Macmillan Co., New York, 1964.

Mirsky, L.

[1] Matrices with prescribed characteristic roots and diagonal elements, *J. London Math. Soc.* **33** (1958), 14–21.

Morf, M., Vieira, A., Kailath, T.

[1]   Covariance characterization by partial autocorrelation matrices. *Ann. Statist.* **6** (1978), 643–648.

Morf, M., Vieira, A., Lee, D., Kailath, T.

[1]   Recursive multichannel maximal entropy spectral estimation, *IEEE Trans. Geosci. Elect.* **16** (1978), 85–94.

Nehari, Z.

[1]   On bounded bilinear forms, *Ann. Math.* **65** (1957), 153–162.

Nikol'skii, N.K.

[1]   Unicellularity and non-unicellularity of weighted shift operators, *Dokl. Akad. Nauk SSSR* **172** (1967), 287–290 (Russian); English Transl., *Soviet Math. Dokl.* **8** (1967), 91–94.

[2]   *Treatise on the Shift Operator*, Springer-Verlag, Berlin, 1986.

Parrott, S.

[1]   On a quotient norm and the Sz.-Nagy-Foias lifting theorem, *J. Funct. Anal.* **30** (1978), 311–328.

Pazy, A.

[1]   *Semigroups of Linear Operators and Applications to Partial Differential Equations*, Applied Mathematical Sciences, Vol. 44, Springer-Verlag, New York, 1983.

Pietsch, A.

[1]   *Eigenvalues and s-numbers*, Cambridge Studies in Advanced Mathematics, Vol. 13, Cambridge University Press, Cambridge, 1987.

Pólya, G., Szegö, G.

[1]   *Aufgaben und Lehrsätze aus der Analysis*, Band I. Grundlehren, Band 19, Springer-Verlag, Berlin, 1964.

Potapov, V.P.

[1]   The multiplicative structure of J-contrative matrix functions, *Trudy Moskov. Math. Obsc.* **4** (1955), 125–136 (Russian); English Transl., *Amer. Math. Soc. Transl.* (Series 2) **15** (1960), 131–243.

Pták, V., Vroba, P.

[1]   On the spectral function of a normal operator, *Czech. Math. J.* **23**(98) (1973), 615–616.

Putnam, C.

[1]   Ranges of normal and subnormal operators, *Mich. Math. J.* **18** (1972), 33–36.

Ran, A.C.M., Rodman, L.

[1]   A matricial boundary value problem which appears in the transport theory, *J. Math. Anal. Appl.* **130** (1988), 200–222.

Read, C.

[1]   Solution to the invariant subspace problem, *Bull. London Math. Soc.* **16** (1984), 337–401.

[2]   Solution to the invariant subspace problem on the space ℓ_1, *Bull. London Math. Soc.* **17** (1985), 305–317.

Rickart, C.E.

[1]   *Banach Algebras*, R.E. Krieger Publ. Co., New York, 1974.

Riesz, F., Sz.-Nagy, B.

[1]   *Leçons d'Analyse Fonctionnelle*, Académie des Sciences de Hongrie, 1955.

Rosenblum, M., Rovnyak, J.

[1]   *Hardy Classes and Operator Theory*, Oxford Mathematical Monographs, Oxford University Press, Oxford, 1985.

Rota, G.C.

[1]   On models for linear operators, *Comm. Pure Appl. Math.* **13** (1960), 468–472.

Sarason, D.

[1]   On spectral sets having connected complement, *Acta. Sci. Math. (Szeged)* **26** (1965), 289–299.

[2]   Generalized interpolation in H^∞, *Trans. Amer. Math. Soc.* **127** (1967), 179–203.

Schäffer, J.J.

[1]   On unitary dilations of contractions, *Proc. Amer. Math. Soc.* **6** (1955), 322.

Schumitzky, A.

[1]   On the equivalence between matrix Ricatti equations and Fredholm resolvents, *J. Computer Systems Sci.* **2**(1) (1968), 76–87.

Showalter, R.E.

[1]   *Hilbert Space Methods for Partial Differential Equations*, Pitman, London, 1977.

Sigal, E.I.

[1] Factor-multiplicity of eigenvalues of meromorphic operator functions, *Mat. Issled* **5**, No. 4(18) (1970), 136–152 (Russian).

Smuljan, Ju.L.

[1] Operators with a degenerate characteristic function, *Dokl. Akad. Nauk* **93** (1953), 985–988 (Russian).

Sobolev, S.L.

[1] Remarks on the numerical solution of integral equations, *Izv. Akad. Nauk SSSR Ser. Mat.* **20** (1956), 431 (Russian); English Transl., *Acad. R. P. Romine. An. Romino-Soviet. Ser. Mat.-Fiz.* **11**(3) (1957), No. 4 (23), 5.

Spitkovsky, I.M., Woerdeman, H.J.

[1] The Carathéodory-Toeplitz problem for almost periodic functions, *J. Funct. Analysis* **114** (1993).

Stampfli, J.G.

[1] A local spectral theory for operators, IV. Invariant subspaces, *Indiana Univ. Math.* **22** (1972), 159–176.

Stone, M.H.

[1] Applications of the theory of Boolean rings to general topology, *Trans. Amer. Math. Soc.* **41** (1937), 375–481.

[2] The generalized Weierstrass approximation theorem, *Math. Mag.* **21** (1947-1948), 167–184, 237–254.

Strang, G.

[1] *Linear Algebra and Its Applications*, Academic Press, New York, 1976.

Stummel, F.

[1] Diskreter konvergenz linearer operatoren, II, *Math. Zeitschr.* **120** (1971), 231–264.

Szegö, G.

[1] Ein Grenzwertsatz über der Toeplitzen Determinanten einer reellen positiven Function, *Math. Ann.* **76** (1915), 490–503.

[2] Beiträge zur Theorie der Toeplitzen Formen, *Math. Z.* **6** (1920), 167–202.

[3] On certain Hermitian forms associated with the Fourier series of a positive function, in: *Commun. du Séminaire Math. de l'Univ. de Lund, Festschrift Marcel Riesz*, Lund, 1952, pp. 228–238.

Sz.-Nagy, B.

[1] Perturbations des transformations autoadjointes dans l'espace de Hilbert, *Comment. Math. Helv.* **19** (1947), 347–366.

[2] Perturbations des transformations linéares fermées, *Acta Sci. Math. Szeged* **14** (1951), 125–137.

[3] *Introduction to Real Functions and Orthogonal Expansions*, Oxford Univ. Press, New York, 1965.

[4] Sur les contractions de l'espace de Hilbert, *Acta Sci. Math.* **15** (1953), 87–92.

[5] Transformations de l'espace de Hilbert, fonctions de type positive sur un groupe, *Acta Sci. Math.* **15** (1954), 104–114.

[6] Sur les contractions de l'espace de Hilbert, II, *Acta Sci. Math.* **18** (1957), 1–15.

[7] Extensions of linear transformations in Hilbert space which extend beyond this space, *Appendix to F. Riesz and B. Sz.-Nagy: Functional Analysis*, New York, 1960. (Transl. of Prolongement des transformations de l'espace de Hilbert qui sortent de cet espace, Budapest, 1955.)

Sz.-Nagy, B., Foiaş, C.

[1] Modèles fonctionnels des contractions de l'espace de Hilbert, La fonction caracteristique, *C.R. Acad. Sci. Paris* **256** (1963), 3236–3239.

[2] Dilation des commutants d'opérateurs, *C.R. Acad. Sci. Paris*, Serie A, **266** (1968), 493–495.

[3] *Harmonic Analysis of Operators on Hilbert Space*, North Holland Publ. Co., Amsterdam-Budapest, 1970.

Weiss, L.

[1] On the structure theory of linear differential systems, *SIAM J. Control* **6** (1968), 659–680.

Whitley, R.W.

[1] Fuglede's commutativity theorem and $\bigcap R(T-\lambda)$, *Canad. Math. Bull.* **33**(3) (1990), 331–334.

Widom, H.

[1] Asymptotic behavior of block Toeplitz matrices and determinants, *Adv. Math.* **13** (1974), 284–322.

[2] On the limit of block Toeplitz determinants, *Proc. Amer. Math. Soc.* **50** (1975), 167–173.

[3] Asymptotic behavior of block Toepliz matrices and determinants, II, *Adv. Math.* **21** (1976), 1–29.

Woerdeman, H.J.

[1]   *Strictly Contractive and Positive Completions for Block Matrices*, Report WS 337,
      Vrije Universiteit, Amsterdam, November 1987.

[2]   *Matrix and Operator Extensions*, Doctoral Dissertation, Department of Mathematics
      and Computer Science, Vrije Universiteit, Amsterdam, 1989 = CWI Tract 68, Centre
      for Mathematics and Computer Science, Amsterdam, 1989.

Wold, H.

[1]   *A Study in the Analysis of Stationary Time Series*, Almqvist and Wiksell, Uppsala,
      1938.

Wolf, F.

[1]   Operators in Banach space which admit a generalized spectral decomposition, *Nederl. Akad. Wetensch. Indig. Math.* **19** (1957), 302–311.

Youla, D.C.

[1]   The synthesis of linear dynamical systems from prescribed weighting patterns, *SIAM J. Appl. Math.* **14** (1966), 527–549.

LIST OF SYMBOLS

$A_\Im$	imaginary part of the operator A, 135
$A_\Re$	real part of the operator A, 146
A^*	(Hilbert space) adjoint of the operator A, 290
A'	conjugate of the operator, 292
$\overline{A}$	minimal closed linear extension of the operator A, 289
$A\|N$	restriction of A to the subspace of N, 8, 326
$A \geq 0$	non-negative operator A, 82
$A \geq B$	$A - B$ is a non-negative operator, 944
$a \geq_\mathcal{R} b$	$a - b$ is positive semidefinite in $\mathcal{R}$, 933
$AC_n(J)$	a space of functions on the interval J, 295
$A(\mathbb{P})$	set of operators that have a diagonal with respect to a chain $\mathbb{P}$, 475
$A_0(\mathbb{P})$	set of operators that are reduced by a chain $\mathbb{P}$, 483
$A_\pm(\mathbb{P})$	triangular algebras, 483
$\mathcal{B}^+$	the set of non-negative elements in a unital C^*-algebra $\mathcal{B}$, 849
$\mathcal{B}^{m\times m}$	the set of $m \times m$ matrices with entries in $\mathcal{B}$, 804
$B(S)$	the space of bounded complex-valued functions on the set S, 789
$\mathbb{C}$	field of complex numbers, 5
$\mathbb{C}_-$	open lower half plane, 777
$\mathbb{C}^n$	complex Euclidean n-space
$\mathbb{C}_\infty$	the extended complex plane, 49
$C\big([a,b]\big)$	the space of complex-valued continuous functions on $[a,b]$, 288
$C(S)$	the space of complex-valued continuous functions on a topological space S, 789
$C_R(S)$	the space of real-valued continuous functions on a topological space S, 845
$C\big[\mathbb{T}[\zeta_1,\ldots,\zeta_k]\big]$	space of functions on the cylinder $\mathbb{T} \times [0,1]$, 628
$C^k(\Omega)$	a set of differentiable functions on Ω, 310
$C^\infty(\Omega)$	the set of C^∞-functions on Ω, 310
$C^\infty(\Omega)$	the set of C^∞-functions with compact support in Ω, 310
$C^k(\overline{\Omega})$	a set of differentiable functions on $\overline{\Omega}$, 310
$\mathbb{D}$	open unit disk in the complex plane
D_A	defect operator $(I - A^*A)^{1/2}$, 665
$\mathcal{D}(A)$	domain of the operator A, 288
D^α	elementary differential operator, 310
$d(A)$	codimension of the range of A, 184
$\Delta_\ell(b),\ \Delta_r(b)$	left, right multiplicative diagonal of b, 932, 933
$\partial\omega$	boundary of the open set Ω, 315
$\det A$	determinant of an operator matrix A, 194
$\det(I + A)$	determinant of $I + A$ with A a trace class operator, 117
$\dim M$	dimension of the linear space M
E_A	a space spanned by eigenvectors and generalized eigenvectors, 31

$\overset{\frown}{\prod} A_j,\ \overset{\frown}{\prod} A_j$	non-commutative products, 493
$\Re\lambda$	real part of the complex number λ
$\mathbb{R}^n$	real Euclidean n-space
$\mathcal{R}_\infty^{m\times m}(\mathbb{R})$	a set of rational $m \times m$ matrix functions, 232
$\operatorname{rank} A$	rank of the operator A
RKHS	reproducing kernel Hilbert space, 496
$r(x)$	spectral radius of a Banach algebra element x, 800
$\rho(A)$	resolvent set of the operator A, 5
$\rho(G, A)$	resolvent set of the pencil $\lambda G - A$, 49
$\rho(x)$	resolvent set of a Banach algebra element x, 797
S_1	the trace class operators, 105
S_2	the Hilbert-Schmidt operators, 143
$s_j(A)$	j-th singular value of the operator A, 96, 212
S^{-*}	the adjoint of S^{-1}, 779, 908
$\mathcal{S}^{\ell\times k}(\mathbb{D})$	the Schur class of contractive-valued analytic $\ell \times k$ matrix function on the open unit disc, 750
$\sigma(A)$	spectrum of the operator A, 5
$\sigma(G, A)$	spectrum of the operator pencil $\lambda G - A$, 49
$\sigma_{\mathrm{ess}}(A)$	essential spectrum of the operator A, 191, 373
$\sigma(x)$	spectrum of a Banach algebra element x, 797
$\Sigma_1,\ \Sigma_2$	Szegö limits, 617
$\mathbb{T}$	unit circle in the complex plane
$\widetilde{\mathbb{T}} = \widetilde{\mathbb{T}}(\zeta_1, \ldots, \zeta_k)$	deformed unit circle, 624
$T_{\max,\tau,J}$	maximal operator, 295
$T_{\min,\tau,J}$	minimal operator, 300
$T_{R,J}$	a differential operator, 300
T_Φ	(block) Toeplitz operator defined by Φ, 575
$\operatorname{tr} A$	trace of the operator A, 110
$\mathcal{T}(C),\ \mathcal{T}_m(C)$	C^*-algebras generated by (block) Toeplitz operators defined by continuous (matrix) functions, 870, 873
$\mathcal{T}_m(PC)$	C^*-algebra generated by block Toeplitz operators defined by piecev continuous matrix functions, 879
$\mathcal{T}_m(PC; \zeta_1, \ldots, \zeta_k)$	C^*-algebra generated by block Toeplitz operators defined by piecev continuous matrix functions with k discontinuities, 875
$\Theta_\Sigma(\cdot)$	transfer function of the unitary system Σ, 705
W_A	numerical range of A, 167
$\mathcal{W},\ \mathcal{W}(\beta)$	Wiener algebras, 789, 790
$^\perp N$	inverse annihilator of N, 292
$S^\perp$	orthogonal complement of S; and also at times, annihilator of S, 2!
$(x_1, \ldots, x_n)^T$	transpose of the row vector $(x_1, \ldots, x_n)$
$\langle\cdot,\cdot\rangle$	inner product
$\|A\|_1$	trace class norm of A, 106
$\|A\|_2$	Hilbert-Schmidt norm of A, 143
$\boxplus$	Hilbert space direct sum, 657, 658
$\bigvee_{n=1}^\infty W_n$	smallest closed linear manifold containing the sets W_n $(n \geq 1)$, 669

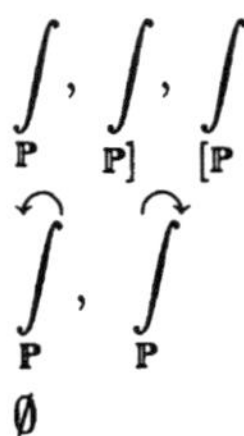

$\int_P, \int_{P]}, \int_{[P}$ integrals along a chain $\mathbb{P}$, 474, 485, 486

$\overset{\frown}{\int}_P, \overset{\frown}{\int}_P$ multiplicative integrals along a chain $\mathbb{P}$, 493

$\emptyset$ empty set

SUBJECT INDEX

TABLE OF CONTENTS OF VOLUME I

Titles previously published in the series

OPERATOR THEORY: ADVANCES AND APPLICATIONS
BIRKHÄUSER VERLAG

3. **K. Clancey, I. Gohberg:** Factorization of Matrix Functions and Singular Integral Operators, 1981, (3-7643-1297-1)

4. **I. Gohberg** (Ed.): Toeplitz Centennial, 1982, (3-7643-1333-1)

5. **H.G. Kaper, C.G. Lekkerkerker, J. Hejtmanek:** Spectral Methods in Linear Transport Theory, 1982, (3-7643-1372-2)

6. **C. Apostol, R.G. Douglas, B. Sz-Nagy, D. Voiculescu, Gr. Arsene** (Eds.): Invariant Subspaces and Other Topics, 1982, (3-7643-1360-9)

7. **M.G. Krein:** Topics in Differential and Integral Equations and Operator Theory, 1983, (3-7643-1517-2)

8. **I. Gohberg, P. Lancaster, L. Rodman:** Matrices and Indefinite Scalar Products, 1983, (3-7643-1527-X)

9. **H. Baumgärtel, M. Wollenberg:** Mathematical Scattering Theory, 1983, (3-7643-1519-9)

10. **D. Xia:** Spectral Theory of Hyponormal Operators, 1983, (3-7643-1541-5)

11. **C. Apostol, C.M. Pearcy, B. Sz.-Nagy, D. Voiculescu, Gr. Arsene** (Eds.): Dilation Theory, Toeplitz Operators and Other Topics, 1983, (3-7643-1516-4)

12. **H. Dym, I. Gohberg** (Eds.): Topics in Operator Theory Systems and Networks, 1984, (3-7643-1550-4)

13. **G. Heinig, K. Rost:** Algebraic Methods for Toeplitz-like Matrices and Operators, 1984, (3-7643-1643-8)

14. **H. Helson, B. Sz.-Nagy, F.-H. Vasilescu, D.Voiculescu, Gr. Arsene** (Eds.): Spectral Theory of Linear Operators and Related Topics, 1984, (3-7643-1642-X)

15. **H. Baumgärtel:** Analytic Perturbation Theory for Matrices and Operators, 1984, (3-7643-1664-0)

16. **H. König:** Eigenvalue Distribution of Compact Operators, 1986, (3-7643-1755-8)

17. **R.G. Douglas, C.M. Pearcy, B. Sz.-Nagy, F.-H. Vasilescu, D. Voiculescu, Gr. Arsene** (Eds.): Advances in Invariant Subspaces and Other Results of Operator Theory, 1986, (3-7643-1763-9)

18. **I. Gohberg** (Ed.): I. Schur Methods in Operator Theory and Signal Processing, 1986, (3-7643-1776-0)

19. **H. Bart, I. Gohberg, M.A. Kaashoek** (Eds.): Operator Theory and Systems, 1986, (3-7643-1783-3)

20. **D. Amir:** Isometric characterization of Inner Product Spaces, 1986, (3-7643-1774-4)

21. **I. Gohberg, M.A. Kaashoek** (Eds.): Constructive Methods of Wiener-Hopf Factorization, 1986, (3-7643-1826-0)
22. **V.A. Marchenko:** Sturm-Liouville Operators and Applications, 1986, (3-7643-1794-9)
23. **W. Greenberg, C. van der Mee, V. Protopopescu:** Boundary Value Problems in Abstract Kinetic Theory, 1987, (3-7643-1765-5)
24. **H. Helson, B. Sz.-Nagy, F.-H. Vasilescu, D. Voiculescu, Gr. Arsene** (Eds.): Operators in Indefinite Metric Spaces, Scattering Theory and Other Topics, 1987, (3-7643-1843-0)
25. **G.S. Litvinchuk, I.M. Spitkovskii:** Factorization of Measurable Matrix Functions, 1987, (3-7643-1883-X)
26. **N.Y. Krupnik:** Banach Algebras with Symbol and Singular Integral Operators, 1987, (3-7643-1836-8)
27. **A. Bultheel:** Laurent Series and their Pade Approximation, 1987, (3-7643-1940-2)
28. **H. Helson, C.M. Pearcy, F.-H. Vasilescu, D. Voiculescu, Gr. Arsene** (Eds.): Special Classes of Linear Operators and Other Topics, 1988, (3-7643-1970-4)
29. **I. Gohberg** (Ed.): Topics in Operator Theory and Interpolation, 1988, (3-7634-1960-7)
30. **Yu.I. Lyubich:** Introduction to the Theory of Banach Representations of Groups, 1988, (3-7643-2207-1)
31. **E.M. Polishchuk:** Continual Means and Boundary Value Problems in Function Spaces, 1988, (3-7643-2217-9)
32. **I. Gohberg** (Ed.): Topics in Operator Theory. Constantin Apostol Memorial Issue, 1988, (3-7643-2232-2)
33. **I. Gohberg** (Ed.): Topics in Interplation Theory of Rational Matrix-Valued Functions, 1988, (3-7643-2233-0)
34. **I. Gohberg** (Ed.): Orthogonal Matrix-Valued Polynomials and Applications, 1988, (3-7643-2242-X)
35. **I. Gohberg, J.W. Helton, L. Rodman** (Eds.): Contributions to Operator Theory and its Applications, 1988, (3-7643-2221-7)
36. **G.R. Belitskii, Yu.I. Lyubich:** Matrix Norms and their Applications, 1988, (3-7643-2220-9)
37. **K. Schmüdgen:** Unbounded Operator Algebras and Representation Theory, 1990, (3-7643-2321-3)
38. **L. Rodman:** An Introduction to Operator Polynomials, 1989, (3-7643-2324-8)
39. **M. Martin, M. Putinar:** Lectures on Hyponormal Operators, 1989, (3-7643-2329-9)
40. **H. Dym, S. Goldberg, P. Lancaster, M.A. Kaashoek** (Eds.): The Gohberg Anniversary Collection, Volume I, 1989, (3-7643-2307-8)
41. **H. Dym, S. Goldberg, P. Lancaster, M.A. Kaashoek** (Eds.): The Gohberg Anniversary Collection, Volume II, 1989, (3-7643-2308-6)
42. **N.K. Nikolskii** (Ed.): Toeplitz Operators and Spectral Function Theory, 1989, (3-7643-2344-2)
43. **H. Helson, B. Sz.-Nagy, F.-H. Vasilescu, Gr. Arsene** (Eds.): Linear Operators in Function Spaces, 1990, (3-7643-2343-4)

44. **C. Foias, A. Frazho:** The Commutant Lifting Approach to Interpolation Problems, 1990, (3-7643-2461-9)

45. **J.A. Ball, I. Gohberg, L. Rodman:** Interpolation of Rational Matrix Functions, 1990, (3-7643-2476-7)

46. **P. Exner, H. Neidhardt** (Eds.): Order, Disorder and Chaos in Quantum Systems, 1990, (3-7643-2492-9)

47. **I. Gohberg** (Ed.): Extension and Interpolation of Linear Operators and Matrix Functions, 1990, (3-7643-2530-5)

48. **L. de Branges, I. Gohberg, J. Rovnyak** (Eds.): Topics in Operator Theory. Ernst D. Hellinger Memorial Volume, 1990, (3-7643-2532-1)

49. **I. Gohberg, S. Goldberg, M.A. Kaashoek:** Classes of Linear Operators, Volume I, 1990, (3-7643-2531-3)

50. **H. Bart, I. Gohberg, M.A. Kaashoek** (Eds.): Topics in Matrix and Operator Theory, 1991, (3-7643-2570-4)

51. **W. Greenberg, J. Polewczak** (Eds.): Modern Mathematical Methods in Transport Theory, 1991, (3-7643-2571-2)

52. **S. Prössdorf, B. Silbermann:** Numerical Analysis for Integral and Related Operator Equations, 1991, (3-7643-2620-4)

53. **I. Gohberg, N. Krupnik:** One-Dimensional Linear Singular Integral Equations, Volume I, Introduction, 1992, (3-7643-2584-4)

54. **I. Gohberg, N. Krupnik:** One-Dimensional Linear Singular Integral Equations, Volume II, General Theory and Applications, 1992, (3-7643-2796-0)

55. **R.R. Akhmerov, M.I. Kamenskii, A.S. Potapov, A.E. Rodkina, B.N. Sadovskii:** Measures of Noncompactness and Condensing Operators, 1992, (3-7643-2716-2)

56. **I. Gohberg** (Ed.): Time-Variant Systems and Interpolation, 1992, (3-7643-2738-3)

57. **M. Demuth, B. Gramsch, B.W. Schulze** (Eds.): Operator Calculus and Spectral Theory, 1992, (3-7643-2792-8)

58. **I. Gohberg** (Ed.): Continuous and Discrete Fourier Transforms, Extension Problems and Wiener-Hopf Equations, 1992, (ISBN 3-7643-2809-6)

59. **T. Ando, I. Gohberg** (Eds.): Operator Theory and Complex Analysis, 1992, (3-7643-2824-X)

60. **P.A. Kuchment:** Floquet Theory for Partial Differential Equations, 1993, (3-7643-2901-7)

61. **A. Gheondea, D. Timotin, F.-H. Vasilescu** (Eds.): Operator Extensions, Interpolation of Functions and Related Topics, 1993, (3-7643-2902-5)

62 **T. Furuta, I. Gohberg, T. Nakazi** (Eds.): Contributions to Operator Theory and its Applications. The Tsuyoshi Ando Anniversary Volume, 1993, (3-7643-2928-9)

63. **I. Gohberg, S. Goldberg, M.A. Kaashoek:** Classes of Linear Operators, Volume 2, 1993, (3-7643-2944-0)